CNR*15 – Fifth International Workshop on Compound-Nuclear Reactions and Related Topics

EPJ Web of Conferences Volume 122 (2016)

Tokyo, Japan
19-23 October 2015

Editors:

T. Kawano
M. W. Paris
S. Chiba
P. Talou

ISBN: 978-1-5108-2605-2

Printed from e-media with permission by:

Curran Associates, Inc.
57 Morehouse Lane
Red Hook, NY 12571

Some format issues inherent in the e-media version may also appear in this print version.

Printed by Curran Associates, Inc. (2016)

For additional information, please contact EDP Sciences – Web of Conferences at the address below.

EDP Sciences – Web of Conferences
17, Avenue du Hoggar
Parc d'Activité de Courtabœuf
BP 112
F-91944 Les Ulis Cedex A
France

Phone: +33 (0) 1 69 18 75 75
Fax: +33 (0) 1 69 28 84 91

contact-edps@webofconferences.org

Additional copies of this publication are available from:

Curran Associates, Inc.
57 Morehouse Lane
Red Hook, NY 12571 USA
Phone: 845-758-0400
Fax: 845-758-2633
Email: curran@proceedings.com
Web: www.proceedings.com

TABLE OF CONTENTS

CNR*15
The 5th International Workshop on Compound-Nuclear Reactions and Related Topics
October 19-23,2015

edp sciences

EPJ Web of Conferences

Statement of Peer review

In submitting conference proceedings to EPJ Web of Conferences, I certify to the Publisher that I adhere to the **Policy on Publishing Integrity** of the journal in order to safeguard good scientific practice in publishing.

1. All articles have been subjected to peer review administered by the proceedings editors.
2. Reviews have been conducted by expert referees, who have been requested to provide unbiased and constructive comments aimed, whenever possible, at improving the work.
3. Proceedings editors have taken all reasonable steps to ensure the quality of the materials they publish and their decision to accept or reject a paper for publication has been based only on the merits of the work and the relevance to the journal.

Title, date and place of the conference

5th International Workshop on Compound-Nuclear Reactions and Related Topics

Oct. 19-23, 2015. Tokyo, Japan

Proceedings editor(s):

Toshihiko Kawano, Satoshi Chiba, Mark W. Paris, Patrick Talou

Date and editor's signature 5/26/2016

A new approach to barrier-top fission dynamics

G.F. Bertsch[a] and J.M. Mehlhaff

[1] *Department of Physics and Institute for Nuclear Theory, University of Washington, Seattle, WA USA*

Abstract. We proposed a calculational framework for describing induced fission that avoids the Bohr-Wheeler assumption of well-defined fission channels. The building blocks of our approach are configurations that form a discrete, orthogonal basis and can be characterized by both energy and shape. The dynamics is to be determined by interaction matrix elements between the states rather than by a Hill-Wheeler construction of a collective coordinate. Within our approach, several simple limits can be seen: diffusion; quantized conductance; and ordinary decay through channels. The specific proposal for the discrete basis is to use the K^π quantum numbers of the axially symmetric Hartree-Fock approximation to generate the configurations. Fission paths would be determined by hopping from configuration to configuration via the residual interaction. We show as an example the configurations needed to describe a fictitious fission decay $^{32}S \rightarrow ^{16}O + ^{16}O$. We also examine the geometry of the path for fission of ^{236}U, measuring distances by the number of jumps needed to go to a new K^π partition.

1 Introduction

In this talk we will advocate a radically different approach to the theory of induced fission. To put this into context, we show in Fig. 1 a schematic view of the fission landscape with the different energy regions indicated by color. In the tunneling region, shown in green, the dynamics is driven by the pairing interaction. High above the barrier, shown in pink, we expect the dynamics to be highly overdamped and amenable to treatment by a diffusion equation. In between is the barrier-top region, shown in blue. This is commonly described by the Bohr-Wheeler theory and its generalization to multiple barriers.

We seek an alternate treatment that does not require well-behaved channels to cross the barrier. The concept of a channel demands that the wave function can be written as a product of an internal part and a one-dimensional function of some collective coordinates. This leads to the Born-Oppenheimer approximation in molecular dynamics, which is well justified by the large separation of electron and nucleus mass scales. But this not at all the case in nuclear physics, where both single-particle and collective structures are on the same 1 MeV energy scale. The tools for microscopic calculations in that framework are just not up to the task.

The present state of the art for modeling induced fission may be seen in the recent calculations of the Los Alamos group [1, 2]. The basic calculational framework was laid out by Bjørnholm and Lynn

[a]e-mail: bertsch@uw.edu

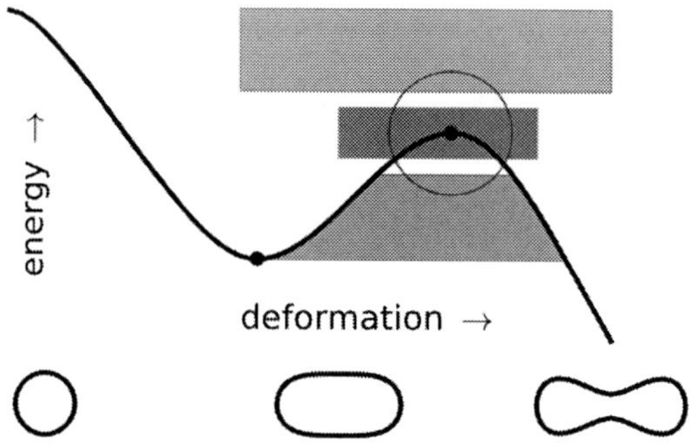

Figure 1. Schematic view of the fission landscape with a simple barrier. The colored regions show the areas of different dynamics: diffusive in pink, tunneling in green, and barrier-top in blue.

[3], who generalized the Bohr-Wheeler theory to deal with a fission landscape having two barriers. The degrees of freedom are depicted in Fig. 2. They are the state densities in wells I and II, indicated by dots in the Figure, and the channels at barriers A and B, indicated by the horizontal lines. The final formula for the fission decay rate is a generalization of the Bohr-Wheeler decay rate formula

$$W = \frac{1}{2\pi\hbar\rho}\sum_c T_c \tag{1}$$

which only has only one compound nucleus density ρ and only transmission coefficients T_c for a single barrier.

To connect this to a microscopic theory based on a nucleonic Hamiltonian, we need to know how to calculate the transmission coefficients T_c between states and channels, and also how calculate mixing between channels. The microscopic framework used up to now is the Generator Coordinate Method (GCM). Unfortunately, it does not have a natural place for the ordinary states and the needed connection to channels seems difficult to incorporate. For the mixing of channels, there is a heroic attempt by Gogny's group to set up a GCM framework for this purpose [5], but it appears to us to be very complicated to carry out in practice.

In our view, the problem is the channel picture itself. Channels are useful if there is clear separation between collective and intrinsic degrees of freedom. That is O.K. for the Born-Oppenheimer framework for molecular physics, but the separation of collective and intrinsic energy scales is simply not present in nuclear physics. Another problem is the non-orthogonality of the channel wave functions in the GCM. In particular, the over-completeness of the basis often gives rise to numerical instabilities that can only be suppressed by ad hoc truncations.

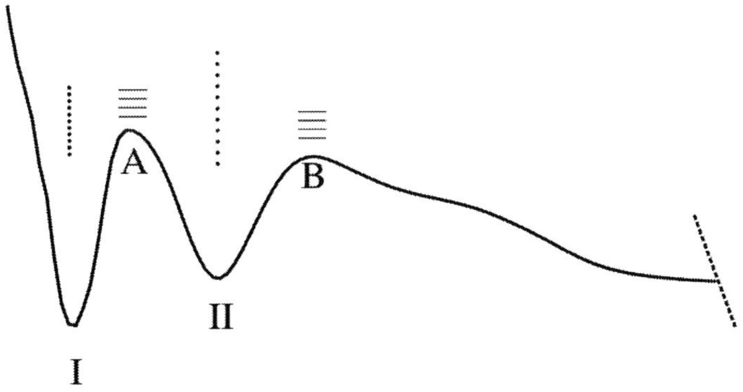

Figure 2. Fission potential energy surface (PES) for ^{236}U, taken from Fig. 7 of Ref. [4]. Dashed line shows energy of separated fission fragments at the same quadrupole deformation coordinate. Superimposed on the PES are the elements needed for the generalized Bohr-Wheeler description of the dynamics. Type I and II state are indicated by dots; channels bridging the barriers are indicated by horizontal lines. In practice, the channels are only needed at the barrier tops.

From the phenomenological side, the dynamics of induced fission may be much closer to a diffusive limit than an inertial limit. One sees from many studies including one presented in this Workshop [6] and a recent one on mass distributions [7] that statistical and diffusive models can describe many features of the fission final state. Furthermore, we also heard in the Workshop a report on a microscopic dynamical model that produced a fission time so large that inertial motion would be highly over-damped [8].

2 Dynamics in a discrete basis

There is an alternative. That is to construct a discrete basis for the Hamiltonian, avoiding completely the introduction of continuous collective degrees of freedom. In this section we show how various limits of the dynamics can emerge, deferring to the next section how we envisage constructing the basis. The basis will be composed of mean-field configurations, allowing one to calculate with well-known methods the matrix elements of Hamiltonians of the usual microscopic form,

$$H = \sum \varepsilon_i a_i^\dagger a_i + \frac{1}{4} \sum v_{ij,kl} a_i^\dagger a_j^\dagger a_l a_k. \tag{2}$$

Besides the close connection to configuration-interaction (CI) computational methods that have been so successful with the nuclear shell model, the discrete-basis framework provides a conceptual bridge to quantum transport in condensed matter physics.

The states need to be characterized by energy and some measure of the shape; these may be determined by the expectation values of the Hamiltonian and some single-particle operator such as the

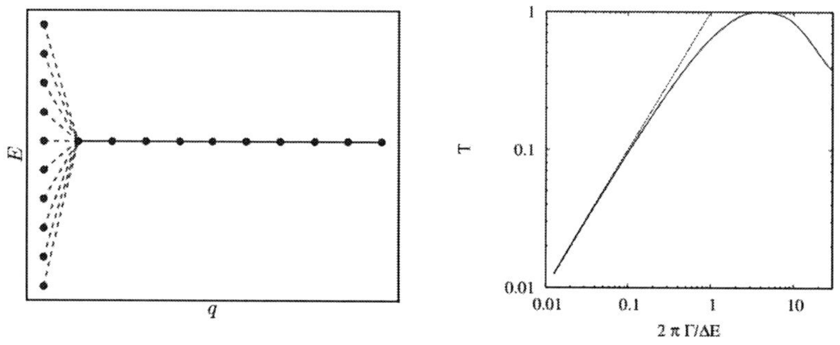

Figure 3. Discrete-state modeling of the compound nucleus coupling to a neutron decay channel. On the left are shown the states as circles and their couplings as lines. On the right is shown the calculated transmission coefficient T compared to the weak-absorption limit $T = 4\pi^2 v_c^2 \rho^2$.

quadrupole moment. It is then easy to set up conditions on Hamiltonian to realize different possibilities for the dynamics.

The first limit is that of a compound nucleus decaying into a single channel. The basis states are set up as shown in Fig. 3 on the left. The compound nucleus states are shown as the tower of states indicated as red points. The channel states are the regularly spaced blue point on the left. The interactions between the compound states and the first channel state are shown as the dashed red lines. They would be taken as Porter-Thomas distributed about some rms average value $\langle v_c^2 \rangle$. The interactions between channel states are shown as the solid blue lines. They would all be equal. Then the entire physics of scattering theory and compound nucleus resonances can be displayed by using standard methods. There has been recent interest in couplings to the continuum that violate Porter-Thomas statistics [9]; perhaps this model would be useful to explore such possibilities.

Another limit is one that approximates diffusive dynamics. By this we mean that wave packets decay according to the diffusion equation with some diffusion coefficient $D(q)$ according to the equation

$$\frac{\partial P}{\partial t} = \frac{\partial}{\partial q} D(q) \frac{\partial P}{\partial q} \tag{3}$$

where q is a shape index. To realize this limit in a discrete basis, we distribute the states in layers according to the shape parameter q and in energy according to Gaussian random matrix ensemble. The layout of states in the (E, q) plane is shown in Fig. 4 on the left. We also assume a Porter-Thomas distribution of interaction matrix elements, limited to states on next-neighbor layers. Evolving the dynamics by the time-dependent Schrödinger, we find probability distributions on the different layers at later times. The results are shown on the right-hand panel of the Figure at some fixed time. The points are an average of different Hamiltonians in the ensemble, with the error bars showing the r.m.s. fluctuations. The connection to macroscopic theory comes if we can relate the statistical properties of the Hamiltonian ensemble to a diffusion coefficent. This is given by the formula [10]

$$D = 2\pi\rho(E)\overline{(q_\alpha - q_\beta)^2 \langle \alpha | v | \beta \rangle^2} \tag{4}$$

where ρ is the level density in a single layer and the bar indicates an average over interactions between states in different layers. The red curve is the corresponding distribution predicted by the diffusion

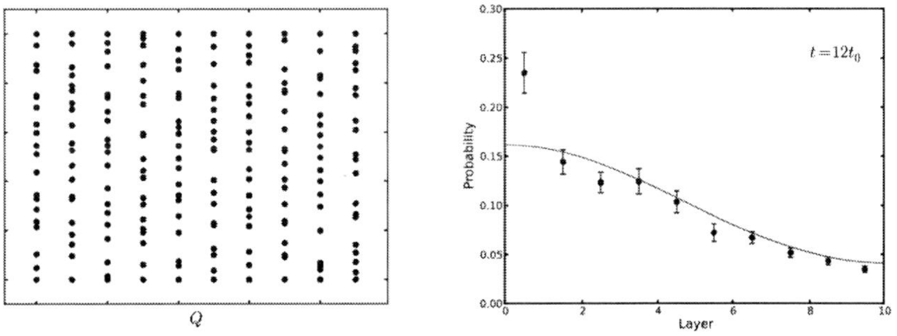

Figure 4. Discrete basis modeling of shape diffusion in a region of high level density. The states are grouped in vertical layers by their deformation, and it is assumed that only nearby layers are coupled by the interaction. The right hand side shows the probability distribution after some time interval, starting from an initial wave function localized on the first layer.

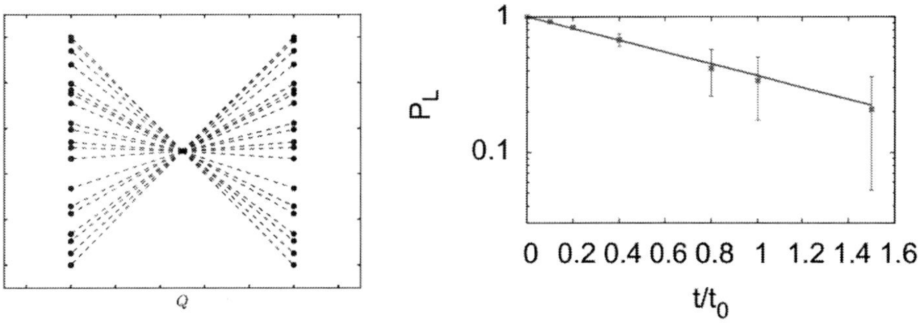

Figure 5. Left-hand panel: discrete state representation of resonance-mediated conductances. The states of the compound nucleus are the red dots, the barrier-top resonance is the state in the middle, and the post-barrier states are in blue. Right-hand panel: probability of remaining in the compound nucleus as a function of time.

equation, the D from Eq. (4). One sees that the classical physics is quite accurate except for some trapping in first layer.

Another interesting limit is what may be called resonance-mediated conductance. It is inspired as a very simplified model for fission dynamics at the barrier top as well as for the conduction of electrons from one conductor to another through a quantum dot. The layout of states is shown in Fig. 5. The transport takes place through the single state in the middle. There is a lot of physics that can be explored with this model [11]. For this talk, we just mention one limit. With suitable restrictions on level densities and interaction strengths, the rate at which the system goes from one side to the other

Table 1. Orbital filling for ^{16}O in the spherical shell model. The columns separate the K quantum number, and the rows separate nucleon type and parity.

K	1/2	3/2	5/2
p^+	2	0	0
p^-	4	2	0
n^+	2	0	0
n^-	4	2	0

can be calculated by Eq. (1) with the transmission coefficient given by the formula [12]

$$T_r = \frac{\Gamma_R \Gamma_L}{E_b^2 + (\Gamma_R + \Gamma_L)^2/4}. \tag{5}$$

Here the decay widths of the resonance to the left and to the right (Γ_L and Γ_R) are calculated by Fermi's Golden Rule. Again, we examine an ensemble of Hamiltonians using random matrix energy spacings, but now with fixed couplings to the resonance to satisfy Eq. (5) with $T_r = 1$. The resulting time-dependent probability for staying on the left is shown in the panel on the right, with error bars showing the fluctuations in the ensemble. The solid line is an exponential decay with the decay rate determined by Eq. (1) with the calculated transmission factor. We see that the average decay rate is well reproduced, but there is considerable fluctuation due to the level-density fluctuations in the random-matrix ensemble.

3 An implementation: the axial basis

A discrete basis should be composed of orthogonal states, and it should be extendible in principle to a complete basis. A good candidate to satisfy these requirements is what we shall call the axial basis. As in traditional theory, the underlying framework is self-consistent mean-field theory. But instead of adding a generator-coordinate field to distinguish states, we use the partition of the particle numbers to orbitals of different K quantum numbers to make a first landscape of the PES. To make this clear, let us consider a very simple example, the shell-model ground state of ^{16}O. The filled shells are $s_{1/2}, p_{3/2}, p_{1/2}$. The K quantum number can be taken as the azimuthal angular momenta j_z of the shells. This produce the fillings shown in Table I, amounting to a partition of the 16 nucleons into 6 K^π sets.

To see how the scheme might work, we examine the partitions for a toy model, the fission of ^{32}S into two ^{16}O nuclei with a Hamiltonian tuned to allow the decay. This model was proposed in Ref. [13]. Tables of the partitions are shown in Fig. 6. The leftmost table gives the K^π partition for ^{32}S as a spherical shell-model configuration. The filled shells are $0s_{1/2}, p_{3/2}, p_{1/2}, d_{5/2}, 1s_{1/2}$. Notice that the nonzero K^π orbitals go up to the $5/2^+$ associated with the $d_{5/2}$ shell. The partition table for ^{16}O+^{16}O can be constructed by a very simple argument. The orbitals of the combined system have the same K quantum numbers as the individual nuclei if the fission is along the z-axis. Each orbital has an extra two-fold degeneracy because of its presence in both O nuclei. The plus and minus linear combinations will have good parity. Thus, each orbital in the individual O gives rise to two orbitals with the same K but opposite parities in the combined system. The resulting partitions are shown on the right hand side of the Figure. Comparing the initial and final partitions, one sees that the decay requires four particles to be moved from $K^\pi = 5/2^+$ orbitals to $K^\pi = 1/2^-$ orbitals. The various configurations involved in the decay are depicted in the energy-versus-shape graph in the bottom panel of the figure. On the left is the ^{32}S ground state. The residual interaction connects it (dotted line) with

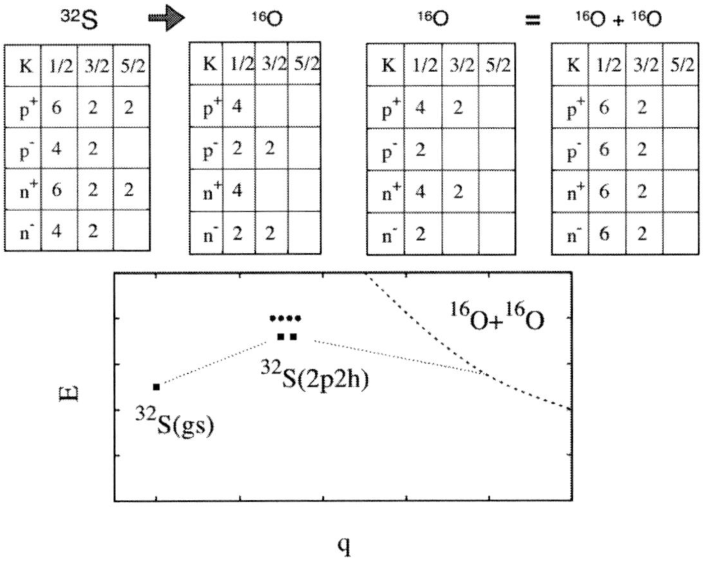

Figure 6. Fission of ^{32}S in a toy model. See text for explanation.

several 2-particle 2-hole states as depicted in the middle. Finally, a second application of the residual interaction connects the intermediate configurations to the final partition. We have indicated the last configuration with a dashed line because it is not clear whether it would have a stable minimum in the Hartree-Fock minimization. It might be that the energy decreases continuously as the distance between centers increases. At that point we cannot avoid dealing with the problem of coupling to the continuum.

The next example exhibits the first steps to building a basis for treating the dynamics in the configuration-interaction framework. We start with a specific Hamiltonian of the usual shell-model form Eq. (2). The goal is to examine all possible partitions and determine the self-consistent minimum of each one. These states will span the range of deformations permitted by the shell-model space. Additional excited states in the same partitions may be constructed as particle-hole excitations with the same mean field as the ground state in that partitions. We have written a code to carry out the fixed-partition ground state minimization, given H of the form Eq. (2). The example we show is the nucleus ^{162}Dy, see Fig. 7. Here we took a Hamiltonian constructed for use in the Shell Model Monte Carlo treatment of level densities [14]. The filled circle shows the ground state of the axial Hartree-Fock approximation. This state is moderately deformed, as is expected for midrange lanthanide nuclei. The open circles show the lowest excited states that arise by changing the partition by a single pair jump. One sees a significant gap in the spectrum, and also that the quadrupole deformations are not very different from that of the ground state. The partial localization in deformation is important to obtain eventually a collective dynamics for a deformation coordinate.

Now we come to the actinide nuclei and realistic fission paths. We start with the two ground states I and II in the PES minima. Fig. 8 shows the difference in the partitions of the two states, as

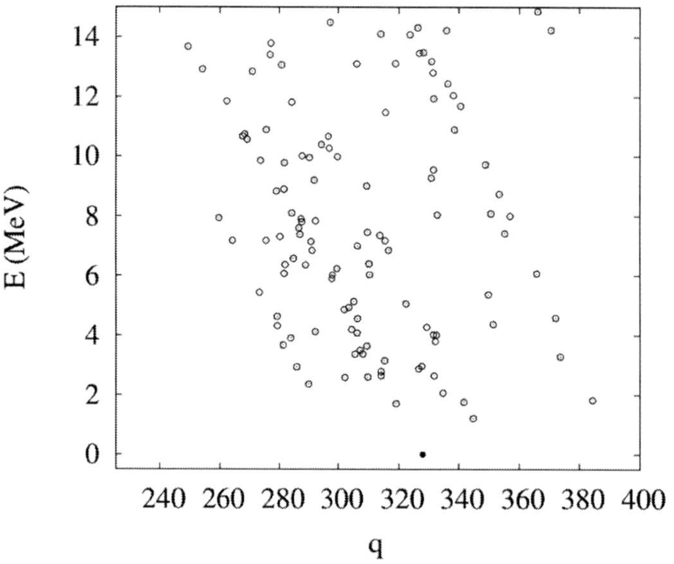

Figure 7. Ground state configuration of ^{162}Dy (filled circle) and excitded state configurations reachable by a single pair jump (open circles).

calculated by Möller in the FRLD model [15, 19]. One sees that particles in higher-K orbitals are moved to lower K in going from the first minimum to the second. The number of pairs to be moved totals 6. To explore the partition space between the two minima, we should have a self-consistent mean field code that allows one to specify the partition, with no other constraints. Lacking that, we can begin to map the distance between configurations with existing codes, using the jump number as the distance measure. We did this for ^{236}U using the HFBaxial code written by Luis Robledo [16]. The configurations we examined are the I and II ground states, the A and B barriers, and a number of configurations beyond the second barrier up to Q = 140 bn. At the largest deformation, the separated fragments have the same energy as the elongated fissioning nucleus. The results are shown in Fig. 9. First note that the shortest path from the I ground state to II is not over the saddle. Past the second saddle, the configurations are labeled by their quadrupole moment. In this region, the system is asymmetric and parity is not a good quantum quantum number. However, for the figure we have reported jump numbers corresponding to symmetric fission. It will be interesting to see the effect on the jump numbers when the parity constraint is released. If they remain substantial, it shows that the saddle-to-scission path requires dynamics beyond that contained in the time-dependent Hartree-Fock approximation.

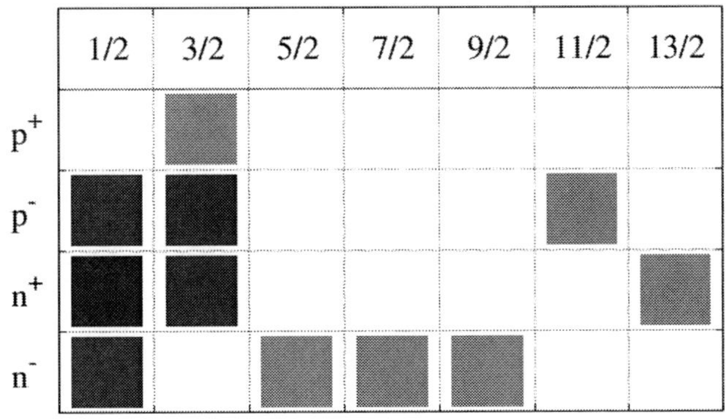

Figure 8. Difference of partitions in ^{236}U for the class I and class II ground states in the FRLD model [19]. The class I partition has one orbital pair more than class II for K^π indicated in green. The red and the blue show K^π's where there is respectively one or two orbital pairs less.

4 The Hamiltonian

While there are many energy density functionals available for self-consistent mean-field theory, there does not yet exist a fully self-consistent effective Hamiltonian for nuclear structure. Still, it should be useful to explore the dynamics that follow from simplified residual interactions.

4.1 Interaction between configurations

The two-particle interaction between configurations can be written schematically as

$$\langle \alpha | v | \beta \rangle = \sum \langle pp | v | pp \rangle \det |\langle \phi_i^\alpha | \phi_j^\beta \rangle| \tag{6}$$

where $\langle pp|v|pp\rangle$ is a two-body matrix element and $\det |\langle \phi_i^\alpha | \phi_j^\beta \rangle|$ is the overlap of spectator orbitals.

From the study of the dynamics as seen for example in Fig. 4, it is clear that an important parameter to approach the diffusion limit is the ratio of the mean-square average of the interaction between configurations $\overline{\langle i|v|j\rangle^2}$ to the local level density $\rho(q)$. If this ratio is large, each state decays exponentially into nearby states and the dynamics will be diffusive. It has been shown in a simple model that this ratio increases with excitation energy [10], so the diffusion limit will be appropriate at high enough excitation energy. In that model, the average interaction strength scaled with excitation energy E as

$$\overline{\langle \alpha | v | \beta \rangle^2} \sim E^{3/2} / \rho(E). \tag{7}$$

We also have analyzed this with a somewhat more sophisticated model, with results shown in Fig. 10. The vertical scale is proportional to the product $\overline{\langle \alpha | v | \beta \rangle^2} \rho(E)$. It is plotted with respect to excitation energy, measured in units of the single-particle energy spacing. The dotted line is a power-law fit to the ratios; it comes out very close to the predicted 3/2-power dependence. Details will be given elsewhere.

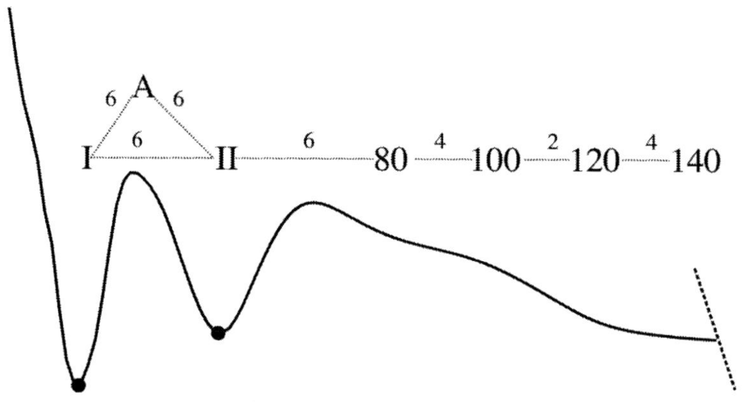

Figure 9. Schematic view of the fission barrier as in Fig. 2. The states I,II,A and B are unconstrained while the states by a quadrupole moment $Q = 80, 100, 120, 140$ are constrained by that operator. The distance between states is shown by the number of pair jumps to get from one to another.

The matrix elements of the two-body interaction are also affected by the changes in the spectator orbitals due to the different mean fields of the partitions. This cuts down the interactions by the determinant of spectator orbital overlaps indicated in Eq. 6. This is a good occasion to mention the early paper by Arima and Yoshida giving analytic expressions for this determinant in the harmonic oscillator basis [17].

So where do we hope to go from here? It seems feasible in the near term to go beyond simple models to sample the CI interactions arising from realistic effective Hamiltonians. We can then use the derived statistical properties to determine where the diffusive limit can be applied.

Acknowledgments

We wish to thank A. Bulgac for discussions, L. Robledo for the use of the HFBaxial code to calculate partitions of ^{236}U, and P. Möller for providing us with the orbital fillings for ^{236}U in the FRLD model. We also acknowledge financial support from the Institute for Nuclear Theory (GFB) and from the DOE under Grant FG02-00ER41132 (JMM).

References

[1] O. Bouland, J.E. Lynn, and P. Talou, Phys. Rev. C **88** 054612 (2013).
[2] O. Bouland, J.E. Lynn, and P. Talou, Nucl. Data Sheets **118** 211 (2014).
[3] S. Bjørnholm and J.E. Lynn, Rev. Mod. Phys. **52** 725 (1980).
[4] R. Rodriguez-Guzman and L.M. Robledo, Phys. Rev. C **89** 054310 (2014).
[5] R. Bernard, H. Goutte, D. Gogny and W. Younes, Phys. Rev. C **84** 044308 (2011)

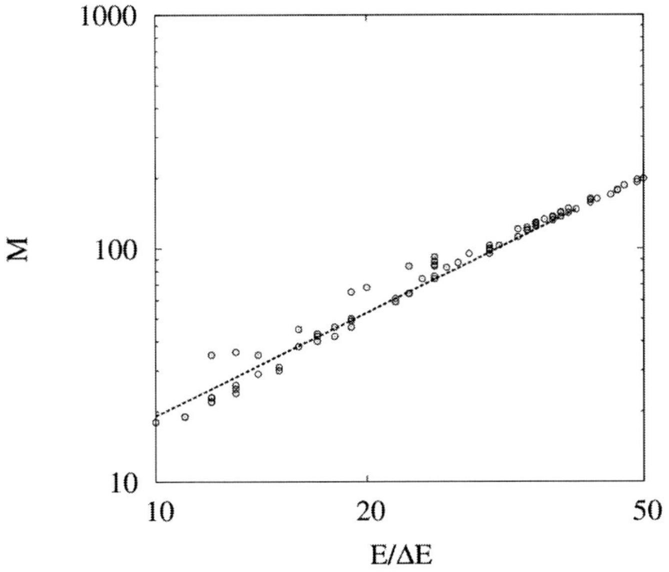

Figure 10. Relative interaction strengths to mix configurations at the same energy, as a function of excitation energy. See text for explanation.

[6] J.-L. Sida, contribution to this workshop and J.-F. Lemaître, et al., Phys. Rev. C **92** 034617 (2015).
[7] J. Randrup and P. Möller, Phys. Rev. Lett. **106** 132503 (2011).
[8] A. Bulgac, contribution to this workshop.
[9] S. Mizutori, contribution to this workshop.
[10] B.W. Bush, G.F. Bertsch, and B.A. Brown, Phys. Rev. Lett. C **45** 1709. (1992).
[11] G.F. Bertsch, arXiv:1407.1899.pdf (2014).
[12] Y. Alhassid, Rev. Mod. Phys. **72** 895 (2000).
[13] J.W. Negele, Nucl. Phys. A **502** 371 (1989).
[14] Y. Alhassid, L. Fang and H. Nakada, Phys. Rev. Lett. **101** 082501 (2008).
[15] P. Möller, et al., Atomic Data and Nuclear Data Tables **59** 185 (1995).
[16] L.M. Robledo, private communication. See also Ref. [18].
[17] A. Arima and S. Yoshida, Nucl Phys. **12** 139 (1959).
[18] M. Warda, J.L. Egido, L.M. Robledo, and K. Pomorski, Phys. Rev. C **66**, 014310 (2002).
[19] P. Möller, private communication.

EPJ Web of Conferences **122**, 01002 (2016) DOI: 10.1051/epjconf/201612201002
*CNR*15*

The scission point configuration of fissioning nuclei

Fedir Ivanyuk[1,2,a]

[1] *Institute for Nuclear Research, Kiev, Ukraine*
[2] *Research Laboratory for Nuclear Reactors, Tokyo Institute of Technology, Tokyo, Japan*

Abstract. We define the optimal shape which fissioning nuclei attain just before the scission and calculate the deformation energy as function of the mass asymmetry at the scission point. The calculated deformation energy is used in quasi-static approximation for the estimation of mass distribution, total kinetic and excitation energy of fission fragments, and the total number of prompt neutrons. The calculated results reproduce rather well the experimental data on the position of the peaks in the mass distribution of fission fragments, the total kinetic and excitation energy of fission fragments. The calculated value of neutron multiplicity is somewhat larger than experimental results. The saw-tooth structure of neutron multiplicity is qualitatively reproduced.

1 Introduction

In the theory of nuclear fission the quasistatic quantities like the potential energy surface, the ground state energy and deformation, the fission barrier height are commonly calculated within the macroscopic-microscopic method [1, 2]. In this method the total energy of the fissioning nucleus consists of two parts, macroscopic and microscopic. Both parts are calculated at a fixed shape of the nuclear surface. In the past a lot of parameterizations of the shape of nuclear surface were proposed and used. A good choice of the shape parameterization is often a key to the success of the theory. Usually, one relies on the physical intuition for the choice of the shape parameterization.

A method to define the shape of the nuclear surface which does not rely on any shape parameterization was proposed by V. Strutinsky in Ref. [3]. In this approach the shape of an axial, left-right symmetric nucleus was defined by the minimization of the liquid-drop energy under the constraints that fix the elongation of the drop and keep the volume constant.

Recently the method was further developed by incorporating the axial and left-right asymmetry, and account of the influence of shell effects on the shape of nuclear surface, see [4] and references therein.

The important result of the Strutinsky procedure [3] is the possibility to definite in a formal way the scission point as the maximal elongation beyond which the solution of the minimization procedure in form of a compact shape does not exist.

Having at one's disposal the shape and the deformation energy at the scission point and assuming the statistical distribution for the population of points in space of deformation parameters we have evaluated the measurables of the fission experiments: the mass distribution, total kinetic and excitation energy of fission fragments, the multiplicity of prompt neutrons.

[a] e-mail: ivanyuk@kinr.kiev.ua

2 The optimal shapes of fissioning nuclei

The shape of an axially symmetric nucleus can be defined by rotation of some profile function $\rho(z)$ around the z-axis. It was suggested in [3] to define $\rho(z)$ looking for the minimum of the liquid-drop energy $E_{\rm LD} = E_{\rm surf} + E_{\rm Coul}$ under the constraint that the volume V and the elongation R_{12} are fixed,

$$\frac{\delta}{\delta\rho}(E_{\rm LD} - \lambda_1 V - \lambda_2 R_{12}) = 0\,, \quad \text{with} \quad R_{12} = \frac{2\pi}{V}\int_{z_1}^{z_2} \rho^2(z)|z|dz\,, \quad V = \pi\int_{z_1}^{z_2} \rho^2(z)dz\,. \tag{1}$$

The elongation parameter R_{12} was chosen in [3] as the distance between the centers of mass of the left and right parts of the nucleus. In (1) λ_1 and λ_2 are the corresponding Lagrange multipliers.

The minimization of $E_{\rm LD} - \lambda_1 V - \lambda_2 R_{12}$ with respect to the profile function $\rho(z)$ leads to an integro-differential equation for $\rho(z)$

$$\rho(z)\frac{\partial^2\rho}{\partial z^2} = 1 + \left(\frac{\partial\rho}{\partial z}\right)^2 - \rho(z)[\lambda_1 + \lambda_2|z| - 10x_{\rm LD}\Phi(z,\rho)]\left[1 + \left(\frac{\partial\rho}{\partial z}\right)^2\right]^{\frac{3}{2}}\,. \tag{2}$$

Here $\Phi(z, \rho(z))$ is the Coulomb potential at the nuclear surface and $x_{\rm LD}$ is the fissility parameter of the liquid drop [5] ,

$$x_{\rm LD} \equiv \frac{E^{(0)}_{\rm Coul}}{2E^{(0)}_{\rm surf}} = \frac{3}{10}\frac{Z^2e^2}{4\pi R_0^3\sigma} \approx \frac{Z^2}{49A}\,, \tag{3}$$

where σ is the surface tension coefficient.

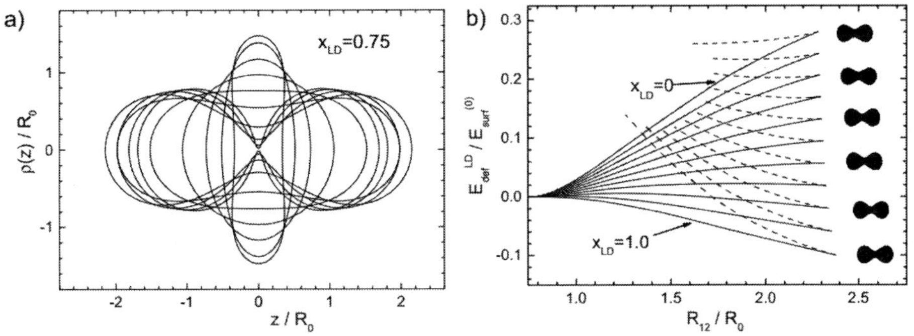

Figure 1. (a) The solutions of Eq. (2) corresponding to different values of Lagrange multiplier λ_2 which fixes the elongation R_{12}; (b) Liquid-drop deformation energy (4) as a function of the parameter R_{12} for different values of fissility parameter $x_{\rm LD}$.

By solving Eq. (2) one obtains the profile function $\rho(z)$ for given x_{LD} and λ_2 (λ_1 is fixed by the volume conservation condition). Varying λ_2 one gets a full variety of shapes from the disk, even with central depression, to the two touching spheres, see Fig. 1a.

The liquid drop deformation energy $E^{LD}_{\rm def} = E_{\rm LD} - E^{(0)}_{\rm LD}$ (in units of the surface energy for a spherical shape)

$$E^{LD}_{\rm def}/E^{(0)}_{\rm surf} = B_{\rm surf} - 1 + 2x_{\rm LD}(B_{\rm Coul} - 1)\,, \tag{4}$$

calculated for the shapes shown in Fig.1a, is presented in Fig.1b. In (4) $B_{\rm Coul} \equiv E_{\rm Coul}/E^{(0)}_{\rm Coul}$, $B_{\rm surf} \equiv E_{\rm surf}/E^{(0)}_{\rm surf}$, where an index $^{(0)}$ refers to the spherical shape.

From Fig.1b one can see that the elongation R_{12} of the shapes shown in these figures is limited by some maximal value $R^{(crit)}_{12}$. Above this deformation mono-nuclear shapes do not exist. This critical deformation was interpreted in [3] as the scission point. Note that, at scission the neck radius is still rather large: at the critical deformation the neck radius is approximately equal to $(0.25 - 0.30)R_0$ for a fissility parameter in the range $0.4 \leq x_{LD} \leq 0.9$

Another peculiarity of Fig. 1b is the upper branch of the deformation energy at large deformation. Along this branch the neck of the drop becomes smaller and smaller until the shape turns into two touching spheres. Both branches are solutions of Eq.(2). It turns out, that the upper branch of $E^{LD}_{\rm def}$ corresponds not to the minimum but to the maximum of the liquid drop energy. Thus, it represents the ridge of the potential energy surface between the fission and fusion valleys.

3 The mass-asymmetric shapes

For the description of fission process the account of mass asymmetry is essential. In Fig. 2 we show the deformation energy of ^{236}U calculated within the two-center shell model. The shape in the two-center shell model is fixed by five deformation parameters. This parametrization allows for the description of both compact and separated shapes. The deformation energy in Fig. 2 was minimized with respect all five deformation parameters keeping fixed the mass asymmetry and the distance R_{12} between the centers of mass of left and right parts of nucleus. One can see, that at some value $R^{(crit)}_{12}$ the potential energy drops suddenly to the potential energy of separated fragments, which is much smaller than the potential energy of compact shape.

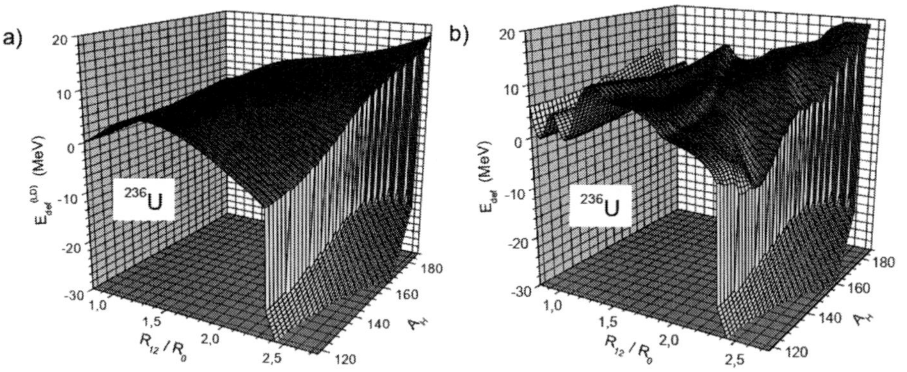

Figure 2. (color online) (a) The liquid-drop deformation energy (4) of ^{236}U as function of the elongation R_{12} and the mass number A_H of heavy fragment calculated within the two-center shell model; (b) The total (liquid-drop plus shell correction) deformation energy of ^{236}U .

In order to give the model-independent definition of $R^{(crit)}_{12}$ one has to generalize the optimal shape approach of [3] to the case of mass-asymmetric shapes. For this aim one has to include into Eq.(1) one more constraint that fix the mass asymmetry δ of the drop. The mass asymmetry δ is commonly defined by the difference of masses M_L and M_R to the left and right of some point z^*. In case that the drop has a neck, z^* coincides with the position of the neck, $z^* = z_{neck}$. By z_{neck} we mean here the point where $\rho(z)$ has a minimum. For the pear-like shape the neck does not exist and z^* could be defined in a different way, see [6]. In the present work we are interested in the scission point configuration for which the neck is well defined. Then Eq. (1) turns into

$$\frac{\delta}{\delta\rho}(E_{\rm LD} - \lambda_1 V - \lambda_2 R_{12} - \lambda_3 \delta) = 0 \quad \text{with} \quad \delta \equiv \frac{M_L - M_R}{M_L + M_R} = \frac{\pi}{V}\int \text{Sign}(z - z^*)\rho^2(z)dz\,. \quad (5)$$

The Euler-Lagrange equation for the variational problem (5) can be solved in the same way as Eq. (2). Some examples of the shapes at the critical point $R_{12}^{(crit)}$ for a few values of the mass asymmetry are shown in Fig. 3a.

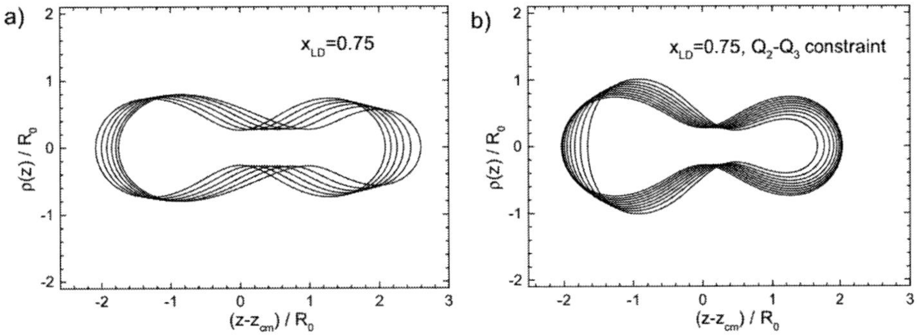

Figure 3. (a) The solutions of Eq. (5) at the maximal elongation $R_{12}^{(sci.)}$ for a few values of the mass asymmetry $\delta = 0, 0.1, ...0.5$; (b) The solutions of Eq. (7) at the maximal elongation $R_{12}^{(crit)}$ for few values of the mass asymmetry $\delta = 0, 0.1, ...0.8$.

Besides R_{12} and δ one could try another popular pair of constraints which are often used in the constrained Hartee-Fock calculations, namely, the quadrupole and octupole moments,

$$Q_2 = \frac{2}{V}\int dVr^2 P_2(\cos\theta)\,, \quad Q_3 = \frac{1}{V}\int dVr^3 P_3(\cos\theta)\,. \quad (6)$$

The moments Q_2 and Q_3 can be defined independently of whether the neck exists or not. The use of quadrupole and octupole moments as constraints leads to the following variational problem

$$\frac{\delta}{\delta\rho}(E_{\rm LD} - \lambda_1 V - \lambda_2 Q_2 - \lambda_3 Q_3) = 0\,. \quad (7)$$

The Euler-Lagrange equation for the problem (7) can be solved in the same way as Eq. (2). The examples of solution of Eq. (7) for few value of the mass asymmetry are shown in the right part of Fig. 3.

Comparing the shapes at the scission point calculated by Eq. (5) and Eq. (7) for the same mass-asymmetry one can see that these two sets of shapes are rather different. The Q_2, Q_3 restrictions lead to scission shapes which are considerably "shorter" as compared with the scission shapes defined with R_{12}, δ restrictions. For an accurate description of the mass-asymmetric nuclear shape at the scission point it is necessary to construct an interpolation between the two sets of definitions for the elongation and mass asymmetry which are applied successfully at small deformations or for separated fragments. Such an interpolation can be achieved by the replacement of the $|z|$ and Sign z functions which appear in (1) and (5) by the *smoothed* absolute value and Sign functions

$$|z| \Longrightarrow f_2(z) \equiv \sqrt{z^2 + (\Delta z)^2}\,, \qquad \text{Sign}\, z \Longrightarrow f_3(z) \equiv \frac{z}{\sqrt{z^2 + (\Delta z)^2}}. \tag{8}$$

The expressions for the elongation and mass asymmetry change then into

$$R_{12} \Longrightarrow \widetilde{R}_{12} \equiv \frac{\pi}{V}\int_{z_1}^{z_2} \sqrt{(z - z^*)^2 + (\Delta z)^2}\rho^2(z)dz\,, \quad \delta \Longrightarrow \widetilde{\delta} \equiv \frac{\pi}{V}\int_{z_1}^{z_2} \frac{(z - z^*)\rho^2(z)}{\sqrt{(z - z^*)^2 + (\Delta z)^2}}dz. \tag{9}$$

The replacement (8) contains an additional parameter - the smoothing width Δz. In principle, one can consider it as an additional collective parameter which has to be taken into account in the dynamical calculations. In the quasi-static limit one could expect that the value of Δz is close to the neck radius in the scission region. In the calculations reported below Δz was chosen as $\Delta z = 0.25R_0$. The calculated results do not change much if Δz varies within the limits $0.1R_0 \lesssim \Delta z \lesssim 0.5R_0$.

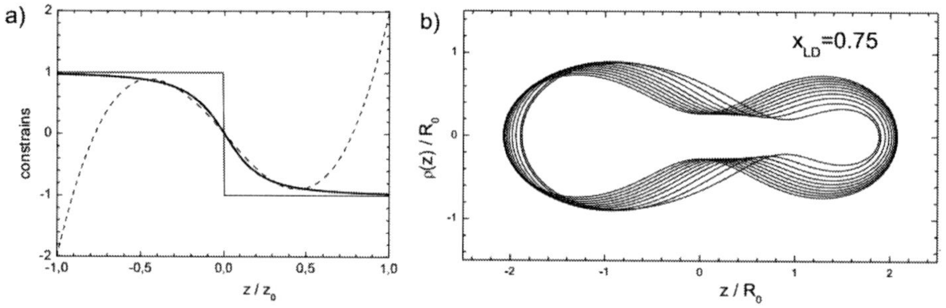

Figure 4. (a) Illustration of the constraints δ (solid), $P_3(z/z_0)$(dash) and smoothed Sign-function (thick solid line); (b) The solutions of Eq. (10) ($\Delta z = 0.25R_0$) at the maximal elongation $R_{12}^{(sci.)}$ for few values of the mass asymmetry $\delta = 0, 0.1, ...0.6$.

The use of the constraints given in (9) leads to the following equation

$$\frac{\delta}{\delta\rho}\left[E_{\rm LD} - \lambda_1 V - \lambda_2\widetilde{R}_{12} - \lambda_3\widetilde{\delta}\right] = 0\,. \tag{10}$$

Eq.(10) can be solved in the same way as Eq.(2). Some examples of the shapes at the scission point $R_{12}^{(crit)}$ (maximal possible value of $\widetilde{R}_{12}$ at fixed λ_3) for a few values of the mass asymmetry $\widetilde{\delta}$ are shown in Fig. 4. The length of these shapes is in between those shown in Fig. 3. One notes that the neck radius at the scision point does not vary much with the mass asymmetry, $R_{neck} \approx 0.3R_0$.

4 The shell effects in the scission point configuration

The optimal shapes given by the solutions of (1, 5, 7, 10) correspond to the minimum of the *liquid drop* energy. However, the shape of fissioning nuclei depends not only on the liquid drop part of the deformation energy but also on the shell effects. It was shown in [8] that due to the shell structure, in the deformation energy of actinide nuclei there exists a deep minimum corresponding to the shape at which one part is almost spherical due to the large shell correction of the double magic fragment ^{132}Sn and another part is very elongated. Clearly, such a shape can not be obtained within the variational approach (10) where the deformation of the fragments is completely fixed by the deformation dependence of the liquid-drop energy.

In order to account for the influence of shell effects on the shape of the scissioning nucleus, it is necessary to include into the optimal shapes procedure more degrees of freedom. For this purpose it was proposed in [7] to include in the variational procedure another two constraints which allow to vary the quadrupole moments of left and right fragments,

$$\frac{\delta}{\delta\rho}[E_{\rm LD} - \lambda_1 V - \lambda_2 \widetilde{R}_{12} - \lambda_3 \widetilde{\delta} - \lambda_5 Q_{2L} - \lambda_6 Q_{2R}] = 0\,. \qquad (11)$$

In (11) Q_{2L} and Q_{2R} are the quadrupole moments of left and right parts of nucleus while the limit between *left* and *right* is defined at the point z_{neck} where the derivative $d\rho(z)/dz$ turns into zero. In what follows we are interested only in shapes at the scission point where the neck position is well defined. Solving (11) for large values of λ_5 and λ_6 one gets very elongated (λ_5 and λ_6 both large positive) or very short shapes (λ_5 and λ_6 both large negative) or shapes where one part is close to spherical and another very elongated (opposite sign of λ_5 and λ_6).

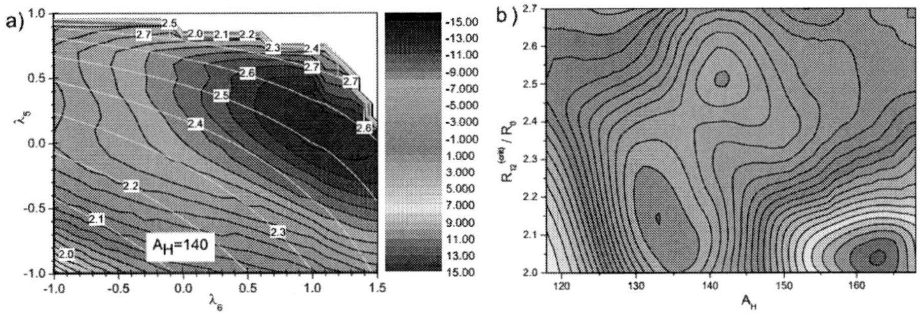

Figure 5. (color online) (a) The total deformation energy at the maximal elongation $R_{12} = R_{12}^{(crit)}$ for the asymmetry corresponding to heavy fragment mass $A_H = 140$ calculated with the solutions of Eq. (11) as function of Lagrange multipliers λ_5 and λ_6, see Eq. (11). The yellow contour lines show the value of $R_{12}^{(crit)}$; (b) The shell component of the scission point deformation energy of ^{236}U as function of the heavy fragment mass number A_H and the maximal elongation $R_{12}^{(crit)}$.

For each value of the mass asymmetry the scission point turns in this case into the "scission surface". In order to visualize this scission surface we reduce it to a "scission line". At each point on the scission surface one can calculate the deformations of left and right fragments Q_L and Q_R and by the critical elongation $R_{12}^{(crit)}$ (which is a function of Q_L and Q_R). For each value of $R_{12}^{(crit)}$ the *total deformation energy* (liquid drop part plus shell correction) was minimized with respect to λ_5 and λ_6, i.e., one fixes the additional parameters λ_5 and λ_6 by the requirement that the total energy at critical elongation $R_{12}^{(crit)}$ is minimal. In this way one gets the deformation energy at the scission point as function of the critical elongation $R_{12}^{(crit)}$. Figs. 5b and 6a show separately the shell correction and the liquid drop energy obtained as the result of such minimization.

One clearly notices two minima in Fig. 5b. One, at *short* scission shape at $A_H \approx 134$ is caused by the large shell correction for ^{132}Sn. The other minimum is caused by the "deformed shell" at $A_H \approx 142$. As will become clear below, these minima have a large influence on the fission fragment mass distribution and the total kinetic energy.

As one could expect, the liquid-drop energy grows with increasing mass asymmetry and decreases with growing critical deformation $R_{12}^{(crit)}$ as seen on Fig. 6a.

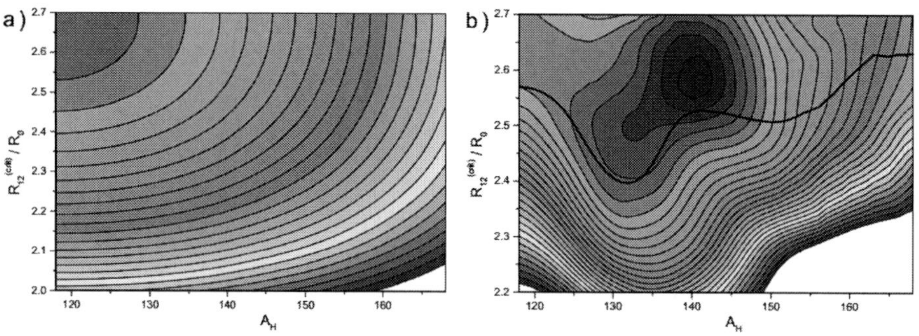

Figure 6. (color online) (a) The liquid drop component of the scission point deformation energy of ^{236}U as function of the heavy fragment mass number A_H and the maximal elongation $R_{12}^{(crit)}$; (b) The total energy (liquid drop plus shell correction) for ^{236}U at the scission point as function of the heavy fragment mass number A_H and the maximal elongation $R_{12}^{(crit)}$. The thick line shows the mean value (14) of $R_{12}^{(crit)}$.

The total scission-point deformation energy for ^{236}U is shown in Fig. 6b. Due to overlay of the liquid drop and shell components of the energy, the minimum around $A_H = 134$ is no longer seen. However, it influences the calculated results for the fission observables (see below). The dependence of the fission fragments kinetic energy on the mass asymmetry is defined mainly by this value.

5 The kinetic and excitation energy of fission fragments

Keeping in mind that fission is a *slow* process, one could assume that during the fission process the state of the fissioning nucleus is close to statistical equilibrium, i.e. each point q_i on the deformation energy surface is populated with a probability given by the canonical distribution,

$$P(\delta_i, q_i) = e^{-(E(\delta_i,q_i)-Z)/T_{coll}}, \qquad Z \equiv -T_{coll} \log \sum_i e^{-E(\delta_i,q_i)/T_{coll}}. \tag{12}$$

Here T_{coll} is a parameter characterizing the width of the distribution (12) in the space of deformation parameters. The energy $E(\delta_i, q_i)$ in (12) is the sum of the liquid-drop deformation energy (4) and of the shell correction δE, shown in Fig. 6b. The δ_i is the mass asymmetry of fissioning system and q_i are the rest of collective parameters (elongation R_{12}, and Lagrange multipliers λ_5 and λ_6).

The distribution (12) is a basic assumption of the scission-point model suggested in [9] and developed further in [10–12], see also [13]. The parameters of this model were fitted in [9] to reproduce the numerous experimental data. The T_{coll} was found to be close to 1 MeV. In the calculations shown below we use somewhat larger value T_{coll}=1.5 MeV.

The comparison of the calculated in [14] TKE for few fission reaction with the available experimental results is shown in Fig. 7. In these calculations we define TKE as the sum of the Coulomb interaction of spherical fragments immediately after scission and the prescission kinetic energy KE_{pre},

$$TKE =< E^{(int)}_{Coul}(\delta) > +KE_{pre}, \text{ with } < E^{(int)}_{Coul}(\delta) >\equiv \frac{1}{2}\sum_i \frac{Z_L Z_H e^2}{R^{(crit)}_{12}(\delta, q_i)} P(\delta, q_i) / \sum_i P(\delta, q_i), \quad (13)$$

where eZ_L and eZ_H are the charges of light and heavy fragments. The summation in the right part of (13) is carried out over λ_{5i} and λ_{6i}. One can see from (13) that the Coulomb interaction energy depends crucially on the mean value of the distance between centers of mass of fragments

$$< R^{(crit)}_{12}(\delta) >\equiv \sum_i R^{(crit)}_{12}(\delta, q_i) P(\delta, q_i) / \sum_i P(\delta, q_i). \quad (14)$$

The small $< R^{(crit)}_{12}(\delta) >$ leads to the large value of $< E^{(int)}_{Coul}(\delta) >$ and, on the contrary, the large $< R^{(crit)}_{12}(\delta) >$ leads to the small value of $< E^{(int)}_{Coul}(\delta) >$.

Like in [4] we define the KE_{pre} from the energy balance, the sum of the prescission kinetic energy and the potential energy just before the scission should be the same as the sum of the ground state energy of mother nucleus and the neutron binding energy:

$$E_{gs}(A_L + A_H) + B_n = E^{(jbs)} + KE_{pre}, \quad (15)$$

i.e. we assume the "complete acceleration": the energy difference between the saddle and scission turns into the kinetic energy of relative motion of fragments, no dissipation takes place. By "just before scission" $E^{(jbs)}$ we mean here the mean value of the total deformation energy

$$E^{(jbs)}(\delta) \equiv \sum_i E_{def}(\delta, \lambda_{5i}, \lambda_{6i}) P(\delta, \lambda_{5i}, \lambda_{6i}) / \sum_i P(\delta, \lambda_{5i}, \lambda_{6i}). \quad (16)$$

The opposite extreme case would be the assumption of overdamped motion: all the energy difference between the saddle and scission turns into the heat, no prescission kinetic energy.

The comparison of the calculated and measured values of TKE shown in Fig. 7 is in favor of "complete acceleration". Without contribution from KE_{pre} the TKE (dash curve in Fig. 7) would be too small.

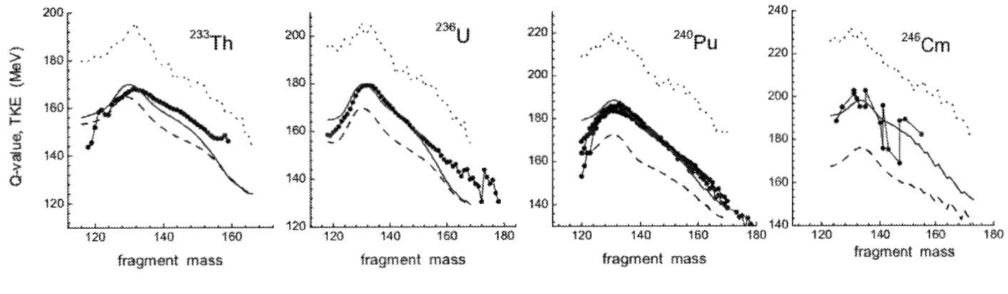

Figure 7. The total kinetic energy (13) calculated with (solid) and without (dash) account of fragments prescission kinetic energy. The experiments results (solid circles) are taken from [15–20]. The Q-values for fission of ^{232}Th, ^{235}U, ^{239}Pu and ^{245}Cm are shown by dotted lines.

Note, that the calculated TKE is rather close to experimental data. The position of maximum of TKE and the drop at symmetric splitting are well reproduced. This can be considered as a confirmation that the potential energy just before scission and the mean value of $< R_{12}^{(crit)} >$ are defined correctly.

The normalized mass distribution of the fission fragments $Y(\delta)$ can be expressed then in terms of the deformation energy at the critical deformation $R_{12}^{(crit)}$,

$$Y(\delta) = \sum_i P(\delta, R_{12}^{(crit)}, \lambda_{5i}, \lambda_{6i}) . \tag{17}$$

The summation in (17) is carried out over the Lagrange multipliers λ_5, λ_6 ($R_{12}^{(crit)}$ is a function of λ_5, λ_6).

The calculated mass distribution (12)-(17) of the fission fragments for the reaction $^{235}U + n_{th}$ is compared with the experimental data [21] in Fig. 8a. The calculated mass distribution is rather close to the experimental results. The presence and the position of two mass asymmetric peaks are well reproduced. At the symmetric splitting the calculated values are somewhat larger as compared with the experiment. The TKE is one of two parts of the total energy release Q, $TKE + TXE = Q$. The

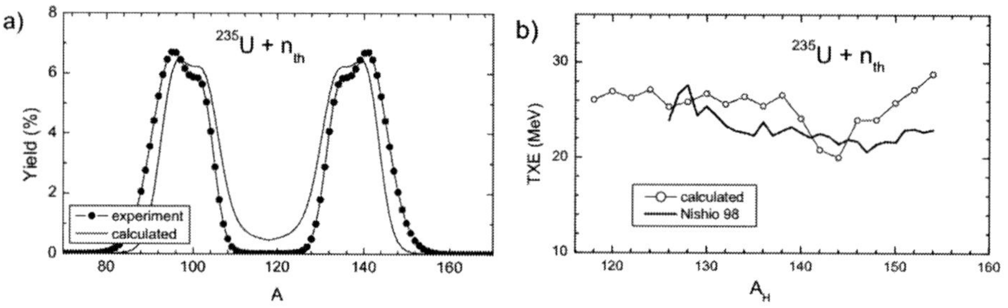

Figure 8. (a) The calculated (12, 17) and experimental [21] values of the mass distribution of fission fragments in reaction $^{235}U + n_{th}$; (b) The calculated (19) and experimental [22] values of the total excitation energy.

Q-value is the difference of the ground state energies

$$Q \equiv E_{gs}(A_L + A_H) + B_n - E_{gs}(A_L) - E_{gs}(A_H) \tag{18}$$

The Q-value does not depend on dynamics, it can be calculated within the macroscopic-microscopic method or taken from the existing databases. Since the experimental value of TKE is rather well reproduced by the present calculations, the calculated total excitation energy TXE

$$TXE = Q - TKE \tag{19}$$

is also quite accurate, see Fig. 8b.

6 The multiplicity of prompt neutrons

After the neck rupture the fragments are de-excited by the emission of neutrons and γ-rays. When the excitation energy is smaller than neutron separation energy S_n the γ-quanta are emitted. The energy available for γ-emission varies from S_n to zero. So, on average the excitation energy available for neutron emission is given by $< TXE > -S_n/2$. The average value of the total number $\bar{\nu}_{tot}$ of prompt neutrons can be estimated by the relation

$$\bar{\nu}_{tot} \approx < TXE > /\bar{S}_n - 1/2, \text{ with } \bar{S}_n = 5.7 \text{ MeV}. \tag{20}$$

In order to calculate the dependence of $\bar{\nu}$ on the fragments mass number A one needs to know how the excitation energy is shared between the fragments. Here we split the last stage of the fission into two steps. On first step the rapid neck rupture takes place. The nucleus turns into two fragments with the same mass asymmetry and the same distance between centers of mass as just before scission. The shape of the fragments immediately after scission is assumed to correspond to the minimum of the liquid drop energy. It was shown in [7] that the optimal shape of the fragments placed at the distance $R_{12}^{(crit)}$ is very close to two spheres. The contribution to energy from quadrupole and higher order deformation can be neglected. So, in present work we assume that immediately after scission the two spherical fragments are formed. The energy immediately after scission consists of the energy of light and heavy spherical fragments plus the energy of their Coulomb interaction,

$$E^{(ias)}(A) = E^{(sph)}(A_L) + E^{(sph)}(A_H) + E^{(int)}_{Coul}(R_{12}, \delta). \tag{21}$$

The primary excitation energy $E^{(ias)} - E^{(jbs)}$ is shared between the fragments. In the recent work [23] it was shown that difference of the excitation energy of light and heavy fragment is not so large, only few MeV. So, for simplicity we will assume here that the primary excitation energy of light and heavy fragments due to the neck rupture is the same. This excitation energy for ^{236}U is shown in Fig. 9b. On

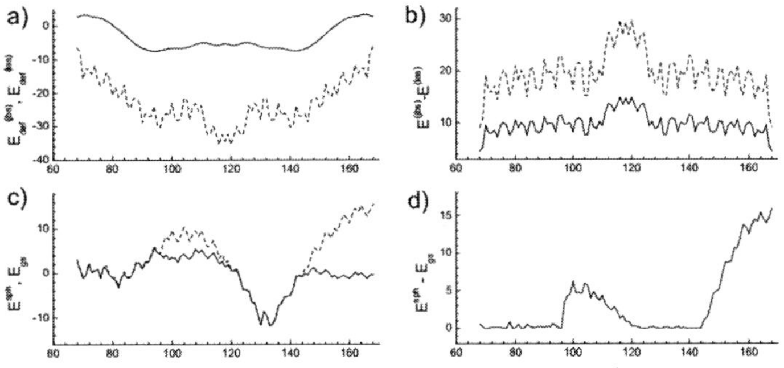

Figure 9. (a) The deformation energy of ^{236}U just before scission (solid) and immediately after scission (dash) as function of fragments mass number A; (b) the primary excitation energy $E^{jbs}_{def} - E^{ias}_{def}$; (c) the deformation energy $E^{(sph)}(A)$ at spherical shape (dash) and $E_{gs}(A)$ at the ground state (solid); (d) the extra deformation energy (22) of fission fragments; e) the excitation energy (23) of fission fragment. All energies in this figure are given in MeV.

the second stage of fission process the fragments relax to the ground state shape and gain some extra deformation energy

$$\Delta E_{def}(A) = E^{(sph)}(A) - E_{gs}(A) \tag{22}$$

So, the total excitation energy of one fragment is given by the sum

$$E_x(A) = [E^{(jbs)}(A) - E^{(ias)}(A)]/2 + \Delta E_{def}(A). \tag{23}$$

In Fig. 10 we compare the excitation energy available for the prompt neutron emission, $E_x(A) - 0.5S_n$ with the experimental value of neutron multiplicity [22, 24] multiplied by the half of two-neutron separation energy S_{2n} (in order to remove the rapid fluctuations due to the odd-even effect in S_n). One can see that there is some discrepancy up to 5 MeV at large mass asymmetries, but on the average the saw-tooth structure is rather well reproduced.

Having at one's disposal the excitation energy $E_x(A)$ one can calculate the total number of prompt neutrons

$$\bar{\nu}_{tot} = \sum_A [E_x(A)/S_n(A) - 1/2]Y(A). \tag{24}$$

The comparison of calculated $\bar{\nu}_{tot}$ with the available experimental results is shown in Fig. 10b. On average the dependence of $\bar{\nu}_{tot}$ on the proton number of fissioning nuclei is qualitatively reproduced. The calculated value of $\bar{\nu}_{tot}$ are however larger than experimental by (0.5-0.9). The source of this discrepancy can be related to the use of very simple estimate (20) for the neutron multiplicity and use of the quasistatic approximation (12) for the mass distribution.

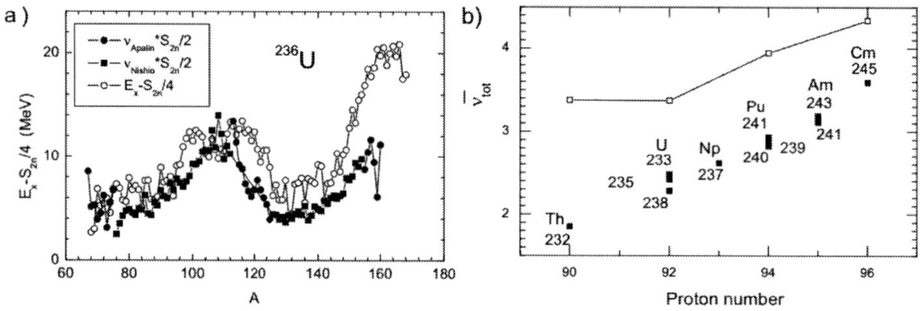

Figure 10. (a) The excitation energy (23) available for prompt neutron emission (open circles) and the experimental results for neutron multiplicity [22, 24] multiplied by half of two-neutron separation energy; (b) The calculated value of total neutron multiplicity (24) (open squares) and the experimental results [25] (solid squares).

For more accurate description of mass distribution the dynamical approach to the fission process seems necessary.

7 Summary

The calculations presented in this work show that the optimal shape prescription offers an accurate definition of the shape of fissioning nuclei just before the scission (the scission point configuration). Within the quasi-static picture the information on the shape and energy at the scission point can be used for the evaluation of the quantities, measured in fission experiments like the mass distribution, the kinetic and excitation energy of fission fragments, the multiplicity of prompt neutrons.

The results of calculations show that the calculated mass distribution and the excitation energy for the reaction ^{235}U + n_{th} are very close to the experimental values. The calculated distributions of the total kinetic energy of fission fragments for the fission of ^{232}Th,^{235}U,^{239}Pu and ^{245}Cm by thermal neutrons are also in rather good agreement with the experimental data.

The sawtooth structure of the neutron multiplicity is qualitatively reproduced. The dependence of total multiplicity of prompt neutrons on the proton number of fissioning nucleus is reproduced only qualitatively. The calculated value of $\bar{\nu}_{tot}$ are larger than experimental by (0.5-0.9).

The reason for the discrepancy between calculated and experimental results for average total prompt neutron multiplicities may be too crude approximation of Eqs. (20, 24) for the relation between the excitation energy and neutron multiplicity. There could be also other reasons, like the use of assumption that each point on the deformation energy surface is populated with a probability given by the canonical distribution (12).

Acknowledgements

The author would like to express his gratitude to the Research Laboratory for Nuclear Reactors, Tokyo Institute of Technology, for the hospitality during his visits to Japan. The contribution includes partly the results of "Comprehensive study of delayed-neutron yields for accurate evaluation of kinetics of high-burn up reactors" entrusted to the Tokyo Institute of Technology by the Ministry of Education, Culture, Sports, Science and Technology of Japan (MEXT).

References

[1] V. M. Strutinsky, Nucl. Phys. **3**, 449 (1966); Nucl. Phys. **A95**, 420 (1967); Nucl. Phys. **A122**, 1 (1968).
[2] M. Brack, J. Damgaard, A.S. Jensen et al, Rev. Mod. Phys. **44**, 320 (1972).
[3] V.M. Strutinsky, N.Ya. Lyashchenko, N.A. Popov, Nucl. Phys.**46**, 639 (1963).
[4] F.A. Ivanyuk, Phys. Scr. **89**, 054012 (2014).
[5] N. Bohr and J. A. Wheeler, Phys. Rev. **56** (1939) 426.
[6] F.Ivanyuk, Int. J. Mod. Phys. E **18**, 130 (2009).
[7] F.A. Ivanyuk, Physics Procedia **47**, 17 (2013).
[8] V.V. Pashkevich, Nucl. Phys. A **169**, 275 (1971).
[9] B.D. Wilkins, E.P. Steinberg, and R.R. Chasman, Phys. Rev. C **14**, 1832 (1976).
[10] J. Moreau, K. Heyde, and M. Waroquier, Phys. Rev. C **28**, 1640 (1983).
[11] A. Ruben, H. Marten, D. Seeliger, Zeit .für Physik A Hadrons and Nuclei **338**, 67 (1991).
[12] S. Panebianco, J.-L. Sida, H. Goutte et al, Phys. Rev. C **86**, 064601 (2012).
[13] H.-J. Krappe and K. Pomorski, *Theory of Nuclear Fission*, Springer Verlag, Heidelberg, 2012.
[14] F.A. Ivanyuk, S. Chiba and Y. Aritomo, Phys. Rev. C **90**, 054607 (2014).
[15] A.I. Sergachev, V.G. Vorob'eva, B.D. Kuz'minov et al, Yad. Fiz. **7**, 778 (1968).
[16] F.-J. Hambsch, H.-H Knitter, C. Budtz-Jorgensen and J. Theobald, Nucl. Phys. A **491**, 56 (1989).
[17] C. Tsuchiya, Y. Nakagome, H. Yamana et al, J. Nucl. Sc. Techn. **37**, 941 (2000).
[18] C. Wagemans, E. Allaert, A. Deruytter et al, Nucl. Phys. A **380**, 61 (1982).
[19] K. Nishio, Y. Nakagome, I. Kanno, I. Kimura, J. Nucl. Sc. Techn. **32**, 404 (1995).
[20] A. Ramaswami, S. Prakash, S. B. Manohar et al, Phys. Rev. C **16**, 716 (1977).
[21] S. Zeynalov, V. Furman, F.-J. Hambsch et al, in *Proceedings of the 13th International Seminar on Interaction of Neutrons with Nuclei (ISINN-13), Dubna, Russia, May 25-28, 2005* (Joint Institute for Nuclear Research, Dubna, 2006), p. 351.
[22] K. Nishio,Y. Nakagome, H. Yamamoto, I. Kimura, Nucl. Phys. A **632**, 540 (1998).
[23] N. Carjan, F.-J. Hambsch, F.A. Ivanyuk, P. Talou, Physics Procedia **59**, 173 (2014).
[24] V.F. Apalin et al., Nucl. Phys. **71**, 553 (1965).
[25] T. Ohsawa, J. Nucl. Radiochem. Sci. **9**, 19 (2008).

Dynamical approach to isotopic-distribution of fission fragments from actinide nuclei

Chikako Ishizuka[1,a], Satoshi Chiba[1], Alexander V. Karpov[2] and Yoshihiro Aritomo[3]

[1] *Research Laboratory for Nuclear Reactors, Tokyo Institute of Technology, Tokyo, Japan*
[2] *Joint Institute for Nuclear Research, Dubna, Russia*
[3] *Kindai University, Osaka, Japan*

Abstract. Measurements of the isotope distribution of fission fragments, often denoted as the primary fission yield (pre-neutron yield) or independent fission yield (post-neutron yield) are still challenging at low excitation energies, so that it is important to investigate it within a theory. Such quantities are vital for applications as well. In this study, fragment distributions from the fission of U isotopes at low excitation energies are studied using a dynamical model. The potential energy surface is derived from the two center shell model including the shell and pairing corrections. In order to calculate the charge distribution of fission fragments, we introduce a new parameter η_Z as the charge asymmetry, in addition to three parameters describing a nuclear shape, z as the distance between two centers of mass, δ as fragment deformation, and η_A as the mass asymmetry. Using this model, we calculated the isotopic distribution of ^{236}U for the n-induced process ^{235}U + n $\rightarrow$ ^{236}U at low excitation energies. As a result, we found that the current model can well reproduce isotopic fission-fragment distribution which can be compared favorably with major libraries.

1 Introduction

In the high burn-up nuclear and light water reactors, information on minor actinides, such as Cm and Am, plays a key role for their safety. However, it is difficult to obtain sufficient accuracy for the concentration of these nuclei by experiments even now. Our aim is to develop a theoretical model, which can provide such high-quality data. From phenomenological models such as superposition of Gaussians to microscopic models, there are so many approaches to nuclear fission. The reasons why we select the Langevin model, are as follows. The Langevin calculations with only a few essential variables can give us an intuitive image of a fissioning nucleus. Such simplicity of the Langevin model enables us to track each physical quantities directly, and to compare them with experiments. However, in general, the Langevin models have been used at high energies where the statistical dynamics is more essential than the quantum effects. Fission reactions in nuclear reactors proceed at low energies, usually 20 MeV at the highest. At such energies, nuclei have their own characteristics. Therefore, we cannot neglect quantum corrections, although these corrections are difficult to treat. Our previous research [1], overcame these difficulties, and successfully reproduced mass distributions of U and

[a] e-mail: chikako@nr.titeich.ac.jp

Pu at low energies. But that model is still not sufficient to apply to nuclear reactor physics. The isotope distributions are required. In this paper, we extend our model in order to treat the charge mode that allows the isotope yield calculation. Difference between the current work and other recent works relating to the fission properties based on the Langevin model [2, 3], is the way of the charge mode. We consider the charge mode simultaneously with other degrees of freedom within unified computational scheme solving a system of coupled equations. Another important feature of the present model is that the microscopic shell effects are essential part of our model that is very important for the considered energy region.

2 Method

We describe the shape of a compound nucleus, using the two-center shell model by three variables. They correspond to distance between the two centers of mass z, deformation δ, and mass asymmetry η_A. In order to treat the isotope distributions, we introduce a new variable η_Z for the charge mode. The Langevin approach treats nuclear fission as a Brownian motion of particle representing by the chosen collective variables. In the equation of motion, the dissipation term and the fluctuation term for the Brownian motion, are added.

$$\begin{aligned} \dot{q}_i &= m_{ij}^{-1} p_j \\ \dot{p}_i &= -\frac{\partial V}{\partial q_i} - \frac{1}{2}\frac{\partial}{\partial q_i} m_{jk}^{-1} p_j p_k - \gamma_{ij} m_{jk}^{-1} p_k + g_{ij} R_j(t) , \end{aligned} \tag{1}$$

where $q_i = z, \delta, \eta_A$, and η_Z $(i = 1, 2, 3, 4)$ are defined as follows.

$$z = \frac{z_0}{BR}, B = \frac{3+\delta}{3-2\delta}, \delta = \frac{3(a-b)}{2a+b}, \eta_A = \frac{A_1 - A_2}{A_1 + A_2}, \eta_Z = \frac{Z_1 - Z_2}{Z_1 + Z_2} . \tag{2}$$

Here, A_1, A_2, Z_1, and Z_2 are the mass and charge numbers of fragment 1 and 2, respectively. We assume that the deformation is common between a left and a right fragment. The random force in Eq. (1) satisfies $< R_i >= 0$ and $< R_i(t_1) R_j(t_2) >= 2\delta_{ij}(t_1 - t_2)$. The Einstein relation between the diffusion coefficient g_{ij} and the heat-bath temperature T is $\sum_k g_{ik} g_{kj} = T\gamma_{ij}$, where γ_{ij} is the friction tensor. The heat-bath temperature for the mass-mode $(i = 1, 2, 3)$ is expressed as $T = \sqrt{E_{int}/a(\mathbf{q})}$ using the level density parameter $a(\mathbf{q})$, while we use the effective temperature $T_Z = \frac{\hbar\omega_Z}{2} \coth \frac{\hbar\omega_Z}{2T}$ for the charge mode $(i = 4)$ instead of the heat-bath temperature T, where ω_Z is the frequency with respect to η_Z. Here, E_{int} is the intrinsic energy defined as follows.

$$E_{int} = E^* - \frac{1}{2} m_{ij}^{-1} p_i p_j - V . \tag{3}$$

E^* is the excited energy, and V is the potential energy.

Basically, for the potential energy, we use a similar potential as in the three dimensional case [1]. The difference between the three dimensional case and the current model is the fluctuation around an expectation value of the charge asymmetry. We have solved the Langevin equations with the charge asymmetry around the expectation value by using the potential as follows. The coefficient of the fluctuation-potential term in Eq. (6) can be derived from the Harmonic approximation. Potential terms relating to the charge asymmetry, are symmetry energy and the Coulomb energy. Then we can obtain the analytical expression for the coefficient $C_{<\eta_Z>}$ with the assumption of homogeneous charge spheres, as derived in [4, 5] because

$$C_{\eta_Z} = \left.\frac{\partial^2 V}{\partial \eta_Z^2}\right|_{\eta_Z = <\eta_Z>} \simeq \left.\frac{\partial^2 V_{Sym}}{\partial \eta_Z^2}\right|_{\eta_Z = <\eta_Z>} + \left.\frac{\partial^2 V_{Coul}}{\partial \eta_Z^2}\right|_{\eta_Z = <\eta_Z>} . \tag{4}$$

The harmonic approximation of the potential is as follows.

$$\begin{aligned} V &= V_{LD} + (V_{Shell}(\mathbf{q}, T=0) + V_{Pair}(\mathbf{q}, T=0))\exp\left(-\frac{aT^2}{E_d}\right) + \frac{l(l+1)\hbar^2}{2I} && (5) \\ &= V(q_i, <\eta_Z>) + \frac{C_{\eta_Z}}{2}(\eta_Z - <\eta_Z>)^2 && (6) \end{aligned}$$

Here, note that the liquid-drop model (the original potential used in [1]) presumes the Unified Charge Distribution (UCD). As already discussed in early 80s [8], the expectation value of the charge asymmetry is not that of the UCD. Deviation from the UCD has been observed.

$$\begin{aligned} V(q_i, <\eta_Z>_{UCD}) &= V_{LD}|_{<\eta_Z>_{UCD}} + (V_{Shell} + V_{Pair})\exp\left(-\frac{aT^2}{E_d}\right)\Bigg|_{<\eta_Z>_{UCD}} + \frac{l(l+1)\hbar^2}{2I}\Bigg|_{<\eta_Z>_{UCD}} && (7) \\ &= V(q_i, <\eta_Z>) + \frac{C_{\eta_Z}}{2}(<\eta_Z>_{UCD} - <\eta_Z>)^2 && (8) \end{aligned}$$

Finally, the potential can be expressed using the potential relating to the liquid-drop potential, the charge fluctuation around the expectation value of the charge asymmetry, and and the charge deviation from the UCD,

$$V = V(q_i, <\eta_Z>_{UCD}) + \frac{C_{\eta_Z}}{2}(\eta_Z - <\eta_Z>)^2 - \frac{C_{\eta_Z}}{2}(<\eta_Z>_{UCD} - <\eta_Z>)^2 \quad (9)$$

As we mentioned in the Introduction, in order to treat low-energy fission, we include quantum corrections in the potential. For the shell correction, we adopt the Strutinsky method from the single-particle levels of the two-center shell model. The BCS pairing effect is also included. The shell correction has a temperature dependence with the shell damping energy of the Ignatyuk's value. This modification of the potential energy changes trajectory, because the derivative of the fluctuation in terms of each variable q_i affects the time evolution. We estimate the $<\eta_Z>$ following [6]. The UCD hypothesis is defined as

$$\begin{aligned} \frac{Z}{A} &= \frac{Z_H^{UCD}}{A'_H} = \frac{Z_L^{UCD}}{Z'_L} && (10) \\ A'_H &= A_H + \nu_H\,, && (11) \\ A'_L &= A_L + \nu_L\,, && (12) \\ A &= A'_H + A'_L\,, && (13) \end{aligned}$$

where the ν_H and ν_L is an averaged number of evaporated neutrons from each fragment, respectively.

Another extension of the Langevin model is the transport coefficients. For the mass mode [1], we use the same one as the three dimensional calculations. The Werner-Wheeler approximation for the inertial mass, and the Wall and Window formula for the friction tensor. For the charge mode [4, 5], we neglect the off-diagonal components. Here, note that we assume the evolution of the charge mode starts just after the neck appearance in a compound nucleus. For the mass tensor of the charge mode, we approximate a nuclear shape near the scission point as two spheres connected by a circular hole with radius r_N. The mass tensor can be described as

$$m_{ij} = \begin{pmatrix} m_{11} & m_{21} & m_{31} & 0 \\ m_{12} & m_{22} & m_{32} & 0 \\ m_{13} & m_{23} & m_{33} & 0 \\ 0 & 0 & 0 & m_{\eta_Z\eta_Z} \end{pmatrix}, \quad (14)$$

where m_{ij} (for i=1, 2, 3) is a matrix element derived from Werner-Wheeler approximation. The indexes i and $j = 1, 2, 3, 4$ correspond to δ, η_A and η_Z, respectively. The charge component of m_{ij} is expressed as follows for a flow of viscous incompressible liquid through that cylindrical neck.

$$m_{\eta_Z\eta_Z} = \frac{m}{3\pi\rho}\frac{ZA^2}{N}\frac{(l+2r_N)}{r_N^2} \tag{15}$$

$$\gamma_{ij} = \begin{pmatrix} \gamma_{11} & \gamma_{21} & \gamma_{31} & 0 \\ \gamma_{12} & \gamma_{22} & \gamma_{32} & 0 \\ \gamma_{13} & \gamma_{23} & \gamma_{33} & 0 \\ 0 & 0 & 0 & \gamma_{\eta_Z\eta_Z} \end{pmatrix}, \tag{16}$$

where the friction parameter for the charge mode,

$$\gamma_{\eta_Z\eta_Z} = \frac{4m}{9\rho}\frac{AZ}{N}[N\bar{v}_p + Z\bar{v}_n]\frac{1}{\Delta\sigma}, \tag{17}$$

is derived in [4, 5] from the one-body dissipation model.

3 Results and Analysis

In this section, we show various properties of obtained fission-fragments using the Langevin model with the charge fluctuation for ^{236}U at $E* = 20$MeV which corresponds to n+^{235}U at $E_n = 14$MeV. The current mass distribution reproduces the peak positions and width of the experimental data as well as the result only with the mass mode. It means that the current modification (including the charge mode simultaneously with the mass mode) keeps the gross feature of the fissioning system. However, the statistics of current calculation is not sufficient compared to the three dimensional case.

Fig. 1 is the JENDL data [7] of ^{236}U thermal fission on the nuclear chart as a reference. Compared to the JENDL data, we found that our result shown in Fig. 2 reproduce the nuclear species on the chart and yield structures, such as the peak positions, the inter-peak structure and the deviation from the UCD line. Unfortunately, it is difficult to see the tail structure in the JENDL (nuclei in blue in Fig. 1) due to the poor statistics, so we expect that it can be reproduced with enough data.

Fig. 3 to Fig. 5 are all isotope distribution of ^{236}U at $E* = 20$MeV. As seen in Fig. 2, the current fission-fragments have narrow distribution width, compared to the experimental data. As a result, the minimum number in the the isotope distribution is larger than the major libraries, while the maximum proton is smaller than them, as seen in Fig. 3) and Fig. 4. However, the peak position and the curvature of our results can reproduce the experimental values, qualitatively and quantitatively.

Symbols are common in Fig. 3 to Fig. 5. Red line with filled circle is our calculation on a log scale. Black line with open circle is the JENDL/FPY-2011 [7] for ^{236}U thermal-fission. Blue dashed-line with open square is the JEFF-3.1.1 [11] for ^{236}U thermal-fission. Purple dashed-line with open triangle is the ENDF/B-VII [12] for ^{236}U thermal fission. Green dashed-line with open inverted-triangle is the GEF [10] for ^{236}U thermal-fission, respectively.

We found that our obtained nuclei with enough yields can reproduce major libraries very well in various mass numbers as shown in Fig. 3 and Fig. 4. Such characteristics of the calculated isotope distribution can be commonly seen in all data between $Z = 39$ to $Z = 52$. Fig. 5 is the same as the previous two figures but for the specific mass number $A = 132$. Even for the fixed mass number, such consistency with the major libraries can be also found.

DOI: 10.1051/epjconf/201612201003

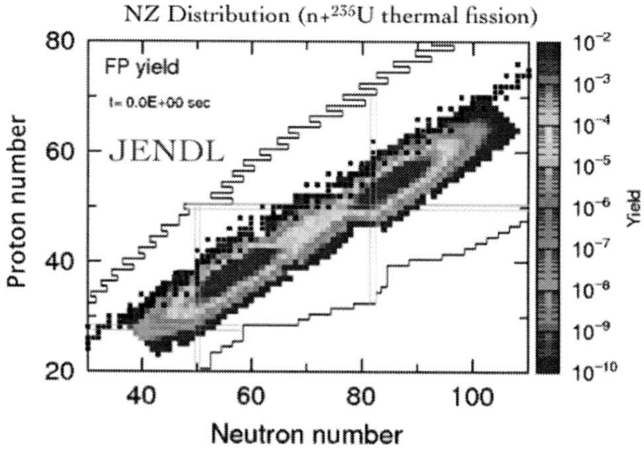

Figure 1. The JENDL data for ^{236}U thermal-fission is plotted in the nuclear chart. Fragment yield is shown in color map.

4 Summary

In this paper, we extended our Langevin model based on the two-center shell-model parametrization by including the contribution from the fluctuation around the most stable charge asymmetry. We performed the Langevin calculation for the compound system of ^{236}U at $E* = 20$ MeV in terms of shape parameters, the mass asymmetry and the charge asymmetry simultaneously. We assumed the expectation value for the charge asymmetry including the charge deviation from the UCD as defined in [6]. As a result, our calculations showed good agreement with the NZ-distributions and isotope distributions of the major libraries, and that our model can provide sufficient accuracy to be compared to them, without any modification.

References

[1] Y. Aritomo and S. Chiba, PRC **88** (2013) 044614 and references therein.
[2] K. Mazurek, C. Schmitt and P. N. Nadtochy, PRC **91** (2015) 041603(R).
[3] M. R. Pahlavani and S. M, Mirfathi, PRC **92** (2015) 024622.
[4] A. V. Karpov and G. D. Adeev, EPJA **14** (2002) 169 and references therein.
[5] A. V. Karpov and G. D. Adeev, *Phys. At. Nucl.* **65** (2002) 1596.
[6] G. D. Adeev, *Fiz. Elem. Chastis At. Yadra* **23** (1992) 1572.
[7] K. Shibata, O. Iwamoto, T. Nakagawa, N. Iwamoto, A. Ichihara, S. Kunieda, S. Chiba, K. Furutaka, N. Otuka, T. Ohsawa, T. Murata, H. Matsunobu, A. Zukeran, S. Kamada, and J. Katakura, *J. Nucl. Sci. Technol.* **48** (2011) 1.
[8] R. K. Gaputa and D. R. Saroha, PRC **30** (1984) 395 and reference therein.
[9] A. C. Wahl, *Atomic Data and Nucl. Data Tables* **39** (1988) 1.
[10] K.-H. Schmidt, B. Jurado, Ch. Amouroux, *JEFF-Report* **24** (2014) 1.

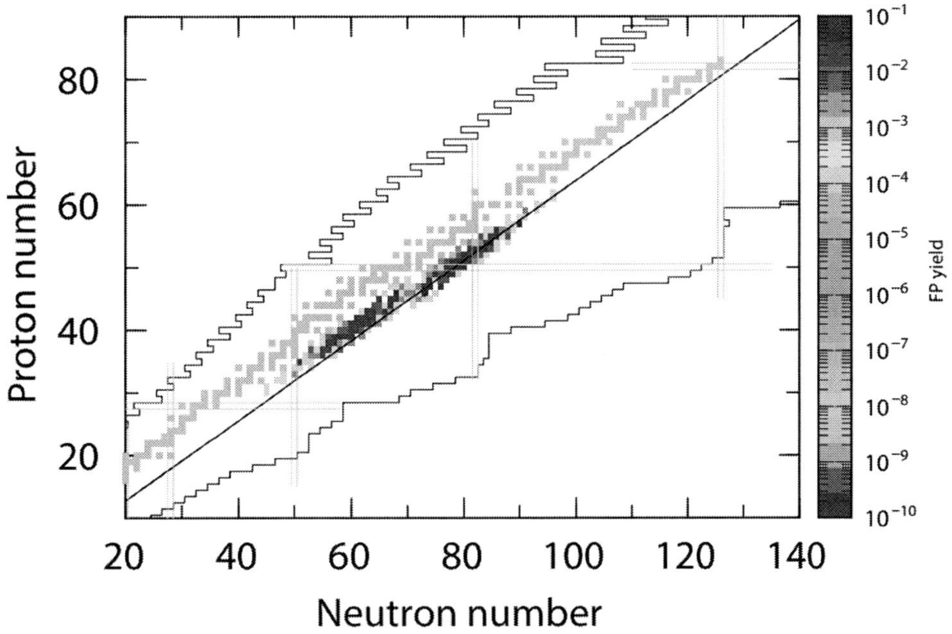

Figure 2. Calculated fragment yields in color map. The 1.5 neutrons are removed from each fragment as emitted prompt neutrons.

[11] A. Santamarina et al., *JEFF-Report* **22** (2009) 1.
[12] M.B. Chadwick et al., *Nucl. Data Sheets* **112** (2011) 2887.

 DOI: 10.1051/epjconf/201612201003

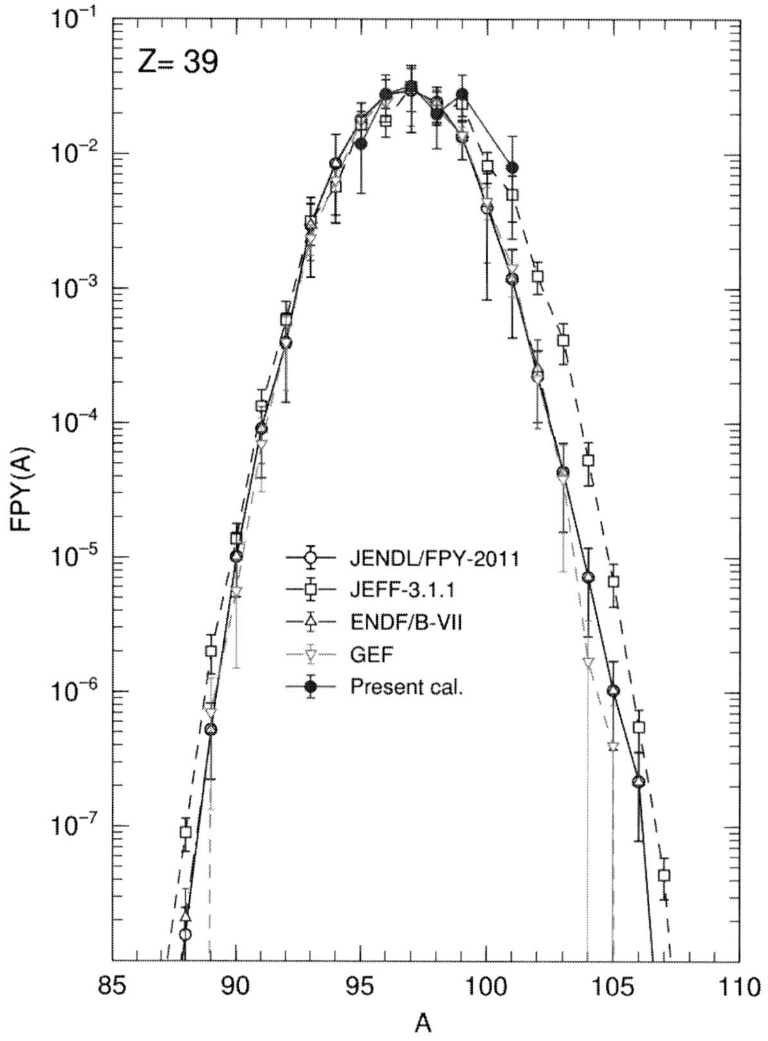

Figure 3. Isotope distribution as a function of mass number A is shown, when $Z = 39$, with major libraries. Red line with filled circle is our calculation on a log scale. Black line with open circle is the JENDL/FPY-2011 [7] for ^{236}U thermal-fission. Blue dashed-line with open square is the JEFF-3.1.1 [11] for ^{236}U thermal-fission. Purple dashed-line with open triangle is the ENDF/B-VII [12] for ^{236}U thermal fission. Green dashed-line with open inverted-triangle is the GEF [10] for ^{236}U thermal-fission.

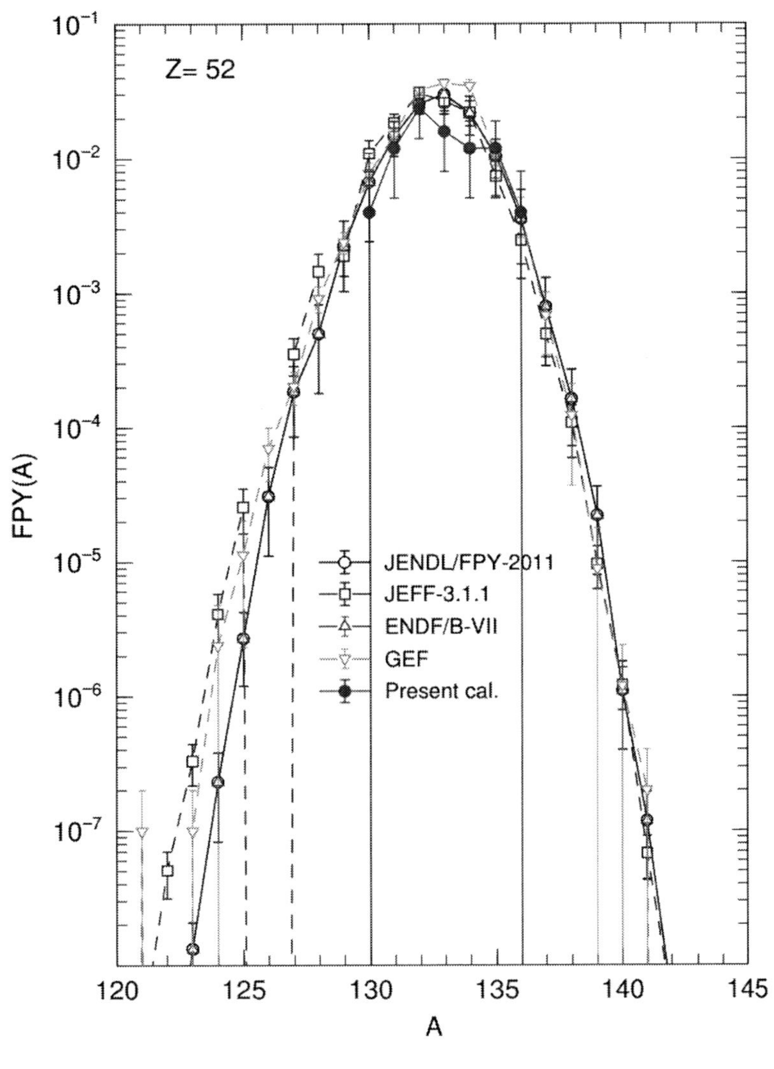

Figure 4. Isotope distribution as a function of mass number A is shown, when $Z = 52$, with major libraries. Red line with filled circle is our calculation on a log scale. Black line with open circle is the JENDL/FPY-2011 [7] for ^{236}U thermal-fission. Blue dashed-line with open square is the JEFF-3.1.1 [11] for ^{236}U thermal-fission. Purple dashed-line with open triangle is the ENDF/B-VII [12] for ^{236}U thermal fission. Green dashed-line with open inverted-triangle is the GEF [10] for ^{236}U thermal-fission.

DOI: 10.1051/epjconf/201612201003

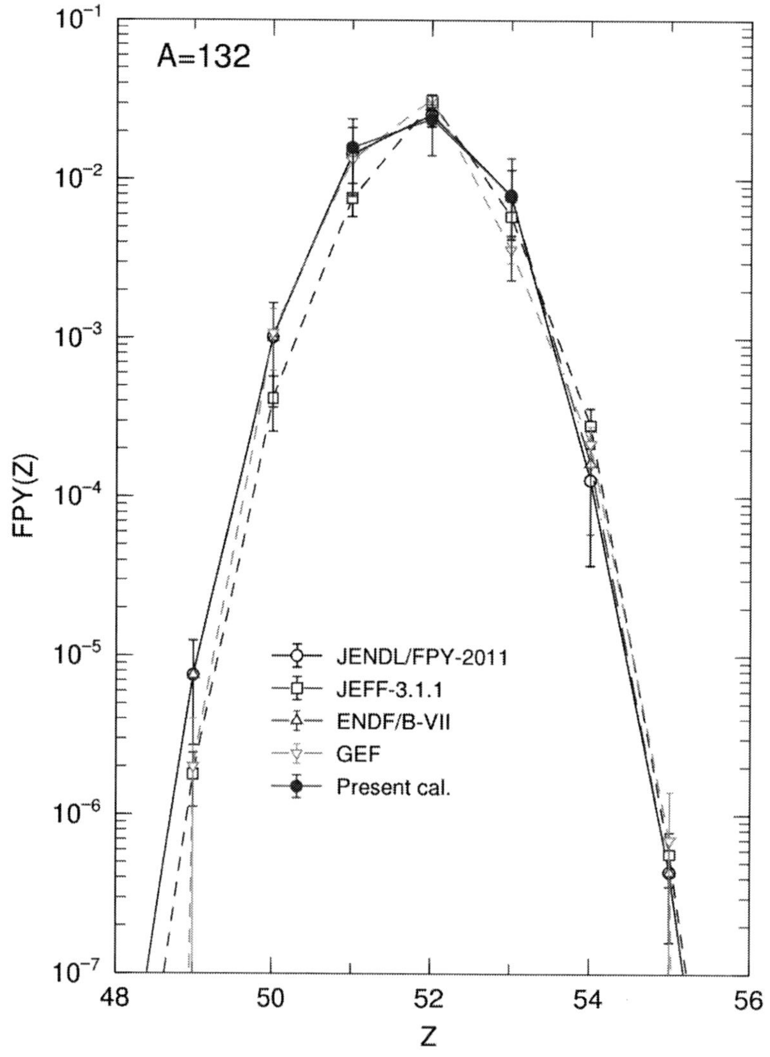

Figure 5. Isotope distribution as a function of charge number Z is shown, when $A = 132$, with major libraries. Red line with filled circle is our calculation on a log scale. Black line with open circle is the JENDL/FPY-2011 [7] for ^{236}U thermal-fission. Blue dashed-line with open square is the JEFF-3.1.1 [11] for ^{236}U thermal-fission. Purple dashed-line with open triangle is the ENDF/B-VII [12] for ^{236}U thermal fission. Green dashed-line with open inverted-triangle is the GEF [10] for ^{236}U thermal-fission.

Fission life-time calculation using a complex absorbing potential

Guillaume Scamps[1,a] and Kouichi Hagino[1,2,b]

[1] *Department of Physics, Tohoku University, Sendai 980-8578, Japan*
[2] *Research Center for Electron Photon Science, Tohoku University, 1-2-1 Mikamine, Sendai 982-0826, Japan*

Abstract.
A comparison between the semi-classical approximation and the full quantum calculation with a complex absorbing potential is made with a model of the fission of ^{258}Fm. The potential barrier is obtained with the constrained Skyrme HF+BCS theory. The life-time obtained by the two calculations agree with each other the difference being only by 25%.

1 Introduction

The description of the quantum tunneling is important to predict the life-time of a spontaneous fission as well as fission product yields. A standard strategy to compute the fission life-time is to first determine a potential energy surface using the liquid drop model with shell correction [1–3] or the self-consistent mean field theory [4–17]. The life-time is then calculated using the semi-classical approximation to determine the fission path that minimizes the action [1, 7, 9, 18–20].

In a previous study [21], we developed a method to obtain the fission life-time using a fully quantum approach with a complex absorbing potential. This method has been tested on a schematic model.

In the present work, we test the semi-classical approximation in a more realistic application, by comparing the full quantum approach to the semi-classical calculation using a potential determined with the constrained Hartree-Fock method with a Skyrme interaction.

2 Complex absorbing potential method

In order to describe the evolution of a quantum system into the continuum spectrum, a possibility is to add an imaginary potential at large distances in order to absorb the wave function, in which the absorbed component of the wave function is assumed to be emitted in the continuum. The modified hamiltonian $\hat{H}' = \hat{H} + iW(R)$ is then composed of the original hamiltonian $\hat{H}$ and the imaginary potential $iW(R)$. The time dependent Shrödinger equation for the system becomes,

$$i\hbar\frac{d}{dt}|\Psi(t)\rangle = \left(\hat{H} + iW(R)\right)|\Psi(t)\rangle. \tag{1}$$

[a] e-mail: scamps@nucl.phys.tohoku.ac.jp
[b] e-mail: hagino@nucl.phys.tohoku.ac.jp

This time-dependent equation is often solved numerically with the Runge-Kutta or other iterative methods. Nevertheless, those iterative methods are limited to the case where the phenomenon of interest takes place in a time scale of the order of magnitude of the characteristic time scale of the system. This is not the case for the spontaneous fission, where the fission life time can be of the order magnitude of years while the characteristic time scale of the nuclear system is the zepto second .

Alternatively, one can integrate the time dependent equation as,

$$|\Psi(t)\rangle = e^{-\frac{i}{\hbar}t\hat{H}'}|\Psi_0\rangle. \tag{2}$$

In order to compute easily the exponential of the matrix, we expand the hamiltonian H' on the biorthogonal basis [22–24] with the left and right eigenfunctions,

$$H'|\varphi_i^r\rangle = E_i|\varphi_i^r\rangle \quad \text{and} \quad \langle\varphi_i^l|H' = E_i\langle\varphi_i^l|, \tag{3}$$

with complex eigenvalue E_i. Using the completeness relation for the bi-orthogonal basis $\sum_i |\varphi_i^r\rangle\langle\varphi_i^l| = \mathbb{1}$, we obtain the simple evolution,

$$|\Psi(t)\rangle = \sum_i e^{-\frac{i}{\hbar}tE_i}\langle\varphi_i^l|\Psi_0\rangle|\varphi_i^r\rangle. \tag{4}$$

It is then possible to describe easily the evolution of the wave packet for a very long time once we compute the overlap $\langle\varphi_i^l|\Psi_0\rangle$.

We note that each eigenstate $|\varphi_i^r\rangle$ is associated with a complex energy, $E_i = E_i^r - i\Gamma_i/2$, with Γ being the resonance width which is related to the life time, τ_i=$\hbar/\Gamma_i$.

3 Results

In the precedent study [21], we tested this method in a model case consisting of a schematic barrier. In the present study, we use a more realistic potential using the constrained Hartree-Fock+BCS theory. The calculation is carried out for the ^{258}Fm with the Skyrme interaction Skm* and a volume type of pairing. The constrained Hartree-Fock BCS equation is solved using the EV8 code [25]. A constraint is applied on the value of the average separation R of the two fragments as in Ref. [26] assuming a symmetric fission with a neck position at z=0.

The obtained potential is shown in fig. 1. The collective hamiltonian is assumed here to have only one dimension,

$$H(R) = -\frac{\hbar^2}{2M}\frac{\partial^2}{\partial R^2} + V(R). \tag{5}$$

In this equation the mass M is assumed to be a constant with a value to reproduce the experimental spontaneous half-life of ^{258}Fm, τ_{exp}=0.37 ms [27], using the semi classical approximation. With this approximation, the life time is determine as,

$$\tau = \frac{2\pi}{\Omega}e^{-2S/\hbar}, \tag{6}$$

with $\Omega = 2(E_0 - V_0)/\hbar$ and the action integral S,

$$S = \int_{R_0}^{R_1} \sqrt{2M(U(R) - E_0)}\,dR, \tag{7}$$

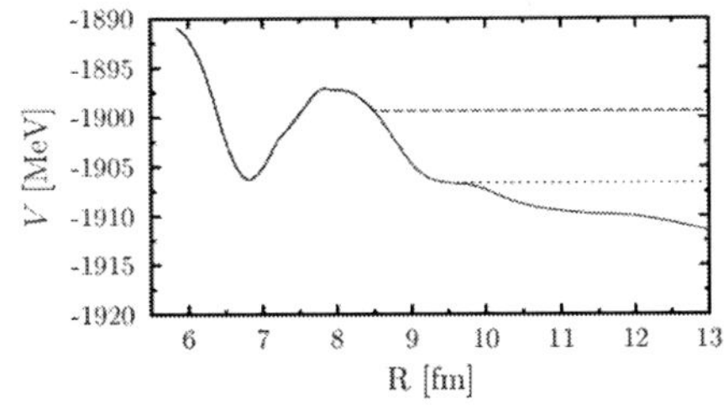

Figure 1. The potential energy as a function of the separation between the two fragments for the symmetric fission of ^{258}Fm. The original potential is shown with the red solid line. The modified potential for the initial wave function and dynamics are shown respectively by the green dashed line and the blue dotted line.

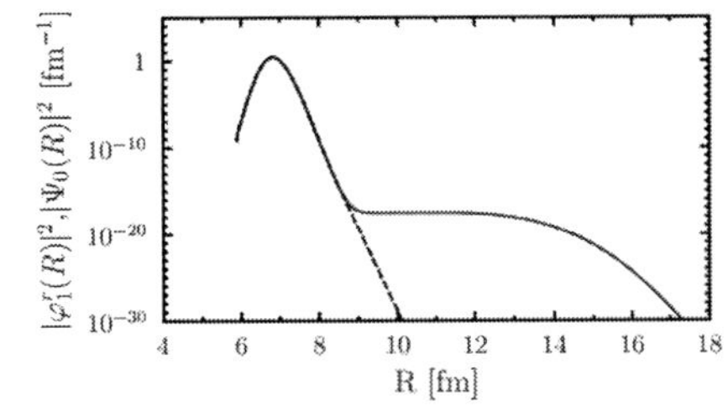

Figure 2. The initial wave function (the blue dashed line) and the resonant wave function (the red solid line).

Here, R_0 and R_1 are the inner and outer turning points, respectively defined as $V(R_0) = V(R_1) = E_0$, E_0 is taken as the energy obtained by solving the static Schrödinger equation by restricting the space before the barrier. The mass that reproduces the experimental life-time is M= 480 MeV $(10^{-22}s)^2$ fm^{-2}. As mentioned earlier, the goal of the present contribution is not to predict the fission life time but to show the feasibility of this approach and to compare the result with the semi-classical approximation.

To solve the tunneling problem, we first determine the initial wave function $|\Psi_0\rangle$ with a modified potential in order to have a bound state inside the barrier. That is the potential is replaced after the barrier by a constant value outside the barrier, after R =8.5 fm. The lowest energy bound state is chosen as the initial wave function and is shown in fig. 2.

To obtain the resonant state, an imaginary potential is then added to the hamiltonian [28, 29],

$$iW(R) = iW_0(R - R_a)^2\theta(R - R_a), \tag{8}$$

with W_0=−0.1 MeV.fm^{-2} and R_a=10.8 fm. We also modified the real potential for the dynamics and is chosen to be constant after R=9.5 fm. We take this prescription in order to stabilize the results with the choice of the imaginary potential. That is, with this prescription, the results are independent of the distance between the barrier and the position of the imaginary potential. Among the eigenfunctions of this Hamiltonian, we select the physical resonance wave function as the state $|\varphi_1^r\rangle$ which has the maximum overlap with the initial state. From the imaginary part of the energy of this state, we determine the decay half-life. The resultant value is 0.290 ms, which agrees well with the semi-classical approximation, τ=0.37 ms. The difference is only by 25%

4 Summary

A comparison between the semi-classical approximation and the full quantum calculation with a complex absorbing potential was made with a model for the fission of ^{258}Fm. We found that those two calculations lead to a similar result to each other. A difference of 25% was found between the two results, that could be attribute to the assumption of harmonicity for the confining potential or to the accuracy of the semi-classical approximation. In this paper, we considered a schematic constant mass for the fission motion, but we plan for a future application to use the mass from a microscopical calculation.

Acknowledgement

G.S. acknowledges the Japan Society for the Promotion of Science for the JSPS postdoctoral fellowship for foreign researchers. This work was supported by Grant-in-Aid for JSPS Fellows No. 14F04769.

References

[1] M. Brack, J. Damgaard, A.J. Jensen, H.C. Pauli, V.M. Strutinsky, and C.Y. Wong, Rev. Mod. Phys. **44**, 320 (1972).
[2] P. Möller, D.G. Madland, A.J. Sierk, and A. Iwamoto, Nature **409**, 785 (2001).
[3] P. Möller, A.J. Sierk, T. Ichikawa, A. Iwamoto, R. Bengtsson, H. Uhrenholt, and S. Aberg, Phys. Rev. C **79**, 064304 (2009).
[4] J. Skalski, Phys. Rev. C **77**, 064610 (2008).
[5] A. Staszczak, A. Baran, J. Dobaczewski, and W. Nazarewicz, Phys. Rev. C **80**, 014309 (2009).
[6] A. Baran, J.A. Sheikh, J. Dobaczewski, W. Nazarewicz, and A. Staszczak, Phys. Rev. C **84**, 054321 (2011).
[7] J. Sadhukhan, K. Mazurek, A. Baran, J. Dobaczewski, W. Nazarewicz, and J.A. Sheikh, Phys. Rev. C **88**, 064314 (2013).
[8] J.D. McDonnell, W. Nazarewicz, J.A. Sheikh, A. Staszczak, and M. Warda, Phys. Rev. C **90**, 021302(R) (2014).
[9] J. Sadhukhan, J. Dobaczewski, W. Nazarewicz, J. A. Sheikh, and A. Baran, Phys. Rev. C **90**, 061304(R) (2014).
[10] M. Warda and J.L. Egido, Phys. Rev. C **86**, 014322 (2012).
[11] R. Rodriguez-Guzman and L.M. Robledo, Phys. Rev. C **89**, 054310 (2014).
[12] S.A. Giuliani, L.M. Robledo, and R. Rodriguez-Guzman, Phys. Rev. C **90**, 054311 (2014).
[13] H. Goutte, J.F. Berger, P. Casoli, and D. Gogny, Phys. Rev. C **71**, 024316 (2005).
[14] R. Bernard, H. Goutte, D. Gogny, and W. Younes, Phys. Rev. C **84**, 044308 (2011).
[15] H. Abusara, A.V. Afanasjev, and P. Ring, Phys. Rev. C **82**, 044303 (2010); Phys. Rev. C **85**, 024314 (2012).
[16] B.-N. Lu, E.-G. Zhao, and S.-G. Zhou, Phys. Rev. C **85**, 011301 (2012); Phys. Rev. C **89**, 014323 (2014).
[17] J. Zhao, B.-N. Lu, D. Vretenar, E.-G. Zhao, and S.-G. Zhou, Phys. Rev. C **91**, 014321 (2015).

[18] A. Schmid, Ann. Phys. **170**, 333 (1986).
[19] T. Kindo and A. Iwamoto, Phys. Lett. **B225**, 203 (1989).
[20] A. Iwamoto, Z. Phys. **A349**, 265 (1994).
[21] G. Scamps and K. Hagino, Phys. Rev. C **91**, 044606 (2015).
[22] P. Morse and H. Feshbach, *Methods of Theoretical Physics* (McGraw-Hill, New York, 1952), p. 884.
[23] M.S. Hussein, M.P. Pato, and A.F.R. de Toledo Piza, Phys. Rev. C **51**, 846 (1995).
[24] K. Hagino and N. Takigawa, Phys. Rev. C **58**, 2872 (1998).
[25] P. Bonche, H. Flocard, and P.-H. Heenen, Comput. Phys. Commun. **171**, 49 (2005).
[26] C. Simenel and A.S. Umar, Phys. Rev. C **89**, 031601(R) (2014).
[27] N. Holden and D. Hoffman, Pure Appl. Chem. **72**, 1525 (2000).
[28] M. Ito and K. Yabana, Prog. Theo. Phys. **113**, 1047 (2005).
[29] R. Otani, R. Kageyama, M. Iwasaki, M. Kudo, M. Tomita, and M. Ito, Phys. Rev. C **90**, 034316 (2014).

Fission cross-sections, prompt fission neutron and γ-ray emission in request for nuclear applications

F.-J. Hambsch[1], P. Salvador-Castiñeira[1,2], S. Oberstedt[1], A. Göök[1], R. Billnert[1]

[1]*EC-JRC-IRMM, Retieseweg 111, B-2440 Geel, Belgium*
[2]*National Physical Laboratory, Hampton Road, Teddington, Middlesex, UK, TW11 0LW*

Abstract. In recent years JRC-IRMM has been investigating fission cross-sections of ^{240,242}Pu in the fast-neutron energy range relevant for innovative reactor systems and requested in the High Priority Request List (HPRL) of the OECD/Nuclear Energy Agency (NEA). In addition to that, prompt neutron multiplicities are being investigated for the major isotopes ^{235}U, ^{239}Pu in the neutron-resonance region using a newly developed scintillation detector array (SCINTIA) and an innovative modification of the Frisch-grid ionisation chamber for fission-fragment detection. These data are highly relevant for improved neutron data evaluation and requested by the OECD/Working Party on Evaluation Cooperation (WPEC). Thirdly, also prompt fission γ-ray emission is investigated using highly efficient lanthanide-halide detectors with superior timing resolution. Again, those data are requested in the HPRL for major actinides to solve open questions on an under-prediction of decay heat in nuclear reactors. The information on prompt fission neutron and γ-ray emission is crucial for benchmarking nuclear models to study the de-excitation process of neutron-rich fission fragments. Information on γ-ray emission probabilities is also useful in decommissioning exercises on damaged nuclear power plants like Fukushima Daiichi to which JRC-IRMM is contributing. The results on the ^{240,242}Pu fission cross section, ^{235}U prompt neutron multiplicity in the resonance region and correlations with fission fragments and prompt γ-ray emission for several isotopes will be presented and put into perspective.

1 Introduction

Data needs for advanced reactor systems and improved nuclear data evaluations have recently been the focus of research at JRC-IRMM. Here mainly the High Priority Request List (HPRL) of the OECD/Nuclear Energy Agency (NEA) is consulted as well as recommendations given in reports of the Working Party on International Evaluation Co-operation (WPEC) of the OECD/NEA. This has resulted in investigations of fission cross sections of the highest priority targets for fast reactors, ^{240,242}Pu(n,f) [1]. The neutron scintillation detector array SCINTIA has been developed to investigate prompt fission neutron emission in resonance neutron induced fission, based on requests from WPEC related investigations [2, 3]. And finally, measurements of γ-ray emission in several actinides have been performed and are planned in the near future in request of the HPRL entries H:3 and H:4 [4]. In the following sections results of the three items mentioned will be highlighted.

2 Fission cross sections

For the next generation of nuclear power plants (GEN-IV), four of the six designs currently under study are based on a fast neutron-energy spectrum instead of a thermal one. Consequently, there are upcoming needs for the nuclear data community to meet more stringent requirements, in order to improve the accuracy on the performance of simulations codes in this energy region. Addressing these aspects, a sensitivity study was performed by the Nuclear Energy Agency (NEA) [1] with the outcome of a list of high priorities of the most important isotopes and their relevant quantities. Not only for isotopes that eventually would be part of the fuel mixture, but, also, for structural materials intended to surround the reactor core. Within the high priority list it is requested to improve the accuracy of the neutron induced fission cross section of ^{240,242}Pu from the current 5%, 20% to a target value of 1-3%, 5%, respectively.

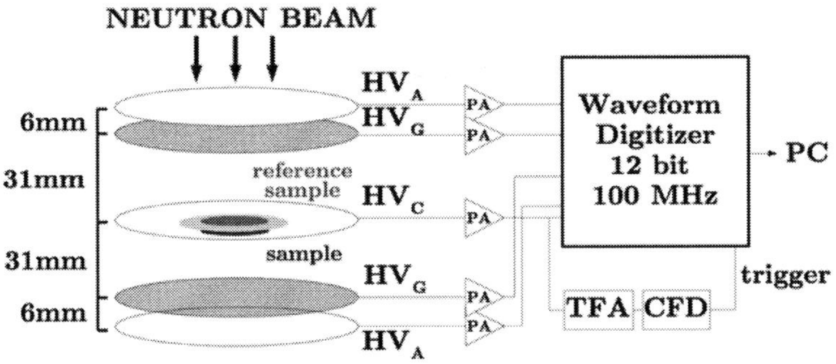

Figure 1. Scheme of the experimental set-up.

2.1 Experimental setup

The respective experiments have been performed at the JRC-IRMM Van de Graaff accelerator during the past years. The main difference to previous works in literature was that three reference targets have been used, namely ^{235,238}U and ^{237}Np. ^{235}U(n,f) is a primary reference standard, ^{238}U(n,f) a secondary standard above 2 MeV and ^{237}Np(n,f) is no standard, but has a lower threshold and higher fission cross section than ^{238}U(n,f). A double Frisch grid ionisation chamber was used in back to back geometry for fission fragment detection. The signals of the chamber were digitised using Waveform digitizers and the full traces of the fission fragments were stored for off-line analysis.

This is the second innovation compared to previous literature works. A schematic scheme of the experimental set-up is given in Fig. 1. Digital signal processing is used to deduce the fission cross-section values. More details are given in Refs. [5, 6]. The incident neutron energy range covered was from 0.3 MeV to 3 MeV. The used targets were prepared at the JRC-IRMM sample preparation group. Since the samples are the most important ingredients to this experiment, a lot of effort has been invested to verify the homogeneity and total mass [7]. Also the spontaneous fission half-lives of ^{240,242}Pu have been re-measured with unprecedented statistical accuracy [8].

2.2 Experimental results

In Figs. 2 and 3 the resulting fission cross section values are given using different colours for the different reference targets. In addition the most recent evaluations are also given. It is obvious that for ^{240}Pu(n,f) our results reproduce the fission threshold very well but in the plateau region the difference to the evaluations is about 5 %. For ^{242}Pu(n,f) our results show a deviation already in the threshold region where the present data tend to a less steep rise of the fission cross section. Furthermore the resonance like structure around 1.1 MeV seen in the evaluations is not present in our data. Furthermore in the plateau region up to about 2.5 MeV our data are 7-9% lower than the evaluations.

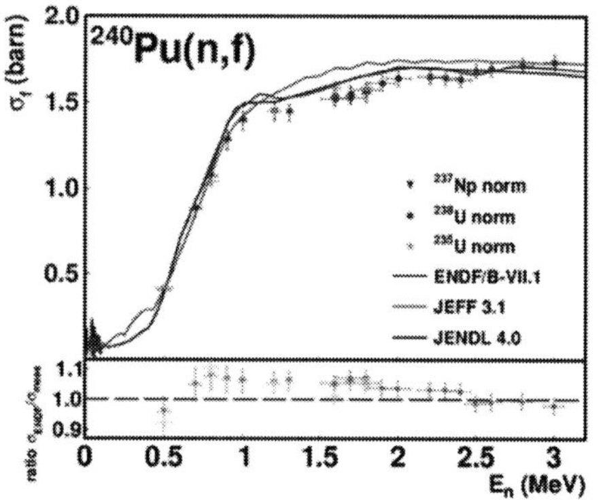

Figure 2. Fission cross section results for ^{240}Pu(n,f) using three different reference samples as given in the legend compared to the most recent evaluations.

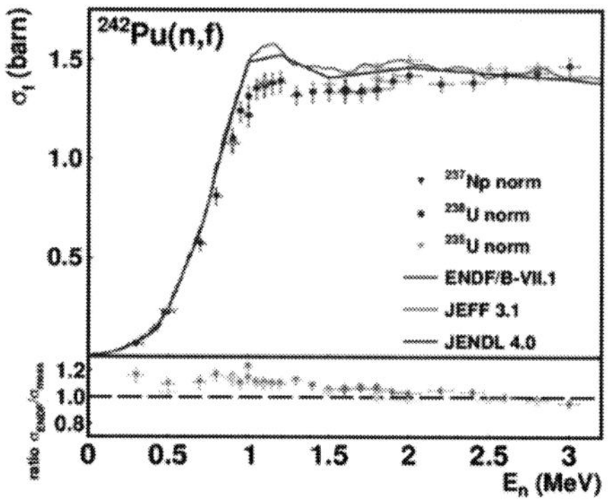

Figure 3. Fission cross section results for ^{240}Pu(n,f) using three different reference samples as given in the legend compared to the most recent evaluations.

Figs. 4 and 5 bring the present results into perspective to selected previous literature results. The weighted average of the present fission cross section results for $^{240,\,242}$Pu(n,f) are shown using three different reference samples as given in the legend and compared to literature data and the most recent evaluations.

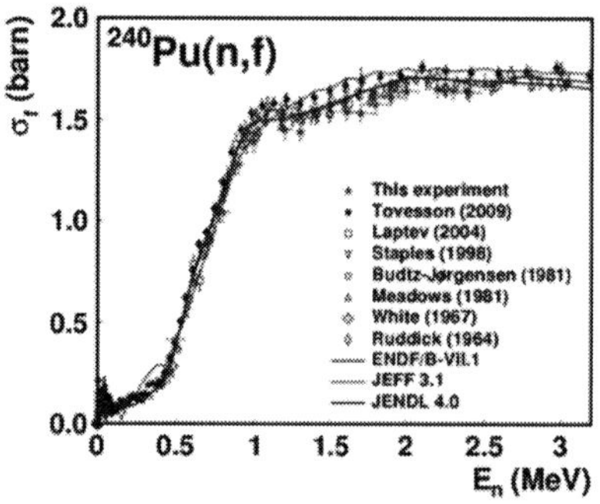

Figure 4. Weighted average of the present fission cross section results for ^{240}Pu(n,f) using three different reference samples as given in the legend compared to literature data and the most recent evaluations.

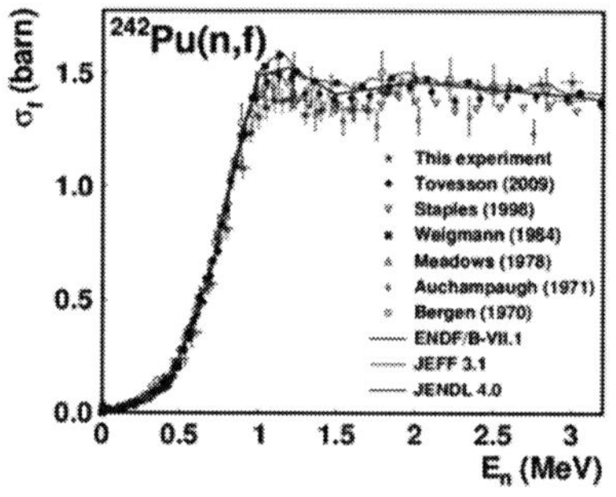

Figure 5. Weighted average of the present fission cross section results for ^{242}Pu(n,f) using three different reference samples as given in the legend compared to literature data and the most recent evaluations.

3 Prompt neutron emission in fission

In understanding the fission process knowledge of prompt fission neutron emission is of particular importance. Information about the properties of prompt fission neutrons, their multiplicities and energy distributions could give answers to questions related not only to the neutron emission itself, but also relevant to the formation of the fission fragments, the sharing of excitation energy among them and the time scale of the process. To improve nuclear modelling and similarly the evaluation of

DOI: 10.1051/epjconf/201612201005

nuclear data the knowledge about fluctuations in the prompt neutron multiplicity as a function of incident neutron energy is requested for the major actinides ^{235}U and ^{239}Pu [2, 3]. Fluctuations in fission fragment mass and total kinetic energy (TKE) for both isotopes have been observed in resonance-neutron induced fission [9, 10]. Independently, the fluctuation in the number of emitted neutrons has been observed, too [11]. In view of the fact that both neutron number and fission fragment properties have been found to vary it is necessary to study the correlations of prompt neutron multiplicity and fission fragments properties in the resonance region [12]. In the most severe cases the relative impact was as large as 20-30 %. Hence an experimental campaign to investigate correlations of prompt fission neutron multiplicity with fragment properties in resonance neutron induced fission on ^{235}U has been started at JRC-IRMM.

3.1 Experimental setup

The present experimental setup for investigating correlations of prompt neutrons with fission fragments in resonance neutron induced fission on ^{235}U is illustrated in Figs. 6 and 7. The fission target is placed inside the ionization chamber at a distance of about 9.2 m from the neutron production target of the GELINA facility. An array of neutron detectors (SCINTIA) is employed to improve the fission-neutron coincidence count rate. The SCINTIA array consists of 7 NE213 equivalent liquid scintillators (Scionix LS-301) and 5 para-therphenyl detectors, a photo is shown in Fig. 6. In this setup with an array of neutron detectors, each detector forms an axis of symmetry around which the orientation of the fission axis needs to be known for the purpose of transforming the emission properties into the fission fragment rest frame. Hence the traditional ionization chamber has been improved with a position sensitive readout structure, which allows determination of all three space coordinates of the fission fragment's direction of travel. For data acquisition a 22 channel fully digital system is used. The digital technique has shown many advantages over the analogue technique for fission fragment spectroscopy with ionization chambers (see e.g. Ref. [13]).

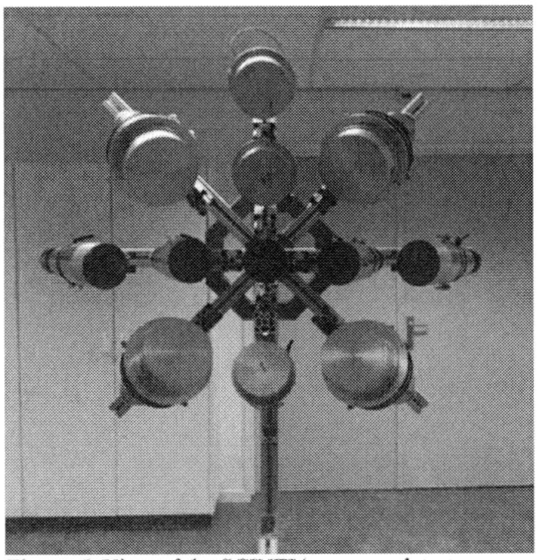

Figure 6. View of the SCINTIA neutron detector array in the laboratory.

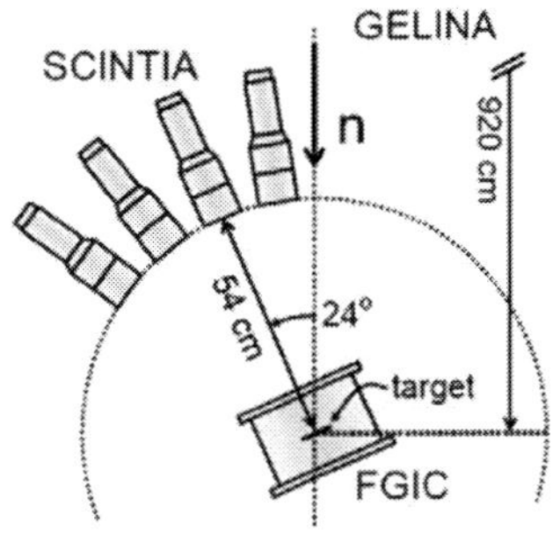

Figure 7. Schematic view of the experimental setup at the GELINA flight path station.

The acquisition is triggered by the current signal from the ionization chamber cathode, giving the instant of a fission event in time with a resolution better than 1 ns FWHM. For each fission event the digital waveforms of all channels, sampled at 400 MS/s with 14-bit resolution, are stored on disk for further off-line processing. The incident neutron time-of- flight is derived from the timestamp of each event. The system has been validated using the spontaneous fission of ^{252}Cf. More detail is given in Refs. [14, 15].

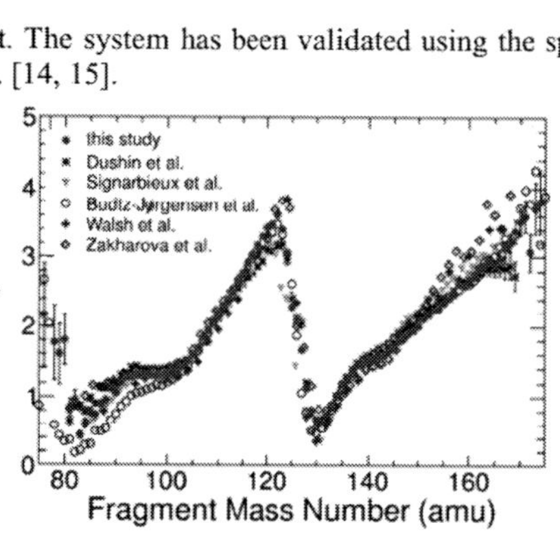

Figure 8 Average total prompt fission neutron multiplicity as a function of the pre-neutron fragment mass number, in comparison with experimental data from literature.

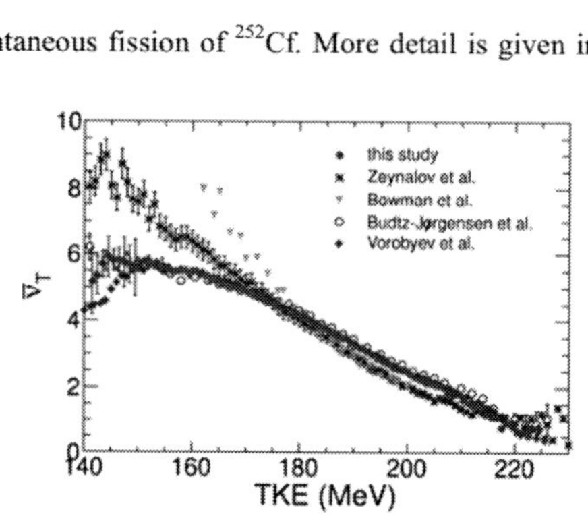

Figure 9. Average total prompt fission neutron multiplicity as a function of TKE, in comparison with experimental data from literature.

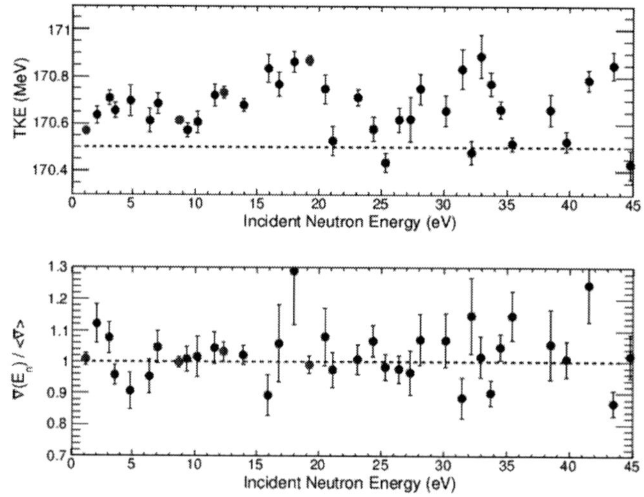

Figure 10. Upper part: Average TKE as a function of incident neutron energy for a given resonance or resonance groups in the energy range up to 45 eV, the dashed line refers to the thermal value. Lower part: Neutron multiplicity for the same resonances or resonance groups relative to the weighted average neutron multiplicity. The red points highlight strong resonances in the fission cross-section of ^{235}U.

3.2 Experimental results

The spontaneous fission of ^{252}Cf was used to verify setup and analysis procedures relevant for studying neutron emission in resonance-neutron induced fission. Prompt fission neutron multiplicity correlations with fission fragment properties were investigated. Beside the purely experimental interest, recent advances in the theoretical description of prompt emission in fission (see e.g. Ref. [16]) require detailed input data for verifying the models. The experiment is essentially a reproduction of the experiment performed by Budtz-Jørgensen and Knitter [17]. Experimental details and results have been published elsewhere [14], a few selected results are repeated in the following.

The mass dependence of the prompt fission neutron multiplicity per fragment from this experiment is compared to literature data in Fig. 8. The data from this study agree quite well with experimental data available in the literature and show in contrast to the data of Ref. [17] a bump at mass number 80 to 100. In Fig. 9 the mean number of neutrons emitted per fission is shown as a function of the TKE of the fission fragments. Except for low TKE the dependence is nearly linear. A least square fit to the data from the present experiment gives an inverse slope of dTKE/dυ = (12.6 +/- 0.2) MeV/n. Due to the correlation between TKE and fragment mass distributions this quantity must not be interpreted as the energy cost to emit one more neutron, as noted already by Nifenecker et al. [18]. A more detailed analysis of dTKE/dυ for separated mass splits results in an average energy cost per neutron of 7.3 MeV/n. Recently, discrepancies in the available experimental data on the behaviour of the total number of neutrons emitted as a function of total kinetic energy have been identified (see e.g. Ref. [19]), therefore efforts were made to identify experimental factors that may influence the result. All of the literature data included in Fig. 9, except one, were extracted using similar techniques as the present study. The factor having the strongest impact on the slope of ν_T(TKE) is the correction for recoil energy imparted to the fission fragment by the detected neutron [20]. This was apparently not properly taken into account in the two deviating data sets in Fig. 9. A more detailed discussion of the results can be found in Ref. [14].

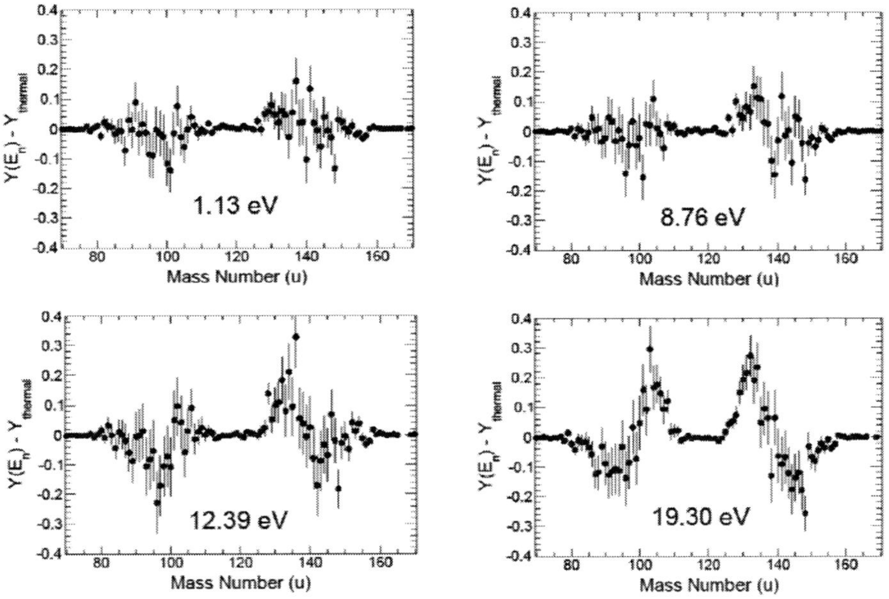

Figure 11. Changes in the mass yield distributions for selected strong resonances of which the energy is given in the inset. The yield differences are calculated to the mass yield distribution at thermal incident neutron energy.

For the prompt fission neutron multiplicity experiment on ^{235}U(n,f), the SCINTIA array (Figs. 6,7) was employed to boost statistics. Preliminary results from a first experimental campaign in the first half of 2015 are shown. In the upper part of Fig. 10 the TKE of fission fragments is plotted as a function of the resonance energy. The TKE shows a strongly fluctuating behaviour as a function of the incident neutron energy, confirming the results of Ref. [9].

In some regions a change in incident neutron energy of only a few eV changes the TKE by more than 10^5 eV. The lower part of Fig. 10 shows relative neutron multiplicities for the same resonance gates as in the upper part. At the present statistical accuracy no statistically significant fluctuations are found.

In Fig. 11 preliminary data on the changes in the fission fragment mass distribution for the strongest resonances are shown. At least the resonances at 12.39 eV and 19.30 eV show a clear change of the fission fragment yield in the inner and other wings of the asymmetric mass distributions. The change amounts to ±0.3 %/u compared to the thermal neutron induced fission yield. Given that the thermal yield is in the order of 6%/u this is a relative change of ±5 %. Again this confirms previous results published in Ref. [9]. Further analysis is in progress.

4 Prompt γ-ray emission in fission

The fact that discrepancies have been observed in various benchmarks for the calculated total γ heating relative to the measured one in the range from 10% to 28% [21], whereas an accuracy of 7.5% is requested, triggered increased interest in new measurements. These new experimental efforts were motivated by requests for new characteristic spectral data, especially for γ -ray multiplicity and mean photon energy per fission in thermal- and fast-neutron induced fission of ^{235}U and ^{239}Pu [22]. Both isotopes are considered the most important ones with respect to the modelling of innovative cores required for fast-spectrum Generation-IV reactors [23]. Instrumental advancements like the development of new γ-ray detectors as well as digital data-acquisition systems made the determination of new and improved prompt fission γ-ray spectral (PFGS) characteristics possible with high precision.

Figure 12. Typical set-up for measuring prompt fission γ-rays at the cold-neutron beam at the KFKI research reactor in Budapest. The fissile actinide sample is mounted in a Frisch-grid ionization chamber.

4.1 Experimental setup

To minimize the uncertainty in determining the γ-ray multiplicity and total released energy, three important detector characteristics need to be considered: (1) energy resolution, (2) intrinsic full peak efficiency and (3) timing resolution. Improved characteristics compared to previously used NaI detectors are nowadays found in cerium-doped lanthanum-halide ($LaBr_3$:Ce) and cerium-halide scintillation detectors ($CeBr_3$). Such detectors have been used in the present experimental campaigns

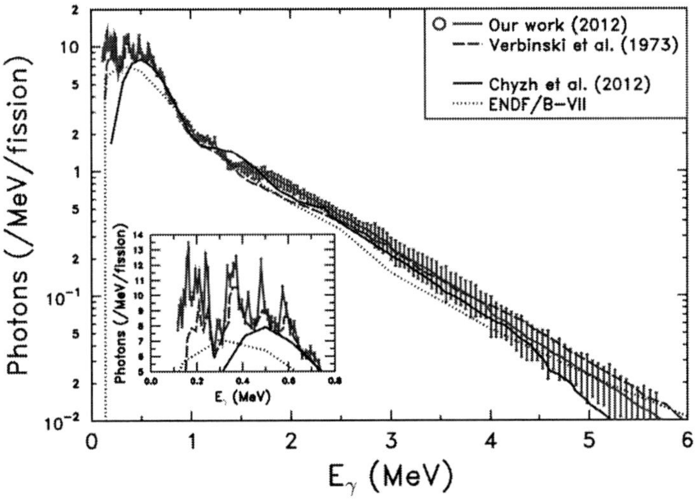

Figure 13. Prompt fission γ-ray spectrum from the spontaneous fission from 252Cf [24] compared to the historical data of Verbinski et al. [25] and recent measurements at LANL [26] (full black line).

together with a twin Frisch grid ionisation chamber as fission fragment detector giving also the trigger for the PFGS measurements. Measurements have been performed for ^{252}Cf(SF) and at thermal beam of the Budapest Research reactor for ^{235}U(n_{th},f and ^{241}Pu(n_{th},f). A summary of the present status is given in Ref. [24]. A few highlights are given in the following.

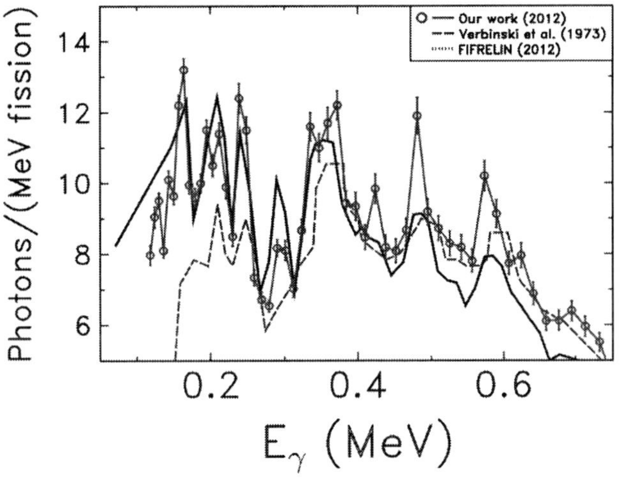

Figure 14. Prompt fission γ-ray spectrum from the spontaneous fission from 252Cf [24] compared to the historical data of Verbinski et al. [25] and recent FIFRELIN model calculations [27]. The dotted line represents the evaluated data ENDF/B-VII.1.

DOI: 10.1051/epjconf/201612201005

In Fig. 13 an example of the PFGS of ^{252}Cf(sf) is given and compared to the historical measurements of Verbinski et al. [25] as well as to measurements using the DANCE detector at Los Alamos National Laboratory (LANL) [26] and the ENDF/B-VII evaluation. Compared to the literature data a more complex structure especially in the low γ-ray energy part is observed as given in the insert of Fig. 13.

In Fig. 14 a zoom of the low energy part of the PFGS is given. In addition to a comparison with Ref. [25] also a comparison with recent model calculations [27] is shown. The agreement is rather remarkable as the model calculation is able to reproduce well the observed structure which rules out the structureless and, hence, unphysical evaluation.

The low-energy structure observed in the PFGS from ^{252}Cf(sf) is also visible in the spectra from thermal-neutron induced fission on ^{235}U and ^{241}Pu with the individual peaks showing different intensities [24].

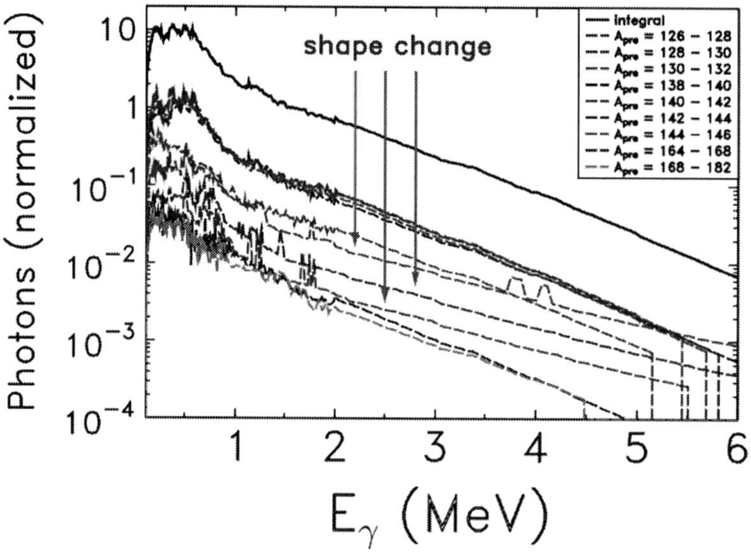

Figure 15. Prompt fission γ-ray spectra as a function of the heavy pre-neutron fission-fragment mass, A_{pre}, shown in logarithmic scale. Each mass cut contains also the contribution from the complementary light fragment, A_{comp}. Indicated is the significant spectral shape change around E = 2.5 MeV for masses between A_{pre} = 126 and A_{pre} = 132 (including A_{comp} = 126 down to A_{comp} = 120).

This may be due to the different shape of the corresponding mass distributions. To study this intriguing feature of the PFGS in more detail correlations of PFGS with fission-fragment properties have been investigated in the spontaneous fission of ^{252}Cf. The focus was on a possible link between the (low-energy) structure and particular mass splits, and on whether the PFGS characteristics change significantly as a function of mass. The PFGS has been analysed as a function of the heavy fragment mass, in Fig. 15 shown for mass cuts of 2 mass units. Apart from the strongly changing structure at γ-ray energies below 1 MeV, a distinct shape change at mass splits with a heavy mass around or smaller than A = 132 is observed. A harder γ -ray spectrum was already observed in the 1990s [28] in an experiment with a 4π NaI-detector array and attributed to non-statistical γ -emission with E > 3.5 MeV from spherical fission fragments. The present data shows that the shape change starts already around

2.5 MeV. Of course, more data is necessary for a quantitative explanation in terms of fission fragment mass and total kinetic energy.

5 Conclusions

The present report gives an overview of the recent research highlights at JRC-IRMM. Progress has been made in recent years on the improvement of important fission cross-sections for innovative reactor system. With the help of digital data acquisition and digital signal processing together with arrays of detector systems for prompt neutron and γ-ray emission more precise measurement capabilities have opened up. For ^{252}Cf(sf) the question on the slope of the neutron multiplicity as a function of TKE has been solved. No strong fluctuations in the prompt neutron multiplicity for ^{235}U(n,f) in resonance induced fission has been found, but fluctuations in mass yield and TKE distributions already mentioned in Ref. [9] have been confirmed. More results concerning this experiment are to be expected. Concerning prompt γ-ray emission in fission, improved detector systems have confirmed, with better energy resolution, the structure at low γ-ray energies observed already in Ref. [25]. In addition a shape change has been found for specific masses around the double magic mass A = 132 confirming results from Ref. [28]. More experiments are planned both at thermal and higher incident neutron energies.

Acknowledgements

Three of the authors (P. S-C, R. B. and A. G.) are indebted to the European Commission for providing a fellowship at EC-JRC IRMM, during which part of this work was carried out. This work was also supported by the European Commission in the frame works of the FP7-ERINDA (agreement number 269499), ANDES (agreement number 249671) and CHANDA (agreement number 605203) programme, which is hereby gratefully acknowledged. The authors are very grateful to the colleagues at the KFKI Budapest for their invaluable support during several measurement campaigns.

References

1. M. Salvatores, *Uncertainty and Target Accuracy Assessment for Innovative Systems Using Recent Covariance Data Evaluations* (NEA/WPEC-26) (Nuclear Energy Agency (OECD), 2008).
2. C. Lubitz, R. Roussin, NEA/WPEC **18** (1999)
3. C. De Saint Jean, R. McKnight, NEA/WPEC **34** (2014)
4. NEA Nuclear Data High Priority Request List https://www.oecd-nea.org/dbdata/hprl/
5. P. Salvador-Castiñeira, T. Brys, R. Eykens, F.-J. Hambsch, A. Göök, A. Moens, S. Oberstedt, G. Sibbens, D. Vanleeuw, M. Vidali, and C. Pretel, Phys. Rev. C **92**, 014620 (2015)
6. P. Salvador-Castiñeira, T. Brys, R. Eykens, F.-J. Hambsch, A. Göök, A. Moens, S. Oberstedt, G. Sibbens, D. Vanleeuw, M. Vidali, and C. Pretel, Phys. Rev. C **92**, 044606 (2015)
7. P. Salvador-Castiñeira, PhD thesis, University Polytecnica de Catalunya, Barcelona (2014).
8. P. Salvador-Castiñeira, T. Brys, R. Eykens, F.-J. Hambsch, A. Moens, S. Oberstedt, G. Sibbens, D. Vanleeuw, M. Vidali, and C. Pretel, Phys. Rev. C **88**, 064611 (2013)
9. F.-J. Hambsch, H.-H. Knitter, C. Budtz-Jørgensen, J.P. Theobald, Nucl. Phys. A**491**, 56.(1989)
10. F.-J. Hambsch, I. Ruskov, L. Dematte, in: Proc. Scientific Workshop on Nuclear Fission Dynamics and the Emission of Prompt Neutrons and Gamma Rays (Theory-1), (2011) EUR 24802, Sinaia, Romania. p. 41.
11. R.E. Howe, T.W. Phillips, C.D. Bowman, Phys. Rev. C **13**, 195 (1976)
12. R. Gwin, R.R. Spencer, R.W. Ingle, Nucl. Sci. Eng. **87**, 381 (1984)
13. A. Al-Adili, F.-J. Hambsch, S. Pomp, S. Oberstedt, Phys. Rev. C **86**, 054601 (2012)
14. A. Göök, F.-J. Hambsch, M. Vidali, Phys. Rev. C **90**, 064611 (2014)
15. A. Göök, F.-J. Hambsch, M. Vidali, Phys. Procedia. **64**, 190 (2015)

16. O. Litaize, O. Serot, Phys. Rev. C **82**, 054616 (2010)
17. C. Budtz-Jørgensen, H.-H. Knitter, Nucl. Phys. A**490**, 307 (1988)
18. H. Nifenecker, C. Signarbieux, R. Babinet, J. Poitou, in: Proc. 3rd Symp. on Physics and Chemistry of Fission, IAEA, Vienna. (1974) p. 51
19. A. Gavron, Nucl. Inst. and Meth. **115**, 99 (1974)
20. A. Tudora, Ann. Nucl. Energy **53**, 507 (2013)
21. A. Lüthi, R. Chawla, and G. Rimpault, Nucl. Sci. Eng. **138**, 3 (2001)
22. Nuclear Data High Priority Request List of the NEA (Req. ID:H.3, H.4), http://www.nea.fr/html/dbdata/hprl/hprlview.pl?ID=421, http://www.nea.fr/html/dbdata/hprl/hprlview.pl?ID=422
23. G. Rimpault, D. Bernard, D. Blanchet, C. Vaglio-Gaudard, S. Ravaux, and A. Santamaria, Phys. Procedia **31**, 3 (2012)
24. S. Oberstedt, R. Billnert, A. Gatera, W. Geerts, P. Halipré, F.-J. Hambsch, M. Lebois, A. Oberstedt, P. Marini, M. Vidali, J.N. Wilson, Phys. Procedia. **64**, 83 (2015)
25. V.V. Verbinski, H. Weber, and R.E. Sund, Phys. Rev C **7**, 1173 (1973)
26. A. Chyzh, C.Y. Wu, E. Kwan, R.A. Henderson, J.M. Gostic, T.A. Bredeweg, R.C. Haight, A.C. Hayes-Sterbenz, M. Jandel, J.M. ODonnell, J.L. Ullmann, Phys. Rev. C **85**, 021601 (2012)
27. D. Regnier, private communication (2013)
28. A. Hotzel, P. Thirolf, Ch. Ender, D. Schwalm, M. Mutterer, P. Singer, M. Klemens, J.P. Theobald, M. Hesse., F. Gönnenwein, H. van den Ploeg, Z. Physik A **356**, 299 (1996).

The SOFIA experiment

Measurement of ^{236}U fission fragment yields in inverse kinematics

L. Grente[1,a], J. Taïeb[1], A. Chatillon[1], J.-F. Martin[1], É. Pellereau[1], G. Boutoux[3], T. Gorbinet[1], G. Bélier[1], B. Laurent[1], H. Alvarez-Pol[6], Y. Ayyad[6], J. Benlliure[6], M. Caamaño[6], L. Audouin[2], E. Casarejos[5], D. Cortina-Gil[6], F. Farget[8], B. Fernández-Domínguez[6], A. Heinz[7], B. Jurado[3], A. Kelić-Heil[4], N. Kurz[4], S. Lindberg[7], B. Löher[4], C. Nociforo[4], C. Paradela[6], S. Pietri[4], D. Ramos[6], J.-L. Rodriguez-Sanchez[6], C. Rodríguez-Tajes[8], D. Rossi[4], K.-H. Schmidt[4], H. Simon[4], L. Tassan-Got[2], H. Törnqvist[4], J. Vargas[6], B. Voss[4], H. Weick[4], and Y. Yan[2]

[1] *CEA, DAM, DIF, 91297 Arpajon, France*
[2] *Institut de Physique Nucléaire d'Orsay, Université Paris-Sud, CNRS, IN2P3, 91406 Orsay, France*
[3] *CNRS, CENBG, 33175 Gradignan, France*
[4] *GSI, Planckstr. 1, 64291 Darmstadt, Germany*
[5] *University of Vigo, 36310 Vigo, Spain*
[6] *Dpt. of Particle Physics, University of Santiago de Compostela, 15782 Santiago de Compostela, Spain*
[7] *Chalmers University of Technology, 41296 Gothenburg, Sweden*
[8] *CNRS, GANIL, Bd H. Becquerel, 14076 Caen, France*

Abstract. The SOFIA (Studies On FIssion with Aladin) experiment aims at measuring fission-fragments isotopic yields with high accuracy using inverse kinematics at relativistic energies. This experimental technique allows to fully identify the fission fragments in nuclear charge and mass number, thus providing very accurate isotopic yields for low-energy fission of a large variety of fissioning systems. This report focuses on the latest results obtained with this set-up concerning electromagnetic-induced fission of ^{236}U.

1 Introduction

The precise knowledge of fission isotopic yields is of great interest in many domains: for fundamental purposes, where it is a unique tool to investigate nuclear structure and collective effects in heavy nucleus and where it has a great influence on our understanding of the r-process nucleosynthesis, as well as for applications where it leads, for example, to a better prediction of the neutron flux in a reactor core and the decay heat after a core shutdown. Despite several decades of experimental investigations on fission, the need for accurate fission-product yields remains.

Indeed, the measurement of isotopic yields is an experimental challenge, and more specifically the measurement of the nuclear charges. Most of the isotopic-yield data originate from neutron-induced fission of actinide targets. Such experiments in direct kinematics are necessarily limited to the charge of the light fragments, because of charge-state fluctuations in the detectors degrading the resolution for the heavier nuclei. Moreover, the use of actinide targets also restrains the study to the

[a] e-mail: lucie.grente@cea.fr

DOI: 10.1051/epjconf/201612201006

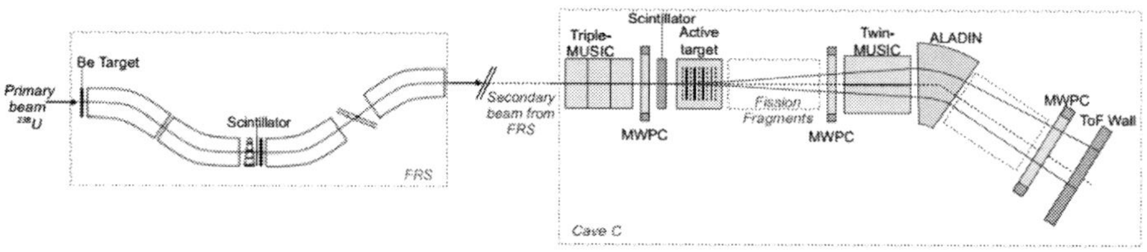

Figure 1. Experimental set-up : production and selection in the FRagment Separator (FRS) of the secondary beam, SOFIA set-up at Cave C.

fission of long-lived nuclei. The alternative method using the inverse kinematics allows to overcome these limitations. The kinematical boost then transmitted to the fission fragments gives access to the charge of the heavier nuclei, with a resolution increasing with their kinetic energy. In addition, this technique offers the possibility of studying a large range of fissioning nuclei thanks to the availability of radioactive beams.

A new type of experiments using this method was introduced at GSI in the 90's [1], representing a breakthrough in the fission-yield measurement : nuclear charges were measured with a good resolution over the whole range of fragments, for the two products of the fission of a large variety of light actinides and pre-actinides. SOFIA (Studies On FIssion with Aladin) falls into the same category of experiments, taking advantage of the inverse kinematics at relativistic energies permitted at GSI, with the objective, this time, of measuring nuclear charges as well as the mass number of the fission fragments. A first set of measurements was performed in 2012 with this set-up where isotopic yields for the low-energy fission of many actinides and pre-actinides were successfully measured [2, 3]. A few extra-shifts were granted in 2014 in order to complete these data, in particular by fission of ^{236}U which is of great interest for the nuclear-reactor community. Neutron-induced fission of ^{235}U occuring in reactors involves indeed the same compound nucleus. First results concerning this recent measurement are presented here.

2 Experimental set-up

SOFIA is a dedicated set-up built around the ALADIN large-acceptance dipole magnet located at GSI (Darmstadt), whose detectors have been specially designed for the study of fission through the simultaneous identification of the nuclear charge and the mass number of both fission fragments, as well as that of the fissioning nuclei. Furthermore, the kinetic energy of the fission fragments is measured, and the total neutron multiplicity is deduced.

2.1 Secondary beam identification

The GSI facility provides an intense primary beam of ^{238}U at 1 A.GeV. The products of the fragmentation reaction of the ^{238}U primary beam on a Be target are selected according to their mass and charge in the FRS (FRagment Separator), a high-resolution recoil spectrometer [4]. The secondary beam of the selected ions of interest is transmitted to the Cave C experimental area where the SOFIA set-up was installed (cf. Figure 1). Those actinides and pre-actinides whose energy is close to 700 A.MeV are fully identified event by event in mass and charge numbers thanks to the simultaneous measurement of their energy loss, magnetic rigidity and time of flight (ΔE - $B\rho$ - ToF technique). A Triple MUSIC

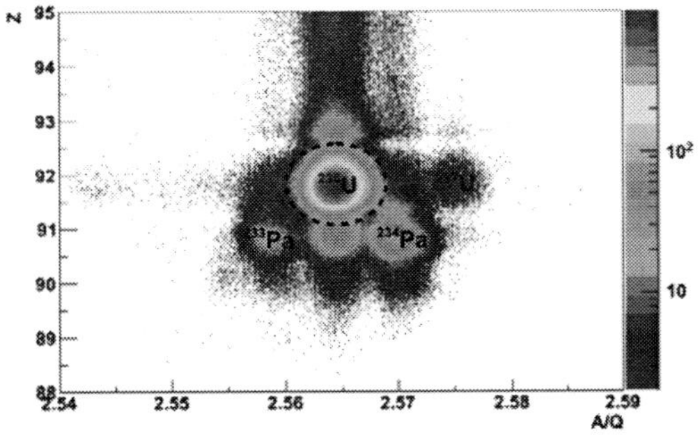

Figure 2. Correlation of the nuclear charge with the ratio of the mass over the charge state for the secondary beam, allowing the identification of the ^{236}U secondary beam.

detector (MUltiple Sampling Ionization Chamber) performs both the energy-loss measurement and the tracking, providing the horizontal angle of the secondary beam. The combination of this angle with the position measurement in a MWPC (Multi-Wire Proportional Chamber) at Cave C gives the magnetic rigidity. The time of flight is measured with two plastic scintillators between the FRS and the Cave C (see figure 1). Figure 2 illustrates the full isotopic identification thus obtained.

2.2 Fission events selection

The fission of the secondary beam nuclei is induced in the active target, a succession of ionization chambers in which four cathodes (two of them made of uranium, one of aluminium and one of lead) act as the targets. The collected anode signals allow to identify in which material the fission takes place. In the frame of this low-energy fission study, one takes an interest only in the electromagnetic-induced fission. This reaction occurs when a target nucleus exchanges a virtual photon with a projectile nucleus going through the target, leading mainly to the excitation of the giant dipole resonance (GDR) of the secondary beam. The latter may then decay through fission. The excitation energy is not measured event by event but its average value is 12 MeV.

Any event corresponding to a nuclear interaction of the secondary beam with a target nucleus has to be discarded, because the fissioning nucleus differs from the one identified before the target, due to a neutron and/or proton removal. Moreover the fragmentation reaction leads, on average, to higher energy fission than the GDR excitation. In the uranium and lead targets, the predominant reaction channel is Coulomb-induced fission and the remaining of the detected fissions follow a nuclear interaction. Conversely, the fissions occuring in the aluminium target are almost only induced by a fragmentation reaction. Based on the limiting-fragmentation hypothesis, the nuclear contribution is deduced from the measurement in the aluminium target and subtracted from the measurement in the uranium and lead targets.

2.3 Fission fragments identification

The fission fragments produced in the active target are emitted in forward direction within a narrow cone of approximately 30 mrad around the secondary beam axis and are deflected in the ALADIN

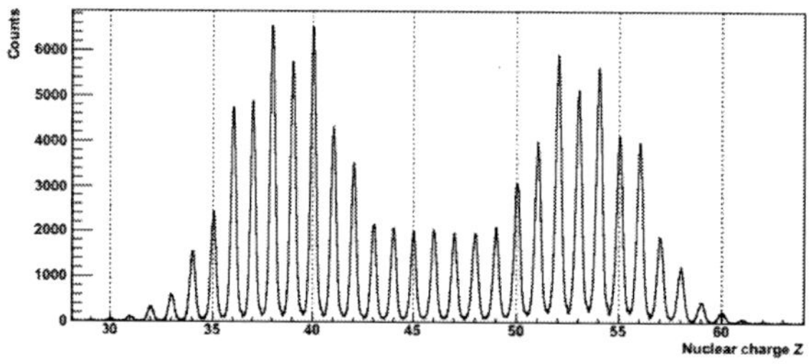

Figure 3. Fission fragments nuclear charge distribution for the electromagnetic-induced fission of ^{236}U.

magnet. The ΔE - Bρ - ToF technique is again applied to each of the two fission fragments. The Twin-MUSIC detector, a double segmented ionization chamber with a common central cathode, performs simultaneously the energy-loss measurement of the two fission products and the tracking, providing their nuclear charge and horizontal angle. The time of flight is measured between the plastic scintillator at the entrance of Cave C and the time-of-flight wall at the very end of the set-up, made of 28 plastic scintillators [5]. Besides the Twin-MUSIC, the complete tracking of the nuclei through the set-up is ensured by two MWPC measuring the coordinates in the plane perpendicular to the spectrometer axis. The mass and the charge numbers of both fission fragments are deduced from these measurements. The analysis of the latest experiment is still ongoing, this report focuses on the nuclear-charge results.

3 Nuclear-charge yields

Figure 3 represents the nuclear-charge distribution of the ^{236}U electromagnetic-induced fission fragments. A very high Z-resolution is reached, slightly reduced for the heavier fragments but at least as good as 0.35 charge unit (FWHM) over the whole range.

Through simulations taking into account the geometry and some cutoff effects of the detectors, the efficiency of the set-up is carefully estimated as a function of the charge and the mass of the detected nuclei, resulting in an average efficiency of 90 % for the last version of the set-up. Elemental yields are then calculated after correction of the fission-fragment charge distributions by this efficiency. Elemental yields for the electromagnetic-induced fission of ^{234,235}U [6] and ^{238}U [7], measured in 2012, and of ^{236}U, measured during the last run in 2014, are compared on figure 4 (top), illustrating the complete consistency between the data from both measurements: whatever the value of Z considered, the yield evolves steadily as a function of the mass of the fissioning uranium isotope.

The bottom of figure 4 highlights the relative statistical uncertainty on the ^{236}U fission yields. The very high statistics gathered for this reaction leads to a relative uncertainty lower than 0.5 % for the most populated fragments and lower than 2 % in the symmetric fission region.

Figure 5 represents the local proton even-odd staggering, expressed following the prescription of [8], as a function of Z for the four reactions ^{234,235,236,238}U (γ,f). A similar trend of an increase of the even-odd effect with the asymmetry of the fission is observed, in accordance with a general phenomenon already observed [9] and explained as the manifestation of the energy sorting mechanism [10] : a common relatively low even-odd effect is observed in the four cases around Z = 46 and is then regularly increasing with the asymmetry.

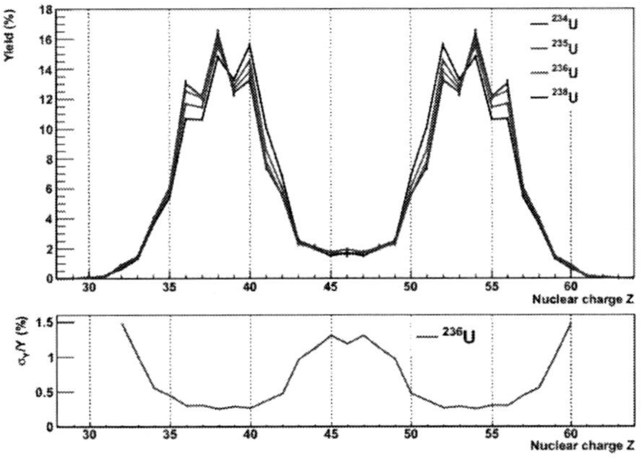

Figure 4. Nuclear charge yields measured with SOFIA for the electromagnetic-induced fission of different Uranium isotopes (top). Relative statistical uncertainty on the elemental yields for the electromagnetic-induced fission of ^{236}U (bottom).

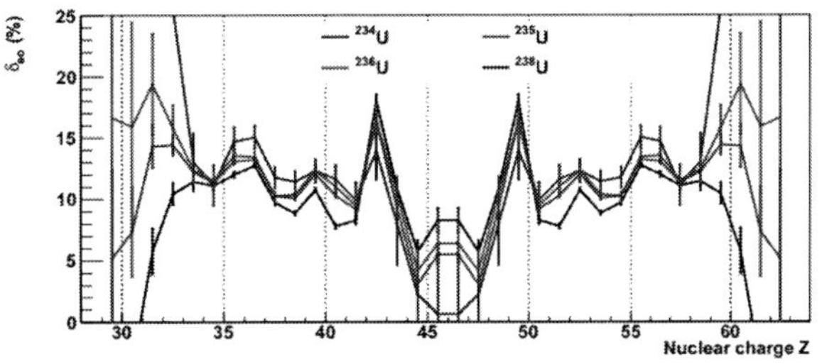

Figure 5. Local even-odd staggering in nuclear-charge yields as a function of the nuclear charge of the fission fragments, for the electromagnetic-induced fission of different U isotopes measured with SOFIA in 2012 for ^{234,235,238}U and in 2014 for ^{236}U.

Finally, elemental yields for the fission of ^{236}U are compared with evaluated data from two libraries, B-VII.1 and JEFF-3.1, for Neutron-induced fission of ^{235}U at different energies on figure 6. A striking feature of this comparison is the very low uncertainty of the SOFIA data compared to evaluated data. One can also note the consistency of the evaluations with an expected mean excitation energy of 12 MeV for the reaction ^{236}U (γ,f), i.e. a neutron energy of approximately 6 MeV for the analog reaction ^{235}U(n,f), by observing the ratio of the yield at the symmetry to the one at the asymmetry which is expected to grow with the excitation energy of the fissioning nuclei.

 DOI: 10.1051/epjconf/201612201006

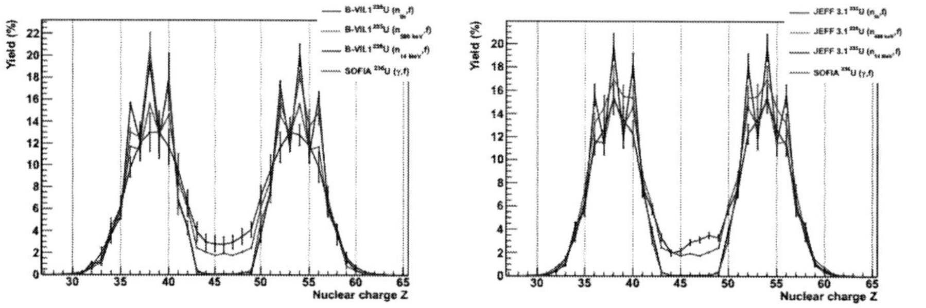

Figure 6. Comparison of the nuclear-charge yields measured with SOFIA for electromagnetic-induced fission of ^{236}U with two different sets of evaluation, B-VII.1 on the right and JEFF-3.1 on the left for the reaction ^{235}U(n,f) at three different neutron energies. Error bars are indicated.

4 Conclusion

Electromagnetic-induced fission of ^{236}U has recently been measured with the SOFIA set-up using inverse kinematics at relativistic energies. The ongoing analysis already reveals the high quality of the data with nuclear-charge yield uncertainties as low as 0.5 % and an unprecedented Z-resolution reaching 0.35 charge unit (FWHM). The comparison with the previous results on the fission of ^{234,235,238}U shows complete consistency.

The ongoing analysis of the fission-fragment masses looks very promising. Very accurate isotopic yields for the ^{236}U fission are thus expected. The kinetic energy and total mean neutron multiplicity, also accessible with SOFIA, still has to be extracted from the data to complete these preliminary results.

Further experimental studies on fission with the SOFIA set-up within the R3B (Reactions with Relativistic Radioactive Beams) collaboration are considered for the near future [2]. The coupling of the SOFIA recoil spectrometer with detectors such as NeuLAND (new Large-Area Neutron Detector) or CALIFA (CALorimeter for the In Flight detection of γ rays and light charged pArticles) would allow to further extend the number of accessible observables of fission in order to get an even more precise insight into the complex phenomenon of fission.

References

[1] K.-H. Schmidt *et al.*, Nucl. Phys. A **665**, 221-267 (2000)
[2] J.-F. Martin *et al.*, Eur. Phys. J. A **51**, 174 (2015)
[3] G. Boutoux *et al.*, Physics Procedia **47**, 166 (2013)
[4] H. Geissel *et al.*, Nucl. Instr. Meth. B **70**, 286-297, (1992)
[5] A. Ebran *et al.*, Nucl. Instr. Meth. A **728**, 40-46 (2013)
[6] J.-F. Martin *et al.*, in preparation
[7] É. Pellereau *et al.*, in preparation
[8] B.L. Tracy *et al.*, Phys. Rev. C **5** 222 (1972)
[9] M. Caamaño *et al.*, J. Phys.G: Nucl. Part. Phys. **38** 035101 (2011)
[10] B. Jurado and K.-H. Schmidt, J. Phys.G: Nucl. Part. Phys. **42** 055101 (2015)

Analysis of prompt fission neutrons in $^{235}U(n_{th},f)$ and fission fragment distributions for the thermal neutron induced fission of ^{234}U

A. Al-Adili[1,a], D. Tarrío[1], F.-J. Hambsch[2,b], A. Göök[2], K. Jansson[1], A. Solders[1], V. Rakopoulos[1], C. Gustafsson[1], M. Lantz[1], A. Mattera[1], S. Oberstedt[2], A. V. Prokofiev[1], M. Vidali[2], M. Österlund[1], and S. Pomp[1]

[1] *Department of Physics and Astronomy, Uppsala University, Sweden*
[2] *EC-JRC - Institute of Reference Materials and Measurements (IRMM), Geel, Belgium*

Abstract. This paper presents the ongoing analysis of two fission experiments. Both projects are part of the collaboration between the nuclear reactions group at Uppsala and the JRC-IRMM. The first experiment deals with the prompt fission neutron multiplicity in the thermal neutron induced fission of $^{235}U(n,f)$. The second, on the fission fragment properties in the thermal fission of $^{234}U(n,f)$.
The prompt fission neutron multiplicity has been measured at the JRC-IRMM using two liquid scintillators in coincidence with an ionization chamber. The first experimental campaign focused on $^{235}U(n_{th},f)$ whereas a second experimental campaign is foreseen later for the same reaction at 5.5 MeV. The goal is to investigate how the so-called saw-tooth shape changes as a function of fragment mass and excitation energy. Some harsh experimental conditions were experienced due to the large radiation background. The solution to this will be discussed along with preliminary results.
In addition, the analysis of thermal neutron induced fission of $^{234}U(n,f)$ will be discussed. Currently analysis of data is ongoing, originally taken at the ILL reactor. The experiment is of particular interest since no measurement exist of the mass and energy distributions for this system at thermal energies. One main problem encountered during analysis was the huge background of $^{235}U(n_{th},f)$. Despite the negligible isotopic traces in the sample, the cross section difference is enormous. Solution to this parasitic background will be highlighted.

1 Introduction

The Uppsala group has a few ongoing projects investigating nuclear fission, e.g. on independent fission yields, high precision cross sections and fission-fragment properties [1]. In recent years the group has also started new collaborations with colleagues from JRC-IRMM to measure prompt fission neutrons. This work discusses the analysis of two ongoing works on $^{235}U(n_{th},f)$ and $^{234}U(n_{th},f)$.

[a] e-mail: ali.al-adili@physics.uu.se
[b] e-mail: franz-Josef.HAMBSCH@ec.europa.eu

1.1 Prompt fission neutron multiplicity

The share of excitation energies between the fission fragments (FF) is of importance for understanding nuclear fission. In particular how the excitation energies (viz. neutron emission) varies as a function of incident neutron energy. The nuclear data community lacks conclusive data to address this issue, although some experimental data supports an increasing $\overline{\nu}(A)$ from the heavy fragments [2]. However, it seems also that theoreticians have agreed to disagree on the explanation of these observations. Certainly, more data are needed to understand the saw-tooth behavior at higher E_n. In addition, systematic studies on a few reactions and correlations between fragment+neutron characteristics could possibly contribute to a better modeling of the neutron emission. The current experimental activity on measuring $\overline{\nu}(A)$, is a feasibility test for a future larger systematic measurement campaign. A first step was to measure ^{235}U(n_{th},f) at the JRC-IRMM Van De Graff accelerator, and this will be compared to a future run at E_n = 5.5 MeV.

1.2 Thermal fission of a fertile system

The second part of the paper concerns data on ^{234}U(n_{th},f) originally measured at the reactor of the Institut Laue-Langevin (ILL) in Grenoble, France. The experiment was performed in 1999 using a conventional Twin Frisch-Grid Ionization Chamber (TFGIC) in conjunction with an analogue data acquisition system. The aim was to measure, for the first time, the fission mass yield, angular- and energy distributions of ^{234}U(n,f) at sub-thermal energy, much below the fission threshold. A cold neutron beam was used to induce fission in the main ^{234}U target and separately in a ^{235}U target for calibration purposes. Unfortunately some problems in the collected data jeopardized the extraction of useful results. Here, some 15 years later, new solutions are attempted to save what can be saved from the data. More details about those challenges will be discussed in sect. 3.

2 Project 1: $\overline{\nu}(A)$ for ^{235}U(n_{th},f)

As commonly known, measuring neutrons is not straightforward. In this work we are forced to detect them indirectly through their recoil to a proton. From the recoil energy the generated light output is extracted and compared to other signal shapes. The shape of the induced signal is related to the sort of particle giving rise to it. In such a way, one could separate gammas from neutrons for instance. The challenges do not stop here, because not all neutrons will have equal chances of getting detected. They even show a dependence on the incoming neutron energy. The internal detector efficiency has to be quantified e.g. by measuring a known standard spectrum which in this case comes from ^{252}Cf(sf). Once a neutron is identified (and successfully distinguished from scattered neutrons), we look for a coincidence with a fission fragment from the ionization chamber. If we found one, a game of kinematics starts to reconstruct the scene to reveal what angle and energy the neutron obtained when ejected. Once all observables are in place we ought to look for correlations that might lead to interesting insights.

2.1 Simulations of the background in the thermal measurement

Prior to the measurement, extensive simulations were performed with the FLUKA code [3] in order to ensure a minimum contribution from beam neutrons. Figure 1 shows the schematic experimental setup with the TFGIC and the neutron detectors (ND). The proton beam is impinging on a Li neutron-converter giving rise to 0.5 MeV neutrons. Paraffin was used to thermalize those neutrons before irradiating a 117 μg/cm^2 ^{235}U target placed in the centre of the chamber cathode. The chamber is

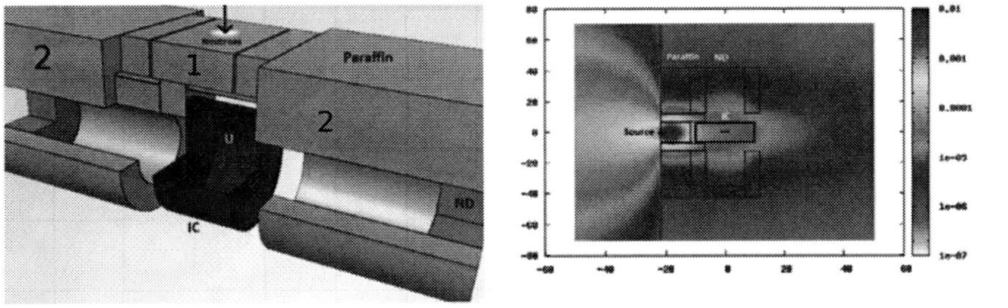

Figure 1. Left: The geometry simulated in FLUKA prior to the first experiment in 2014. Paraffin was used to thermalize the neutron beam. Right: The neutron flux density for the thermalized neutron beam.

tilted 90° to the neutron beam. The fission fragments are detected when they ionize the chamber gas (P-10 at 1.05 bar pressure) and the liberated electrons are collected on the anode plates. The neutrons are anisotropically emitted in the laboratory frame with preference towards the motion of the fully accelerated FF. The neutron detectors are placed on each side of the chamber in order to detect the FF with the highest mass-, energy and angular resolution (due to minimum energy losses in the target). Paraffin and Lead were planned to be used to protect the detectors from unwanted background from the neutron field. The neutron fluxes in the target and neutron detectors were estimated (See Fig. 1). A balance had to be achieved between the n-FF coincidence rate and the acceptable level of neutron background in the detectors. The best combination found was to use a low energetic neutron beam of 0.5 MeV together with large amounts of paraffin, because the NE213-based scintillator is practically transparent to thermal neutrons. The neutron flux at the U target was estimated to be roughly 10^4 n/cm^2 s, of which 93% are below 100 keV. In total this corresponds to about $\approx$ 50 ff/s. The solid angle of the neutron detectors, at 35 cm distance, is roughly 1% which gives a coincidence rate of about 0.5 coincidences/s. About 2.4 neutrons are emitted on average per fission event, but this is compensated by the neutron detector intrinsic efficiency which is about 30% on average. Thus, about half a million neutron-FF coincidences were to be expected in 15 days of stable beam-time at 5 μA. The background level in the neutron detectors are in the order of a million n/s. However, only a small fraction are above the detection threshold of the liquid scintillator. Because the chamber provided the fission trigger and considering the time of flight of the fission neutrons, the background level in the detectors was regarded to be of no concern for the thermal measurement.

2.2 Experiments and analysis

The experimental data was collected using a digital data acquisition system. The fission fragment energies and angles were obtained from the chamber signals. The neutron energy is derived from the time-of-flight (ToF) calculated as the time difference between the cathode signal and the neutron detector pick-off signal. Hence, the signal from both chamber and neutron detectors were digitized and stored for off-line analysis. The neutron multiplicity is estimated by counting the number of times a coincidence was registered between the FF and the neutron detector. The experimental and analysis procedure is very much similar to the work of Ref. [4]. First, ^{252}Cf(sf) was measured using the identical experimental setup. By utilizing the Mannhart evaluation for the neutron spectrum, one can determine the detector efficiency function. In order to estimate the level of neutron scattering, the

 DOI: 10.1051/epjconf/201612201007

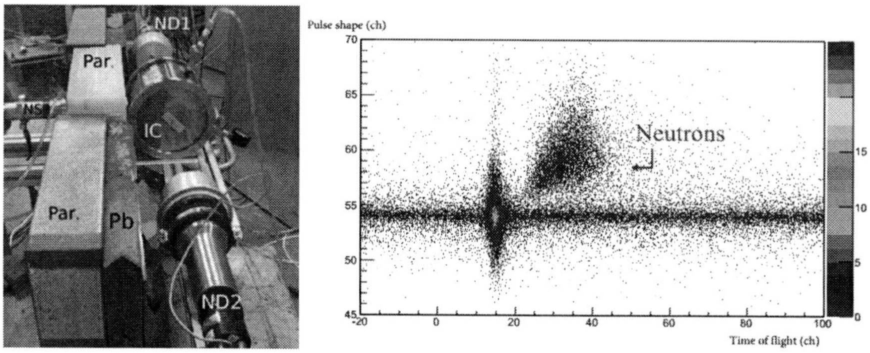

Figure 2. Left: The experimental setup. The neutron source (NS) can be seen along with the shielding material (Pb), moderator (Paraffin), the IC and the two ND. Right: The raw data from the ^{235}U run neutron detector showing gammas and neutrons.

surrounding material was removed in one run and ^{252}Cf was remeasured. In this way one can later study the impact of the shielding material on the neutron spectrum and to understand the scattering component from the Prompt Fission Neutron Spectrum (PFNS). An additional reduction to the effect of scattered neutrons, is done by analyzing the neutron detector pulse height versus ToF and looking for the maximum transfer of energy to the recoil protons in the scintillator.

During the measurement, up to 10^5 γ/s were registered in the ND. Most of which came from the beam or from neutron capture in hydrogen. Luckily, due to the possibility of neutron gamma separation based on pulse shape (PS), one may get rid of the gamma component. Some raw data are shown in Fig. 2 and the gammas are appearing uncorrelated with time and have a particular signal fall-time (zero-crossing method). However neutrons are correlated with time since they are in coincidence with the trigger from the fission chamber. Note also the prompt fission gammas are also in coincidence with the fission fragments in the chamber. By means of Gaussian fitting as a function of signal height, one may retrieve the border line between gammas and neutrons as seen in Fig. 3. The timing resolution can be looked for in the prompt fission gamma peak and is slightly above 1 ns. Despite the resolution limit from the sampling rate (2.5 ns/sample) digital interpolations allow for obtaining a better timing resolution. The lower plots of Fig. 3 show the ToF spectra before and after PS discrimination. The green line shows the neutron energy after discrimination against gammas and a ToF cut. From the tof spectrum, the neutron energy is calculated following non-relativistic kinematics. The resulting neutron spectra for the two detectors are shown in the left plot of Fig. 4 together with the Mannhart evaluation. The agreement between the two detectors is very good. However the energy calibration is still under investigation so the two detectors might not have the same threshold. The neutron detector efficiency (shown as an example in the right plot of Fig. 4) is determined from the Cf measurement by dividing the calculated neutron energy spectrum with the Mannhart evaluation. The solid angle has to be accounted to estimate the detection efficiency. The detector efficiency is dependent on the threshold in the detector as shown in a simulation in the figure for different pulse height [5]. Ongoing work is now focused on the energy calibration to ensure the same threshold in both detectors.

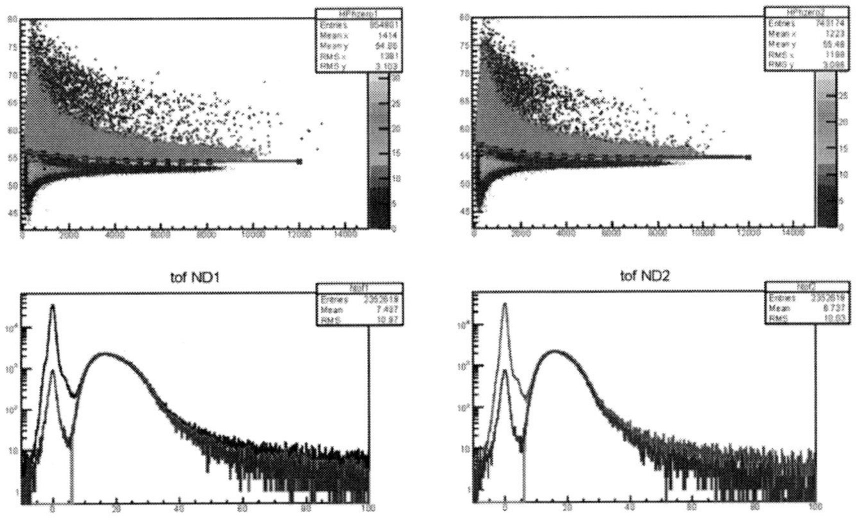

Figure 3. Upper :The neutron gamma separation by use of the pulse shape discrimination technique. Gaussian fits as a function of pulse height in the ND give the separation line between prompt gamma and prompt neutrons. Lower: The ToF spectra before and after pulse shape (PS) discrimination. The green lines show the neutron energy after discrimination against gammas and a ToF cut.

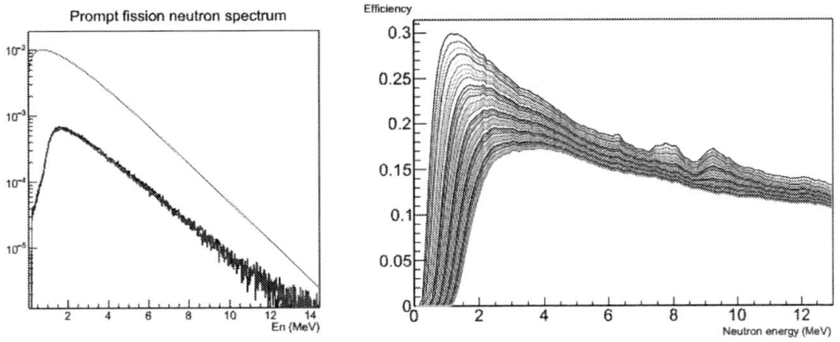

Figure 4. Left: The ^{252}Cf neutron spectra from the two detectors together with the Mannhart evaluation. Right: The detector efficiency as a function of detector threshold (GEANT simulation from Ref. [5]).

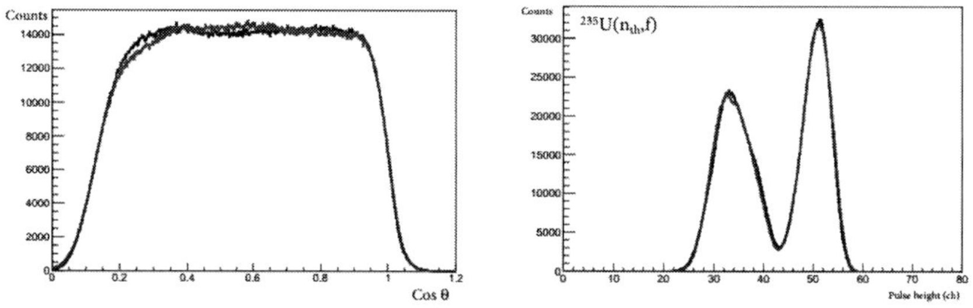

Figure 5. Left: The cosine distribution of the fission fragments from both sample and backing sides. Right: The pulse-height (energy) distributions from sample and backing sides.

2.3 Preliminary results

We are still in the process of ensuring the quality of the neutron data. Once the data are in optimum shape the fission fragment correlation study can be finalized. However preliminary, the provisional distributions have been studied to obtain an overall check. Some rough analysis on the ^{235}U fragment data was done and the distributions show a decent behavior. The cosine distribution is plotted for the two chamber sides in Fig. 5 together with the pulse height distributions (proportional to the fragment energy). A decent matching is seen, despite improvements will enhance the agreement between the two sides. Figure 6 shows a preliminary plot on the coincidence condition between the two detectors and the chamber. To the left and in the middle, are the two provisional mass distributions for ^{235}U(n$_{th}$,f) with (green, pink) and without (blue, red) coincidence condition. The asymmetric distributions are from the coincidence condition where the light fragments seem to emit more neutrons. The ratio between those two distributions give a rough idea on the saw-tooth trend (see right plot of Fig. 6). Albeit the roughness of the provisional plots and the discrepancy between the two detectors we are optimistic that the data look reasonable. Final data analysis is ongoing and preparations for the next campaign is also under exploration.

2.4 Testing and preparations for measuring at E_n = 5.5 MeV

A test run was performed later in 2015 on the 5.5 MeV case, also with two liquid scintillators and a Frisch-Grid ionization chamber. The aim was to investigate the background conditions and to compare roughly with the thermal measurement. MCNP and FLUKA simulations were performed before the experiment to estimate the neutron fields and to optimize the experimental setup. Due to problems with the beam, statistics became an issue for the 5.5 MeV test run and focus was instead put on benchmarking the simulations with different shielding conditions/materials. Shielding materials such as Pb, Cu and Paraffin were all used in different configurations and thicknesses. A throughout benchmarking is ongoing between the different simulation codes and the experimental data.

3 Project 2: Fission properties of ^{234}U(n$_{th}$,f)

3.1 Experiments

In 1999, neutrons from the ILL reactor were irradiating the Uranium sample inside a Frisch-grid ionization chamber. Two different targets were used separately during the experimental campaign; one

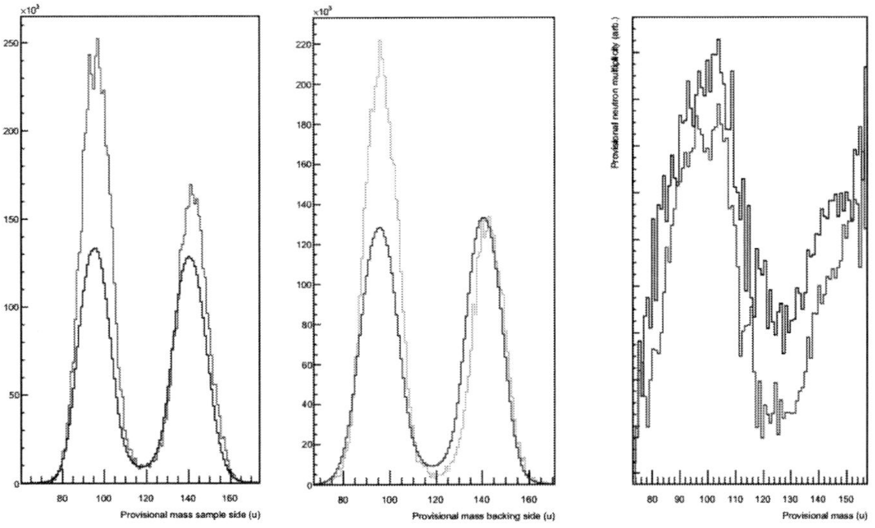

Figure 6. Left and middle: The provisional mass distributions from the two chamber sides. The more symmetric distributions show the non-coincidence mode whereas the asymmetric show the coincident events between neutrons and fission fragments. The ratio between the distributions is seen in the right plot and comprised a preliminary saw-tooth trend.

^{234}U sample for the main measurement and another natural uranium target for the reference measurement with ^{235}U. Unfortunately, two problems affected the data and have implied a rather cumbersome analysis of these data:

1. **Fission events from the reaction ^{235}U(n$_{th}$,f).** The ^{234}U sample used for irradiation has a contamination of 0.07% of ^{235}U residues. Despite the negligible trace, the thermal fission cross section is much larger (584 barn instead of 0.067 barn). Due to this difference in fission cross section a larger ^{235}U(n$_{th}$,f) background was contaminating the desired ^{234}U data (roughly 87%). The problem arises in the kinematics calculations since the mass of the fissioning compound nucleus (m$_{cn}$) is unknown yet an important parameter. With a large background of ^{235}U(n,f), m$_{cn}$ is in fact to a large extent 236 u rather than 235 u. Proper treatment of this background was necessary. One possible solution is to randomize m$_{cn}$, event-by-event wise, between the two values (based on the ratio of cross sections) and analyze the data in a Monte Carlo manner. Another solution is to utilize the reference measurement to subtract for the unwanted background. The latter approach was found better suited for this application. The ^{235}U data was treated in the same way as the background data in the ^{234}U data set, namely by using m$_{cn}$=235 and the neutron multiplicity of $\overline{\nu}_{234}(A)$.

2. **Fission product range in the gas.** Due to some restricting conditions, a CF_4 counting gas was used at a reduced gas pressure. Unfortunately, the high energetic ions were not fully stopped (light fission products). As a result, a clear artefact can be seen in the plots of energy versus emission angle. This problem implied that the angle determination of the light ions could not be extracted with confidence and had to be determined otherwise. The solution was to use only the emission angle from the complementary heavy fragment in each fission event, since these have lower kinetic energy and have obtained full stopping range in the gas. In addition, the data with smaller emission angles, corresponding to non-fully measured pulse height had to be excluded from the full analysis.

In the following section, the ongoing search for solutions to these problems is explained.

3.2 Analysis

The analysis of the fission fragment data followed the procedure described in Ref. [6]. The conservation of mass and momentum provides the means of calculating the pre-fission neutron emission distributions. In order to address the first issue of sect. 3.1, the analysis procedure was as follows:

1. The absolute energy calibration was performed by analyzing the ^{235}U(n$_{th}$,f) data. The experimental parameters were fine-tuned to reach the literature values on the average total kinetic energy (TKE = 170.5 MeV) and average heavy mass of 139.6 u. m_{cn} = 236 is used along with $\bar{\nu}_{235}(A)$ [7].
2. The ^{234}U(n$_{th}$,f) data was analyzed with m_{cn} = 235 and $\bar{\nu}_{234}(A)$ [8].
3. The ^{235}U(n$_{th}$,f) data was now re-analyzed assuming it is ^{234}U(n$_{th}$,f), i.e. with m_{cn} = 235 and $\bar{\nu}_{234}(A)$.
4. The background subtraction is performed on item (2), by subtracting item (3) from it after scaling with the fission cross section and mass. The fission rate is given by the sum of all fissioning isotopes:

$$FR = (\sigma_{234} \times N_{234} + \sigma_{235} \times N_{235} + \sigma_{other} \times N_{other})\,\Phi \quad , \tag{1}$$

where N denotes the number of atoms, σ is the fission cross section and Φ is the neutron flux. The main contributions come from the two first isotopes, whereas the other are negligible. Therefore, the fraction of fissions coming from ^{235}U(n,f) is:

$$FR_{235} = \frac{\sigma_{235} \times N_{235}}{\Sigma\{\sigma \times N\}} \approx \frac{584b \times 0.077\%}{0.067b \times 99.923\% + 584b \times 0.077\%} \approx 87\% \tag{2}$$

Although, the 99.923 % includes all other isotopes but ^{235}U, using σ_{234} is a fairly good approximation.

3.2.1 α Pile-up

The ^{234}U sample had a pile-up rate of about 10^5 α/s. These alpha particles affect also the background data (i.e both ^{234,235}U(n$_{th}$,f)). Therefore, in view of the needed background subtraction stressed in point 1, sect. 3.1 the background has to be affected by the same pile-up signals. The response of the

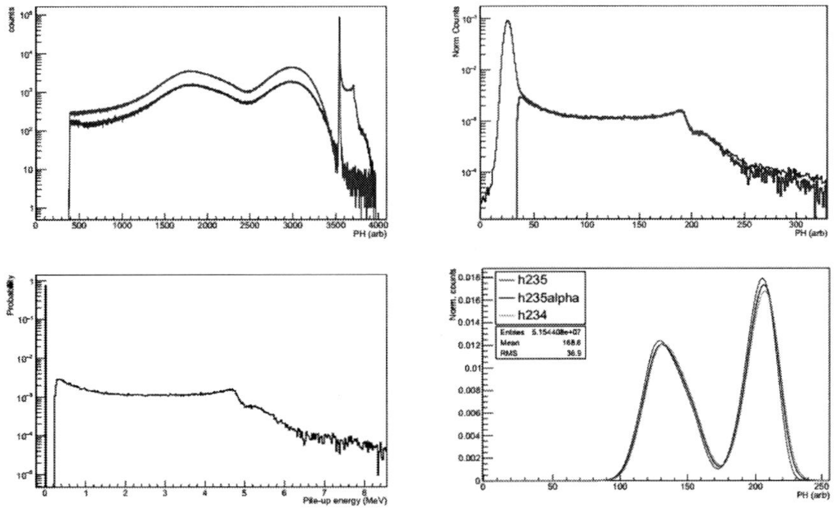

Figure 7. a) The α pile up in ^{234}U (blue) in contrast to the pulse height from ^{235}U (red), together with the signal from the precision pulse generator (at ch 3500). b) The α pile up contribution (red) deduced from the affected pulser signal. c) The probability density function (PDF) of α pile up signals, generated from the pulser response function. 30% is the estimated α pile up rate. d) The pulse height distributions for ^{234}U (green), the background run ^{235}U (red) and the simulated α pile up affected ^{235}U background (blue).

detector to the pile-up can be studied on the precision pulse-generator, originally used for electronic-drift monitoring. Since the pulser gives a well-defined signal, the impact of the alpha pile-up is nicely visible. Figure 7a shows the pulse-height distributions of ^{234,235}U(n$_{th}$,f) and the pulser at channel 3500 (note the pile up on the pulser in the case of ^{234}U (blue line)). The pulser signal was used as a probability density function (PDF) to simulate the α pile up in the background data. The α pile up contribution is seen in Fig. 7b. About 70 % of the events are free from pile-up. The PDF was created by converting the α contribution to energy (see Fig. 7c) and adding a zero channel with the amplitude 1-Integral(α contribution). In the last step, a random generator was used to sample from the generated PDF and thus emulate the pile-up process by adding the signal to the anode signals stochastically event-by-event. The result is shown in Fig. 7d where the pulse height distribution is shown for ^{234}U (green line) and ^{235}U (red line). By adding the simulated α pile up, the ^{235}U distribution (blue line) is closer to the naturally affected ^{234}U data (which in turn is dominated by the background of ^{235}U).

3.2.2 Emission angle and energy losses

The two signals used from the ionization chamber are the anode and sum (anode+grid) signals. After the grid inefficiency correction [9, 10] the angular distributions were determined via the summing method [11].

A zoomed plot is shown on the light fragments in Fig. 8 where the effect of the second issue of sect. 3.1 is clearly visible. The fragments do not acquire their full stopping range and thus accumulate

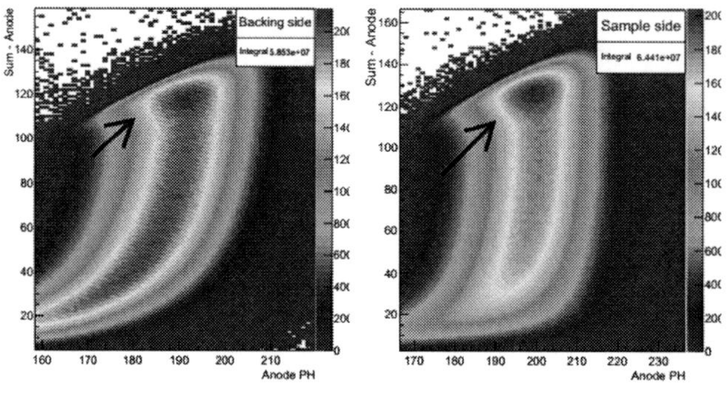

Figure 8. a) $PH_{\Sigma} - PH_A$ versus PH_A, used to extract the emission angle (See text for details). Note the range issue at larger PH_{Σ} values (sect. 3.1).

for smaller emission angles and high energies. The $\overline{X}/D$ distribution for all light fragments will be affected by a too small range. The resulting $\cos(\theta)$ distributions from the two chamber exhibits an overshoot at angles close to $\cos(\theta)=1$ especially for the sample side. In addition the slope of the backing side seems to be too steep. The angular resolution in terms of Full Width at Half Maximum (FWHM) is 0.084. Because of the second issue of sect. 3.1, the angular resolution is determined only for $0.5 \leq \cos(\theta) \leq 0.85$. The best solution found to the range problem was to use the emission angle from the heavy fragment solely. Since the angular resolution seems to be fairly good, the uncertainty from doing so is relatively small.

The energy-loss correction is done on the mean anode pulse heights from both chamber sides as a function of $1/\cos(\theta)$. A linear fit is extrapolated to the intersection at the y-axis corresponding to an assumed ideal energy-loss free channel to which the pulse heights are then corrected to, depending on the emission angle. Due to the degraded angular resolution at higher angles, the final fragment mass calculation includes only events with $\cos(\theta) \geq 0.5$. In addition, due to the issue 2 of sect. 3.1, the angle is further restricted to $0.5 \leq \cos(\theta) \leq 0.85$.

3.2.3 Fragment-mass determination

The "pulse-height defect" (PHD) accounts normally for about 4 MeV reduction in the detected FF energy. It depends mainly on the fragment mass and energy, and can be corrected for by the parametrization from Ref. [12]. In this work, the PHD had to be larger (about 6-7 MeV) in order to obtain proper reference distributions. Probably this is due to the different conditions of this experiment compared to standard measurements performed at the JRC-IRMM (i.e. properties of the counting gas CF_4 compared to P-10 gas).

One critical step in the analysis is to estimate the neutron emission on an event-by-event basis. The neutron multiplicity of $^{235}U(n_{th},f)$ was obtained from Ref. [7]. The neutron multiplicity of ^{234}U was obtained as the average of ^{233}U and ^{235}U [13]. In addition to the mass dependence, the neutron multiplicity is dependent on the TKE. Both sawtooth shapes used in this work are shown in Fig. 9a.

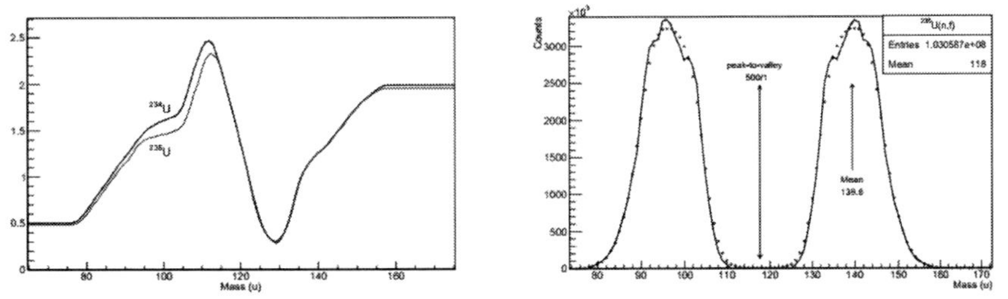

Figure 9. a) The neutron multiplicity for both systems. ^{235}U was obtained from Ref. [7]. ^{234}U was obtained from an average of ^{233}U and ^{235}U. b) the experimental mass distribution of ^{235}U(n,f) (red dots) compared to literature data from Ref. [14] (solid line).

The full kinematics necessary to determine the fragment masses are described in Ref. [8]. In this work, the equations are simplified since the neutron energy is about 0 MeV. The pre-neutron emission energy in the CM system can be calculated by:

$$E_{\text{pre}}^{\text{CM}} = \frac{A_{\text{pre}}}{A_{\text{pre}} - \nu(A, \text{TKE})} E_{\text{post}}^{\text{LAB}} \tag{3}$$

The final pre-neutron emission fragment masses are calculated as:

$$A_{1,\text{pre}} = A_{\text{CN}} \cdot \frac{E_{2,\text{pre}}^{\text{CM}}}{E_{1,\text{pre}}^{\text{CM}} + E_{2,\text{pre}}^{\text{CM}}} \quad \text{and} \quad A_{2,\text{pre}} = A_{\text{CN}} \cdot \frac{E_{1,\text{pre}}^{\text{CM}}}{E_{1,\text{pre}}^{\text{CM}} + E_{2,\text{pre}}^{\text{CM}}} \tag{4}$$

The calculations above are repeated in an iterative loop where the newly calculated masses are always compared with the earlier determined. When the mass difference, $A_{\text{new}} - A_{\text{old}}$, in two sequential iterations is less than 1/16 amu the calculation is finished.

3.3 Preliminary results

In sect. 3.2 the analysis scheme was presented. In the first step, the calibration is done using ^{235}U(n_{th},f). The resulting mass distribution of the reference run is shown in Fig. 9b (full line), in comparison to the one of Ref. [14] (red dots). The peak-to-valley ratio between the peaks and the symmetric yield region is around 500 and the mass distributions are in relatively good agreement. Due to the limited mass resolution in the present experiment the finer structure is not reproduced.

In the second step,^{234}U(n,f) was analyzed and the resulting mass vs TKE is shown in Fig. 10a. In the third step and in order to subtract the background, the reference ^{235}U was re-analyzed with m_{cn} = 235 and $\bar{\nu}_{234}$. Fig. 10b shows the mass vs TKE. Note also that to the reference also a simulated α pile up (see sect. 3.2.1 for details) was added. Finally the background was subtracted from the 2-dimensional mass and TKE plot after scaling the background to the calculated fission rate (Fig. 10c).

Figure 11a shows the fission yields for ^{234}U(n,f) plus the background, the background normalized to the fission cross section, ^{235}U, and the background-subtracted ^{234}U(n,f). The difference between the two distributions is the assumed clean ^{234}U(n,f) mass yield. Figure 11b shows all cases compared to

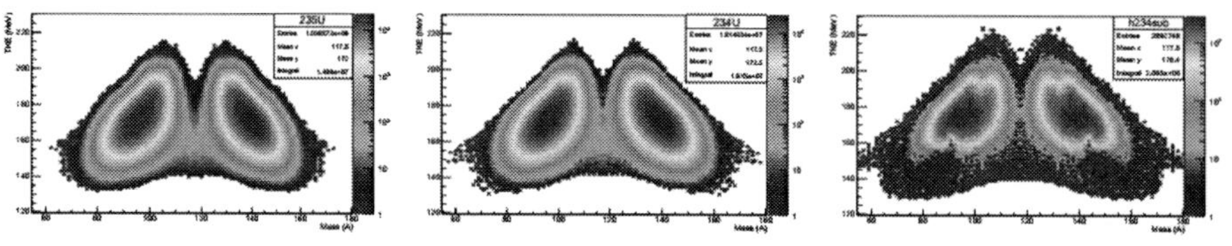

Figure 10. a) ^{234}U(n,f) including the 85% background of ^{235}U(n,f). b) ^{235}U, analyzed as ^{234}U(n,f) and normalized to the cross section. Note the number of entries which is the actual statistics and the integral which is the normalized number of events. c) The difference between the two distributions is the assumed clean ^{234}U(n,f) yield.

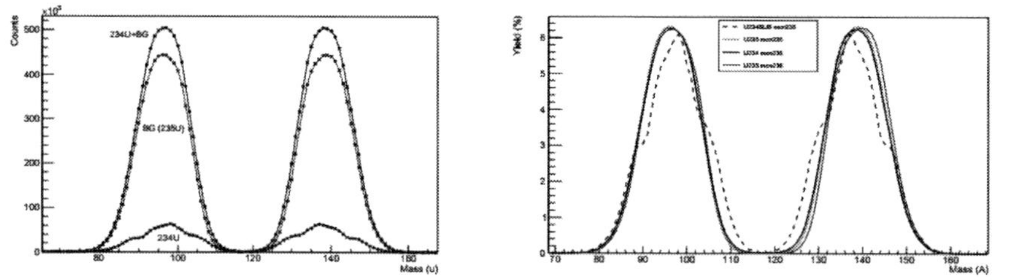

Figure 11. a) ^{234}U+background (BG) shown together with ^{235}U (Bg), analysed as ^{234}U(n,f) and normalized to the cross section. The difference between the two distributions is the assumed clean ^{234}U(n,f) mass yield. b) The mass distribution for ^{235}U analysed as ^{235}U(n,f) (red) and ^{234}U(n,f) (green). In addition ^{234}U(n,f) with (dashed) and without background subtraction (black).

each other, namely: The mass distribution for ^{235}U analyzed as m_{cn}=236 (red), ^{235}U(n,f) as m_{cn}=235 (green), ^{234}U(n,f) with (dashed) and without background subtraction (black). As seen already in Fig. 11c, unfortunately the final mass distribution of ^{234}U(n,f) has some problems and the solution to this is still under investigation.

Acknowledgements

The authors would like to thank CHANDA and EUFRAT for their financial support to perform these scientific works. The Van De Graff staff of the JRC-IRMM are acknowledged for their technical support and for providing the beam.

References

[1] Al-Adili, et al. Phys. Proc. **64**, 145 (2015).
[2] A. A. Naqvi, et al. Phys. Rev. **C34**, 218 (1986).
[3] T.T. Böhlen, et al., Nucl. Data Sheets **120**, 211 (2014).
[4] A. Göök, et al. Phys. Rev. **C90**, 064611 (2014).

[5] A. Göök, Private communication.
[6] C. Budtz-Jørgensen, et al. Nucl. Meth. Instr. **A258**, 209 (1987).
[7] A.C. Wahl, At. Data and Nucl. Data Tables **39**, 1 (1988).
[8] A. Al-Adili, et al. Phys. Rev. **C86**, 054601 (2012).
[9] A. Göök, et al. Nucl. Meth. Instr. **A664** 289 (2012).
[10] A. Al-Adili, et al. Nucl. Meth. Instr. **A673**, 116 (2012).
[11] A. Al-Adili, et al. Nucl. Meth. Instr. **A671**, 103 (2012).
[12] F.-J Hambsch, et al. Nucl. Meth. Instr. **A361**, 257 (1995).
[13] A. Al-Adili, et al. Phys. Proc. **47**, 131 (2013).
[14] G. Simon, et al. Nucl. Meth. Instr. **A286**, 220 (1990).

Fission yield measurements at IGISOL

M. Lantz[1,a], A. Al-Adili[1], D. Gorelov[2], A. Jokinen[2], V.S. Kolhinen[2], A. Mattera[1], I. Moore[2], H. Penttilä[2], S. Pomp[1], A.V. Prokofiev[1], V. Rakopoulos[1], S. Rinta-Antila[2], V. Simutkin[1,2], A. Solders[1], and the IGISOL group[2]

[1] *Division of Applied Nuclear Physics, Uppsala University, Sweden*
[2] *Department of Physics, University of Jyväskylä, Finland*

Abstract. The fission product yields are an important characteristic of the fission process. In fundamental physics, knowledge of the yield distributions is needed to better understand the fission process. For nuclear energy applications good knowledge of neutron-induced fission-product yields is important for the safe and efficient operation of nuclear power plants. With the Ion Guide Isotope Separator On-Line (IGISOL) technique, products of nuclear reactions are stopped in a buffer gas and then extracted and separated by mass. Thanks to the high resolving power of the JYFLTRAP Penning trap, at University of Jyväskylä, fission products can be isobarically separated, making it possible to measure relative independent fission yields. In some cases it is even possible to resolve isomeric states from the ground state, permitting measurements of isomeric yield ratios. So far the reactions U(p,f) and Th(p,f) have been studied using the IGISOL-JYFLTRAP facility. Recently, a neutron converter target has been developed utilizing the Be(p,xn) reaction. We here present the IGISOL-technique for fission yield measurements and some of the results from the measurements on proton induced fission. We also present the development of the neutron converter target, the characterization of the neutron field and the first tests with neutron-induced fission.

1 Introduction

The IGISOL facility at University of Jyväskylä has for several decades been used for experiments with exotic nuclei far from the valley of β stability. A useful method for producing these highly unstable nuclei is through particle-induced fission of heavy nuclei.

Besides performing experiments on the resulting fission products and their decay products, there are a number of reasons to study the yields of the fission process itself. For a more structured planning of experiments with neutron-rich nuclei, it is essential to have better knowledge of the production cross sections of those nuclei that can be obtained through particle-induced fission.

From a theoretical point of view, accurate knowledge of fission yields is of importance for a better understanding of the fission process. Fission yields also have an impact on the result of the astrophysical rapid neutron capture process, which is responsible for the formation of a large fraction of the nuclei heavier than iron [1]. During the r-process the region of fissionable nuclei is reached, which terminates the r-process path and generates fission products. These have an impact on the resulting

[a] e-mail: Mattias.Lantz@physics.uu.se

stable isotope distribution. The fissioning nuclei are so far from stability that their fission yields have to be calculated based on theoretical models. To improve these models, accurate experimental data on the fission of reachable nuclei, such as ^{238}U, is needed.

For nuclear energy applications good knowledge of neutron induced fission product yields is crucial in many aspects, including criticality and reactivity calculations for reactor design, dosimetry and fission gas production for reactor safety and improved burn-up predictions. Good knowledge of the composition of the spent fuel is required for the management of the nuclear waste, i.e. depositories, reprocessing, transmutation and so on. It is also needed for Generation IV reactor concepts, including fast reactors where fast neutrons change the fission yield distribution compared to thermal neutrons. The inventory of fission products in a reactor determines the decay heat, both residual heat after reactor shut down and the decay heat from spent fuel outside of the reactor. Typically, the decay heat is dominated by fission products for the first 50-80 years after extraction of spent nuclear fuel from a reactor. The successful operation of nuclear power plants shows that the current knowledge of the underlying nuclear physics processes is generally sufficient. Predictions of macroscopic reactor parameters with model codes, as well as calculations of the isotopic composition of spent nuclear fuel, are in reasonable agreement with reality. Nevertheless, different evaluations display significant discrepancies for some fission products of relevance [2]. Thus more accurate nuclear data would improve the predictions of fuel compositions and hence both safety and fuel economy.

In view of these needs the objective of the present project is to measure fission yields of high accuracy for different systems, in particular

- the energy dependence of neutron-induced fission yields by varying the incident neutron field,
- the fission yield dependence on the initial system by measuring on different fissile nuclei, and with different incident particles (neutrons, protons, deuterons, ...), and
- other observables such as isomeric yield ratios.

The project is a collaboration between the IGISOL group at University of Jyväskylä and the applied nuclear physics division at Uppsala University.

2 Experimental method of IGISOL

IGISOL is an acronym for Ion Guide Isotope Separator On-Line and is a technique that has been developed in Jyväskylä since the early 1980's [3]. The facility has been extensively explained elsewhere [4–8] but some key features will be repeated here. Figure 1 gives a principle view of the steps being described below, part a) of Fig. 4 may also be useful.

The IGISOL technique is based on stopping the nuclear reaction products recoiling out of the reaction target in a helium gas. Here the reaction of interest is fission and thus the reaction products are fission products. The helium gas flows out of the reaction chamber and transports the reaction products towards an electrostatic field that accelerates them through a SextuPole Ion Guide (SPIG). After the SPIG follows further accelerating steps in vacuum, sending the reaction products for analysis or for further treatment.

The reaction products are ionised as they are formed in the nuclear reaction and the high ionisation potential of helium causes a majority of the extracted reaction products to end up as singly charged ions. This significantly simplifies the subsequent mass analysis, where the ions are separated based on their charge to mass (q/m) ratios by sending them through a dipole magnet.

In earlier versions of IGISOL the nuclides in the isobaric chain selected by the dipole magnet would be identified through γ-spectroscopy. The introduction of JYFLTRAP, a Penning trap with 7 Tesla magnetic field strength, has significantly improved the experimental method. Before the trap

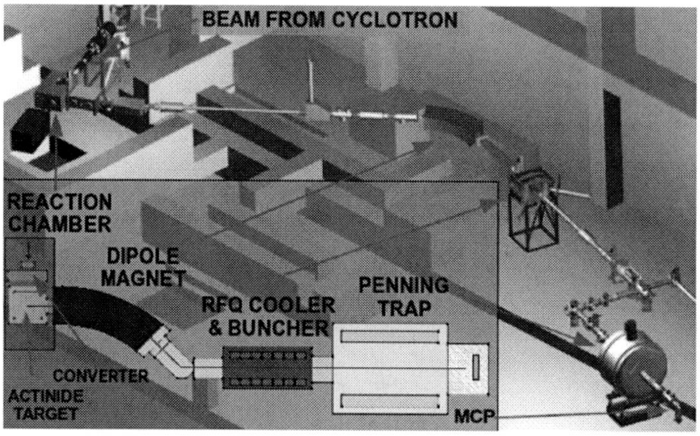

Figure 1. Layout of the IGISOL-JYFLTRAP facility (schematic view in the insert). A particle beam enters from the top left impinging either directly on a fissile target, or alternatively on a neutron converter resulting in a neutron field causing fission in the target. The fission products are accelerated and transported through a dipole magnet for mass selection, followed by an RFQ cooler and buncher for improvement of the beam properties. The cooled ions are injected into the Penning trap for high precision mass separation and finally the count rate is determined by sending the ions to a MCP detector. The acronyms are explained in the text. CAD drawing by Tommi Eronen.

the ions are sent to a radio-frequency quadrupole (RFQ) cooler and buncher, from which they are sent as a bunch to the Penning trap.

The Penning trap can be used as a high-power mass-resolving filter to allow unambiguous identification of the nuclei based on their atomic mass, through the relation between cyclotron frequency, f_c, and mass of the nuclide, m,

$$f_c = \frac{1}{2\pi}\frac{q}{m}B\,, \tag{1}$$

where q is the ion charge (+1) and B is the magnetic field strength of JYFLTRAP. The filtering, usually referred as to mass purification, can be performed in less than 100 ms, however, a longer purification time is required for the highest mass resolving power. In the experiments performed so far the total purification time has typically been 0.6-1.3 s. The achieved mass resolution after the Penning trap is of the order of $\frac{A}{\Delta A} = 10^5$. In some cases it is possible to separate isomeric states from the ground state down to about 200 keV [9].

For the most neutron rich fission products it is necessary to know their half-lives to be able to correct for the losses due to radioactive decay. In addition to the atomic masses and the half-lives no further information about the properties of the fission products are needed, since they are observed by ion counting with multichannel plate (MCP) detectors after JYFLTRAP. It may also be noted that since the detection of the fission products does not require their decay, the yield of very long-lived or even stable fission products is determined equally fast and effectively as that of radioactive nuclides.

The frequency of the Penning trap is scanned to generate a high resolution mass spectrum. One isotope of each element is selected as a reference, against which the yields of the other isotopes are determined. In the measurements of independent fission yields, reference mass spectra are collected

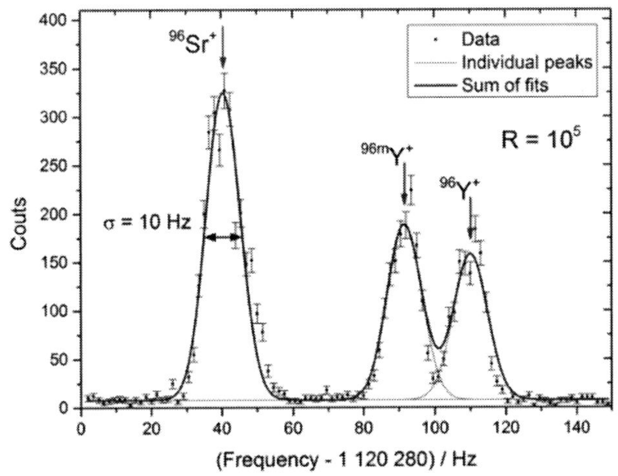

Figure 2. The quadrupole frequency spectrum collected after the Penning trap, showing three identified nuclides of mass 96. Black dots with error bars are the number of counts at each measured frequency. The red lines are gaussian fits to each individual peak (for the ^{96}Sr peak the red line is superimposed by the sum of the fits) and the blue line is the sum of the fits. The isomer ^{96m}Y at 1.14 MeV excitation energy is easily distinguished from the ground state.

before and after each measurement of the studied isotope. Reference runs may take as long as 20 minutes, while the measurement of the studied isotope can last up to two hours for the lowest production rates. The yield may fluctuate over time in which case it has to be corrected for. After determining the intensity ratio between the mass peaks of the reference and the measured isotopes, applying the appropriate decay corrections and, when necessary, correction for the yield fluctuations, one obtains isotopic yield distributions.

3 Proton and deuteron induced fission yields

At IGISOL-JYFLTRAP fission yield experiments have been performed with 25 MeV protons on Th [10] and ^{238}U [11], measuring the isobaric chains for more than 20 different elements. There have also been measurements for a number of nuclei with 50 MeV protons on ^{238}U and with 25 MeV deuterons on ^{238}U [4, 5]. The IGISOL method has been developed over time, gradually improving the experimental method and analysis [12]. Comparisons with older data obtained through γ-spectroscopy give good agreement and show the advantages of the method.

Figure 2 shows an example of the experimental resolution of the method, where the isomer ^{96m}Y at 1.14 MeV excitation energy can be clearly distinguished from the ground state.

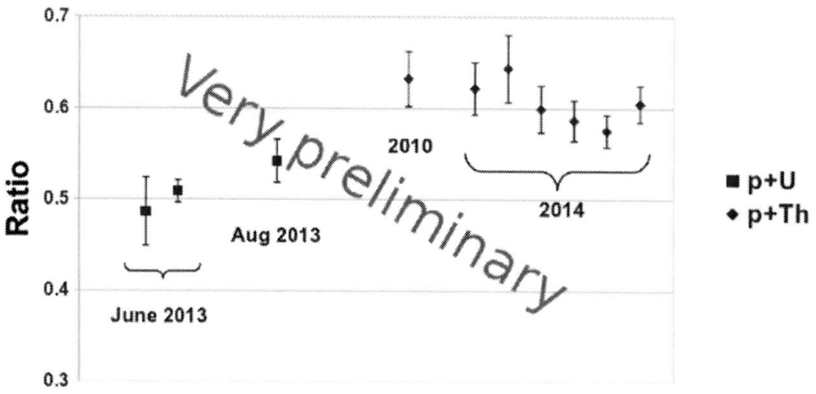

Figure 3. Comparison of isomeric yield ratios for ^{96}Y from p+^{238}U (blue) and p+Th (red) induced fission from four different experiments. For the June 2013 and 2014 experiments there is more than one measurement of ^{96}Y. The error bars only indicate the statistical uncertainty, other corrections have not been implemented yet an may affect both the values and the size of the error bars.

3.1 Isomeric yield ratios

One interesting observable to measure is isomeric yield ratios. As there is no need for absolute normalization they are relatively easy to measure, provided that the isomer has long enough half life and an excitation energy large enough to be distinguished with the mass resolution given by JYFLTRAP. Data on isomeric yield radios provide valuable information for theoretical development. Here the isomeric yield ratio is defined as

$$Ratio = \frac{Yield(^{96m}Y)}{Yield(^{96m}Y) + Yield(^{96}Y)} . \tag{2}$$

A variable to consider is whether or not the isomeric yield ratios have a dependence on the initial system, i.e. will the ratio be the same for isomers produced from fission of different actinides, or from fission on a given actinide induced by different projectiles? The comprehensive evaluation and compilation by England and Rider [13] assumes no dependence on the initial system and we are unaware of any published experimental results indicating such an effect. From a series of IGISOL measurements with 25 MeV p+Th and 25 MeV p+^{238}U performed between 2010 and 2014 there are several isomeric pairs to compare [14].

Figure 3 shows preliminary results from the analysis from four different experiments of isomeric yield ratios for the ground state of ^{96}Y and its isomer. The results seem to indicate such a dependence on some isomeric pairs, though all corrections have not been implemented yet to the dataset, and the error bars are likely to increase [15]. For one of the experiments there are also data obtained through γ-spectroscopy, which will be compared with the results from JYFLTRAP. Since the 2010 experiment the IGISOL-JYFLTRAP facility has been upgraded and moved within the laboratory, so the 2010 data are essentially taken at a different experimental setup.

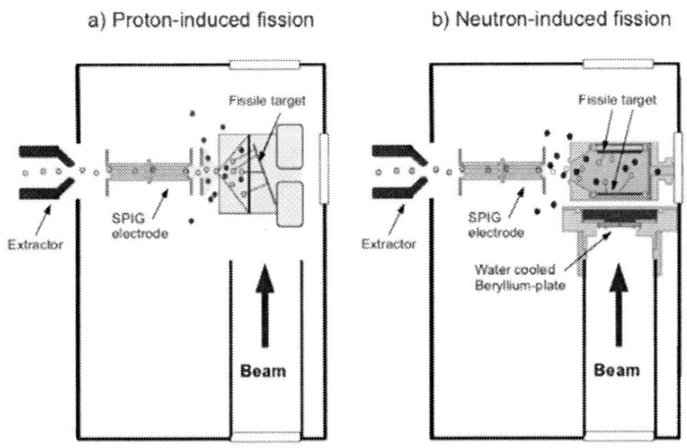

Figure 4. Comparison of the IGISOL chamber for a) experiments with proton/deuteron-induced fission, and b) with a neutron converter for neutron-induced fission. The reaction chamber has a helium flow passing through from the right to left, transporting stopped ions towards the SPIG electrode and an electrostatic field that accelerates the ions towards the RFQ cooler (see Fig. 1). The position of the proton-neutron converter is adjustable and can be mounted further upstream in order to allow the insertion of a moderator for measurements with thermal neutrons.

Future experiments should include further measurements of isomeric yield ratios with proton induced fission. It will also be of interest to measure the same isomeric yield ratios with neutron induced fission.

4 Neutron induced fission yields

A project was initiated in 2010 to design a proton-neutron converter in order to measure neutron induced fission yields. Besides the need for nuclear data in view of nuclear energy applications, neutron induced fission enables fundamental research on even more neutron rich nuclei. When the IGISOL-JYFLTRAP facility was moved to a new location it was also supplied with a high intensity MCC30/15 cyclotron [16], which would make it feasible to obtain a high enough neutron flux for measurements of relevance. The ambition was to have a design that allows 10^{12} fast neutrons on a large fissile target [17].

After studying different materials and geometries the decided approach was to use a 5 mm thick Beryllium (Be) plate, slightly thinner than the stopping range for 30 MeV protons. The back side of the plate is cooled by water, and as the majority of the protons stop in the water the cooling requirements are easier to fulfil with a large safety margin [18]. Part b) of Fig. 4 shows the proton-neutron converter assembly with the water cooled Be plate, followed by a cylindrical ion chamber that enables a large area of fissile material, e.g. ^{238}U, to be mounted close to the neutron source. The close proximity of the converter allows the ambition for 10^{12} fast neutrons to be fulfilled.

A characterisation measurement of the neutron flux from a prototype of the Be converter was performed at the TSL facility in Uppsala by means of two different measurement techniques, time-

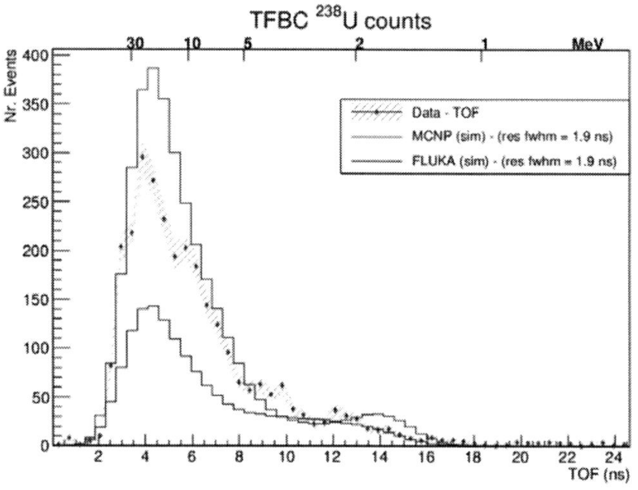

Figure 5. Neutron spectra obtained through ToF with TFBCs at IGISOL. Comparisons with simulations performed with the Monte Carlo codes MCNP and FLUKA.

of-flight measurement (ToF) and Bonner sphere spectroscopy [19–22]. In 2014 the first tests with a prototype at IGISOL were performed. Measurements of the neutron flux within the IGISOL reaction chamber has also been performed [23] with Thin Film Breakdown Counter detectors (TFBCs) [24] and the first attempts to extract ions out of the chamber were performed in 2015 [10].

Simulations of the expected neutron fluxes have been done using the Monte Carlo codes MCNPX [25] and FLUKA [26]. Figure 5 shows a comparison of simulated ToF neutron spectra with experimental data obtained with TFBCs within the IGISOL chamber. The MCNPX simulations agree somewhat better with the experimental values. The discrepancy in absolute magnitude between the two codes is a bit surprising and requires further investigations.

5 Status and outlook

The IGISOL-JYFLTRAP facility is very promising for measurements of independent fission yields to complement experimental data from other experimental techniques. Proton and deuteron induced fission yields may be further explored. For nuclear energy applications the possibility to use different neutron fields will be very important.

Initial test runs have been performed with the neutron converter and we foresee actual data taking runs during 2016. In order to fully utilize the new MCC30/15 cyclotron a number of projects have been initialized, for instance

- simulations in order to verify that the neutron field resulting from proton beams with intensity of about 100 μA will not cause any issues with respect to radiation protection,
- further characterizations of the neutron fields in order to plan measurements and assess the experimental uncertainties, and

- modeling and simulation of ion guide geometries and the related physics aspects, in order to optimize and better characterize the yields extracted from the IGISOL reaction chamber [27].

A challenge, but also a possibility, is the relatively long time between the fission event and the actual identification of the fission products. The fastest possible measurement is about 300 ms, though around 1 second is more realistic when accounting for the RFQ cooler and buncher, and the purification time in JYFLTRAP. In comparison with γ-spectroscopy this is very fast, but it is long when considering that during that time there may have been a number of decays and even beta delayed neutron emission. It is therefore fair to ask what it is that is actually being measured. As some of the upstream decay branching ratios are not known when measuring isomeric pairs, it will be difficult to perform relevant corrections in order to claim a certain isomeric yield ratio. In order to make the data useful for theoretical development it is therefore very important to explain what has been measured and at what time after the fission event. This limitation in the experimental method may also be an opportunity. The time can be increased, which gives a possibility to vary the time for some isomeric pairs with short half lives, and thereby obtain more information about the upstream branching ratios.

The dual purpose of using IGISOL-JYFLTRAP to measure nuclear data for fundamental physics research and simultaneously for applications for neutron induced fission yields lead to some constraints, but also many possibilities. The combination of both purposes strengthen the motivation for the existance of the facility, and as a long-term commitment we foresee many new challenges and valuable nuclear data.

Acknowledgements

Part of this work has been financed by the Swedish Radiation Safety Authority (SSM), the Swedish Nuclear Fuel and Waste Management Company (SKB), and experimental work through the EU funded research framework programmes "European Research Infrastructures for Nuclear Data Applications" (ERINDA) and "solving CHAllenges in Nuclear DAta" (CHANDA). We also want to acknowledge support from the Academy of Finland under the Finnish Center of Excellence Programme 2012–2017 (Nuclear and Accelerator Based Physics Programme at JYFL) and the project 139382 (Precision fission studies for practical needs).

References

[1] I.V. Panov, et al., Nucl. Phys. A **747**, 633-654 (2005)

[2] B. Lourdel, *Research on fission yields evaluation and measurements within the ALFONS project* (MSc thesis, Chalmers university of technology and Uppsala university, Göteborg, 2010)

[3] J. Äystö, Nucl. Phys. A **693**, 477–494 (2001)

[4] H. Penttilä et al., Eur. Phys. J. A **44**, 147-168 (2010)

[5] H. Penttilä et al., Eur. Phys. J. A **48**, 43 (2012)

[6] V.S. Kolhinen et al., Nucl. Instrum. Meth. B **317**, 506-509 (2013)

[7] H. Penttilä et al., Nuclear Data Sheets **119**, 334-337 (2014)

[8] A. Solders et al., Nuclear Data Sheets **119**, 338-341 (2014)

[9] K. Peräjärvi et al., Appl. Radiat. Isotopes **68**, 450-453 (2010)

[10] D. Gorelov, *Nuclear Fission Studies with the IGISOL Method and JYFLTRAP* (PhD thesis, University of Jyväskylä, Jyväskylä, 2015)

[11] H. Penttilä et al., manuscript

[12] P. Karvonen, *Fission yield studies with SPIG-equipped IGISOL: A novel method for nuclear data measurements* (PhD thesis, University of Jyväskylä, Jyväskylä, 2010)

[13] T.R. England and B.F Rider, *Fission product yields per 100 fissions for thermal neutron induced fission decay* (Report LA-UR-94-3106, ENDF-349, Los Alamos National Lab, Los Alamos, 1993)

[14] D. Gorelov et al., Acta Physica Polonica B **45**, 211-216 (2014)

[15] V. Rakopoulos et al., manuscript

[16] P. Heikkinen, Proc. 18th International Conference On Cyclotrons And Their Applications, 128 (2007)

[17] M. Lantz et al., Phys. Scr. **T150**, 014020 (2012)

[18] M. Lantz et al., *Compendium of Neutron Beam Facilities for High Precision Nuclear Data Measurements*, (IAEA-TEKDOC-1743, IAEA, Vienna, 2014) sec 3.2

[19] A. Mattera et al., Nuclear Data Sheets **119**, 416-418 (2014)

[20] R. Bedogni et al., manuscript

[21] V. Rakopoulos et al., EPJ Web of Conferences **66**, 11032 (2014)

[22] A. Mattera *Characterization of a Neutron Source for Fission Yield Studies* (PhLic thesis, Uppsala university, Uppsala, 2014)

[23] B. Eriksson, *Neutron field characterization using TFBCs and comparison to Monte Carlo simulations* (BSc thesis, Uppsala university, Uppsala, 2015)

[24] A.N. Smirnov et al., Nucl. Instrum. Meth. A **687**, 14–22 (2012)

[25] G.W. McKinney et al., *MCNPX 2.7.0 - New Features Demonstrated*, (Report LA-UR-12-25775, Los Alamos National Lab, Los Alamos, 2012)

[26] T.T. Böhlen et al., Nuclear Data Sheets **120**, 211-214 (2014), A. Ferrari et al., *FLUKA: a multi-particle transport code* (Report CERN-2005-10 (2005), INFN/TC_05/11, SLAC-R-773, CERN, Geneva, 2005)

[27] A. Al-Adili et al., Eur. Phys. J. A **51**, 59 (2015)

EPJ Web of Conferences **122**, 01009 (2016) DOI: 10.1051/epjconf/201612201009
*CNR*15*

Inversion of the Odd-Even Effect in Cold Fission from the Time-Dependent Pairing Equations

M. Mirea[1,a]

[1] *Horia Hulubei National Institute for Physics and Nuclear Engineering, P.O. Box MG-G, Bucharest, Romania*

Abstract. A peculiar phenomenon was observed experimentally in cold fission: the odd partition yields are favored over the even ones for excitations energies of the fragments smaller than 4 MeV. In this contribution, a microscopic model is proposed for the explanation of this odd-even effect in cold fission. This explanation is based on a mixing configuration mechanism that is produced during the fission process. This configuration mixing mechanism is obtained dynamically by solving a the generalized system of time-dependent pairing equations, which include a pair-breaking effect. The time dependent equations of motion for the pair breaking effect were corroborated with a condition that fixes dynamically the number of particles on the two fission fragment. The single particle level scheme was calculated with the Woods-Saxon superasymmetric two center shell model, providing a continuous variation of the single particle energies and of the wave functions from one nucleus up to two separated fragments. A first rule can be extracted from this model. The even-even fission products cannot be obtained at zero excitation energies because of the existence of dynamical excitations produced in the avoided- level-crossing regions when the nuclear system deforms slowly.

1 Introduction

There are some strange experimental evidences in low energy fission. In the thermal neutron induced fission, the mass and charge distributions of fragments exhibit a fine structure that show a preference for even neutrons and protons fission products [1–5]. We expect that this phenomenon will be more evidenced in the case of cold fission, where the excitation energy of the fragments is so small that no neutrons can be emitted. This is not true as evidenced experimentally [6, 7]. For example, in the case of the mass fragmentation (132/104) for the reaction ^{235}U(n$_{th}$,f) in the cold fission regime, the even-even charge split (50/42) can be observed up to a total kinetic energy several MeV lower than the Q-value (206 MeV). That means, the final fission products are released with at least several MeV excitation energy. In the same time, the odd-odd fragmentation (51/41) has yields that amount a total kinetic energy close to the Q-value (202 MeV). It is strange, but at very low excitation energy an inversion of the odd-even effect is produced and the odd-odd fragmentation dominates. The even-even fragmentation dominates at larger excitation energies while the odd-odd one dominates at very low excitation energies. This behavior is typical, being observed also for the fission of Cf [8, 9]. This inversion can be explained statistically by involving the level densities of odd and even nuclei

[a] e-mail: mirea@ifin.nipne.ro

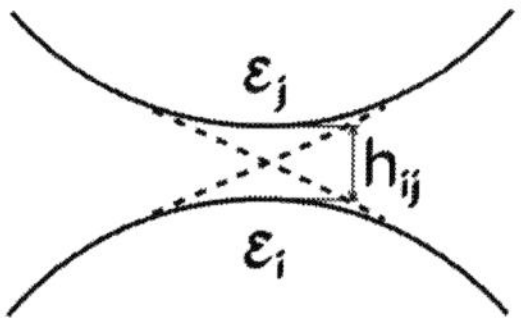

Figure 1. Interaction h_{ij} between two adiabatic levels ϵ_i and ϵ_j in an ideal avoided level crossing region.

[10, 11]. This effect is also linked to the important question of intrinsic excitation (quasiparticle excitations) during the dynamical evolution of the fissioning nucleus on ints way from ground state to scission. Another explanation can be based on the Landau-Zener promotion mechanism during fission [12] or at scission [13].

2 The Landau-Zener effect

The single particles or potential levels characterized by the same good quantum numbers associated to some symmetries of the system cannot in general intersect and exhibit avoided level crossing regions. An interaction h_{ij} is always present in these regions between two single particle levels ϵ_i and ϵ_j. Such an ideal avoided level crossing is plotted in figure 1 as function of a deformation parameter of the nuclear system, that could be the internuclear distance between the two nascent fragments. If the system deforms slowly in time and a nucleon is located on the orbital ϵ_i, the nucleon will remain on the same adiabatic level after the passage of the avoided crossing region. If the deformation parameter varies rapidly, the nucleon will skip with a large probability from one adiabatic level to the other. It is possible to know the final probabilities of occupation of a single particle level after the passage through the avoided crossing level region by solving a set of time dependent coupled channel equations. These equations are derived from the time dependent Schrödinger equation. In general the Landau-Zener effect describes the non adiabatic transitions in avoided level crossing regions between potential curves or energy levels.

3 Mixing of seniority zero and seniority two configurations

It will be interesting to investigate the behavior of this effect in the case of superfluid systems, where each level is filled with some probabilities by a Cooper pair. Due to the interaction in the avoided crossing regions, it is possible that only one nucleon of the pair skips to the another level producing a dynamical quasiparticle excitation. Three situations depicted in figure 2 are possible. In the first case, the pairs remain on the same single particle level after the passage through the avoided level crossing region. In the second case, a pair destruction is illustrated. In the last case, two fermions generates a pair. A formalism is elaborated to describe these situations by using quasiparticle creation α_i^+ and annihilation operators α_i. Within such operators, we construct some interactions able to break or to

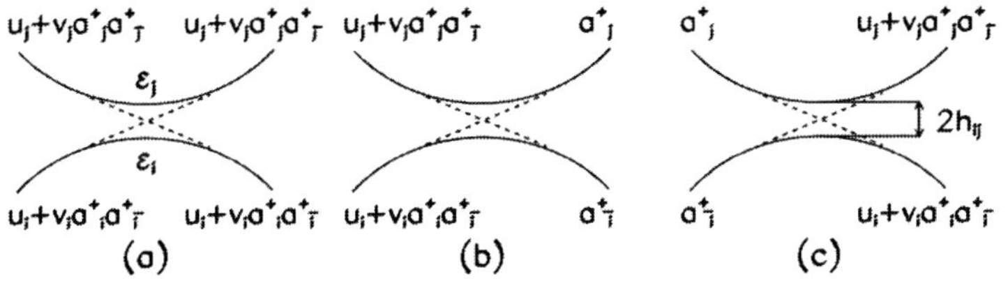

Figure 2. (a) The configuration remains unchanged after the passage through the avoided crossing region; (b) A pair is broken; (c) A pair is created.

create a pair when the system traverses an avoided level crossing region. The second situation can be obtained with an operator of the type $\alpha_i\alpha_j$ that acts on a Bogoliubov wave function, two particles being obtained on the levels i and j. The third situation is obtained with operators $\alpha_i^+\alpha_j^+$ acting on a seniority two configuration, a pair being initially distributed on the levels i and j. This kind of perturbations were postulated in a previous work where the Landau-Zener effect was generalized in the superfluid seniority one systems [14]. Using creation and annihilation operators, the equations that governs the pair breaking mechanism are obtained from the variational principle [15, 16]. A residual perturbation in the avoided level crossing region is postulated as follows:

$$H'(t) = \sum_{i,j\neq i}^{n} h_{ij}(t) \quad \Big[\alpha_{i(0)}\alpha_{\bar{j}(0)} \prod\nolimits_{k\neq i,j} \alpha_{k(0)}\alpha_{k(ij)}^{+} + \alpha_{i(0)}^{+}\alpha_{\bar{j}(0)}^{\pm} \prod\nolimits_{k\neq i,j} \alpha_{k(ij)}\alpha_{k(0)}^{+}\Big],$$

where h_{ij} is the interaction energy between levels.

Another problem is that the two fission products must be characterized by integer numbers of neutrons and protons. As a consequence, the sums of the occupation probabilities of single particle levels of both fragments must give integer mass and charge numbers. By solving the equations of motion, unfortunately, the sum of single-particle densities (BCS occupation probabilities) of the single particle levels belonging to each fission fragment don't give exactly their numbers of nucleons at scission. A recipe can be implemented to fix dynamically these numbers of particles in the two final fragments by using the operators for the number of particles $\hat{N}_i$ $(i = 1, 2)$ that act on each fission product. At scission, the two fission fragments must be characterized by a supplementary condition

$$\mid N_2\hat{N}_1 - N_1\hat{N}_2 \mid = 0, \tag{1}$$

where N_1 and N_2 are the number of particles in the final fragments labeled 1 and 2, respectively, and

$$\hat{N}_1 = \sum_{k1}(a_{k_1}^{+}a_{k_1} + a_{\bar{k}_1}^{+}a_{\bar{k}_1}); \quad \hat{N}_2 = \sum_{k2}(a_{k_2}^{+}a_{k_2} + a_{\bar{k}_2}^{+}a_{\bar{k}_2}) \tag{2}$$

are the corresponding operators. Here, k_1 and k_2 run over the pairing active level states that are located in the potential wells of the final fragment 1 and of the final fragment 2, respectively. The condition (1) can be introduced in the equations of motion by means of the Lagrange multipliers [17, 18].

DOI: 10.1051/epjconf/201612201009

All the previous ingredients could be used to obtain the microscopic equations of motion. These equations are obtained from the variational principle by minimizing the following energy functional

$$\delta\mathcal{L} = \delta\left\langle\varphi\left|H + H' - \lambda \mid N_2\hat{N}_1 - N_1\hat{N}_2 \mid -i\hbar\frac{\partial}{\partial t}\right|\varphi\right\rangle. \tag{3}$$

The trial many-body function φ is a superposition of Bogoliubov seniority zero and seniority two wave functions

$$\begin{aligned} \mid \varphi(t)\rangle = \; & \Big[c_0 \textstyle\prod_k \left(u_{k(0)}(t) + v_{k(0)}(t)a_k^+ a_{\bar{k}}^{\pm}\right) \\ & + \textstyle\sum_{j,l\neq j} c_{jl}(t)a_j^+ a_{\bar{l}}^{\pm} \prod_{k\neq j,l}\left(u_{k(jl)}(t) + v_{k(jl)}(t)a_k^+ a_{\bar{k}}^{\pm}\right)\Big] \mid 0\rangle. \end{aligned}$$

c_0 is the amplitude of the seniority zero wave function while c_{jl} are amplitudes for the seniority two wave functions for configurations in which the single particle orbitals j and l in the active pairing space are each blocked by only one unpaired nucleon. Here, the vacancy u_k and occupation v_k amplitudes are not the adiabatic solutions of the BCS equations and depend on the variation in time of the generalized parameters and the history of the nuclear system. λ is a Lagrange multiplier. The next coupled channel equations are obtained, eventually:

$$i\hbar\dot{\rho}_{k(0)} = \kappa_{k(0)}\Delta^*_{k(0)} - \kappa^*_{k(0)}\Delta_{k(0)}, \tag{4}$$

$$i\hbar\dot{\rho}_{k(jl)} = \kappa_{k(jl)}\Delta^*_{k(jl)} - \kappa^*_{k(jl)}\Delta_{k(jl)}, \tag{5}$$

$$\begin{aligned} i\hbar\dot{\kappa}_{k(0)} = & \left(2\rho_{k(0)} - 1\right)\Delta_{k(0)} + 2\kappa_{k(0)}\left(\epsilon_k - sN_{i_k}\lambda\right) \\ & -2G_{kk}\rho_{k(0)}\kappa_{k(0)}, \end{aligned} \tag{6}$$

$$\begin{aligned} i\hbar\dot{\kappa}_{k(jl)} = & \left(2\rho_{k(jl)} - 1\right)\Delta_{k(jl)} + 2\kappa_{k(jl)}\left(\epsilon_k - sN_{i_k}\lambda\right) \\ & -2G_{kk}\rho_{k(jl)}\kappa_{k(jl)}, \end{aligned} \tag{7}$$

$$i\hbar\dot{P}_0 = \sum_{l,j\neq l} h_{lj}(S^*_{0jl} - S_{0jl}) \tag{8}$$

$$i\hbar\dot{P}_{jl} = h_{lj}(S_{0jl} - S^*_{0jl}) \tag{9}$$

$$\begin{aligned} i\hbar\dot{S}_{0jl} = & S_{0jl}(\bar{E}_0 - \bar{E}_{jl}) + S_{0jl}\left(\textstyle\sum_{k\neq j,l} T_{k(jl)} - \sum_k T_{k(0)}\right) \\ & + \textstyle\sum_{\{mn\}\neq\{jl\}} h_{mn}S_{mnjl} + h_{jl}(P_{jl} - P_0) \end{aligned} \tag{10}$$

$$\begin{aligned} i\hbar\dot{S}_{mnjl} = & S_{mnjl}(\bar{E}_{mn} - \bar{E}_{jl}) + S_{mnjl}\left(\textstyle\sum_{k\neq m,n} T_{k(mn)} - \sum_{k\neq j,l} T_{k(jl)}\right) \\ & + h_{mn}S_{0jl} - h_{jl}S^*_{0mn} \end{aligned} \tag{11}$$

where j, k, l, m, n label the single particle levels in the active pairing space. The sign $s = \pm 1$ ensures that the matrix element of the expression (1) is positive. $N_{i_k} = N_2$ or $N_{i_k} = -N_1$ if the state k will be located in the fragment 1 or in the fragment 2 after the scission, respectively. T_γ are state dependent energy terms

$$\begin{aligned} T_{k(0)} = & \tfrac{i\hbar}{2}(v^*_{k(0)}\dot{v}_{k(0)} - \dot{v}^*_{k(0)}v_{k(0)}) = 2 \mid v_{k(0)} \mid^2 (\epsilon_k - sN_{i_k}\lambda) - 2G_{kk} \mid v_{k(0)} \mid^4 \\ & + \tfrac{\Delta^*_{k(0)}}{2}\left(\tfrac{|v_{k(0)}|^4}{(v_{k(0)}u_{k(0)})^*} - v_{k(0)}u_{k(0)}\right) + \tfrac{\Delta_{k(0)}}{2}\left(\tfrac{|v_{k(0)}|^4}{v_{k(0)}u_{k(0)}} - (v_{k(0)}u_{k(0)})^*\right); \end{aligned} \tag{12}$$

with similar equations for the seniority two configurations. The following notations are used:

$$\begin{array}{cc} \Delta_{k(0)} = \sum_{k'} \kappa_{k'(0)} G_{kk'}; & \Delta_{k(jl)} = \sum_{k' \neq j,l} \kappa_{k'(jl)} G_{kk'}; \\ \kappa_{k(0)} = u_{k(0)} v_{k(0)}; & \rho_{k(0)} = \mid v_{k(0)} \mid^2; \\ \kappa_{k(jl)} = u_{k(jl)} v_{k(jl)}; & \rho_{k(jl)} = \mid v_{k(jl)} \mid^2; \\ P_0 = \mid c_0 \mid^2; & P_{jl} = \mid c_{jl} \mid^2; \\ S_{0jl} = c_0 c^*_{jl}; & S_{mnjl} = c_{mn} c^*_{jl}. \end{array} \tag{13}$$

The symbol Δ_γ gives the gap parameter. (The label γ denotes here generically a specific configuration.) The variables that depend on the time through the generalized coordinates are the single particle densities ρ_γ, the pairing moment components κ_γ, the probabilities to have a given seniority configuration P_γ, and the moment components between configurations $S_{\gamma\gamma'}$. The relations (4)-(7) are the well known time dependent paring equations for one seniority configuration previously deduced in Refs. [19, 20]. These formulas are identical to the time dependent Hartree-Fock-Bogoliubov equations [21–23]. The symbol h_γ denotes the Landau-Zener interaction, while E_γ and T_γ are energy terms. The condition that $\sum_\gamma P_\gamma=1$ is implicitly ensured through Eqs. (8) and (9) because $\dot{P}_0 + \sum_\gamma \dot{P}_\gamma = 0$.

4 The dissipation

The energy of the nuclear system in the seniority zero state is

$$\begin{aligned} E_0 &= \langle \prod_k (u_{k(0)}(t) + v_{k(0)}(t) a_k^+ a_{\bar{k}}^+) | H | \prod_k (u_{k(0)}(t) + v_{k(0)}(t) a_k^+ a_{\bar{k}}^+) \rangle \\ &= 2 \sum_k \rho_{k(0)} \epsilon_k - \sum_k \kappa_{(0)} \sum_{k'} \kappa^*_{k'(0)} G_{kk'} - \sum_k \rho^2_{k(0)} G_{kk}; \end{aligned} \tag{14}$$

and in the seniority two state is

$$\begin{aligned} E_{jl} &= \langle a_j^+ a_{\bar{l}}^+ \prod_{k \neq j,l} (u_{k(jl)}(t) + v_{k(jl)}(t) a_k^+ a_{\bar{k}}^+) | H | a_j^+ a_{\bar{l}}^+ \prod_{k \neq j,l} (u_{k(jl)}(t) + v_{k(jl)}(t) a_k^+ a_{\bar{k}}^+) \rangle \\ &= 2 \sum_{k \neq j,l} \rho_{k(jl)} \epsilon_k - \sum_{k \neq j,l} \kappa_{k(jl)} \sum_{k' \neq j,l} \kappa^*_{k'(jl)} G_{kk'} - \sum_{k \neq j,l} \rho^2_{k(jl)} G_{kk} + \epsilon_j + \epsilon_l. \end{aligned}$$

The corresponding lower energy states E_0^0 and E_{jl}^0 of the nuclear system in a given configuration are obtained with the previous relations by replacing the densities ρ_γ and the pairing moment components κ_γ with the adiabatic values obtained in the BCS approximation. Consequently, as defined in Ref. [19], along the fission path the average dissipated energies E_γ^* will be

$$E_0^* = E_0 - E_0^0; \quad E_{jl}^* = E_{jl} - E_{jl}^0, \tag{15}$$

in the seniority zero and the seniority two configurations, respectively. We subtracted from the total potential energy of the nuclear system, its adiabatic value. It was already shown in Ref. [24] that the mean value of the dissipated energy becomes larger when the velocities of the generalized coordinates increase.

5 Results

It is possible to solve the equations of motion if the fission trajectory or the potential barrier are provided. In order to determine the fission barriers within the least action principle, two ingredients are required: the deformation energy and the inertia. The total energy of the nuclear system is computed in the framework of the macroscopic-microscopic method [25, 26]. Within this method, the whole system is characterized by some collective coordinates that determine approximately the behavior of many other intrinsic variables. The generalized coordinates vary in time leading to a split of the system. The procedure is analogue to that used in more sophisticated models as the Hartree-Fock, where external constraints simulate the deformation [27, 28]. The essential idea of our approach is that a macroscopic model, as the liquid drop one, describes quantitatively the smooth trends of the potential energy with respect the particle number and deformation whereas a microscopic approach as the shell model describes local fluctuations. The combined macroscopic-microscopic method should reproduce both smooth trends and local fluctuations. The basic ingredient in such an analysis is the shape parametrization that depends on several macroscopic degrees of freedom. The macroscopic deformation energy is calculated within the liquid drop model. A microscopic potential must be constructed to be consistent with this nuclear shape parametrization. A microscopic correction is then evaluated using the Strutinsky procedure [29].

The basic ingredient of the model is the nuclear shape parametrization. In the following, an axial symmetric nuclear shape surface during the deformation process from one initial nucleus to the separated fragments is obtained by smoothly joining two spheroids of semi-axis a_i and b_i (i=1,2) with a neck surface generated by the rotation of a circle of radius R_3 around the axis of symmetry. By imposing the condition of volume conservation we are left by five independent generalized coordinates $\{q_n\}$ (n=1,5) that can be associated to five degrees of freedom: the elongation R given by the distance between the centers of the spheroids; the necking parameter $C = S/R_3$ related to the curvature of the neck, the eccentricities ϵ_i associated with the deformations of the nascent fragments and the mass asymmetry parameter $\eta = V_1/V_2$, V_i (i =1,2) denoting the volumes of the virtual ellipsoids characterized by the semi-axis a_i and b_i. The nuclear shape parametrization is displayed in figure 3. The entire model can be considered valid as long as the generalized coordinates and their variations in time make sense.

The many-body wave function and the single particle energies are provided by the Woods-Saxon two-center shell model [14]. The Woods-Saxon potential, the Coulomb interaction and the spin orbit term must be diagonalized in an eigenvector basis. The asymmetric two center shell oscillator provides an orthogonal eigenvector basis for only one Hermite space [30–32]. In this Hermite space the behavior of both fragments can be described. When the elongation R is zero, the eigenvector basis becomes that of the anisotropic oscillator. When R tends to infinity, a two oscillator eigenvector system is obtained in the same Hermite space, centered in the two fragments. In the intermediate situation, each eigenfunction has components in the two subspaces that belong to the fragments. So, the two center shell model provides permanently the wave functions associated to the lower energies of the single particle states pertaining to a major quantum number N_{max}. Therefore, molecular states formed by two fragments at scission could be precisely described. Another feature of the two center shell model is related to the localization of the single particle wave function in one of the two potential wells after the scission. As evidenced in Ref. [17], it is possible to predict this localization for a given fragmentation before that the scission is produced. This feature helps us to fix the number of particles in each fragment when Eqs. (4)-(11) are resolved. This model was widely used by the Bucharest group in the calculations addressing the cluster [33, 34] and alpha decay [35–37], the fission [38–40], or the heavy element synthesis [41–43]. For example, the model was able to describe two tangent nuclei in a wide range of mass asymmetries. The half-lives for cluster decay were reproduced. A mechanism for the formation of an α-particle on the nuclear surface was supplied. Fission barriers that agree with the evaluated ones were calculated. It was also shown that the two center parametrization is a good approximation for the optimal shapes [44]. Due to these advantages, versions based

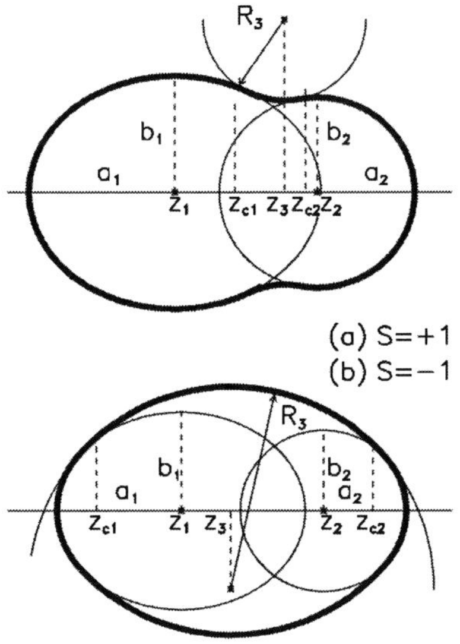

Figure 3. Nuclear shape parametrization. The elongation is defined as $R = z_2 - z_1$. The curvature of the neck parameter is $C = S/R_3$, where $S = 1$ for necked shapes in the median surface and $S = -1$ otherwise. The eccentricities of the fragments are $\epsilon_i = \sqrt{1-(b_i/a_i)^2}$ $(i = 1, 2)$. The mass asymmetry parameter can be defined as $\eta = a_1 b_1^2/(a_2 b_2^2)$.

on a generalized Nilsson model are widely used in investigating nuclear disintegrations or collisions [45, 46].

The fission yields as function of the total excitation energy for the fragmentation ^{90}Kr+^{144}Ba (even-even), ^{90}Rb+^{144}Cs (odd-odd) will be investigated in the framework of our model. Experimental data are available for the reaction [7] ^{233}U(n_{th},f) where these partitions are investigated. So, two fission channels should be investigated. In the cases of the odd-odd partition, the barrier must be increased with approximately the value of the excitation energy of the unpaired nucleons. In this case, the penetrability is decreased exponentially. The amount of which the barrier is modified can be estimated by accounting the specialization energy. Wheeler defined this specialization energy [47] as the excess of the energy of a nucleon with a given spin over the energy for the same spin nucleon in the state of lowest energy. The least action principle was used to determine the best fission trajectory in our configuration space spanned by our five generalized coordinates. We tested several models for the inertia: the non-adiabatic cranking model [48], the Gaussian overlap approximation [49, 50], and the cranking approach [12]. The final results were similar within these different models for the inertia. By minimization, we obtained the most probable fragmentation, that corresponds to a heavy fragment with mass around 138. In our work, the fission path for the 144/90 fragmentation is required. Sophisticated models are used to describe the Z distributions, as for example in Ref. [51]. To overcome these difficulties, some simple assumptions are made. We rely on the results of Refs. [18, 38] where the mass distribution of the fragments was relatively well reproduced by considering that the variation of the mass asymmetry is linear from the saddle configuration of the outer barrier up to the exit point. So, it is considered that under the rapid descent from the top of the second barrier different mass partitions are obtained. To obtain these different mass partitions, the simplest way is to vary the mass asymmetry parameter η and the averaged deformations of fragments as in Ref. [18].

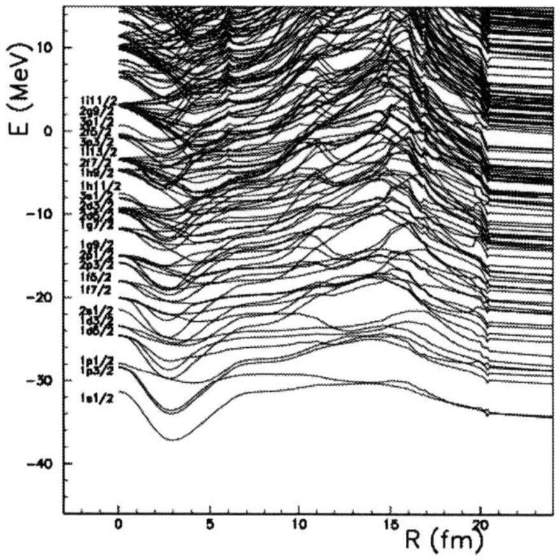

Figure 4. Proton single particle level scheme. The levels with spin projection Ω=3/2 that give the lower energy configuration for the unpaired fragments is plotted with a full line. The Fermi energy of the compound nucleus is displayed with a thick dashed line. Four avoided level crossing regions were identified for $R \approx 9.1$, 14.7, 16 and 20 fm.

The penetrabilities for the odd-odd channel is obtained within the same path as that for the even-even channel, but the specialization energy was taken into account.

Once the fission path determined, we need the single particle energies, the variations of the pairing matrix elements G_{kl}, the interactions in the avoided crossing regions, and the velocity of passage through these regions in order to solve the equations of motion. In figure 4, the single particle levels obtained within the two center Woods-Saxon model are displayed along the fission trajectory for protons, as an example. The Fermi level of the parent nucleus is plotted with a dashed thick line. The pair of levels with the same spin projection that gives the lower single particle seniority two excitation (or specialization energy) are also displayed with thick full lines. It can be observed that the levels that pertain to the lower excitation are close to the Fermi energy. In the case of protons the spin projection Ω of these levels is 3/2 while in the case of neutrons it is Ω=5/2. These values of Ω give also the final spin of the partners. Our pairing active space is constructed with 58 single particle levels around the Fermi energy. An important question is the identification of the avoided crossing regions that arise from the strong energy fluctuations of the single particle levels observed along the fission path. For this purpose, we selected all the pairs of adjacent levels with the same spin projection in the active pairing space. From an analyze of the rearrangement of these levels in a manner similar to that of Ref. [52], the avoided crossing regions were identified. In these regions, the excitations are possible between the seniority two and the seniority zero configurations. In other words, these regions are like gates for mixing the configurations, or for changing the seniority. Using the interpolation method described in Ref. [53], the magnitude of the interactions was calculated. Finally, we selected 32 seniority two configurations for protons and 31 configurations for neutrons that are coupled to the seniority zero configuration through avoided levels crossings regions. In the frame of the adiabatic BCS model [12], a mean value of the pairing interaction G can be associated to a given active pairing space, by using a renormalization procedure. So, the values of G for all seniority states were computed for the parent and for the two fragments. A linear interpolation between the values of G from those of the parent to those of the fragments is also realized around the scission point. So, all the most important quantities required to solve the equations of motion are provided: the single particle energies ϵ_k, the perturbations h_{ij} and the pairing interactions G. The dependence in time is introduced by means of the variations of the collective coordinates. In this respect, several values of the internuclear velocity

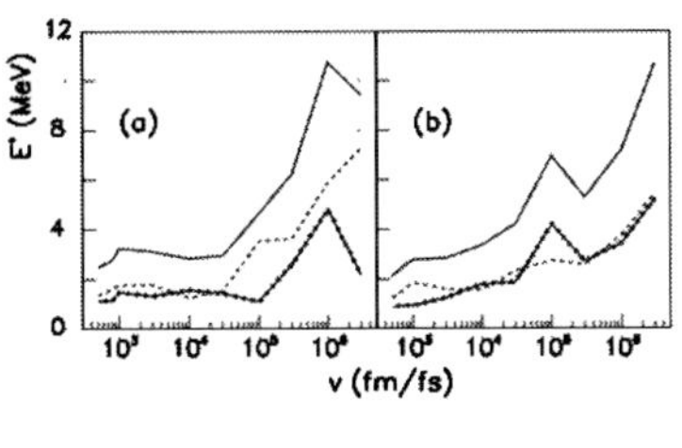

Figure 5. a) Total dissipation energy after the scission E^* as function of the internuclear velocity υ for the seniority zero configuration. The neutron and proton components of the dissipated energy are plotted with a dot-dashed and a dashed line respectively. (b) Total dissipation for the seniority two configuration with lower single particle excitation. The neutron and proton components are displayed with the same line types as in panel (a).

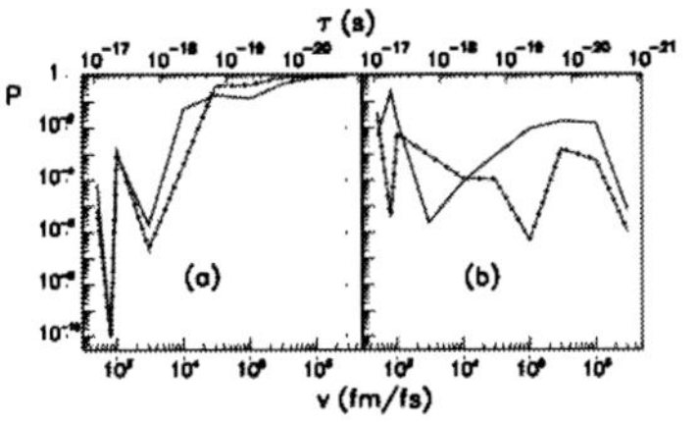

Figure 6. (a) Probability of realization of the adiabatic seniority zero configuration at scission as function of the internuclear velocity for neutrons (full line) and protons (dot-dashed line). (b) Probability of the realization of the seniority two configuration at scission with the lower single particle excitation for neutrons (full line) and protons (dot-dashed line). The upper axis corresponds to the saddle to scission time τ.

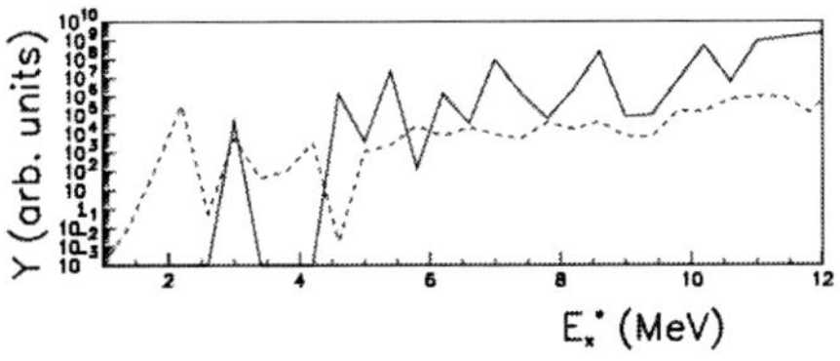

Figure 7. Fission yields for even-even and odd-odd partitions. Full line: dependence of even-even yield in arbitrary units Y_0 as function of the final excitation E_x^* of the fission fragments. Dashed line: dependence of the odd-odd yields Y_2 as function of the excitation energy.

$\upsilon = \dot{R}$ are tested in order to solve the equations: 5×10^2, 8×10^2, 10^3, 3×10^3, 10^4, 3×10^4, 10^5, 3×10^5, 10^6, and 3×10^6 fm/fs.

The Equations of motion were solved by taking as initial conditions the BCS ground state values. The final excitation and the probabilities of realizations are plotted in figure 5. The general trend exhibited in both cases is an increase of the final dissipated energy when the internuclear velocity becomes larger. The same behavior is typical for all seniority configurations involved. So the general rule that assesses that the dissipation is proportional to the velocity is retrieved. Another way to introduce the dissipation in quantum systems is to consider forces proportional to the velocity in analogy with the friction forces in classical mechanics [54], leading to a generalized Schrödinger non-linear equation for an open system.

The initial condition for the equations (8)-(11) that describe the configuration mixing is P_0=1, all the other values being zero. The final probabilities of realization of two configurations for neutrons and protons as function of the internuclear velocities are displayed in Fig. 6. This figure evidences the fact that the probability of realization of adiabatic seniority zero states are close to zero for small velocities, exactly in the energy region in which the dissipation is smaller. So, the main features concerning the relation between the final excitation energy and the probability of realization of a given channel presented in Ref. [15] are retrieved and are valid even if the method for dynamical projection on final atomic and mass numbers is used. The saddle to scission time τ depends on the intrinsic velocity υ. This scission time τ labels the upper axis of Fig. 6.

The results concerning the final excitation and the probabilities of realization presented above are coupled through the collective velocity parameter. In order to compare the theoretical findings with the experimental data, this velocity must be eliminated. Therefore, a simple model in accordance to the experimental arrangement is conceived in order to relate the dissipation and the yields. By bombarding the ^{233}U with thermal neutrons, using mass evaluations [55] we found that an excitation of at least B_n=6.85 MeV is accumulated in the compound ^{234}U nucleus. This energy is shared between a potential part U and a kinetic one $T = B_0 v^2/2$, where B_0 is the inertia in the ground state configuration. A constant population of all values of the kinetic energy is assumed. The penetrability $P_b(v,\gamma_n,\gamma_p)$ of each channel depends on the turning points of the barrier, that is, on v through the relation $B_n = U+T$. That leads to a larger penetrability when the velocity decreases. The penetrability depends also on the excitation channel $\{\gamma\}$ that can be constructed with the specialization energies of the configurations $\{0\}$ or $\{ij\}$ of the two isospins. The final excitation energy of the two fragments is $E_x^*(v,\gamma_n,\gamma_p) = E^*(v,\gamma_n,\gamma_p) + E_{sp}(\gamma_n,\gamma_p)$, that is, it corresponds to a sum between the dissipated energy E^* and the single particle excitations E_{sp} of the two fragments. The dependence of the yield as function of the final excitation E_x^* in the interval $[E_{x1}^*, E_{x2}^*]$ exhibits the following proportionality:

$$\begin{aligned}Y_0(E_x^*) \propto \frac{1}{E_{x2}^* - E_{x1}^*}\int_0^{v_{max}} & [P_{0(n)}(v)P_{0(p)}(v)P_b(v,0,0) \\ & +P_{0(n)}(v)\sum_{\gamma_p} P_{\gamma_p(p)}(v)P_b(v,0,\gamma_p) \\ & +P_{0(p)}(v)\sum_{\gamma_n} P_{\gamma_n(n)}(v)P_b(v,\gamma_n,0) \\ & +\sum_{\gamma_n,\gamma_p} P_{\gamma_n(n)}(v)P_{\gamma_p(p)}(v)P_b(v,\gamma_n,\gamma_p)] \\ \times w(v)\theta(E_{x1}^* - E_x^*(v,\gamma_n,\gamma_p))\theta(E_x^*(v,\gamma_n,\gamma_p) - E_{x2}^*))dv, & \end{aligned} \tag{16}$$

for the even-even channel. Here P_b are the penetrabilities that depend on a specific channel and P_0 are the probabilities of realization given by the time dependent equations for neutrons (index (n)) and protons (index (p)). For all velocities that give an excitation in the interval $[E_{x1}^*, E_{x2}^*]$, the previous formula reflects the fact that the yields are proportional to these penetrabilities and probabilities. Of course, at scission it is possible to obtain even-even partitions even if a Cooper pair is broken. In this case, one of the fission products picks this broken pair and will carry a very large excitation energy. Therefore, the sums that run over the channels γ in the expression (16) take into consideration the fact that some configurations are formed with a broken pair in only one partner. The probabilities of realization of a given seniority configuration $P_{\gamma_n(p)}(v)$ depend on the internuclear velocity. The penetrability of the fission barrier $P_b(v,\gamma_n,\gamma_p)$ at the velocity v in the channel $\{\gamma_n,\gamma_p\}$ depends on the variation of the probabilities of realization along the fission path. The factor $w(v) = B_0 v$ is a weighting that reflects the dependence of the kinetic energy of the velocity, because $dT = w(v)dv$, and θ is the step Heaviside distribution used to select only events in the interval $[E_{x1}^*, E_{x2}^*]$. B_0 is considered to be the inertia in the ground state of the parent nucleus. The maximal value of the the velocity is obtained from the boundary $T_{max} = \frac{1}{2}B_0 v_{max}^2 = B_n$. The penetrability $P_b(v,\gamma_n,\gamma_p)$ is calculated by considering the turning points at the energy $U = B_n - \frac{1}{2}B_0 v^2$. The yields for odd-odd partitions are proportional with the expression:

$$\begin{aligned}Y_2(E_x^*) \propto \frac{1}{E_{x2}^* - E_{x1}^*}\int_0^{v_{max}} & \sum_{\gamma_n}\sum_{\gamma_p} P_{\gamma_n(n)}(v,)P_{\gamma_p(p)}(v,)P_b(v,\gamma_n,\gamma_p)w(v) \\ & \times\theta(E_{x1}^* - E_x^*(v,\gamma_n,\gamma_p))\theta(E_x^*(v,\gamma_n,\gamma_p) - E_{x2}^*)dv. \end{aligned} \tag{17}$$

In the previous relation, the sums run over all the seniority two configuration taken into consideration for neutrons (index n) and protons (index p) that give unpaired nucleons in the two fragments.

The results obtained for the folded distributions $Y_0(E_x^*)$ and $Y_2(E_x^*)$ are displayed in figure 7. These distributions are proportional to the even-even and the odd-odd yields obtained for the fission fragments. The averaging interval is $E_{x2}^* - E_{x1}^*$=0.5 MeV. The observed experimental trends exhibited in Fig. 4 of Ref. [7] for cold fission yields were reproduced. The trends are same for the three approaches used for the inertia. At low excitation energy the odd-odd yields surpass the even-even ones. The even-even yields become larger for excitation energies larger that 3-4 MeV, in accordance with the experimental findings.

6 Discussion and conclusions

In this work, a microscopic model is proposed for the explanation of the odd-even effect in cold fission. This explanation is based on a mixing configuration mechanism that is produced during the fission process. This configuration mixing mechanism is obtained dynamically by solving a the generalized system of time dependent pairing equations, that include a pair-breaking effect. A first rule can be extracted from this model. The even-even fission products cannot be obtained at zero excitation energies because of the existence of dynamical excitations produced in the avoided level crossing regions when the nuclear system deforms slowly.

The magnitudes of the interactions and the location of the the avoided crossing regions are fixed along the fission path and are independent of the velocity of passage through these regions. If this velocity is large, the perturbation will act onto the Cooper pair a small fraction of time. So, the chance to break a pair will be small. If the velocity is low, the pairs will traverse the regions in larger time durations. So, the probability to break a pair increases. On another hand, high velocities lead to large dissipation energies. Another characteristic that was not exploited in this work can be featured from the model. The lower excitation energies of a combination of two odd-odd partners can be obtained only if their spin are the same for neutrons and protons. If the spins of the partners are not the same, the model predicts that at least two pair ruptures are produced for neutrons or protons and additional single particle excitations must be taken into consideration.

The possibility to jump from one level to another in a large scale amplitude motion was predicted by Hill and Wheeler in Ref. [12]. Dissipation in terms of Landau-Zener crossings during fission was first proposed in Ref. [56] where excitations were considered only for time-reversed pairs, neglecting the possible existence of unpaired nucleons. As evidenced in Ref. [57], many studies were performed in order to exploit this mechanism in different type of processes. It was also shown in Ref. [20] that the Landau-Zener mechanism is cached in the time dependent pairing equations (4) and (7). Pairs undergo Landau-Zener transitions on virtual levels with coupling strengths given by the value of the magnitude of the gap parameter. Anyhow, it is the first time that the dynamical pair breaking effect was used to explain the odd-even effect in fission. It must be mentioned that a time-dependent microscopic approach to the scission process was described in Ref. [58]. They observed that for scission time larger than 5×10^{-21} s, the single particle excitations are negligible. This time that characterizes the neck rupture corresponds to a velocity of 10^6 fm/fs in our calculations. A typical value for saddle to scission time could be considered as $\tau_s \approx 1.3\times10^{-21}$. As plotted in Fig. 6, this value of the scission time is consistent with a total excitation energy of the even-even fragments of about 9 MeV (10 MeV for odd-odd ones). For this value τ_s, the probability to have fully paired partners is close to unity.

The density of single particle levels increases in the region of the second barrier. Therefore, from the outer saddle to the scission, many avoided crossing regions are produced and the chance to break a pair is enhanced. Up to the second saddle, the number of avoided crossing regions is small an the system evolves merely in the seniority zero configuration. That gives a large penetrability for all channels, the mixing of configurations being produced especially in the outer barrier region.

In this paper, the scission configuration takes in account values for the fragment deformations close to the fundamental ones while the fission path corresponds to cold fission. Therefore, small values of the excitation energy are expected. However, if other deformations of the fragment at scission and a path for more elongated shapes are provided, the same prescriptions can be used for the odd-even effect description. The dissipated energy depends on the modality in which the microscopic levels are rearranged along the specific fission trajectory. If the deformations at scission differ from the ground state ones, the specialization energy and the deformation energy of each fragment increase in magnitude giving finally larger excitation energies.

In conclusion, by solving the dynamical microscopic equations of motion for an fissioning even-even system it is found that the probability to obtain an odd-odd partition overcomes the probability of an even-even one at excitation energies smaller than 4 MeV, for the same division in mass numbers. The theoretical results are in accordance with the experimental behavior of the odd-even distributions at high kinetic energies. It is the first time that this behavior was explained within a quantum mechanical approach.

Acknowledgements Work supported by CNCS-UEFISCDI, project number PN-II-ID-PCE-2011-3-0068.

References

[1] W.N. Reisdorf, J. P. Unik, and L. E. Glendenin, Nucl. Phys. A **205**, 348 (1973)
[2] H. Wohlfarth, W. Lang, H.-G. Clerc, H. Schrader, K.-H. Schmidt, and H. Dann, Phys. Lett. B 63, **275** (1976)
[3] C. Signarbieux, M. Montoya, M. Ribrag, C. Mazur, C. Guet, P. Perrin, and M. Maurel, J. Phys. Lett. **42**, 437 (1981)
[4] M. Caamano, F. Rejmund, and K.-H. Schmidt, J. Phys. G **38**, 035101 (2011)
[5] B. Jurado and K.-H. Schmidt, EPJ Web of Conferences **62**, 07003 (2013)
[6] H.-G. Clerc, in Heavy Elements and Related New Phenomena, edited by W. Greiner and R. K. Gupta (World Scientific, Singapore, 1999), p. 471
[7] E. Schwab, H.-G. Clerc, M. Mutterer, J. P. Theobald, and H. Faust, Nucl. Phys. A **577**, 674 (1994)
[8] F.-J. Hambsch, H.-H. Knitter, and C. Budtz-Jorgensen, Nucl. Phys. A **554**, 209 (1993)
[9] H.-H. Knitter, F.-J. Hambsch, and C. Budtz-Jorgensen, Nucl. Phys. A **536**, 221 (1992)
[10] V.Avrigeanu, A. Florescu, A. Sandulescu, and W. Greiner, Phys. Rev. C **52**, R1755 (1995)
[11] B. Jurado and K.-H. Schmidt, J. Phys. G. **42**, 055101 (2015)
[12] D. L. Hill and J. A. Wheeler, Phys. Rev. **89**, 1102 (1953)
[13] F. Gonnenwein, Phys. Procedia **47**, 107 (2013)
[14] M. Mirea, Phys. Rev. C **78**, 044618 (2008)
[15] M. Mirea, Phys. Lett. B **680**, 316 (2009)
[16] M. Mirea, Phys. Rev. C **89**, 034623 (2014)
[17] M. Mirea, Phys. Rev. C **83**, 054608 (2011)
[18] M. Mirea, Phys. Lett. B **717**, 252 (2012)
[19] S.E. Koonin, J.R. Nix, Phys. Rev. C **13**, 209 (1976)
[20] J. Blocki, H. Flocard, Nucl. Phys. A **273**, 45 (1976)
[21] B. Avez, C. Simenel and Ph. Chomaz, Phys. Rev. C **78**, 044318 (2008)

[22] S. Ebata, T. Nakatsukasa, T. Inakura, K. Yoshida, Y. Hashimoto and K. Yabana, Phys. Rev. C **82**, 034306 (2010)
[23] J.W. Negele, Rev. Mod. Phys. **54**, 913 (1982).
[24] M. Mirea, L. Tassan-Got, C. Stephan, C.O. Bacri, P. Stoica, R.C. Bobulescu, Nucl. Phys. A **735**, 21 (2004).
[25] J.R. Nix, Ann. Rev. Nucl. Sci. **22**, 65 (1972).
[26] W.J. Swiatecki, S. Bjornholm, Phys. Rep. **4**, 325 (1972).
[27] C. Simenel, A.S. Umar, Phys. Rev. C **89**, 031601(R) (2014).
[28] G. Scamps, C. Simenel, D. Lacroix, Phys. Rev. C **92**, 011602(R) (2015).
[29] M. Brack, J. Damgaard, A.S. Jensen, H.C. Pauli, V.M. Strutinsky, C.Y. Wong, Rev. Mod. Phys. **44**, 320 (1972).
[30] J. Maruhn, W. Greiner, Z. Phys. **251**, 431 (1972).
[31] M. Mirea, Phys. Rev. C **54**, 302 (1996).
[32] M. Mirea, Nucl. Phys. A **780**, 13 (2006).
[33] M. Mirea, A. Sandulescu, D.S. Delion, Nucl. Phys. A **870**, 23 (2011).
[34] M. Mirea, A. Sandulescu, D.S. Delion, Eur. Phys. J. A **48**, 85 (2012).
[35] A. Sandulescu, M. Mirea, D.S. Delion, EPL **101**, 62001 (2013).
[36] M. Mirea, Eur. Phys. J. A **51**, 36 (2015).
[37] M. Mirea, Rom. J. Phys. **60**, 156 (2015).
[38] M. Mirea, D.S. Delion, A. Sandulescu, Phys. Rev. C **81**, 044317 (2010).
[39] M. Mirea, L. Tassan-Got, Centr. Eur. J. Phys. **9**, 116 (2011).
[40] R. Budaca, A. Sandulescu, M. Mirea, Mod. Phys. Lett. A **30**, 1550129 (2015).
[41] M. Mirea, D.S. Delion, A. Sandulescu, EPL **85**, 121001 (2009).
[42] A. Sandulescu, M. Mirea, Eur. Phys. J. A **50**, 110 (2014).
[43] D. Aranghel and A. Sandulescu, Rom. J. Phys. **60**, 147 (2015).
[44] F.A. Ivanyuk, S. Chiba, A. Arimoto, Phys. Rev. C **90**, 054607 (2014).
[45] A. Diaz-Torres, W. Scheid, Nucl. Phys. A **757**, 373 (2005).
[46] Y. Arimoto, S. Chiba, Phys. Rev. C **88**, 044614 (2013).
[47] J.A. Wheeler, *Niels Bohr and the Development of Physics*, edited by W. Pauli, L. Rosenfeld, and W. Weisskopf (Pergamon Press, London, 1955) p. 163.
[48] M. Mirea, R.C. Bobulescu, J. Phys. G **37**, 055106 (2010).
[49] A. Gozdz, K. Pomorski, M. Brack, E. Werner, Nucl. Phys. A **442**, 26 (1985).
[50] A. Gozdz, K. Pomorski, M. Brack, E. Werner, Nucl. Phys. A **442**, 50 (1985).
[51] J. Randrup, P. Moller, Phys. Rev. Lett. **106**, 132503 (2011).
[52] M. Mirea, L. Tassan-Got, C. Stephan, C.O. Bacri and R.C. Bobulescu, Phys. Rev. C **76**, 064608 (2007).
[53] W. Greiner, J.Y. Park, and W. Scheid, *Nuclear Molecules* (World Scientific, Singapore, 1995).
[54] A. Sandulescu, H. Scutaru, Ann. Phys. **173**, 277 (1987).
[55] G. Audi, A.H. Wapstra, C. Thibault, Nucl. Phys. A **729**, 337 (2003).
[56] G. Schutte, L. Wilets, Z. Phys. A **286**, 313 (1978).
[57] R.W. Hasse, Rep. Prog. Phys. **41**, 1027 (1978).
[58] M. Rizea, N. Carjan, Proc. Rom. Acad. Series A **16**,176 (2015).

Studies for γ-ray emission in the fission process with LICORNE

M. Lebois[1,a], J. Wilson[1], Q. Liqiang[1], P. Halipré, R. Shearman[2,3], I. Matea[1], A. Oberstedt[4], S. Oberstedt[5], R. J. Carroll[2,], P. H. Reagan[2,3], and D. Verney[1]

[1] *Institut de Physique Nucléaire d'Orsay, Bât. 100, 15 rue G. Clémenceau, 91406 Orsay Cedex, France*
[2] *Department of Physics, University of Surrey, Guildford, GU 7XH, UK*
[3] *National Physical Laboratory, Teddington, Middlesex, TW11 0LW, UK*
[4] *Fundamental Fysik, Chalmers Tekniska Högskola, 41296 Göteborg, Sweden*
[5] *European Commission, Joint Research Centre (IRMM), 2440 Geel, Belgium*

Abstract. The LICORNE neutron source is a new device at the ALTO facility. Its use of inverse kinematics makes the production of naturally focused neutron beams possible with an energy range of 0.5 to 4 MeV. This is perfect for any studies concerning fast neutron induced reaction based on detection setup that require to be placed in a compact geometry – such as Ge based γ detection setup.
In this paper, neutron production with LICORNE is described. The development of a gas cell target and the extension of the energy range up to 7 MeV with the p(^{11}B,n)^{11}C are presented. An overview of the major research fields studied with LICORNE is given and two types of experiment are presented. The first one dedicated to prompt fission γ-ray emission in fission, as a function of incident neutron energy, is described. Some preliminary results are shown. The second, in the context of the MINORCA campaign, is detailed. The most recent outcomes in the data analysis process are also presented.

1 Introduction

The LICORNE project started in 2012 with the test of a new technique to produce neutrons using the p(^{7}Li,n)^{7}Be reaction in inverse kinematics. A commissioning experiment clearly showed the capability of LICORNE, to naturally produce fast neutrons, with energy of 0.5 - 4 MeV, in narrow cones (< 20°).
The development of this neutron source using an hydrogen gas target has opened up possibilities for new experiments about the γ spectroscopy of fast neutron induced reactions. Since then, two research programs have been initiated. The first one concerns the emission of prompt fission γ-rays (PFG) from major actinide nuclei. This is of interest for understanding the emission of neutral particles in the fission process. Indeed, the competition between neutrons and gammas is still a puzzle that theoretical descriptions need to understand. At the same time, these measurements are directly related to nuclear energy and reactor safety, since such data are used as source terms in gamma heating calculations in reactor cores. The other axis of research of the LICORNE physics program concerns the γ spectroscopy of individual exotic fission fragments at the moment of their formation. This gives access to collective and highly excited states which will provide new interesting information

[a] e-mail: lebois@ipno.in2p3.fr

on nuclear structure for fission fragments that are still unknown.To achieve this goal, LICORNE was coupled in March 2015 with the Miniball high efficiency gamma ray spectrometer on loan from ISOLDE at CERN. This coupling was only made possible by careful elimination of all secondary isotropic neutron fields generated by the LICORNE directional neutron source.
In the following, a description of the LICORNE neutron source and its recent developments will be given. Then, an overview of the most recent results obtained for PFG and the MINIBALL experiment will be given.

2 Presentation of the LICORNE neutron source

In 2012, at the ALTO facility [1, 2], a new project was started. The aim was to add neutron beam production capability to this facility in prevision of the NFS project [3] start at GANIL. In Europe, several neutron facilities were already running at that time: GELINA in Geel [4], ILL in Grenoble [5], or KFKI research reactor [6] in Budapest, ... It was necessary for LICORNE to be different from these installations in order to be able to explore new ways of using neutrons. In many facilities, a light primary beam impinging an heavy target is used to induce nuclear reactions which produce neutrons. In this case, the neutron emission is isotropic. As a consequence, it is mandatory, for any detection setup, to be at a great distance from the source. Thus, the neutron fluxes are quite small compare to the total amount of produced neutrons, making almost 99% of them useless and participating to the neutron background. In addition, any detector, placed in the neighbourhood of the sample, still needs to be heavily shielded to not be blinded by the source neutrons.

In order to be complementary with the other sources, LICORNE also had to be able to produce decent fluxes of fast neutrons without the use of a very intense ($< \mu$A), or very energetic ($>$ 100 MeV), primary beams. It also required to provide in the ALTO environment, where experimental caves are small compare to typical installations, a very low neutron background.

2.1 Description of the p(^{7}Li,n)^{7}Be reaction for neutron production

The LICORNE neutron source meets those criteria thanks to the clever use of specific reactions in inverse kinematics. In his work, M. Drosg [7] have listed a wide range of reactions that produce neutrons with energies from keV to 500 MeV. To satisfy our needs (see §.3), fast neutrons with energy close to fission neutrons were required. For this reason, the p(^{7}Li,n)^{7}Be reaction was chosen. The results of 2-body relativistic kinematics calculations are shown on the left side of figure 1. With beam energy in the range of 13.5 to 16.5 MeV, it is possible to produce neutron beams in the laboratory frame with two different energies corresponding to forward and backward emission of neutrons in the centre of mass frame. The relative size of the principal (high energy) peak and satellite (low energy) peak is governed by both the relativistic kinematics of the focusing and the angular distribution of emission in the centre of mass frame, which changes drastically as a function of ^{7}Li bombarding energy. The gain from the focusing and natural collimation can be expressed in terms of neutron flux enhancement over the non-inverse reaction (see right part of fig. 1). To obtain this curve, the reaction rates have been calculated for different ^{7}Li kinetic energies. The neutron flux per steradian has been normalized to the ^{7}Li beam intensity. At threshold the enhancement factor is maximal, since the emitted neutrons move with the centre of mass of the system, which follows the ^{7}Li beam direction. As a consequence, close to the threshold energy, it is possible to produce very narrow ($< 5°$) conical neutron field. With the increasing ^{7}Li bombarding energy, the cone broadens and the number of neutrons in a given solid angle decreases, which causes the enhancement factor to drop. For a beam energy lower than 16.5 MeV, the maximum opening of the cone is smaller than 29°. In this case, an enhancement factor of

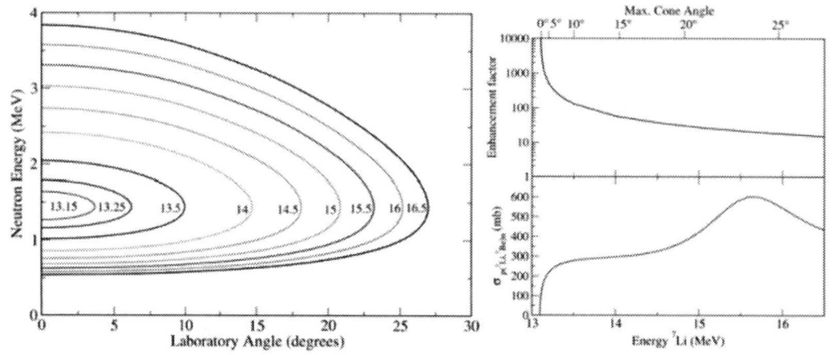

Figure 1. On the left: kinematic curves relating the angle of neutron emission to neutron energy in the laboratory frame for different ^{7}Li bombarding energies from 13.15 to 16.5 MeV, calculated using two-body relativistic kinematics.
On the right: the top panel shows the enhancement factor of the neutron flux between the inverse kinematic and the direct kinematic reaction as a function of ^{7}Li bombarding energy. The bottom panel shows the p(^{7}Li,n)^{7}Be reaction cross section over the same energy range.

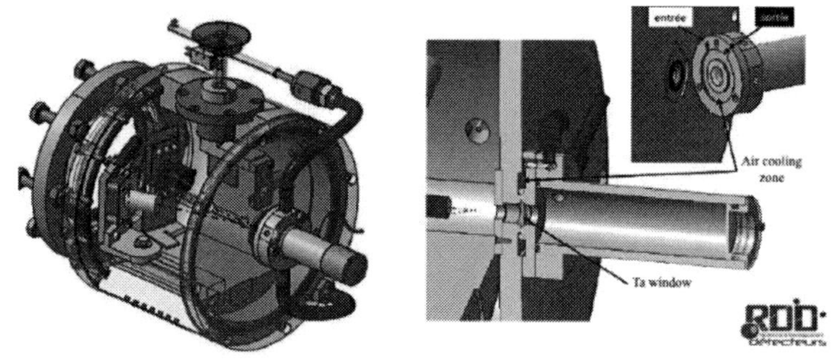

Figure 2. Technical scheme of the LICORNE neutron source chamber. The right hand figure is a zoom of the gas cell.

30 can still be expected over the non-inverse reaction.

The maximum available fluxes of up to 10^7 neutrons/s/sr (assuming ^{7}Li beam currents of 100 nA) are comparable to those available at other conventional quasi-monoenergetic neutron sources. However, the natural collimation and neutron beam directionality provides at least two distinct advantages: (i) the placement of gamma detectors adjacent to the sample being irradiated becomes feasible; (ii) the scattered neutron background in the room will be reduced by up to a factor of 100.

2.2 Facts about the LICORNE neutron source

To produce neutrons with the Tandem ^{7}Li beam, a dedicated reaction chamber had to be built. At first, the reaction chamber is a cylindrical aluminum chamber of 17 cm. The exit window is made of aluminum and is only 0.3 mm thick. Inside, a very proton rich target ($\sim 10^{19}$ p/cm^2) had to be placed. A mini-camera and an illuminating LED are included for beam tuning and inspecting the target from inside. At first, a simple layer of polypropylene of 4.4 μm thickness was chosen. However, the carbon content allowed parasitic fusion-evaporation reactions that produce isotropic neutrons. Added to this, this material cannot stand high intensity beam exposure. For these reasons, the choice of a gas target was made: it was possible to limit the presence of low Z material in the beam path. The exit flange was modified to be able to attach a cylindrical gas cell made of steel. A 5 mm diameter hole was drilled in the flange. The hole's contour were coated with a 50 μm thick tantalum foil to prevent any fusion-evaporation reaction with the flange material. The gas is separated from the beam line vacuum by a 4 μm tantalum foil. In case of a high intensity beam, the tantalum window is air cooled. Inside the gas cell, pure H_2 gas flows with pressure above 1 bar and up to 1.8 bar. The bottom of the gas cell is coated with 100 μm lead foil to form a beam dump. Figure 2 shows a technical drawing of the design.

Different sizes of the gas cell are available depending on the needs. The longer the gas cell, the greater is the ^{7}Li beam energy loss and the wider is the neutron energy dispersion. Four sizes are now available: 2, 3.5, 5.5 and 7.5 cm. The bigger ones are used when a high neutron flux is wanted: the ^{7}Li enter with the maximum energy and slows down to threshold energy. All the neutron energies between 0.5 and 4 MeV are produced in a broad spectrum. On the contrary, short cells are used when a precise neutron energy is required: ^{7}Li beam looses little energy and the neutron energy spread is small. The minimal energy spread measured was 300 keV.

Nowadays, only gas cells are used for the neutron production with LICORNE. In the near future, some upgrades are foreseen to increase the energy range.

2.3 The LICORNE energy upgrade

Getting to higher energies requires a different primary beam. By comparing M. Drosg's work [7] an the available beams at the Tandem [10], ^{11}B was considered. It is a good candidate because it can be produced with a rather good intensity ($\sim$ 100 nA) and the kinematics allow the extension of LICORNE energy range up to 12 MeV – see fig.3.a – with features similar to ^{7}Li kinematics. However, the cross-section (see fig.3.b) for the p(^{11}B,n)^{11}C reaction is, on average, four times smaller than for the ^{7}Li induced reaction. This beam has been used for the first time, in may 2015, during an experiment dedicated to test the PARIS detector [14] response to neutrons with energies of 4 to 12 MeV. During this experiment the reaction kinematics at 0° was the expected one. Neutrons, up to 12 MeV, were produced. However, another test experiment was conducted in november to make quantitative measurements of fluxes and source background. We observed an opening of fusion-evaporation channels when beam energy reaches 44.5 MeV. From that energy, as many neutrons are produced outside the cone as inside and the natural collimation provided by the inverse kinematics is no longer an advantage.

This test experiment demonstrated the new capabilities of LICORNE to produce neutrons, in a narrow cone, up to 7 MeV with fluxes of 10^{5-6} n/s/sr.

3 Studies of Statistical properties of Prompt fission γ-rays

As mentioned previously (see 1), the LICORNE physics program follows two major axes. The first one is dedicated to the measurement of prompt fission γ-ray properties. Prompt γ emission in nuclear

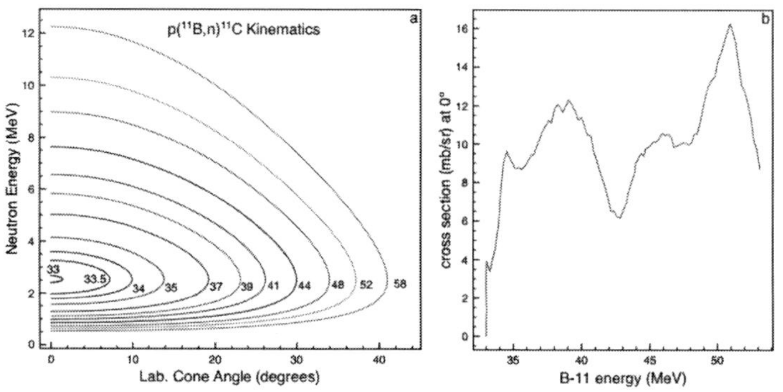

Figure 3. a. kinematic curves relating the angle of neutron emission to neutron energy in the laboratory frame for different ^{11}B bombarding energies from 33 to 58 MeV, calculated using two-body relativistic kinematics.
b. Measured cross section for the p(^{11}B,n)^{11}C reaction taken from [11, 12].

fission is one of the least well-understood parts of the fission process. It has important applications for nuclear reactors. Prompt fission γ emission contributes up to 40% to the reactor core γ-ray spectrum. Induced γ heating effects, particularly at the core periphery are closely linked to reactor safety. These are still largely underestimated [13] and measurement of spectral shapes, average multiplicities and distributions, average energy per quantum and total energy are at the top of OECD/NEA high priority request list [15, 16]. Besides its importance for more detailed databases for nuclear reactors physics, understanding γ/neutron competition in the fission process is important for the modeling of post-scission phases of the nuclear fission. Unlike neutrons, prompt γ-rays remove most of the fission fragment spin. Studying the spectrum can help understanding the deexcitation process of fission fragments and the angular momentum generation. So, this type of studies can provide information on the partition of energy and angular momentum, and also on the competition between neutron and γ emission during fission fragment de-excitation.

In 2013, comparative prompt fission γ-rays characteristics for fast neutron induced fission of two key nuclei, ^{235}U and ^{238}U, have been measured for the first time. In this experiment, a broad neutron spectrum with an average energy of 1.5 MeV was produced using LICORNE. Fourteen BaF_2 crystals from the Château de Cristal were placed 30 cm away from an ionisation chamber. After a careful selection of γ-rays, by time-of-fight, relative shapes of the PFG spectrum from fast neutron induced fission of ^{235}U and ^{238}U were observed to be similar within the experimental error bars [17].

In february 2015, we participated in our CEA/DAM collaborators' experiment with LICORNE. At first, it was designed to measure prompt neutron emission for ^{238}U fast neutron induced fission. For a week, they bombarded an ionisation chamber containing ~ 300 mg of ^{238}U with the LICORNE neutron beam. Two energies were selected: 2.4 and 3.3 MeV. In the setup, three 2"x2" $LaBr_3$ crystals were placed 40 cm above the ionisation chamber to measure prompt fission γ-rays at the same time. Off beam, a similar ionization chamber containing ^{252}Cf sample replaced the one containing ^{238}U. For two days, the prompt γ-rays of spontaneous fission of ^{252}Cf were measured. Analysis on this smaller data set was performed. First, fission fragments were selected in the fission chamber. They can be easily separated from α and β thanks to the higher charge collected on the anodes. With this fission

DOI: 10.1051/epjconf/201612201010

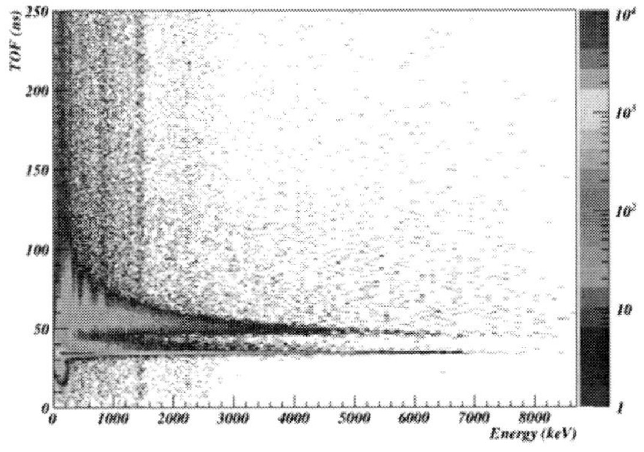

Figure 4. TOF vs measured energy for one $LaBr_3$ crystal. Two major horizontal structures can be distinguished. The lower one corresponds to the γ. The triangular one just above corresponds to neutrons.

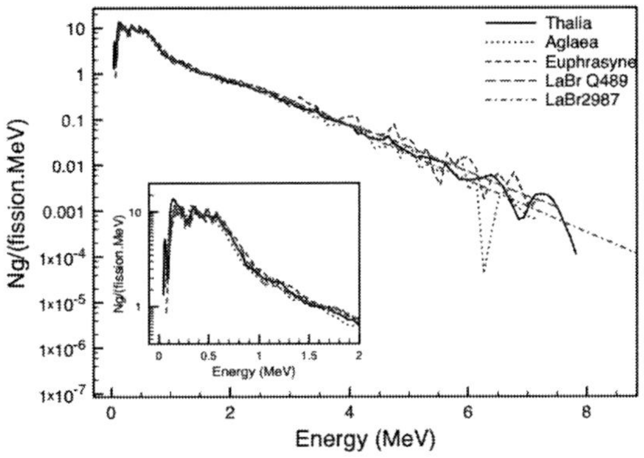

Figure 5. Prompt fission γ-ray spectra for each of the $LaBr_3$ crystals: Thalia, Aglaea and Euphrasyne. The two other spectra are taken from [19]. The subview in the bottom left corner is a zoom of the low energy region.

tag, we measured times of flight of particles between ionization chamber and $LaBr_3$ crystals. Then, a matrix representing time of flight as a function of the energy measured in one crystal is generated (see fig. 4). In such matrix several structure can be distinguished. The lowest horizontal one corresponds to γ energy deposit in the crystal. A projection of this structure on the X-axis will give the experimental prompt fission γ-ray spectrum. The second triangular structure, just above, is created by neutrons. The vertical lines in this triangle correspond to (n,γ) reaction with La or Br isotopes in the crystal.

After projection, and background substraction, the obtained spectrum needs to be unfolded according to the response function of the crystal, which have been simulated using GEANT4 [18]. Up to 200 energies of incident γ-rays were simulated in the range of 10 keV to 10 MeV, with an energy interval determined by the detector FWHM function. The simulation takes into account the detector geometry, its environment, as the different photons interactions that occur in the crystal. Then, a deconvolution algorithm, detailed in [19], is used to extract the spectrum emitted during the fission process. This analysis technique is then applied to each crystal separately to obtain three independent PFGS for ^{252}Cf. Then, the PFGS have to be normalised to the number of fission detected during the experiment. After these different steps, our first need was to validate the data analysis process. For this, our spectra were compared to R. Billnert's work[19]. Results are presented in figure 5. One can

Table 1. Statistical properties of prompt fission γ-rays measured for ^{252}Cf and compared with R. Billnert's *et al.* works

Detector Name	Total Energy (MeV)	Average Energy (keV)	Average Multiplicity
Aglaea	6.04 ± 0.25	742 ± 23	8.14 ± 0.24
Thalia	6.44 ± 0.26	761 ± 22	8.47 ± 0.24
Euphrasyne	7.08 ± 0.29	842 ± 25	8.41 ± 0.25
Average	6.47 ± 0.15	777 ± 13	8.34 ± 0.14
Billnert *et al.*	6.65 ± 0.04	800 ± 10	8.29 ± 0.03

notice the perfect agreement of our three spectra with our collaborator's work. The lower left corner subview enlarge the energy region between 60 keV and 2 MeV. Again, all the structure at low energy are also observed. From these spectral data, we can extract average energy, multiplicity and total energy emitted under the form of γ-rays in the fission process. The results are shown in table 1 and compare with R. Billnert's work. Again, the obtained results are in perfect agreement with previous measurements. As a consequence, our data analysis on ^{238}U PFGS for two neutron energies of 2.4 MeV and 3.3 MeV can be continued.

In 2016, two experiments are planned to measure prompt fission γ-ray. First, we will use higher neutron energies (up to 7 MeV) to understand γ emission when second chance of fission channel is opened. In the second one, we will induce fission of ^{239}Pu to study influence of the fissionning system on prompt fission γ properties.

4 Spectroscopy of Prompt fission γ-rays of ^{238}U

4.1 Presentation of the LICORNE experiment in the MINORCA campaign

In parallel to our PFG research activity, we started in 2014 to study the feasibility of performing fine γ spectrometry of neutron induced reaction – radiative capture, fission, ... The possibility to study the nuclear γ emission of fission fragments at the moment of their creation could provide important information about their structure. The ratio of neutron/proton is a very interesting quantity, which can be used to determine the predicted properties of a nucleus. The extreme of this ratio define the limits of existence for nuclear matter. For neutron rich nuclei, which are close to the limits of existence, specific parts of the nuclear interaction are strongly amplified or tend to vanish, which depict our understanding of the nuclear interaction between nucleons.
In the past, major campaigns of measurement, such as EXILL [20], were performed. Compared to these, fission of fertile nuclei can be induced with LICORNE. With, from start, three more neutrons and a rather cold process, one should expect to produce more exotic species. A test experiment with the ORGAM detector [21] was performed in january 2014 to prepare the MINIBALL campaign (MINORCA) in 2014-15 at ALTO. This experiment was used to understand γ background and neutron exposure of germanium crystals due to scattering in the target. After making clear there was no risk for the Ge crystals, during the campaign, LICORNE was coupled to the MINIBALL array [22] as can be seen in picture 6. A γ detection efficiency of ~ 10% was achieved. A peak-to-total ratio of ~ 20% was expected. The uranium target was placed 1.5 cm away from gas cell. It was a half cylinder of uranium metal with 1.2 cm diameter and a length of 3 cm. A pulsed ^{7}Li beam from the IPN tandem was used to bombard hydrogen gas in gas cell to generate neutrons. The beam pulses were 2 ns wide and separated by 400 ns with an average beam current of approximately 4 nA. The beam energy and

Figure 6. Picture of the MINIBALL setup. In the center, the LICORNE gas cell is visible. It is surrounded with lead and copper shielding. Around, the LICORNE chamber the eight MINIBALL cluster are placed around a sphere of 14 cm radius. At the center a metallic target of ^{238}U is placed to be irradiated.

gas cell length were chosen to maximize the flux for a cone of neutrons with an opening angle of not greater than 20 degrees. The length and pressure of the gas chamber (3.5 cm and 1.5 atm) assured that the beam energy slowed from 15 to 13 MeV producing maximum number of neutrons per second (order of 10^6n/s).

A limited fission rate of 50 kHz was achieved in order to maintain MINIBALL single crystal counting rate about 8 – 9 kHz. In order to tag the fission process a trigger was implemented in the data acquisition system. Three clusters had to be fired in a time window of 1.8 μs. Among those three γ-rays firing clusters, two had to be prompt ($\Delta t < 50$ns with a beam pulse) and the third one could be either prompt, or delayed: $\Delta t < 1.8\mu$s with a beam pulse. A trigger rate of 5 – 6 kHz was achieved. Data were acquired for two weeks and up to 10^9 triple coincidences, and higher folds, were stored to disk.

4.2 Preliminary Results

A preliminary analysis has been performed on this data set. After calibration and alignement in time of the different crystals, add-back algorithm has been implemented. It improved peak-to-total ratio from 16% up to 21%. The result of this first step is shown on figure 7. It represents the γ-ray energy measured *versus* the time difference between a γ hit and a beam pulse in a 1.8 μs window. Because of the 400 ns beam period, several pulses can be seen on this matrix. Horizontal lines can also be spotted. They belong to isomeric de-excitation in some fission fragment isomers.

4.2.1 Prompt γ-γ matrix

The first method to extract interesting data from this matrix is to take all the coincidences between γ-rays belonging to the first pulse. These two γ-rays are considered as prompt. This technique allow the selection of a fission fragment and its partner. However, the γ-γ matrix contains many lines. Figure 8 shows the spectrum obtained with a 814 keV energy gate on the matrix. This γ line belongs to ^{96}Sr. The fission partners are Xe isotopes with masses around 141. On the spectrum, many lines are identified and belong to the expected Xe isotope. The nuclei identified are up to eight neutrons away

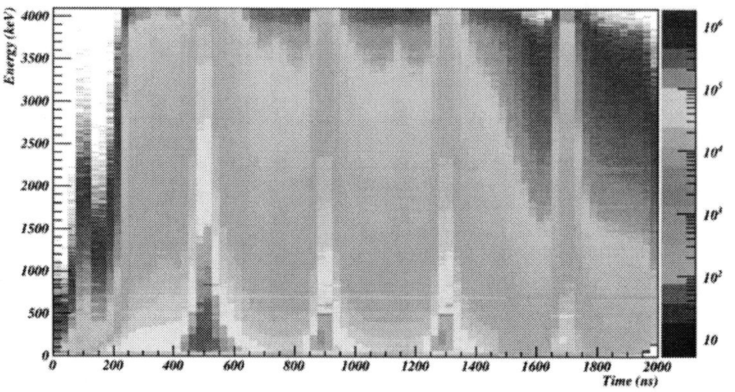

Figure 7. Two dimensional plot of the energy measured in one of the cluster as a function of the time difference between hit and beam pulse. The total time window is 1.8 μs. The vertical structures correpond to different beam pulse.

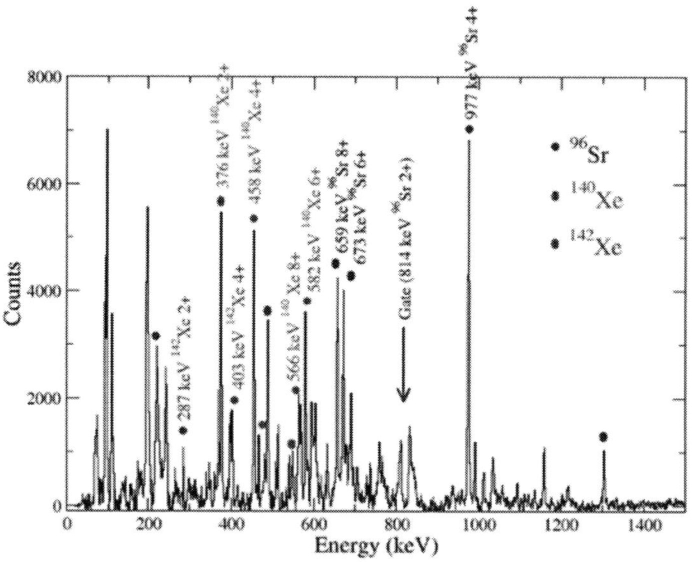

Figure 8. Spectrum conditioned by a 814 keV γ-ray extracted for the γ-γ matrix obtained from fig.7. ^{96}Sr and ^{140,142}Xe γ lines are visibles.

DOI: 10.1051/epjconf/201612201010

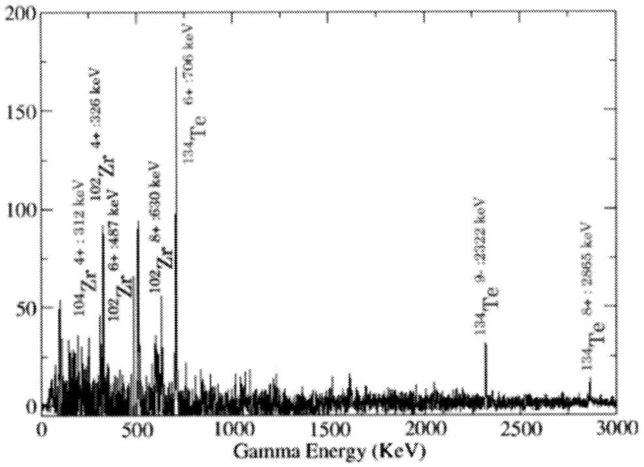

Figure 9. Total projection of the γ-γ matrix conditioned by the isomeric decay of the 164 ns state in ^{134}Te. ^{134}Te and ^{102,104}Kr γ-lines are visibles.

from the last stable isotope. This preliminary result is a good proof of concept for this experimental technique dedicated to the study of fission fragment at the moment of their formation. Excited states with angular momentum up to 8^+ were identified. If one can expect higher spins, this technique is already complementary to actual β-decay measurements which are usually pioneers in accessing structure data very far from stability. Unfortunately, the low peak-to-total ratio is a major drawback making only the most produced fission fragment γ lines visible in the matrix. This way of analysing the data is not perfectly suited to explore the most exotic regions.

4.2.2 Delayed-Prompt γ-γ-γ coincidences.

As seen previously, the peak visibility in the fold two matrix needs improvement to give access to more neutron rich fission fragment. Thus, a higher selectivity is required. Based on coincidence technique, a good way to access to exotic nucleus is to select one that is close to stability. For example, a gate on a transition on very long lived Sn isotopes could give access to Mo nuclei which are beyond the actual knowledge on structure in this region. A simple energy gate might not be selective enough to clean the conditioned spectrum. For this reason, the use of isomeric transitions was considered.
On figure 7, an horizontal line at 1279 keV can been observed between the pulses. It corresponds to the de-excitation of the isomeric state ($T_{1/2} \sim$ 164 ns) decay of ^{134}Te. In addition, a projection of the region between beam pulses on the vertical axis would allow the identification of up to 20 isomeric transitions in different nuclei. If we select any two γ in the pulse in coincidence with the isomeric line, we obtain the figure 9. In this spectrum, γ lines from ^{102,104}Kr are visible. These nuclei are 6 and 8 neutrons away from the last stable Kr isotope. This spectrum is also very "clean" compared to figure 8.
If using triple coincidences can end in a lack of statistics, this result demonstrates that isomers can

provide a very high selectivity which is crucial to this technique. We expect that future data analysis will allow us to cross the actual frontier on structure knowledge for very neutron rich fission fragments.

5 Conclusion and Perspectives

In this paper an overview of the research activity with the LICORNE neutron source has been given. Well suited for the study of neutron induced reactions, LICORNE is now providing important results about prompt fission γ-rays. In the near future, the outcomes of the data analysis about the ^{238}U experiment with 2.4 and 3.3 MeV neutrons will be compared with theoretical calculations. Furthermore, this experimental program will be completed in 2016 with two another experiments.
In parallel, MINORCA data will be extensively analysed. The information extracted will prepare us for our future measurement campaign. In the next two years, a Ge/$LaBr_3$ hybrid array is expected to be mounted in Orsay. The very good time resolution of $LaBr_3$ will increase the selectivity of the triple coincidences technique described previously. Added to a greater γ detection efficiency and a better peak-to-total ratio, we intend to explore even exotic regions. In addition, ambitious experiments dedicated to fission isomer studies will be considered.
As a matter of fact, in the next years until NFS starts, LICORNE will pave the way by testing bench for many new experimental techniques and by producing many interesting data.

References

[1] F. Ibrahim *et al.*, Nucl. Phys. A **787**, p. 110-117 (2007)
[2] S. Essabaa *et al.*, Nucl. Inst. Meth. B **317**, p. 218-222 (2013)
[3] C. Gustavsson, Phys. Scr. T **150**,014017 (2012)
[4] D. Ene *et al.*, Nucl. Inst. Meth. A **618**, p. 54-68 (2010)
[5] J. Winter, Physica B+C **127**, p. 385-387 (1984)
[6] L. Rosta *et al.*, Physica B **234-236**, p. 1196-1198 (1997)
[7] M. Drosg, *Monoenergetic Neutron Production by Two Body Reactions in the Energy Range from 0.0001 to 500 MeV* (TCM-Meeting of IAEA, 1999)
[8] M. Drosg, *The* $^{1}H(^{7}Li,n)^{7}Be$ *Reaction as a Neutron Source in the* MeV *Range* (Los Alamos National Laboratory, 1981)
[9] M. Lebois *et al.*, Nucl. Inst. Meth A **735**, p. 46 (2014)
[10] http://ipnwww.in2p3.fr/Available-beams
[11] J.C. Overley and R.R. Borchers, Nucl. Phys. **65**, p. 156-166 (1965)
[12] L. Van Der Zwan and K. W. Geiger, Nucl. Phys. A**306**, p. 45-52 (1978)
[13] G. Rimpault, *Proceedings of the workshop on nuclear data needs for generation IV* (Belgium World Scientific ISBN 981-256-830-1, Antwerp, 2005) p. 46
[14] A. Maj *et al.*, Acta Phys. Pol. B**40**, p. 565 (2009)
[15] www.oecdnea.org/dbdata/hprl/hprlview.pl?ID=422 (2006).
[16] www.oecdnea.org/dbdata/hprl/hprlview.pl?ID=421 (2006).
[17] M. Lebois *et al.*, Phys. Rev. **C92** 034618 (2015).
[18] S. Agostinelli *et al.*, Nucl. Inst. Meth. A**506** p. 250-303 (2003).
[19] R. Billnert *et al.*, Phys. Proc. **59** p. 17-23 (2014).
[20] G. De France *et al.*, Pramana Journal of Physics **85** p. 467-472 (2002).
[21] M. Ferraton, PhD Thesis (2011).
[22] P. Reiter *et al.*, Nucl. Phys. A**701** p. 209-212 (2002).

 DOI: 10.1051/epjconf/201612201011

Total prompt γ-ray emission in fission

C.Y. Wu[1,a], A. Chyzh[1,2], E. Kwan[1,3], R.A. Henserson[1], T.A. Bredeweg[4], R.C. Haight[4], A.C. Hayes-Sterbenz[4], H.Y. Lee[4], J.M. O'Donnell[4], and J.L. Ullmann[4]

[1]*Lawrence Livermore National Laboratory, Livermore, California 94550 USA*
[2]*North Carolina State University, Raleigh, North Carolina 27695 USA*
[3]*National Superconducting Cyclotron Laboratory, East Lansing, Michigan, 48824 USA*
[4]*Los Alamos National Laboratory, Los Alamos, New Mexico 87545 USA*

Abstract. The total prompt γ-ray energy distributions for the neutron-induced fission of ^{235}U, ^{239,241}Pu at incident neutron energy of 0.025 eV – 100 keV, and the spontaneous fission of ^{252}Cf were measured using the Detector for Advanced Neutron Capture Experiments (DANCE) array in coincidence with the detection of fission fragments by a parallel-plate avalanche counter. DANCE is a highly segmented, highly efficient 4π γ-ray calorimeter. Corrections were made to the measured distribution by unfolding the two-dimension spectrum of total γ-ray energy vs multiplicity using a simulated DANCE response matrix. The mean values of the total prompt γ-ray energy, determined from the unfolded distributions, are ~ 20% higher than those derived from measurements using single γ-ray detector for all the fissile nuclei studied. This raises serious concern on the validity of the mean total prompt γ-ray energy obtained from the product of mean values for both prompt γ-ray energy and multiplicity.

1 Introduction

The total prompt γ-ray emission in fission accounts for about 40% of the total energy released by γ-ray emission that makes up about 10% of the total energy released in reactor core. The heating in nuclear reactors attributed to the total γ-ray emission in fission is underestimated up to 28% using the evaluated data for the main reaction channels, ^{235}U(n,f) and ^{239}Pu(n,f) [1]. This discrepancy is significantly greater than 7.5%, an upper bound of the uncertainty deemed necessary to adequately model the heat deposit in the fuel core [2,3]. Therefore, efforts are needed to improve the experimental data on the γ-ray emission in fission. As a matter of fact, the request for the new data on the prompt fission γ rays at thermal energy and above for those two isotopes has been categorized as the high-priority by the Nuclear Energy Agency under the Organization for Economic Co-operation and Development [4]. The majority of measurements made for the prompt γ-ray emission in fission always employed a single or a few γ-ray detectors. For example, a single NaI detector was used by Verbinski *et al.* [5] more than 40 years ago and the cerium-doped $LaBr_3$, $CeBr_3$, and $LaBr_3$ detectors were used recently by Billnert *et al.* [1] and Oberstedt *et al.* [6,7].

[a] Corresponding author: wu24@llnl.gov

Below we describe the analysis and results on the total γ-ray emission in fission measured by the DANCE array [8,9]. DANCE consists of 160 equal-volume, equal-solid-angle BaF_2 detectors, covering a 4π geometry space, and is located at the Los Alamos Neutron Science Center (LANSCE). Several unique features exhibited by DANCE are particularly attractive for those measurements, such as the nearly γ-ray energy independence for the detection efficiency, the multiplicity response, and the peak-to-total ratio, all of which are described in detail in Refs. [10-12]. For example, it enables one to measure the total γ-ray energy as a function of multiplicity. The only limitation is the energy resolution, which is about 14% for the measured total γ-ray energy. A series of measurements of the prompt γ rays in the neutron-induced fission of ^{235}U and ^{239,241}Pu, and the spontaneous fission of ^{252}Cf has been carried out recently using DANCE in coincidence with the detection of fission fragments by a compact parallel-plate avalanche counters (PPAC) [13]. The results on the measured and unfolded fission prompt γ-ray energy and multiplicity distributions for those isotopes have been published [12,14]. An independent analysis of the same data for ^{239}Pu, by assuming a general parameterized correlation between E_γ and M_γ, was presented in Ref. [15]. We also reported the total prompt γ-ray energy distributions for those isotopes, obtained by unfolding the measured two-dimensional spectrum of total γ-ray energy vs multiplicity [16]. This unfolding procedure and the implication on the γ heating in nuclear reactors are described.

2 Experiments and data analysis

The measurements of the prompt γ emission in the neutron-induced fission of ^{235}U and ^{239,241}Pu as well as the spontaneous fission in ^{252}Cf were performed at the Lujan Center of LANSCE. The experimental setup and the data analysis have been described in detail in our early publications [12,14-16]. A brief summary of the experiments is given here. For the neutron-induced fission experiment, neutrons with energies from thermal up to several hundred keV were produced first by bombarding an 800-MeV proton beam at a repetition rate of 20 Hz on a tungsten target then moderated by water. The prompt γ rays emitted in fission were detected by the DANCE array in coincidence with the detection of fission fragments by a compact PPAC [13]. More than 10^6 fission events with at least one γ ray detected by DANCE were collected for all isotopes studied. The threshold for detecting γ-ray energy by DANCE was set to 150 keV. The summed energy of all γ rays detected by DANCE within a time window of 40 ns was defined as the total prompt γ-ray energy ($E_{\gamma,tot}$) in fission for a given event. With this time window extended to 100 ns, little change was observed for the $E_{\gamma,tot}$ spectrum [15]. The possible background contribution to $E_{\gamma,tot}$ is due to capture of thermalized prompt fission neutrons by Ba isotopes, which is on the order of μs and too long in the time scale for prompt γ rays. Additional suppression of neutron contribution is made by placing a gate on the pulse height spectrum of PPAC in addition to the 8-ns gate on the time spectrum between PPAC and DANCE, show in Fig. 1. All the offline data analysis was carried out using the code, FARE [17]. Note that both DANCE and PPAC have a similar intrinsic time resolution of ~ 1.2 ns [13]. The total γ-ray multiplicity (M_γ) in fission is established not according to the number of detectors observing the γ ray, but instead according to the number of clusters by grouping adjacent detectors catching the γ ray in the same time window. This counting method for M_γ is closer to the simulated results using the γ-ray calibration sources [10-12]. In addition, the nearly γ-ray energy independence of the DANCE response to M_γ, indicated by the numerical simulations, enables one to unfold approximately the measured M_γ distribution in fission for the first time [12,14].

Corrections have to be made to the measured $E_{\gamma,tot}$ distribution to obtain the physical one, which would be useful for the applications. This can be accomplished by unfolding the two-dimensional spectrum of $E_{\gamma,tot}$ vs M_γ. The two-dimensional unfolding is necessary because of the strong dependence of $E_{\gamma,tot}$ on M_γ. It is numerically implemented by adopting the iterative Bayesian method [18-20]. The DANCE response matrix for $E_{\gamma,tot}$ vs M_γ is simulated using the GEANT4 [21] geometrical model of both DANCE and PPAC [12,14,22]. To make sure this two-dimensional response matrix has a sufficient coverage of the phase space beyond the measured one, the value of M_γ up to 25 and $E_{\gamma,tot}$ up to 40 MeV are included. The $E_{\gamma,tot}$ has a bin size of 200 keV and an energy threshold of 150 keV. So the response matrix has a size of 200 × 25.

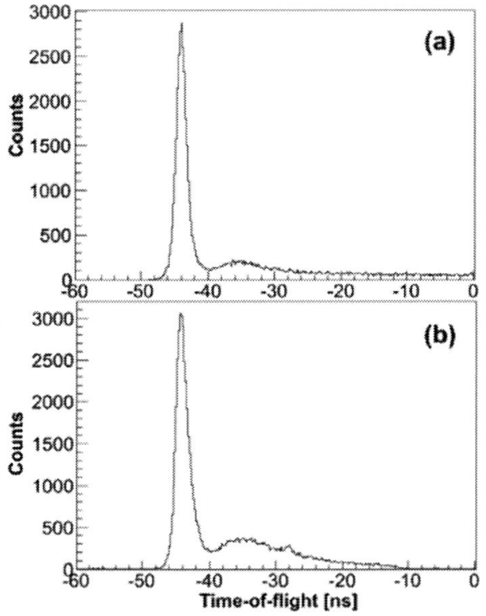

Figure 1. Time difference between γ rays detected by DANCE and fission fragments detected by PPAC for (a) ^{235}U and (b) ^{241}Pu experiments with an achieved time resolution of ~ 1.7 ns. The bump next to the peak is related to events with ambiguous correlation between DANCE and PPAC.

For any given grid point ($E_{\gamma,tot}$, M_γ) in the response matrix, a two-dimensional DANCE response matrix of a size of 200 × 25 is generated using GEANT4 with a given assembly of no more than 20,000 samples. Note that the DANCE response to the total prompt γ-ray is relatively insensitive to the content of γ rays for a given sample since the γ-ray detection efficiency (84 to 88%) and the peak-to-total ratio (~ 55%) remain nearly constant for the γ-ray energy ranging from 150 keV to 10 MeV [10-12]. Each sample has a matching number of γ rays to M_γ, selected randomly according to the unfolded γ-ray energy distributions [12,14] with the condition on the total γ-ray energy that is equal to $E_{\gamma,tot}$ ± 100 keV. This simulation is repeated for all the grid points within the lower and upper bound of $E_{\gamma,tot}$ for a given M_γ, established by this random sampling technique.

The resulting ($E_{\gamma,tot}$, M_γ) DANCE response matrix consists of ~ 3300 two-dimensional matrices with a size of 200 × 25 each. This numerically simulated DANCE response matrix is unique for each isotope studied, and was used to unfold the measured two-dimensional spectrum of $E_{\gamma,tot}$ vs M_γ into a physical one using the iterative Bayesian method. During the iteration stage, a single factor was applied to and varied for the response matrix at any given grid point.

3 Results and discussions

Typically it takes about 30 iterations to reach the convergence in the unfolding of the two-dimensional spectrum of $E_{\gamma,tot}$ vs M_γ using the Bayesian method. The results for the neutron-induced fission in ^{239}Pu are shown in Fig. 2 where the unfolded $E_{\gamma,tot}$ vs M_γ spectrum together with the measured one are given. In addition, the comparisons of the projected $E_{\gamma,tot}$ and M_γ distributions between the unfolded and measured ones are also given. The general trend of the results is that the mean value and the width of projected $E_{\gamma,tot}$ and M_γ distributions increases noticeably after the unfolding.

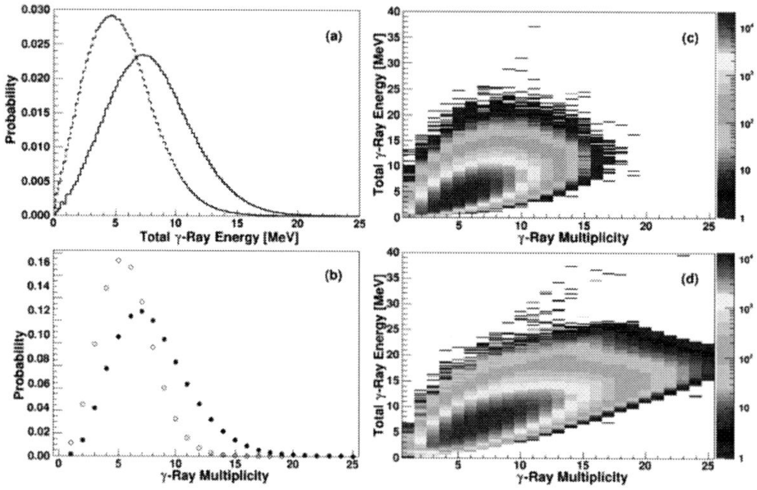

Figure 2. Shown in panel (c) and (d), respectively, are the measured and unfolded total prompt γ-ray energy vs. multiplicity distribution for the neutron-induced fission of ^{239}Pu. Comparison of the projected total γ-ray energy and multiplicity distributions between measured (dashed line, open circles) and unfolded one (solid line, filled circles) are given in panels (a) and (b), respectively.

Given in Table 1 is the comparison of $\langle M_\gamma \rangle$ derived from the unfolded M_γ distribution between the recent work (2-D) and the early one using the one-dimension unfolding technique [14] for all isotopes studied. For ^{235}U, the recent (2-D) mean value of 7.35 is 0.37 higher than 6.98 in the earlier 1-D work. However, the latter value is known to be underestimated by about 0.30. Since these values were derived from the same data set, this consistency in the derived mean M_γ from both the one- and two-dimensional unfolding techniques gives us a certain confidence in the validity of our work. This trend is the same for ^{239,241}Pu and ^{252}Cf. The comparison with other measurements and evaluations also is given in Table 1. Our measured $\langle M_\gamma \rangle$'s for all isotopes studied are consistently higher than the weighted-average of earlier measurements [23] by ~ 10% except for the most recent measurements [1,6,7], where their measured $\langle M_\gamma \rangle$ is ~ 11% greater than ours for ^{235}U but near in agreement with ours for both ^{241}Pu and ^{252}Cf. Moreover, ours are consistent with the evaluated data listed in ENDF/B-VII.1 [24]. The uncertainty for our derived $\langle M_\gamma \rangle$ has an upper bound of about 0.3-0.4 or ~ 5%.

The comparison of $\langle E_{\gamma,tot} \rangle$ between our measurements and previous ones is given in Table 2. For ^{235}U, the recent (2-D) derived mean $E_{\gamma,tot}$ of 8.35 MeV is higher than 6.53(20) MeV, the weighted average of previous measurements [23], and 6.60 MeV, the evaluated data listed in ENDF/B-VII.1. It also is higher than 6.92(9) MeV, the most recent measurement [6]. The same comparisons are also made for the neutron-induced fission in ^{239,241}Pu and the spontaneous fission in ^{252}Cf. Our measured $\langle E_{\gamma,tot} \rangle$ are consistently higher than the previous ones [1,6,7,23] by ~ 20% for all isotopes studied. The uncertainty for our derived $\langle E_{\gamma,tot}\rangle$ is dominated by the systematic error and roughly estimated to be better than 5%, assuming a similar uncertainty to that of the derived $\langle M_\gamma\rangle$.

Table 1. Comparison of the mean M_γ between our recent measurements and previous ones for the neutron-induced fission of ^{235}U and ^{239,241}Pu as well as the spontaneous fission of ^{252}Cf.

Isotope	2-D	1-D	Ref. 15	ENDF/B-VII.1	Ref. 23	Refs. 1, 6, 7
^{235}U	7.35	6.95		7.04	6.60(10)	8.19(11)
^{239}Pu	7.93	7.50	7.15	7.78	7.06(20)	
^{241}Pu	7.97	7.50		8.18		8.21(9)
^{252}Cf	8.75	8.16			7.98(40)	8.30(8)

An independent analysis of the same DANCE data for ^{239}Pu by assuming a very general parameterized correlation between E_γ and M_γ has been carried out by Ullmann et al. [15], which yields the $\langle E_{\gamma,tot}\rangle$ = 7.46 MeV and $\langle M_\gamma\rangle$ = 7.15. The $\langle E_{\gamma,tot}\rangle$, derived from the $E_{\gamma,tot}$ distribution, agrees within 6% of that obtained by using the 2-D unfolding technique. This agreement is significant and indicates the importance of the correlation between E_γ and M_γ to be considered in the determination of $\langle E_{\gamma,tot}\rangle$. It raises serious concern on the validity of the equation, $\langle E_{\gamma,tot}\rangle = \langle E_\gamma\rangle \times \langle M_\gamma\rangle$, which ignores the correlation between E_γ and M_γ exhibited in Fig. 2.

Table 2. Comparison of the mean $E_{\gamma,tot}$ (MeV) between our recent measurements and previous ones for the neutron-induced fission of ^{235}U and ^{239,241}Pu as well as the spontaneous fission of ^{252}Cf.

Isotope	2-D	Ref. 15	ENDF/B-VII.1	Ref. 23	Refs. 1, 6, 7
^{235}U	8.35		6.60	6.53(20)	6.92(9)
^{239}Pu	7.94	7.46	6.74	6.78(10)	
^{241}Pu	8.01		7.26		6.41(6)
^{252}Cf	8.52			6.95(30)	6.64(8)

4 Summary

A systematic study of the total prompt γ-ray emission in the neutron-induced fission of ^{235}U and ^{239,241}Pu as well as the spontaneous fission of ^{252}Cf has been carried out using the DANCE array together with a compact PPAC to select the fission event by detecting its fission fragments. The total γ-ray energy vs multiplicity spectrum for all fissile nuclei studied was constructed and unfolded using a two-dimensional unfolding technique, numerically implemented by adopting the iterative Bayesian method. The $\langle E_{\gamma,tot}\rangle$ derived from the projected $E_{\gamma,tot}$ distribution of the unfolded $E_{\gamma,tot}$ vs M_γ spectrum is about 20% higher than the previous measurements for all fissile nuclei studied. However, it agrees reasonably well with the result derived from the analysis by considering the correlation between E_γ and M_γ in a very general parameterization manner. In addition, the measured total prompt γ-ray energy vs multiplicity spectrum in fission enables one to evaluate the variance in addition to the average value of the energy deposited in a reactor core by the prompt fission γ rays. This may improve our understanding of the γ heating in many applications involving nuclear fission.

Acknowledgement

This work benefited from the use of the LANSCE accelerator facility as was performed under the auspices of the US Department of Energy by Lawrence Livermore National Security, LLC under contract DE-AC52-07NA27344 and by Los Alamos National Security, LLC under contract DE-AC52-06NA25396. Partial funding is gratefully acknowledged from the U.S. DOE/NNSA Office of Defense Nuclear Nonproliferation Research and Development. All isotopes used in the measurements were obtained from Oak Ridge National Laboratory.

References

1. R. Billnert, F.-J. Hambsch, A. Oberstedt, and S. Oberstedt, Phys. Rev. C **87,** 024601 (2013)
2. G. Rimpault, in Nuclear Data Needs for Generation IV, edited by P. Rullhusen (World Scientific, Antwerp, 2006), p. 46
3. G. Rimpault, A. Courcelle, and D. Blanchet, comment to HPRL: ID H.3 and H.4
4. Nuclear Data High Priority Request List of the NEA (Req. ID: H.3, H4) http://www.oecd-nea.org/dbdata/hprl/hprlview.pl?ID=421 and http://www.oecd-nea.org/dbdata/hprl/hprlview.pl?ID=422
5. V.V. Verbinski, H. Weber, and R.E. Sund, Phys. Rev. C **7,** 1173 (1973)
6. A. Oberstedt *et al.*, Phys. Rev. C **87**, 051602(R) (2013)
7. A. Oberstedt et al., Phys. Rev. C **90**, 024618 (2014)
8. M. Heil, R. Reifarth, M. M. Fowler, R. C. Haight, F. Kappeler, R. S. Rundberg, E. H. Seabury, J. L. Ullmann, and K. Wisshak, Nucl. Instrum. Methods Phys. Res. A **459**, 229 (2001)
9. R. Reifarth *et al.*, IEEE Transactions on Nuclear Science **53**, 880 (2006)
10. R. Reifarth *et al.*, Los Alamos National Laboratory, LA-UR-01-4185 (2001)
11. R. Reifarth *et al.*, Los Alamos National Laboratory, LA-UR-03-5560 (2003)
12. A. Chyzh, C. Y. Wu *et al.*, Phys. Rev. C **85**, 021601(R) (2012)
13. C.Y. Wu, A. Chyzh, E. Kwan, R. Henderson, J. Gostic, D. Carter, T. Bredeweg, A. Couture, M. Jandel, J. Ullmann, Nucl. Instrum. Methods Phys. Res. A **694**, 78 (2012)
14. A. Chyzh, C.Y. Wu *et al.*, Phys. Rev C **87**, 034620 (2013)
15. J.L. Ullmann *et al.* Phys. Rev. C **87**, 044607 (2013)
16. A. Chyzh, C. Y. Wu *et al.*, Phys. Rev. C **90**, 014602 (2014)
17. M. Jandel, T.A. Bredeweg, A. Couture, J.M. O'Donnel, and J.L. Ullmann, Los Alamos National Laboratory, LA-UR-12-21171 (2012)
18. G. D'Agostini, Nucl. Instrum. Methods Phys. Res. A **362**, 487 (1995)
19. G. D'Agostini, arXiv:1010.0612 (2010)
20. T.J. Adye, arXiv:1105.1160 (2011)
21. S. Agostinelli *et al.*, Nucl. Instrum. Methods Phys. Res. A **506**, 550 (2003)
22. M. Jandel *et al.*, Nucl. Instrum. Methods Phys. Res. B **261**, 1117 (2007)
23. T.E. Valentine, Ann. Nucl. Energy **28**, 191 (2001) and references therein
24. M.B. Chadwick *et al.*, Nucl. Data Sheets **112**, 2887 (2011)

Neutron-induced fission: properties of prompt neutron and γ rays as a function of incident energy

I. Stetcu[1,a], P. Talou[1], and T. Kawano[1]

[1] *Theoretical Division, Los Alamos National Laboratory, Los Alamos, New Mexico, 87545, USA*

Abstract. We have applied the Hauser-Feshbach statistical theory, in a Monte-Carlo implementation, to the de-excitation of fission fragments, obtaining a reasonable description of the characteristics of neutrons and gamma rays emitted before beta decays toward stability. Originally implemented for the spontaneous fission of ^{252}Cf and the neutron-induced fission of ^{235}U and ^{239}Pu at thermal neutron energy, in this contribution we discuss the extension of the formalism to incident neutron energies up to 20 MeV. For the emission of pre-fission neutrons, at incident energies beyond second-chance fission, we take into account both the pre-equilibrium and statistical pre-fission components. Phenomenological parameterizations of mass, charge and TKE yields are used to obtain the initial conditions for the fission fragments that subsequently decay via neutron and γ emissions. We illustrate this approach for ^{239}Pu(n,f).

1 Introduction

Applications often require the knowledge of prompt neutron and γ-ray spectra and other observables as a function of neutron incident energy for neutron-induced fission reactions. In this contribution, we describe the implementation of the incident neutron energy dependence into the CGMF (Cascading Gamma-ray and Multiplicity for Fission) code developed at Los Alamos for simulating prompt neutron and γ observables.

In CGMF, the excited fission fragments are treated as compound nuclei, whose de-excitation is modeled in a Monte-Carlo implementation of the Hauser-Feshbach statistical decay model [1]. The Hauser-Feshbach framework provides the most detailed description of the de-excitation process. This approach requires knowledge of a large number of properties of far-from-stability nuclei, from density levels to neutron transmission coefficients or electromagnetic decay properties, in contrast to the Los Alamos model [2], which is based on a small number of parameters. However, in addition to the average neutron multiplicity and spectrum calculated in the Los Alamos model, CGMF provides many other observables, like neutron multiplicity distribution, γ-ray observables (average multiplicity and spectrum, multiplicity distribution), and spatial and energy correlation between prompt neutrons and/or γ rays. Other codes that treat the de-excitation fission fragments via emission of neutrons and γ rays are available [3–5].

In addition to the parameters that describe properties of the fission fragments, other quantities that directly relate to the fission itself are needed. CGMF does not simulate the dynamics of fission, and

[a] e-mail: stetcu@lanl.gov

thus the production of fission fragments. Hence, the mass, charge and total kinetic energy (TKE) distributions are taken from available experimental data or models. As data are scarce for other incident energies than thermal, we are forced to rely on systematics for neutron incident energies up to 20 MeV. The initial angular momentum distribution and the mechanism of the available total excitation energy (TXE) are quantities that are also necessary to perform Hauser-Feshbach simulations of the fission fragments. For these, experimental data provide only indirect evidence. Multi-chance fission probabilities (fission after the emission of one or more neutrons) are also required above about 5.5 MeV. We calculate the multi-chance fission probability by using the code CoH3 [6], and, depending on the fission barrier and nuclear structure parameters, the results can be closer or farther from the ENDF/B-VII.0 [7] evaluation. Only one set of multi-chance fission probabilities are considered here. The neutron emission is modeled by considering a mixture of pre-equilibrium (obtained using the exciton model [8–10]) and compound nucleus contributions. In this contribution, we concentrate in Sec. 2 on the evolution of mass, charge and TKE distribution parameters with the energy of the incident neutron, and leave the spin and sharing mechanism for a detailed discussion in future publications. The first results of the current implementation in CGMF for ^{239}Pu(n,f) for incident energies up to 20 MeV are presented in Sec. 3, while a summary and outlook are presented in Sec. 4.

2 Energy dependence of mass and TKE distributions

The main assumptions in CGMF are that (i) the neutrons and γ rays are only emitted from the fully accelerated fission fragments, (ii) no emission occurs during the evolution of the system between saddle and scission, or at the neck rupture, and (iii) the fission fragments are compound nuclei. Thus, in order to obtain the properties of neutrons and γ rays, we start with a collection of fission fragments, which are sampled from a mass and charge distribution. Initially, CGMF has been developed only for the description of thermal neutron induced fission of ^{235}U and ^{239}Pu and the spontaneous fission of ^{252}Cf [11, 12]. For these system the yield in mass and TKE, $Y(A, \mathrm{TKE})$, is directly available from experiment, or a host of data exist separately on mass yields, $Y(A)$, and average TKE and width as a function of primary fragment mass, so that $Y(A, \mathrm{TKE})$ can be constructed. In both cases, the charge distribution is taken from the Wahl systematics [13].

For the incident energy dependence of the fission fragments yields, we adopt the same five-Gaussian model as in Ref. [5], where the mass distribution is decomposed in fission modes whose parameters depend on the incident neutron energy E_n:

$$Y(A, E_n) = \frac{N_0}{\sqrt{2\pi}\sigma_0} e^{-\frac{(A-A_0)^2}{2\sigma_0^2}} + \sum_{i=1,2} \frac{N_i(E_n)}{\sqrt{2\pi}\sigma_i(E_n)} \left[e^{-\frac{(A-A_0-D_i(E_n))^2}{2\sigma_i^2(E_n)}} + e^{-\frac{(A-A_0+D_i(E_n))^2}{2\sigma_i^2(E_n)}} \right]. \quad (1)$$

We take the same incident energy dependence for $D_i(E_n)$, $\sigma_i(E_n)$ and $N_i(E_n)$ as in Ref. [5]. For the symmetric mode, much less constrained than the asymmetric modes, the width of the distribution σ_0 does not vary with energy, and the only energy dependence comes from the normalization condition

$$N_0 + 2N_1(E_n) + 2N_2(E_n) = 2. \quad (2)$$

While we preserve the same energy dependence, we have modified the parameters of Ref. [5] at thermal energy for ^{235}U(n,f) and ^{239}Pu(n,f) to better agree with experimental data.

The fission yields parameterized in Eq. (1) are sampled and a list of possible fission fragments is generated. In the case of incident neutron energy above the threshold for multi-chance fission, we first simulate the emission of a number of neutrons according to the multi-chance fission probabilities. Both compound and preequilibrium emission components of the pre-fission neutron(s) are taken into account. The emission has an important effect: not only it reduces the mass of the fissioning system, but also reduces the excitation energy available in the system. If the excitation energy in the fissioning system is below the fission barrier, we reject that event. A more detailed description will be presented in a future publication.

The TKE distribution is also needed in order to obtain the initial conditions for the fission fragments. A similar decomposition to Eq. (1) can be done for TKE, but the fit should include at the same time both the mass and TKE data. In this work, we assume that for each fragment fission A, the TKE distribution is a Gaussian characterized by the average $\langle \mathrm{TKE}\rangle(A, E_n)$ and width $\sigma_{\mathrm{TKE}}(A, E_n)$. This data is readily available for incident thermal neutron, and much more scarce for higher incident energies. However, in Fig. 1(a) we plot $\langle \mathrm{TKE}\rangle(A, E_n)$ scaled to $\langle \mathrm{TKE}\rangle(E_n)$ for a large number of fissioning systems and several neutron incident energies below the threshold for second-chance fission. This plot suggests that most of the energy dependence in $\langle \mathrm{TKE}\rangle(A, E_n)$ comes from the energy dependence of the average TKE, with little dependence on the particular isotope as well. Hence, we chose to relate the average TKE as a function of mass at an arbitrary incident neutron energy E_n to be proportional to $\langle \mathrm{TKE}\rangle(A, E_{\mathrm{th}})$, i.e.,

$$\langle \mathrm{TKE}\rangle(A, E_n) = C(E_n)\,\langle \mathrm{TKE}\rangle(A, E_{\mathrm{th}}) \tag{3}$$

with the scaling constant determined so that the average TKE at that particular incident energy is reproduced:

$$\langle \mathrm{TKE}\rangle(E_n) = C(E_n) \sum_A Y(A, E_n) \times \langle \mathrm{TKE}\rangle(A, E_{\mathrm{th}}). \tag{4}$$

Here the summation runs only over the heavy fragment masses. Equation (3) is a first-order approximation, as a small, but systematic change in $\langle \mathrm{TKE}\rangle(A, E_n)$ is observed with increasing E_n. Taking into account that for large enough energies neutrons can be emitted before scission, $\langle \mathrm{TKE}\rangle(E_n)$ depends on the actual system that fission. We denote the mass of the fissioning system by $\mathcal{A} - n_{\mathrm{pre}}$, where $\mathcal{A}$ is the mass of the initial compound system (e.g., $\mathcal{A} = 240$ for the ^{239}Pu(n,f) reaction) and n_{pre} is the number of neutrons emitted before fission. The Madland systematics [14] assumes the linear dependence of the average TKE with the neutron incident energy:

$$\langle \mathrm{TKE}\rangle(E_n) = E_0(\mathcal{A} - n_{\mathrm{pre}}) + s\,E_n, \tag{5}$$

where s is a slope determined from experimental data below second-chance fission threshold, while E_0 is the average TKE at (equivalent) thermal neutron incident energy. The Madland parameterization is only valid below the second-chance fission, and, indeed, in the experimental data an increase in $\langle \mathrm{TKE}\rangle(E_n)$ is observed above the threshold for second-chance fission. However, we extend the same linear energy dependence of $\langle \mathrm{TKE}\rangle(E_n)$ up to 20 MeV, assuming that the slope s is independent of the fissioning system, while $E_0(\mathcal{A})$ is determined so that we reproduce $\bar{\nu}$ at thermal energy for the $\mathcal{A}$ system. If neutrons are emitted prior to fission, we parameterize $E_0(\mathcal{A} - n_{\mathrm{pre}})$ as a function of the Coulomb parameter $\mathcal{Z}^2/(\mathcal{A} - n_{\mathrm{pre}})^{1/3}$, where $\mathcal{Z}$ is the charge of the compound fissioning system. For example, for ^{239}Pu(n,f), we have obtained from fits to isotopes in the Pu chain

$$E_0(\mathcal{A} - n_{\mathrm{pre}}) = 178.6 - 0.4498\left(\frac{94^2}{(\mathcal{A} - n_{\mathrm{pre}})^{1/3}} - 1423.8\right). \tag{6}$$

Because the pre-scission neutrons carry energy, the energy of an equivalent incident neutron that would produce the fission of the $\mathcal{A} - n_{\mathrm{pre}}$ compound would be smaller. This explains the increase in TKE above the threshold for second-chance fission. Around those energies, the observed fissioning system is a superposition of the $\mathcal{A}$ and $\mathcal{A} - 1$ compounds, with the latter at a lower excitation energy.

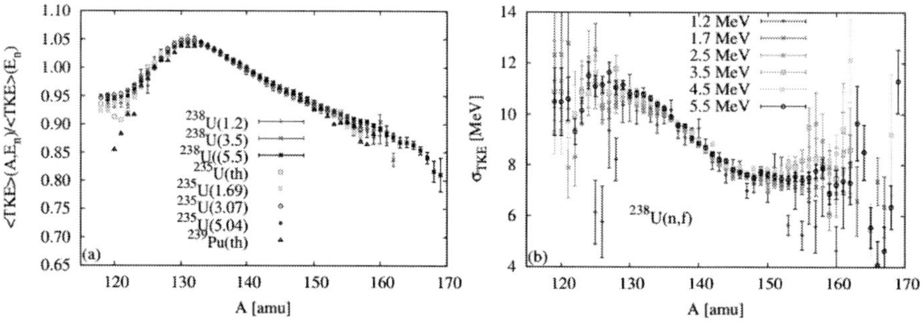

Figure 1. (a) The average TKE as a function of mass, scaled to the total average TKE for the neutron induced fission of ^{235}U, ^{238}U and ^{239}Pu at several incident energies below the second-chance fission threshold. (b) The TKE width as a function of mass for several incident neutron energies in the case of the ^{238}U(n,f) reaction.

The mixing is given by the fission probabilities and the result is an increase in TKE. With the simple extension of the linear behavior of $\langle \mathrm{TKE} \rangle$ with E_n for each fissioning system, we can reproduce the features of the experimental data on $\langle \mathrm{TKE} \rangle (E_n)$ by only fitting $\bar{\nu}$ below the second-chance fission threshold. Thus, we can predict a large number of observables above the threshold to produce a neutron energy before fission without further adjustments. Finally, we assume no energy dependence of $\sigma_{\mathrm{TKE}}(A)$, as suggested by Fig. 1(b).

For a set of fission fragments generated by Eq. (1) we obtain the TKE by sampling the Gaussian distribution with the parameters $\langle \mathrm{TKE} \rangle (A)$ and $\sigma_{\mathrm{TKE}}(A)$ computed through the procedure outlined above. Once TKE is known, from the energetics of the reaction one can compute the total excitation energy (TXE) available to both fragments. The sharing of this energy is a complicated mechanism, and at this moment we utilize a parameterization that ensures a good reproduction of the average number of neutrons as a function of mass at thermal incident energy. Since more neutrons are emitted from the heavy fragments at higher incident energies, our model will not describe this feature correctly. However, this assumption does not change significantly the average of total neutrons emitted per fission event, $\bar{\nu}$. It also does not affect the γ-ray properties, since the residual energy available for γ emission is controlled to a large extent by the average neutron separation energy. The implementation of a more sophisticated energy sharing mechanism is in progress.

$\bar{\nu}$ is extremely sensitive to TKE, as this quantity determines the energy available for neutron emissions. On the other hand, the competition between the neutron and γ-ray emissions also influences $\bar{\nu}$, but to a much smaller degree. In this implementation, we use the γ-neutron competition to fine tune $\bar{\nu}$ below the second-chance fission, via the parameter α that controls the initial spin distribution of the fission fragments (see Refs. [11, 12, 15] for a comprehensive discussion of the spin distributions). Thus, as in the case of TKE distribution parameters, we fit an energy dependence for α below the second-chance fission threshold only, which is then extended to 20 MeV incident neutron energy using multi-chance fission probabilities. More details regarding this parameterization will be given in a longer publication.

 DOI: 10.1051/epjconf/201612201012

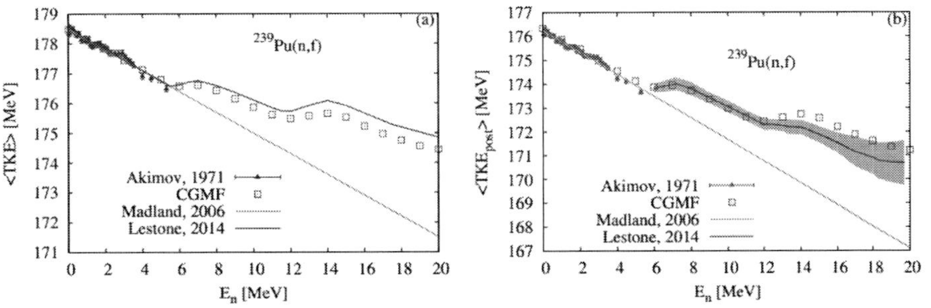

Figure 2. Pre-neutron emission (a) and post-neutron emission (b) average TKE of the fission fragments as a function of the neutron incident energy. We compare or results to the Madland systematics [14], valid up to 5.5 MeV, and the Lestone parameterization [16]. By construction, the Leston parameterization reproduces the Madland systematics below 5.5 MeV. For the post-neutron emission average TKE, we also show the uncertainty band for the Lestone parameterization.

3 Results for ^{239}Pu(n,f)

In this section, we present our results in the case of ^{239}Pu(n,f) reaction for $\langle \mathrm{TKE}\rangle(E_n)$, $\langle \mathrm{TKE}_{post}\rangle(E_n)$ (average fission fragment TKE after neutron emission), $\bar{\nu}(E_n)$, and $\langle E_\gamma^{tot}\rangle(E_n)$, where $\langle E_\gamma^{tot}\rangle$ is the total average γ-ray energy emitted per fission event. As discussed in Sec. 2, below the second-chance fission threshold, $\langle \mathrm{TKE}\rangle(E_n)$ is fitted to the Madland systematics [14], while $\bar{\nu}(E_n)$ to the ENDF values [7]. However, $\langle E_\gamma^{tot}\rangle(E_n)$ and $\langle \mathrm{TKE}_{post}\rangle(E_n)$ are predictions from 0 to 20 MeV, while $\langle \mathrm{TKE}\rangle(E_n)$ and $\bar{\nu}(E_n)$ are predictions above the threshold for the second-chance fission.

In Fig. 2, we present the predicted values for average TKE, pre- and post-neutron emissions. At thermal incident energy, the average pre-neutron emission TKE is shifted within recommended errors from the Madland value in order to obtain a good agreement with $\bar{\nu}$ (also fine tuned with α). The emission of the pre-fission neutrons beyond the threshold for second-chance fission decreases the excitation energy in the fissioning system, and thus TKE, both pre- and post-neutron emission, increases. The final value of TKE depends on the multi-chance fission probabilities, which are not directly observable. We note, however, a reasonable agreement between the CGMF predictions and the Lestone parameterization [16] for both pre- and post-neutron emission TKE. Experimental data for the ^{239}Pu(n,f) reaction for $\langle \mathrm{TKE}_{post}\rangle(E_n)$ exist from measurements with the SPIDER detector [17] at the Los Alamos Neutron Science Center and will be published shortly.

Figure 3(a) shows the dependence of $\bar{\nu}$ with the neutron incident energy. The good agreement between CGMF and ENDF suggests that our simple assumptions regarding the pre-neutron emission TKE are reasonable. We also obtain reasonable agreement between CGMF and available data for $\langle E_\gamma^{tot}\rangle$ up to 12 MeV, which is around the threshold for third-chance fission. Around this energy, CGMF starts to deviate with respect to the evaluated data. We note that the data below 5.5 MeV suggests an increase of the initial angular momentum with the neutron incident energy. At the second-chance fission threshold, the emission of a neutron decreases the excitation energy of the fissioning system and the initial angular momentum decreases (the probability to emit a neutron first and then fission increases quickly after the threshold). Hence, a small drop appears in the CGMF results for $\langle E_\gamma^{tot}\rangle$, and a plateau in the data, which remains consistent with the CGMF values. After the drop,

DOI: 10.1051/epjconf/201612201012

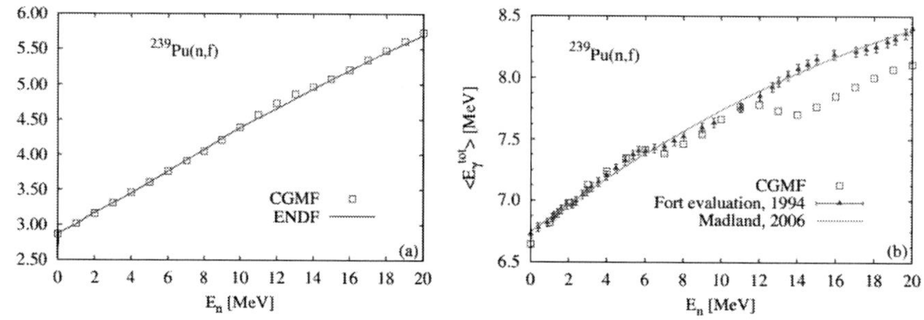

Figure 3. (a) CGMF and ENDF values of $\bar{\nu}$ as a function of incident neutron energy. (b) Comparison between CGMF and available data for $\langle E_\gamma^{tot} \rangle$ as a function of incident energy.

one expects that the angular momentum to increase, and a second drop to appear when the energetics will allow for a second neutron to be emitted before fission (third-chance fission threshold). This is where the CGMF results start to disagree with the data. This either suggest that our interpretation of the behavior of the initial spin distribution with increasing the incident neutron energy is incorrect and something happens around third-chance fission threshold, or the data, which have not been published in a peer-reviewed journal, are suspicious. Future measurements of this quantity would be thus quite useful in order to guide the models.

4 Summary and outlook

For neutron-induced fission reactions we have extended the range of applicability of CGMF from thermal energies up to 20 MeV incident neutron energy. The mass yields and charge are taken from systematics, while above the threshold for second-chance fission, we simulate the emission of one or more neutrons before the fission process occurs, considering the competition between pre-equilibrium and compound nucleus emission. The fragments are considered compound nuclei, whose de-excitation is described within the Hauser-Feshbach formalism. We have assumed that the same $\langle \mathrm{TKE} \rangle (A)$, measured at thermal incident energy, can be used over the full range of incident energies, scaled to reproduce the TKE systematics of Madland, extended to 20 MeV for each fissioning system. The widths $\sigma_{\mathrm{TKE}}(A)$ of the distribution have been taken the same over for all the incident energies. After we fit the average neutron multiplicity below 5.5 MeV to the ENDF values, we predicted several observables up to 20 MeV. Thus, we obtain a good reproduction of $\bar{\nu}$ up to 20 MeV, and of the total average γ-ray energy up to 12 MeV. Average TKE, pre- and post-neutron emission is also consistent with available data and current models.

Other available theoretical models for the mass [18, 19] and TKE [20] yields can be incorporated into our approach. Moreover, we work currently on the implementation of a more accurate mechanism for the sharing of the excitation energy between the fission fragments. This will allow us to more realistically predict the mass-dependent average neutrons emitted per fission event. In parallel with these developments, we will study the sensitivity of our results with respect to the input parameters as some observables will be more sensitive to some parameters than others. Finally, the current implementation can be extended to more isotopes as well as photo-fission reactions.

Acknowledgements

We thank R. Vogt and J. Randrup for useful discussions and for providing us with the parameters for the mass yields. This work was performed at Los Alamos National Laboratory, under the auspices of the National Nuclear Security Administration of the U.S. Department of Energy at Los Alamos National Laboratory under contract No. DE-AC52-06NA25396. We acknowledge partial support from the Office of Defense Nuclear Nonproliferation Research & Development (DNN R&D), National Nuclear Security Administration, US Department of Energy.

References

[1] W. Hauser, H. Feshbach, Phys. Rev. **87**, 366 (1952)
[2] D.G. Madland, J.R. Nix, Nucl. Sci. Eng. **81**, 213 (1982)
[3] K.H. Schmidt, B. Jurado, Tech. rep. (2015), `http://www.khs-erzhausen.de/GEF.html`
[4] D. Regnier, Ph.D. thesis, Université de Grenoble (2013)
[5] R. Vogt, J. Randrup, D.A. Brown, M.A. Descalle, W.E. Ormand, Phys. Rev. C **85**, 024608 (2012)
[6] T. Kawano, P. Talou, M.B. Chadwick, T. Watanabe, J. Nucl. Sci. Tech. **47**, 462 (2010)
[7] M. Chadwick, P. Obložinský, M. Herman, N. Greene, R. McKnight, D. Smith, P. Young, R. MacFarlane, G. Hale, S. Frankle et al., Nuclear Data Sheets **107**, 2931 (2006), evaluated Nuclear Data File ENDF/B-VII.0
[8] C. Kalbach, Phys. Rev. C **33**, 818 (1986)
[9] C. Kalbach, Phys. Rev. C **71**, 034606 (2005)
[10] A. Koning, M. Duijvestijn, Nuclear Physics A **744**, 15 (2004)
[11] B. Becker, P. Talou, T. Kawano, Y. Danon, I. Stetcu, Phys. Rev. C **87**, 014617 (2013)
[12] I. Stetcu, P. Talou, T. Kawano, M. Jandel, Phys. Rev. C **90**, 024617 (2014)
[13] A.C. Wahl, Report LA-13928, Los Alamos National Laboratory (2002)
[14] D. Madland, Nuclear Physics A **772**, 113 (2006)
[15] I. Stetcu, P. Talou, T. Kawano, M. Jandel, Phys. Rev. C **88**, 044603 (2013)
[16] J. Lestone, T. Strother, Nuclear Data Sheets **118**, 208 (2014)
[17] F. Tovesson, C.W. Arnold, T. Bredeweg, M. Jandel, A.B. Laptev, K. Meierbachtol, A. Sierk, W. M., A.A. Hecht, D. Mader et al., *SPIDER: A new instrument for fission yield measurement*, in *Proceedings of the Fifth International Conference on ICFN5*, edited by J.H. Hamilton, A.V. Ramayya (World Scientific, 2013), p. 361
[18] J. Randrup, P. Möller, Phys. Rev. Lett. **106**, 132503 (2011)
[19] J. Randrup, P. Möller, A.J. Sierk, Phys. Rev. C **84**, 034613 (2011)
[20] A. Sierk, Report LA-UR-14-27056, Los Alamos National Laboratory (2014)

EPJ Web of Conferences **122**, 01013 (2016) DOI: 10.1051/epjconf/201612201013
*CNR*15*

Assessing the role of the $(n, \gamma f)$ process in the low-energy fission of actinides

Patrick Talou[1,a], J. E. Lynn[1], T. Kawano[1], S. Mosby[2], A. Couture[2], and O. Bouland[3]

[1] *Nuclear Physics Group, Theoretical Division, Los Alamos National Laboratory, Los Alamos, USA*
[2] *P-27, Physics Division, Los Alamos National Laboratory, USA*
[3] *CEA, DEN, DER, SPRC, LEPh, Cadarache, F-13108 Saint-Paul-lez-Durance, France*

Abstract. We review the role of the $(n, \gamma f)$ process in the low-energy neutron-induced fission reaction of ^{239}Pu. Recent measurements of the average total γ-ray energy released in this reaction were performed with the Detector for Advanced Neutron Capture Experiments (DANCE) at Los Alamos. Significant fluctuations of this quantity in the resonance region below 100 eV can be interpreted by invoking the presence of the indirect $(n, \gamma f)$ process. Modern calculations of the probability for such an event to occur are presented.

1 Introduction

The possibility for a heavy compound nucleus formed through low-energy neutron-induced reactions to emit a γ ray before fissioning was postulated theoretically long ago by Stavinsky and Shaker [1] and Lynn [2]. In this scenario, a compound nucleus is first formed with excitation energy $E^* = \epsilon_n + B_n$, angular momentum $J = I \pm \frac{1}{2}$, and parity π. ϵ_n is the kinetic energy of the incident neutron, B_n the neutron binding energy for the target, and I the spin of the target in its ground-state. Since we are considering low-energy reactions, only s-wave ($l = 0$) neutrons are taken into account. In the case of n+^{239}Pu reaction, and considering the target nucleus in its ground-state with $I^\pi = \frac{1}{2}^+$, only 0^+ and 1^+ compound states can be populated. At those energies, only the neutron elastic, fission and capture channels are open. Because the excitation energy E^* is higher than the fission barrier height by $\sim$1.5 MeV, γ rays can be emitted leaving the nucleus with enough residual excitation energy to overcome the fission barrier, leading to an additional $(n, \gamma f)$ channel.

Beyond the rather esoteric interest for this particular fission decay, there are practical implications to the existence of this process. The emitted γ ray carries with it an excitation energy of about 1 MeV, which then represents a deficit for the total excitation energy available for the emission of prompt neutrons. Because the average neutron multiplicity $\overline{\nu}$ plays such an important role in nuclear technologies, as for instance in the very accurate simulation of critical assembly and/or nuclear reactor benchmarks, a small change in $\overline{\nu}$ has important practical consequences, as noted in [3, 4].

Such fluctuations of $\overline{\nu}$ have been observed already, as shown in Fig. 1 for incident energies up to 100 eV in the neutron-induced fission of ^{239}Pu. To a good approximation, the neutron multiplicity reflects how much excitation energy is available in the fission fragments shortly after scission. In turn, this energy depends on the total kinetic energy of the complementary fragments and on the Q value of

[a] e-mail: talou@lanl.gov

the fission reaction. The latter is a function of the specific fragments (A, Z) produced in the reaction. Besides the $(n, \gamma f)$ process, fluctuations in the particular fragment yields in mass and charge could also lead to fluctuations in the number of emitted neutrons [5]. However, such an interpretation would not explain fluctuations in the observed total γ-ray energy, as will be explained below.

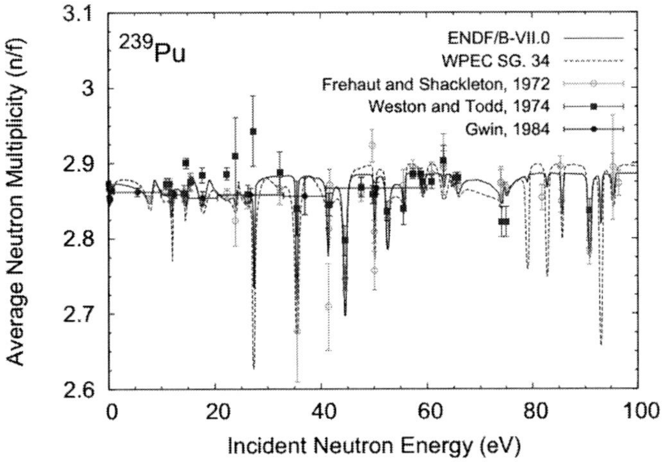

Figure 1. Fluctuations in the average prompt fission neutron multiplicity $\bar{\nu}_n$ for the ^{239}Pu (n, f) reaction.

2 Theoretical Framework

Here, we briefly review the theoretical framework used to compute the pre-fission γ-ray emission probability $\Gamma_{\gamma f}$ and spectrum. This probability can be written as follows

$$\Gamma_{\gamma f}(E^*, J^\pi) = \sum_{Xl} \sum_{J'=|J-l|}^{J+l} \int_0^{E^*} d\epsilon_\gamma \rho(E^* - \epsilon_\gamma, J', (-)^{Xl}\pi)\Gamma_{\gamma,Xl}(\epsilon_\gamma) P_f(E^* - \epsilon_\gamma, J', (-)^{Xl}\pi), \quad (1)$$

where $\rho(E, J, \pi)$ is the level density in energy, spin and parity; $\Gamma_{\gamma,Xl}$ is the γ-ray strength function, and $P_f(E, J, \pi)$ is the fission probability for a given energy, spin and parity. The level density is calculated at the residual excitation energy after γ-ray emission, $E^* - \epsilon_\gamma$. The residual spin and parity depend on the particular electromagnetic transition– denoted by Xl, leading to the final configuration. We calculate the compound nucleus level density following the combinatorial model described in [6]. Single-particle states are combined with vibrational and rotational enhancements to infer the total level density. The neutron and proton pairing gap parameters are adjusted to reproduce the known average resonance spacing for ^{240}Pu.

Only $E1$ and $M1$ electromagnetic transitions are considered in this work, with higher-order transitions assumed to be negligible. The γ-ray strength functions are obtained using the giant dipole resonance formalism. The Kopecky-Uhl generalized Lorentzian expression is used to compute $E1$ transitions, while a standard Lorentzian is used for $M1$ transitions.

Finally, the fission probabilities entering in Eq. 1 are calculated in the cross-section modeling formalism discussed in [6, 7]. This model is especially suited to the study of fission reactions near or below the fission barrier, as coupling between the two wells along the fission path lead to significant corrections to the simpler statistical model [6]. Since ^{239}Pu is a fissile nucleus, its fission barrier height lies below its neutron separation energy, and the direct estimation of its barrier height and width using neutron-induced reactions is not possible. Instead, surrogate reactions such as ^{238}Pu (t, pf) transfer-induced fission reactions [8] were used to infer the needed input parameters for the fission channel.

More details on the model and each of these quantities will be provided in a longer publication, in preparation.

3 DANCE Experiment and Simulations

The Detector for Advanced Neutron Capture Experiments (DANCE) at the Los Alamos Neutron Science Center (LANSCE) is a 4π-calorimeter that consists of 160 BaF_2 crystals, each 734 cm^3 in volume with faces located 19 cm from the beam center. Such an array has a very high efficiency and is very well suited to measure the total energy and multiplicity released by prompt γ rays in a fission event. Located on the Lujan flight path, incident neutrons are produced by spallation on a tungsten target and moderated to energies ranging from thermal up to a few hundred keV. Of particular interest to the present study is the region below 100 eV where well-resolved neutron resonances appear. Those resonances are also well characterized in energy and spin, providing an ideal test ground for our calculations.

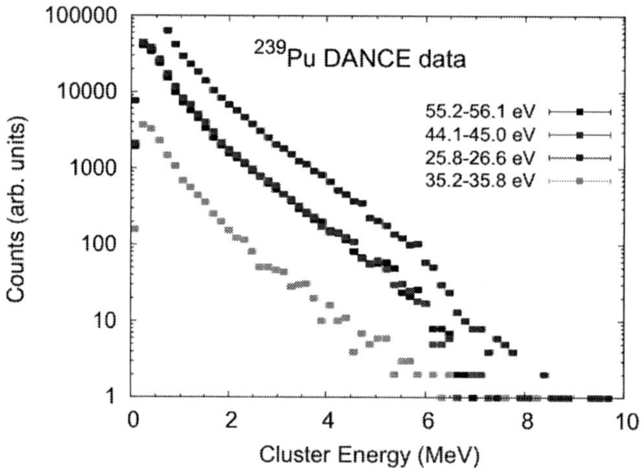

Figure 2. Total γ-ray energy measured with the DANCE detector for different energy bins.

Measurements of the total γ-ray energy and total γ-ray multiplicity for different neutron resonances are shown in Figs. 2 and 3, respectively. Due to the complexity of the DANCE detector, the unfolding of the raw signal into a useful and physical γ-ray spectrum is no easy task. Instead, we chose to forward propagate simulated γ rays into a transport simulation of the DANCE detector

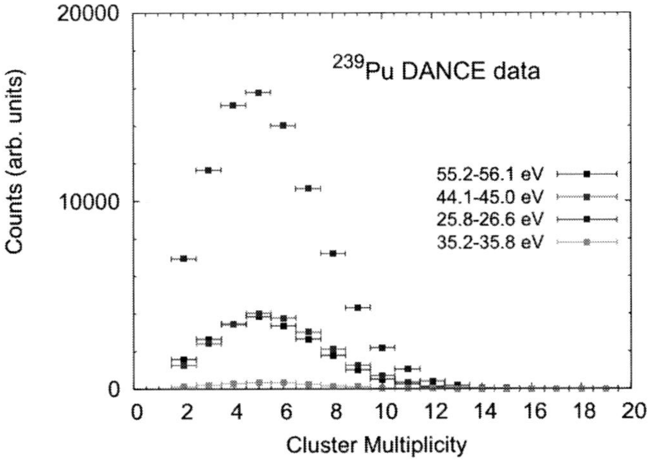

Figure 3. Total γ-ray multiplicity measured for the same energy bins as in Fig. 2.

response. The GEANT4 transport code was used, leveraging an already built model of the complete DANCE array. As mentioned above, we are not able to differentiate between γ rays emitted right before fission from those emitted right after, so we have to calculate the contribution from the fission fragments as well. The prompt fission γ rays evaporated from the fission fragments are calculated using our CGMF code [9], which computes the de-excitation of the primary fragments through neutron and γ emissions in a Monte Carlo implementation of the Hauser-Feshbach statistical nuclear reaction theory. The calculated prompt fission γ-ray spectrum, excluding the contribution from the $(n,\gamma f)$ events, for the thermal neutron-induced fission of ^{239}Pu is shown in Fig. 4. The agreement between the CGMF calculations and the measurement by Chyzh [10] and by Verbinski [11] is quite good.

The total γ-ray spectrum is obtained by combining the CGMF-calculated spectrum of γ rays evaporated from the fragments with the γ-ray spectrum from the $(n,\gamma f)$ process, weighted by the ratio $\Gamma_{\gamma f}/\Gamma_f$

$$N_\gamma^{tot}(\epsilon_\gamma) = \frac{\Gamma_{\gamma f}}{\Gamma_f} N_{\gamma f}(\epsilon_\gamma) + N_{FF}(\epsilon_\gamma), \tag{2}$$

where Γ_f is the sum of the width for the "direct" fission process and $\Gamma_{\gamma f}$. To obtain this equation, we are making the reasonable assumption that the spectrum of prompt γ rays emitted from the fragments is mostly insensitive to a modest change in the initial excitation energy of the fragments.

We then perform random samplings of the total spectrum N_γ^{tot} to produce random fission events that are used as input in the GEANT4 transport simulations. The final propagated results can then be compared to raw experimental data from DANCE.

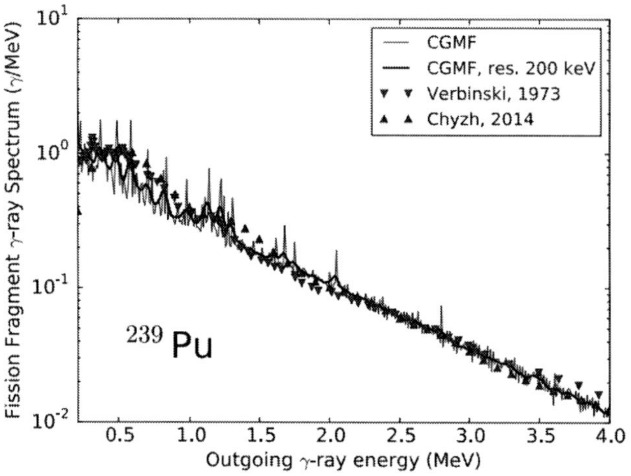

Figure 4. Prompt fission γ-ray spectrum calculated with the CGMF code for the thermal neutron-induced fission of ^{239}Pu. Experimental data sets from Chyzh *et al.* [10] and Verbinski *et al.* [11] are plotted for comparison.

4 Results and Discussion

Using the ^{238}Pu (t, pf) transfer-induced fission reactions [8], the characteristics in height and width of the double-hump fission barrier for ^{240}Pu* were obtained and are in good agreement with the values given in [7]. Of course, those values are somewhat model-dependent and not directly observable.

The fission probabilities calculated for ^{240}Pu* are shown in Fig. 5 for several spin and parity states. At low excitation energy, the spectrum of an even-even nucleus such as ^{240}Pu is dominated by the K=0 ground-state band (0^+, 2^+, ...) and at slightly higher energy, the mass asymmetry band (1^-, 3^-, ...). With s-wave neutrons ($l = 0$) impinging on ^{239}Pu in its ground-state ($1/2^+$), only 0^+ and 1^+ compound nuclear states can be formed. Only $M1$ transitions can then lead to 0^+ and 2^+ fission transition states, hence explaining the stronger than expected contribution of $M1$ relative to $E1$ transitions in the $(n, \gamma f)$ process.

In Eq. 1, it is the convolution of the fission probabilities P_f with the widths $\Gamma_{\gamma,Xl}$ that determines the magnitude of the $(n, \gamma f)$ process. Contrary to what is commonly observed in capture reactions, the $M1$ component is strongly enhanced in this case, due to the higher values of the fission probabilities for the 0^+ and 2^+ states, which again, can only be accessed through $M1$ transitions.

As noted earlier, disentangling γ rays emitted in the $(n, \gamma f)$ process from all prompt γ rays emitted in coincidence with a fission event is not possible. Due to the large number of compound nucleus states available, the width $\Gamma_{\gamma f}$ can be assumed to be rather constant across resonances. Due to the large fluctuations of the total fission widths Γ_f, the impact of the $(n, \gamma f)$ process can only be revealed for resonances with very small total fission width. In the n+^{239}Pu reaction, only a few resonances satisfy this criterion. Below 100 eV, of particular interest are the 1^+ resonances at energies 27.24, 35.50, 41.42, 44.48, 50.08, 52.60, 90.75, and 92.97 eV. The most promising candidate is the 41.42

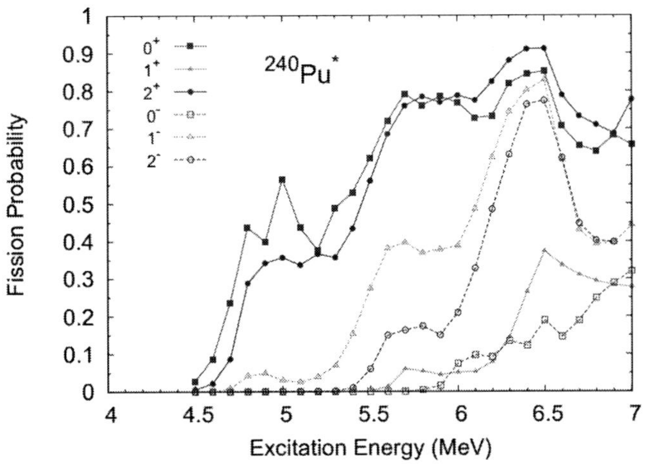

Figure 5. Spin and parity-dependent fission probabilities calculated for $^{240}Pu^*$. Positive-parity states are depicted with full symbols and solid lines, while negative-parity states are shown with empty symbols and dashed lines.

eV resonance with a total fission width of only 6.4 meV [12]. All 0^+ states have much larger fission widths.

One way to reveal the contribution of the pre-fission γ rays to the total fission γ-ray spectrum is to study the ratio of γ-ray spectra obtained for two different resonances, with two very different total fission widths. As an illustrative example, we shown in Fig. 6 the results obtained for the 35.5 eV resonance (Γ_f=3.5 meV) vs. the 14.6 eV resonance (Γ_f=30 meV). The observed (black) and simulated (red) γ-ray spectra clearly show a "bump" in the 1-1.5 MeV range, where the contribution from the $(n, \gamma f)$ process is expected to be strongest. The agreement between the simulated and experimental data is quite good in this ratio plot. Note that the addition of a $M1$ "scissors" mode contribution worsens the agreement by predicting too many pre-fission γ rays. However, other parameterizations of this process could lead to a different conclusion. This work is in progress.

Finally, to address the question of the role of the fluctuations of the mass and the total kinetic energy yields of the fission fragments in the resonance region as observed by Hambsch *et al.* [5], we performed CGMF calculations of the decay of the fission fragments by modifying artificially the average total kinetic energy of the fragments and study the sensitivity of the calculated neutron and γ-ray multiplicities on TKE. The result is shown in Fig. 7. As can be seen, the average neutron multiplicity (red) is extremely sensitive to TKE. On the other hand, the average γ-ray multiplicity is almost constant over a wide range, ±2 MeV, of TKE values. Such a result confirms that only the $(n, \gamma f)$ mechanism can explain the type of fluctuations observed with DANCE and in previous experiments. This is not to say that the yield fluctuations in mass and TKE do not play a role in the fluctuation of $\overline{\nu}$ observed in the resonance region (see Fig. 1). The exact contributions of the two phenomena for each resonance would be difficult to assess though.

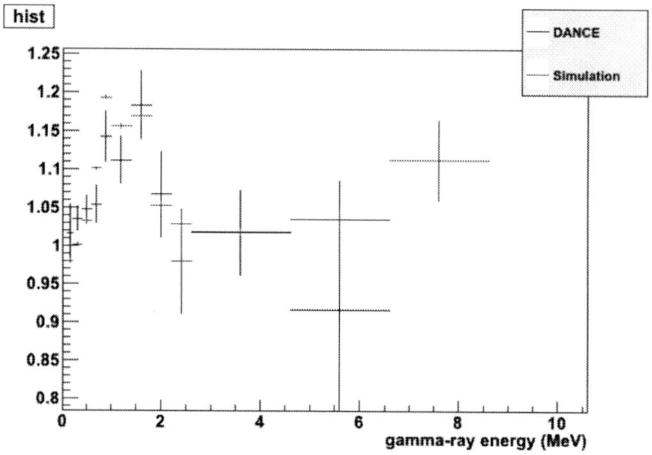

Figure 6. Ratio of γ-ray energy spectrum for two resonances, 35.5 eV (Γ_f=5 meV) and 14.6 eV (Γ_f=33 meV), as observed (black) and simulated (red) in the DANCE detector. A clear "bump" appears in the 1-1.5 MeV range, corresponding to the predicted average γ-ray energy emitted in the $(n,\gamma f)$ process.

5 Conclusion

In this contribution, we reviewed an old problem with fresh eyes, fresh model calculations, and fresh experimental data. While the possibility for the emission of a γ ray prior to fission was postulated theoretically 50 years ago, rather conclusive experimental evidence for such a process came only in the seventies. Very little information on the pre-fission γ-ray spectrum is available however, even though it is important to infer the particular nature of the γ transitions preceding fission. We have applied a modern theoretical framework to compute this spectrum, along with the spectrum of γ rays from the fission fragment themselves. Propagating these results into a GEANT4 model of the DANCE detector array, we were able to directly compare the calculated results with recent DANCE experimental data. Because of the intrinsic difficulties in disentangling pre- and post-fission γ rays, we studied the ratios of various spectra for different resonances. The contribution of the $(n,\gamma f)$ process is best observed for resonances with very small total fission widths. In the case of ^{239}Pu (n,f), we were able to show and explain the important role of $M1$ transitions in the pre-fission γ-ray spectrum. More work is in progress to complete calculations and experimental data analysis for neutron-induced fission reactions on both ^{239}Pu and ^{235}U.

Acknowledgments

We would like to thank C. Y. Wu for providing data and explanations on his recent measurement of the prompt fission γ-ray spectrum for thermal neutron-induced fission of ^{239}Pu. This work was performed under the auspices of the National Nuclear Security Administration of the U.S. Department of Energy at Los Alamos National Laboratory under Contract No. DE-AC52-06NA25396.

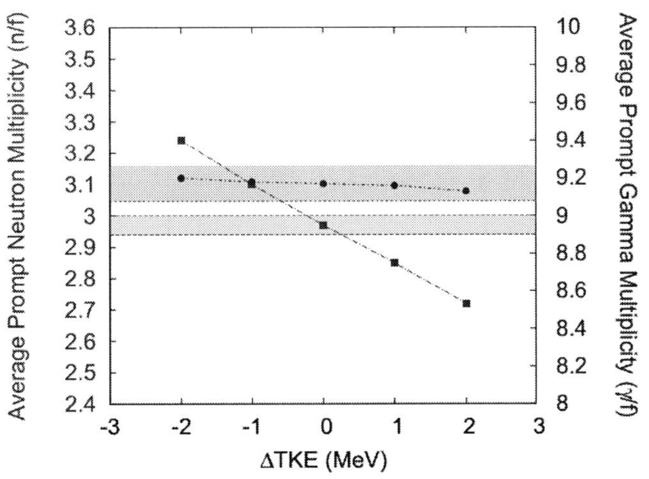

Figure 7. Sensitivity of the calculated average prompt neutron and γ-ray multiplicities on a change in the total kinetic energy TKE of the fission fragments.

References

[1] V. Stavinsky and M. O. Shaker, Nucl. Phys. **62**, 667 (1965).
[2] J. E. Lynn, Phys. Lett. **18**, 31 (1965).
[3] C. De Saint Jean, R.D. McKnight, et al., OECD NEA/NSC/WPEC/DOC 447 (2014).
[4] R. Capote *et al.*, Nucl. Data Sheets **131**, 1 (2016).
[5] F. -J. Hambsch, H. -H. Knitter, C. Budtz-Jørgensen, and J. P. Theobald, Nucl. Phys. **A491**, 56 (1989).
[6] O. Bouland, J. E. Lynn, and P. Talou, Phys. Rev. C **88**, 054612 (2013).
[7] S. Bjørnholm and J. E. Lynn, Rev. Mod. Phys. **52**, 725 (1980).
[8] J. D. Cramer and H. C. Britt, Phys. Rev. C **2**, 2350 (1970).
[9] I. Stetcu, P. Talou, T. Kawano and M. Jandel, Phys. Rev. C **90**, 024617 (2014).
[10] A. Chyzh, C. Y. Wu, E. Kwan, R. A. Henderson, T. A. Bredeweg, R. C. Haight, A. C. Hayes-Sterbenz, H. Y. Lee, J. M. O'Donnell, and J. L. Ullmann, Phys. Rev. C **90**, 014602 (2014).
[11] V. V. Verbinski, H. Weber and R. E. Sund, Phys. Rev. C **7**, 1173 (1973).
[12] S. F. Mughabghab, "Atlas of Neutron Resonances," Elsevier, 2006.

Level densities of heavy nuclei in the shell model Monte Carlo approach

Y. Alhassid[1,a], G.F. Bertsch[2,3], C.N. Gilbreth[3], H. Nakada[4], and C. Özen[5]

[1] *Center for Theoretical Physics, Sloane Physics Laboratory, Yale University, New Haven, Connecticut 06520, USA*
[2] *Department of Physics, University of Washington, Seattle, Washington 98915, USA*
[3] *Institute for Nuclear Theory, Box 351560, University of Washington, Seattle, Washington 98915, USA*
[4] *Department of Physics, Graduate School of Science, Chiba University, Inage, Chiba 263-8522, Japan*
[5] *Faculty of Engineering and Natural Sciences, Kadir Has University, Istanbul 34083, Turkey*

Abstract. Nuclear level densities are necessary input to the Hauser-Feshbach theory of compound nuclear reactions. However, the microscopic calculation of level densities in the presence of correlations is a challenging many-body problem. The configuration-interaction shell model provides a suitable framework for the inclusion of correlations and shell effects, but the large dimensionality of the many-particle model space has limited its application in heavy nuclei. The shell model Monte Carlo method enables calculations in spaces that are many orders of magnitude larger than spaces that can be treated by conventional diagonalization methods and has proven to be a powerful tool in the microscopic calculation of level densities. We discuss recent applications of the method in heavy nuclei.

1 Introduction

Level densities play an important role in the Hauser-Feshbach theory [1] of compound nuclear reactions, but are not always accessible by direct measurements. The calculation of level densities in the presence of correlations is a challenging many-body problem. Most approaches are based on empirical modifications of the Fermi gas formula or on mean-field approximations [2] that can often miss important correlations. The configuration-interaction (CI) shell model approach accounts for correlations, but conventional diagonalization methods are limited to spaces of dimensionality $\sim 10^{11}$. The dimension of the many-particle model space increases combinatorially with the numbers of valence nucleons and single-particle orbitals, hindering applications of the CI shell model approach to mid-mass and heavy nuclei.

The auxiliary-field quantum Monte Carlo method, also known in nuclear physics as the shell model Monte Carlo (SMMC) method [3–6], enables microscopic calculations in spaces that are many orders of magnitude larger (e.g., $\sim 10^{30}$ in recent applications to rare-earth nuclei) than those that can be treated by conventional methods.

[a] e-mail: yoram.alhassid@yale.edu

The outline is as follows. In Sec. 2 we briefly review the SMMC method. In Sec. 3 we discuss the emergence of collectivity in heavy nuclei within the framework of the CI shell model. In Sec. 4 we present results for the state densities in families of lanthanide isotopes and find them to be good agreement with experimental data. In Sec. 5 we benchmark finite-temperature mean-field approximations to level densities by comparing them with SMMC state densities. In Sec. 6 we discuss nuclear deformation in the CI shell model. Deformation is an important concept for understanding heavy nuclei but it is usually introduced in a mean-field approximation that breaks rotational symmetry. Here we discuss how the intrinsic quadrupole deformation distributions can be derived in a framework that preserves rotational symmetry. We conclude in Sec. 7.

2 Shell model Monte Carlo method

2.1 Hubbard-Stratonovich transformation

The SMMC method is based on the Hubbard-Stratonovich (HS) transformation [7], in which the Gibbs ensemble $e^{-\beta\hat{H}}$ at inverse temperature $\beta = 1/T$ for a nucleus described by an Hamiltonian $\hat{H}$ can be viewed as an imaginary-time propagator and written as a functional integral over propagators of non-interacting particles in external time-dependent auxiliary fields $\sigma(\tau)$

$$e^{-\beta\hat{H}} = \int \mathcal{D}[\sigma] G_\sigma \hat{U}_\sigma \,. \tag{1}$$

Here G_σ is a Gaussian factor and $\hat{U}_\sigma$ is a many-particle propagator for non-interacting particles for a given configuration of the auxiliary fields. The thermal expectation value of observable $\hat{O}$ is given by

$$\langle \hat{O} \rangle = \frac{\mathrm{Tr}\,(\hat{O} e^{-\beta\hat{H}})}{\mathrm{Tr}\,(e^{-\beta\hat{H}})} = \frac{\int \mathcal{D}[\sigma] W_\sigma \langle \hat{O} \rangle_\sigma \Phi_\sigma}{\int \mathcal{D}[\sigma] W_\sigma \Phi_\sigma} \,, \tag{2}$$

where $W_\sigma = G_\sigma |\mathrm{Tr}\,\hat{U}_\sigma|$ is a positive-definite function, $\Phi_\sigma = \mathrm{Tr}\,\hat{U}_\sigma / |\mathrm{Tr}\,\hat{U}_\sigma|$ is the Monte Carlo sign function and $\langle \hat{O} \rangle_\sigma \equiv \mathrm{Tr}\,(\hat{O}\hat{U}_\sigma)/\mathrm{Tr}\,\hat{U}_\sigma$. The grand canonical traces in the integrands of Eq. (2) can be calculated in terms of the single-particle representation $\mathbf{U}_\sigma$ (an $N_s \times N_s$ matrix, where N_s is the number of single-particle orbitals) of $\hat{U}_\sigma$. The trace of $\hat{U}_\sigma$ is given by

$$\mathrm{Tr}\,\hat{U}_\sigma = \det(\mathbf{1} + \mathbf{U}_\sigma) \,, \tag{3}$$

while the grand canonical expectation value of a one-body operator $\hat{O} = \sum_{i,j} O_{ij} a_i^\dagger a_j$ is given by

$$\langle a_i^\dagger a_j \rangle_\sigma \equiv \frac{\mathrm{Tr}\,(a_i^\dagger a_j \hat{U}_\sigma)}{\mathrm{Tr}\,\hat{U}_\sigma} = \left[\frac{1}{\mathbf{1} + \mathbf{U}_\sigma^{-1}} \right]_{ji} \,. \tag{4}$$

The grand canonical expectation value of a two-body observable can be calculated using Wick's theorem.

2.2 Canonical projection

In the finite nucleus it is important to consider the canonical ensemble of fixed number of protons and neutrons. The particle-projected partition function for $\mathcal{A}$ particles is given by a discrete Fourier transform [8]

$$\mathrm{Tr}_{\mathcal{A}} U_\sigma = \frac{e^{-\beta\mu\mathcal{A}}}{N_s} \sum_{m=1}^{N_s} e^{-i\varphi_m \mathcal{A}} \det\left(\mathbf{1} + e^{i\varphi_m} e^{\beta\mu} \mathbf{U}_\sigma\right) \,, \tag{5}$$

where $\varphi_m = 2\pi m/N_s$ $(m = 1, \ldots, N_s)$ are quadrature points and μ is a chemical potential introduced to stabilize the numerical evaluation of the Fourier sum. Similarly for a one-body observable $\hat{O} = \sum_{i,j} O_{ij} a_i^\dagger a_j$

$$\mathrm{Tr}_{\mathcal{A}}\left(\hat{O}\hat{U}_\sigma\right) = \frac{e^{-\beta\mu\mathcal{A}}}{N_s} \sum_{m=1}^{N_s} e^{-i\varphi_m \mathcal{A}} \det\left(\mathbf{1} + e^{i\varphi_m + \beta\mu}\mathbf{U}_\sigma\right) \mathrm{tr}\left(\frac{1}{\mathbf{1} + e^{-i\varphi_m - \beta\mu}\mathbf{U}_\sigma^{-1}}\mathbf{O}\right), \tag{6}$$

where $\mathbf{O}$ is the matrix with elements O_{ij}.

2.3 Monte Carlo sampling

In SMMC we choose a set of uncorrelated auxiliary-field configurations $\sigma^{(k)}$ from the positive-definite weight function W_σ. The expectation value of an observable in Eq. (2) is then estimated from

$$\langle \hat{O} \rangle \approx \frac{\sum_k \langle \hat{O} \rangle_{\sigma^{(k)}} \Phi_{\sigma^{(k)}}}{\sum_k \Phi_{\sigma^{(k)}}} \tag{7}$$

The statistical error of $\langle \hat{O} \rangle$ can be estimated from the variance of the "measurements" $\langle \hat{O} \rangle_{\sigma^{(k)}}$.

3 Collectivity in heavy nuclei in the CI shell model

The single-particle model space we use for rare-earth nuclei is composed of the orbitals $0g_{7/2}, 1d_{5/2}, 1d_{3/2}, 2s_{1/2}, 0h_{11/2}, 1f_{7/2}$ for protons and $0h_{11/2}, 0h_{9/2}, 1f_{7/2}, 1f_{5/2}, 2p_{3/2}, 2p_{1/2}, 0i_{13/2}, 1g_{9/2}$ for neutrons. The single-particle levels are taken from a Woods-Saxon potential plus a spin-orbit interaction. The interaction includes attractive monopole pairing plus attractive multipole-multipole terms with quadrupole, octupole and hexadecupole components. The interaction parameters are determined empirically as discussed in Refs. [9] and [10].

Quantum Monte Carlo methods are often limited by the so-called sign problem, which leads to large statistical errors. However, the dominant components of effective nuclear shell model interactions [11] have a good sign (in the grand canonical ensemble) and are usually sufficient for realistic calculations of statistical and collective properties of nuclei. Small bad sign components of the interaction can be treated using the extrapolation method of Ref. [4]. The interaction discussed above has a good Monte Carlo sign in the grand canonical ensemble and the sign remains good in the canonical ensemble for even-even nuclei. For even-odd nuclei there is a moderate sign problem at high and intermediate temperatures that becomes more severe at low temperatures [12].

Heavy nuclei are known to exhibit various types of collectivity that are well described by empirical models. An important question is whether such collectivity, and, in particular, rotational collectivity, can be described in a truncated spherical shell model approach. Various types of collectivity are usually identified by their characteristic spectra of low-lying states. However, it is a challenge to obtain detailed spectroscopic information in SMMC. We found that the type of collectivity can be characterized by the low-temperature behavior of the thermal observable $\langle \mathbf{J}^2 \rangle_T$, where $\mathbf{J}$ is the total angular momentum of the nucleus. For an even-even nucleus, we find (at low temperatures)

$$\langle \mathbf{J}^2 \rangle_T \approx \begin{cases} 30\frac{e^{-E_{2^+}/T}}{\left(1-e^{-E_{2^+}/T}\right)^2} & \text{vibrational band} \\ \frac{6}{E_{2^+}}T & \text{rotational band} \end{cases}, \tag{8}$$

where E_{2^+} is the excitation energy of the lowest 2^+ level.

In Fig. 1, we show the low-temperature behavior of $\langle \mathbf{J}^2 \rangle_T$ vs. T for ^{162}Dy (left panel) and ^{148}Sm (right panel). The results are consistent with ^{162}Dy being a rotational nucleus, and ^{148}Sm a vibrational nucleus (see figure caption for details).

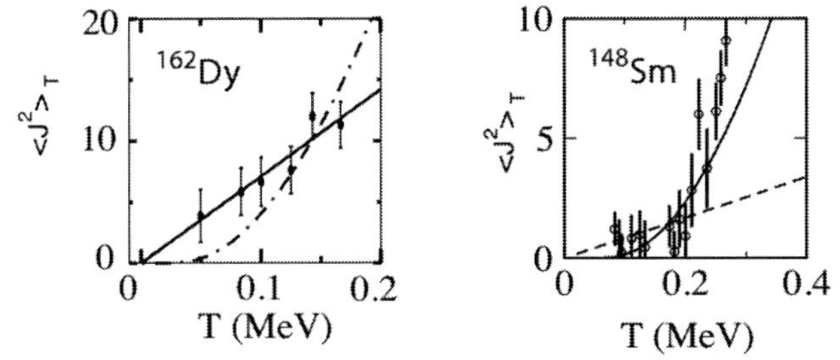

Figure 1. Left panel: $\langle \mathbf{J}^2 \rangle_T$ vs. T for ^{162}Dy. The SMMC results (solid circles) fit well the rotational model (solid line) but not the vibrational model (dashed-dotted line). Adapted from Ref. [9]. Right panel: $\langle \mathbf{J}^2 \rangle_T$ vs. T for ^{148}Sm are in better agreement with the vibrational model (solid line) than the rotational model (dashed line). Taken from Ref. [13].

3.1 Crossover from vibrational to rotational collectivity

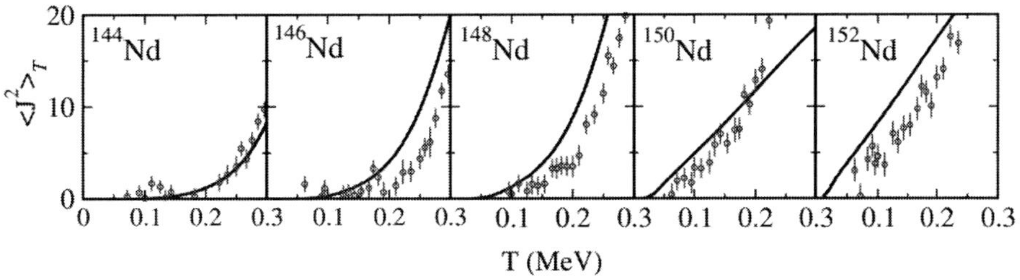

Figure 2. $\langle \mathbf{J}^2 \rangle_T$ vs. T for a family of even-even neodymium isotopes $^{144-152}$Nd. The SMMC results (open circles) are compared with results extracted from experimental data (solid lines, see text).

The observable $\langle \mathbf{J}^2 \rangle_T$ can also be used to describe the crossover from vibrational to rotational collectivity a family of isotopes. Fig. 2 shows this observable vs. T in a family of even-even neodymium isotopes. We observe a crossover from a soft response to temperature in the vibrational nucleus ^{146}Nd to a rigid behavior in the rotational nucleus ^{152}Nd. The solid lines are determined from experimental data using

$$\langle \mathbf{J}^2 \rangle_T = \frac{1}{Z(T)} \left(\sum_i^N J_i(J_i+1)(2J_i+1)e^{-E_i/T} + \int_{E_N}^{\infty} dE_x\, \rho(E_x)\, \langle \mathbf{J}^2 \rangle_{E_x}\, e^{-E_x/T} \right) , \quad (9)$$

DOI: 10.1051/epjconf/201612202001

where the partition function $Z(T)$ is

$$Z(T) = \sum_i^N (2J_i + 1)e^{-E_i/T} + \int_{E_N}^{\infty} dE_x \rho(E_x) e^{-E_x/T} . \quad (10)$$

The summations in Eqs. (9) and (10) are over a complete set of experimentally measured levels with energies E_i and spins J_i up to an energy of E_N. Above E_N we use a back-shifted Fermi gas level density $\rho(E_x)$ with parameters determined from level counting at low excitation energies and neutron resonance data at the neutron separation energy.

4 State densities in lanthanide isotopes

SMMC has been a powerful method for calculating state densities in the presence of correlations [9, 10, 14–18]. In SMMC we calculate the canonical energy vs. inverse temperature β as an observable $E_c(\beta) = \langle \hat{H} \rangle$. The canonical partition function $Z_c(\beta)$ is then determined by integrating the thermodynamic relation $E_c(\beta) = -\partial \ln Z_c / \partial \beta$. The level density is related to the partition function by an inverse Laplace transform

$$\rho(E) = \frac{1}{2\pi i} \int_{-i\infty}^{i\infty} d\beta \, e^{\beta E} Z_c(\beta) . \quad (11)$$

The average state density is obtained by calculating Eq. (11) in the saddle-point approximation [19]

$$\rho(E) \approx \left(2\pi \left| \frac{\partial E}{\partial \beta} \right| \right)^{-1/2} e^{S_c(\beta)} , \quad (12)$$

where $S_c(E) = \ln Z_c + \beta E_c$ is the canonical entropy. The inverse temperature β is determined as a function of E from the saddle-point condition $E_c(\beta) = E$.

In Fig. 3 we show the SMMC state densities (open circles) in families of even-mass samarium and neodymium isotopes. They are in good agreement with level counting data at low excitation energies (histograms) and neutron resonance data (triangles) [21] when available.

5 Mean-field approximations to level densities

In SMMC we take into account all correlations within the CI shell model space. However, these calculations are computationally intensive and it is interesting to determine the accuracy of mean-field approximations to level densities. Recently we used SMMC level densities to benchmark finite-temperature mean-field approximations to level densities in a strongly deformed nucleus ^{162}Dy and in a spherical nucleus ^{148}Sm with a strong pairing condensate [22].

A finite-temperature mean-field theory such as the Hartree-Fock (HF) and the Hartree-Fock-Bogoliubov (HFB) approximations work in the grand canonical ensemble while the SMMC calculations are canonical with fixed numbers of protons and neutrons. It is therefore necessary to carry out an approximate particle-number projection in the mean-field theory and this is usually done in the saddle-point approximation. The saddle-point approximation leads to a three-dimensional Jacobian $|\partial(E, N_p, N_n)/\partial(\beta, \alpha_p, \alpha_n)|$ (where α_p/β and α_n/β are chemical potentials for N_p protons and N_n neutrons) as a pre-exponential factor to the level density. We carried out this calculation in two steps; in the first step we evaluated the particle-number projection by a saddle-point approximation with respect to α_p and α_n, and in a second step we evaluated the canonical to micro-canonical integration

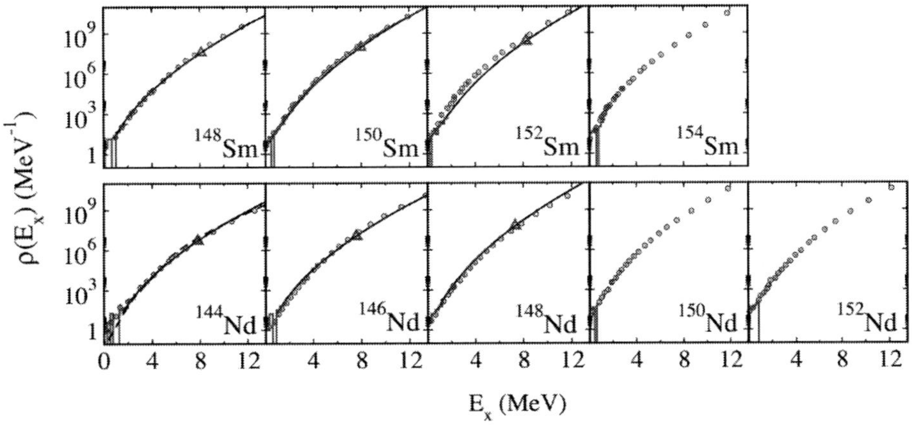

Figure 3. State densities in even-mass samarium $^{148-154}$Sm and neodymium isotopes $^{144-152}$Nd. The SMMC results (solid circles) are compared with experimental data: level counting at low excitation energies (histograms) and neutron resonance data (triangles) [21]. The solid lines were computed using the empirical back-shifted Fermi gas formula with parameters determined from the experiments. Adapted from Refs. [10] and [20]

in (11) by a saddle point approximation with respect to β. The level density is then approximated by Eq. (12) in which the canonical entropy is

$$S_c(\beta, N_p, N_n) = S_{gc} - \ln\zeta + \beta\delta E\ . \tag{13}$$

with S_{gc} being the grand canonical entropy. Here ζ is given by

$$\zeta = 2\pi\left|\frac{\partial(N_p, N_n)}{\partial(\alpha_p, \alpha_n)}\right|^{1/2}\ , \tag{14}$$

and α_p, α_n are determined by the 2-D saddle-point conditions $N_i = \partial \ln Z_{gc}/\partial\alpha_i$ $(i = p, n)$. The correction $\beta\delta E$ with $\delta E = -d\ln\zeta/d\beta$ is absent in the usual 3-D saddle-point approximation.

5.1 Entropies

^{162}Dy is a deformed nucleus with a weak pairing condensate and the appropriate mean-field theory is the finite-temperature HF approximation. ^{148}Sm on the other hand is a spherical nucleus with strong pairing condensate, and the appropriate mean-field theory is the finite-temperature HFB approximation. In Figs. 4 and 5 we show the various entropy functions vs. β for ^{162}Dy and ^{148}Sm, respectively.

The saddle-point approximation to particle-number projection breaks down when the particle-number fluctuations are small. This happens in ^{162}Dy using the HF approximation for β above $\sim$ 5 MeV^{-1}. We then replace the saddle-point expression by a discrete Gaussian approximation discussed in Ref. [22]. In this approximation, $\zeta \to 1$ at large β values, and the entropy (13) vanishes approximately in this limit. In SMMC, the entropy at low temperatures (that are still above $\sim$ 0.05 MeV) is finite and is well described by the contribution from the ground-state rotational band $S_{\rm rot} = 1 + \ln\left(\mathcal{I}_{\rm gs}T/\hbar^2\right)$.

 DOI: 10.1051/epjconf/201612202001

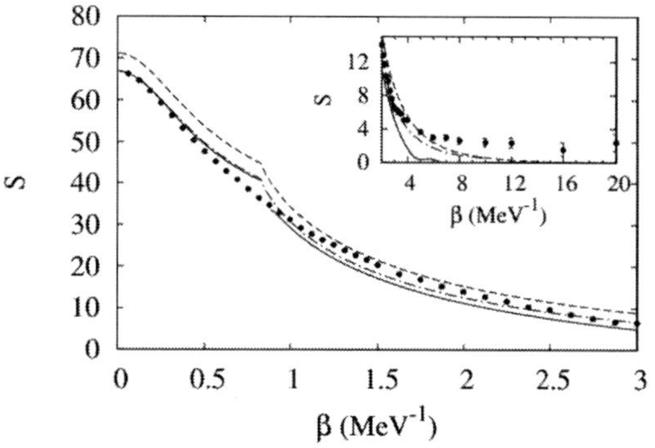

Figure 4. Entropy vs. inverse temperature β in ^{162}Dy. The SMMC entropy (solid circles) is compared with the grand canonical HF entropy (dashed line), the approximate canonical entropy defined in Eq. (13) (solid line), and the approximate canonical entropy obtained in the 3-D saddle-point approximation without the $\beta\delta E$ correction (dashed-dotted line). The calculations use the discrete Gaussian approximation [22]. The inset shows the large β values. Adapted from Ref. [22].

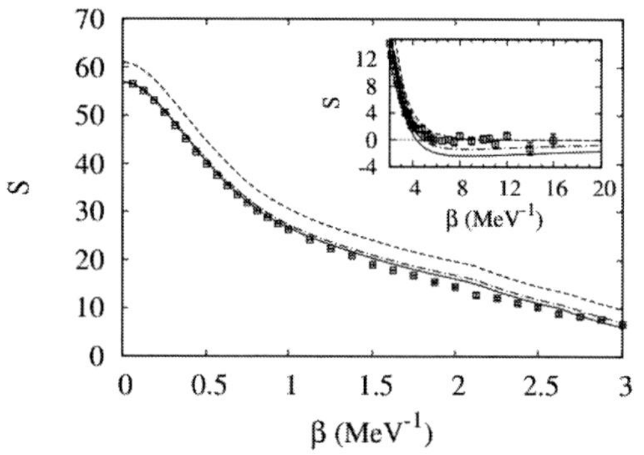

Figure 5. Entropy vs. inverse temperature β in ^{148}Sm. The SMMC entropy (open squares) is compared with entropies obtained in the HFB approximation. The convention for the lines is as in Fig. 4 except that we use the finite-temperature HFB approximation. Taken from Ref. [22].

In ^{148}Sm we use the HFB approximation, in which the pairing condensate violates particle-number conservation and we find a negative approximate canonical entropy at low temperatures.

5.2 Level densities

Here we benchmark the finite-temperature mean-field level densities in comparison with the SMMC densities. The SMMC results are exact (up to statistical errors) for our model Hamiltonian except for the saddle-point approximation (12), which is also used in the finite-temperature mean-field approximations. In Fig. 6 we show the SMMC state density of ^{162}Dy in comparison with the approximate canonical HF level density. We observe a large enhancement of the SMMC state density in comparison with the HF density that eventually disappears in the vicinity of the shape transition energy. The HF density measures the density of intrinsic states while SMMC includes also rotational bands that are built on top of the intrinsic band heads, hence the observed enhancement. A simple model that estimates the contribution of rotational bands [23], leads to overestimating the SMMC density around the neutron separation energy. Assuming a spin cutoff model [24] for the spin distribution with rigid-body moment of inertia gives a mean level spacing of $D = 0.5$ eV for the s-wave resonances. The SMMC result of $D = 2.4 \pm 0.3$ eV is consistent with the experimental result of $D = 2.4$ eV [21].

The single-particle HF spectrum at the neutron separation energy is similar to the ground-state HF spectrum and a frozen potential model in which the single-particle spectrum is fixed (and taken from HF at $T = 0$) gives a good approximation for the HF density at the neutron separation energy.

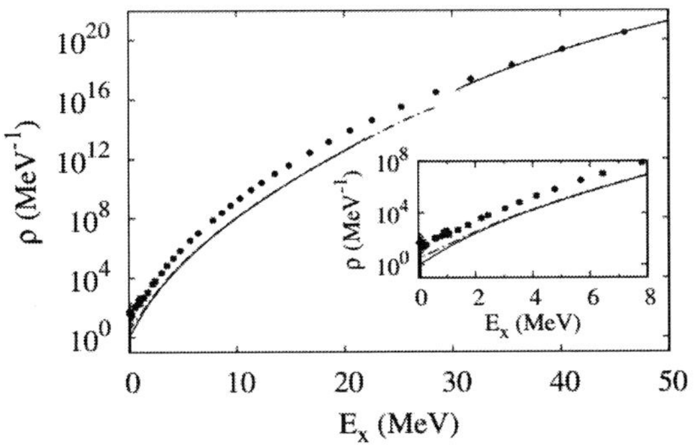

Figure 6. State densities vs. excitation energy E_x in ^{162}Dy. The SMMC state density (solid circles) is compared with the HF density calculated from (12) using the approximate canonical entropy defined in Eq. (13) (solid line), and the approximate level density obtained without the δE correction in the approximate energy and canonical entropy (dashed-dotted line). The inset shows lower values of the excitation energy. Taken from Ref. [22].

In Fig. 7 we compare the SMMC state density of ^{148}Sm with the HFB results. For the neutron resonance spacing we find the HFB value of $D = 4.1$ eV to be in good agreement with SMMC value of $D = 3.7$ eV and the experimental value of $D = 5.7$ eV. The neutron separation energy of ~ 8.2 MeV is above the pairing transition where the HFB approximation reduces to the HF approximation and the HFB just resets the scale of the ground-state energy by the pairing correlation energy. This provides some justification to the back-shifted Fermi gas model.

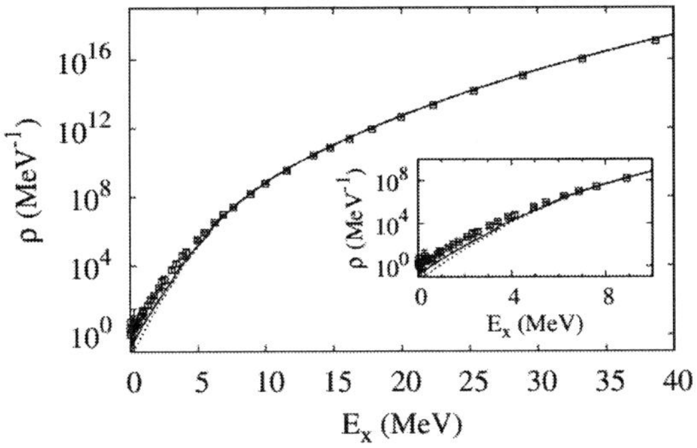

Figure 7. State densities vs. excitation energy E_x in ^{148}Sm. The SMMC state density (open squares) is compared with the HFB density calculated from (12) using the approximate canonical entropy defined in Eq. (13) (solid line), and the approximate level density obtained when ζ is calculated from particle-number fluctuations (dotted line) [22]. The inset shows lower values of the excitation energy. Taken from Ref. [22].

6 Nuclear deformation in the CI shell model

The modeling of dynamical nuclear processes, such as fission, often requires knowledge of the level density as a function of deformation [25]. Nuclear deformation, however, is a concept introduced in the framework of a mean-field approximation that breaks rotational invariance of the nuclear Hamiltonian. The challenge is to study nuclear deformation and calculate statistical properties of nuclei as a function of deformation in an approach that preserves rotational invariance, such as the CI shell model.

Here we focus on the quadrupole deformation. In Sec. 6.1 we discuss the distribution of the axial quadrupole operator in the laboratory frame and show that it provides a model-independent signature of nuclear deformation [26]. In Sec. 6.2 we describe how we can calculate the distribution of the intrinsic quadrupole deformation within a framework that preserves rotational symmetry.

6.1 Quadrupole distributions in the laboratory frame

The axial quadrupole operator $\hat{Q}_{20} = \sum_i \left(2z_i^2 - x_i^2 - y_i^2\right)$ does not commute with the Hamiltonian $\hat{H}$. Its distribution $P_T(q)$ in the laboratory frame at temperature T is given by

$$P_T(q) = \sum_n \delta(q - q_n) \sum_m \langle q, n|e, m\rangle^2 e^{-\beta e_m} , \tag{15}$$

where $|q, n\rangle$ are eigenstates of $\hat{Q}_{20}$ satisfying $\hat{Q}_{20}|q, n\rangle = q_n|q, n\rangle$ and similarly for $|e, m\rangle$. The spectrum of $\hat{Q}_{20}$ in the truncated CI shell model is discrete.

6.1.1 Axial quadrupole projection

The distribution $P_T(q)$ can be calculated in SMMC using its Fourier representation

$$P_T(q) \equiv \frac{\mathrm{Tr}\left[\delta(\hat{Q}_{20} - q)e^{-\beta\hat{H}}\right]}{\mathrm{Tr}\, e^{-\beta\hat{H}}} = \frac{\int_{-\infty}^{\infty} \frac{d\varphi}{2\pi} e^{-i\varphi q}\, \mathrm{Tr}\left(e^{i\varphi\hat{Q}_{20}} e^{-\beta\hat{H}}\right)}{\mathrm{Tr}\, e^{-\beta\hat{H}}} \tag{16}$$

together with the HS representation (1). In practice we use a discretized version of the Fourier transform, in which the interval $[-q_{\max}, q_{\max}]$ is divided into $2M+1$ intervals of length $\Delta q = 2q_{\max}/(2M+1)$. For a given configuration σ of the auxiliary fields, we have

$$\mathrm{Tr}\left[\delta(\hat{Q}_{20} - q)\hat{U}_\sigma\right] \approx \frac{1}{2q_{\max}} \sum_{k=-M}^{M} e^{-i\varphi_k q_m} \mathrm{Tr}(e^{i\varphi_k\hat{Q}_{20}}\hat{U}_\sigma) , \tag{17}$$

where $q_m = m\Delta q$ $(m = -M, \ldots, M)$ and $\varphi_k = \pi k/q_{\max}$ $(k = -M, \ldots, M)$ are quadrature points.

Since $\hat{Q}_{20}$ is a one-body operator we can use the group property to represent the Fock space operator $e^{i\varphi_k\hat{Q}_{20}}\hat{U}_\sigma$ in the single-particle space by an $N_s \times N_s$ matrix $e^{i\varphi_k\mathbf{Q}_{20}}\mathbf{U}_\sigma$, where $\mathbf{Q}_{20}$ is the axial quadrupole matrix in the single-particle space. In analogy with Eq. (3), the grand canonical traces in Eq. (17) are given by

$$\mathrm{Tr}(e^{i\varphi_k\hat{Q}_{20}}\hat{U}_\sigma) = \det\left(1 + e^{i\varphi_k\mathbf{Q}_{20}}\mathbf{U}_\sigma\right) . \tag{18}$$

6.1.2 Results

In Fig. 8 we show the distributions $P_T(q)$ calculated with SMMC for ^{154}Sm at three temperatures. At a low temperature (T = 0.1 MeV) close to the ground state, the distribution is skewed and it becomes Gaussian like at high temperatures (see, e.g., at T = 4 MeV). At T = 0.1 MeV the distribution is in good qualitative agreement with that of a prolate rigid rotor with an intrinsic quadrupole moment equal to the value found in the ground-state HFB solution (dashed line). The intermediate temperature of T = 1.2 MeV is close to the shape transition temperature in HFB from a deformed to spherical shape. This distribution is still skewed, suggesting that deformation effects survive above the mean-field shape transition temperature.

In Fig. 9 we show such distributions $P_T(q)$ for ^{148}Sm. In HFB, this nucleus is spherical in its ground-state solution, and we observe a Gaussian-like distribution already at the low temperature of T = 0.1 MeV. We conclude that $P_T(q)$ is a model-independent signature of deformation.

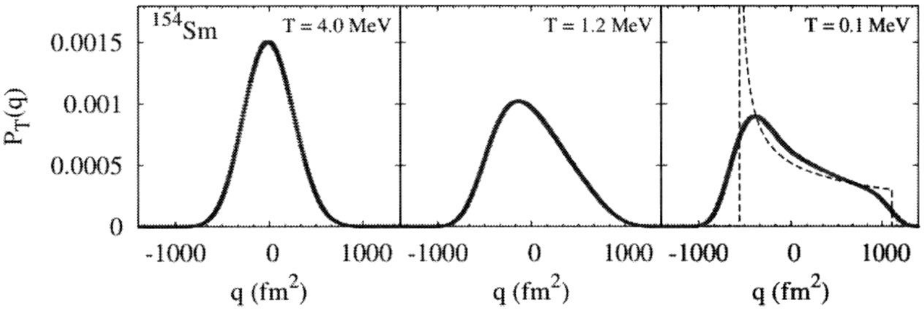

Figure 8. Axial quadrupole distributions $P_T(q)$ in the laboratory frame for ^{154}Sm at three temperatures.

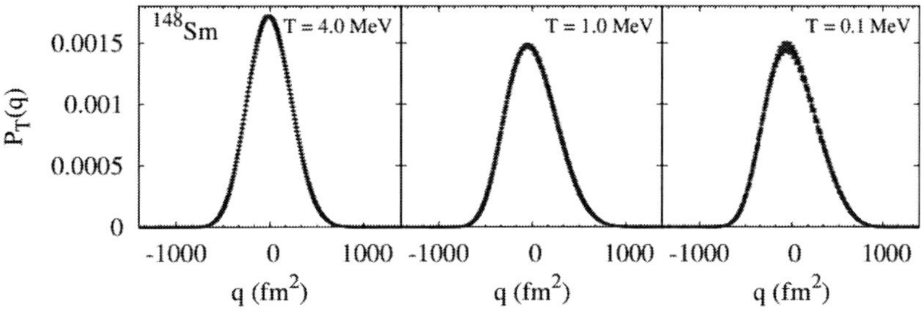

Figure 9. Axial quadrupole distributions $P_T(q)$ in the laboratory frame for ^{148}Sm at three temperatures.

6.2 Quadrupole distributions in the intrinsic frame

The quadrupole deformation is described by a second-rank tensor with spherical components $q_{2\mu}$ ($\mu = -2, \ldots, \mu$). In the intrinsic frame, characterized by a set of Euler angles Ω, the deformation parameters are denoted by β, γ. Information on β, γ can be obtained from the expectation values of rotationally invariant combinations of the quadrupole tensor, known as quadrupole invariants [27]. The distribution P_T is invariant under rotations and therefore in the intrinsic frame it depends only on β, γ. We expand $-\ln P_T(\beta, \gamma)$ in the quadrupole invariants. These invariant are unique up to fourth order and are given by $\beta^2, \beta^3\cos(3\gamma)$ and β^4. To this order we have

$$-\ln P_T(\beta, \gamma) = N + A\beta^2 - B\beta^3 \cos 3\gamma + C\beta^4 + \ldots \,, \tag{19}$$

where A, B, C are temperature-dependent parameters and N is a normalization constant. We can determine A, B, C from the expectation values of the above three invariants, which in turn can be calculated from the moments of $\hat{Q}_{20}$ in the laboratory frame [26]

$$\langle \hat{Q} \cdot \hat{Q} \rangle = 5\langle \hat{Q}_{20}^2 \rangle \; ; \; \langle (\hat{Q} \times \hat{Q}) \cdot \hat{Q} \rangle = -5\sqrt{\frac{7}{2}}\langle \hat{Q}_{20}^3 \rangle \; ; \; \langle (\hat{Q} \cdot \hat{Q})^2 \rangle = \frac{35}{3}\langle \hat{Q}_{20}^4 \rangle \,. \tag{20}$$

The moments of $\hat{Q}_{20}$ in Eq. (20) can be directly calculated from the SMMC distributions $P_T(q)$. The expressions for the expectation values of the invariants in terms of $P_T(\beta, \gamma)$ in Eq. (19) requires the metric $\prod_\mu dq_{2\mu}$, which in the intrinsic frame parameters is given by

$$\prod_\mu dq_{2\mu} = \frac{1}{2}\beta^4 |\sin 3\gamma|\, d\beta\, d\gamma\, d\Omega \tag{21}$$

with $\beta \geq 0$ and $0 \leq \gamma \leq \pi/3$. To test the accuracy of the expansion (19), we expressed the invariants in terms of $q_{2\mu}$ in the laboratory frame and integrated over $q_{2\mu}$ for all $\mu \neq 0$ to find the marginal distribution $P_T(q_{20})$. We found this marginal distribution to be essentially indistinguishable from the SMMC distribution $P_T(q)$.

In Fig. 10 we show $-\ln P_T(\beta, \gamma = 0)$ as a function of β for ^{154}Sm at three temperatures. We observe that these curves mimic the deformed to spherical shape transition in the Landau mean-field theory of shape transitions [28, 29], in which the quadrupole tensor $q_{2\mu}$ is the order parameter. However, the curves here are calculated within the CI shell model approach that preserves rotational symmetry.

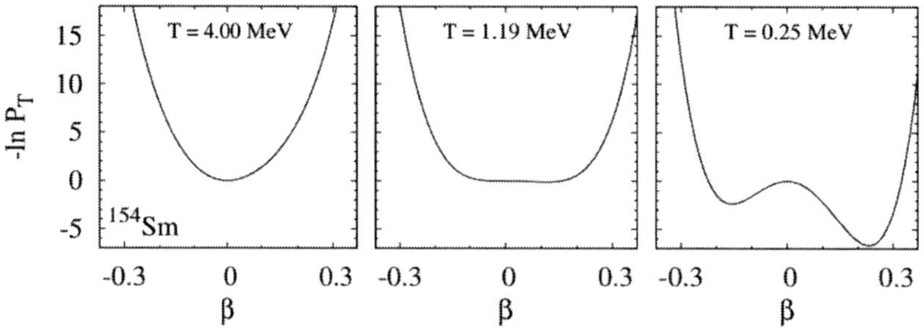

Figure 10. $-\ln P_T(\beta, \gamma = 0)$ in Eq. (19) as a function of axial deformation β for ^{154}Sm at three temperatures. The parameters A, B, C are determined from the expectation values of the three lowest-order quadrupole invariants.

The quadrupole distributions $P_T(\beta, \gamma)$ can in principle be converted to level densities $\rho(E_x, \beta, \gamma)$ as a function of intrinsic deformation by using the saddle-point approximation.

7 Conclusion

The SMMC method is a powerful method to calculate statistical and collective properties of nuclei in very large model spaces in the presence of correlations. We discussed recent applications of the method in heavy nuclei for identifying various types of collectivity and for calculating state densities. We also studied the SMMC distributions of the axial quadrupole operator in the laboratory frame and showed that they provide model-independent signatures of deformation. Finally we outlined a method to determine the quadrupole distributions in the intrinsic frame using the rotationally invariant framework of the CI shell model. This method can be used to determine level densities as a function of intrinsic deformation.

Acknowledgements

This work was supported in part by the DOE grant Nos. DE-FG-0291-ER-40608 and DE-FG02-00ER411132, by Grant-in-Aid for Scientific Research (C) No. 25400245 by the JSPS, Japan, and by the Turkish Science and Research Council (TÜBITAK) grant No. ARDEB-1001-110R004 and ARDEB-1001-112T973. The research presented here used resources of the National Energy Research Scientific Computing Center, which is supported by the Office of Science of the U.S. Department of Energy under Contract No. DE-AC02-05CH11231. It also used resources provided by the facilities of the Yale University Faculty of Arts and Sciences High Performance Computing Center.

References

[1] W. Hauser and H. Feshbach, Phys. Rev. **87**, 366 (1952).
[2] S. Hilaire and S. Goriely, Nucl. Phys. A **779**, 63 (2006).
[3] G.H. Lang, C.W. Johnson, S.E. Koonin, and W.E. Ormand, Phys. Rev. C **48**, 1518 (1993).
[4] Y. Alhassid, D.J. Dean, S.E. Koonin, G.H. Lang, and W.E. Ormand, Phys. Rev. Lett. **72**, (1994) 613.
[5] S.E. Koonin, D.J. Dean, and K. Langanke, Phys. Rep. **278**, 2 (1997).
[6] Y. Alhassid, Int. J. Mod. Phys. B **15**, 1447 (2001).
[7] J. Hubbard, Phys. Rev. Lett. **3**, 77 (1959); R.L. Stratonovich, Dokl. Akad. Nauk. S.S.S.R. **115**, 1097 (1957).
[8] W.E. Ormand, D.J. Dean, C.W. Johnson, G.H. Lang, and S.E. Koonin, Phys. Rev. C **49**, 1422 (1994).
[9] Y. Alhassid, L. Fang and H. Nakada, Phys. Rev. Lett. **101**, 082501 (2008).
[10] C. Özen, Y. Alhassid and H. Nakada, Phys. Rev. Lett. **110**, 042502 (2013).
[11] M. Dufour and A.P. Zuker, Phys. Rev. C **54**, 1641 (1996).
[12] C. Özen, Y. Alhassid and H. Nakada, Phys. Rev. C **91** (2015) 034329.
[13] Y. Alhassid, M. Bonett-Matiz, A. Mukherjee, H. Nakada, and C. Özen, Journal of Physics: Conference Series **580**, 012009 (2015).
[14] H. Nakada and Y. Alhassid, Phys. Rev. Lett. **79**, 2939 (1997).
[15] Y. Alhassid, S. Liu and H. Nakada, Phys. Rev. Lett. **83**, 4265 (1999).
[16] Y. Alhassid, G. F. Bertsch, S. Liu, and H. Nakada, Phys. Rev. Lett. **84**, 4313 (2000).
[17] Y. Alhassid, G.F. Bertsch, and L. Fang, Phys. Rev. C **68**, 044322 (2003).
[18] Y. Alhassid, S. Liu and H. Nakada, Phys. Rev. Lett. **99**, 162504 (2007).
[19] A. Bohr and B. R. Mottelson, *Nuclear Structure*, Vol. 1 (Benjamin, New York, 1969).
[20] Y. Alhassid Y, C. Özen, and H. Nakada, Nuclear Data Sheets **118**, 233 (2014).
[21] R. Capote, M. Herman, P. Obložinský, P.G. Young, S. Goriely, T. Belgya, A.V. Ignatyuk, A.J. Koning, S. Hilaire, V.A. Plujko, M. Avrigeanu, O. Bersillon, M.B. Chadwick, T. Fukahori, Zhigang Ge, Yinlu Han, S. Kailas, J. Kopecky, V.M. Maslov, G. Reffo, M. Sin, E.Sh. Soukhovitskii, and P. Talou, Nuclear Data Sheets, **110**, 3107 (2009).
[22] Y. Alhassid, G.F. Bertsch, C.N. Gilbreth, and H. Nakada, arXiv:1512.03773.
[23] S. Bjørnholm, A. Bohr and B.R. Mottelson, *Physics and Chemistry of Fission 1973* (IAEA, Austria, 1974), p. 367.
[24] T. Ericson, Adv. Phys. **9**, 425 (1960).
[25] G.F. Bertsch and J.M. Mehlhaff, arXiv:1511.01936.
[26] Y. Alhassid, C.N. Gilbreth, and G.F. Bertsch, Phys. Rev. Lett. **113**, 262503 (2014).
[27] K. Kumar, Phys. Rev. Lett. **28**, 249 (1972); D. Cline, Ann. Rev. Nucl. Part. Sci. **36**, 683 (1986).
[28] S. Levit and Y. Alhassid, Nucl. Phys. A **413**, 439 (1984).
[29] Y. Alhassid, S. Levit and J. Zingman, Phys. Rev. Lett. **57**, 539 (1986).

EPJ Web of Conferences **122**, 02002 (2016) DOI: 10.1051/epjconf/201612202002
*CNR*15*

Persistence of Vibrational Collectivity in Nuclear Level Densities

H. Nakada[1,a], E. Matsuyama[1], and C. Özen[2]

[1] *Department of Physics, Graduate School of Science, Chiba University, Inage, Chiba 263-8522, Japan*
[2] *Faculty of Engineering and Natural Sciences, Kadir Has University, Cibali 34083, Istanbul, Turkey*

Abstract. By applying the particle-number projection to the finite-temperature BCS theory to the state densities in the rare-earth nuclei and comparing its results to the SMMC ones, we investigate effects of the particle-number conservation on the collective enhancement factor for the state densities. Once we restore the particle-number conservation, the rapid decay of vibrational enhancement disappears. This suggests that the vibrational collectivity may survive up to higher energy and the relevant enhancement factor is insensitive to the excitation energy.

1 Introduction

Nuclear level densities are one of the key quantities in understanding compound nuclear reactions. Widely used conventional approaches to nuclear level densities are based on the mean-field (MF) theories including the Fermi-gas model [1], which are suitable for global description covering almost the whole region of the nuclear chart. However, it has been known that collective excitations are missed in the MF approaches, and so-called collective enhancement factor (CEF) has been phenomenologically introduced as complements. Although the CEF is considered to represent the vibrational and the rotational collective degrees of freedom, microscopic understanding of the CEF has not been sufficient.

The shell model gives us a microscopic framework suitable for studying nuclear level densities, because it is a full quantum theory that can describe both collective and non-collective states in the model space. Although the model space required for reasonable description of level densities is much larger than tractable in the conventional diagonalization methods, the shell model Monte Carlo (SMMC) methods [2] enables us to overcome this difficulty. Indeed the SMMC methods have successfully been applied to the state densities of medium- to heavy-mass nuclei [3–9]. With respect to the collective excitations, the SMMC methods have successfully described crossover from the vibrational to the rotational collectivity in the rare-earth nuclei with a single class of Hamiltonians [10, 11]. On this basis, the energy-dependence of the CEF has been investigated from microscopic standpoint, by comparing the SMMC densities to the corresponding finite-temperature (FT) Hartree-Fock-Bogolyubov (HFB) densities [10]. It has been found that the decays of the CEF seem to correlate to the phase transitions in HFB; the decay of the vibrational enhancement to the pairing transition and that of the rotational enhancement to the shape transition. However, in MF approaches such as the HFB, certain symmetries are broken due to phase transitions, although those symmetries should hold in actual nuclei. Therefore, restoration of the broken symmetry is important to investigate genuine effects of the collectivity.

[a] e-mail: nakada@faculty.chiba-u.jp

We here focus on the breaking and restoration of the global gauge symmetry, which is connected to the particle-number conservation. The FT-MF approaches are formulated in the grand-canonical ensembles, in which the thermal average is taken in the entire Fock space. In contrast, the nuclear level densities are argued for a specific nucleus with fixed proton and neutron numbers. Therefore the canonical ensembles are more appropriate for describing level densities, as adopted in the SMMC calculations. Reduction from the grand-canonical ensembles to the canonical ensembles in the FT-MF theories is inspected recently [12]. However, in finite quantum systems such as nuclei, the phase transitions observed in the MF regime are washed out in truth [3], and full restoration is often desired to investigate influence of the existing symmetry. In the present study, we apply the particle-number projection [13–15] to the FT Bardeen-Cooper-Schrieffer (BCS) theory, and investigate effects of the symmetry restoration.

2 Particle-number projection in FT-BCS theory

2.1 Thermodynamical quantities

For canonical ensembles with a given Hamiltonian H, the statistical density operator at temperature $T = 1/\beta$ is defined by

$$w^{(\mathrm{C})} = \frac{e^{-\beta H}}{\mathrm{Tr}_{\mathrm{C}}(e^{-\beta H})}, \tag{1}$$

by which the thermal expectation value of an observable O is calculated as

$$\langle O\rangle_{\beta}^{(\mathrm{C})} = \mathrm{Tr}_{\mathrm{C}}(O w^{(\mathrm{C})}). \tag{2}$$

Here Tr_{C} denotes the canonical trace.

The statistical density operator for grand-canonical ensembles is defined by

$$w^{(\mathrm{GC})} = \frac{e^{-\beta(H-\sum_\tau \mu_\tau N_\tau)}}{\mathrm{Tr}_{\mathrm{GC}}(e^{-\beta(H-\sum_\tau \mu_\tau N_\tau)})}, \tag{3}$$

where $\mathrm{Tr}_{\mathrm{GC}}$ denotes the grand-canonical trace, N_τ is the number operator and μ_τ represents the chemical potential, with $\tau\,(= p, n)$ distinguishing protons and neutrons. The thermal expectation value of an observable is obtained by

$$\langle O\rangle_{\beta}^{(\mathrm{GC})} = \mathrm{Tr}_{\mathrm{GC}}(O w^{(\mathrm{GC})}). \tag{4}$$

In the FT-MF theories, the Hamiltonian in $w^{(\mathrm{GC})}$ is approximated by the MF Hamiltonian H_0, giving

$$w_{\mathrm{MF}}^{(\mathrm{GC})} = \frac{e^{-\beta(H_0-\sum_\tau \mu_\tau N_\tau)}}{\mathrm{Tr}_{\mathrm{GC}}(e^{-\beta(H_0-\sum_\tau \mu_\tau N_\tau)})}, \tag{5}$$

The MF Hamiltonian has the form

$$H_0 = E_0 + \sum_k \varepsilon_k \alpha_k^\dagger \alpha_k, \tag{6}$$

where E_0 is a constant, k stands for a quasiparticle (q.p.) state and $\alpha_k^\dagger$ (α_k) the corresponding creation (annihilation) operator. The thermal expectation value in the FT-MF approximations is

$$\langle O\rangle_{\beta,\mathrm{MF}}^{(\mathrm{GC})} = \mathrm{Tr}_{\mathrm{GC}}(O w_{\mathrm{MF}}^{(\mathrm{GC})}). \tag{7}$$

In any case above, the energy expectation value is calculated by $E(\beta) = \langle H \rangle_\beta = \mathrm{Tr}(Hw)$, and the heat capacity by $C = dE/dT = -\beta^{-2} dE/d\beta$. The entropy can be represented as

$$S = -\mathrm{Tr}(w \ln w) . \tag{8}$$

Within the FT-MF framework, this is reduced to

$$S_{\mathrm{MF}}^{(\mathrm{GC})} = -\sum_k [f_k \ln f_k + (1 - f_k) \ln(1 - f_k)] , \tag{9}$$

where (assuming the q.p. state k belongs to the particle type τ)

$$f_k = \langle \alpha_k^\dagger \alpha_k \rangle_{\beta,\mathrm{MF}}^{(\mathrm{GC})} = \frac{1}{e^{\beta(\varepsilon_k - \mu_\tau)} + 1} . \tag{10}$$

At thermal equilibrium, the free energy $F = E - TS$ takes minimum for a given temperature T in the canonical ensembles, while the grand potential $\Omega = F - \sum_\tau \mu_\tau N_\tau$ does in the grand-canonical ensembles. H_0 in the FT-MF theories is determined so as to minimize $\Omega_{\mathrm{MF}}^{(\mathrm{GC})}$.

2.2 Particle-number projection

In Ref. [12] an approximate method of reducing the grand-canonical ensemble to the canonical ensemble is proposed and applied to the FT-MF theories. To restore the particle-number conservation exactly, it is desired to apply the projection operator on the proton and neutron numbers (n_p, n_n),

$$P_{(n_p,n_n)} = \prod_{\tau=p,n} \int_0^{2\pi} \frac{d\varphi_\tau}{2\pi} e^{i\varphi_\tau(N_\tau - n_\tau)} . \tag{11}$$

We can then handle $w^{(\mathrm{C})}$ and $\langle O \rangle_\beta^{(\mathrm{C})}$ in the grand-canonical ensembles,

$$w^{(\mathrm{C})} = \frac{e^{-\beta(H - \sum_\tau \mu_\tau n_\tau)}}{\mathrm{Tr_{GC}}(P_{(n_p,n_n)} e^{-\beta(H - \sum_\tau \mu_\tau N_\tau)} P_{(n_p,n_n)})} = \frac{e^{-\beta H}}{\mathrm{Tr_{GC}}(P_{(n_p,n_n)} e^{-\beta H} P_{(n_p,n_n)})} , \tag{12}$$

$$\langle O \rangle_\beta^{(\mathrm{C})} = \mathrm{Tr_{GC}}(P_{(n_p,n_n)} O w^{(\mathrm{C})} P_{(n_p,n_n)}) . \tag{13}$$

In practical calculations the integral in Eq. (11) is implemented by a finite sum of the φ variables. Particularly when we work in a finite model space as below, the integral operator is identical to the operator represented by the discrete sum,

$$\int_0^{2\pi} \frac{d\varphi}{2\pi} e^{i\varphi(N-n)} = \frac{1}{D+1} \sum_{m=0}^{D} e^{i\varphi_m(N-n)} ; \quad \varphi_m = \frac{2m\pi}{D+1} , \tag{14}$$

where D is the number of the single-particle states in the model space.

In the FT-MF theories, H_0 is determined by variation as mentioned above. In the variation-before-projection (VBP) method, H_0 is determined by the variation of $\Omega_{\mathrm{MF}}^{(\mathrm{GC})}$ in the grand-canonical ensemble, and the projection is applied when computing physical quantities ($\langle O \rangle_\beta$). It should be noted, however, the VBP method violates the thermodynamical consistency [12]. Although the variation-after-projection (VAP) is a more desirable method, in which H_0 should be determined by minimizing the free energy $F_{\mathrm{MF}}^{(\mathrm{C})}$ in the canonical ensemble, the VAP has not been well formulated for the following reasons.

The nuclear Hamiltonian satisfies $[H, N_\tau] = 0$, and therefore $[H, P_{(n_p,n_n)}] = 0$. Suppose that this also holds for the FT-MF Hamiltonian H_0; $[H_0, N_\tau] = 0$. It is then straightforward to apply $P_{(n_p,n_n)}$ to the FT-MF theory, by defining the canonical density operator by

$$w_{\mathrm{MF}}^{(\mathrm{C})} = \frac{e^{-\beta(H_0-\sum_\tau \mu_\tau n_\tau)}}{\mathrm{Tr_{GC}}(P_{(n_p,n_n)} e^{-\beta(H_0-\sum_\tau \mu_\tau N_\tau)} P_{(n_p,n_n)})} = \frac{e^{-\beta H_0}}{\mathrm{Tr_{GC}}(P_{(n_p,n_n)} e^{-\beta H_0} P_{(n_p,n_n)})} . \tag{15}$$

The VAP is formulated by minimizing the free energy $F = E - TS$, with the entropy

$$\begin{aligned} S_{\mathrm{MF}}^{(\mathrm{C})} &= -\mathrm{Tr_{GC}}([P_{(n_p,n_n)} w_{\mathrm{MF}}^{(\mathrm{C})} P_{(n_p,n_n)}] \ln [P_{(n_p,n_n)} w_{\mathrm{MF}}^{(\mathrm{C})} P_{(n_p,n_n)}]) \\ &= \beta \frac{\mathrm{Tr_{GC}}(P_{(n_p,n_n)} H_0 e^{-\beta H_0} P_{(n_p,n_n)})}{\mathrm{Tr_{GC}}(P_{(n_p,n_n)} e^{-\beta H_0} P_{(n_p,n_n)})} + \ln \mathrm{Tr_{GC}}(P_{(n_p,n_n)} e^{-\beta H_0} P_{(n_p,n_n)}) . \end{aligned} \tag{16}$$

However, the FT-MF Hamiltonian does not satisfy $[H_0, N_\tau] = 0$ when the pairing correlation exists. Without this commutability, it is prohibitively difficult to use $S_{\mathrm{MF}}^{(\mathrm{C})}$ [16], as the equality of the second line of Eq. (16) is lost. Whereas an approximate VAP method has been proposed by respecting the Pierls inequality [16], assuming an approximate entropy

$$S_{\mathrm{MF}}^{\prime(\mathrm{C})} = \beta \frac{\mathrm{Tr_{GC}}(P_{(n_p,n_n)} H_0 e^{-\beta H_0} P_{(n_p,n_n)})}{\mathrm{Tr_{GC}}(P_{(n_p,n_n)} e^{-\beta H_0} P_{(n_p,n_n)})} + \ln \mathrm{Tr_{GC}}(P_{(n_p,n_n)} e^{-\beta H_0} P_{(n_p,n_n)}) , \tag{17}$$

it again violates the thermodynamical consistency, leading to negative entropy at $T \approx 0$.

Despite the problems mentioned above, the particle-number projection is still useful in investigating influence of the particle-number conservation qualitatively. As mentioned above, the sharp transitions in the FT-MF theories do not remain in actual nuclei, closely connected to restoration of the broken symmetry. To investigate behavior around the critical temperature, the quantum-number projection provides us with a indispensable tool. In the present study, we shall apply the approximate VAP, which we shall abbreviate as VAP′, as well as the VBP methods, keeping in mind their defects and limitations.

2.3 State density in the saddle-point approximation

The state density ρ is quite relevant to the thermodynamics. It can be represented by the inverse Laplace transform of the partition function. By applying the saddle-point approximation, ρ is approximately expressed in terms of the thermodynamical quantities. From the canonical partition function we reach [3]

$$\rho^{(\mathrm{C})} \simeq \frac{e^{S^{(\mathrm{C})}}}{\sqrt{2\pi\beta^{-2}C^{(\mathrm{C})}}} , \tag{18}$$

while from the grand-canonical partition function

$$\rho^{(\mathrm{GC})} \simeq \frac{e^{S^{(\mathrm{GC})}}}{\sqrt{(2\pi)^3 (\Delta N_p)^2 (\Delta N_n)^2 \beta^{-2} C^{(\mathrm{GC})}}} , \tag{19}$$

with $(\Delta N_\tau)^2 = \langle (N_\tau - n_\tau)^2 \rangle_\beta^{(\mathrm{GC})}$.

It is recently argued that there could be an additional contribution to the denominator of Eq. (19) in the FT-MF framework, due to the proton- and neutron-number violation [12], because of H_0 itself depends on μ_τ. In the present study, we do not take it into account for the sake of simplicity.

We compute the heat capacity C by numerical differentiation of $E(\beta) = \langle H \rangle_\beta$, either in the canonical or the grand-canonical ensemble. The entropy $S^{(\mathrm{GC})}$ is given by Eq. (9) in the usual FT-MF theories in the grand-canonical ensembles. In the SMMC, $S^{(\mathrm{C})}$ is calculated by integrating $E(\beta)$, using the relation of the thermodynamical consistency

$$S^{(\mathrm{C})} = \beta E + \ln Z^{(\mathrm{C})}\,; \quad \ln Z^{(\mathrm{C})}(\beta) = \ln Z^{(\mathrm{C})}(0) - \int_0^\beta d\beta'\, E(\beta')\,. \tag{20}$$

In the particle-number projected MF approaches, the incommutability of H_0 and N_τ causes a problem again. If $[H_0, N_\tau] = 0$, Eq. (16) provides $S^{(\mathrm{C})}_{\mathrm{MF}}$. If $[H_0, N_\tau] \neq 0$, Eq. (20) loses its ground. Even in that case, we again apply $S'^{(\mathrm{C})}_{\mathrm{MF}}$ in Eq. (17) to $\rho^{(\mathrm{C})}_{\mathrm{MF}}$ as an approximation.

The state density will be presented as a function of the excitation energy E_x, which is obtained by $E_x = E(\beta) - E_{g.s.}$ with $E_{g.s.} = \lim_{\beta\to\infty} E(\beta)$.

2.4 Model space and effective Hamiltonian

In this study we take several rare-earth nuclei as examples. For them the SMMC state densities are in good agreement with the experimental data [8, 10, 11]. The model space in the SMMC calculations is the full $Z = 50 - 82$ shell plus $1f_{7/2}$ for protons, and the full $N = 82 - 126$ shell plus $0h_{11/2}$, $1g_{9/2}$ for neutrons. The Hamiltonian is comprised of the monopole pairing and the quadrupole, octupole, hexadecupole interactions, by which dominant collective features are well taken into account. See Ref. [10] for details of the Hamiltonian. This SMMC setup is a good starting point to investigate CEF.

In Ref. [10], the CEF was argued by comparing the state densities obtained in the FT-HFB framework to those obtained from the SMMC calculations. The model space and the Hamiltonian in those FT-HFB calculations is identical to those in the SMMC. For most nuclei the FT-HFB approaches gives three phase transitions; the proton and neutron pairing transitions which occur at $E_x \lesssim 5\,\mathrm{MeV}$, and the deformed-to-spherical transition which takes place at $E_x \gtrsim 8\,\mathrm{MeV}$. To the best of our knowledge, there are no FT-HFB plus particle-number projection computer codes available for the shell model Hamiltonian. For this reason, we introduce the following effective Hamiltonian, to which the particle-number projection is applicable on top of the FT-BCS:

$$H' = \sum_k \varepsilon'_k a^\dagger_k a_k - \sum_{\tau=p,n} g'_\tau P^\dagger_\tau P_\tau\,; \quad P^\dagger_\tau = \sum_{k>0(\in\tau)} a^\dagger_k a^\dagger_{\bar{k}}\,. \tag{21}$$

Here ε'_k is fixed by the HF results at $T = 0$. Although this ignores energy-dependence of the nuclear shape, it is not significant at energies below the critical temperature of the shape transition. g'_τ is adjusted so as to reproduce the FT-HFB state densities. In practice, it is enhanced from the pairing strength g_τ in the original shell model Hamiltonian as $g'_\tau = c_\tau g_\tau$, with $c_p = 1.4$ and $c_n = 1.6$. The enhancement factor c_τ should partially represent contribution of the multipole interactions to the pairing. We implement the FT-BCS calculations with H', and investigate influence of the restoration of the particle-number conservation by applying the VBP or the VAP′ method.

3 Collective enhancement factor for state densities

It has been known that the nuclear level densities calculated from MF theories underestimate the experimental data. This originates from the collective excitations which is not well described within the MF theories. To correct the influence of the collective excitations, the collective enhancement factor (CEF) K is customarily introduced [1],

$$\rho(E_x) = K(E_x)\rho_{\mathrm{MF}}(E_x)\,. \tag{22}$$

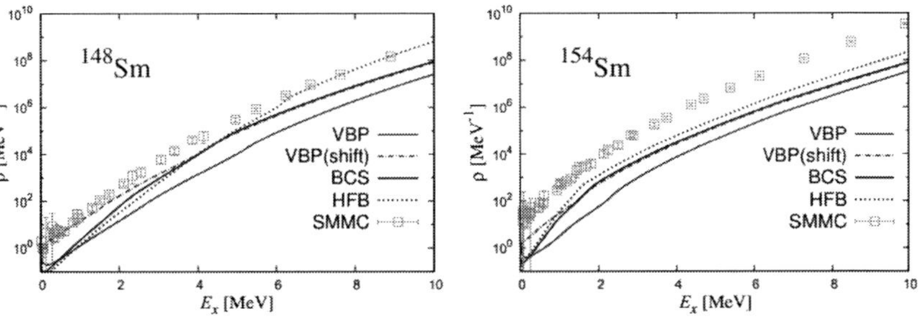

Figure 1. State densities $\rho(E_x)$ for ^{148}Sm and ^{154}Sm. See text for details.

The MF density $\rho_{\mathrm{MF}}(E_x)$ is calculated by the FT-MF theory or by the combinatorial methods of counting the q.p. excitations [17]. The CEF contains effects of the collective vibration and the rotation, and may be factorized as

$$K(E_x) = K_{\mathrm{vib}}(E_x)\, K_{\mathrm{rot}}(E_x)\,. \tag{23}$$

For stable nuclei, experimental data on the state densities are obtained from direct counting of levels at low E_x and from the neutron resonance around the neutron separation energy. The SMMC densities $\rho^{(\mathrm{C})}_{\mathrm{SMMC}}(E_x)$, which are available in the Nd-Sm region [10, 11, 18], are in good agreement with them. Since $\rho^{(\mathrm{C})}_{\mathrm{SMMC}}(E_x)$ gives state densities in wider energy range than the data, we shall use it as $\rho(E_x)$ in the lhs of Eq. (22).

It has been suggested in Ref. [10], by defining the CEF by $K(E_x) = \rho^{(\mathrm{C})}_{\mathrm{SMMC}}(E_x)/\rho^{(\mathrm{GC})}_{\mathrm{HFB}}(E_x)$, that the vibrational enhancement decays rapidly, in well correlation to the pairing phase transitions in the FT-HFB theory. Whereas it is difficult to separate the vibrational and rotational effects by comparing the MF density to the SMMC density, violation and restoration of the particle-number conservation should mainly be connected to the vibrational enhancement. At $E_{\mathrm{pair}} < E_x < 10\,\mathrm{MeV}$ in the rare-earth nuclei, where E_{pair} is the pairing energy, we have $K_{\mathrm{vib}} \lesssim 3$ from an empirical formula [1].

4 Numerical results of particle-number projection

4.1 State densities

We take ^{148}Sm and ^{154}Sm as typical examples. ^{148}Sm is spherical with strong pairing correlation at $T = 0$, while ^{154}Sm has quadrupole deformation with weak pairing [10]. In Fig. 1 several types of MF densities are compared with one another and with the SMMC densities. Among the MF densities, $\rho^{(\mathrm{GC})}_{\mathrm{HFB}}$ (blue dotted lines) is the FT-HFB results with the same Hamiltonian as in the SMMC. In the FT-HFB calculations, the pairing transition occur at $E_x \approx 6\,\mathrm{MeV}$ (1 MeV) for protons and $E_x \approx 4\,\mathrm{MeV}$ (2 MeV) for neutrons in ^{148}Sm (^{154}Sm). $\rho^{(\mathrm{GC})}_{\mathrm{BCS}}$ (blue solid lines) is the results of the FT-BCS calculation with H' in Sec. 2.4. We observe $\rho^{(\mathrm{GC})}_{\mathrm{BCS}} \approx \rho^{(\mathrm{GC})}_{\mathrm{HFB}}$ to reasonable precision, at $E_x \lesssim 6\,\mathrm{MeV}$ ($E_x \lesssim 10\,\mathrm{MeV}$) for ^{148}Sm (^{154}Sm). On this basis we use the FT-BCS results in place of the FT-HFB ones to investigate influence of the particle-number conservation. Red solid lines in Fig. 1 represent $\rho^{(\mathrm{C})}_{\mathrm{BCS}}$, which are obtained by applying the VBP when calculating $E(\beta)$ in the FT-BCS. We do not show the VAP′ results because they are very close to the VBP results.

In ^{148}Sm, $\rho^{(GC)}_{\mathrm{HFB}}$ agrees well with $\rho^{(C)}_{\mathrm{SMMC}}$ above the pairing transitions, indicating no rotational enhancement in this spherical nucleus. In the deformed nucleus ^{154}Sm, most of the HFB levels should correspond to the lowest states of the rotational bands, not including other members of each band. Thereby $\rho^{(GC)}_{\mathrm{HFB}}$ underestimates the state densities even above the pairing transitions, and the discrepancy stands for the rotational enhancement.

A notable effect of the particle-number projection is the suppression of the kink in the state densities at the pairing transition. This suggests much weaker energy-dependence of the CEF than reported in Ref. [10], if we properly take into account the conservation laws in the MF theories. We shall return to this point in the subsequent subsection.

If absolute values of $\rho^{(C)}_{\mathrm{BCS}}(E_x)$ is compared to $\rho^{(GC)}_{\mathrm{BCS}}(E_x)$, the particle-number conservation seems to reduce the state densities at all excitation energies under observation. This reduction might be attributed to elimination of the spurious states that have wrong particle numbers. In particular, the one-q.p. states give dominant contribution to the lowest-lying densities in the grand-canonical ensembles, but should not be present in practice. However, we should also keep in mind the negative entropy ($S'^{(C)}_{\mathrm{MF}} \approx -2$ at $T \approx 0$), which may reduce the densities fictitiously. This problem obscures effect of the particle-number conservation on the absolute values of the state densities. It is noted as well that the MF densities have been shown to agree with the SMMC densities fairly well after appropriate corrections [12].

The particle-number projection lowers $E_{g.s.}$, by which the projected densities have shifted toward rhs in Fig. 1. To view influence of this energy shift, we also plot $\rho^{(C)}_{\mathrm{BCS}}$ by the VBP but as a function of $E'_x = E^{(C)}_{\mathrm{BCS}}(\beta) - E^{(GC)}_{\mathrm{BCS}}(\infty)$ (red dot-dashed line), namely by taking $E_{g.s.}$ to be the value without the projection. Quite interestingly, we have $\rho^{(C)}_{\mathrm{BCS}}(E'_x) \approx \rho^{(GC)}_{\mathrm{BCS}}(E_x)$ above the pairing transitions. Below the transitions, $\rho^{(C)}_{\mathrm{BCS}}(E'_x)$ is greater than $\rho^{(GC)}_{\mathrm{BCS}}(E_x)$, in complete opposite to $\rho^{(C)}_{\mathrm{BCS}}(E_x)$. $\rho^{(C)}_{\mathrm{BCS}}(E'_x)$ is close to $\rho^{(C)}_{\mathrm{SMMC}}$ below the transitions in ^{148}Sm. Thus the particle-number projection gives improvement on the state densities, if the ground-state energy is properly shifted. These results seem consistent with those in Ref. [12].

4.2 Collective enhancement factors

The CEFs defined by $K(E_x) = \rho^{(C)}_{\mathrm{SMMC}}(E_x)/\rho_{\mathrm{MF}}(E_x)$ are shown in Fig. 2, for a number of even-even nuclei in the Nd-Sm region. To confirm effects of the particle-number conservation, we take $\rho^{(GC)}_{\mathrm{BCS}}(E_x)$ and $\rho^{(C)}_{\mathrm{BCS}}(E_x)$ (with the VBP) for the MF densities and compare the resultant CEFs.

As is expected from the state densities, the energy-dependence of the CEFs comes weaker after the particle-number projection. This indicates that the vibrational collectivity around the ground state remains up to high energy region. $\rho^{(C)}_{\mathrm{BCS}}(E_x)$ significantly lower than $\rho^{(C)}_{\mathrm{SMMC}}(E_x)$ gives high CEF. However, the problems for the absolute values of $\rho^{(C)}_{\mathrm{BCS}}(E_x)$ is inherited for the absolute values of $K(E_x)$, which prohibit us from making quantitative predictions based on the present calculations. The present results suggest that the CEF evaluated around the ground state will be important, not varying much up to relatively high energy.

The combinatorial methods based on the MF state at $T = 0$ are sometimes applied to global description of the level densities. The combinatorial methods give a reasonable basis in evaluating the number of low-lying states. They could be suitable for estimate of the CEF around the ground state. Although they may miss energy-dependence of the nuclear mean fields as typified by the phase transitions, effects of the energy-dependence seem milder than predicted by the FT-MF theories. This may account for a part of the reasons why the combinatorial methods work moderately well.

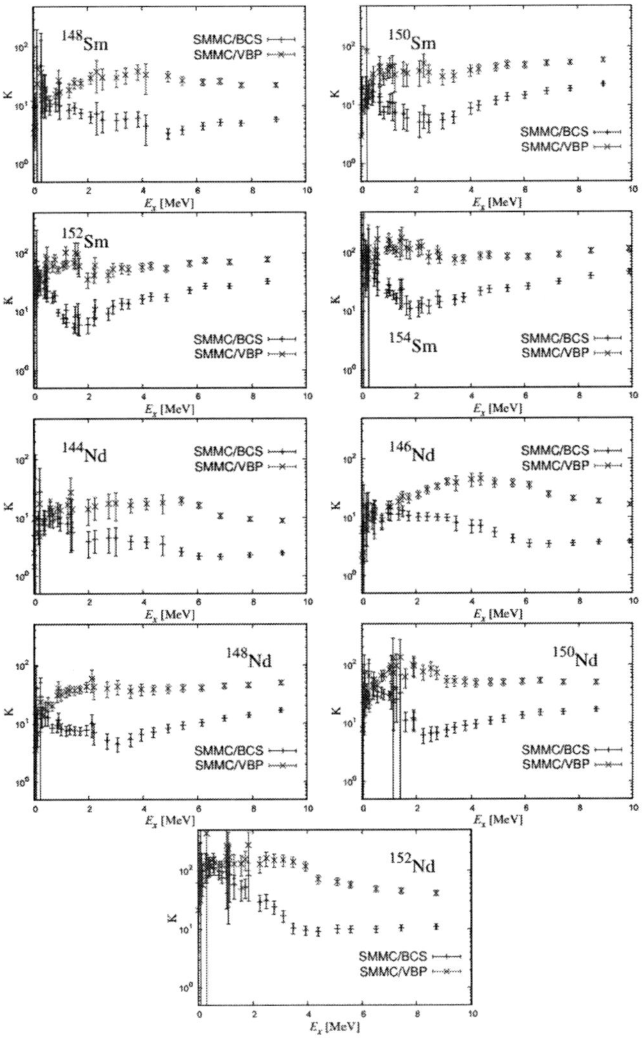

Figure 2. CEFs for Nd-Sm nuclei. Blue plus symbols and red crosses represent $\rho^{(C)}_{\rm SMMC}(E_x)/\rho^{(GC)}_{\rm BCS}(E_x)$ and $\rho^{(C)}_{\rm SMMC}(E_x)/\rho^{(C)}_{\rm BCS}(E_x)$ (with the VBP), respectively.

5 Summary and outlook

We have investigated effects of the particle-number conservation on the nuclear state densities, particularly on their collective enhancement, by applying the projection to the finite-temperature (FT) mean-field (MF) theories. Although there remain problems in consistency in the projected FT-MF theories, qualitative effects can be observed from the projected results. It is found that the energy-dependence of the vibrational enhancement comes significantly weaker than the usual FT-MF results, almost washing out the decaying features of the collective enhancement factor (CEF) at the pairing transitions. This indicates that the vibrational collectivity around the ground state persists up to relatively high energy, in contrast to the unprojected FT-MF results.

Energy-dependence of the pairing correlation, which is manifested by the pairing transitions in the FT-MF theories, substantially affects the spin distribution of the level densities, as supported by the SMMC studies [19]. Even if the pairing transitions do not influence the state densities significantly after the particle-number conservation is restored, it is still of interest how they affect the spin distribution. Simultaneous implementation of the angular-momentum projection and the particle-number projection on the FT-MF theories [15] will enable us to investigate this point, though it is beyond scope of this paper and we leave it as a future subject.

The loss of the thermodynamical consistency has been a serious obstacle against quantitative arguments. It seems impractical to restore the consistency in the projected FT-MF theories. Still, there is a room of improvement. In the present study, we have adopted a simple effective Hamiltonian which is fitted to the FT-HFB state densities at low energy. This restricts the energy region of investigation. By applying the projection to the FT-HFB observables, it will be possible to extend the energy region and to investigate effects of the rotational symmetry in a similar manner, at least in a qualitative respect.

As mentioned above, the present results indicates significance of evaluating the CEF around the ground state. The combinatorial methods seem suitable for this purpose, and may help quantitative arguments, complementary to the FT-MF approaches, although they are unable to incorporate the energy-dependence of the pairing correlation appropriately.

Acknowledgment

One of the authors (H. N.) is grateful to G. F. Bertsch and Y. Alhassid for discussions. This work was supported by Grant-in-Aid for Scientific Research (C) No. 25400245 by the JSPS, Japan, and by the Turkish Science and Research Council (TÜBİTAK) grant No. ARDEB-1001-110R004 and ARDEB-1001-112T973.

References

[1] R. Capote *et al.*, Nucl. Dat. Sheets **110**, 3107 (2009).
[2] G. H. Lang, C. W. Johnson, S. E. Koonin and W. E. Ormand, Phys. Rev. C **48**, 1518 (1993).
[3] H. Nakada and Y. Alhassid, Phys. Rev. Lett. **79**, 2939 (1997).
[4] W. E. Ormand, Phys. Rev. C **56**, R1678 (1997).
[5] H. Nakada and Y. Alhassid, Phys. Lett. B **436**, 231 (1998).
[6] Y. Alhassid, S. Liu, and H. Nakada, Phys. Rev. Lett. **83**, 4265 (1999).
[7] C. Özen, K. Langanke, G. Martinez-Pinedo and D. J. Dean, Phys. Rev. C **75**, 064307 (2007).
[8] Y. Alhassid, L. Fang and H. Nakada, Phys. Rev. Lett. **101**, 082501 (2008).
[9] H. Nakada and Y. Alhassid, Phys. Rev. C **78**, 051304(R) (2008).

[10] C. Özen, Y. Alhassid and H. Nakada, Phys. Rev. Lett. **110**, 042502 (2013).
[11] C. Özen, Y. Alhassid and H. Nakada, Phys. Rev. C **91**, 034329 (2015).
[12] Y. Alhassid, G. F. Bertsch, C. N. Gilbreth and H. Nakada, arXiv:1512.03773.
[13] C. Esebbag and J. L. Egido, Nucl. Phys. A **552**, 205 (1993).
[14] R. Rossignoli and P. Ring, Ann. Phys. (N.Y.) **235**, 350 (1994).
[15] K. Tanabe and H. Nakada, Phys. Rev. C **71**, 024314 (2005).
[16] H. Nakada and K. Tanabe, Phys. Rev. C **74**, 061301(R) (2006).
[17] A. J. Koning, S. Hilaire and S. Goriely, Nucl. Phys. A **810**, 13 (2008).
[18] C. Özen, Y. Alhassid and H. Nakada, EPJ Web Conf. **69**, 00011 (2014).
[19] Y. Alhassid, S. Liu, and H. Nakada, Phys. Rev. Lett. **99**, 162504 (2007).

EPJ Web of Conferences **122**, 02003 (2016) DOI: 10.1051/epjconf/201612202003
*CNR*15*

Stochastic estimation of level density in nuclear shell-model calculations

Noritaka Shimizu[1,a], Yutaka Utsuno[2,1], Yasunori Futamura[3], Tetsuya Sakurai[3,4], Takahiro Mizusaki[5], and Takaharu Otsuka[6,1,7,8]

[1] *Center for Nuclear Study, the University of Tokyo, Hongo Tokyo 113-0033, Japan*
[2] *Advanced Science Research Center, Japan Atomic Energy Agency, Tokai, Ibaraki 319-1195, Japan*
[3] *Faculty of Engineering, Information and Systems, University of Tsukuba, Tsukuba, 305-8573, Japan*
[4] *CREST, Japan Science and Technology Agency, Kawaguchi, 332-0012, Japan*
[5] *Institute of Natural Sciences, Senshu University, Tokyo 101-8425, Japan*
[6] *Department of Physics, the University of Tokyo, Hongo Tokyo 113-0033, Japan*
[7] *National Superconducting Cyclotron Laboratory, Michigan State University, East Lansing, MI 48824, USA*
[8] *Instituut voor Kern- en Stralingsfysica, Katholieke Universiteit Leuven, B-3001 Leuven, Belgium*

Abstract. An estimation method of the nuclear level density stochastically based on nuclear shell-model calculations is introduced. In order to count the number of the eigenvalues of the shell-model Hamiltonian matrix, we perform the contour integral of the matrix element of a resolvent. The shifted block Krylov subspace method enables us its efficient computation. Utilizing this method, the contamination of center-of-mass motion is clearly removed.

1 Introduction

Nuclear level density is a key ingredient of the Hauser-Feshbach calculations [1] to understand neutron-capture processes and compound nucleus microscopically. Theoretically, some phenomenological formulas for level density are well known such as the backshifted Fermi gas model [2]. However, microscopic calculation of the nuclear level density is prerequisite for the precise prediction of r-process nuclei and for further microscopic understanding of the statistical properties of nuclei.

Nuclear shell-model calculation is one of the most powerful theoretical frameworks to evaluate the level density. However, the rapid growth in the dimension of the shell-model Hamiltonian matrix remains an inevitable obstacle. An efficient numerical method to solve the eigenvalue problem of the huge sparse matrix and to estimate its eigenvalue distribution has been demanded. Much effort has been paid to develop efficient methods to estimate nuclear level densities based on shell model, *e.g.* the shell model Monte Carlo (SMMC) [3], the moment-based methods [4, 5], and the Lanczos-based method [6]. In this proceedings, we review stochastic estimation of the eigenvalue distribution based on shifted Krylov-subspace method [7] and its application to nuclear shell model calculations [8]. We also discuss its validity and feasibility.

[a] e-mail: shimizu@cns.s.u-tokyo.ac.jp

2 Nuclear shell model and its Hamiltonian matrix

The Hamiltonian in nuclear shell-model calculations is written in second quantization as

$$H = \sum_i t_i c_i^\dagger c_i + \sum_{i<j,k<l} v_{ijkl} c_i^\dagger c_j^\dagger c_l c_k, \tag{1}$$

where $c_i^\dagger$ denotes the creation operator of the single particle state i.

The many-body wave function is described as a linear combination of a vast number of Slater determinants, which are the antisymmetrized products of the single-particle wave functions. The simplest representation for the Slater determinant is called "M-scheme" basis state:

$$|M_i\rangle = \prod_{\alpha=1}^{A} c^\dagger_{M_i^{(\alpha)}} |-\rangle \tag{2}$$

where A and $|-\rangle$ are the number of active nucleons and an inert core, respectively. $M_i^{(\alpha)}$ denotes the single-particle state which is occupied by the α-th active nucleon. Thus, $M_i = \{M_i^{(1)}, M_i^{(2)}, ..., M_i^{(A)}\}$ is called "configuration" and means that 1st, 2nd, ..., A-th particles occupy $M_i^{(1)}, M_i^{(2)}, ..., M_i^{(A)}$ single-particle states, respectively.

Since the model space is fully spanned by the M-scheme basis states, the shell-model wave function is expressed as a linear combination of them:

$$|\Psi\rangle = \sum_{i=1}^{D} v_i |M_i\rangle, \tag{3}$$

where D is the number of the M-scheme basis states. The Schrödinger equation is transformed into the eigenvalue problem of the Hamiltonian matrix H_{ij},

$$\sum_{j=1}^{D} H_{ij} v_j = E v_i, \tag{4}$$

where the matrix $H_{ij} = \langle M_i | H | M_j \rangle$ is real symmetric.

There are two symmetries which the Hamiltonian matrix in the M-scheme basis explicitly has: invariant under rotational symmetry around z-axis and parity inversion. The operators concerning these symmetry are referred by J_z and Π and the corresponding quantum numbers are M and π. Owing to these symmetries, the Hamiltonian matrix is a block-diagonal form. We need to solve only a block matrix specified by the M and π due to the degeneracy. The largest dimension of this block matrix is often called the M-scheme dimension. Moreover, there is an additional symmetry to be conserved, namely total angular momentum J^2. However, it does not contribute to the block form of the matrix since the M-scheme state is not a good eigenstate of J^2.

Furthermore, this block matrix is quite sparse, namely the fraction of the non-zero matrix elements is quite small, since the shell-model Hamiltonian in Eq.(3) consists only of one-body and two-body interactions. Figure 1 shows the non-zero matrix elements of the Hamiltonian matrix of ^{44}Ti with pf shell and $M^\pi = 0^+$. Its M-scheme dimension is 4000. In this case, the fraction of the non-zero matrix elements is around 5%. The fraction is expected to be small in a larger system.

Owing to this sparsity, the matrix-vector product is performed efficiently with explicitly avoiding the treatment of the zero matrix elements, and the Krylov-subspace method is quite advantageous to

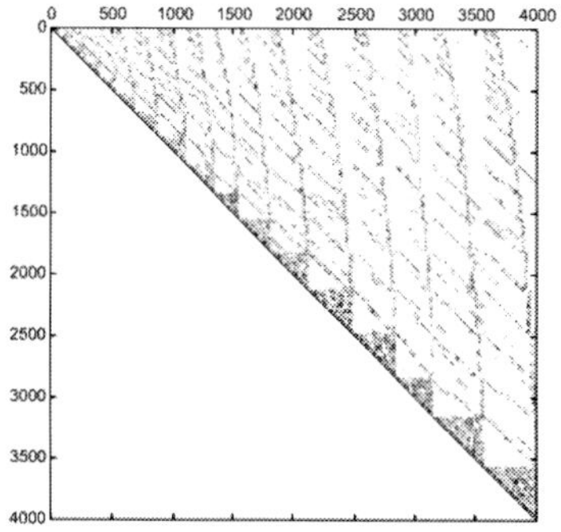

Figure 1. M-scheme Hamiltonian matrix of ^{44}Ti with pf shell and $M = 0$. The points represent non-zero matrix elements. Only upper-right part is shown since it is a real symmetric matrix.

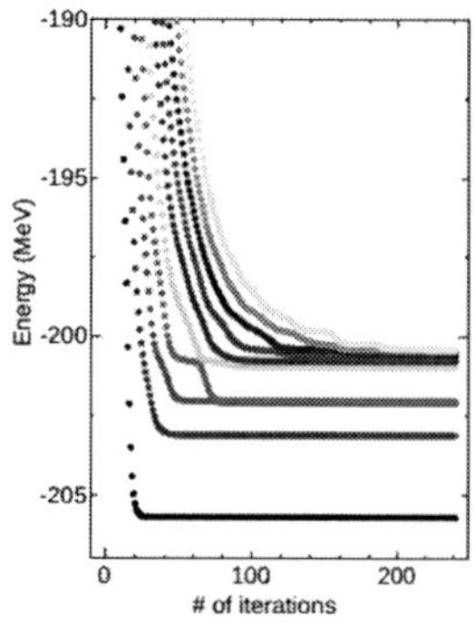

Figure 2. Convergence of the Lanczos method against the number of iteration, n. The symbols denote the shell-model energies of $J^\pi = 0^+_1$ (black circles) to $J^\pi = 0^+_{10}$ (yellow circles) states of ^{56}Ni with pf shell and GXPF1A interaction [9].

solve the eigenvalue problem. The Krylov subspace is spanned by an initial vector, $\boldsymbol{v}$, and its products with the matrix H to the power:

$$K_n(H, \boldsymbol{v}) = \mathrm{span}\{\boldsymbol{v}, H\boldsymbol{v}, H^2\boldsymbol{v}, ..., H^{n-1}\boldsymbol{v}\}. \tag{5}$$

In usual shell-model calculations the Lanczos method, one of the Krylov-subspace methods, is used to obtain low-lying eigenvalues. The convergence of the eigenvalue as a function of the number of Lanczos iteration, n, is shown in Fig. 2. The y axis shows the eigenvalues of the Hamiltonian matrix in the Krylov subspace. While the lowest eigenvalue converges quite fast, the high-lying eigenvalue converges slower as the eigenvalue goes higher. At a rough guess, the current limitation of the Lanczos method is that the product of the M-scheme dimension and the number of eigenvectors does not exceed $O(10^{10})$. While the largest dimension for a few low-lying states are $O(10^{10})$, the limitation for the calculation of the level density is much strict: since it requires $O(10^3)$ eigenvectors, the largest dimension becomes $O(10^7)$. The largest dimension is extended up to $O(10^{10})$ by utilizing our proposed method of stochastic estimation, the detail of which is described in the following section.

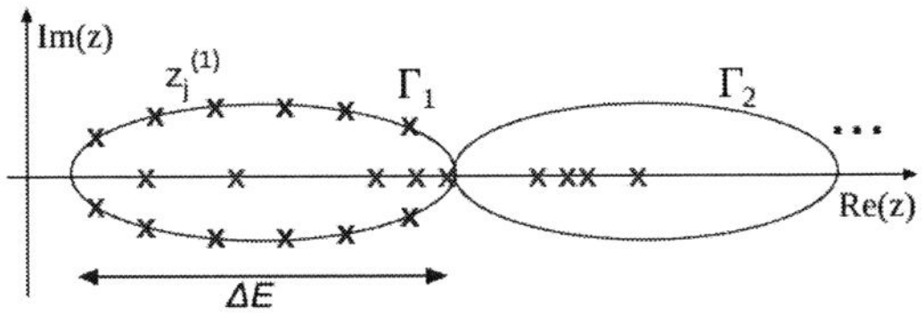

Figure 3. Schematic view of the contour line in complex plane z and its discretized points for the numerical calculations. The red crosses show eigenvalues of H.

3 Framework of the stochastic estimation of the eigenvalue count

We briefly review the stochastic estimation of eigenvalue density of sparse matrix in this section. This method was introduced and demonstrated in condensed matter physics [7].

For the estimation of the level density, we count the number of the eigenvalues in a specified energy region, Γ. It is given as $\mu = \mathrm{Tr}(P^\Gamma)$, where P^Γ is a projection operator to the subspace spanned by the eigenvectors whose eigenvalues are inside Γ. It is realized by the Cauchy integral as

$$P^\Gamma = \frac{1}{2\pi i}\oint_\Gamma dz \frac{1}{z-H} \simeq \sum_j w_j(z_j - H)^{-1} \tag{6}$$

where Γ is a contour to surround a certain energy region, or an oval Γ_1 in Fig. 3. Although we set Γ_i for each energy bin for evaluating the level density, we omit the index of energy bin i for simplicity. The contour integral is approximated by a summation of the discretized points z_j shown as blue symbols in Fig. 3 with the corresponding weights w_j. Note that this projector also appears in the Sakurai-Sugiura method, which is an eigenvalue solver [10, 11].

However, since the trace of Eq.(6) is difficult to calculate directly, we estimate it utilizing Hutchinson's estimator [12] as the following equation:

$$\hat{\mu} = \frac{1}{N_s}\sum_i^{N_s} \boldsymbol{v}_s^T P^\Gamma \boldsymbol{v}_s, \tag{7}$$

where $\boldsymbol{v}_s$ is a sample vector whose matrix elements are taken as -1 or 1 randomly. N_s is the number of sample vectors. While this estimation contains stochastic error, the magnitude of the error depends on the property of P^Γ. For example, if P^Γ is a diagonal matrix, the stochastic error becomes zero. In preceding works, this estimator gives sufficiently small error even with a small number of sample vectors [7]. In this work, we take $N_s = 16$.

The next step is to evaluate $\boldsymbol{v}_s^T \frac{1}{z-H}\boldsymbol{v}_s$ by solving a linear equation $(z-H)\boldsymbol{x} = \boldsymbol{v}_s$. Since we have multiple $\boldsymbol{v}_s$, we adopt the block conjugate gradient (BCG) method to solve the multiple linear equations $(z-H)X = V$ where V is the $D \times N_s$ matrix whose column is $\boldsymbol{v}_s$. We adopt block CG-rQ (BCG-rQ) method [14] for efficient computation.

With $A = z - H$, the BCG-rQ algorithm is shown in Algorithm 1 where qr(V) denotes QR decomposition of V. The X_k, P_k, Q_k, are also $D \times N_s$ matrices. The residual matrix Q_k is orthonormalized by QR decomposition every iteration to improve the numerical stability [13]. The block algorithm accelerates the convergence and also helps to use a CPU efficiently by increasing contiguous memory access.

However, it requires enormous computational resources to perform the BCG-rQ calculation for each z_j individually. This problem is overcome by utilizing the shift invariance of the Krylov subspace. This invariance means that the Krylov subspace of the matrix H is the same as that of $z - H$:

$$K_n(z-H, \boldsymbol{v}) = \mathrm{span}\{\boldsymbol{v}, (z-H)\boldsymbol{v}, (z-H)^2\boldsymbol{v}, ..., (z-H)^{n-1}\boldsymbol{v}\} = K_n(H, \boldsymbol{v}). \tag{8}$$

Algorithm 1 Algorithm of the Block CG-rQ method.

$X = O,\ Q_0\rho_0 = \text{qr}(V),\ \Delta_0 = \rho_0,\ P_0 = Q_0$
for $k = 0, 1, ...$ until $\|\Delta_k\| < \varepsilon\|V\|$
  $\alpha_k = (P_k^H A P_k)^{-1}$
  $X_{k+1} = X_k + P_k\alpha_k\Delta_k$
  $Q_{k+1}\rho_{k+1} = \text{qr}(Q_k - AP_k\alpha_k)$
  $\Delta_{k+1} = \rho_{k+1}\Delta_k$
  $P_{k+1} = Q_{k+1} + P_k\rho_{k+1}^H$
end for

The solution of $(z - H)\boldsymbol{x} = \boldsymbol{v}$ by the CG method is expressed in this subspace, if the number of iterations, n, is large enough. Owing to the shift invariance of the Krylov subspace, the solution of $(z_j - H)\boldsymbol{x} = \boldsymbol{v}$ for any j is also expressed in the same subspace. Thus, by considering this shift invariance, once we solve $(z_{\text{ref}} - H)X = V$ for a reference z_{ref}, we can obtain the solution of $(z_j - H)X = V$ for any j without further time-consuming matrix-vector products if n is large enough.

The block Krylov subspace is also invariant with the shift z:

$$K_n(z-H, \{v_1, ..., v_{N_s}\}) = \text{span}\{v_1, ..., v_{N_s}, (z-H)v_1, ..., (z-H)v_{N_s}, ..., (z-H)^{n-1}v_{N_s}\} = K_n(H, \{v_1, ..., v_{N_s}\}). \tag{9}$$

Therefore, the shift algorithm can also be used for the block CG-rQ method. In practice, shifted block CG-rQ (SBCG-rQ) method is adopted for efficient computations [14]. Finally, the level density is obtained as $\rho = \frac{\mu}{\Delta E}$, where δE is the length of Γ along the real axis shown in Fig. 3.

The "level density" means a summation of the numbers of levels without counting the degeneracy of z-component of angular momentum, namely the factor $2J + 1$. Because of this, especially for even (odd) nuclei, the level density is equal to the eigenvalue density of the Hamiltonian matrix with $M = 0$ $(M = \frac{1}{2})$. In order to obtain the spin-dependent level density, we replace the sample vector v_s in Eq.(7) by the angular-momentum projected vector, $P^J v_s$.

The procedure is summarized as follows:

1. Prepare z_j and w_j for the integral shown in Fig. 3 and Eq.(6).
2. Prepare sample vectors v_s, whose elements are taken as -1 or 1 randomly. If spin-dependent level density is required, those sample vectors are replaced by the angular-momentum projected vectors $P^J v_s$.
3. Solve linear equations $(z_j - H)\boldsymbol{x}_{s,j} = \boldsymbol{v}_s$ and obtain $\boldsymbol{v}_s^T\boldsymbol{x}_{s,j}$ by the SBCG-rQ method.
4. Level density is obtained by $\rho = \frac{1}{\Delta E}\frac{1}{N_s}\sum_{s=1}^{N_s}\sum_j w_j \boldsymbol{v}_s^T \boldsymbol{x}_{s,j}$.

In the present framework, the most time-consuming part is the matrix-vector product in the SBCG-rQ method. The dimension of the vector, the M-scheme dimension, often reaches quite huge and high-performance computing is required. We combine the nuclear shell-model code "KSHELL" [15] and the eigenvalue-solver library "z-Pares" [16], which enable us to utilize state-of-the-art supercomputers efficiently.

4 Benchmark test and level density of ^{56}Fe

In this section, we discuss the level density of ^{56}Fe as a demonstration of the present estimation method being compared with the exact Lanczos method and experimental results. For the shell-model

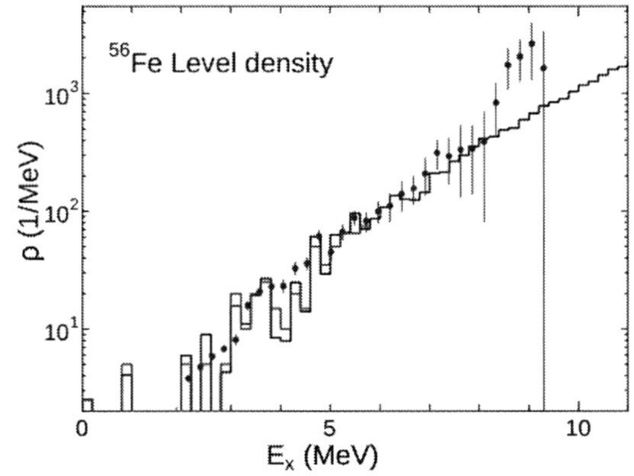

Figure 4. Level densities of ^{56}Fe against the excitation energy. These are obtained by stochastic estimation (black line) and by direct counting of the Lanczos results (red line). The experimental values (blue symbols with error bars) are taken from [17]. The energy bin is 0.2 MeV.

calculations, the model space is taken as the pf shell and the GXPF1A interaction [9] is adopted for the realistic effective interaction. This interaction was constructed based on G-matrix interaction and chi-square fitted for the experimental data of low-lying states. The M-scheme dimension of ^{56}Fe reaches 501,113,392.

The blue symbols with error bars in Fig. 4 show the experimental level density of ^{56}Fe obtained by the Oslo method [17]. The red line shows the exact shell-model level density of ^{56}Fe. The exact values are obtained by the direct counting of the 100 lowest states. The energy bin is $\Delta E = 0.2$ MeV, and it is the same as the experimental one. The exact Lanczos value shows good agreement the experimental one up to 6 MeV. However, it becomes rapidly difficult to obtain the exact value in higher-energy region since the level density, and hence the number of eigenvalues to be evaluated, increase exponentially.

The black line in Fig. 4 is obtained by the present stochastic estimation with $N_s = 16$ and 1550 BCG iteration. This estimation reproduces well the exact value within a certain stochastic error. The estimation is feasible beyond the Lanczos limitation and it reproduces well the experimental value up to 10 MeV. The deviation around $E_x \sim 9$ MeV is considered to be the contribution of negative-parity states, which is beyond the present model space. As the excitation energy increases, the ratio of the stochastic error to the level density would decrease since the level density increases exponentially and its error is expected to be proportional to the square root of the level density. Further discussion of the error estimation remains for future work.

B. A. Brown and A. C. Larsen also showed the shell-model results of the level density of ^{56}Fe by the direct counting of the Lanczos results in Ref.[18]. They used the same model space and interaction, but the truncation scheme which they applied in order to save computational resources brought about modest underestimation.

5 Removal of the contamination of the center-of-mass motion

Since an atomic nucleus is an isolated system and translational invariance should be taken into account the contamination of center-of-mass motion should be removed in nuclear shell model calculations beyond $0\hbar\omega$ model space [19]. In order to remove this contamination, we replace the Hamiltonian by

$$H' = H + \beta(H_{\rm CM} - \frac{3}{2}\hbar\omega) \tag{10}$$

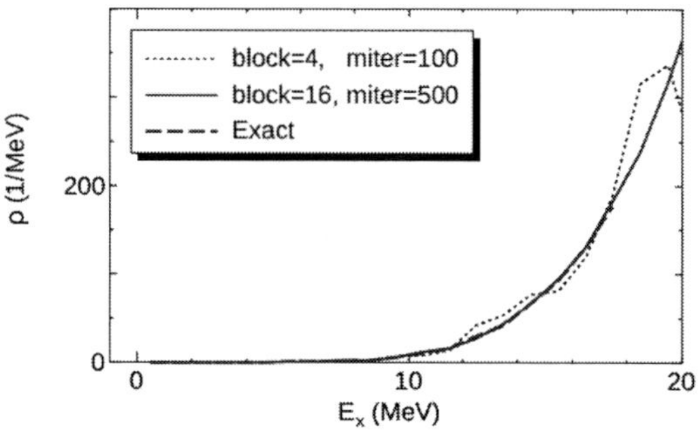

Figure 5. Level densities of $J^\pi = 1^-$ states of ^{48}Ca against the excitation energy. These are obtained by the stochastic estimation with $N_s = 4$ and 100 CG iterations (black dashed line), that with $N_s = 16$ and 500 CG iterations (blue solid line), and the exact values obtained by direct counting of the Lanczos results (red dashed line). The energy bin is 1 MeV.

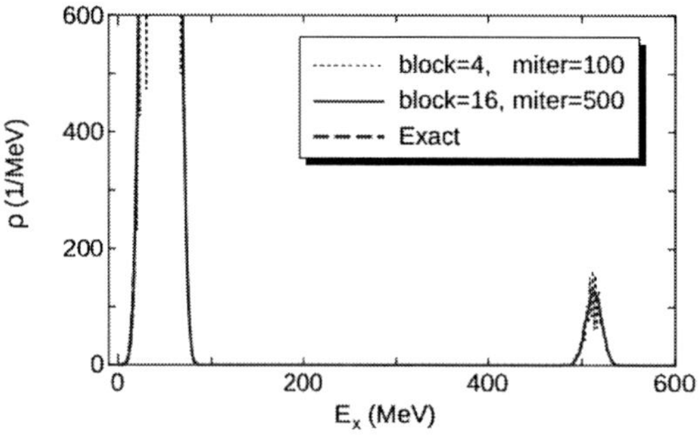

Figure 6. High-energy region of Fig. 5. Small peak at $E_x \sim 520$ MeV corresponds to the spurious density lifted up by the Lawson method.

with $\beta\hbar\omega/A = 10$ MeV. $H_{\rm CM}$ is the Hamiltonian of the center-of-mass motion in a harmonic oscillator potential whose energy quanta is $\hbar\omega$. We take β large enough so that the expectation value of $H_{\rm CM} - \frac{3}{2}\hbar\omega$ is close to zero and the excitation of the center-of-mass motion is suppressed. The contaminated states are lifted up to high-energy region. This prescription is called the Lawson method [20].

Figure 5 shows the level density of $J^\pi = 1^-$ states of ^{48}Ca. The stochastic estimation in shell model is performed with *sd-pf-sdg* shells and $1\hbar\omega$ truncation. The level densities are obtained in 1 MeV energy bin. The realistic interaction presently used was introduced in Ref.[21] and provides us with a good description of collective states such as the 3^-_1 state and the giant dipole resonances of Ca isotopes [22]. The exact value obtained by the Lanczos method is well reproduced by the present estimation with $N_s = 16$ and 500 BCG iterations. The estimation with $N_s = 4$ and 100 BCG iterations is shown by the dotted line. Since the number of iterations is small and the BCG does not converge well, it shows pseudo oscillation and disagreement with the exact one.

Figure 6 shows the high-energy region of Fig. 5. The spurious states are lifted up to $E_x \sim 520$ MeV, corresponding a small peak in Fig. 6, thereby separated clearly from the main peak. Such separation is also clearly seen even in the dotted line, which is obtained before the convergence.

6 Summary

We proposed a novel stochastic estimation method for calculation level densities in nuclear shell model. This method enables us to estimate level density in a rather small energy bin, such as 200 keV as demonstrated in the level density of ^{56}Fe. It is also demonstrated that the contamination of the center-of-mass motion is clearly removed by the Lawson prescription. The present method is feasible up to $O(10^{10})$ M-scheme dimension, which is a similar order to the limitation to obtain some low-lying states by the Lanczos method. By adopting this framework, we successfully reproduced the parity equilibration of 2^+ and 2^- level densities of ^{58}Ni in low-energy region in Ref.[8]. Further applications of this method are in progress.

Acknowledgments

This work has been supported by the HPCI Strategic Program from MEXT, CREST from JST, the CNS-RIKEN joint project for large-scale nuclear structure calculations, and KAKENHI grants (25870168, 23244049, 15K05094) from JSPS, Japan. The numerical calculations were performed on K computer at RIKEN AICS (hp140210, hp150224), FX10 supercomputer at the University of Tokyo, and COMA supercomputer at the University of Tsukuba.

References

[1] W. Hauser and H. Feshbach, Phys. Rev. **87**, 366 (1952).
[2] H. A. Bethe, Phys. Rev. **5**, 332 (1936); J. A. Holmes, S. E. Woosley, W. A. Fowler, and B. A. Zimmerman, At. Data Nucl. Data Tables **18**, 305 (1976); J. J. Cowan, F.-K. Thielemann, and J. W. Truran, Phys. Rep. **208**, 267 (1991).
[3] Y. Alhassid, G.F. Bertsch, S. Liu and H. Nakada, Phys. Rev. Lett. **84** 4313 (2000); H. Nakada, and Y. Alhassid, Phys. Rev. Lett. **79** 2939 (1997).
[4] R. A. Sen'kov and V. Zelevinsky, arXiv:1508.03683 [nucl-th]; R. A. Sen'kov and M. Horoi, Phys. Rev. C **82** 024304 (2010).
[5] E. Teran and C. W. Johnson, Phys. Rev. C **73** 024303 (2006).
[6] B. Strohmaier, S. M. Grimes, and S. D. Bloom, Phys. Rev. C 32 1397 (1985); S. M. Grimes, S. D. Bloom, R. F. Hausman Jr., and B. J. Dalton, Phys. Rev. C **19** 2378 (1979).
[7] Y. Futamura, H. Tadano and T. Sakurai, JSIAM Lett. **3**, 61 (2010).
[8] N. Shimizu, Y. Utsuno, Y. Futamura, T. Sakurai, T. Mizusaki and T. Otsuka, Phys. Lett. B **753**, 13 (2016).
[9] M. Honma, T. Otsuka, B. A. Brown, and T. Mizusaki, Eur. Phys. J. A **25**, Suppl. 1, 499 (2005).
[10] T. Sakurai and H. Sugiura, J. Comput. Appl. Math. **159**, 119 (2003).
[11] T. Mizusaki, K. Kaneko, M. Honma and T. Sakurai, Phys. Rev. C **82**, 024310 (2010).
[12] M. F. Hutchinson, Commun. Stat., Simulation Comput., **19** 433, (1990).
[13] A. A. Dubrulle, Elect. Trans. Numer. Anal. **12** 216 (2001).
[14] Y. Futamura, T. Sakurai, S. Furuya, and J. Iwata, VECPAR 2012, LNCS 7851 226 (2013).
[15] N. Shimizu, arXiv:1310.5431 [nucl-th]
[16] Y. Futamura and S. Sakurai, http://zpares.cs.tsukuba.ac.jp/
[17] E. Algin *et al.*, Phys. Rev. C **78**, 054321 (2008).
[18] B. A. Brown and A. C. Larsen, Phys. Rev. Lett. **113**, 252502 (2014).
[19] E. Caurier, G. Martinez-Pinedo, F. Nowacki, A. Poves, and A. P. Zuker, Rev. Mod. Phys. **77** 427 (2005).
[20] D. H. Gloeckner and R. D. Lawson, Phys. Lett. **53B** 313 (1974).
[21] Y. Utsuno, T. Otsuka, B. A. Brown, M. Honma, T. Mizusaki, and N. Shimizu, Prog. Theor. Phys. Suppl. **196**, 304 (2012).
[22] Y. Utsuno, N. Shimizu, T. Otsuka, S. Ebata and M. Honma, Prog. Nucl. Ener. **82**, 102 (2015).

Effective photoexcitation cross section of $^{115}\mathrm{In}(\gamma,\gamma')^{115\mathrm{m}}\mathrm{In}$ from photoactivation data

M. Versteegen[1,a], D. Denis-Petit[2,1], V. Méot[2], T. Bonnet[1,2], M. Comet[1,2], F. Gobet[1], F. Hannachi[1], P. Morel[2], M. Martini[3], S. Péru[2], M. Tarisien[1], and M-M. Aléonard[1]

[1] *Université de Bordeaux, CNRS-IN2P3, CENBG F-33175 Gradignan, France*
[2] *CEA, DAM, DIF F-91297 Arpajon, France*
[3] *CEA, DSM/IRFU/SPhN/ESNT, F-91191 Gif sur Yvette Cedex*

Abstract. Photoexcitation yields of the ^{115m}In metastable state were measured with Bremsstrahlung γ beams over a range of endpoint energies between 4.5 and 18 MeV. An effective cross section of the $^{115}\mathrm{In}(\gamma,\gamma')^{115m}\mathrm{In}$ photoexcitation cross section was determined to reproduce the data. This cross section is built from a cross section calculated with the TALYS code, to which an enhancement is added at about 8.5 MeV.

1 Introduction

The production of radioactive nuclei is the basic principle of nuclear activation diagnostics. The yield of activated nuclei is measured after irradiation and is used to extract the energy distribution and absolute number of particles of the incident beam, provided the cross section of the activation reaction is well known [1].

The ^{115}In is a perfect candidate for nuclear activation diagnostics. It is a stable and easy to handle isotope, with a first excited isomeric state lying 336 keV above the ground state with a half-life of 4.49 h. It is already used for the characterization of irradiation facilities [2], and as an activation standard in the context of inertial fusion experiments [3, 4]. The cross section of the $^{115}\mathrm{In}(n,n')^{115m}\mathrm{In}$ neutron inelastic scattering reaction is well known [5], and the production yield of ^{115m}In is measured to determine the yield of the 2.45 MeV fusion neutrons, emitted in DD fusion reactions. In the context of laser-matter experiments, high intensity Bremsstrahlung γ sources are also produced [6]. These sources are characterized by large numbers of particles emitted in bunches of the order of a few ps or fs, with continuous energy distributions. We investigate the possibility of using the production of the ^{115m}In isomer by photoexcitation to constrain the low energy part of the absolute energy distributions of these γ beams.

For such purposes, the $^{115}\mathrm{In}(\gamma,\gamma')^{115m}\mathrm{In}$ photoexcitation cross section has to be well known. However, surprisingly little and contradictory information is available. Early studies from the mid-1950s show a peak-shaped cross section, with a maximum of a few mb at about 8-10 MeV but disagree on the exact value of the maximum as well as on its position [7, 8]. A renewed interest in photoactivation and nuclear resonance fluorescence experiments on isomers led to several studies in the late 1980s and

[a] e-mail: versteeg@cenbg.in2p3.fr

early 1990s on ^{115}In [9–11]. Contradictory results about a second maximum in the $^{115}\text{In}(\gamma,\gamma')^{115m}\text{In}$ cross section at 22 MeV, well beyond to the Giant Dipole Resonance, are reported [12, 13], but have recently been contradicted [14].

New experimental data are needed to clarify these results. Irradiations of natural In samples were performed at the ELSA electron linac of CEA/DAM in Arpajon, France [15]. The measured photoexcited yields of ^{115m}In are compared with calculated yields obtained using different models of the cross section. As all calculated yields fail to reproduce the data, an effective cross section is built. The cross section is enhanced at low energies by means of an additionnal lorentzian contribution, the parameters of which are determined by minimizing a χ^2 between measured and calculated yields.

Section 2 describes the set up of the experiment as well as the measured data. Section 3 summarizes the data analysis, and presents the measured yields of photoexcited ^{115m}In. Section 4 shows the available photoexcitation cross sections and presents the effective cross section which allows to reproduce the data.

2 Experiment Set Up

The mono-energetic electron beam of the radio-frequency electron linac ELSA is delivered with energies up to 19 MeV with very low dispersion (0.05% RMS) using four RF cavities. In this experiment, beam intensities between 1.6 and 3.3 μA were set depending on beam energy, which varied between 4.5 and 18 MeV.

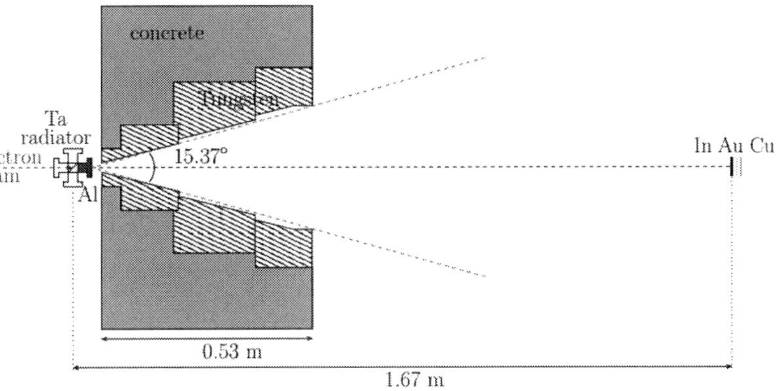

Figure 1: Set up of the experiment.

As presented on Fig.1, the electron beam is focused onto a 1.2 mm thick tantalum radiator with an angle of incidence of 45° to produce Bremsstrahlung photons. A cylindrical aluminum hardener 40 mm in length is placed against the radiator to absorb low energy photons as well as high energy electrons passing through. The photon beam is collimated by four cone-shaped collimators mainly composed of tungsten and embedded in the 0.53 m thick concrete wall separating the accelerator cave from the experimental area. The resulting opening angle of the photon beam is about 15°.

Natural indium, gold and copper samples were irradiated simultaneously at 1.67 m from the radiator, during 20 min to 2 h depending on beam energy. The gold and copper samples serve as references for the determination of the photon flux. The targets were disk-shaped, 2.5 cm in diameter. The natural In (95.71% of ^{115}In and 4.29% of ^{113}In) targets were 3 mm thick, with typical masses of 11 g, while the ^{197}Au and natural Cu (69.15% of ^{63}Cu and 30.85% of ^{65}Cu) were 0.1 mm thick, with typical masses of 0.9, respectively 0.4 g.

Sample activation was measured offline. The γ-ray decay spectra of activated In and Au samples were each measured using high-purity coaxial, respectively broad range, HPGe detectors, with relative efficiencies of 80%, respectively 85%. Irradiated In targets were placed at a distance of 6.5 cm from the detector entrance window for counting, and the Au targets at 15.1 cm. Each detector was protected from ambient radioactivity by an 11 cm thick shielding, composed of 10 cm of Pb and 1 cm of Cu. The acquisition dead time was under 0.1% and will be neglected in the following analysis.

The β^+ activity of the Cu targets corresponding to the ^{64}Cu and ^{62}Cu daughter nuclei of the Cu(γ,n) reactions was measured using the multidetector NATALIE [1]. The two 511 keV annihilation photons are detected in coincidence by two NaI detectors, with an efficiency of 5.6±0.2% or 5.9±0.2% for ^{62}Cu or ^{64}Cu respectively. The two radioactive isotopes are identified in the time decay spectrum by their half-lives, equal to 762.06(2) min for ^{64}Cu and 9.673(8) min for ^{62}Cu.

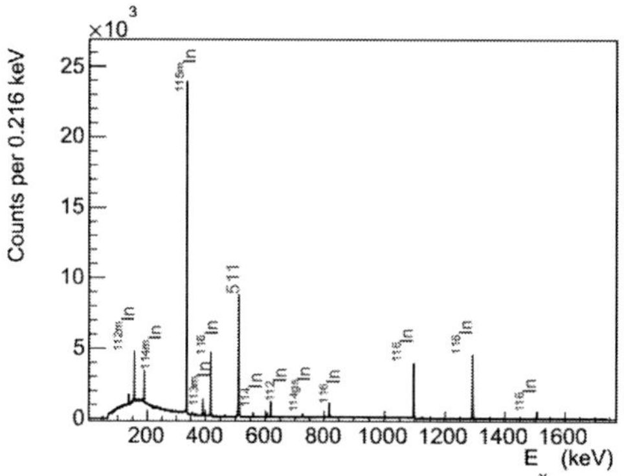

Figure 2: Decay spectrum of a natural In target irradiated at 17 MeV, measured offline during 15h of counting. The main decay lines are directly identified to the In isotope they originate from.

Figure 2 shows an example of a γ-ray decay spectrum from an In sample irradiated at the electron beam energy of 17 MeV, measured offline during 15h of counting. The photopeaks at 156, 190, 336 and 392 keV correspond to the isomeric states of ^{112m}In, ^{114m}In, ^{115m}In and ^{113m}In respectively. The ^{112}In and ^{114}In isotopes are produced by (γ,n) reactions on ^{113}In and ^{115}In present in the sample. Photopeaks at 417, 818, 1096, 1292 and 1506 keV are also observed, and correspond to the decay products of excited ^{116}In, produced by (n,γ) reactions on ^{115}In. The photopeaks at 559 and 725 keV sign the decay of ^{114gs}In, while the ones at 607 and 617 keV sign the decay of ^{112gs}In.

3 Data Analysis

3.1 Activation yields

The measured activation yield is determined from the measured number of counts in the full-energy peak corresponding to the radionuclide decay, N_γ, by:

$$Y_{act} = \frac{t_{\rm irr}}{Q} \frac{\lambda}{1 - e^{-\lambda t_{\rm irr}}} e^{\lambda t_{\rm loss}} \frac{N_\gamma - N_{bg}}{I_\gamma \epsilon (1 - e^{-\lambda t_{\rm meas}})} \tag{1}$$

if the radionuclide is populated by only one process. In Eq.(1), $t_{\rm meas}$ is the duration of counting, I_γ is the absolute γ-ray intensity, λ is the decay constant, ϵ is the detector photopeak efficiency, and Q is the total beam charge delivered during irradiation. The values of I_γ and λ are taken from Refs. [16–18] depending on the radionuclide. The background underlying the full-energy peak, N_{bg}, is taken into account, as well as decay losses during irradiation ($t_{\rm irr}$) and between the end of irradiation and the beginning of counting ($t_{\rm loss}$). The obtained yields are determined in each target sample and normalized per μC of incident electron beam.

The measured yields of ^{115m}In are compared to expected yields determined by:

$$Y_{act}(E_e) = N_{tgt} \int_{E_{\rm thres}}^{E_e} \frac{dN_\gamma}{dE_\gamma}(E_\gamma, E_e)\, \sigma(E_\gamma)\, dE_\gamma \tag{2}$$

where dN_γ/dE_γ is the Bremsstrahlung distribution of the impinging photons, with endpoint energy E_e corresponding to the electron beam energy, $\sigma(E_\gamma)$ the activation cross section at the photon energy E_γ and $E_{\rm thres}$ the reaction threshold energy. N_{tgt} is the target areal density. The calculated yield can be expressed per μC of incident electron beam when the Bremsstrahlung spectrum is normalized to the total electron beam charge.

3.2 Detector efficiencies

As shown in Eq.(1), the absolute photopeak efficiencies ϵ of the counting set-ups are necessary to determine the absolute activation yields obtained in the experiment. These efficiencies are obtained using Monte Carlo simulations to take the extended source geometry into account. The exact dimensions of the broad range and the coaxial HPGe detectors were determined using X-ray radiographies to ascertain crystal dimensions, endcap-to-crystal distances and endcap thickness. Efficiencies were calculated in the 0.12 to 1.4 MeV energy range using GEANT4 v.9.6.p03 [19]. They were compared to measured data from point-like calibration ^{152}Eu sources at four different source-to-detector distances : 15.1, 20.1, 25.1 and 30.1 cm for the broad range HPGe, and 6.5, 11.5, 16.5 and 21.5 for the coaxial HPGe.

In the case of the broad range HPGe detector, measured and calculated efficiencies were found in agreement over the whole energy range when the source-to-detector distance was beyond 20 cm. At a distance of 15.1 cm, which corresponds to the ^{196}Au activity measurement, a constant correction of about 20% had to be applied to the simulation to render measured point-like source efficiencies.

In the case of the coaxial HPGe detector, the simulated efficiencies were overestimated compared to measurements by a constant factor of 1.37 ± 0.02 over the whole energy range and at all source-to-detector distances. As the correction factor showed no dependence with energy or source-to-detector distance, it was applied to simulated efficiencies in the extended source geometry for the In samples at a distance of 6.5 cm.

Table 1. summarizes the obtained efficiencies for the decay peaks pertinent to this work. They are of the order of 1.8%, and 0.7%, for the coaxial, respectively broad range, HPGe, with uncertainties of the order of 0.02% related to statistics and to the correction factor described above.

Table 1: Absolute photopeak efficiencies in extended source geometry of the HPGe detectors used in this work.

reaction or excitation process	E_γ (keV)	ϵ (%)
^{113}In$(\gamma, n)^{112m}$In	156.61 (3)	1.82 (3)
^{115}In$(\gamma, n)^{114m}$In	190.27 (3)	1.87 (2)
^{115}In$(\gamma, \gamma')^{115m}$In	336.244(17)	1.55 (2)
^{197}Au$(\gamma, n)^{196}$Au	355.73 (5)	0.70 (1)

3.3 Neutron production yields

The total measured yield of ^{115m}In is the sum of the contributions from excited nuclei produced by photoexcitation and by neutron inelastic scattering reactions. These contributions are undistinguishable and the latter needs to be evaluated. To do so, all possible sources of impinging neutrons are identified.

To evaluate the excitation yields due to photo-neutrons produced by (γ, n) in the beam structures, irradiation configurations were modified. At each electron beam energy, irradiations were performed with and without shielding the targets by a 5 cm thick block of polyethylene (CH), acting as a neutron moderator, and a 1 cm thick foil of cadmium, acting as an absorber of the moderated neutrons. This shielding was placed in the beam path in front of the In sample. The expected impact on the photon spectrum at E_e = 18 MeV was determined by Monte Carlo simulations using GEANT4 and was shown to be a constant transmission of 90 ± 7% over all photon energies of the Bremsstrahlung spectrum. The uncertainty corresponds to the standard deviation. Similar simulations were performed for the expected neutron spectrum using MCNPx. The ratio between calculated neutron spectra with and without the shielding is 30 ± 10% over all neutron energies.

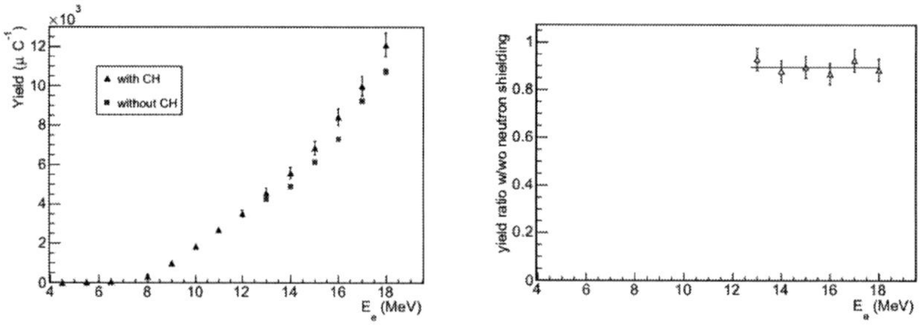

(a) Total measured excitation yields of ^{115m}In with and without the polyethylene (CH) and cadmium neutron shielding in front of the target during irradiation.

(b) Ratio between measured photoexcitation yields of ^{115m}In with and without the neutron shielding. The red line is a fit by a constant, which gives a mean ratio of 89±1%.

Figure 3

Figure 3a shows the obtained total yields of ^{115m}In with and without the neutron shielding, while Fig. 3b shows the ratio between these yields at electron beam energies above the neutron emission threshold energy. The mean ratio obtained over all electron beam energies is of the order of 90%, which is in agreement with the transmission of the photon spectra. The (n, n') cross section is of the order of 300 mb over the 2 to 10 MeV neutron energy range [20], which is about two orders of magnitude larger than the maximum value of the (γ, γ') cross sections, which is of the order of a few mb. The attenuation of the measured yields is thus not compatible with the expected attenuation of an impinging neutron spectrum, and the contribution of (n, n') excitations induced by neutrons coming from the beam structures is neglected in the following.

The yields of photoneutrons produced within the targets themselves are obtained from the measured yields of all (γ, n) reaction products. The (γ, n) reactions in the Au and Cu samples produce the ^{196}Au, ^{63}Cu and ^{65}Cu isotopes, the yields of which are already specifically measured to serve as reference for the calculation of the Bremsstrahlung spectra. The yields of ^{112m}In and ^{114m}In are determined from the measured number of counts in the 156 and 190 keV full-energy peaks in the measured decay spectrum of the irradiated In sample, using Eq.(1). The yield of ^{112gs}In is deduced from the time decay analysis of the 617 keV γ-ray line, which depends on the ^{112m}In and ^{112gs}In production yields. The production yields ^{114gs}In can not directly be extracted from measured data, as ^{114gs}In mainly de-

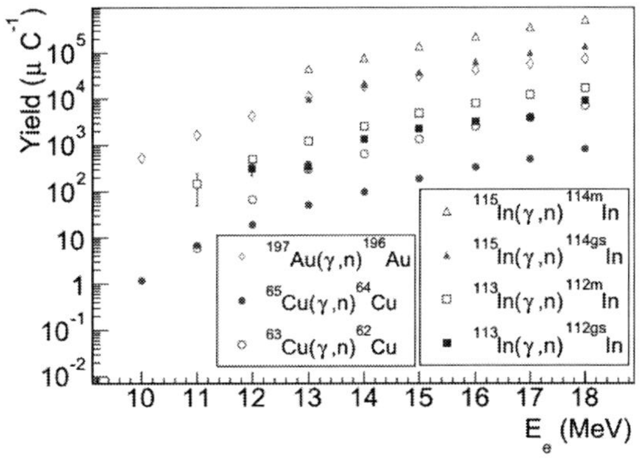

Figure 4: Production yields of photoneutrons in the In, Au and Cu samples.

cays onto the stable ^{114gs}Sn with a half life of about 71 ns [17]. An expected ratio between ^{114gs}In and ^{114m}In production yields is thus calculated using Eq.(2). The Bremsstrahlung photon fluxes are calculated, using Monte Carlo simulations, and are validated by the Au and Cu data, which are activation standards [21–23]. Cross sections of the ^{115}In$(\gamma, n)^{114gs}$In and ^{115}In$(\gamma, n)^{114m}$In reactions are calculated using the TALYS code [24], with input parameters which will be discussed in Section 4. The calculated ratio between ^{114gs}In and ^{114m}In yields is of the order of 27% over the 10 to 18 MeV energy range. This ratio is applied to the measured yields of ^{114m}In to obtain the ones of ^{114gs}In. Figure 4 presents the obtained photoneutron production yields, corresponding to the above-mentioned reactions. A total of 7×10^5 neutrons per μC of electron beam is produced in the samples in the worst case of E_e = 18 MeV. The (n, n') cross section and Monte Carlo simulations are used to determine the

DOI: 10.1051/epjconf/201612203001

corresponding yield of ^{115m}In produced by $(n\ n')$ excitation processes in the In sample, as described in the following.

3.4 ^{115m}In photoexcitation yields

In order to evaluate the yield of ^{115m}In produced by neutron inelastic scattering, Monte Carlo simulations using the GEANT4 code are performed. Neutrons are emitted isotropically within each sample volumes following the energy distributions of photoneutrons produced by (γ, n) reactions predicted by the TALYS code. The energy of the each neutron interacting in the In sample by neutron inelastic scattering is retrieved to build the energy distributions of interacting neutrons. These distributions

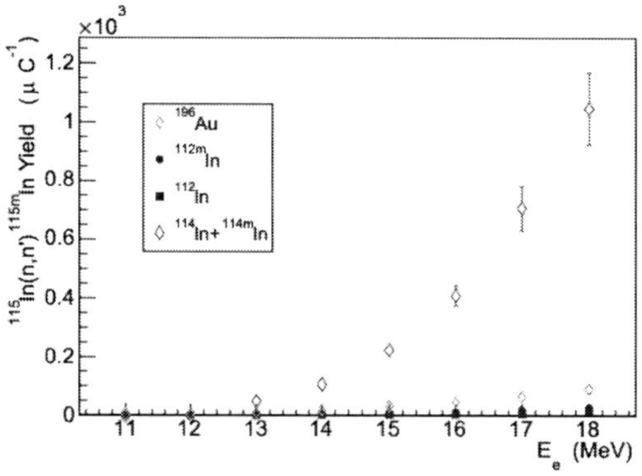

Figure 5: (Color online) Evaluated yields of ^{115m}In excited by (n, n') neutron inelastic scattering process, from photoneutrons produced in the Au and In samples.

are used to calculate the yield of ^{115m}In produced by the ^{115}In$(n, n')^{115m}$In process using an equation similar to Eq. (2), with the yields of photoneutrons presented in Fig.4 and the cross section from [20]. Details of the procedure will be published elsewhere [25]. The obtained yields are presented in Fig.5. The contribution from photoneutrons produced by the ^{115}In$(\gamma, n)^{114gs}$In and ^{115}In$(\gamma, n)^{114m}$In reactions within the In sample is the major one. The contribution from photoneutrons produced in the Au sample are one order of magnitude lower at $E_e = 18$ MeV.

These contributions are subtracted to the total measured yield of ^{115m}In, determined from the measured number of counts in the 336 keV full-energy peak (Eq.(1)), as presented on Fig.6. At 18 MeV, the contribution due to (n, n') reactions corresponds to about 10% of the total measured yield of ^{115m}In. This contribution decreases with decreasing electron beam energy, and is negligible below 11 MeV.

4 ^{115m}In photoexcitation cross section

The measured yields of photoexcited ^{115m}In are compared with calculated yields, obtained following Eq.(2) with the cross sections presented in Fig.7 and the calculated Bremsstrahlung spectra.

Figure 7a presents the photoexcitation cross sections from Refs. [7, 8, 26], in a figure extracted from Ref. [14]. These cross sections were obtained from activation data and show large discrepancies.

 DOI: 10.1051/epjconf/201612203001

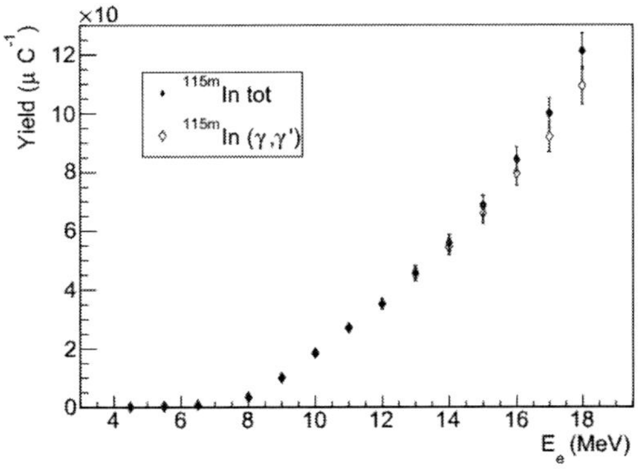

Figure 6: Total and (γ, γ') yields of ^{115m}In excited in the sample per μC of electron beam.

Authors from Ref. [8] estimate the uncertainty on σ_2 in the peak region at ~30%, and the uncertainty on the position and full width at half maximum of the peak to be ±1 MeV. The cross section σ_1 from

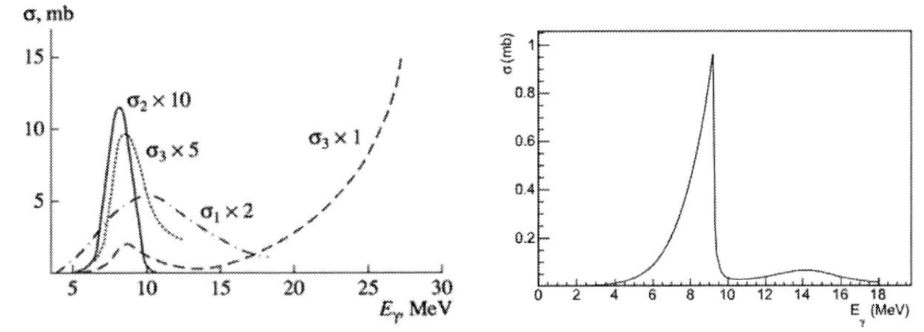

(a) Experimental cross sections, extracted from [14], with σ_1 from Ref. [7], σ_2 from Ref. [8] and σ_3 from Ref. [26].

(b) Calculated cross section, obtained with the TALYS code [24], using the GLO photon strength function. See text for details.

Figure 7: ^{115}In$(\gamma, \gamma')^{115m}$In cross sections.

Ref. [7] presents a maximum value at 4.4 mb positioned at 9 MeV with a width at half maximum of 9 MeV. The authors mention that the uncertainty associated with their method of extraction of the cross section from the activation data can be up to 50% [27]. Authors from Ref. [8] discuss the data from Ref. [7] and argue that they are overestimated. The peak value should be at 2.2 mb [8] and the width of the peak should be narrower (see Ref. [8] for details). In Fig. 7a, the maximum value of σ_1 is neither 4.4 mb as published in Ref. [7], nor 2.2 mb as advised by the authors of Ref. [8], but 2.5 mb. As this could not be explained, the σ_1 cross section used in the following is the one directly extracted from Ref. [7], with a maximum value of 4.4 mb. Finally, the cross section from Ref. [26] exhibits a second large peak above the Giant Dipole Resonance, which is located about 15.5 MeV in ^{115}In. This second peak is largely discussed in Ref. [14] and is contradicted.

Figure 7b presents a calculated photoexcitation cross section obtained with the TALYS 1.4 code. TALYS 1.4 performs the calculation with the global dispersive optical potential of Morillon and Romain [28]. The phenomenological constant temperature model of Gilbert and Cameron [29] is used for level densities, and standard options based on the RIPL database [30] are used for nuclear structure parameters. The generalized Lorentzian model (GLO) of Kopecky and Uhl [31] is used for the E1 γ strength function. This model is fully parametrized so as to reproduce (γ, n) data. All radiations other than E1 are treated under the Brink Axel hypothesis [32, 33] and the corresponding γ strength function takes the form of a standard Lorentzian function. The calculated cross section exhibits a maximum value of about 1 mb at 9.2 MeV, comparable to the experimental cross section of Ref. [8]. The sharp drop at energies beyond 9 MeV corresponds to the opening of the neutron emission decay channel.

The photoexcitation yields measured in this work are compared to calculated yields obtained with the experimental cross sections of Fig.7a and the calculated one presented on Fig.7b. These yields are presented in Fig.8. The lower energy points are presented in logarithmic scale in the inset. As there

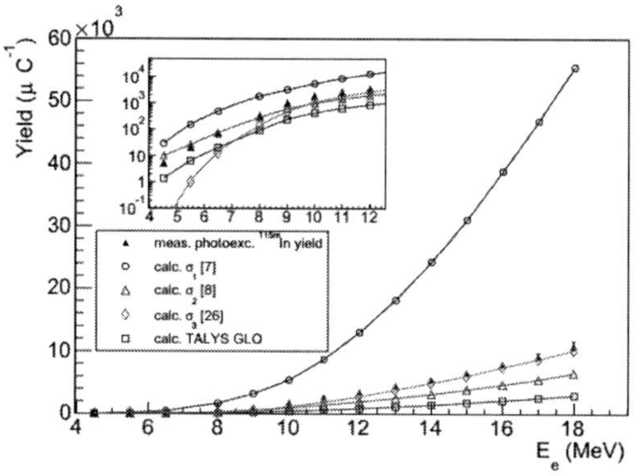

Figure 8: Comparison between photoexcitation yields of ^{115m}In measured in this work (closed black triangles) and calculated yields obtained using the (γ, γ') cross sections presented in Fig.7.

is no experimental value below 3.8 and 5.4 MeV in the σ_1 and σ_3, these cross sections supposed to drop smoothly to zero from their lowest energy point for the calculation of the yields. This hypothesis is supported by the fact that no extra photon strength is expected in the energy region below 5 MeV in ^{115}In, as none is observed in the Cd and Sn neighboring nuclei [34–36]. The yields calculated with σ_3 from Ref. [26] reproduce well our data at high energies, but fail to do so below the neutron emission threshold energy. On the contrary, the yields calculated with σ_2 from Ref. [8] reproduce the data below ~8 MeV, and are underestimated at higher energies. The cross section σ_1 from Ref. [7] was argued to be overestimated by the authors of Ref. [8], and the corresponding yields corroborate this observation, as they also are overestimated compared to the data. Even a factor of 2 on the cross section σ_1 would not allow to reproduce the measured yields. Finally the yields obtained with the cross section from TALYS are underestimated by a factor of 4 at all electron beam energies.

In order to determine an effective photoexcitation cross section which can be used in the context of laser experiments, the calculated cross section from TALYS is modified to reproduce the data at all energies. A lorentzian function with the following expression:

$$L(E_\gamma) = \frac{1}{2\pi} \frac{A\,\Gamma}{(E_\gamma - E_0)^2 + \Gamma^2/4} \tag{3}$$

is introduced and directly added to the TALYS cross section. The parameters A, Γ and E_0 represent the normalization constant, the width and mean energy of the lorentzian function. They are adjusted to minimize a χ^2 between calculated and measured yields. Figure 9 shows the measured and calculated yields for which the minimum χ^2 value is obtained ($\chi^2/NDF = 1.2$), with the lower energy points in logarithmic scale in the inset. The calculated yields obtained using the modified cross section are in open pink circles. Measured and calculated photoexcitation yields are in very good agreement both at

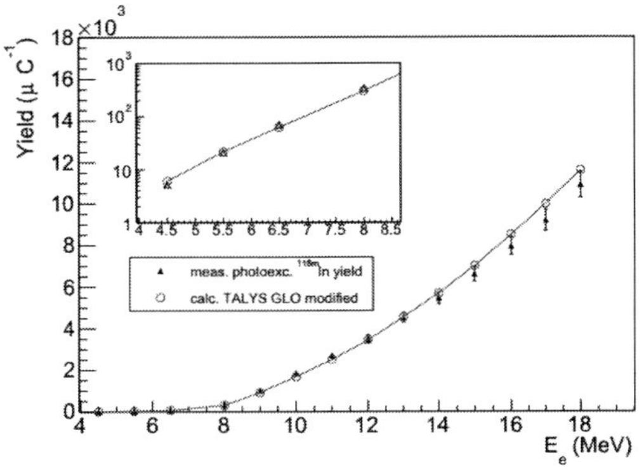

Figure 9: (Color online) Measured photoexcitation yields of ^{115m}In and calculated yields obtained with the modified cross section from TALYS. See text for details.

low and high electron beam energies. The parameters of the added lorentzian function are :

$$A = 4.3 \text{ mb} \qquad \Gamma = 0.9 \text{ MeV} \qquad \text{and} \qquad E_0 = 8 \text{ MeV}$$

Figure 10 shows the corresponding effective cross section (pink dashed line). The added lorentzian function increases the maximum of the cross section from ~1 mb at ~9.5 MeV to ~4 mb at about 8.5 MeV, compared to the cross section calculated with TALYS alone. The experimental cross sections already presented on Fig. 7a are also reported, with the extrapolation to zero introduced in this work. The σ_3 cross section from Ref. [26] which reproduced the photoexcitation yields at energies above the neutron emission threshold is comparable to the effective cross section, but falls more rapidly to zero below 5 MeV, thus explaining the underestimation of the yields at low energies. The σ_2 cross section from Ref. [8] is very similar to the effective cross section at low energies, but is lower by about a factor of four at the maximum, which is enough to underestimate the yields at high energies. The σ_1 cross section from Ref. [7] is largely overestimated, as already noted.

DOI: 10.1051/epjconf/201612203001

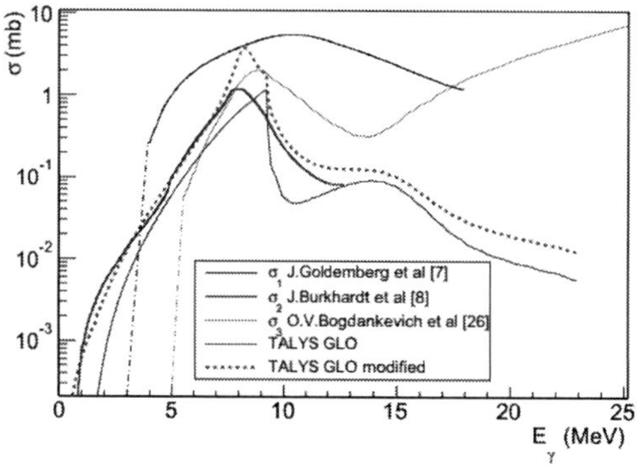

Figure 10: Cross sections of $^{115}In(\gamma,\gamma')^{115m}In$. Extrapolations of experimental cross sections σ_1 and σ_3 in dash-dotted line were performed in this work. See text for details.

5 Conclusion

Photoexcitation yields of ^{115m}In have been measured on an energy range between 4.5 and 18 MeV using a Bremsstrahlung photon source. These yields are used to test the experimental cross sections of the (γ,γ') excitation process available in the literature, as well as a cross section obtained by calculation using the TALYS code. As none of these cross sections allows to reproduce the data on the whole energy range, an effective cross section is built by adding a lorentzian contribution to the calculated one. The parameters of this lorentzian are determined from the minimization of a χ^2 between measured and calculated yields.

The effective cross section allows to reproduce the data, but no fundamental physics motivation supports the direct modification of the cross section. However, it has been shown that additionnal strength is missing in the GLO model in neighboring nuclei, at energies around the neutron emission threshold [34–36]. Rather than directly modifying the cross section, a systematic study on the γ strength function should be performed. Such a study was carried out and will be published elsewhere [25].

References

[1] M. Tarisien, C. Plaisir, F. Gobet, F. Hannachi, M.M. Aléeonard, A. Rebii, Rev.Sci.Instrum. **82** (2011)

[2] U. Kneissl, Prog. Part. Nucl. Phys. **46**, 79 (2001)

[3] C. Ruiz et al., Rev.Sci.Instrum. **63** (1992)

[4] O. Landoas et al., Rev.Sci.Instrum. **82** (2011)

[5] N. Otuka et al., Nuclear Data Sheets **120**, 272 (2014)

[6] C. Courtois, A. Compant La Fontaine, O. Landoas, G. Lidove, V. Méot, P. Morel, R. Nuter, E. Lefebvre, A. Boscheron, J. Grenier et al., Phys. Plasmas **16**, 013105 (2009)

[7] J. Goldemberg, L. Katz, Phys. Rev. **90**, 308 (1953)

[8] J. Burkhardt, E.J. Winhold, T.H. Dupree, Phys. Rev. **100**, 199 (1955)
[9] J.J. Carroll et al., Phys. Rev. C **43**, 1238 (1991)
[10] P. von Neumann-Cosel et al., Phys. Lett. B **266**, 9 (1991)
[11] P. von Neumann-Cosel, V.Y. Ponomarev, A. Richter, C. Spieler, Z. Phys. A **350**, 303 (1995)
[12] L. Lakosi, J. Sáfár, A. Veres, T. Sekine, H. Kaji, K. Yoshihara, J. Phys. G: Nucl. Part. Phys. **19**, 1037 (1993)
[13] V. Bokhinyuk, A. Guthy, A. Parlag, M. Sabolchy, I. Sokolyuk, I. Khimich, Ukr. J. Phys. **51**, 657 (2006)
[14] L.Z. Dzhilavyan, Phys. Atom. Nucl. **78**, 624 (2015)
[15] A. Chauchat, V.L. Flanchec, A. Binet, J. Flament, P. Balleyguier, J. Brasile, S. Muller, J. Ortega, Nucl. Instrum. Meth. A **608**, S99 (2009)
[16] S. Lalkovski, F. Kondev, Nuclear Data Sheets **124**, 157 (2015)
[17] J. Blachot, Nuclear Data Sheets **113**, 515 (2012)
[18] J. Blachot, Nuclear Data Sheets **113**, 2391 (2012)
[19] S. Agostinelli et al. (GEANT4), Nucl. Instrum. Meth. **A506**, 250 (2003)
[20] M. Bhike, A. Saxena, B.J. Roy, R.K. Choudhury, S. Kailas, S. Ganesan, Nucl. Sci. Eng. **163**, 175 (2009)
[21] C. Nair, M. Erhard, A.R. Junghans, D. Bemmerer, R. Beyer, E. Grosse, J. Klug, K. Kosev, G. Rusev, K.D. Schilling et al., Phys. Rev. C **78**, 055802 (2008)
[22] R.E. Sund, M.P. Baker, L.A. Kull, R.B. Walton, Phys. Rev. **176**, 1366 (1968)
[23] C. Plaisir, F. Hannachi, F. Gobet, M. Tarisien, M. Aléonard, V. Méot, G. Gosselin, P. Morel, B. Morillon, Eur. Phys. J. A **48**, 68 (2012)
[24] S.H. A.J. Koning, M. Duijvestijn, *TALYS-1.0*, in *Proceedings of the International Conference on Nuclear Data for Science and Technology April 22-27, 2007, Nice, France* (editors O.Bersillon, F.Gunsing, E.Bauge, R.Jacqmin, and S.Leray, EDP Sciences, 2008), pp. 211–214
[25] M. Versteegen et al., to be published
[26] O.V. Bogdankevich, L.E. Lazareva, F. Nikolaev, Sov. Phys. JETP **4**, 320 (1956)
[27] H.E. Johns, L. Katz, R.A. Douglas, R.N.H. Haslam, Phys. Rev. **80**, 1062 (1950)
[28] B. Morillon, P. Romain, Phys. Rev. C **70**, 014601 (2004)
[29] A. Gilbert, A.G.W. Cameron, Can. J. Phys. **43**, 1446 (1965)
[30] R. Capote et al., Nuclear Data Sheets **110**, 3107 (2009)
[31] J. Kopecky, M. Uhl, Phys. Rev. C **41**, 1941 (1990)
[32] D. Brink, Nucl. Phys. **4** (1957)
[33] P. Axel, Phys. Rev. **126** (1962)
[34] A.C. Larsen, I.E. Ruud, A. Bürger, S. Goriely, M. Guttormsen, A. Görgen, T.W. Hagen, S. Harissopulos, H.T. Nyhus, T. Renstrøm et al., Phys. Rev. C **87**, 014319 (2013)
[35] H.K. Toft, A.C. Larsen, A. Bürger, M. Guttormsen, A. Görgen, H.T. Nyhus, T. Renstrøm, S. Siem, G.M. Tveten, A. Voinov, Phys. Rev. C **83** (2011)
[36] H.K. Toft, A.C. Larsen, U. Agvaanluvsan, A. Bürger, M. Guttormsen, G.E. Mitchell, H.T. Nyhus, A. Schiller, S. Siem, N.U.H. Syed et al., Phys. Rev. C **81**, 064311 (2010)

Extended Intranuclear Cascade model for pickup reactions induced by 50-MeV-range protons

Yusuke Uozumi[1a], Taiki Mori[1], Akifumi Sonoda[1] and Masahiro Nakano[2]

[1]*Kyushu University, Faculty of Engineering, 819-0395 Fukuoka, Japan*
[2]*Junshin Gakuen Univeersity, Faculty of Health Sciences, 815-8150 Fukuoka, Japan*

Abstract. The intranuclear cascade model was investigated to explain (*p*, *dx*) and (*p*, *αx*) reactions at incident energies of around 50 MeV. Since these reactions are governed mainly by the direct pickup process, the model was expanded to include exclusive pickup processes leading to hole-state-excitations. The energy of the outgoing clusters is determined with single-particle energies of transferred nucleons, the reaction *Q*-value, and the recoil of the residual nucleus. The rescattering of the produced cluster inside the nucleus is treated within the intranuclear cascade model. The emission angle is given by the sum of momentum vectors of transferred nucleons in addition to the deflection at the nuclear surface, which was introduced to explain angular distributions of elastic scattering. Double differential cross sections of reactions were calculated and compared with experimental data. The proposed model showed a high predictive power over the wide range of emission energies and angles. The treatment of the cluster transport inside the nucleus was also verified.

1 Introduction

The intranuclear cascade (INC) model is a powerful tool for predicting the double-differential cross sections (DDXs) of nuclear reactions. Many studies have been devoted to improving the accuracy of the INC model. As a consequence, the INC model has high predictive power at energies above a few hundreds of MeV, and plays an essential role in transport code systems for a variety of applications of engineering, medicine and fundamental sciences. In our previous works, the INC model has been successfully extended toward proton productions [1], and cluster productions [2,3] induced by protons of energies above 200 MeV or higher. More recently, we have successfully made extension of the INC model to the lower beam energies, around 50 MeV [4,5] by including collective excitations, trajectory deflections and barrier transmission coefficients. However, this improvement was carried out for only (*p*, *p'x*) reactions. For broadening the applicability of transport codes, further extension toward cluster production reactions below 100 MeV is crucial.

The INC model assumes that nuclear reactions can be described by sequential nucleon-nucleon collisions inside the nucleus, and that all nucleons other than the colliding pair behave as spectators. To describe the cluster production reactions, we have developed [2,3] the INC model to describe the knockout and the indirect pickup process by including nucleon correlations. These processes are responsible for reaction of incident energies above 100 MeV, and are suppressed strongly at the lower

[a] Corresponding author: uozumi@nucl.kyushu-u.ac.jp

energies. INC models developed elsewhere [6,7] incorporate the coalescence model, in which emitting particles within a certain phase space can be combined to form a cluster. The coalescence model is similar to the indirect pickup model, and cannot explain the high energy part of spectra where clusters have energies close to the incident beam energy. None of these models can be relied on to show a high predictive power at low beam energies. Since experimental observations of (p, d) reactions reveal discrete levels, which are attributed to the direct pickup process, up to at least 10 MeV of excitation energies, the inclusion of the direct pickup should therefore be a key to extending the INC model. It bears emphasis that all existing INC models lack explicit treatment of the direct pickup process. In the present work, we extend the INC model to (p, dx) and $(p, \alpha x)$ reactions of around 50 MeV.

The exclusive (p, d) reaction is explained as the direct single-neutron pickup process. The angular distribution of each lj transfer is calculated reasonably by the Distorted Wave Born Approximation (DWBA). Such discrete levels are observed up to about 10 MeV of excitations of residual nucleus. The continua at the higher excitations are not understood well. We assume that the higher excitation corresponds to the pickup from the deeply-lying shell orbit and the continua are the overlap of levels populated by intrinsic lj transfer.

The mechanism of the (p, α) reaction is unclear. The transitions to the ground state regions have been explained as the triton-pickup process in terms of DWBA analysis [8]. Although the mechanism of transitions to the excitation region higher than 5 MeV was concluded as the knockout process [8], the incident energies of 22 and 72 MeV are supposed to be too low to allow kinematically. Due to our INC prediction, α knockout is not observed at 70 MeV, and deuteron knockout appears barely. In the present study, we assume that these reactions are governed by the direct pickup process.

The rescattering of the cluster inside nucleus is treated within the INC framework. In order to check the particle transport inside the nucleus, DDXs for $(d, d'x)$ and $(\alpha, \alpha'x)$ reactions are also investigated. To describe the process, major modification of the INC model is necessary in order to include physical aspects neglected in previous studies. The evaporation stage is beyond the scope of this work.

2 Model

We assumed a two-stage model for nuclear reactions: the cascade and the evaporation processes. The cascade process is the first stage and described by the INC model, which assumes sequential *NN* collisions take place inside the target nucleus. The second stage is the evaporation process, which is the slow and statistical process. In the present work, we used the code GEM [9] for this stage. After the cascade process finishes, information of the residue nucleus is forwarded from INC to GEM.

2.1 Intranuclear cascade model

The prescription of the present INC model calculation is detailed in [1,3,4]; here we describe sole the essential points. The present model describes the time-development of the many-particle system. The target nucleus is considered to be a sphere with the nuclear radius R_0 and diffuseness a. They were given by $R_0 = \left(0.978 + 0.0206A^{1/3}\right)A^{1/3}$, and $a = 0.54$, respectively. Each nucleon is treated as spinless particles, and is given position and momentum randomly. However, target nucleons stay at the initial location before being struck out over the Fermi sea by the *NN* collision. The nucleon above the Fermi sea is assumed to travel inside the nucleus freely.

When two nucleons i and j approach to each other within a distance equivalent to the nucleon–nucleon cross section σ_{NN} [10]:

$$r_{ij} < \sqrt{\frac{\sigma_{NN}}{\pi}}, \quad (1)$$

they have a chance to undergo a collision. Pauli blocking is introduced to judge the collision phenomenologically using an operator $\hat{Q}$:

$$\hat{Q}|ij\rangle = [1 - \Theta(E_i - E_F)][1 - \Theta(E_j - E_F)]|ij\rangle, \tag{2}$$

where E_i and E_F are the energy of the nucleon after the collision and the Fermi energy, respectively. Θ denotes the unit step function. The scattering angle is determined stochastically to follow angular distributions of *NN* collisions in a way that energy and momentum are conserved.

In the INC model, DDX is assumed [4,5] to be

$$\frac{d^2\sigma}{dEd\Omega}(\theta,\varepsilon) = \sigma_{total}\frac{1}{2\pi\Delta E\Delta\cos\theta}P(\theta,\varepsilon), \tag{3}$$

where ΔE, $\Delta\cos\theta$, and σ_{total} are the bin width of the outgoing energy ε, the angle θ and proton-nucleus total reaction cross section, respectively. The form of proton emission probability P for (p, $p'x$) reactions, for instance, is given by

$$P(\theta,\varepsilon) = P_{def}^{in}(1 + P_{CE})(G + GP_{NN}G + GP_{NN}GP_{NN}G + \cdots)P_{tr}P_{def}^{out}\big|_{\theta,\varepsilon}, \tag{4}$$

where G is the space development operator for energetic particles inside the target nucleus. The deflection function P_{def} has been determined to reproduce the angular distribution of elastic scattering, and gives deflection angle at the nuclear surface for both the entrance and the exit channels. P_{CE} is probability of the energy loss due to collective excitations of which cross sections are given in [5] as function of target mass number and incident proton energy. P_{NN} is *NN* collision probability and provides the energy and direction of scattered nucleon after the in-medium collision. P_{tr} gives barrier transmission probability [5] for the escaping proton.

2.2 Neutron pickup model for (p, d) reaction

To describe the direct (p, d) reaction process, the single-neutron pickup probability P_{np} was introduced. Since the knockout and the indirect pickup processes are suppressed strongly at incident energies around 50 MeV, the deuteron production is attributed to the direct pickup process. We assumed the pickup takes place when the incident proton enters the nucleus, and the deuteron emission probability is written by:

$$P^{pd}(\theta,\varepsilon) = P_{def}^{in} P_{np}(G + GP_{dN}G + GP_{dN}GP_{dN}G + \cdots)P_{tr}^{d}P_{def}^{out}\big|_{\theta,\varepsilon}, \tag{5}$$

where P_{tr}^{d} is the deuteron transmission probability, which is given by the unit step function at the Coulomb potential at a distance of $R_0 + 1.5$ fm from the center of the target nucleus. P_{dN} is a probability of the in-medium deuteron-nucleon collision. To transport the deuteron inside the nucleus, the deuteron is assumed to be a single elementary particle. The *d-N* cross section and the deuteron potential depth are assumed to be double for that of protons.

The deuteron energy E_d of the exclusive (p, d) reaction is given by

$$E_d = E_p - Ex + Q, \tag{6}$$

where E_p is the incident proton energy, Ex is the excitation energy of residual nucleus, and Q is the Q-value of the reaction, respectively. The kinematical correction is applied according to the emission angle and the excitation energy of residue. To introduce this relation into the INC model, we rewrite as

$$E_d = E_p - (E_F - E_n) + Q, \qquad (7)$$

where E_F and E_n are the Fermi energy and the single-particle energy of the transferred neutron, respectively.

When the incident proton energy is 30 MeV or lower, the deuteron spectra are dominated by transitions to the ground state region. This means that the neutron in the outer shell orbit is picked up predominantly. With increasing the incident energy, pickup from inner shells occurs more frequently. It is supposed whether the transferred neutron is picked up from the surface orbit or not is the essential point to reproduce observed spectra over a wide incident energy range. Therefore, we introduce a weighting factor to emphasis the surface pickup contribution. The weighting factor was determined depending on the beam energy and the mass of target nucleus to fit the experimental energy spectra. The illustrations are shown in Figs. 1 (A) and (B) to explain the weighting factor for the choice of picked-up neutron from the two classes.

Angular distributions of *(p, d)* reactions depend on the transferred angular momenta. In the present model, we assume that the momentum direction of the produced deuteron is given by the added vector of the incident proton and the transferred neutron. When the deuteron leaves the nucleus, it undergoes deflection.

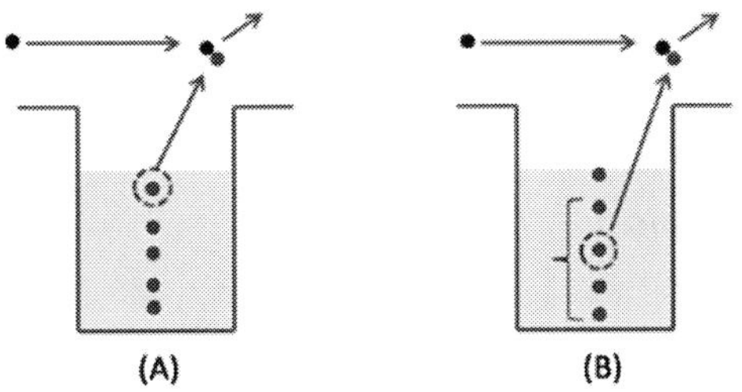

Figure 1. The two classes of the picked-up neutron in direct *(p, d)* reactions; (A) the surface one, and (B) one from the inner orbit.

2.3 Triton pickup model for (p, α) reaction

The pickup process for the exclusive *(p, α)* reaction is treated in a similar way to that for the *(p, d)* reaction. The α particle emission probability is given in the same way as eq.(5), but is displaced by triton pickup probability, α-*N* collision, and the α transmission efficiency. We assume that the residual nucleus is in a (ν^{-2}, π^{-1}) state and the α energy is given by

$$E_\alpha = E_p - (E_F - E_{n,1} - E_{n,2} - E_p) + Q, \qquad (8)$$

where E_n and E_p is the single-particle energy of transferred neutron and proton, respectively. These energies of transferred nucleons, two neutrons and one proton, are chosen with a weighting factor, which was determined to fit the energy spectra depending on incident energy and mass of target nucleus. The class of the weighting factor is four as being illustrated in Fig. 2; (A) is the case to choose the most-surface-particles. (B) and (C) are the cases that include one or two particles lying in the deeper shell orbits, respectively. The case (D) includes no surface particles.

It is assumed that the momentum direction of the produced α particle is given by the added vector of the incident proton and the transferred nucleons. When the α particle leaves the nucleus, it undergoes deflection.

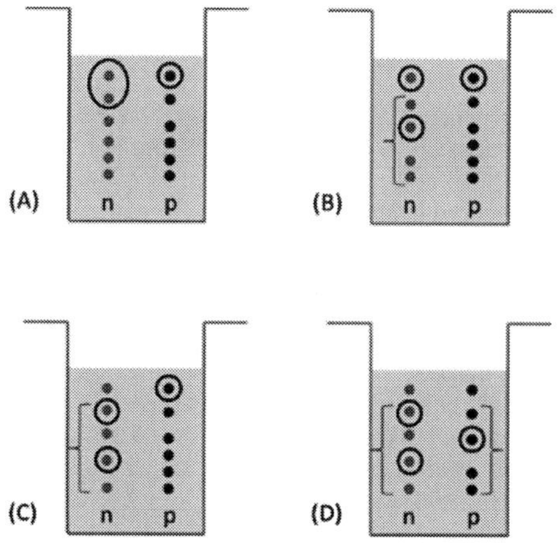

Figure 2. The class of the picked-up nucleons in direct (p, α) reactions; (A) the surface one, and (B) one neutron is from an inner orbit, (C) two neutrons are from inner orbits, and (D) all nucleons are chosen from inner orbits.

3 Results and discussion

To validate the proposed model, DDXs of (p, dx) and $(p, \alpha x)$ reactions were calculated and compared with experimental data. All the data used presently were obtained from the EXFOR data base [11].

3.1 (p, d) reactions

The solid and the dashed lines in Fig. 3 show the INC contributions of pickup from the surface and the inner orbits, respectively for the 42-MeV $^{27}Al(p, dx)$ reaction at laboratory angles of 30° and 60°.

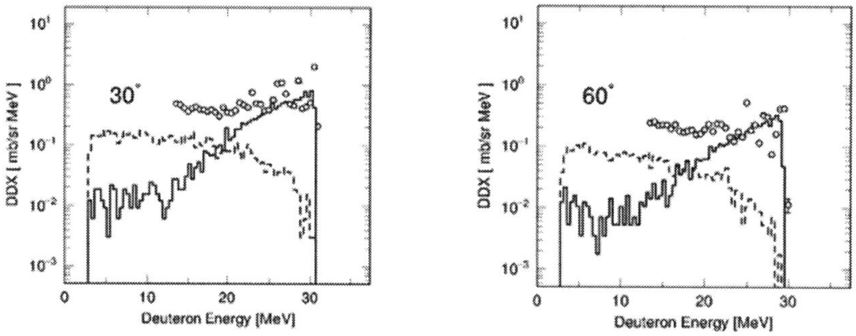

Figure 3. The INC contributions of the pickup from the surface orbit (solid line) and the inner orbits (dashed line) are shown for the $^{27}Al(p, dx)$ reaction at 42 MeV at laboratory angles of 30° (left) and 60° (right).

Open circles are experimental data. Since the experiments were conducted with detectors of poor energy resolutions, discrete levels were not very clear. Although data points distribute largely in the low excitation regions of these figures, average tendency is reproduced by solid lines, the surface pickup. The low-energy regions shown by the solid lines are due to rescattering of deuteron inside the nucleus. The pickup from inner-shells contributes to the flat shape spectra in the low-energy range.

In Fig. 4, spectral DDXs are compared between present calculation results and experiments for the same reaction as Fig. 3, in which solid lines and open circles indicate the calculation results and the experiments, respectively. Each calculation result is the sum of present INC (pickup from the surface and the inner orbits) and GEM for the evaporation. In experiments, evaporation regions were not observed due to the dead layers of detectors. Since experimental points show rising and falling due to the discrete level spectra reflected by the shell structure, there exist some discrepancies. However, the present calculations are in reasonable agreements with experimental data at four angles.

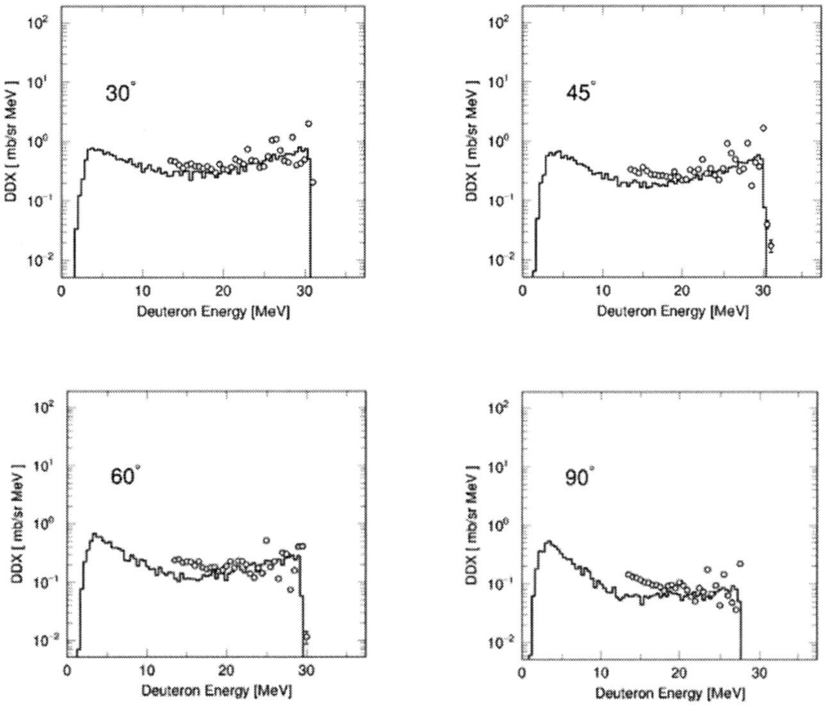

Figure 4. Comparison of double-differential cross sections (DDX) between present calculation (solid lines) and experiment (open circles) for the ^{27}Al(p, dx) reaction at 42 MeV. The calculation results include the evaporation components in addition to the total of the present INC (pickup from the surface and the inner orbits).

Figure 5 is the same as Fig. 3, but for the deuterons from the 42-MeV ^{58}Ni(p, dx) reaction. The experimental spectra in the low excitation regions are somehow sharper than those for ^{27}Al in Fig. 3. These experimental spectra are interpreted well by the calculation with the surface pickup model.

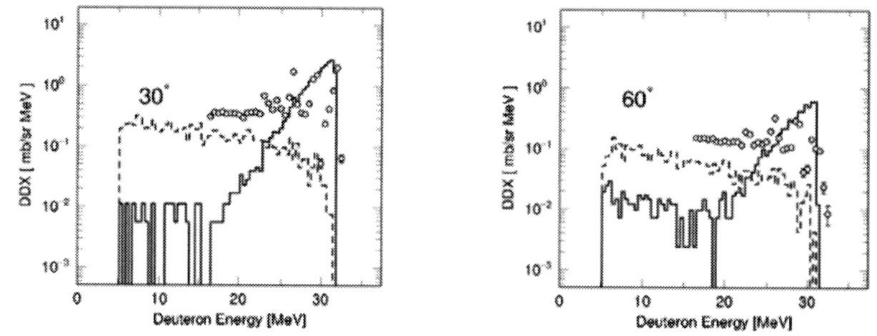

Figure 5. The contributions of the pickup from the surface orbit (solid line) and the inner orbits (dashed line) are shown for the ^{58}Ni(p, dx) reaction at 42 MeV at laboratory angles of 30° (left) and 60° (right).

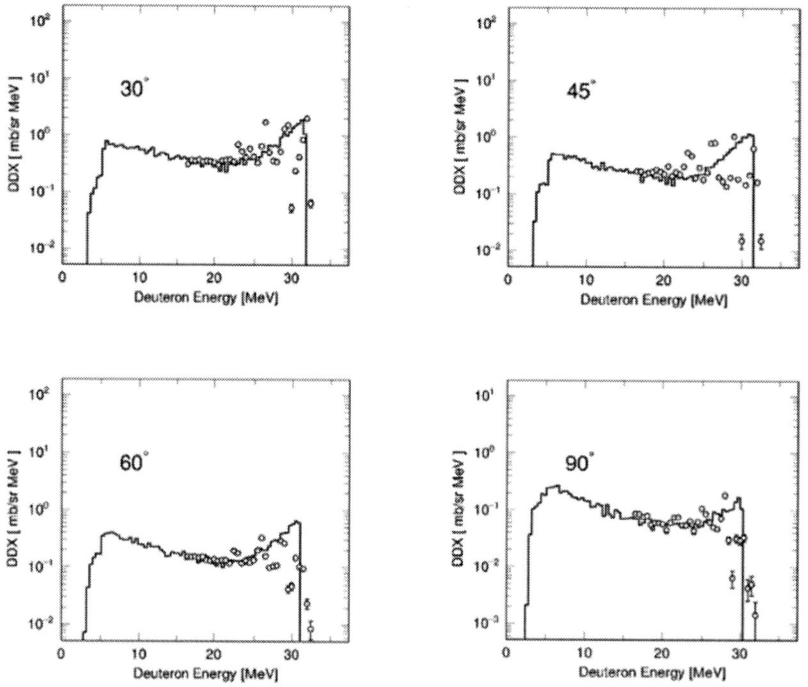

Figure 6. Comparison of double-differential cross sections (DDX) between present calculation (solid lines) and experiment (open circles) for the ^{58}Ni(p, dx) reaction at 42 MeV. The calculation results include the evaporation components in addition to the total of the present INC (pickup from the surface and the inner orbits).

Spectral DDXs for the ^{58}Ni(p, dx) reaction at 42 MeV are compared between present calculation and experiment in Fig. 6, which is the same as Fig. 4. As the case of the ^{27}Al target, reasonable agreements are observed between the present calculation and the experiment.

The present model was verified at the higher energy. Figure 7 is the same as Fig. 4 and 6, but for the ^{58}Ni(p, dx) reaction at 68 MeV. As in the case of the lower incident energy, present calculations are in reasonable agreements with the experiment.

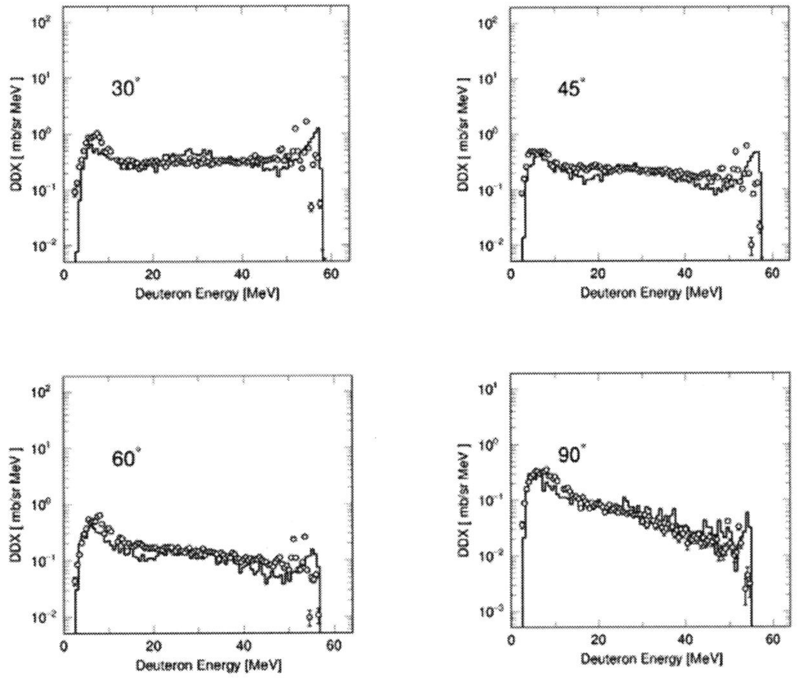

Figure 7. Comparison of double-differential cross sections (DDX) between present calculation (solid lines) and experiment (open circles) for the ^{58}Ni(p, dx) reaction at 68 MeV. The calculation results are the sum of the evaporation components and the present INC (pickup from the surface and the inner orbits).

3.2 (p, α) reactions

In Fig. 8, INC contributions of pickup from the surface and the inner orbits, which are discussed in Fig. 2, are displayed for the ^{58}Ni(p, αx) reaction at incident energies of 42 MeV (left) and 68 MeV (right) at laboratory angles of 30°. At the higher incident energy, the contribution from the inner shell becomes more important. The surface pickup is responsible to the low excitation region and the inner-orbit pickup appears in the high excitation region. These results are consistent with understanding of the direct pickup mechanism.

Spectral DDXs of the ^{58}Ni(p, αx) reaction at 42 MeV are displayed in Fig. 9 at angles from 30° to 90°, in which solid lines are the total of INC and GEM results, and dashed lines indicate GEM results only. As shown in Fig. 8, INC includes only two classes of pickup combinations at this incident

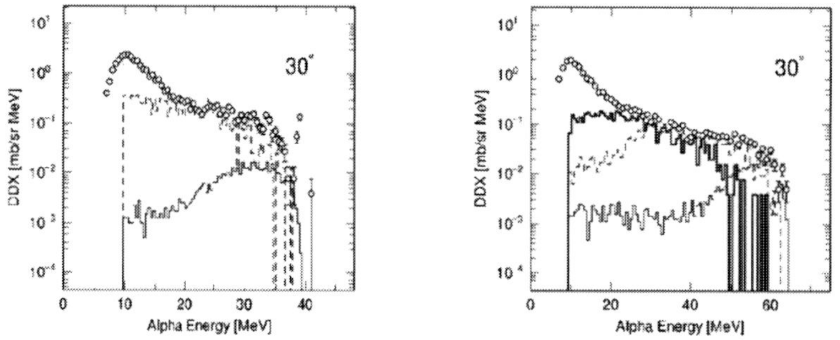

Figure 8. Contribution of each class of pickup of the present INC for ^{58}Ni(p, αx) reaction at incident energies of 42 MeV (left) and 68 MeV (right) at laboratory angles of 30°. Thine, broken, and thick lines correspond to (A)-(C) of Fig. 2, respectively.

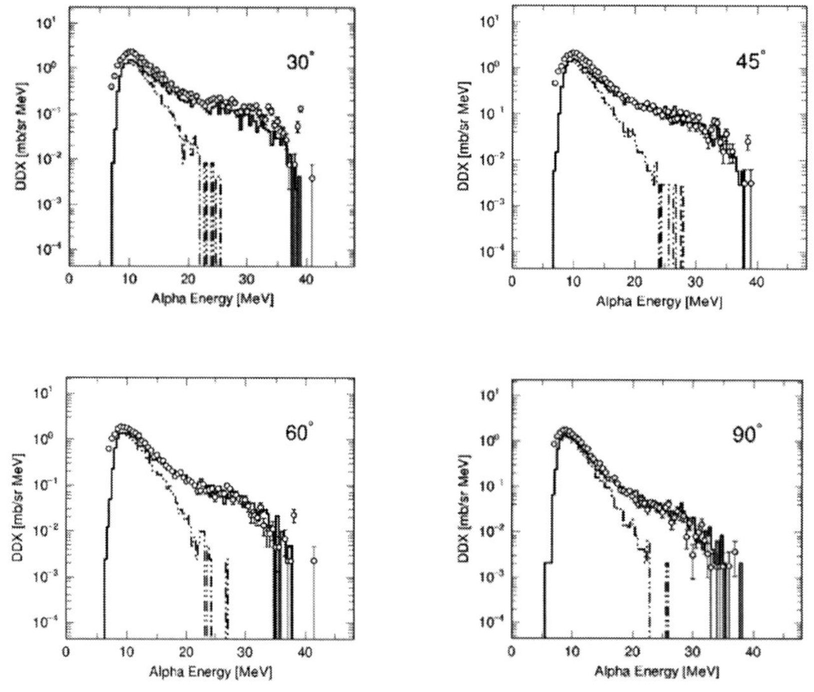

Figure 9. Comparison of double-differential cross sections (DDX) between present calculation (solid lines) and experiment (open circles) for the ^{58}Ni(p, αx) reaction at 42 MeV. The calculation results with solid lines include the evaporation components in addition to the total of the present INC (pickup from the surface and the inner orbits). Dashed lines indicate only the evaporation results.

energy. The experimental results show significant peaks, which come from the evaporation process, at around 10 MeV. With increasing the α energy, the experimental DDXs decrease gradually. This tendency is explained well by the present model. At the highest energy region, experiments show a few peaks and falling, which reflect the shell model structure of residual nuclei. Although the calculation model ignores the shell structure, calculation results are in reasonable agreements with the experiment. It should be stressed that the overall agreements are shown over the three-order changes in DDX magnitude at 90°.

In Figs. 10 and 11 are shown DDXs for the $(p, \alpha x)$ reactions at 68 MeV on ^{58}Ni and ^{27}Al, respectively. At this incident energy, the INC calculations employed four classes of pickup combinations, which are shown in Fig. 2. Experimental spectra are very similar to those at 42 MeV except that the peak structures are not observed. Although slight discrepancies are indicated, overall close agreements are shown between the present model calculations and experiments.

The triton pickup probabilities were determined to fit experimental data in terms of pickup classes defined in Fig. 2. The resultant probabilities are shown in Fig. 12 on the target of ^{27}Al and ^{58}Ni as function of incident energy.

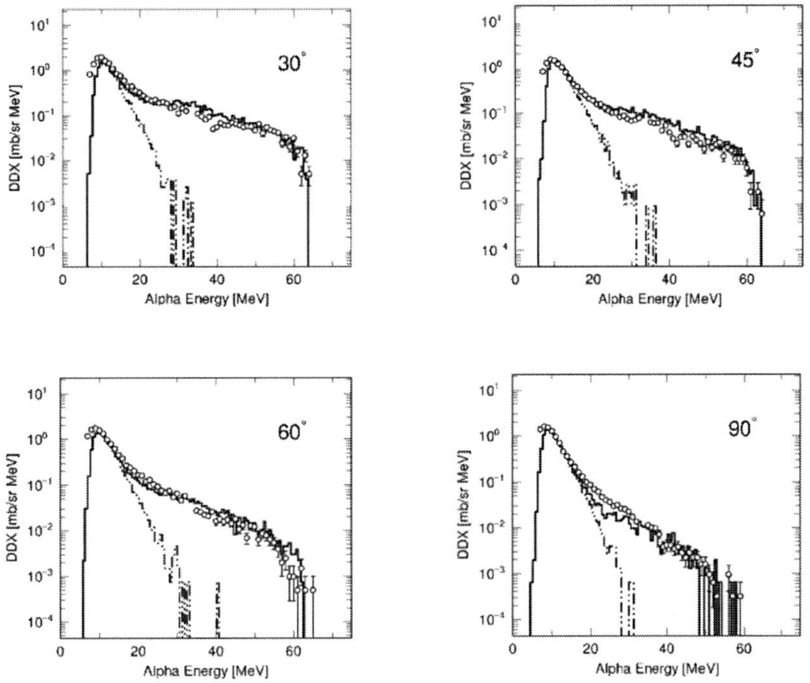

Figure 10. Comparison of double-differential cross sections (DDX) between present calculation (solid lines) and experiment (open circles) for the ^{58}Ni$(p, \alpha x)$ reaction at 68 MeV. The calculation results with solid lines include the evaporation components in addition to the total of the present INC (pickup from the surface and the inner orbits). Dashed lines indicate only the evaporation results.

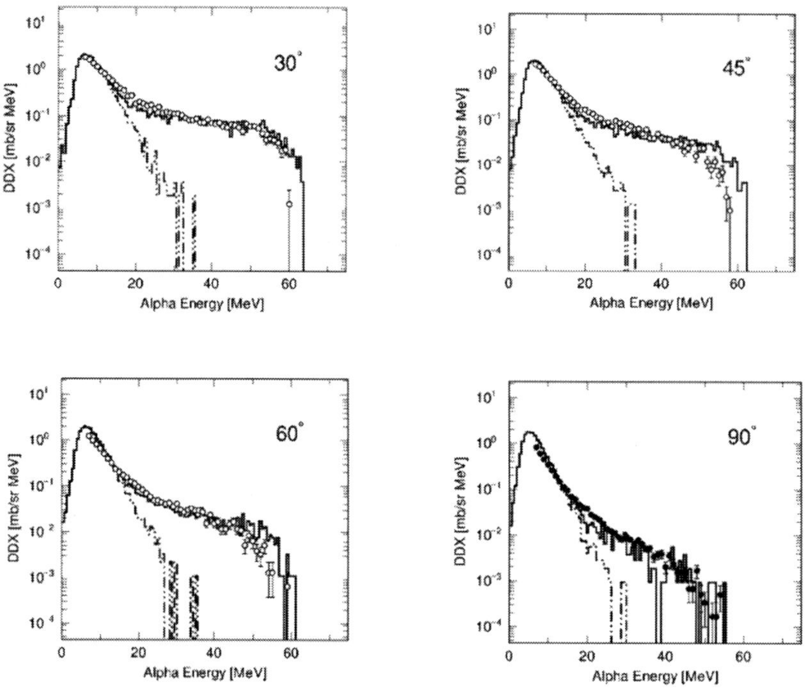

Figure 11. Same as Fig. 10, but for the ^{27}Al$(p, \alpha x)$ reaction at 68 MeV.

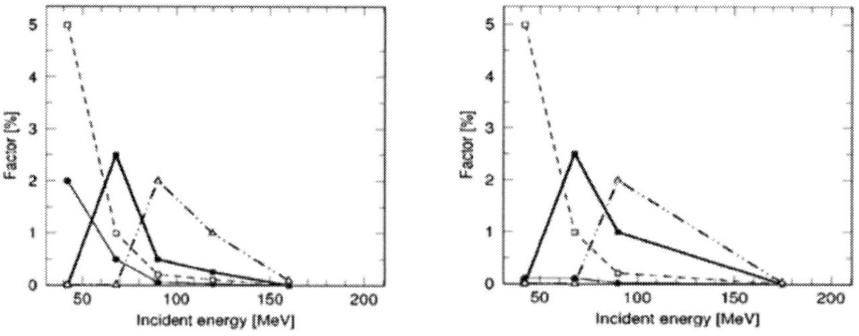

Figure 12. Factors of pickup probabilities for pickup classes of $(p, \alpha x)$ reactions on ^{27}Al (left) and ^{58}Ni (right) as function of incident energy. Dots, open squares, closed squares and open triangles show pickup classes shown in Fig. 2 (A) to (D), respectively.

3.3 (d, d'x) and (α, α'x) reactions

The present model to transport deuterons and α–particles inside the target nucleus was verified. Calculated DDXs for 80-MeV ^{27}Al (*d, d'x*) reactions are compared with experiments in Fig. 13, and the comparison for 140-MeV ^{27}Al(α, α'x) reactions were displayed in Fig. 14. The observed angles are 30° and 90° for the both cases. Disagreements appear at the forward angle, where collective excitation process contributes significantly. At the large angle, the experimental magnitude variations over three-orders are interpreted successfully by the present model. Therefore, it is validated that the present model reproduces experimental data reasonably.

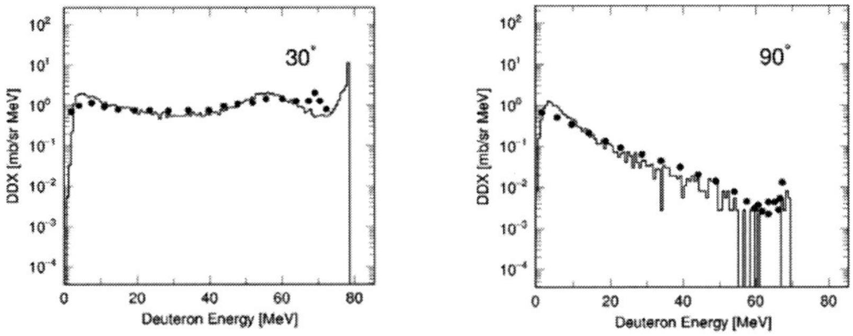

Figure 13. Comparison of double-differential cross sections (DDX) for the ^{27}Al(d, $d'x$) reaction at 80 MeV, at laboratory angles of 30° (left) and 90 ° (right). Solid lines are the calculation results including the evaporation components. Dots are experimental results.

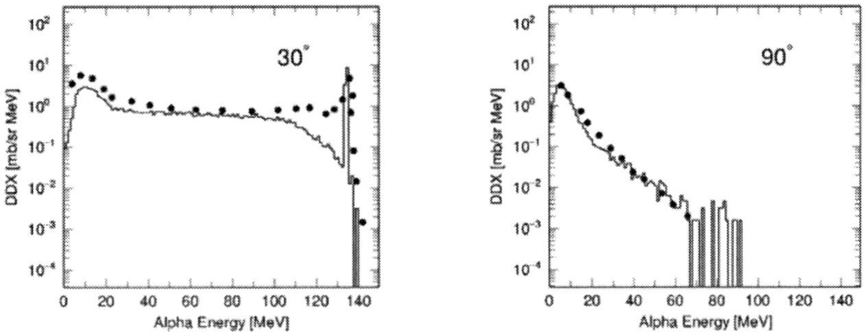

Figure 14. Comparison of double-differential cross sections (DDX) for the ^{27}Al(α, $\alpha'x$) reaction at 140 MeV, at laboratory angles of 30° (left) and 90 ° (right). Solid lines are the calculation results including the evaporation components. Dots are experimental results.

4 Conclusion

The Intranuclear Cascade (INC) model was extended to incorporate the direct pickup process to interpret (p, dx) and (p, αx) reactions at incident energies below 100 MeV, where the standard INC model has poor predictive ability. The proposed pickup model is based on a semi-classical interpretation of the nucleon transfer processes. In order to reproduce the whole energy range of emitted particles, possible classes of picked-up particles were introduced.

Calculations of the proposed model followed by the generalised evaporation model GEM were performed for (p, dx) and (p, αx) reactions on ^{27}Al and ^{58}Ni at incident energies of 40 and 68 MeV to be compared with experiments. Some of the experimental data distribute largely due to the reflection of the shell model structure, which was ignored in the model. However, the overall close agreements are confirmed.

References

1. H. Iwamoto, M. Imamura, Y. Koba, Y. Fukui, G. Wakabayashi, Y. Uozumi, T. Kin, Y. Iwamoto, S. Hohara, M. Nakano, Phys. Rev. C **82**, 034604 (2010).
2. Y. Uozumi, *Proceedings of the 2010 Symposium on Nuclear Data*, JAEA-Conf 2011-02, 65 (2011).
3. Y. Uozumi, Y. Sawada, A. Mzhavia, S. Nogamine, H. Iwamoto, T. Kin, S. Hohara, G. Wakabayashi, M. Nakano, Phys. Rev. C **84**, 064617 (2011).
4. Y. Uozumi, T. Yamada, S. Nogamine, M. Nakano, Phys. Rev. C **86**, 034610 (2012).
5. Y. Uozumi, T. Yamada, M. Nakano, Jour. Nucl. Sci. Tech. **52**, 264 (2015).
6. G. Montarou, J. P. Alard, J. Augerat, L. Fraysse, M. J. Parizat, R. Babinet, Z. Fodor, et al., Phys. Rev. C **44**, 365 (1991).
7. A. Boudard, J. Cugnon, J.-C. David, S. Leray, D. Mancusi, Phys. Rev. C **87**, 014606 (2013).
8. R. Bonetti, F. Crespi, K.-I.Kubo, Nucl. Phys. A **499**, 381 (1989).
9. S. Furihata and T. Nakamra, Jour. Nucl. Sci. Tech. **39**, Suppl. 2, 758 (2001).
10. J. Cugnon, D. L'Hote, J. Vandermeulen, Nucl. Instrum. Methods Phys. Res. B**11**, 251 (1996).
11. EXFOR Experimental Nuclear Reaction Data, https:// www-nds.iaea.org/exfor/exfor.htm.

Feasibility study of nuclear transmutation by negative muon capture reaction using the PHITS code

Shin-ichiro Abe[1a], and Tatsuhiko Sato[1]

[1]*Japan Atomic Energy Agency, Nuclear Science and Engineering Center, 2-4 Shirakata, Tokai-mura, Naka-gun, Ibaraki 319-1195, Japan*

Abstract. Feasibility of nuclear transmutation of fission products in high-level radioactive waste by negative muon capture reaction is investigated using the Particle and Heave Ion Transport code System (PHITS). It is found that about 80 % of stopped negative muons contribute to transmute target nuclide into stable or short-lived nuclide in the case of ^{135}Cs, which is one of the most important nuclide in the transmutation. The simulation result also indicates that the position of transmutation is controllable by changing the energy of incident negative muon. Based on our simulation, it takes approximately 8.5×10^{8} years to transmute 500 g of ^{135}Cs by negative muon beam with the highest intensity currently available.

1 Introduction

Partitioning and transmutation (P&T) have been investigated to reduce the radiotoxicity and storage time of the high-level radioactive waste (HLW) [1-5]. In the present reprocessing, U and Pu are recovered from spent fuel and HLW is vitrified. In the P&T process, HLW is separated into below groups; minor metals (e.g., Ru, Rh, Pd), minor actinides (MA), Sr and Cs, and the other fission products (FPs). Minor metals are utilized as catalysts, material for electronic device and so on. MA are transmuted into stable of short-lived nuclei by neutron irradiation with nuclear fuels in the thermal reactor, fast reactor or accelerator driven sub-critical system (ADS). Sr and Cs are stabilized as a calcified waste form because they have high heat load, and the other FPs are vitrified.

After transmutation of MA, residual FPs become major cause for long-term repository of radioactive wastes. Therefore, transmutation of FP contributes further reduction of storage time. Several studies of the transmutation of FP have been performed [4, 5]. Even though transmutation by thermal neutrons can be applied to some nuclides (e.g., ^{99}Tc, ^{129}I), it is difficult to transmute FPs having small cross section for neutron capture reaction. Table 1 shows the thermal neutron capture cross section, content in spent fuel and half-life of typical FPs.

Negative muon has possibility to apply transmutation of FP through capture reaction. As a result of the weak interaction, a proton in the nucleus changes to a neutron and nucleus gains excitation energy:

$$\mu^{-} + \mathrm{X}(N,Z) \rightarrow \nu_{\mu} + \mathrm{Y}(N+1, Z-1)^{*} + (m_{\mu} + m_{\mathrm{X}} - m_{\mathrm{Y}})\mathrm{MeV} \quad (1)$$

[a] Corresponding author: abe.shinichiro@jaea.go.jp

Table 1. Thermal neutron capture cross section [6], content in spent fuel and half-life [7] of typical FPs.

Radio isotope	Cross section of (n_{th}, γ) [barn]	Content in spent fuel [kg/ton]	Half-life [year]
^{79}Se	50	0.006	2.95e+5
^{90}Sr	0.0101	0.6	28.79
^{93}Zr	2.239	1	1.61e+6
^{99}Tc	23.59	1	2.11e+5
^{107}Pd	9.241	0.3	6.5e+6
^{126}Sn	0.09	0.03	2.30e+5
^{129}I	30.31	93.3	1.57e+7
^{135}Cs	8.3	223	2.3e+6
^{137}Cs	0.27	659	30.08

where m_μ, m_X and m_Y are the rest mass of negative muon and nucleus before and after negative muon capture reaction, respectively. The unique feature of negative muon capture reaction is the controllability of capture position. A negative muon is captured by nucleus just before it stops.

Recently, negative muon capture process has been implemented into the Particle and Heavy Ion Transport code System (PHITS) [8]. In the present work, the production ratio of secondary nucleus from negative muon capture reaction on typical FPs are calculated using PHITS. The controllability of the position of transmutation is demonstrate. After that, the feasibility of nuclear transmutation by negative muon capture reaction for FPs is discussed.

2 Model of negative muon capture reaction in PHITS

At first, a negative muon is captured by an atom and the muonic atom is formed. The captured negative muon cascades down to the 1s orbit while emitting characteristic X-rays. After that, the negative muon is decayed in the 1s orbit or captured by the nucleus. As shown in eq. (1), a proton in the nucleus captures the negative muon and it becomes a neutron.

In PHITS, the muonic atom cascade program [9] is employed to simulate the cascade of the captured negative muon in the atomic orbits. The decay rate of negative muon is derived by using the Huff factor taken from [10] or calculated by fitting function of [11]. The nuclear capture rate of negative muon is calculated by the Goulard-Primakoff formula [12]. For light atoms (Z < 10), the measured nuclear capture rate taken from [10] are tableted and used.

When a negative muon is captured by a nucleus, the excitation energy of the nucleus is sampled from excitation function proposed in [13]. A proton in the nucleus is selected randomly and is replaced to a neutron added the excitation energy. After that, the time evolution of nucleons is calculated by JAERI Quantum Molecular Dynamics (JQMD) model [14] and evaporation process is calculated by Generalized Evaporation Model (GEM) [15], respectively.

Figure 1 shows the differential cross section of secondary neutron from negative muon capture on ^{40}Ca and ^{208}Pb. The experimental data [16, 17] are also plotted. PHITS is in good agreement with experimental data at low emission energies for ^{40}Ca target, although it tends to overestimate for ^{208}Pb. At emission energies above 30 MeV, PHITS underestimate neutron production for both targets. It was reported that high energy neutron is generated from $\mu \rightarrow q\bar{q}\nu_\mu$ decay mechanism [11], which does not considered in PHITS. Figure 2 compares the neutron multiplicity from negative muon capture on ^{nat}Ag and ^{208}Pb with the experimental data [18]. PHITS describes the neutron multiplicity well over the whole range for ^{nat}Ag, whereas PHITS overestimate the number of emitted neutron from ^{208}Pb. It is expected that PHITS describes the production of secondary nuclei from negative muon

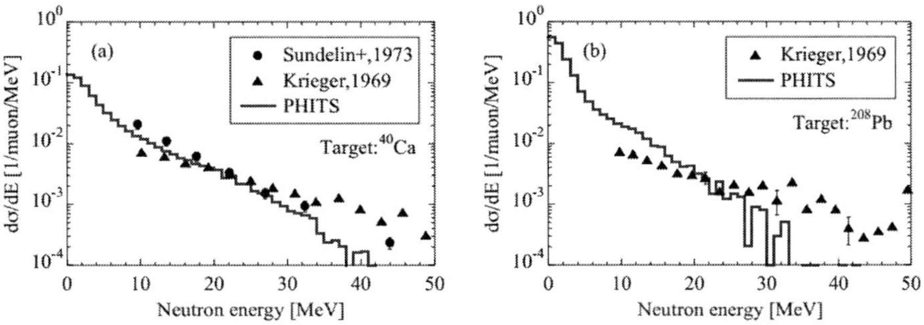

Figure 1. Differential cross section of secondary neutron from negative muon capture on (a) ^{40}Ca and (b) ^{208}Pb. The experimental data are taken from [16, 17].

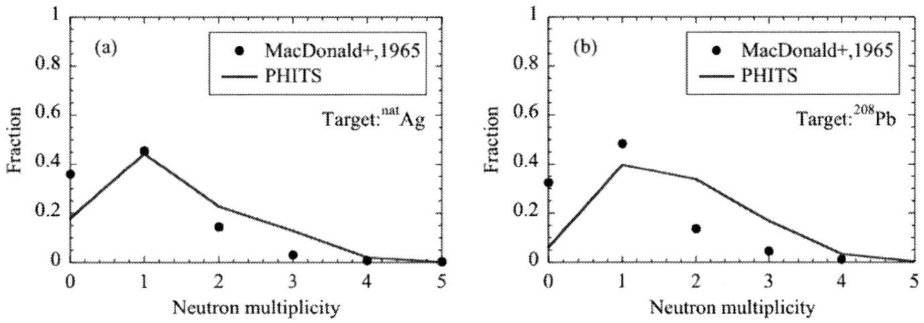

Figure 2. Neutron multiplicity from negative muon capture on (a) ^{nat}Ag and (b) ^{208}Pb. The experimental data is taken from [18].

capture on FPs within reasonable range because the atomic number of target FPs is closer to Ag rather than Pb.

3 Simulation of nuclear transmutation for FPs

Nuclear transmutation by negative muon capture is simulated for ^{79}Se, ^{90}Sr, ^{93}Zr, ^{99}Tc, ^{107}Pd, ^{126}Sn, ^{129}I, ^{135}Cs and ^{137}Cs. Table 2 shows the capture rate, the decay rate, and mean productions of neutron and proton for each target nuclide. About 5 % of stopped negative muons decay in orbit for all target nuclide and hence they do not contribute to transmutation. The number of event for proton emission is less than 8 %, thus the target nucleus with an atomic number of Z is mostly transmuted to nucleus with an atomic number of Z–1. Table 3 and 4 show the production ratio and decay chain of secondary nuclides produced by negative muon capture reactions on ^{135}Cs and ^{137}Cs as examples. It is found that 12.5 % of ^{135}Cs become ^{135}Xe and they return to ^{135}Cs by β- decay. As a result, 83.4 % of ^{135}Cs become to stable or short-lived nuclides. In the case of ^{137}Cs, 4.6 % of ^{137}Cs return to the original nuclide and 26.7 % of ^{137}Cs are transmuted to ^{135}Xe and they become to ^{135}Cs having longer half-life than original nuclide. Consequently, 61.5 % of ^{137}Cs become to stable or short-lived nuclides. For the other FPs, approximately 66 % to 87 % of FPs are transmuted to stable or short-lived nuclides by negative muon capture reaction.

In order to demonstrate the controllability of the position of transmutation, irradiation of negative muon beam on ^{135}Cs target are simulated by PHITS. The momentum of incident negative muon is 60

Table 2. Capture rate, decay rate, and mean productions of neutron and proton for each target nuclide.

Target nuclide	Capture rate [1/muon]	Decay rate [1/muon]	Mean production [1/muon]	
			neutron	proton
^{79}Se	0.925	0.075	1.197	0.059
^{90}Sr	0.935	0.065	1.540	0.048
^{93}Zr	0.948	0.052	1.413	0.068
^{99}Tc	0.949	0.051	1.432	0.070
^{107}Pd	0.962	0.038	1.325	0.081
^{126}Sn	0.940	0.060	1.622	0.008
^{129}I	0.959	0.041	1.436	0.062
^{135}Cs	0.959	0.041	1.401	0.063
^{137}Cs	0.954	0.046	1.632	0.052

Table 3. Production ratio and decay chain of secondary nuclides from negative muon capture reactions on ^{135}Cs

Product [1/muon]	0.129	0.224	0.393	0.125
Decay chain (Half-life)	^{132}Xe (stable)	^{133}Xe (5.25 d) ▼ ^{133}Cs (stable)	^{134}Xe (stable)	^{135}Xe (9.14 h) ▼ ^{135}Cs (2.3e+6 y) ▼ ^{135}Ba (stable)

Table 4. Production ratio and decay chain of secondary nuclides from negative muon capture reactions on ^{137}Cs

Product [1/muon]	0.163	0.267	0.388	0.046
Decay chain (Half-life)	^{134}Xe (stable)	^{135}Xe (9.14 h) ▼ ^{135}Cs (2.3e+6 y) ▼ ^{135}Ba (stable)	^{136}Xe (stable)	^{137}Xe (3.18 m) ▼ ^{137}Cs (30.1 y) ▼ ^{137}Ba (stable)

MeV/c with the deviation of 5 % (i.e., it corresponds to the incident energy of 16 ± 1.5 MeV) which desired momentum for muon beam of MuSIC [19] at RCNP. The shape of target is cylinder with a radius of 1 cm and a length of 4 cm. A lead moderator with a thickness of 1.5 mm is put in front of the target. The energy loss of the incident negative muon in the moderator is about 6 MeV. The energy straggling and the angular straggling are considered in the simulation. Figure 3 shows the depth distribution of produced nucleus with or without the moderator. The central depth of produced nucleus is 0.7 cm and 1.5 cm, which correspond to the range of the negative muon in ^{135}Cs target having the energy of 10 MeV and 16 MeV, respectively. The spatial deviation of produced nucleus is 0.35 cm

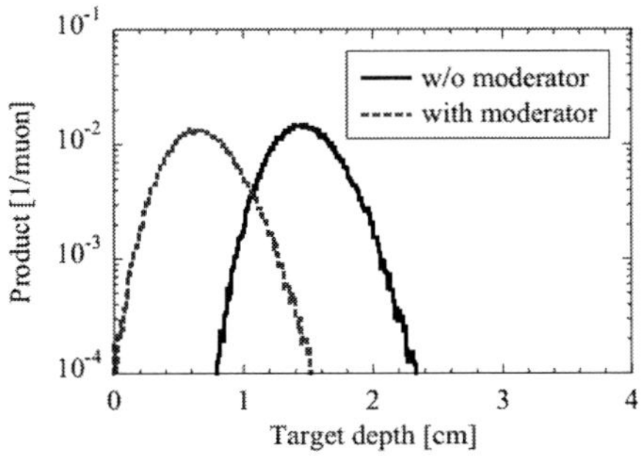

Figure 3. Depth distribution of produced Xe with or without the lead moderator.

due to the deviation of the incident momentum. This result indicates that the energy distribution of incident negative muon is directly reflected on the spatial distribution of secondary nuclei. The position of transmutation is controllable by using moderator or changing beam energy.

Finally, the condition of negative muon source required for transmutation is discussed. The time required to transmute of all FPs can be estimated briefly as follows:

$$t = N_A \times \frac{g}{M} \times \frac{\alpha}{\phi} \tag{2}$$

where N_A is the Avogadro's number, g is the content of the FP in spent fuel, M is the mass number of the target FP, α is the transmutation ratio to stable or short-lived nucleus, and ϕ is the intensity of negative muon. Even though the intensity of negative muon reaches 10^8 muons/sec which corresponds to the muon intensity of MuSIC at RCNP, it takes about 8.5×10^8 year to transmute 500 g of ^{135}Cs. Therefore, much higher intensity of negative muon source is required for transmutation by negative muon capture reaction.

4 Conclusions

Nuclear transmutation of FPs by negative muon capture reaction was simulated by PHITS. It was found that target nuclide with an atomic number of Z is mostly transmuted to nuclide with an atomic number of Z–1. As a result, for our investigated case, 61 % to 87 % of stopped negative muons contribute to transmute target nuclide into stable or short-lived nuclide. In the case of ^{137}Cs, 26.7 % of ^{137}Cs is transmuted into ^{135}Cs which have longer life-time than original nuclide. The irradiation of negative muon beam was also simulated. The result indicates that the position of transmutation by negative muon capture is controllable by changing incident energy of negative muon. From a rough estimate, it takes about 8.5×10^6 year to transmute 500 g of ^{135}Cs with the negative muon intensity of 10^8 muons/sec. We conclude that much higher intensity of negative muon source is required for transmutation by negative muon capture reaction.

References

1. OECD/NEA, NEA Report No.6894 (2011).
2. K. Nishihara et al., J. Nucl. Sci. Technol., **45**, 84 (2012).
3. K. Tsujimoto et al., J. Nucl. Sci. Technol., **41**, 21 (2004).
4. T. Wakabayashi et al., Prog. Nucl. Energy, **32**, 555 (1998).
5. W. S. Yang et al., Nucl. Sci. Eng., **146**, 291 (2004).

6. K. Shibata et al., J. Nucl. Sci. Technol., **48**, 1 (2011).
7. Japanese Nuclear Data Committee, Chart of the Nuclides (2014).
8. T. Sato et al., J. Nucl. Sci. Technol., **50**, 913 (2013).
9. V. R. Akylas et al., Comp. Phys. Comm., **15**, 291 (1978).
10. T. Suzuki et al., Phys. Rev. C, **35**, 2212 (1987).
11. M. V. Kossov, Eur. Phys. J. A, **33**, 7 (2007).
12. B. Goulard et al., Phys. Rev. C, **10**, 2034 (1974).
13. P. Singer et al., Il Nuovo Cimento, **23**, 669 (1962).
14. K. Niita et al., Phys. Rev. C, **52**, 2620 (1995).
15. S. Furihata, Nucl. Instr. and Meth. in Phys. Res. B, **171**, 251 (2000).
16. R. M. Sundelin et al., Phys. Rev. C, **7**, 1037 (1973).
17. M. H. Krieger, Ph.D Thesis, Columbia Univ. (1969).
18. B. MacDonald et al., Phys. Rev., **139**, B1253 (1965).
19. Y. Hino et al., Nucl. Phys. B Proc. Suppl., **253-255**, 206-207 (2014).

Pion Production from Proton Synchrotron Radiation under Strong Magnetic Field in a Relativistic Quantum Approach

Tomoyuki Maruyama[1,a], Myung-Ki Cheoun[2], Toshitaka Kajino[3], and Grant J. Mathews[4]

[1] *College of Bioresource Sciences, Nihon University, Fujisawa 252-8510, Japan*
[2] *Department of Physics, Soongsil University, Seoul, 156-743, Korea*
[3] *National Astronomical Observatory of Japan, 2-21-1 Osawa, Mitaka, Tokyo 181-8588, Japan*
[4] *Center of Astrophysics, Department of Physics, University of Notre Dame, Notre Dame, IN 46556, USA*

Abstract. We study pion production from proton synchrotron radiation in the presence of strong magnetic fields by using the exact proton propagator in a strong magnetic field and explicitly including the anomalous magnetic moment. Results in this exact quantum approach do not agree with those obtained in the semi-classical approach. Then, we find that the anomalous magnetic moment of the proton greatly enhances the production rate by about two orders magnitude, and that the decay width satisfies a robust scaling law.

1 Introduction

It is widely accepted that soft gamma repeaters (SGRs) and anomalous X-ray pulsars (AXPs) correspond to magnetars [1], and that the associated strong magnetic fields may have a significant role in the production of high energy photons.

The magnetars have also been proposed [2, 3] as an acceleration site for ultra high-energy (UHE) cosmic rays (UHECRs), and a possible association between magnetar flares [4] and UHECRs has also been observed. Synchrotron radiation can be produced by high-energy protons accelerated in an environment containing a strong magnetic field. This process has been proposed as a source for high-energy photons in the GeV – TeV range [5–10]. The meson-nucleon couplings are about 100 times larger than the photon-nucleon coupling, and the meson production process is expected to exceed photon synchrotron emission in the high energy regime.

However, theoretical calculations were performed approximately in a semi-classical way [11–15] and each model gave a different result. In addition, these model could not give a momentum-distribution of a final pion.

In the semi-classical approach for the production of synchrotron radiation, the magnetic field strength is characterized by the curvature parameter, $\chi = E_i^3/(M^3 R_c)$ given in terms of the incident particle mass M, its energy E_i, and the curvature radius R_c. For protons propagating in a strong magnetic field, the value of χ can be written as

$$\chi = \frac{E_i^2}{M_p^3 R_c} = \frac{eBE_i}{M_p^3}, \tag{1}$$

[a] e-mail: maruyama.tomoyuki@nihon-u.ac.jp

where M_p is the proton mass. Pion production is the dominant process compared to direct photon emission when $\chi \sim 0.01 - 1$ [15].

There are various semi-classical calculations [11–15] which give different results, but all of those models suggest that the decay width depends only on the parameter χ, not on the initial energy E_i and the strength of the magnetic field B individually, though this has not been confirmed.

In this work, we examine the scaling relation in the quantum calculation. Based upon this we can realistically estimate for the first time the results for a much larger Landau numbers from the results obtained from a smaller number of Landau levels.

Furthermore, we compare the results with those in the semi-classical approaches [14, 15]. In addition, we show the energy and angular distributions of emitted pions.

2 Formalism

Here, we briefly explain our formalism.

We assume a uniform magnetic field along the z-direction, $\boldsymbol{B} = (0, 0, B)$, and take the electromagnetic vector potential A^μ to be $A = (0, 0, xB, 0)$ at the position $\boldsymbol{r} \equiv (x, y, z)$. The relativistic proton wave function ψ is obtained from the following Dirac equation:

$$\left[\alpha_z p_z - i\alpha_x \partial_x + \alpha_y (p_y - eB_x) + M_p \beta - \frac{\kappa_p e}{M_p} B\Sigma_z\right] \psi(x, p_z, s) = E\psi(x, p_z, s), \tag{2}$$

where M_p is the proton mass, κ_p is the proton anomalous magnetic moment (AMM), e is the elementary charge, and E is the single particle energy written as with

$$E(n, p_z, s) = \sqrt{p_z^2 + (\sqrt{2eBn + M_p^2} - s\kappa_p eB/M_p)^2}. \tag{3}$$

To calculate the pionic decay width, furthermore, we use the pseudo-vector coupling interaction Lagrangian density as

$$\mathcal{L} = \frac{if_\pi}{m_\pi} \bar{\psi} \gamma_5 \gamma_\mu \tau_a \psi \partial^\mu \phi_a, \tag{4}$$

where f_π is the pseudo-vector pion-nucleon coupling constant, m_π is the pion mass, and ϕ is the pion field. Then, we obtain the differential decay width of the proton as

$$\frac{d^3\Gamma_{p\pi}}{dq^3} = \frac{1}{8\pi^2 E_\pi} \left(\frac{f_\pi}{m_\pi}\right)^2 \sum_{n_f, s_f} \delta(E_f + E_\pi - E_i) W_{if}, \tag{5}$$

with

$$\begin{aligned} W_{if} &= \mathrm{Tr}\left\{\rho_M(n_i, s_i, q_z) O_\pi \rho_M(n_f, s_f, p_z - q_z) O_\pi^\dagger\right\}, \\ \rho_M &= \frac{\left[E\gamma_0 + \sqrt{2eBn}\gamma^2 - p_z\gamma^3 + M_p + (\kappa_p eB/M_p)\Sigma_z\right]}{4E} \left[1 + \frac{s\left(\kappa_p eB/M_p + p_z \gamma_5 \gamma^0 - E\gamma_5 \gamma^3\right)}{\sqrt{(2eBn + M_p^2)}}\right], \end{aligned}$$

where $q \equiv (q_0; \boldsymbol{q})$ is a momentum of a emitted pion. In the above equation, f_n is the harmonic oscillator wave function with the principle quantum number n, and

$$\begin{aligned} O_\pi &= \gamma_5 \left\{\left[\mathcal{M}\left(n_i, n_f\right) \frac{1+\Sigma_z}{2} + \mathcal{M}\left(n_i - 1, n_f - 1\right) \frac{1-\Sigma_z}{2}\right]\left[\gamma_0 q_0 - \gamma^3 q_z\right] \right. \\ &\quad \left. - \left[\mathcal{M}\left(n_i, n_f - 1\right) \frac{1+\Sigma_z}{2} + \mathcal{M}\left(n_i - 1, n_f\right) \frac{1-\Sigma_z}{2}\right] \gamma^2 q_T\right\}, \end{aligned} \tag{6}$$

where $q_T \equiv \sqrt{q_x^2 + q_y^2}$, and $\mathcal{M}(n_1, n_2)$ is the harmonic oscillator (HO) overlap integral defined as

$$\mathcal{M}(n_1, n_2) = \sqrt{eB} \int dx f_{n_1}\left(\sqrt{eB}x + \frac{q_T}{2\sqrt{eB}}\right) f_{n_2}\left(\sqrt{eB}x - \frac{q_T}{2\sqrt{eB}}\right). \tag{7}$$

3 Results

Now, we show numerical results for pion emission.

In Fig.1, we show the total pionic decay widths of the proton with $p_{iz} = 0$ when $B = 10^{18}$G (a) and $B = 5 \times 10^{17}$G (b) as functions of the Landau level n_i. These decay widths are averaged over the proton spin, $\Gamma = [\Gamma(n_i, +1) + \Gamma(n_i, -1)]/2$. The solid and dot-dashed lines represent the decay widths of the proton with and without AMM, respectively. We see that the decay width with AMM is about 50 times larger than that without AMM.

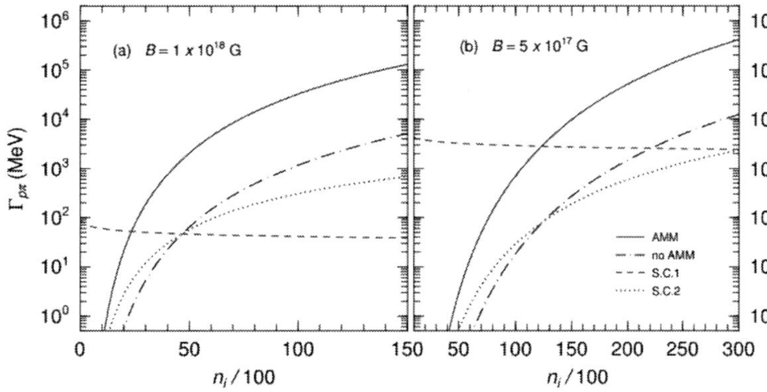

Figure 1. Pionic decay width of protons versus the initial Landau level at $p_{iz} = 0$ for $B = 5 \times 10^{17}$G. The decay width is averaged over the proton spin. The solid and dot-dashed lines represent the decay widths of the proton with and without the AMM, respectively. The dashed and dotted lines indicate the results in the semi-classical approaches of Refs. [14] and in [15], respectively.

For comparison, furthermore, we also plot results in the semi-classical approaches of Ref. [14] (dashed line) and Ref. [15] (dotted line). The results in the semi-classical calculations do not agree with those in the quantum calculation, particularly when the AMM is included.

In order to study the AMM effect more clearly, we next examine the HO overlap integral $M(n_i, n_f)$ defined in Eq. (7). In Fig. 2 we show the n_f-dependence of $M(n_i, n_f)$ when $n_i = 46$. The solid circle and triangle in the inset represent the peak positions in the cases with $s_i = -s_f = -1$ and $s_i = -s_f = 1$ with the AMM included. The open diamond indicates the case of a spin-flip $s_i = -s_f$ without including AMM.

$M(n_i, n_f)$ shows an oscillating behavior, but only the strength after the last peak contributes to the results. In this region the strength rapidly decreases with increases of $n_i - n_f$. A pion cannot be produced in the free kinematics, $B = 0$, because of energy momentum conservation. Under the influence of a magnetic field, momentum conservation is not satisfied, so that a pion can be produced in the kinematical condition far from the free kinematics, where $M(n_1, n_2)$ rapidly decreases.

The AMM gives a repulsive potential for $s_i = -1$ and attractive for $s_i = 1$. The transition with $s_i = s_f = -1$ introduces an additional energy to be consumed in the pion production. Thus, the difference between the initial and final Landau-levels is shifted toward a smaller number. In addition, the transition with $s_i = s_f = 1$ reduces the production energy. In this kinematical region a small difference between the initial and final Landau-levels significantly changes the transition strength. Therefore, the AMM plays the important roles of increasing greatly the pionic decay width when $s_i = s_f = -1$ and to decrease it $s_i = s_f = 1$.

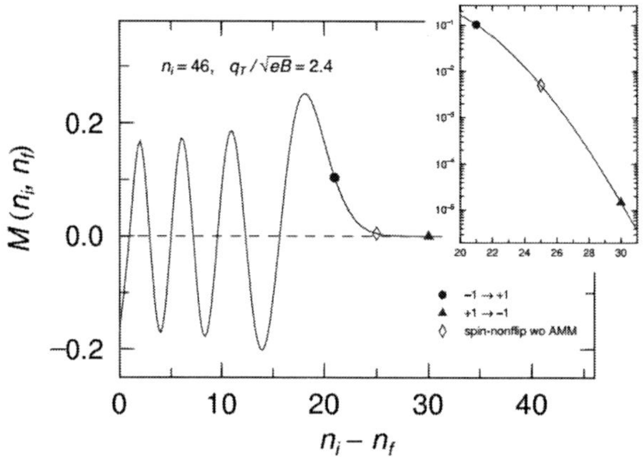

Figure 2. The transition strength function $M(n_i, n_f)$, Eq. (7), when $n_i = 46$. The solid circle and triangle show the peak positions in the cases $s_i = -s_f = -1$ and $s_i = -s_f = 1$, and the opened diamond indicates the case of the spin-flip $s_i = -s_f$ when $\kappa_p = 0$.

In Fig. 3 we present the decay widths, $\Gamma_{p\pi}(n_i, n_f)$ as functions of $(n_i - n_f)/n_i$ when the parameter χ is fixed. The upper panels show the results with the AMM included for $\chi = 0.02$ (a1) and $\chi = 0.07$ (a2), and the lower panels show the results without the AMM for $\chi = 0.02$ (b1) and $\chi = 0.07$ (b2). The dashed, dot-dashed, solid and dotted lines show the results with initial Landau numbers, $n_i = 5 \times 10^3$, 2×10^4, and 6×10^4, respectively. In all results the initial and final proton spins are set to be $s_i = -s_f = -1$ because this is the dominant contribution.

First we note that the dependence of $(n_i - n_f)/n_i$ for the different cases are almost completely in agreement when χ is fixed. There are however various semi-classical calculations [11–15] which give different results, but all of the models indicate that the decay width depends only on the parameter χ, not on the initial energy E_i and the magnetic field B. Thus, we confirm that this scaling relation is satisfied in the quantum calculations, independently of the AMM.

In the all results the peak positions are $(n_i - n_f)/n_i \approx 0.3 - 0.4$. As χ increases, the peak position is shifted to a larger value, and the peak width becomes broader.

When comparing the results with the AMM included (a1, a2) and those without the AMM (b1, b2) we can see that an AMM slightly shifts the peak position to smaller values of $(n_i - n_f)/n_i$, and that the peak width is slightly broader than in the case without the AMM.

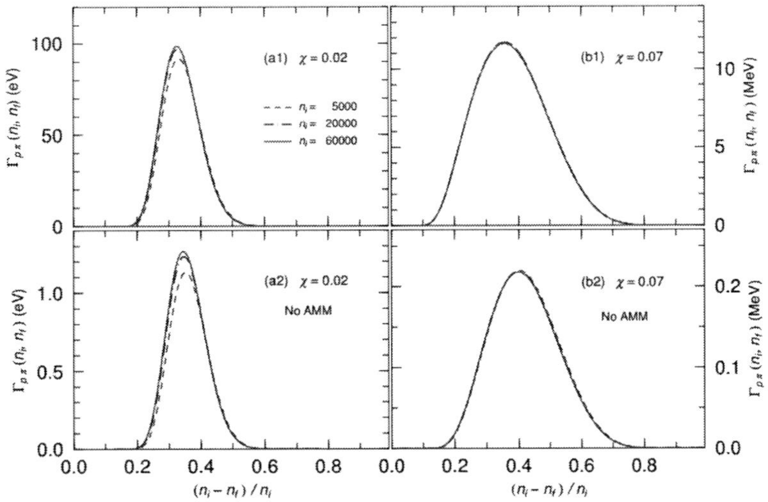

Figure 3. (Color online) Pionic decay widths of protons when $s_i = -s_f = -1$ as a function of $(n_i - n_f)/n_i$ for $\chi = 0.02$ (a1) and $\chi = 0.07$ (b1). The bottom panels (a2) and (b2) show the results without the AMM for $\chi = 0.02$ and $\chi = 0.07$, respectively. The dashed, dotted-dashed and solid lines represent the results with initial Landau number, $n_i = 5 \times 10^3$, 2×10^4 and 6×10^4, respectively.

Moreover, we see that the peak position is shifted by including the AMM with the scaling relation being satisfied. This result indicates that the AMM remains important for any magnetic field strength and proton energy.

As written in Ref. [16], the very large effect of the AMM comes from the shift of this peak position, i.e. where the HO overlap integral $\mathcal{M}(n_i, n_f)$ changes rapidly. For $n_i \approx 48$ and the shift of $n_i - n_f$ to 2, the absolute value of $\mathcal{M}$ increases by about a factor of 100. When the magnetic field is weaker or the initial energy becomes larger, n_i increases, and the AMM effect is expected to be smaller. When χ is fixed, however, the AMM effect remains and plays an important role in any regime.

Furthermore, we can see that $(n_i - n_f)/n_i \approx 0.3 - 0.4$, namely the Landau level difference in the pion emission between the initial and final states is of the same order of the initial and final Landau levels, $(n_i - n_f) \sim n_i \sim n_f$.

In the adiabatic limit it can be assumed that the relative momentum between the final proton and the pion is zero, and that the two particles have the same velocity. In this limit the ratio between these two energies is the same as the mass ratio: $E_\pi / E_f \approx m_\pi / M_p$, and the final proton and pion energy become

$$E_f \approx \frac{M_p}{M_p + m_\pi} E_i, \quad E_\pi \approx \frac{m_\pi}{M_p + m_\pi} E_i. \tag{8}$$

The initial proton energy is very large $E_i \gg M_p$, $E_{i,f} \approx \sqrt{2n_{i,f}eB}$, so that the following relation is given in the adiabatic and high energy limit as $\sqrt{n_i} - \sqrt{n_f} \approx (m_\pi/M_p)\sqrt{n_i}$. This leads to

$$n_i - n_f \approx \frac{M_p^2 - (M_p - m_\pi)^2}{M_p^2} n_i \approx 0.28 n_i. \tag{9}$$

The actual πN-interaction is the p-wave, and the relative momentum is not zero even in the adiabatic limit, so that the actual value of $n_i - n_f$ is larger than the above value; this argument is consistent with the present results.

In the usual semi-classical approximations it is assumed that $n_i - n_v \ll n_i$. This assumption is equivalent to the adiabatic limit when a produced particle is massless such as a photon. Thus, the assumption, $n_i - n_v \ll n_i$, is not satisfied when the emitted particle is massive, so that its results do not agree with the results in the quantum calculations.

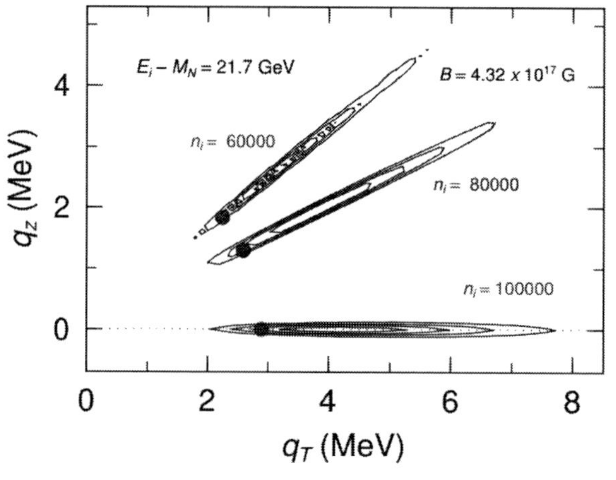

Figure 4. (Color online) The contour plot of the differential pionic luminosity at the initial spin state $s_i = -1$, the initial kinetic energy $E_i - M_p = 21.7$GeV and the magnetic field, $B = 4.32 \times 10^{17}$G ($\chi = 0.07$) with initial Landau number, $n_i = 100,000$ ($p_{iz} = 0$), 80,000 and 60,000. Each lines show the relative strength. The vertical and horizontal axises are the z-component and the transverse component of the emitted pion momentum.

In Fig. 4, we show luminosity distribution of the emitted pion, $d^3I/dq_\pi^3 = E_\pi d^3\Gamma/dq_\pi^3$, with the contour plots, when the initial proton has the initial spin $s_i = -1$ and the initial energy $E_i - M_p = 21.7$GeV and $n_i = 100{,}000$, 80,000, 60,000 when $B = 4.32 \times 10^{17}$G; the state with $n_i = 100,000$ corresponds to that with $p_{iz} = 0$ and $\chi = 0.07$. The pion momentum is distributed narrowly for the polar angle, but is distributed widely in the absolute value.

The solid circles indicate the pion momentum in the adiabatic limit, $q_z = r_\pi p_{iz}$ and $q_z = r_\pi \sqrt{2n_i eB}$ with $r_\pi = m_\pi/(M_p + m_\pi)$. We see that the absolute value of the emitted pion momentum is almost larger than that in the adiabatic limit, and that the direction of the emitted pion momentum is almost same as that of the initial nucleon momentum.

As the magnetic field decreases, and the initial Landau number increases, the width of the angular distribution becomes narrower, and this width for the polar angle must become negligible in the zero limit of the magnetic field.

4 Summary

We have calculated the pion synchrotron radiation from high energy protons propagating in strong magnetic fields in a microscopic quantum field theoretical framework. We solved the Dirac equation in a strong magnetic field and obtained the proton propagator from its solution. Then, we derived the pionic decay width of the propagating proton in a fully relativistic and quantum mechanical way.

We find out that the anomalous magnetic moment has a very large effect which enlarges the emission rate by about 50 times, and that the polar angles of the emitted pion and the final pion are almost the same as that of the initial proton, when the proton energy is 10 GeV and the magnetic field is 5×10^{17}G. One can not know these effects before performing the quantum calculations.

Furthermore, we have shown that the results in our quantum approach do not agree with those in the usual semi-classical approximations. In the usual semi-classical approximation it is assumed that $n_i - n_\upsilon \ll n_i$, where the HO overlap function $\mathcal{M}$ can be approximately written with the Airy function. This assumption is not satisfied when a produced particle is massive,

In actual magnetars the surface magnetic field is known to be of order $B \sim 10^{15}$G . In the present method we did not perform a calculation for such a magnetic field strength because of the large number of Landau levels involved: $n_i \sim 10^{12} - 10^{13}$.

The present work suggests a better way to treat pion production. The result of $\Gamma(n_i, n_f)$ depends only on χ and $\Delta n_{if}/n_i$ demonstrates that one can calculate $\Gamma(n_i, n_f)$ for values of $n_i \sim 10^{4-5}$ and scale those results to more realistic conditions. Also, when $p_{iz} \neq 0$, the decay width can be obtained from a Lorenz transformation along z-direction.

References

[1] S. Mereghetti, Annu. Rev. Astron. Astrophys. **15**, 225 (2008).
[2] M.A. Hillas, Annu. Rev. Astron. Astrophys. **22**, 425 (1984).
[3] J. Aarons, ApJ **589**, 871 (2003).
[4] K. Ioka, S. Razzaque, S. Kobayashi and P. Mészáros, ApJ **633**, 1013 (2005).
[5] N. Gupta and B. Zhang, MNRAS **380**, 78 (2007).
[6] M. Böttcher, C. D. Dermer, ApJL **499**, L131 (1998).
[7] T. Totani, ApJL **502**, L13 (1998).
[8] P.C. Fragile, G.J. Mathews, J. Poirier and T. Totani, APh **20**, 591 (2004).
[9] K. Asano and S. Inoue, ApJ **671**, 645 (2007) .
[10] K. Asano, S. Inoue and P. Mészáros, ApJ **699**, 953 (2009).
[11] V.L. Ginzburg and S.I. Syrovatskii, UsFiN **87**, 65 (1965).
[12] V.L. Ginzburg and S.I. Syrovatskii, Annu. Rev. Astron. Astrophys. **3**, 297 (1965).
[13] G.F. Zharkov, Sov. J. Nucl. Phys., 1, 17314 (1965).
[14] V. Berezinsky, A. Dolgoy and M. Kachelriess, Phys. Lett. **B 351**, 261 (1995)
[15] A. Tokushita and T. Kajino, ApJ **515**, L117 (1999).
[16] T. Maruyama, M.-K. Cheoun, T. Kajino Y. Kwon, G.J. Mathews, Phys. Rev. **D91**,123007 (2015)

Modelling and analysis of nucleon emission from deuteron-induced reactions at incident energies up to 100 MeV

Shinsuke Nakayama[1a], Hiroshi Kouno[2], Yukinobu Watanabe[2], Osamu Iwamoto[1], Tao Ye[3], and Kazuyuki Ogata[4]

[1]*Nuclear Science and Engineering Center, Japan Atomic Energy Agency, 2-4, Shirakata, Tokai-mura, Ibaraki, 319-1195, Japan*
[2]*Department of Advanced Energy Engineering Science, Kyushu University, 6-1, Kasuga-koen, Kasuga-shi, Fukuoka 816-8580, Japan*
[3]*Institute of Applied Physics and Computational Mathematics, Beijing, 100094, China*
[4]*Research Center for Nuclear Physics, Osaka University, 10-1, Mihogaoka, Ibaraki-shi, Osaka, 567-0047, Japan*

Abstract. We have so far developed a computational code system dedicated to deuteron-induced reactions in combination with some theoretical models. In our previous works, the code system was successfully applied to systematic analyses of double-differential cross sections (DDXs) of (*d,xp*) reactions for ^{12}C, ^{27}Al, and ^{58}Ni at incident energies up to 100 MeV. In the present work, we apply the code system to neutron emission from deuteron-induced reactions. Since there is few experimental data of DDXs of (*d,xn*) reactions, double-differential thick target neutron yields (TTNYs) are calculated and compared with experimental data instead of DDXs. The calculation using the code system reproduces the measured TTNYs for carbon at incident energies up to 50 MeV.

1 Introduction

In recent years, research and development of accelerator neutron sources has led to renewed interest in the study of deuteron-induced reactions. In these facilities, the (*d,xn*) reaction on light nuclei (Li, Be, C, etc.) is considered as a promising reaction to generate intensive neutron beams. Fig.1 shows experimental thick target neutron yields (TTNYs) from (*p,xn*) and (*d,xn*) reactions on ^{9}Be [1, 2]. This figure shows some advantages of a (*d,xn*) neutron source over a widely-used (*p,xn*) neutron source. First, the amount of generated neutrons is large. Second, the neutron spectrum has a broad energy peak around half the deuteron incident energy. This means that the most probable neutron energy can be changed by adjusting incident deuteron energy. In addition, the (*d,xn*) reaction has strongly forward-peaked angular distribution, which is an additional advantage from the point of view of shielding. From these favourable features, intensive neutron sources using deuteron accelerator have been proposed for various applications such as production of radioisotopes for medical use [3, 4] and irradiation testing of fusion reactor materials [5].

Engineering design of such (*d,xn*) neutron sources requires understanding of the interaction of deuteron with various materials used in the system components, e.g., a neutron converter, a beam collimator, a vacuum duct, etc. Therefore, we have so far developed a computational code system

[a] Corresponding author: nakayama.shinsuke@jaea.go.jp

dedicated to deuteron-induced reactions in combination with some theoretical models [6-9]. Elastic breakup and stripping reactions to continuum are calculated using the codes based on the Continuum-Discretized Coupled-Channels method (CDCC) [10] and the Glauber model [6], respectively. In addition, DWUCK4 [11], which is the calculation code based on conventional zero-range Distorted Wave Born Approximation (DWBA), is used to calculate stripping reactions to bound states in the residual nuclei. Finally, preequilibrium and evaporation components are estimated using the CCONE code [12], which was successfully applied to neutron nuclear data evaluation for the latest version of Japanese Evaluated Nuclear Data Library (JENDL-4.0) [13]. In our previous works, the code system has been successfully applied to systematic analyses of double-differential cross sections (DDXs) of (*d,xp*) reactions for ^{12}C, ^{27}Al, and ^{58}Ni at incident energies up to 100 MeV [7-9]. The calculation is in good agreement with the experimental data. This result shows that the neutron stripping process to continuum is dominant in forward proton emission in the high incident energy region.

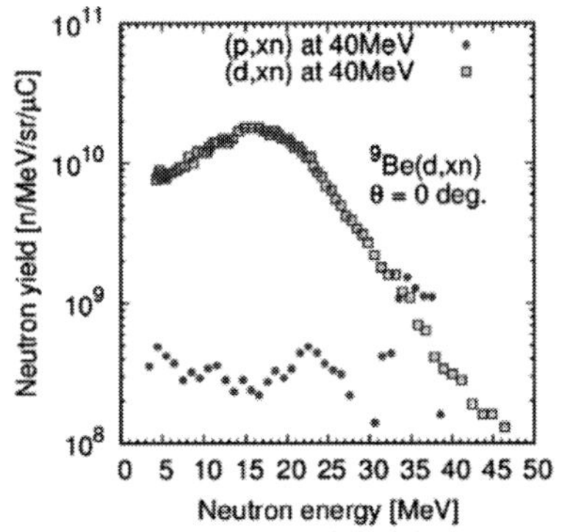

Figure 1. Experimental double defferential thick target neutron yields from (*p,xn*) and (*d,xn*) reactions on ^{9}Be at 40 MeV. The data are taken from Refs. [1,2].

In this study, we apply the code system to neutron emission from deuteron-induced reactions. Since there is few experimental data of DDXs of (*d,xn*) reactions, TTNYs for carbon are calculated and compared with the corresponding experimental data instead of DDXs. The measured TTNYs for carbon are available in the wide incident energy range, thus they enable us to investigate the incident energy dependence of individual reaction processes in detail. Including the results of our previous works for the (*d,xp*) reactions, validation of the present modelling for both proton and neutron emissions from deuteron-induced reactions is discussed through comparison with available experimental data.

2 Analysis of stripping reactions by DWBA

First of all, we analyze the stripping reaction process to discrete states by DWBA. These reactions make a large contribution to nucleon emission in the low incident energy region. In TTNYs calculation, we assume that natural carbon consists of only ^{12}C isotope whose natural abundance is 98.89%. Therefore, we analyze the ^{12}C(*d,n*)^{13}N reactions to the ground and some excited states.

2.1 DWBA calculation

The DWBA differential cross section for the (*d,n*) reaction leading to the bound state *i* is given by

$$\frac{d\sigma_{DWBA,i}}{d\Omega}(E_d) = \frac{D_0^2}{10^4}\frac{2J_i+1}{2J_A+1}\frac{S_i}{2j+1}\frac{d\sigma_{DWUCK4,i}}{d\Omega}(E_d) \tag{1}$$

where E_d is the incident deuteron energy, D_0 is a constant used in zero-range approximation, J_A and J_i are the spins of target nucleus ^{12}C and *i*-th state of residual nucleus ^{13}N, respectively, S_i is the spectroscopic factor for each state, j is the spin of transferred proton, and $d\sigma_{DWUCK4,i}/d\Omega$ is the differential cross section calculated with the DWUCK4 code. Input parameters used in DWUCK4 calculations are listed in Table 1. These parameters are based on our previous work [9].

Table 1. Input parameters used in the DWUCK4.

Neutron potential	Koning-Deraloche (KD) [14]
Deuteron potential	Adiabatic potential [15] from KD
Proton binding potential	Woods-Saxon form r_0 = 1.25 (fm), a_0 = 0.65 (fm)
Finite range correction factor (fm)	0.7457
Zero range constant D_0^2 (MeV^{2}fm^3)	1.5 × 10^4
Nonlocality parameters	neutron:0.85, deuteron:0.54

In the DWBA analyses for the ^{12}C(d,n)^{13}N reactions, we consider the four states of ^{13}N where experimental differential cross sections are available [16-20], namely the ground state (J^π=1/2$^-$), the first excited state (E_x= 2.37 MeV, J^π= 1/2$^+$), the second excited state (E_x= 3.51 MeV, J^π= 3/2$^-$), and the third excited state (E_x= 3.55 MeV, J^π= 5/2$^+$). The excited states of ^{13}N are unbound since the proton separation energy of ^{13}N is 1.94MeV. Following Ref. [18], we have calculated the DWBA cross section for these state using a very loosely bound state wave function by setting the binding energy to 0.01MeV in DWUCK4 calculations. In addition, the statistical decay contribution from compound nuclei is calculated using the CCONE code.

Figure 2 shows the comparison of calculated and experimental differential cross sections for the ^{12}C(d,n)^{13}N reactions in the incident energy range from 7 to 18 MeV. The experimental data are taken from Refs. [16–20]. The values of spectroscopic factor (SF) are extracted by fitting the calculated DWBA cross section to the corresponding experimental one at forward angles where proton stripping process is dominant. It should be noted that the experimental values for the second and the third excited states are unresolved. From the analyses for the ^{12}C(d,p)^{13}C, which is the isobaric analog state of ^{13}N [9], we estimate the SF values of the third excited state which is four times as large as that of the second excited state. The sum of the direct stripping component calculated by DWBA and the contribution from statistical decay show good agreement with the experimental data at small angles.

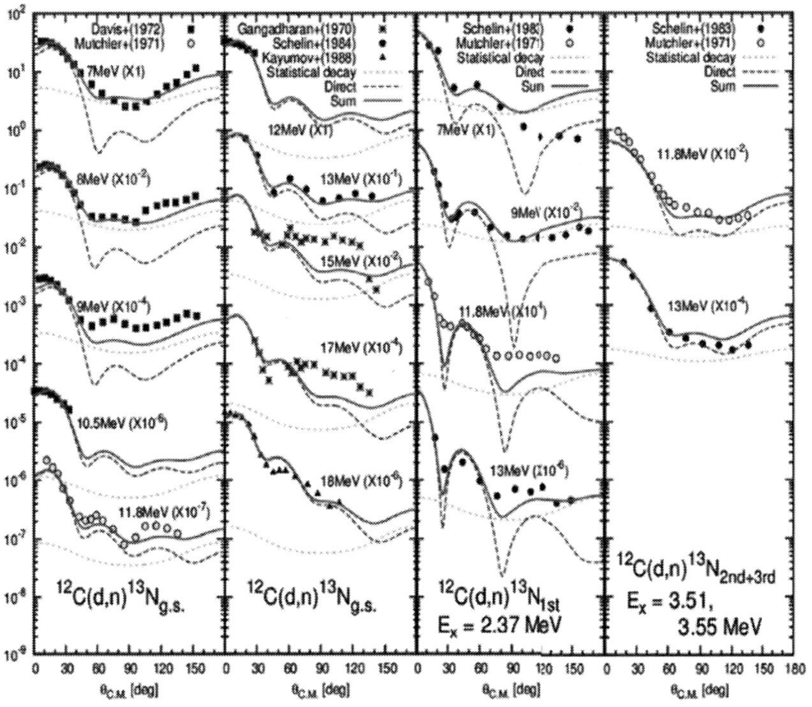

Figure 2. Calculated and experimental differential cross sections for the ^{12}C(d,n)^{13}N reaction.

2.2 Spectroscopic factor

The experimental SF values for the ^{12}C(d,n)^{13}N reaction extracted from the present analyses are shown in Table 2. They are also presented in solid squares in Fig.3. Fluctuation among the experimental SF values can be seen in the figure. This is due to the effect of resonance structures as described in Ref. [16].

Table 2. List of spectroscopic factors for the ^{12}C(d,n)^{13}N reaction extracted from the present analysis

i	E_x (MeV)	J^π	l	$S_i(E_d)$									
				7	8	9	10.5	11.8	12	13	15	17	18
0	0 (g.s.)	$\frac{1}{2}^-$	1	0.50	0.46	0.42	0.93	0.40	1.00	0.28	0.30	0.31	0.69
1	2.37	$\frac{1}{2}^+$	0	0.29	-	0.43	-	0.54	-	0.39	-	-	-
2	3.51	$\frac{3}{2}^-$	1	-	-	-	-	0.66	-	0.64	-	-	-
3	3.55	$\frac{5}{2}^+$	2	-	-	-	-	0.17	-	0.16	-	-	-

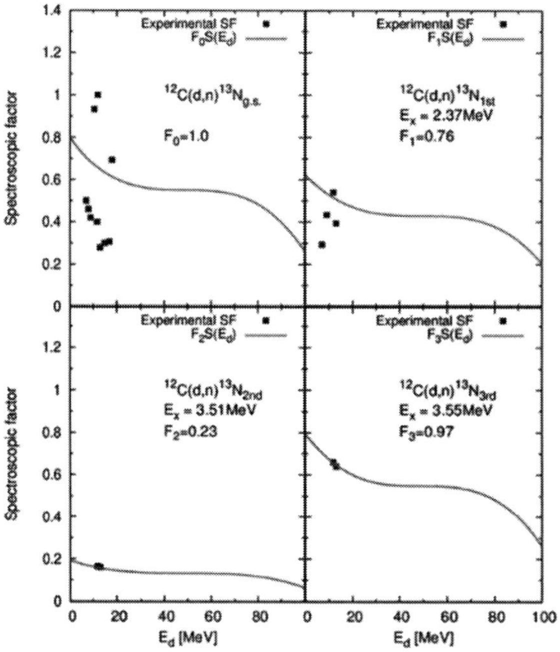

Figure 3. Experimental and empirical spectroscopic factors for the $^{12}C(d,n)^{13}N$ reaction.

In our systematic DWBA analyses for the (d,p) reactions on ^{12}C, ^{27}Al, ^{40}Ca, and ^{58}Ni, the extracted SF values have shown similar incident energy dependence for all the target nuclei. Therefore, we proposed the following empirical expression to express the energy dependence of the experimental SF values [9] :

$$S(E_d) = -2.18\times10^{-6}E_d^3 + 3.19\times10^{-4}E_d^2 - 1.56\times10^{-2}E_d + 8.20\times10^{-1}. \qquad (2)$$

We assume that the SF value for each $^{12}C(d,n)^{13}N$ reaction has the same energy dependence as $S(E_d)$. Since $S(E_d)$ is re-normalized for the case of the $^{12}C(d,p)^{13}C_{g.s.}$ reaction, it is necessary to fix a scaling factor F_i depending on each i-th state of ^{13}N. We average the effect of resonance structures and fix the each F_i as follows. For the ground and the first excited states of ^{13}N, we adopt the same F_i as the ones for the corresponding states of ^{13}C derived in Ref. [9]. For the second and the third excited state of ^{13}N, we fix each F_i so as that $F_iS(E_d)$ pass through the center of experimental SF values. The empirical SF values $F_iS(E_d)$ for the each reaction are shown by solid line in Fig. 3.

3 Analysis of TTNYs

In the engineering design of deuteron accelerator neutron sources, DDXs for (d,xn) reactions are critically important. However, the experimental data are sparse over wide ranges of target mass number and incident energy. On the other hand, experimental data of TTNYs for carbon are available in the wide incident energy range. To validate the present modelling for neutron emission from deuteron-induced reactions, we derive TTNYs from DDXs for (d,xn) reactions calculated with the present code system and compare the resultant TTNYs with the experimental data.

 DOI: 10.1051/epjconf/201612204004

3.1 Calculation method

In TTNYs calculation, we assume that natural carbon consists of only ^{12}C. TTNYs for the ^{12}C target can be derived as

$$\frac{d^2Y}{dEd\Omega}(E_d) = N\int_0^{E_d} dE_i \frac{d^2\sigma}{dEd\Omega}(E_i)\left[\frac{dE_i}{dx}\right]^{-1} \exp\left(-N\int_{E_i}^{E_d} dE'\sigma_{rea}(E')\left[\frac{dE'}{dx}\right]^{-1}\right) \tag{3}$$

where E_d is the incident deuteron energy, N is the atomic density of the target, E_i is the deuteron energy in the target, $d^2\sigma/dEd\Omega$ is the DDXs for the ^{12}C(d,xn) reaction, dE/dx is the deuteron stopping power of deuteron in the ^{12}C target, and σ_{rea} is the deuteron total reaction cross section on ^{12}C.

We calculate the stopping power and the total reaction cross section with the SRIM-2010 code [21] and the CCONE code, respectively. DDXs for the ^{12}C(d,xn) reaction are calculated in the same way as described in Refs. [8, 9].The input parameters in DWBA calculations are the same as those given in Table 1. In the CDCC and Glauber models, nucleon optical potentials (OPs) of KD [14] are used for proton and neutron. The OPs used in the CCONE code are the global nucleon OPs of KD for proton and neutron and the adiabatic OP [15] for deuteron in a consistent way as in DWBA calculations. Default values in the CCONE code are used for other input parameters such as the level density parameter. The SF values for the ^{12}C(d,n)^{13}N reactions are given by each $F_iS(E_d)$ shown in Fig. 2.

3.2 Results and discussion

Figure 4 shows the comparisons between the calculated and experimental TTNYs for carbon at angles around 0°. The incident deuteron energies are from 9 to 50 MeV. The experimental data are taken from Refs. [22-27].

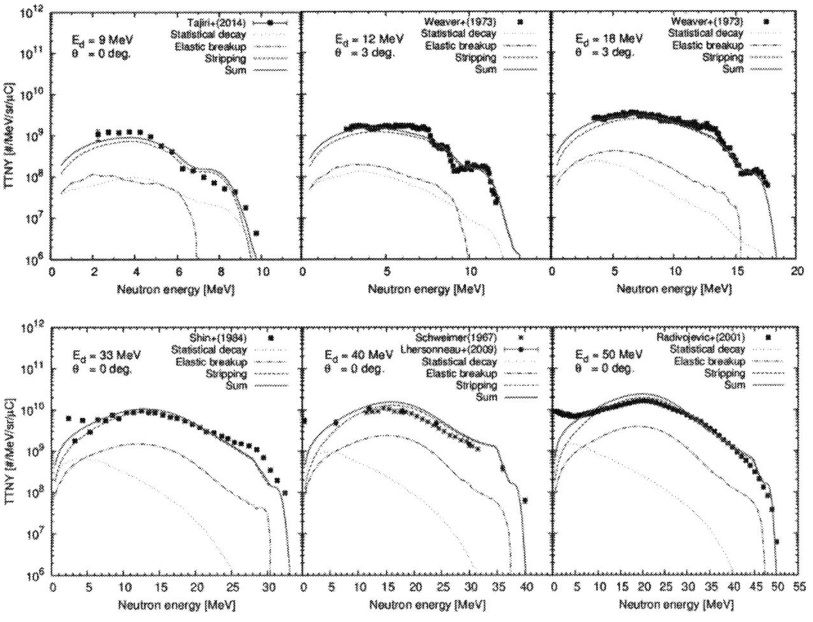

Figure 4. Calculated and experimental TTNYs for carbon in the incident energy from 9 to 50 MeV.

In the wide incident energy range, the present calculation reproduces both the shape and magnitude of the experimental TTNY spectra at forward angles around 0°. As shown in Fig.4, the proton stripping process is dominant in the incident energy below 50 MeV. In our previous works [6-9], the calculation results have shown that the neutron stripping process is dominant in forward proton emission at 56 and 100 MeV. These results indicate that the stripping processes make a dominat contribution to neucleon emission in the whole incident energy range below 100 MeV and they should be taken into account using reliable theoretical models. However, the present calculation fails to reproduce the TTNY spectra in the low emission energy region. This discrepancy may be due to insufficiency of the statistical decay calculation.

Figure 5 shows the comparisons between the calculated and experimental TTNYs for carbon for several angles at 18 MeV. The present calculation reproduces the experimental data at several forward angles. It should be noted that the step structure seen in the high energy region arises from the stripping reaction to the ground and some excited states of ^{13}N calculated by DWBA. The Glauber model cannot deal with individual transitions to bound states but can describe the sum of stripping contributions to both continuum and bound states as a continuous spectrum. Thus, if we use the Glauber model alone to calculate the stripping component, these step structure in TTNYs cannot be reproduced well.

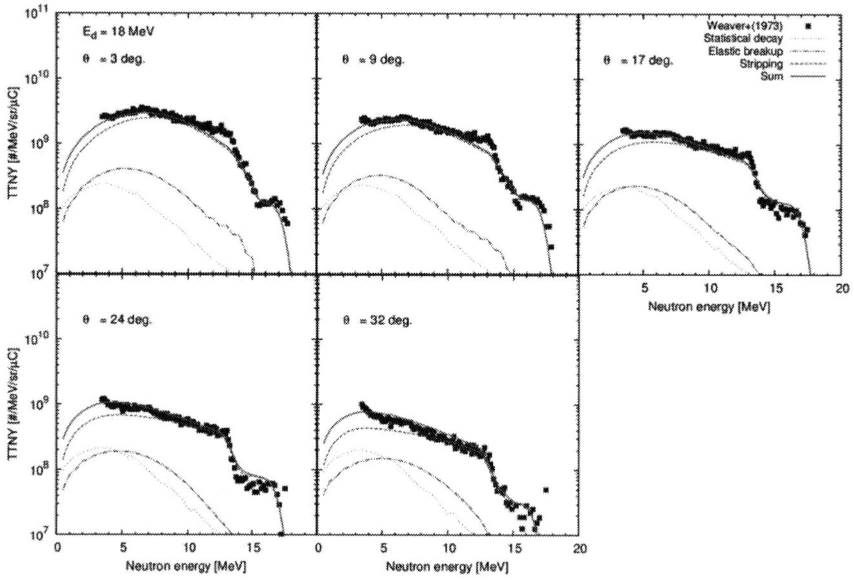

Figure 5. Calculated and experimental TTNYs for carbon at 18 MeV.

4 Summary and outlook

A computational code system dedicated to deuteron-induced reactions has been applied to neutron emission in deuteron-induced reactions. The code system consists of the codes based on the CDCC for elastic breakup and the Glauber model for the stripping reaction to continuum, DWUCK4 based on conventional zero-range DWBA for the stripping reaction to bound states in the residual nuclei, and the CCONE code for preequilibrium and evaporation components. The calculations with the code system reproduce successfully the experimental thick target neutron yields (TTNYs) for carbon at incident energies from 9 to 50 MeV. Along with the results of our previous works for double differential (*d,xp*) cross sections, these results demonstrate the applicability of the present modelling for both proton and neutron emissions from deuteron-induced reactions.

In the future, we plan to apply the present modelling to analyses of TTNYs for other nuclei, such as Be, Li, Al, etc., which are required in the engineering design of deuteron accelerator neutron sources.

Acknowledgement

The authors wish to thank S. Hashimoto for fruitful discussion and comments on DWBA analyses.

References

1. M.J. Saltmarsh, C.A. Ludemann, C.B. Fulmer, R.C. Styles, Nucl. Instr. Meth. **145**, 81-90 (1977).
2. Y. Uwamino, T. Okubo, A. Torii, T. Nakamura, Nucl. Instrum. Methods Phys. Res. A **271**, 546 (1988).
3. Y. Nagai, K. Hashimoto, Y. Hatsukawa, H. Saeki, S. Motoishi, N. Sato, M. Kawabata, H. Harada, T. Kin, K. Tsukada, T. Sato, F. Minato, O. Iwamoto, N. Iwamoto, Y. Seki, K. Yokoyama, T. Shiina, A. Ohta, N. Takeuchi, Y. Kawauchi, N. Sato, H. Yamabayashi, Y. Adachi, Y. Kikuchi, T. Mitsumoto, T. Igarashi, J. Phys. Soc. Jpn. **82**, 064201 (2013).
4. T. Kin, Y. Nagai, N. Iwamoto, F. Minato, O. Iwamoto, Y. Hatsukawa, M. Segawa, H. Harada, C. Konno, K. Ochiai, K. Takakura, J. Phys. Soc. Jpn. **82**, 034201 (2013).
5. A. Moeslang, V. Heinzel, H. Matsui, M. Sugimoto, Fusion Eng. Des. **81**, 863 (2006).
6. T. Ye, S. Hashimoto, Y. Watanabe, K. Ogata, M. Yahiro, Phys. Rev. C **84**, 054606 (2011).
7. S. Nakayama, S. Araki, Y. Watanabe, O. Iwamoto, T. Ye, K. Ogata, Nucl. Data Sheets **118**, 305 (2014).
8. S. Nakayama, S. Araki, Y. Watanabe, O. Iwamoto, T. Ye, K. Ogata, Energy Procedia **71**, 219 (2015).
9. S. Nakayama, Y. Watanabe, J. Nucl. Sci. Technol. (2015) *in press*.
10. M. Yahiro, K. Ogata, T. Matsumoto, K. Minomo, Prog. Theor. Exp. Phys. **2012**, 01A206 (2012) and references therein.
11. P.D. Kunz, computer code DWUCK4 (1974), http://spot.colorado.edu/~kunz/DWBA.html.
12. O. Iwamoto, J. Nucl. Sci. Technol. **44**, 687 (2007).
13. K. Shibata, O. Iwamoto, T. Nakagawa, N. Iwamoto, A. Ichihara, S. Kunieda, S. Chiba, K. Furutaka, N. Otuka, T. Ohsawa, T. Murata, H. Matsunobu, A. Zukeran, S. Kamada, J. Katakura, J. Nucl. Sci. Technol. **48**, 1 (2011).
14. A.J. Koning, J.P. Delaroche, Nucl. Phys. A **713**, 231 (2003).
15. D.G. Madland, Proc. Specialists' Meeting on Preequilibrium Nuclear Reactions, 103, Semmering, Austria, 10-12 Feb (1988).
16. J.R. Davis, G. U. Din, Nucl. Phys. A **179**, 101 (1972).
17. H.R. Schelin, E. Farrelly Pessoa, W.R. Wylie, E.W. Cybulska, K. Nakayama, L.M. Fagundes, R.A. Douglas, Nucl. Phys. A **414**, 67 (1984).
18. G.S. Matchler, D. Rendic, D.E. Velkley, W.E. Sweeney jr., G.C. Phillips, Nucl. Phys. A **172**, 469 (1971).
19. S. Gangadharan, R.L. Wolke, Phys. Rev. C **1**, 1333 (1970).
20. M.A. Kayumov, S.S. Kayumov, S.P. Krekoten, A.M. Mukhamedzhanov, K.D. Razikov, K. Khamidova, R. Yarmukhamedov, Sov. J. Nucl. Phys. **48**, 629 (1988).
21. J. F. Ziegler, M. D. Ziegler, and J. P. Biersack, Nucl. Instrum. Methods Phys. Res. B **268**, 1818 (2010).
22. Y. Tajiri, Y. Watanabe, N. Shigyo, K. Hirabayashi, T. Nishizawa, K. Sagara, Prog. Nucl. Sci. Technol. **4**, 582 (2014).
23. K.A. Weaver, J.D. Anderson, H.H. Barschall, J.C. Davis, Nucl. Sci. Eng. **52**, 35 (1973).

24. K. Shin, K. Hibi, M. Fujii, Y. Uwamino, T. Nakamura, Phys. Rev. C **29**, 1307 (1984).
25. G. Lhersonneau, T. Malkiewicz, K. Kolos, M. Fadila, H. Kettunen, M.G. Saint-Laurent, A. Pichard, W.H. Trzaska, G. Tyurin, L. Cousin, Nucl. Instrum. Methods Phys. Res. A **603**, 228 (2009).
26. G.W. Schweimer, Nucl. Phys. A **100**, 537 (1967).
27. Z. Radivojevic, A. Andrighetto, F. Brandolini, P. Dendooven, A. Honkanen, J. Aysto, V. Lyapin, V. Rubchenya, W.H. Trzaska, D. Vakhtin, G. Walter, Nucl. Instrum. Methods Phys. Res. B **183**, 212 (2001).

Application of JAERI quantum molecular dynamics model for collisions of heavy nuclei

Tatsuhiko Ogawa[1,a], Shintaro Hashimoto[1], Tatsuhiko Sato[1], and Koji Niita[2]

[1]*Japan Atomic Energy Agency, 2-4 Shirakata, Tokai-mura, Naka-gun, Ibaraki-ken, Japan, 319-1195;*
[2]*Research Organization for Information Science and Technology, 2-4 Shirakata, Tokai-mura, Naka-gun, Ibaraki-ken, Japan, 319-1195;*

Abstract. The quantum molecular dynamics (QMD) model incorporated into the general-purpose radiation transport code PHITS was revised for accurate prediction of fragment yields in peripheral collisions. For more accurate simulation of peripheral collisions, stability of the nuclei at their ground state was improved and the algorithm to reject invalid events was modified. In-medium correction on nucleon-nucleon cross sections was also considered. To clarify the effect of this improvement on fragmentation of heavy nuclei, the new QMD model coupled with a statistical decay model was used to calculate fragment production cross sections of Ag and Au targets and compared with the data of earlier measurement. It is shown that the revised version can predict cross section more accurately.

1 Introduction

Heavy ion beam irradiation on heavy nuclei is common in the heavy ion cancer therapy and fundamental science studies on nuclear physics. Accurate prediction of reactions induced by heavy ions with general-purpose radiation transport simulation codes is important for such applications to estimate detector performance and evaluate radiological safety of the experiment. JQMD (Jaeri Quantum Molecular Dynamics code) [1] incorporated into PHITS (Particle and Heavy Ion Transport code System) [2] is one of such codes and used at some facilities. Calculation of neutron production for heavy ion irradiation was well benchmarked so far in various target-projectile-energy combinations [3, 4]. To predict fragment production from light nuclei and medium-mass nuclei, JQMD was recently upgraded to JQMD-2.0 [5]. Its performance was tested in various conditions [6, 7] and it was confirmed that fragment production owing to peripheral collisions, underestimated by previous version of JQMD, was predicted for better accuracy.

In the revision of JQMD to JQMD-2.0, stability of nuclei before reactions, description of the interactions between nucleons, and in-medium effect on nucleon-nucleon interactions were modified. However, it is likely that ground-state stability and in-medium effect in light to medium-mass nuclei and those in heavy nuclei are different. Therefore, JQMD-2.0 was applied to interactions of heavy nuclei such as Ag and Au and its accuracy was benchmarked.

[a]e-mail: ogawa.tatsuhiko@jaea.go.jp

2 Method

Description of reaction mechanism in JQMD and JQMD-2.0 is explained briefly. Since complete description of simulation scheme in JQMD-2.0 is provided elsewhere [5], the aspects of JQMD-2.0 important for simulation of heavy nuclei are given here. By randomly choosing the coordinates of each nucleon in the 6-dimension phase space and adjusting them to reproduce the total binding energy, the initial states of the target and projectile are configured. In JQMD, thus configured ground states were directly used for collision event simulation; however, time evolution of the configured states was followed and only the states which stayed stable for 150 fm/c was used for collision simulation in JQMD-2.0. After configuring the ground states, the target and the projectile were transferred to their center-of-mass frame. In the frame, nucleons are transported for 150 fm/c based on their initial momenta and interactions between nucleons. Interaction between nucleons is one of the most important issues in JQMD because nucleon-nucleon interaction was not described in a Lorentz-covariant form in JQMD, therefore nuclei configured in their rest frame were occasionally excited or disintegrated during time evolution in the center-of-mass frame. Lorentz-covariant equation of motion adopted in JQMD-2.0 [8] is described as follows;

$$\dot{\mathbf{r}}_\mathbf{i} = \frac{\mathbf{p}_\mathbf{i}}{2p_i^0} + \sum_j^N \frac{m}{p_j^0}\frac{\partial\langle\hat{V}_j\rangle}{\partial\mathbf{p}_\mathbf{i}},$$

$$\dot{\mathbf{p}}_\mathbf{i} = -\sum_j^N \frac{m}{p_j^0}\frac{\partial\langle\hat{V}_j\rangle}{\partial\mathbf{r}_\mathbf{i}}.$$

$$p_i^0 = \sqrt{\mathbf{p}_\mathbf{i}^2 + m^2 + 2m\langle\hat{V}_i\rangle},$$

where $\mathbf{r}_\mathbf{i}$ is the spatial coordinate of the centroid of the i-th nucleon, $\mathbf{p}_\mathbf{i}$ is the momentum of the i-th nucleon, m is the rest mass of nucleons, $\langle\hat{V}_j\rangle$ is the potential of j-th particle, and N is the number of particles in the system. In both JQMD and JQMD-2.0, the potential term of the Hamiltonian V is a sum of the Skyrme-type force term, Coulomb interaction term, and symmetry term. The potential V_i is described by the following formula;

$$V_i = \frac{1}{2}\frac{A}{\rho_s}\langle\rho_i\rangle + \frac{1}{1+\tau}\frac{B}{\rho_s^\tau}\langle\rho_i\rangle^\tau + \frac{1}{2}\sum_j \frac{c_i c_j e^2}{|\mathbf{R_i}-\mathbf{R_j}|}\mathrm{erf}\left(\frac{|\mathbf{R_i}-\mathbf{R_j}|}{\sqrt{4L}}\right) + \frac{C_s}{2\rho_s}\sum_j(1-2|c_i-c_j|)\rho_{ij},$$

where A is a Skyrme force parameter (= -219.4 MeV), ρ_s is the saturation density (=0.168 fm^{-3}), $\langle\rho_i\rangle$ is the overlap integral of wave packets between the i-th nucleon and all the other nucleons, B is another Skyrme force parameter (= 165.3 MeV), τ is 4/3, c_i is 1 for protons and 0 for neutrons, e is the elementary charge, $\mathbf{R_i}$ denotes the position of i-th nucleon, L is the square of the width of wave packet representing nucleons (= 2 fm^2), C_s is the symmetry energy parameter (= 25 MeV), and ρ_{ij} is the overlap integral of wave functions of the i-th and j-th nucleons. The potential description of JQMD-2.0 was inherited from previous JQMD without any modifications.

In addition, in-medium effects on nucleon-nucleon scattering cross sections were modified. Previously, the modified version of the Cugnon's formula [9, 10], which assumed that scattering is suppressed by the Pauli blocking effect, was always used to calculate the cross sections. In JQMD-2.0, reduced cross sections or free-space cross sections were used in central and peripheral collisions, respectively. Thus in-medium effect on scattering was considered.

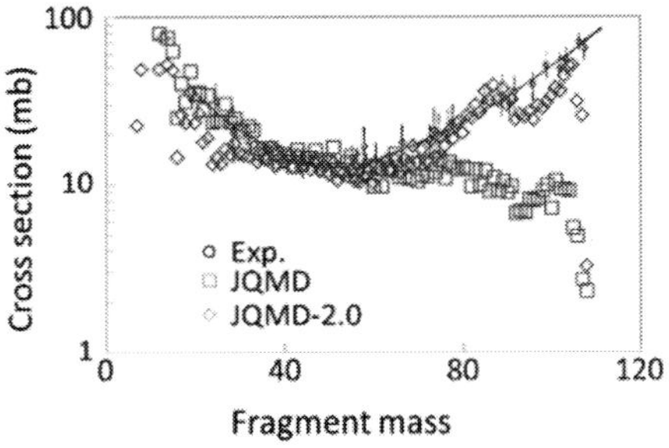

Figure 1. Mass distribution of fragments produced in 3650 MeV/u $^{Nat}Ag(^{12}C,x)$ reactions.

After the time evolution of 150 fm/c, nucleons close to each other are bound to finally form clusters. This process was inherited from JQMD without any modifications. Energy balance was checked after the clusterization. Initial energy, which is the sum of projectile kinetic energy, projectile rest mass and target rest mass, was compared with final energy, which is the sum of kinetic energy, excitation energy and the ground-state rest mass of all the clusters. The excitation energies of the clusters were scaled to satisfy the energy conservation. If the energy at the end deviates from that at the beginning so largely that energy cannot be balanced by excitation energy scaling, the simulated reaction event was rejected and event simulation was started afresh.

Thus developed JQMD-2.0 was combined with the statistical decay model GEM (Generalized Evaporation Model) [11] and the statistical multi-fragmentation model (SMM) [12] to calculate fragmentation yields of heavy nuclei such as Ag and Au.

3 Results and discussion

In Fig. 1, mass distribution of the fragments produced in $^{Nat}Ag(^{12}C,x)$ reaction by 3650 MeV/u C ions was calculated and compared with the measurement data [13]. Yields above A=70 was underestimated by JQMD but the simulation by JQMD-2.0 is in better agreement with the measurement data because the improvement in this study is particularly effective for peripheral collisions. It is also reasonable that the calculation by JQMD and that by JQMD-2.0 are mostly the same in the range from A=35 to A=70 because these nuclei are produced by intermediate collisions. The measured mass distribution exhibits steady declining trend down to A=65 whereas the yield calculated by JQMD-2.0 has a dip near A=95. The cause of this dip is discussed later.

Fig. 2 shows the comparison of calculated and measured [13] mass distribution of fragments produced in $^{197}Au(^{12}C,x)$ reaction by 3650 MeV/u C ions. The cross section is increased in a wider mass range in Fig.2 because Au has a larger nuclear surface than Ag and the improvement in this study is particularly effective in peripheral collisions. Similar to the case of Ag, the mass distribution calculated by JQMD and JQMD-2.0 has a dip aroud A=170, which is not seen in the measurement data.

To clarify the cause of the dip in the calculated mass distribution, fragment mass distributions for $^{Nat}Ag(^{12}C,x)$ reaction were calculated with changing the statistical decay model. The comparison is shown in Fig.3,

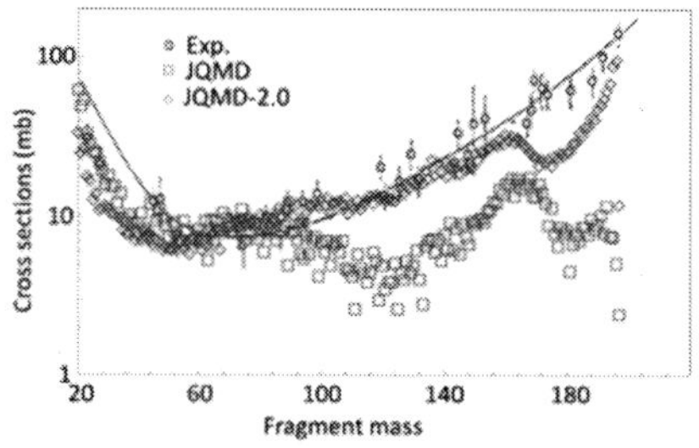

Figure 2. Mass distribution of fragments produced in 3650 MeV/u ^{197}Au(^{12}C,x) reactions.

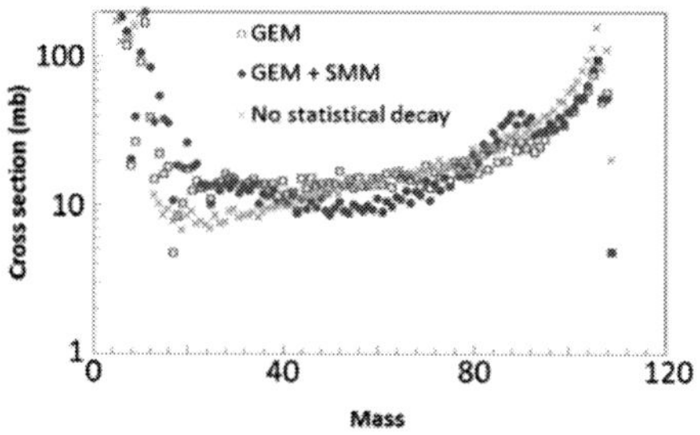

Figure 3. Comparison of fragment mass distribution in 3650 MeV/u ^{Nat}Ag(^{12}C,x) reactions.

The dip was not observed if SMM was turned off though the yields below A=20 become insufficient. The comparison of GEM+SMM, GEM, and no statistical decay (i.e., just after JQMD) suggests that the yield bump at the left of the dip is attributed to the statistical decay and particularly to the SMM stage. It may be explained by the asymmetric binary decay in SMM, which takes place if the pre-fragment has excitation energy about 4 MeV/nucleon [14]. Both GEM and GEM+SMM underestimate the yield above A=90; therefore nucleon knock-out and residual nucleus exitation in peripheral collisions should be suppressed in JQMD-2.0 for more precise prediction.

4 Conclusion

It is shown that mass distribution of the fragments produced by interactions of heavy nuclei such as Au and Ag with heavy ions was predicted by JQMD-2.0 better than the old JQMD. The yield of fragment with mass close to the original mass was significantly improved. Particularly the improvement was pronounced for Au than for Ag because of larger cross sections corresponding to peripheral collisions.

The fragmentation of heavy nuclei is significantly affected by multifragmentation, dip of yield attributed to multifragmentation is observed only in calculated result. This fact indicates that revision of reaction mechanism description relevant to estimate of nucleon loss and excitation energies in peripheral collisions is the key for more accurate prediction of fragment yield.

JQMD-2.0 is incorporated to PHITS Version 2.76 and later and is generally applicable for calculation of residual nuclide production and secondary particle production.

Acknowledgments

We wish to thank Dr. Davide Mancusi for providing the source code of R-JQMD. We also thank the operation team of the Center for Computational Science and E-systems (CCSE) in the Japan Atomic Energy Agency. Monte-Carlo simulations reported in this paper were executed on a PC-cluster system of CCSE. This work was partly supported by JSPS KAKENHI Grant Numbers 24860072 (Grant-in-Aid for Research Activity Start-up), and 26790072 (Grant-in-Aid for Young Scientists B).

References

[1] K. Niita, S. Chiba, T. Maruyama, T. Maruyama, H. Takada, T. Fukahori, Y. Nakahara, A. Iwamoto, Phys. Rev. C **52**, 2620 (1995)
[2] T. Sato, K. Niita, N. Matsuda, S. Hashimoto, Y. Iwamoto, S. Noda, T. Ogawa, H. Iwase, H. Nakashima, T. Fukahori et al., Journal of Nuclear Science and Technology **50**, 913 (2013), `http://dx.doi.org/10.1080/00223131.2013.814553`
[3] D. Satoh, T. Kurosawa, T. Sato, A. Endo, M. Takada, H. Iwase, T. Nakamura, K. Niita, Nuclear Instruments and Methods in Physics Research Section A: Accelerators, Spectrometers, Detectors and Associated Equipment **583**, 507 (2007)
[4] D. Satoh, D. Moriguchi, T. Kajimoto, Y. Koba, Y. Nakamura, N. Shigyo, M. Ueyama, U. Y., Y. M., J. Korean Phys. Soc. **59**, 1741 (2011)
[5] T. Ogawa, T. Sato, S. Hashimoto, D. Satoh, S. Tsuda, K. Niita, Phys. Rev. C **92**, 024614 (2015)
[6] T. Ogawa, T. Sato, S. Hashimoto, K. Niita, journal of italian physical society **to be published**, (2016)
[7] T. Ogawa, T. Sato, S. Hashimoto, K. Niita, CERN Conference series **to be published**, (2016)
[8] D. Mancusi, K. Niita, T. Maruyama, L. Sihver, Phys. Rev. C **79**, 014614 (2009)
[9] J. Cugnon, T. Mizutani, J. Vandermeulen, Nuclear Physics A **352**, 505 (1981)
[10] J. Cugnon, Phys. Rev. C **22**, 1885 (1980)
[11] S. Furihata, Nuclear Instruments and Methods in Physics Research Section B: Beam Interactions with Materials and Atoms **171**, 251 (2000)
[12] T. Ogawa, T. Sato, S. Hashimoto, K. Niita, Nuclear Instruments and Methods in Physics Research Section A: Accelerators, Spectrometers, Detectors and Associated Equipment **723**, 36 (2013)
[13] P. Kozma, C. Damdinsuren, Czechoslovak Journal of Physics **40**, 38 (1990)
[14] J. Bondorf, R. Donangelo, I. Mishustin, H. Schulz, Nuclear Physics A **444**, 460 (1985)

Angular distribution of evaporated protons from 50-MeV-range proton-nucleus reactions

Yuji Yamaguchi[1], Yusuke Uozumi[1], and Masahiro Nakano[2]

[1]*Kyushu University, Department of Applied Quantum Physics and Nuclear Engineering, 819-0395 Fukuoka, Japan*
[2]*Junshin Gakuen University, Faculty of Health Sciences, 815-8510 Fukuoka, Japan*

Abstract. The angular distribution of compound reactions at bombarding energies lower than 10 MeV is known to be 90° symmetry. At the higher incident energies, 50-MeV range, the quantization axis tilts from the beam axis due to the particle emission in the cascade or the pre-equilibrium process. Therefore, it is necessary to know the tilted quantization axis for the angular distribution calculation of the evaporated protons from (p, p'x) reactions. In the present work, we applied the intranuclear cascade (INC) model to determine the tilted quantization axis by a classical vector analysis. The proton evaporation was calculated by the generalized evaporation model (GEM). By fitting calculations to experimental angular distributions, we deduced the angular momentum transfer from the equilibrium state.

1 Introduction

The nuclear reactions at bombarding energies higher than 10 MeV/nucleon are interpreted by a two-stage model. The first stage is the cascade or the multi-step direct process, where several energetic particles are emitted during series of intranuclear NN collisions. When the cascade process is finished, the nucleus reaches equilibrium and forms a highly-excited state. The second stage is the particle emission via the decay of the excited nucleus. Since the emitted particles show energy distributions of the Maxwellian shape, it is called the nuclear evaporation process.

Extensive efforts in both theory and experiment were made to understand the nuclear evaporation process for compound reactions at lower energies than 10 MeV/nucleon. A lot of experimental observations show[1] symmetrical angular distributions with respect to 90° in the CM system. This symmetry has been proved theoretically [2] to be ascribable to the angular momentum, which is brought into target nucleus by the incident particle. The angular momentum determines the nuclear level density and the centrifugal barrier, which are essential to interpret the energy distributions of the evaporation process.

In contrast, no theoretical study was made for the evaporation from a highly-excited equilibrium state produced in medium-energy (20-80 MeV) reactions. Since energetic particles are emitted during the first stage, the angular momentum will deviate from the initial value that is brought in by an incident particle. The quantitative analysis of angular momentum is crucial. However, there was no microscopic theory that explains the first stage reasonably with the angular momentum vector, and therefore no quantitative interpretation was performed in the past. The output of previous studies is only some empirical formulations presenting the shapes of angular distributions. Model calculation codes developed recently for general applications ignore the angular momenta and employ an assumption of the isotropic angular distributions. Indeed, some disagreements have been pointed out between experimental and calculated values. Since we succeeded in development of a reliable model [3,4] for the first stage, the intranuclear cascade (INC) model, it is now possible to estimate the angular momentum vector in a classical way.

The object of the present work is to apply the classical vector analysis to obtain angular momentum of evaporated protons in nuclear reactions induced by protons of medium-energy i.e., around 50-MeV-range protons. The tendency of angular momenta is discussed in terms of the target mass.

2 Model and analysis

2.1 Outline of the first-stage model

The angular momentum vector of cascade recidue was calculated by the INC model for the first-stage. The detailes of the present INC model is given in Refs. [3,4]; here sole the outline is described. The target nucleus is considered to be a sphere containing constituent nucleons with complete isospin degeneracy. Each nucleon has position and momentum and travels inside the nucleus freely. When a nucleon comes close to the other within a distance of the nucleon–nucleon cross section, nucleon-nucleon collision takes place. The inmedium scattering angle is chosen randomly by using experimental angular distributions of nucleon-nucleon cross sections [5]. When all energetic particles are emitted and no nucleons have chance to escape from the target nucleus, the cascade process finishes and the angular momentum vector $\boldsymbol{L}_{res}$ is calculated for residual nucleus in a classical many-body mechanics:

$$\boldsymbol{L}_{res} = \boldsymbol{L}_0 - \sum_{i-emitted} \boldsymbol{r}_i \times \boldsymbol{p}_i, \tag{1}$$

where $\boldsymbol{L}_0$ is the angular momentum brought by an incident particle, and $\boldsymbol{r}_i$, $\boldsymbol{p}_i$ are position and momentum of the emitted particle i, respectively.

2.2 Evaporation model

The Generalized Evaporation Model (GEM) code [6,7] was used to calculate the energy and emission angle of evaporated protons in the present work. It is assumed in GEM that the total emission width Γ_j of a particle j escaping from the parent nucleus i is expressed by

$$\Gamma_j = \frac{(2S_j + 1)m_j}{\pi^2 \hbar^2} \frac{1}{\rho_i(E)} \int_V^{E-Q} \varepsilon \sigma_{inv}(\varepsilon) \rho_d (E - Q - \varepsilon) d\varepsilon, \tag{2}$$

where the symbols ε, m_j, S_j are kinetic energy, mass and spin of the emitted particle j, respectively. Q is the Q-value, V is the Coulomb barrier, and d denotes the daughter nucleus. The inverse reaction cross section σ_{inv} for charged particles is given by

$$\sigma_{inv}(\varepsilon) = \sigma_g \alpha \left(1 - \frac{V}{\varepsilon}\right), \tag{3}$$

where σ_g is the geometric cross section, and α is a parameter. The level density function ρ for the parent and daughter nucleus is calculated based on the Fermi-gas model.

The total emission width of an ejectile j is summed over its ground state and all its excited states. The ejectile j is selected according to the probability distribution p_j calculated by

$$p_j = \frac{\Gamma_j}{\sum_n \Gamma_n}. \tag{4}$$

The original code of GEM employes the assumption that the angular distribution of emitted particles is isotoropic in the CM system. In the present work, we modify calculation method for the angular distribution.

2.3 Angular distribution

The angular distribution $W(\theta)$ of evaporated particles from compound system formed by a complete fusion is expressed by

$$W^{com}(\theta) = \sum_{L_{even}} C_L P_L(\cos\theta), \tag{5}$$

where C_L is constants depending on L which is the transferred angular momentum. In the case of emission of particles from a highly-excited nucleus formed in medium-energy reactions, the quantization axis should tilt from the beam axis. Then, we assume the form:

$$W(\theta) = \Lambda_{\theta' \to \theta} \left| \sum_{L_{even}} C_L P_L(\cos\theta') \right|, \tag{6}$$

where θ' is the emission angle in a new scattering plane defined by the angular momentum vector of residual nucleus, and Λ is a projection operator to the coordinate system of the laboratory frame, where the observation is made(Fig. 1). Since the evaporated protons have low energies, possible transferred angular momenta should be L = 0 and 2 in general. We determined the ratio of these transferred angular momenta to fit experimental angular distributions of (p, p'x) reactions at around 50 MeV of incident energies.

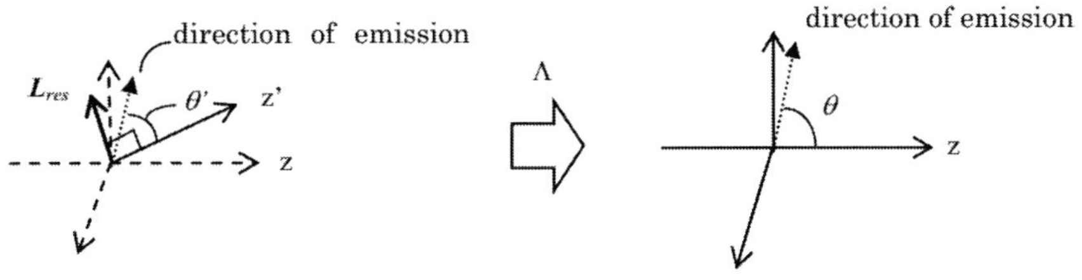

Figure 1. The relation between the emission angle in a new scattering plane and that in the system of the laboratory frame, where the observation is made .

3 Results and discussion

The calculations were carried out for (p, p'x) reactions on target nuclei of ^{27}Al, ^{56}Fe, ^{89}Y, and ^{197}Au at 60-MeV of an incident energy. Angular distributions were made from experimental energy-angle double-differential cross section (DDX) data for emitted protons of energies where evaporation is dominant. The experimental data were taken from EXFOR [8]. Calculated and observed spectral DDXs are shown in Fig. 2 for the 61.5-MeV ^{197}Au(p, p'x) reaction by solid lines and circles, respectively. For the GEM calculation, an assumption of isotropic angular distribution is employed. The evaporation regions are focused and spectra of higher energies than 20 MeV are not displayed in

Fig. 2. The calculated evaporation contribution is also indicated by broken lines. It is found that the energy region of 5 to 7 MeV is governed by the evaporation process.

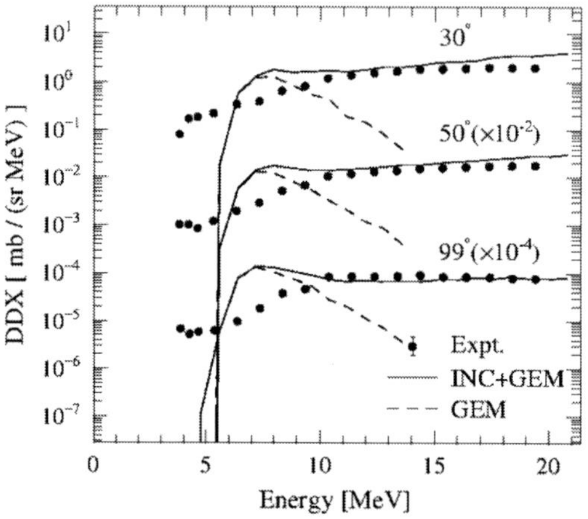

Figure 2. Proton energy spectra from inclusive ^{197}Au(p, p'x) reaction at 61.5 MeV. The angular distribution of GEM calculation is isotropic. The higher-energy parts are not shown, but only the evaporation regions are displayed.

The angle deviation $\Delta\theta$ from the initial quantized axis was calculated by the INC model for the ^{27}Al(p, p'x) and ^{197}Au(p, p'x) reactions at 60-MeV. The resulted $\Delta\theta$ distribution is shown in Fig. 3. The distributions show monotonic decrease for both ^{27}Al and ^{197}Au targets. The average $\Delta\theta$ as a function of incident energy on the ^{197}Au target nucleus is shown in Fig. 4. With increasing incident energy, the average $\Delta\theta$ slightly increases. In Fig. 5, the average $\Delta\theta$ as a function of target mass is shown at 60 MeV. With increasing target mass, $\Delta\theta$ slightly increases and becomes constant for heavier targets.

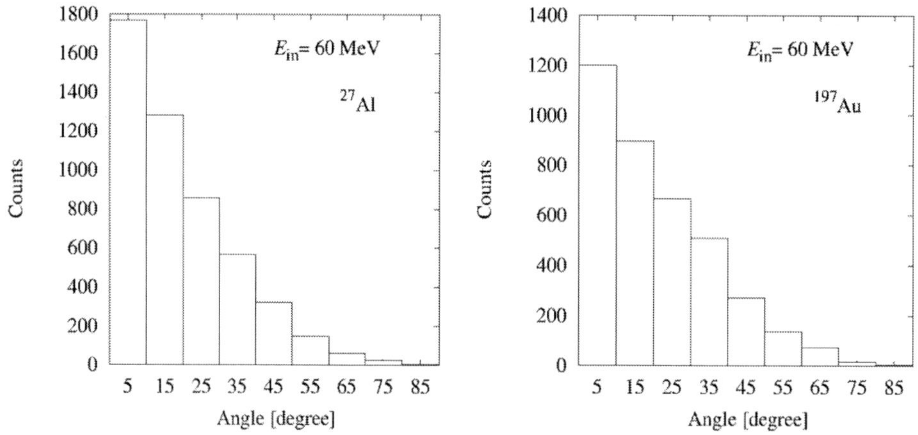

Figure 3. Angle difference of angular momenta from the initial direction for targets of ^{27}Al (left) and ^{197}Au (right).

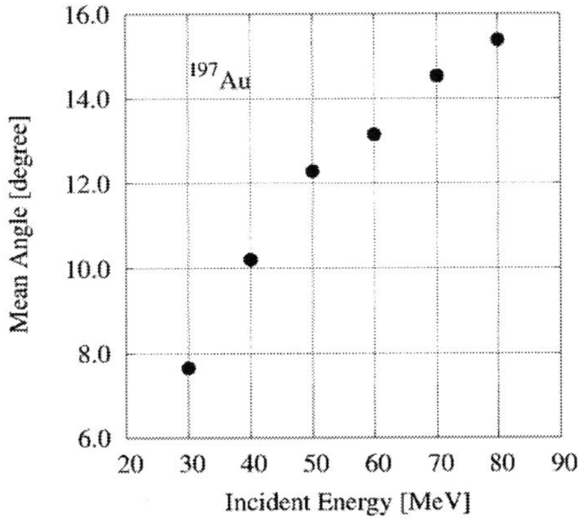

Figure 4. Mean angle difference of angular momenta from the initial direction as function of incident energy.

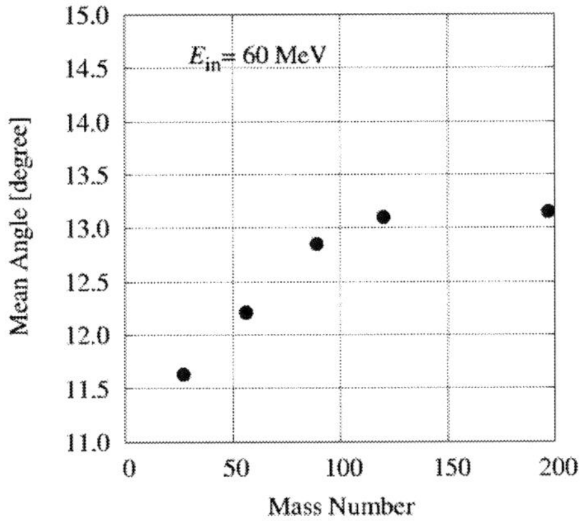

Figure 5. Mean angle difference of angular momenta from the initial direction as function of target mass.

Angular distributions obtained by calculations and experiments are shown in Fig. 6 and Fig. 7 for targets of ^{27}Al, ^{56}Fe in 60-MeV (p, p'x) reactions, respectively. Since transferred angular momentum $L = 0$ is dominant for targets of ^{27}Al and ^{56}Fe, only isotropic results are displayed. The results of $L = 0$ calculations are in good agreements with experiments.

In Fig. 8 and Fig. 9, angular distributions are shown for targets of ^{89}Y and ^{197}Au in 60-MeV (p, p'x) reactions, respectively. Although no experimental data were obtained at backward angles, the experimental data on the ^{197}Au target show much smaller values at 100° than those at forward angles. Thus, the isotropic distribution cannot explain the data of around 90° as illustrated by the dashed curves. In contrast, calculations with $L = 2$ reproduce the observation closely. For the case of the ^{89}Y target, $L = 0$ and 2 ratio of 8:2 was the best to fit the experiments.

From the present results, mass dependence of the ratio of $L = 0$ and 2 was observed. In Fig. 10, target mass dependence of the ratio is shown. For the light targets such as ^{27}Al and ^{56}Fe, angular distributions are explained by $L = 0$ only. With increasing mass, the ratio of $L = 2$ increases. Since the angular momentum L_{res} shows larger values for heavier targets, the angular momentum L of evaporeted protons, also, may show larger values.

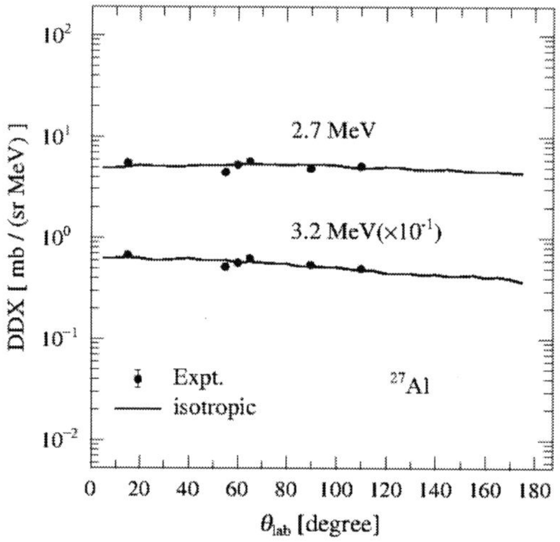

Figure 6. Angular distributions of 2.7- and 3.2-MeV protons from 61.7-MeV $^{27}Al(p,p'x)$ reactions. Experimental data are shown with closed circles. GEM calculation assuming isotropic angular distribution is shown with the solid line.

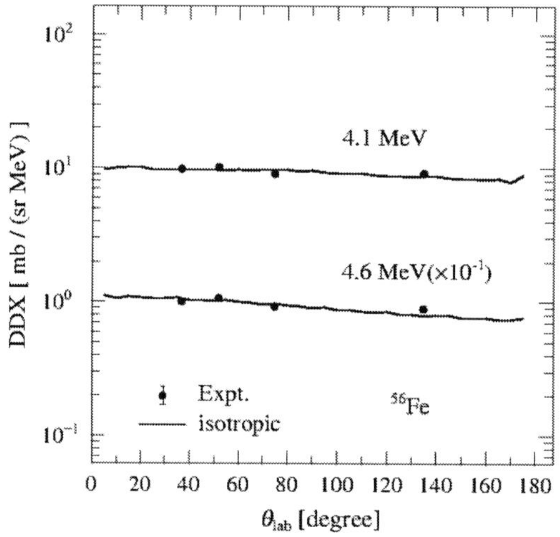

Figure 7. Angular distributions of 4.1- and 4.6-MeV protons from 61.5-MeV $^{56}Fe(p,p'x)$ reactions. See Fig. 6 caption.

DOI: 10.1051/epjconf/201612204006

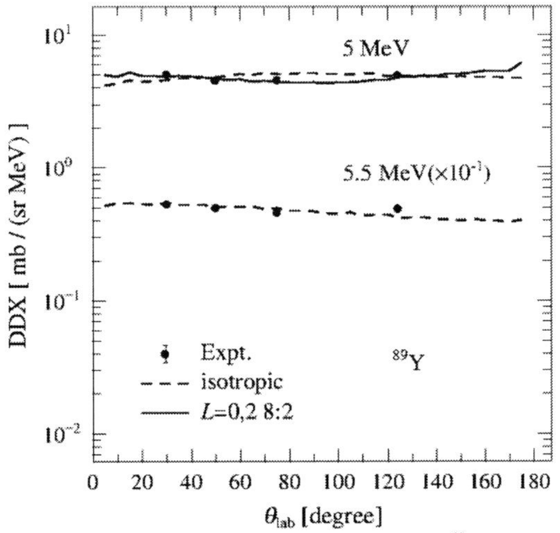

Figure 8. Angular distributions of 5- and 5.5-MeV protons from 61.5-MeV $^{89}Y(p,p'x)$ reactions. Experimental data are shown with closed circles. GEM calculation assuming isotropic angular distribution is shown with the broken line and present calculation with the solid line.

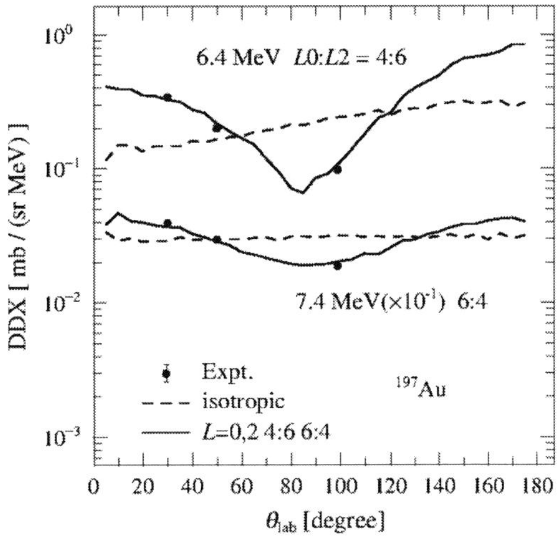

Figure 9. Angular distributions of 6.4- and 7.4-MeV protons from 61.5-MeV $^{197}Au(p,p'x)$ reactions. See Fig. 8 caption.

 DOI: 10.1051/epjconf/201612204006

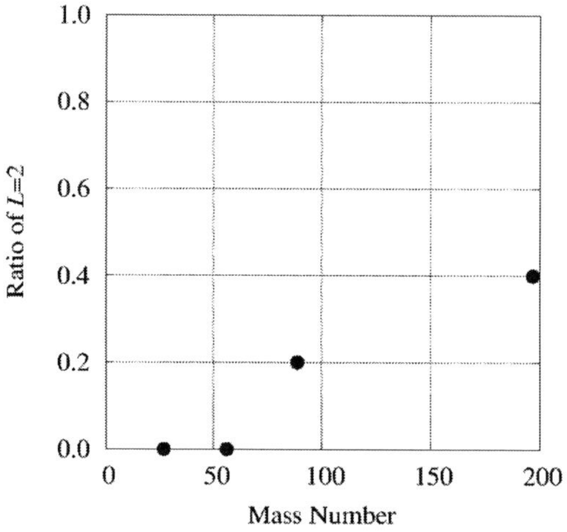

Figure 10. Target mass dependence of the ratio of $L = 2$ to $L = 0$.

4 Conclusion

The classical vector analysis, where the angular momentum is calculated in a classical many-body mechanics in terms of intranuclear cascade model was performed to deduce angular momenta of evaporated protons from medium-energy proton-nucleus reactions. The INC model is applied to extract the angular momentum vector after the cascade process. The mixing ratios of transferred angular momenta in the expression of angular distribution were determined to fit experimental angular distributions. Resultant angular momenta for targets over the periodic table show reasonable mass dependence ; angular momenta of evaporated protons show a rising tendency with mass number. Since the angular momentum L_{res} shows larger values for heavier targets, the angular momentum of evaporeted protons, also, shows larger values.

References

1. W. J. Knox, A. R. Quinton, C. E. Anderson, Phys. Rev. **120**, 2120–2128, (1960) (therein)
2. T. Ericson, V. Strutinski, Nucl. Phys. **8**, 284–293, (1958)
3. Y. Uozumi, T. Yamada, S. Nogamine, M. Nakano, Phys. Rev. C, **86**, 1–7, (2012)
4. Y. Uozumi, T. Yamada, M. Nakano, Jour. Nucl. Sci. Tech. **52**, 264–273, (2015)
5. J. Cugnon, D. L'Hote, J. Vandermeulen, Nucl. Instr. and Meth. B, **111**, 215–220, (1996)
6. S. Furihata, Adv. Monte Carlo Rad. Trans. Simul and Appli., 1045–1050, (2001)
7. S. Furihata, T. Nakamura, Jour. Nucl. Sci. Tech. **39**, Suppl.2, 758–762, (2002)
8. EXFOR Experimental Nuclear Reaction Data, https:// www-nds.iaea.org/exfor/exfor.htm.

The CERN n_TOF facility: a unique tool for nuclear data measurement

F. Mingrone[1,2,a], O. Aberle[2], J. Andrzejewski[3], L. Audouin[4], V. Bécares[5], M. Bacak[6], J. Balibrea-Correa[5], M. Barbagallo[7], S. Barros[8], F. Bečvář[9], C. Beinrucker[10], E. Berthoumieux[11], J. Billowes[12], D. Bosnar[13], M. Brugger[2], M. Caamaño[14], F. Calviño[15], M. Calviani[2], D. Cano-Ott[5], R. Cardella[2], A. Casanovas[15], D. M. Castelluccio[16,1], F. Cerutti[2], Y. Chen[4], E. Chiaveri[2], N. Colonna[7], M. A. Cortés-Giraldo[17], G. Cortés[15], L. Cosentino[18], L. Damone[7], M. Diakaki[11], C. Domingo-Pardo[19], R. Dressler[20], E. Dupont[11], I. Durán[14], B. Fernández-Domínguez[14], A. Ferrari[2], P. Ferreira[8], P. Finocchiaro[18], V. Furman[21], S. Ganesan[22], A. A. Garcia-Rios[5], A. Gawlik[3], I. Gheorghe[23], T. Glodariu[23], I. F. Gonçalves[8], E. González[5], A. Goverdovski[24], E. Griesmayer[6], C. Guerrero[17], F. Gunsing[11,2], K. Göbel[10], H. Harada[25], T. Heftrich[10], S. Heinitz[20], J. Heyse[26], G. Jenkins[27], E. Jericha[6], F. Käppeler[28], Y. Kadi[2], T. Katabuchi[29], P. Kavrigin[6], V. Ketlerov[24], V. Khryachkov[24], A. Kimura[25], N. Kivel[20], M. Kokkoris[30], M. Krtička[9], E. Leal-Cidoncha[14], C. Lederer[31,10], H. Leeb[6], J. Lerendegui[17], S. Lo Meo[16,1], S. Lonsdale[31], R. Losito[2], D. Macina[2], J. Marganiec[3], T. Martínez[5], C. Massimi[1,32], P. Mastinu[33], M. Mastromarco[7], F. Matteucci[34], E. A. Maugeri[20], E. Mendoza[5], A. Mengoni[16], P. M. Milazzo[34], M. Mirea[23], S. Montesano[2], A. Musumarra[18], R. Nolte[35], A. Oprea[23], N. Patronis[36], A. Pavlik[37], J. Perkowski[3], J. Praena[17], J. M. Quesada[17], K. Rajeev[22], T. Rauscher[38,39], R. Reifarth[10], A. Riego-Perez[15], P. Rout[22], C. Rubbia[2], J. A. Ryan[12], M. Sabaté-Gilarte[2,17], A. Saxena[22], P. Schillebeeckx[26], S. Schmidt[10], D. Schumann[20], P. Sedyshev[21], A. G. Smith[12], A. Stamatopoulos[30], G. Tagliente[7], J. L. Tain[19], A. Tarifeño-Saldivia[19], L. Tassan-Got[4], A. Tsinganis[30], S. Valenta[9], G. Vannini[1,32], V. Variale[7], P. Vaz[8], A. Ventura[1], V. Vlachoudis[5], R. Vlastou[30], A. Wallner[40], S. Warren[12], M. Weigand[10], C. Weiss[2,6], C. Wolf[10], P. J. Woods[31], T. Wright[12], and P. Žugec[13,2]

[1]*Istituto Nazionale di Fisica Nucleare, Sezione di Bologna, Italy*
[2]*European Organization for Nuclear Research (CERN), Switzerland*
[3]*University of Lodz, Poland*
[4]*Institut de Physique Nucléaire, CNRS-IN2P3, Univ. Paris-Sud, Université Paris-Saclay, F-91406 Orsay Cedex, France*
[5]*Centro de Investigaciones Energeticas Medioambientales y Tecnológicas (CIEMAT), Spain*
[6]*Technische Universität Wien, Austria*
[7]*Istituto Nazionale di Fisica Nucleare, Sezione di Bari, Italy*
[8]*Instituto Superior Técnico, Lisbon, Portugal*
[9]*Charles University, Prague, Czech Republic*
[10]*Johann-Wolfgang-Goethe Universität, Frankfurt, Germany*
[11]*CEA Saclay, Irfu, Gif-sur-Yvette, France*
[12]*University of Manchester, United Kingdom*
[13]*University of Zagreb, Croatia*
[14]*University of Santiago de Compostela, Spain*
[15]*Universitat Politècnica de Catalunya, Spain*

[a]e-mail: federica.mingrone@cern.ch

 DOI: 10.1051/epjconf/201612205001

[16]*Agenzia nazionale per le nuove tecnologie (ENEA), Bologna, Italy*
[17]*Universidad de Sevilla, Spain*
[18]*Istituto Nazionale di Fisica Nucleare, Sezione di Catania, Italy*
[19]*Instituto de Física Corpuscular, Universidad de Valencia, Spain*
[20]*Paul Scherrer Institut (PSI), Villingen, Switzerland*
[21]*Joint Institute for Nuclear Research (JINR), Dubna, Russia*
[22]*Bhabha Atomic Research Centre (BARC), India*
[23]*Horia Hulubei National Institute of Physics and Nuclear Engineering*
[24]*Institute of Physics and Power Engineering (IPPE), Obninsk, Russia*
[25]*Japan Atomic Energy Agency (JAEA), Tokai-mura, Japan*
[26]*European Commission JRC, IRMM, Geel, Belgium*
[27]*University of York, United Kingdom*
[28]*Karlsruhe Institute of Technology, Germany*
[29]*Tokyo Institute of Technology, Japan*
[30]*National Technical University of Athens, Greece*
[31]*University of Edinburgh, United Kingdom*
[32]*University of Bologna, Italy*
[33]*Istituto Nazionale di Fisica Nucleare, Sezione di Legnaro, Italy*
[34]*Istituto Nazionale di Fisica Nucleare, Sezione di Trieste, Italy*
[35]*Physikalisch Technische Bundesanstalt, Braunschweig, Germany*
[36]*University of Ioannina, Greece*
[37]*Universität Wien, Austria*
[38]*University of Hertfordshire, United Kingdom*
[39]*Department of Physics and Astronomy, University of Basel, Switzerland*
[40]*Australian National University, Canberra, Australia*

Abstract. The study of the resonant structures in neutron-nucleus cross-sections, and therefore of the compound-nucleus reaction mechanism, requires spectroscopic measurements to determine with high accuracy the energy of the neutron interacting with the material under study.

To this purpose, the neutron time-of-flight facility n_TOF has been operating since 2001 at CERN. Its characteristics, such as the high intensity instantaneous neutron flux, the wide energy range from thermal to few GeV, and the very good energy resolution, are perfectly suited to perform high-quality measurements of neutron-induced reaction cross sections. The precise and accurate knowledge of these cross sections plays a fundamental role in nuclear technologies, nuclear astrophysics and nuclear physics.

Two different measuring stations are available at the n_TOF facility, called EAR1 and EAR2, with different characteristics of intensity of the neutron flux and energy resolution. These experimental areas, combined with advanced detection systems lead to a great flexibility in performing challenging measurement of high precision and accuracy, and allow the investigation isotopes with very low cross sections, or available only in small quantities, or with very high specific activity.

The characteristics and performances of the two experimental areas of the n_TOF facility will be presented, together with the most important measurements performed to date and their physics case. In addition, the significant upcoming measurements will be introduced.

 DOI: 10.1051/epjconf/201612205001

1 Introduction

The time-of-flight facility of CERN, called n_TOF, became operative in 2001 based on an idea by Rubbia et al. [1], and since then it occupies a major role in the field of neutron cross-section measurements. In particular, thanks to the broad energy spectrum covered and the possibility to use the time-of-flight (TOF) technique to precisely select the incident neutron energy, high-resolution neutron spectroscopic measurements are possible. The pulsed neutron beam at n_TOF is produced by spallation of 20 GeV/c protons from the CERN Proton Synchrotron accelerator on a water-cooled Pb target [2]. The pulsed neutron source is used together with a moderation system, so that the n_TOF neutron beam covers about eleven orders of magnitude in energy from thermal to GeV. A scheme of the facility is given in Fig. 1. The spallation mechanism is a remarkably powerful source of neutrons, and at the proton energy of 20 GeV about 300 neutrons per proton are produced in the n_TOF target. The instantaneous intensity of the n_TOF neutron output is therefore one of the highest among the TOF facilities.

After a phase of design and feasibility studies in the late 90s, the commissioning of the facility started in 2000 and the first phase of measurements ran from 2001 to 2004. In this period of time, called n_TOF Phase-1, 25 (n,γ) and 12 (n,f) measurements have been successfully performed. In 2004 a technical problem related with the spallation target interrupted the data-taking for four years, during which the target was re-designed and replaced, and improvements such as the upgrade of EAR1 to a Class-A laboratory were implemented. From the end of 2008 till the end of 2012 the facility resumed operation in its Phase-2, during which 14 (n,γ) and 3 (n,f) measurements have been performed, together with the first two (n,cp) measurements. During the Long Shutdown 1 (LS1) of CERN [3] a second short flight-path [4], complementing the existing 185 m one, has been constructed and completed on the 25th of July 2014 when the n_TOF Phase-3 started. Since then, in this new experimental area (EAR2) one (n,γ), two (n,f) and two (n,cp) measurements have been successfully

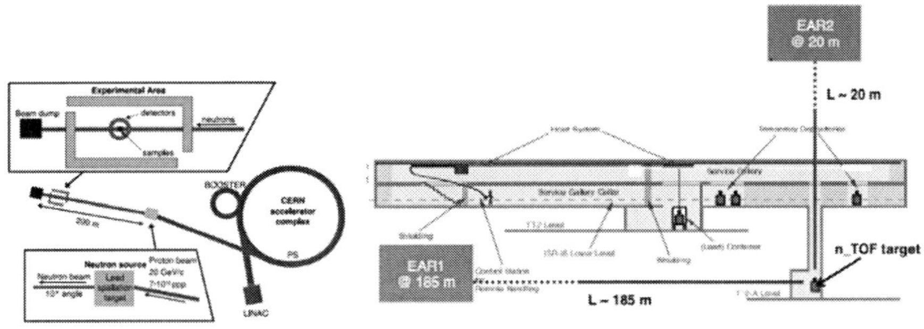

Figure 1. Left: Layout of the n_TOF facility within the CERN accelerator complex [5]. The LINAC feeds the PS-Booster, which provides the PS with protons of 1.4 GeV/c for acceleration up to 20 GeV/c. This beam is extracted and sent to the n_TOF lead spallation target in bunches of 7×10^{12} protons. The experimental hall is located near the end of the 200 m long neutron beam line. Right: Schematic of the two different beam lines of the n_TOF facility. The first experimental area (EAR1) is located horizontally at a distance of 185 m from the neutron source, while the new experimental area (EAR2) is located at 90° with respect to the proton beam direction with a flight-path of about 20 m.

completed, together with the seven (n,γ) and one (n,f) measurements performed in parallel at the first measuring station (EAR1).

2 The n_TOF neutron beam

Two different spallation targets have been used at n_TOF as neutron-producing targets. The first one, composed of Pb blocks rectangular in shape and cooled by a 5 cm water layer acting also as the moderator of the neutron spectrum, had to be replaced after four years of operation due to damages caused in some spots by inefficient cooling. The second target, installed in 2008, was equipped with a more efficient cooling system based on recirculating water, and with a separate moderator circuit to permit the use of different moderating materials. This target is made of a monolithic cylindrical block of lead, 40 cm in length and 60 cm in diameter. On the neutron exit face of the target opposite to the proton entrance, cooling is ensured by a water layer of 1 cm thickness, while moderation of the neutron spectrum is performed with a 4 cm thick layer of either normal, heavy or borated water. Borated water is now mostly being used to minimize the production of 2.2 MeV in-beam γ rays from n+ H $\rightarrow {}^2$H $+\gamma$ reactions, that constitutes the main source of background in measurements of capture cross-sections in the keV neutron-energy region.

The innovative features of the n_TOF neutron beam derive from the characteristics of the pulsed proton-beam from the CERN Proton-Synchrotron (PS), consisting of bunches with a high momentum of 20 GeV/c, a high peak current of 7×10^{12} protons per bunch and a low duty cycle of 0.5 Hz, thus avoiding pile-up of events coming from consecutive bunches and resulting in the very high intensity of the n_TOF neutron source, of the order of 10^4 neutrons/cm^2/pulse in EAR1 and 10^6 neutrons/cm^2/pulse in EAR2.

Two different beam lines originate from the spallation target: one horizontal, 200-m long, which leads to the first experimental area, and one vertical, 20-m long, which ends in the second experimental area. Along both the beam lines a sweeping magnet is placed to deflect the remaining charged particles in the beam, and the tube has to cross various shielding elements that stop particles travelling around the beam. Two collimators are used to shape the neutron beam, and a filter station allows to insert black-resonance materials in the beam to accurately evaluate the background in the experimental areas. The complete descriptions of the two beam lines can be found in Ref. [6, 7].

At the beginning of the second experimental campaign (n_TOF Phase-2), the first experimental area was upgraded to a *Work Sector Type A*, with a series of safety and monitoring systems, in order to allow measurements of high-activity samples without certified sealing. Moreover, the second experimental area has been built as a Class-A laboratory. This key modification was essential to exploit the full potential of the facility.

A fundamental piece of the puzzle of high precision measurements of reaction cross-sections is the point-wise knowledge of the incident neutron flux as a function of energy. The knowledge of its energy dependence is as much important as knowing its absolute value, considering that often cross sections are measured relative to some well known quantity in a particular energy range, as for example a standard cross section, the cross section value at thermal energies or the capture yield at the top of a well isolated saturated resonance.

At n_TOF the neutron flux was experimentally determined using several detection systems based on different principles and reactions. A large effort was devoted to minimize all possible uncertainties, both systematic and statistical, in particular on the energy dependence of the neutron flux. All details can be found in Ref. [8].

Starting from 2010, the neutron flux in EAR1 has been moderated by borated water, while for the previous years normal water has been used. Since the separated moderator circuit is present only on

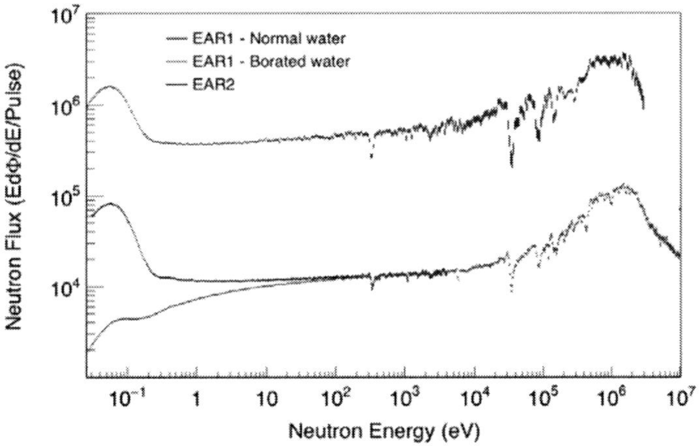

Figure 2. n_TOF neutron flux at EAR1 with normal (black) and borated (red) water as moderator compared with the neutron flux at EAR2 (blue).

the neutron exit face of the target opposite to the proton entrance, the neutron beam arriving in EAR2 can be moderated only by normal water. The effect of the ^{10}B-loaded moderator is a strong reduction of the thermal peak, which changes also with time depending on the ^{10}B concentration. Above a few hundred eV, the flux does not depend on the moderator anymore and, as expected, is found to be constant over time. In Fig. 2 the flux in EAR1 with different moderators is compared with the flux in EAR2. The intensity of the neutron flux in EAR2 is about 40 times higher than in EAR1 for neutron energies above 10 eV, while for lower energies, where the borated water is strongly reducing the flux in EAR1, it is about 500 times higher. On the other hand, it is important to point out that, because of the kinematics within the spallation target, a notable amount of neutrons arrive in EAR2 only for energies below about 300 MeV, while in EAR1 (the beam line in the direction of the incoming proton beam) neutron energies up to 1 GeV can be reached.

Another cardinal characteristic of the n_TOF facility is its energy resolution. Being a time-of-flight facility, the neutron kinetic energy is calculated from the time taken by the neutrons to travel the distance between the spallation target and the sample under analysis. Because of the neutron transport in the spallation target and in the moderator circuit, and the pulse width of the primary proton beam, the relation between TOF and E_n is represented by distributions of a particular shape that depends on the neutron energy. These distributions form the resolution function, i.e. the probability that a neutron with energy E_n is detected with a corresponding time of flight t, of the n_TOF spectrometer.

The resolution function changes significantly from EAR1 and EAR2, because of its strong dependance on the neutron transport within the target and on the neutron propagation to the experimental areas. Since these effects cannot be experimentally measured, the two resolution functions for the spallation target-moderator assembly have been simulated with FLUKA [9] and MCNP [10] Monte Carlo codes for both the configuration with normal and borated water as moderator material, and neutrons have been propagated through the two different beam lines. In Table 1 the comparison between the resolution broadening as function of neutron energy in EAR1 and EAR2 shows that the energy resolution in EAR2 is limited in resolving individual resonances at high energies compared to EAR1.

Table 1. The energy resolution as function of neutron energy for EAR1 with borated water as moderator [6] and EAR2 [11].

E_n	$\Delta E/E$	
	EAR1	EAR2
1 eV	3.2×10^{-4}	4.8×10^{-3}
10 eV	3.2×10^{-4}	5.7×10^{-3}
100 eV	4.3×10^{-4}	8.1×10^{-2}
1 keV	5.4×10^{-4}	1.4×10^{-2}
10 keV	1.1×10^{-3}	2.3×10^{-2}
100 keV	2.9×10^{-3}	4.6×10^{-2}
1 MeV	5.3×10^{-3}	5.6×10^{-2}

With the characteristics listed above n_TOF represents a unique facility for measurements of radioactive isotopes and very low cross-sections, as for example capture cross section of s-process branching-point isotopes, or capture and fission cross-sections of actinides. Depending on the requirements of the reaction under study (e.g. the need of high energy resolution or rather of high neutron flux), the two experimental areas of the n_TOF facility provide a remarkable flexibility to select the best configuration for successful measurement. To fully exploit this potential, different detection systems are available and will be described in the next Section.

3 Detection Systems

Several detection systems can be used for neutron induced reaction measurements at the n_TOF facility. Among them, the most significant ones for neutron induced capture, fission and (n,cp) reaction measurements will be here presented.

For capture measurements, two different detection systems have been set up: an array of deuterated liquid scintillator detectors (C_6D_6) and a 4π BaF_2 Total Absorption Calorimeter (TAC), both shown in Fig. 3. The first apparatus is characterized by a low sensitivity to background signals induced by scattered neutrons, and it is particularly suitable for measurements of light isotopes for which the elastic scattering cross-section could be several orders of magnitude higher than for capture. An array of four custom made C_6D_6 scintillators at the cutting edge of present technology [12] are used at n_TOF in both experimental areas. For highly radioactive and fissile isotopes, in particular for minor actinides, capture measurements are performed in EAR1 with the TAC, which permits the identification of capture events from competing reactions by reconstructing the total energy of the γ-ray cascade. The relatively large neutron sensitivity of the apparatus is reduced by hardware expedients, as described in Ref. [13]. Recently, the simultaneous measurement of neutron-induced capture and fission reactions have been tested using a MicroMegas [14] detector in combination with the TAC, in order to accurately disentangle the two types of reactions [15]. This method, which has been tested with a measurement of ^{235}U in 2012, allows accurate capture cross-section measurements of fissile isotopes.

Different detection systems are exploited also for fission cross-section measurements. A multi-stack Fission Ionization Chamber (FIC), used in the first campaign, has been replaced in Phase-2 by a high-performance MicroMegas detector [14, 16] characterized by a better signal-to-noise ratio and has been successfully used in both experimental areas. A second method used at n_TOF relies on the detection of both fission fragments in coincidence. To this purpose, a stack of position-sensitive Parallel Plate Avalanche Counters (PPACs) is used, which also determines the angular distribution

Figure 3. Total absorption calorimeter (TAC) in the n_TOF-EAR1 (left panel) and C_6D_6 scintillators set-up in n_TOF-EAR2 (right panel). The horizontal and vertical beam lines are visible as well.

of the fission fragments [17, 18]. In all fission measurements, the ratio method is applied, where the cross section of a given isotope is determined relative to the established standard ^{235}U. To this purpose, reference samples of ^{235}U are always mounted in the detector in use and measured simultaneously.

Another important set of measurements deals with the (n,cp) reactions, which has become particularly notable thanks to the characteristics of EAR2 that allows one to measure very low cross sections. These measurements can be performed with MicroMegas or Silicon-based detectors. Recently, a new concept of Si-based detector has been developed with the sample sandwiched between two silicon detectors that exploits the emission of back-to-back reaction products together with the coincidence technique to reject the background [19].

4 The n_TOF facility for science and technology

As previously pointed out, nuclear reactions induced by neutrons play a fundamental role in nuclear technology, being at the basis of reactor physics, and are of great importance in other fields such as Nuclear Astrophysics and fundamental Nuclear Physics. Many of the existing experimental data have been evaluated and made available through nuclear data libraries and databases. Nevertheless, new developments in emerging nuclear technologies, as well as the newest astrophysical and nuclear models, require more precise and accurate data together with new measurements for a large number of isotopes.

For nuclear technologies, the prediction of the behaviour of reactor cores depends strongly on capture and fission cross-section data. In particular, the R&D of new reactor concepts, as Generation-IV [20] and Accelerator Driven System (ADS) [21], require precise and accurate measurements of neutron capture and fission cross sections of actinides in the thermal (meV), epithermal (eV-keV) and fast (MeV) energy regions. At n_TOF different measurements of reaction cross-sections addressed as urgent nuclear data requirements by the Nuclear Energy Agency [22] have been performed. As an example, the measurement of the ^{238}U radiative capture cross-section [23] and of the ^{240,242}Pu(n,*f*) cross-section [24, 25], with the purpose of obtaining a very high accuracy to solve present inconsistencies and improve the performances of new reactors. All these measurements have been performed in EAR1, but the ^{240}Pu(n,f) one has suffered from several problems due to the damage of the MicroMegas detector caused by the very high α-activity of the sample. For this reason the measurement has been repeated in 2014 exploiting the n_TOF EAR2, which has allowed to reduce the measurement time, thus preventing the damage of the detector.

The experimental campaign devoted to Nuclear Astrophysics is focused mostly on neutron magic nuclei, which act as bottle neck for the flow of s-process. Branching point isotopes for the s-process nucleosynthesis chain, i.e. those isotopes for which the time scale of neutron capture reactions and β decays are comparable, have also been investigated. Among them, in 2012 the capture cross-section of ^{63}Ni, which is the first branching point of the s-process nucleosynthesis chain, has been measured for the first time in a wide energy range from thermal neutron energies up to 200 keV [26]. In total the capture kernels of 12 (new) resonances were determined, and stellar model calculations show that the new data have a significant effect on the s-process production of ^{63}Cu, ^{64}Ni, and ^{64}Zn in massive stars.

In addition, isotopes of special interest such as the osmium involved in the so-called cosmic clock have been studied at n_TOF. The clock is based on the extremely long half-life of ^{187}Re($\tau_{1/2}$ = 43.3 Gyr), decaying to ^{187}Os, and on the fact that ^{186}Os and ^{187}Os are shielded against direct r-process production. Thanks to the well established s-process abundances of the ^{186}Os and ^{187}Os, the age of the Universe can be inferred, in the the Re/Os clock, by the enhancement in the abundance of ^{187}Os due to ^{187}Re $\rightarrow^{187}$Os decay. The neutron capture cross sections of ^{186,187,188}Os have been measured at the n_TOF facility with improved accuracy and over a wide energy range of neutron energies from 1 eV to 1 MeV. Using the n_TOF results within a schematic model that assumes an exponentially decreasing production rate for ^{187}Re, an age of the Universe of 15 $\pm$ 2 Gyr was obtained from the Re/Os cosmo-chronometer. More details can be found in Ref. [27–29].

With the construction of EAR2 the possibility to measure highly radioactive isotopes has been opened. In particular, for the first time the ^{7}Be(n,α) cross section has been measured in a wide energy range. This enters in the so-called "cosmological ^{7}Li problem" in Big Bang Nucleosynthesis (BBN), with the aim of lowering the uncertainty related to nuclear physics inputs in the BBN calculations. The challenges of the measurement, related to the low (n,α) cross section and the very high specific activity of ^{7}Be, have required a comprehensive study on its feasibility together with the development of a new Si-based detector.

The characteristics of the n_TOF facilities could be exploited also for measurements providing valuable information on basic nuclear physics quantities, such as levels densities, γ-ray strength functions and angular distributions. As an example, the study of (n,f) reactions with PPAC detectors provides valuable information on the angular distribution of the fission fragments [30]. Moreover, the high energy resolution of the facility allows a statistical analysis of the resonances obtained in capture measurements,, which can result in important constraints both for Astrophysics and Nuclear Physics models. In the next paragraph the ^{25}Mg(n,γ) measurement is illustrated as a significant example of the study of the level structure of the compound nucleus.

4.1 The study of ^{25}Mg+n reaction

Elements heavier than iron are produced in stellar environments through the balance of neutron capture and β–decay processes. About half of these elements appear through the nucleosynthesis chain of the so called s-process (i.e. slow neutron capture process) [31]. Although most of the astrophysical sites of stellar nucleosynthesis and the nuclear processes involved have been identified, important physical and nuclear details are still largely unexplained, thus hindering a comprehensive understanding of the origin of the elements. In this view, an accurate knowledge of the nuclear reaction rates

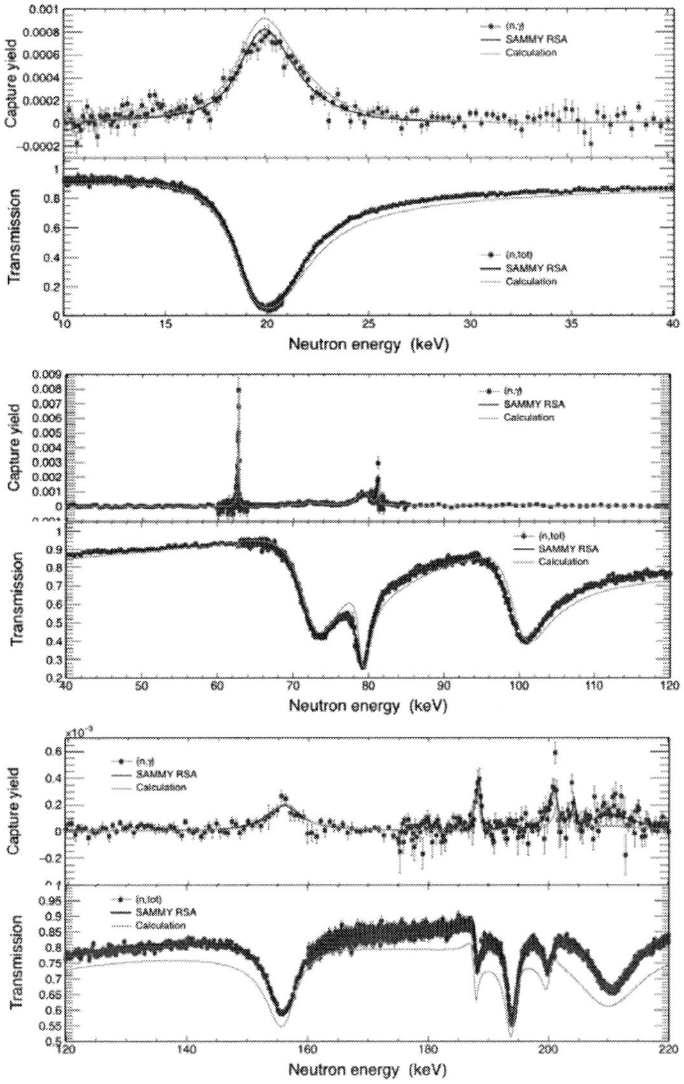

Figure 4. Comparisons between capture and transmission data in different energy ranges. The SAMMY RSA best fit curve is shown (blue line) together with the calculation performed using the parameters from the evaluation by Koehler [36] (green line).

at stellar energy (i.e. keV energy region) is fundamental to constrain theoretical predictions of stellar models.

The (α,n) reaction on ^{22}Ne is the major astrophysical neutron source of the s process in massive stars and in intermediate AGB stars, while it is partially activated in low mass AGBs. Despite lots of attempts (see Ref. [32] and reference therein), direct measurements in the energy range of astrophysical sites are extremely difficult, mostly because of the extremely low cross section of the reaction at these energies and of cosmic-ray induced background. Data present in literature are therefore not accurate or even lacking, with the consequence that the ^{22}Ne(α,n)^{25}Mg reaction rate is very uncertain. This reaction rate at s process temperature depends on the level structure of the compound nucleus ^{26}Mg above the α threshold (Q = 10.615 MeV), close to the neutron threshold (S_n = 11.093 MeV). Information on the properties of these poorly known states, especially on their spin-parity J^π, can be obtained from the ^{25}Mg(n,γ) reaction by selecting only those states which can be populated by ^{22}Ne+α reaction. For ^{22}Ne and α particles $J^\pi = 0^+$, and therefore only natural parity states $(0^+, 1^-, 2^+, \ldots)$ can participate in the ^{22}Ne(α,n)^{25}Mg reaction, which correspond to a subset of ^{26}Mg states populated in the ^{25}Mg(n,γ) reaction. In this context, it is crucial to have a precise assigment of the spin-parity numbers to each state studied in ^{25}Mg+n reaction.

To this aim, the measurement of the ^{25}Mg(n,γ) cross section has been performed at n_TOF in 2012, and in addition the total cross section on ^{25}Mg has been measured at the EC-JRC-IRMM facility GELINA in Belgium [33].

For the capture measurement, a low solid angle detection system composed of two C_6D_6 scintillators has been used. As previously pointed out, these detectors have a very low sensitivity to γ-rays induced by scattered neutrons, being therefore optimized to investigate light isotopes such as the ^{25}Mg. The total energy detection technique has been used, in combination with the *pulse-height weighting method* [34] to obtain the required proportionality between the the efficiency of the detection system and the total radiative energy emitted by the capture event. The measurement has been carried out in EAR1, where single resonances could be resolved up to about 500 keV.

The experimental yield has been analyzed with the resonance shape analysis (RSA) code SAMMY [35] to obtain the parametrization of the capture cross-section in terms of resonance parameters. A simultaneous resonance shape analysis of both capture and transmission data is performed, which leads to much more precise resonance parameters as if the two data sets would have been analyzed separately. In Fig. 4 comparisons between capture and transmission data in different energy ranges are shown. The calculation performed using the parameters from the evaluation by Koehler [36] is shown as well on top of the data, together with the SAMMY best fit. It is evident from the figures that the new capture and transmission measurements call for a major revision of the evaluations, which suffered from several problems mainly due to impurities in the samples analyzed (see Ref. [37] and references therein).

Thanks to the new high-quality data provided at n_TOF and GELINA, a preliminary analysis of the statistical properties of the resonances has pointed out that for the level at E_R = 79 keV the spin parity should change from $J^\pi = 3^+$ to $J^\pi = 3^-$. If this would be confirmed, this state will become a natural-parity one and must be added to the number of states that can be populated by the ^{22}Ne(α,n) reaction, affecting therefore its rate.

5 Conclusion

The n_TOF facility became operative at CERN in 2001, and since then it is at the cutting edge for neutron induced reaction measurements. Since July 2014 a second experimental area (EAR2) has been built 20 m above the spallation target, with unique characteristics which made it a powerful tool for measurements of highly radioactive and of low-mass samples, and of very low cross sections. This new measuring station is complementary to the existing first experimental area (EAR1), and the characteristics of both are extremely competitive in terms of flux intensity, energy range covered and energy resolution. Depending on the requirements of the reaction under study precise and accurate results can be achieved, and several detection systems are available to investigate (n,γ), (n,cp) and (n,f) reactions. These cross sections play a fundamental role in nuclear technologies, nuclear astrophysics and fundamental nuclear physics, and significant results have been attained, relating to new reactor designs, to stellar and primordial nucleosynthesis models and to the investigation of the compound nucleus. Associated to the last field, a recent measurement on the ^{25}Mg capture cross-section has allowed an accurate study of the level structure of the ^{26}Mg compound nucleus, providing important constraints on levels which can be populated by the ^{22}Ne(α,n) reaction, one of the two main neutron source of the s process.

References

[1] C. Rubbia et al., CERN/LHC/98-02, CERN (1998).
[2] The n_TOF Collaboration, CERN INTC-2002-037 (2003).
[3] http://home.web.cern.ch/about/updates/2013/02/long-shutdown-1-exciting-times-ahead.
[4] The n TOF Collaboration, CERN-INTC-2012-029 / INTC-O-015 (2012).
[5] The CERN Accelerator Complex, http://public.web.cern.ch/public/en/research/AccelComplex-en.html.
[6] C. Guerrero *et al.*, Eur. Phys. J. A **49** (2013) 27.
[7] C. Weiß *et al.*, Nucl. Instr. and Meth. A **799** (2015) 90.
[8] M. Barbagallo *et al.*, Eur. Phys. J. A **49** (2013) 156.
[9] A. Ferrari and P.R. Sala, *Intermediate and High Energy Physics Models in FLUKA: Improvements, Benchmarks and Applications*, Proc. Int. Conf. on Nuclear Data for Science and Technology, NDST-97, ICTP, Miramare, Trieste, Italy, (19-24 May 1997).
[10] J. Briesmeister, *MCNP-A General Monte Carlo N-Particlke Transport Code-Version 4c2*, LA-13709-M (2000).
[11] J. Lerendegui-Marco *et al.*, submitted to Eur. Phys. J. A.
[12] P.F. Mastinu *et al.*, n_TOF-PUB-2013-002, CERN-n_TOF-PUB-2013-002 (2013).
[13] C. Guerrero *et al.*, Nucl. Instr. and Meth. A **608** (2009) 424-433.
[14] Y. Giomataris *et al.*, Nucl. Instr. and Meth. A **376** (1996) 29.
[15] C. Guerrero *et el.*, Eur. Phys. J. A **48** (2012) 29.
[16] S. Andriamonje *et al.*, *Development and performance of Microbulk Micromegas detectors*, Proceedings of the 1st International Conference on Micro-Pattern Gaseous Detectors, Crete, Greece, 2009.
[17] C. Paradela *et al.*, Physical Review C **82** (2010) 034601.
[18] L. Tassan-Got *et al.*, CERN-INTC-2006-016, INTC-P-209.
[19] M. Barbagallo, A. Musumarra *et al.*, CERN-INTC-2012-029 / INTC-O-015 (2012).
[20] Generation IV International Forum (GIF), https://www.gen-4.org/gif/jcms/c_9260/Public.
[21] OECD Nuclear Energy Agency, *Accelerator-Driven Systems (ADS) and Fast Reactor (FR) in advanced nuclear fuel cycles. A comparative study*, 2002.
[22] NEA-HPRL, High Priority Request List, hwww.nea.fr/html/dbdata/hprli.
[23] F. Mingrone *et al.*, Nuclear Data Sheets **119** (2014) 18.
[24] M. Calviani, E. Berthoumieux *et al.*, CERN-INTC-2010-042 / INTC-P-280 (2010).

[25] A. Tsinganis *et al.*, Nuclear Data Sheets **119** (2014) 58.
[26] C. Lederer *et al.*, Phys. Rev. Lett. **110** (2013) 022501.
[27] M. Mosconi *et al.*, Phys. Rev. C **82** (2010) 015802.
[28] M. Mosconi *et al.*, Phys. Rev. C **82** (2010) 015803.
[29] K. Fujii *et al.*, Phys. Rev. C **82** (2010) 015804.
[30] D. Tarrio *et al.*, Nucl. Instr. and Meth. A **743** (2014) 79-85.
[31] E.M. Burbidge, G.R. Burbidge, W.A. Fowler and F. Hoyle, Rev. Mod. Phys. **29** (1957) 547.
[32] M. Jaeger *et al.*, Phys. Rev. Lett. **87** (2001) 202501.
[33] W. Mondelaers and P. Schillebeeckx, Notiziario **11** (2006) 19.
[34] U. Abbondanno *et al.*, Nuclear Instruments and Methods in Physics Research A **521** (2004) 454.
[35] N. M. Larson, ORNL/TM-9179/R8, Oak Ridge National Laboratory, Oak Ridge, TN, USA (2008). Also ENDF-364/R2.
[36] P. E. Koehler, Phys. Rev. C 66, 055805 (2002).
[37] C. Massimi *et al.*, Phys. Rev. C **85** (2012) 044615.

Neutron capture cross section and capture gamma-ray spectra of ^{89}Y

Tatsuya Katabuchi[1,a], Tohomohiro Okamiya[1], Shotaro Yanagida[1], Motoharu Mizumoto[1], Kazushi Terada[2], Atsushi Kimura[2], Nobuyuki Iwamoto[2], and Masayuki Igashira[1]

[1]*Research Laboratory for Nuclear Reactors, Tokyo Institute of Technology, Meguro ku, 152-8550, Tokyo, Japan*
[2]*Japan Atomic Energy Agency, Tokai, Ibaraki 319-1195, Japan*

Abstract. The neutron capture cross section of ^{89}Y was measured by the time-of-flight method in an energy range from 15 to 100 keV. A pulse-height weighting technique was applied to derive the capture yield. The absolute cross section was determined based on the standard reaciotn ^{197}Au$(n,\gamma)^{198}$Au reaction. The neutron capture γ-ray spectrum was derived by unfolding the pulse-height spectrum with detector response functions.

1 Introduction

In the slow neutron capture process (s-process) model, heavier nuclides than ^{56}Fe are synthesized through the neutron capture process and successive β-decay in stellar environments. The s-process occurs along the stability line of the nuclear chart. The s-process model calculations are performed based on nuclear data of neutron capture cross sections and β-decay half-lives. Among the nuclear data inputs, the neutron capture cross sections of neutron magic nuclides (N = 50 and 82) are particularly important because the neutron magic nuclides form bottlenecks of the s-process reaction network due to their small neutron capture cross sections.

Yttrium-89 is an $N = 50$ neutron magic nuclide. The reliable neutron capture cross sections of ^{89}Y and other $N = 50$ isotones (^{88}Sr and ^{90}Zr) are necessary in the s-process model calculation. In our previous works, we have already measured the neutron capture cross sections of ^{88}Sr and ^{90}Zr in the Tokyo Institute of Technology [1, 2]. The present work was carried out to measure the cross section of the ^{89}Y$(n,\gamma)^{90}$Y reaction. The neutron capture γ-ray spectrum that provides more information on the reaction mechanism was also measured.

2 Experiments and Data Analysis

Experimental details of the Tokyo Tech experiments were described elsewhere [3]. Only a brief description is given here. Neutrons were generated through the ^{7}Li$(p,n)^{7}$Be reaction, bombarding a lithium target with a proton beam from a Pelletron accelerator at an incident proton energy of 1.902 MeV, 22 keV above the reaction threshold. The neutron energy distributes from a few keV

[a]e-mail: buchi@nr.titech.ac.jp

to 100 keV. The time-of-flight (TOF) method was employed to measure the incident neutron energy. Neutrons were detected with a small ^{6}Li glass scintillator located at a flight distance of 30 cm from the neutron source. The yttrium sample was Y_2O_3 powder cased in a graphite container. An isotopically enriched sample was not needed since Yttrium is a monoisotopic element. The net yttrium mass was 18.5 g. The size was 55 mm in diameter and 7.5 mm in thickness. A gold sample with the same diameter was also used as a standard sample. The sample was placed at a flight distance of 12 cm. Gamma-rays emitted from the sample were detected with an NaI(Tl) detector with an anti-Compton shield [4]. Runs for ^{89}Y, Au and blank were repeated to average out change of experimental conditions. Both the pulse height (PH) and the TOF of signals from the NaI(Tl) detector were recorded event-by-event in a list-data format.

For data analysis, we set eight TOF gates: 15-20, 20-23, 23-26, 26-29, 29-39, 39-46, 46-54 and 54-100 keV. Two-dimensional PH-TOF data were sorted with the eight TOF gates, thereby giving gated foreground PH spectra. Another sorting with a background gate that is a fast-neutron absent TOF region provided background PH spectra. After subtracting the background PH spectra from the foreground PH spectra, net PH spectra were obtained.

A pulse-height wighting technique was employed to derive the neutron capture yields [5]. The weighting function was made from detector response functions calculated by the Monte Carlo simulation method. The absolute $^{89}Y(n,\gamma)^{90}Y$ cross sections were determined in reference to the standard $^{197}Au(n,\gamma)^{198}Au$ cross section from JENDL-4.0. In addition, corrections for neutron self-shielding and multiple scattering, chemical impurities, and discriminated γ-rays below 620 keV were made.

Capture γ-ray spectrum was derived by unfolding the net PH spectrum with the detector response functions. The detector response functions were calculated from Monte Carlo simulations. Spectrum unfolding was done with a unfolding code FERDOR [6].

3 Results and Discussion

Figure 1 shows the obtained cross section. Previous measurements and evaluated cross sections from JENDL-4.0 and ENDF/B-VII.1 are also plotted for comparison. The errors of the present data include the statistical and systematic uncertainties. Except for Gate 1 (15-20 keV), the total errors are 5 - 6%. The error in Gate 1 is 9%. As shown in Fig. 1, errors are smaller than previous experiments. The evaluated cross sections agree with the present results in the high energy region but deviate below 30 keV.

The unfolding capture γ-ray spectrum is shown in Fig. 2. The bars above the spectrum indicate the energy positions of primary γ-rays calculated from known excited levels of ^{90}Y. Transitions to the ground and the first excited states were observed clearly.

4 Summary

We measured the neutron capture cross section and capture γ-ray spectrum of ^{89}Y in the energy region from 15 to 100 keV. The capture cross section was determined with a smaller uncertainties than previous measurements. Evaluated cross sections from JENDL-4.0 and ENDF/B-VII.1 agree with the present work in high energy region but disagreement was found in low energy region. The neutron capture γ-ray spectrum was derived by unfolding the PH spectrum with detector response functions. We plan to measure the capture cross section of ^{89}Y at a higher energy point (550 keV). Combining with the upcoming high energy data, the present results must be included in future evaluation of nuclear data libraries.

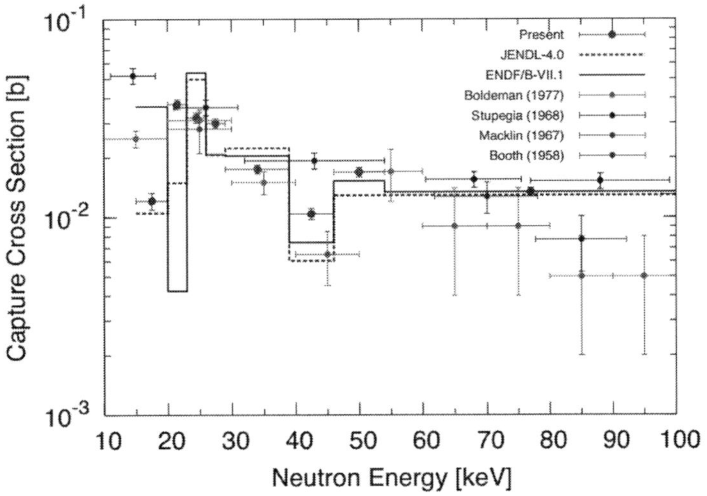

Figure 1. Neutron capture cross section of ^{89}Y compared with previous measurements and evaluated cross sections.

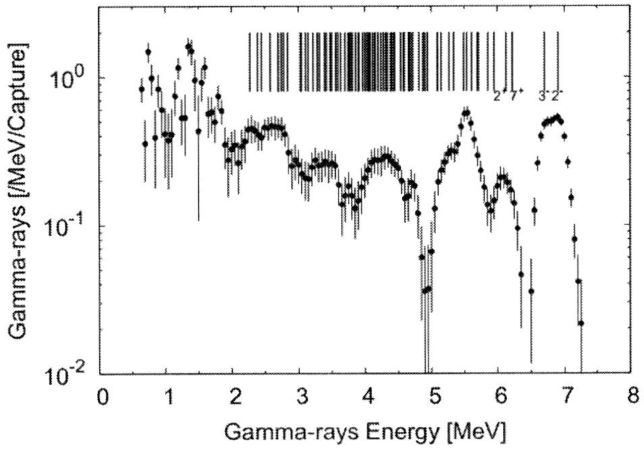

Figure 2. Neutron capture γ-ray spectrum of ^{89}Y. The bars indicate primary γ-ray positions calculated from known excited levels of ^{90}Y.

References

[1] T. Katabuchi *et al.,* J. Korean Physical Society, **59**, 1844-1847 (2011).
[2] K. Ohgama, *Doctoral Dissertation*, Tokyo Institute of Technology, (2005), [in Japanese].
[3] S. Mizuno, M. Igashira, K. Masuda, J. Nucl. Sci. Technol., **36**[6], 493-507 (1999).
[4] M. Igashira, H. Kitazawa, N. Yamamuro, Nucl. Instrum. Methods, **A245**, 432-437 (1986).
[5] R. L. Macklin, J. H. Gibbons, Phys. Rev., **159**, 1007-1015 (1967).
[6] H. Kendrick and S. M. Speering, GA-9882 (1970).

Study for s-process using neutron beam provided from ANNRI of J-PARC

Takehioto Hayakawa[12], Yosuke Toh[1], Akira Kimura[1], Shoji Nakamura[1], Toshiyuki Shizuma[1] and Hideo Harada[1]

[1]*Japan Atomic Energy Agency, Tokai, Ibaraki, Japan*
[2]*Natonal Astronomical Observatory of Japan, Mitaka, Tokyo, Japan*

Abstract. Most isotopes heavier than iron are synthesized by the slow neutron capture reaction process (s-process) in stars. Isomers in stable isotopes have sometimes an important role as a branching point in nucleosynthesis flow in the s-process. An isomer with a half-life of 14.1 y in ^{113}Cd is a branching point from which a nucleosynthesis flow reaches to a rare isotope ^{115}Sn. The astrophysical origin of ^{115}Sn has remained still an open question. The s-process abundance of ^{115}Sn depends on the ratio of the ^{112}Cd(n, γ)^{113}Cd^{m} reaction cross section to the ^{112}Cd(n, γ)^{113}Cd^{gs} reaction cross section. However, the isomer production ratio following the neutron capture reaction has not been measured in the energy region higher than the thermal energy. An intense neutron beam experimental system, ANNRI, in J-PARC has a high purity germanium (HPGe) detector system consisting of two cluster detectors. We have measured γ-rays decaying to the ground state and the isomer using the HPGe detectors in conjunction with a time-of-flight method at ANNRI.

1 Introduction

Nuclides heavier than iron were predominantly synthesized by two stellar neutron induced reactions of slow neutron capture reactions (s-process) and rapid neutron capture reaction process (r-process). The astrophysical origin of rare isotopes in Cd-In-Sn region, ^{113}In, ^{114}Sn and ^{115}Sn, have been a long-standing problem [1-3]. These isotopes are classified to "p-nuclei" which have following features. Their isotopic fractions are typically lower than 1%. They are neutron-deficient isotopes. They are located on the outside of the main path of the s-process and shielded against the β-decay after the freezeout of the r-process by stable isobars. Hayakawa et al. [4] found an empirical relationship between a p-nucleus and an s-nucleus with the same atomic number in the solar abundances. This relationship is evidence that the most probable origin of 27 p-nuclei is photodisintegration reactions in supernova explosions (γ-process). However, the origin of odd-N isotope, ^{115}Sn, has remained still an open question, since the relationship cover only the even-N nuclei.

Isomers in stable isotopes have sometimes an important role as a branching point in nucleosynthesis flow in the s-process (for examples, see refs. [3,5]). As shown in Fig. 1, β-unstable isomers in stable isotopes have important role to synthesis these three p-nuclei. The p-nuclei, ^{113}In and ^{115}Sn, may be produced through isomers in ^{113}Cd and ^{115}In, respectively, by the β-decay after the r-process and/or weak branches of the s-process. The nucleus ^{115}Sn may be produced by a weak branch of the s-process (see the dotted arrows in Fig. 1). The neutron capture reaction cross-sections were systematically measured and the resonance parameters were also evaluated [6]. However, the partial neutron capture cross section to $^{113}Cd^{m}$ has not been measured except thermal energy. The abundance of ^{115}Sn depends on the population ratio of $^{113}Cd^{m}$ to $^{113}Cd^{gs}$ following neutron capture reactions on ^{112}Cd. It was suggested that the population of the isomer in ^{113}Cd should follow by thermal-equilibrium conditions in s-process environments of kT=8-30 keV. In such cases, the population ratio, m/(m+gs), depend only on the temperature of the environment where the s-process occurs.

Hayakawa et al. [3] measured the neutron capture cross-section to $^{113}Cd^{m}$ using thermal neutrons provided by a nuclear research reactor JRR-3. To estimate the ratio of $^{112}Cd(n, \gamma)^{112}Cd^{m}$ to $^{112}Cd(n, \gamma)^{112}Cd^{m}$ + $^{112}Cd(n, \gamma)^{112}Cd^{gs}$ at the s-process energies of 5-50 keV the ratio as a function of the neutron incident energy was calculated by using a statistical model. It was found that the m/(m+gs) ratio depends strongly on branching ratios of γ-decay of excited states. The calculated ratios are almost constant over a wide range of E < 100 keV. The s-process contribution to ^{115}Sn using a classical "steady flow" model with the evaluated ratio at stellar temperatures was evaluated. It was obtained that 3-4% of the ^{115}Sn solar abundance may originate from the s-process. However, the m/(m+gs) ratios depend on the detailed nuclear structure and thus m/(m+gs) ratios in an energy region of 5-50 keV should be experimentally measured. Thus, we have measured γ-rays using the HPGe detectors in conjunction with a time-of-flight method at an intense neutron beam experimental system, ANNRI, in J-PARC.

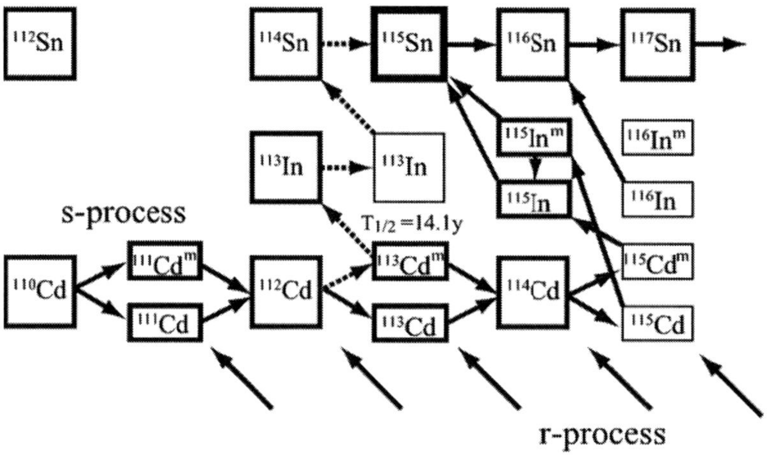

Figure 1. Partial nuclear chart around ^{115}Sn and nucleosynthesis flows.

2 Experimental producers

The experiment was performed using Accurate Neutron-Nucleus Reaction measurement Instrument (ANNRI), installed at a neutron beam line, BL04, in the Materials and Life Science Experimental Facility (MLF) in the Japan Proton Accelerator Research Complex (J-PARC) [7]. A pulsed proton beam with an energy of 3 GeV was injected into a mercury target of the spallation source in MLF. The proton beam was accelerated with a beam power of 200 kW and at a repetition rate of 25 Hz. Neutrons were emitted from the mercury target by spallation reactions. The proton beam was operated with the double bunch mode; A single bunch includes two proton pulses with pulse widths of 100 ns and with an time interval between the two pulses of 600 ns. Thus, the time resolution was determined by the double pulse mode. The neutron beam collimator with a diameter of 7 mm was used, which was located upstream from the target position. The diameter of the neutron beam on the target is 7 mm.

A lead absorber with a thickness of 37.5 mm was located downstream from the neutron source to absorb prompt γ-rays generated by the spallation reaction.

The ANNRI has a large highly pure germanium (HPGe) detector array, 4π Ge spectrometer, which consists of two cluster HPGe detectors, eight coaxial HPGe detectors, and BGO Compton-suppression detectors. In the present experiment, the two cluster HPGe detectors were used. This detector system was located 21.5 m downstream from the neutron source in the MLF. The γ-rays were recorded when at least two HPGe crystals detected photons. The energy of a neutron was measured using a Time-of-Flight (TOF) method by recording the time when a detector is fired. The energies of the γ-rays and a neutron were recorded in event by event as the list mode. A ^{112}Cd foil enriched to 98.27% was used as the target.

The energy difference between the ground state and the isomer is only 264 keV and thus it is difficult to distinguish two decays to the ground state or the isomer following neutron capture reactions using standard large volume scintillation detectors such as NaI(Tl). Thus, in the present experiment, we measured relative ratios of γ-rays decaying to the ground state to that to the isomer using the HPGe detectors.

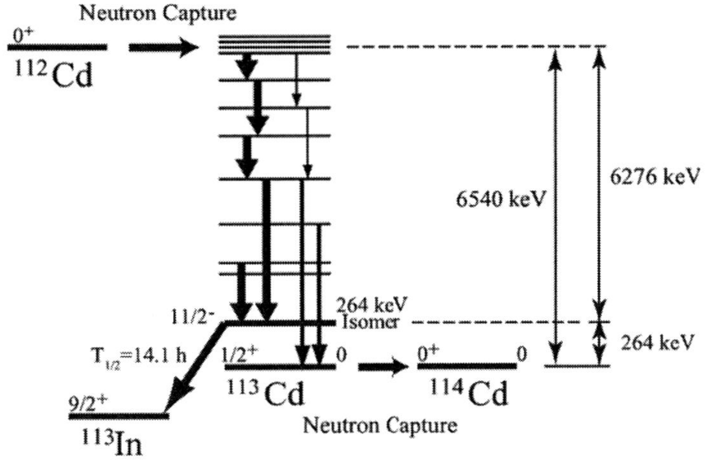

Figure 2. Partial level scheme of ^{115}Sn.

3 Results and Discussion

Figure 3 shows a TOF spectrum higher than the thermal energy. In the thermal energy region, the large peak originated from ^{113}Cd is observed. This ^{113}Cd is inferred to originate from the contamination to the ^{112}Cd enriched target. The fraction of 113Cd is only 0.6%. The thermal neutron capture cross-section of ^{113}Cd is 20650 barn, whereas that of ^{112}Cd is considered to be lower than 2 barn. As discussed in the previous paper [3], the thermal neutron capture cross section of ^{112}Cd has not been measured yet. As shown in Fig. 3, the relatively strong resonances like 66.8 eV, 227 eV, and 440 eV, are observed. This suggests that γ-rays of ^{112}Cd is dominant in the energy region higher than eV.

Figure 4 shows a γ-ray energy spectrum gated by the 66.8-eV neutron energy in TOF. The γ-rays decaying to the ground state with energies of 299 and 316 keV are clearly observed. A γ-ray of 259 keV is also observed, which decays to the isomer on ^{113}Cd. In this way, the present experiment

 DOI: 10.1051/epjconf/201612205003

suggests that it is possible to evaluate the isomer production ratio to the ground state as a function of neutron energy with high statics.

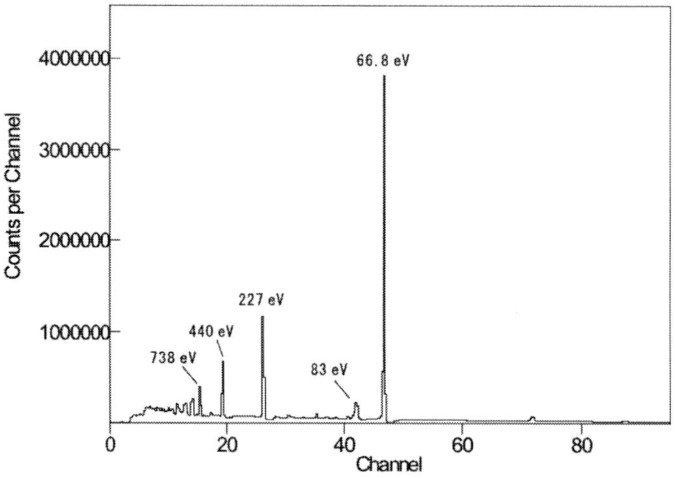

Figure 3. TOF spectrum on the ^{112}Cd enriched target.

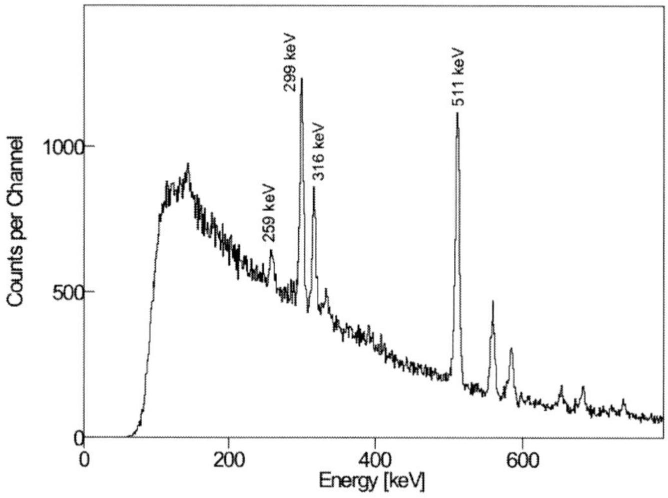

Figure 4. Partial nuclear chart around ^{115}Sn and nucleosynthesis flows.

4 Summary

The astrophysical origin of ^{115}Sn has remained still an open question. An isomer with a half-life of 14.1 y in ^{113}Cd is a branching point from which a nucleosynthesis flow reaches to a rare isotope ^{115}Sn. The s-process abundance of ^{115}Sn depends on the ratio of the ^{112}Cd(n, γ)^{113}Cd^{m} reaction cross section to the ^{112}Cd(n, γ)^{113}Cd^{gs} reaction cross section. However, the isomer production ratio following the neutron capture reaction has not been measured in the energy region higher than the thermal energy.

We have measured γ-rays following neutron capture reactions on ^{112}Cd using the HPGe detectors in conjunction with a time-of-flight method at ANNRI. In the neutron energy region higher than eV, we observed decay γ-rays to the ground state and the isomer. The present result suggests that it is possible to evaluate the isomer production ratio to the ground state as a function of neutron energy higher than eV with high statics.

References

1. R. A. Ward and H. Beer, Astrophy. Astron. **103**, 189A (1981).
2. Ch. Theis, et al., Astrophy. J. **500**, 1039M (1998).
3. T. Hayakawa et al., Astrophy. J. **707**, 859 (2009).
4. T. Hayakawa et al., Phys. Rev. Lett. **93**, 161102 (2004).
5. T. Hayakawa et al., Astrophy. J. **628**, 533 (2005).
6. K. Volev, et al., Nucl. Instr. Method Phys. Res. B **300**, 11 (2013).
7. K. Kino, et al., Nucl. Instr. Method Phys. Res. A **626**, 58 (2011).

$^{16}O(n,\alpha)$ cross section investigation using LENZ instrument at LANSCE

H.Y. Lee[1,a], S. Mosby[1], R.C. Haight[1], and M.C. White[2]

[1] *P-27, Physics Division, Los Alamos National Lab, Los Alamos, New Mexico, U.S.A.*
[2] *XCP-5; X-Computational Physics Division, Los Alamos National Lab, Los Alamos, New Mexico, U.S.A.*

Abstract. Importance of studying the $^{16}O(n,\alpha)$ reaction is motivated by multiple nuclear applications. The Los Alamos Neutron Science Center (LANSCE) produces a white neutron spectrum ranging from thermal to several hundreds of MeV energies. We have recently developed the LENZ (Low Energy NZ-neutron induced charged particle detection) capability to measure high-precision (n,α) cross sections. In order to provide more reliable data, we have enhanced solid angle coverage, and improved signal-to-noise ratios and time-of-flight resolution by implementing digitizer waveform analysis. The LENZ was commissioned by studying the $^{59}Co(n,\alpha)$ reaction with neutron beams in early 2015. For the $^{16}O(n,\alpha)$ reaction, we investigate solid oxygen targets and make a relative measurement to a better known cross section, such as the $^{6}Li(n,\alpha)$ reaction in order to further reduce systematic uncertainty. We will discuss the progress of the $^{16}O(n,\alpha)$ study at LANSCE and the outlook for improving Hauser-Feshbah prediction on (n,p) reaction cross sections.

1 Introduction : New Nuclear Data needs for the $^{16}O(n,\alpha)$ reaction

Oxygen is present in many materials - water, oxides, concrete and elsewhere - and the uncertainties in its nuclear data can have a significant impact on many applications. Several nuclear applications require precise nuclear reaction cross sections on water. Naval reactors are light water reactors for cooling and moderating, which require neutron reactivity input for reactor designs. Data testing with "solution critical", which contains a lot of water, is to benchmark critical assemblies. Experiments that use a manganese bath measure the total number of neutrons through the activation of manganese, therefore precise knowledge on neutron-induced reactions on water is necessary. Radio-biology takes advantage of better understanding of reactions related to oxygen, due to the bio-system's oxygen content. In particular, neutron-absorption reactions cause reduction of available neutrons in applications. Therefore, improving the precision of the evaluated $^{16}O(n,\alpha)$ reaction cross section is in order and for that we need to provide a new and independent set of experimental data.

This has been recognized and is the motivation for its inclusion as part of the Collaborative International Evaluated Library Organization (CIELO) project [1] looking at improving the nuclear data for key isotopes. At the heart of the questions on the oxygen-16 data is a 30-50% discrepancy between various (n,α) cross section measurements. Reconciling these discrepancies and settling on a best value are key deliverables for the CIELO project and requires new measurements for confirmation.

[a] e-mail: hylee@lanl.gov

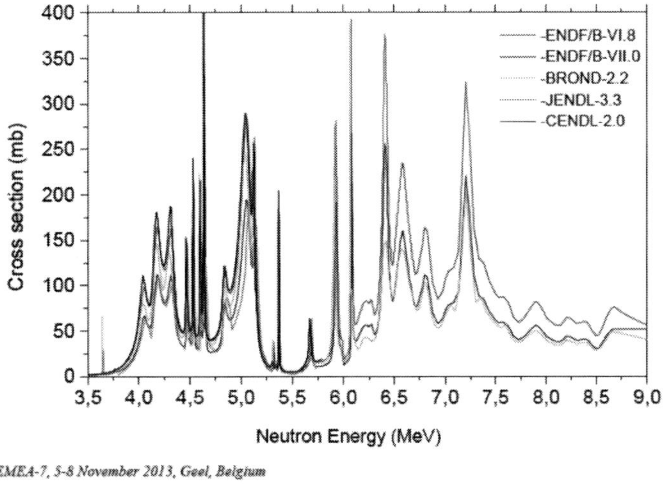

Figure 1. Summary of available evaluations on the $^{16}O(n,\alpha)$ reaction taken from the NEMEA-7/CIELO meeting presentation [2]

2 Status of the $^{16}O(n,\alpha)$ cross section evaluations : experimentalist point of view

Evaluated cross sections on the $^{16}O(n,\alpha)$ reaction have shown discrepancy on normalizations and energy-dependent shapes, as shown in Figure 1. Six different evaluations are shown and there are no evaluations with good agreement in the whole energy range. In particular, the ENDF/B-VII.0 evaluation [3] was released in 2006 after including the experiment result from Ref. [4] and the whole normalization was reduced by about 30 % compared with the previous evaluation of ENDF/B-VI.8 [5]. Since then, a new normalization factor to apply to the ENDF/B-VII.0 evaluation has been suggested, based on additional new measurements and the recent development of the LANL R-matrix study on ^{17}O-nucleus system [6].

Previously measured cross sections (The subsets are shown in Figure 2) were performed by two different approaches . One is to measure the $^{16}O(n,\alpha)$ reaction directly [2, 7–10]. Gas detectors like a Frisch-gridded ionization counter or a proportional counter were used in a gas mixture, which contained oxygen in order to measure $^{16}O(n,\alpha)^{13}C$ reaction cross section. Since the target gas plays as a counting gas as well, identifying α signals from background signals is crucial in data analysis. Most measurements were performed by thorough data reduction in order to remove background contribution, however there could be incomplete reduction of non-linear behaviors to the foreground. Since neutrons are produced from charged particle beams impinging on a gas target, there was in-built significant energy broadening, which will be discussed in detail in the following Section 4.

The other is the time-reversed reaction of $^{13}C(\alpha,n)^{16}O$ using the principle of detailed balance [4, 11, 12]. It is measured by impinging an alpha beam on a ^{13}C target and followed by neutron detection. In order to compensate decreasing cross sections at low energy, a thick target is used. However, self absorption of alpha particles in a thick target has to be corrected in order to disentangle an effective

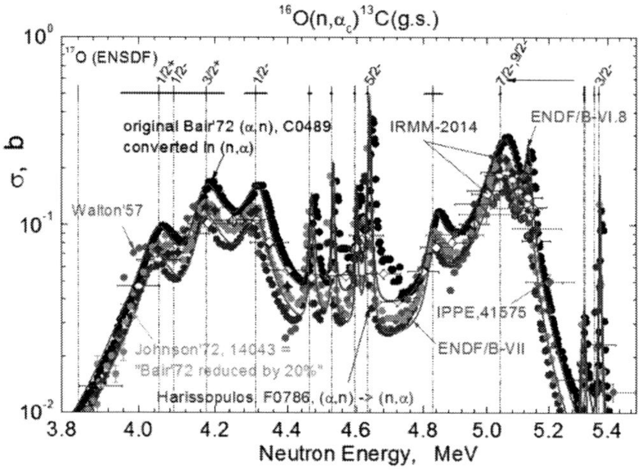

Figure 2. Subset of available experimental data on the $^{16}O(n,\alpha)$ reaction taken from the NEMEA-7/CIELO meeting presentation [2]

energy using a realistic stopping power for straggling particles. The neutron detector efficiency is determined by simulating neutron transport in the detector system, which is then calibrated with the measured yield relative to a known reference. Bair and Hass [12] used a Sb-Be source to calibrate their neutron detectors. Harissopulos *et al.* [4] used a ^{252}Cf spontaneous fission source, which has a shape of Maxwellian distribution with the temperature of 1.42 MeV and the peak around 1 MeV. The prompt fission neutron spectrum might not be optimal for calibrating neutrons above 4 MeV, where the number of available neutrons from fissions drops rather quickly. On other hand, the neutron energy populated from this $^{13}C(\alpha,n)$ reactions ranges from 3.5 MeV to 9.0 MeV. The neutron detectors used here, like BF_3 counters or ^3He counters, can not provide neutron angular information, which has to be corrected from a Monte Carlo simulation using predicted angular distribution as input.

In order to enhance the quality of a new measurement, the requirements are (a) a large number of target atoms and a detection system with high efficiency, due to small cross sections, (b) a large signal-to-background ratio and low detection threshold, in order to capture all the low energy alphas from the reaction, (c) a good energy resolution, since some resonances have their resonance width less than 100 keV, (d) an improved systematic uncertainty for resolving the discrepancy among current data sets, and (e) additional angular distributions would improve data analysis for using advanced R-matrix codes.

3 LENZ (Low Energy NZ) capability

At LANSCE, neutrons are produced in the energy range of thermal and several hundreds MeV [13]. A 800 MeV proton impinges a bare tungsten target through spallation process, then produces neutrons in white spectrum to be delivered to 6 different flight paths at Weapons Neutron Research (WNR). Protons are pulsed to have the time structure of 1.8 μs repetition, so neutron energy is determined by

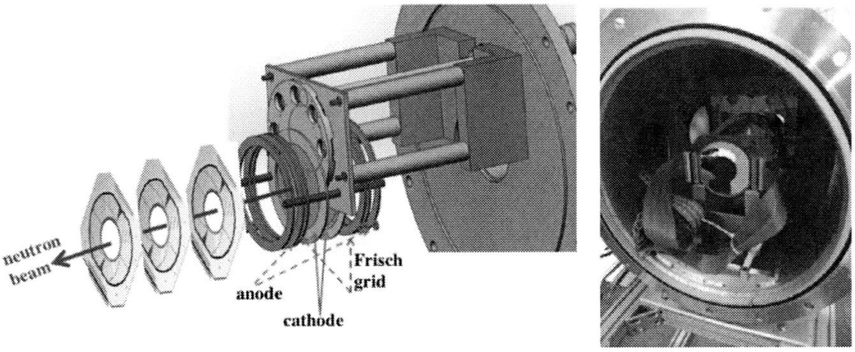

Figure 3. (left) The drawing of LENZ detector diagram (right) The picture of LENZ chamber

measuring the time stamp at the location of neutron detection relative to the proton time stamp right before bombarding the tungsten target. This broad neutron energy range and neutron flux at LANSCE are ideal to study the double differential cross sections for these (n,p) and (n,α) reactions.

The LENZ instrument is an update of the preceding NZ chamber [14] at LANSCE, to measure cross sections of neutron-induced charged particle reactions. The LENZ instrument is designed to have a large solid angle and low detection threshold in particular measuring low-energy alpha particles. It is composed of twin Frisch-gridded ionization counter [15] as a delta-E detector and double-sided silicon strip detectors (DSSD) as E detector to be used as "telescope", as shown in Figure 3.

All the electrodes in an ionization counter are 63.5 mm in radius and the distances for grids and anodes relative to cathodes can be easily adjusted, in order to be flexible for detecting different types and/or energy range of charged particles. The Frisch grid signal allows to correlate emitting angles through the arrival time at the grid and the fast timing of the cathode signal is specifically optimized to provide a good neutron energy resolution, which was deduced from the time of flight method. The target wheel can hold up to 8 different targets, including a backing material, targets of interest, a calibration target, a normalization target, etc. The DSSDs include two of 1000 micron thick S1 type detectors (an active area is defined by a disc of the 48 mm outer radius with a hole of the 24 mm inner radius) and one 1500 micron thick S3 detector (an active area is defined by a disc of the 35 mm outer radius with a hole of the 11 mm inner radius). The location of these DSSDs can be configured to optimize each experiment's goal, depending on either largest solid angle coverage or largest proton energy coverage. For the maximum solid angle coverage, the combination of the twin ionization counter and DSSDs can cover up to 70 degrees. The gas of P10 was chosen due to expected neutron-beam induced backgrounds on gas elements. Based on previous studies in the literature [16], the prominent background peaks in the energy spectrum are reported, due to the reactions like O(n,α), Ar(n,α), and recoil protons. Therefore, depending on the reaction of interest, meaning the potential overlap of the pulse heights with these background reactions, the beam-induced background reactions on gas elements need to be investigated case by case.

Since LANSCE provides a white neutron energy spectrum, the beam energy information is deduced from a time of flight at a target, in this case the time stamp measured in cathode. For reducing the energy resolution, the timing information needs to be preserved as precise as possible. We used waveform digitizers, which allowed us store partial wavelet information to obtain the best timing and

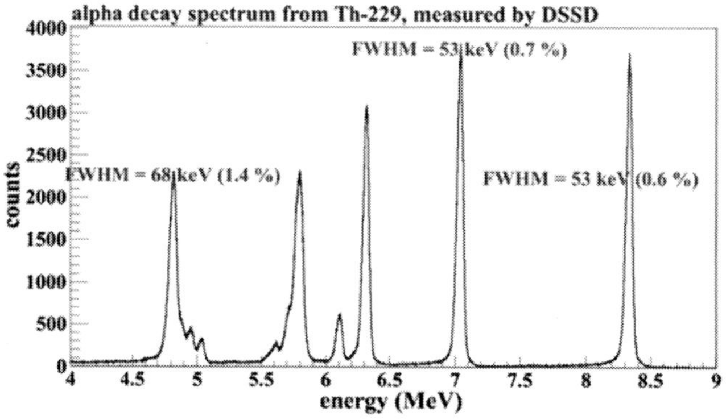

Figure 4. Energy of a DSSD detector is calibrated with ^{229}Th α-emitting source using a CAEN V1724 digitizer.

energy resolution, via post-processing with the use of various digital filters. Digital waveform analysis was implemented to achieve better signal-to-noise ratios, lower alpha detection threshold, and better timing resolution of the cathode signal. In order to match the characteristics of each detector preamplifier signal, two different types of CAEN digitizers [17] were tested. The preamplifier outputs of ionization counter signals have about 500 ns rise time with about 50 μs fall time, and the pulse height is up to 200 mV, with a 550 Torr. of P10 gas pressure. The preamplifier outputs of DSSD signals have about 40 ns rise time with about 35 μs fall time. When the ionization counter signals were used with the CAEN 1720 digitizer, which is a 12 bit and 250 MS/s sampling rate digitizer, the

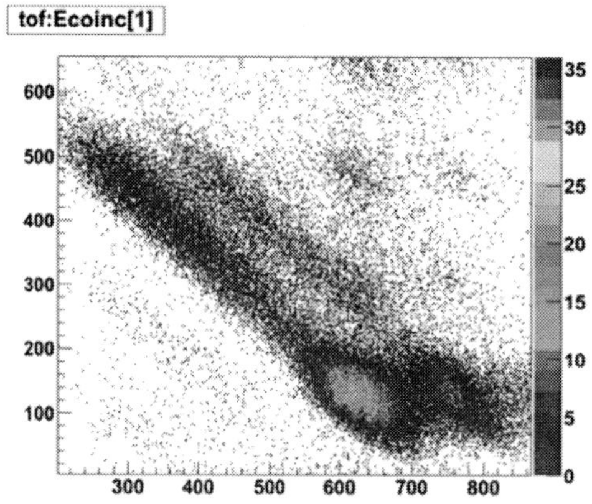

Figure 5. 2d plot of Time of Flights between T0 - Cathode (x-axis) and DSSD - Cathode (y-axis). Both axes are in arbitrary units.

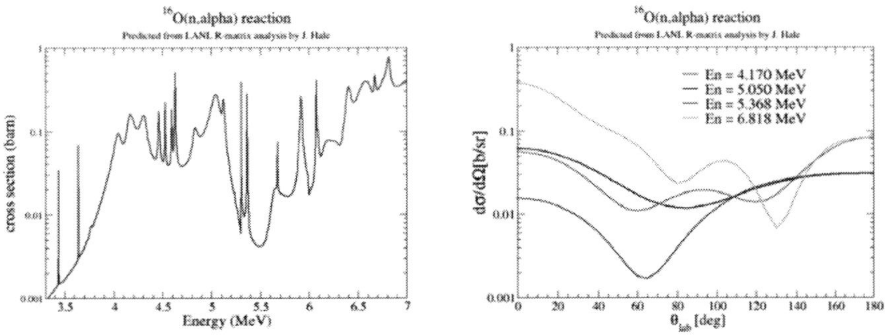

Figure 6. (left) The calculated reaction cross section of $^{16}O(n,\alpha)$. (right) Calculated angular distributions for several strong resonances between 4 and 7 MeV of neutron energies. Both are performed using the Los Alamos developed R-matrix analysis by G. Hale [18].

time information was taken by using a double-derivative filter and peak height was charge integration. For DSSD detector signals, the CAEN 1724 digitizer with a 14 bit and 100 MS/s sampling rate was used to preserve most peak height information by fitting to an exponential decay from saved wavelets.

The parameter optimization to be used in these digital filters was performed for both detector signals using a ^{229}Th source. Figure 4 demonstrates the energy resolution in Full Width Half Maximum obtained for DSSD detectors in ~30 mTorr. vacuum pressure, without P10 gas being present. The energy resolution is determined to be as good as 53 keV above 6 MeV, which is better than the specification, 75 keV, provided by the Microsemiconductor manufacturer [19].

The in-beam commissioning of the LENZ instrument at LANSCE was performed by repeating the ^{59}Co(n,p) and (n,α) reaction, which were measured previously using the NZ chamber [20]. This is in particular to investigate the beam-induced background reactions on gas elements and to determine the timing resolution of detectors relative to the incoming neutron time stamp. For this run, the P10 gas pressure at 550 Torr. was chosen to generate enough energy loss for protons and alphas in the gas volume and one S1 DSSD was coupled to the ionization counter at forward angles. As shown in Figure 5, the different time of flight groups were measured to correlate the information on different reaction channels. With higher incoming neutron energy, different types of charged particles were populated through energetically allowed different reaction channels. The detailed analysis, including optimization of best timing resolution in ionization counter signals, is still on going.

4 The $^{16}O(n,\alpha)$ measurement at LANSCE

Regarding the discrepancy on the $^{16}O(n,\alpha)$ cross sections, which is known to be 30-50 % among currently available experimental data sets, we plan to improve the quality of this measurement by achieving the uncertainty to be less than 20 %. For absolute cross section measurements, the major uncertainty is caused by estimating the detection efficiency including a dead time in data acquisition system, the beam flux, the amount of target material, and the target uniformity and the beam profile if beam has a sizable area. A relative measurement to the reaction with a better known cross-section knowledge, called a "ratio method", could be performed to reduce the uncertainties associated with

DOI: 10.1051/epjconf/201612205004

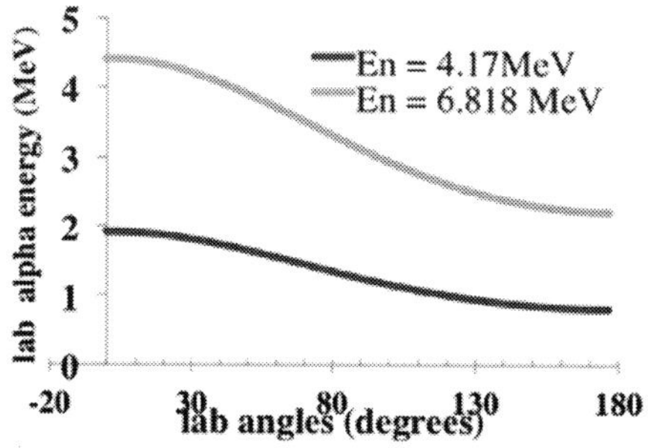

Figure 7. Calculated kinematics of alpha from the $^{16}O(n,\alpha)$ reaction. For the resonances of 4.17 MeV and 6.818 MeV, the forward angles in LAB system are preferable for detecting alphas due to higher LAB energies than that at the backward angles.

beam intensity and detection efficiency estimation. For (n,α) reactions, the $^{6}Li(n,\alpha)$ reaction can be used as a reference reaction. Target characterization still needs to be taken into account for the uncertainty estimation, although this ratio method is used.

As Figure 6 (left) shows multiple resolved resonances, improving the neutron energy resolution, i.e. timing resolution of the cathode signal in the ionization counter is critical. R-matrix analysis in Figure 6 (right) shows larger angular distributions at forward angles, where more DSSDs are positioned to cover larger angles. Based on the kinematics calculations shown Figure 7, it confirms to locate more DSSDs at forward angles for detecting kinematically favored alpha energies. In order to detect alpha above 1.5 MeV, the gas pressure in the ionization counter and the preamplifier parameters are optimized for small pulse-height signals. The gas type is chosen to be P10, because any oxygen contained gas mixture is excluded for potential background. The pressure and the distance between the cathode and anode are set to be 300 Torr. and 12 mm to generate least energy loss and energy straggling from low-energy alphas in the gas volume. For the same reason, the Frisch grid is not utilized and the angle information is taken from the DSSDs, which has annular segmentations. The DSSDs is located in such a way to provide the angular resolution of about one degree.

Since the amount and uniformity of the target material was dominant contribution to systematic uncertainties in the previous measurements, solid oxygen targets were investigated for this experiment.

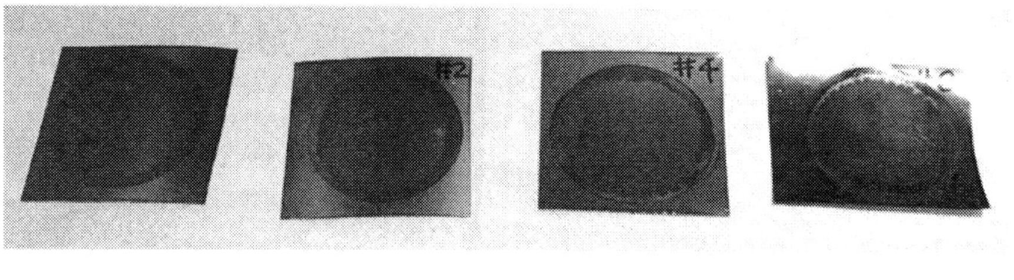

Figure 8. Picture of Ta_2O_5 targets manufactured by the anodizing technique

We have investigated two methods for manufacturing solid oxygen targets. One is the anodizing technique by applying an electric field in ^{16}O-enriched water, via etching oxygen ions onto the backing material (tantalum) surface with relatively small variation in thickness [22]. The target was manufacture by applying 250 V, which corresponds to about 4000 Å in thickness. The pictures of the Ta_2O_5 targets in different thicknesses are shown in Figure 8 and the color on the deposit could be used to identify a specific oxygen thickness. The other method was to evaporate lithium carbonate (Li_2CO_3) powder on a tantalum backing, which can be used as the in-built normalization factor correction for neutron beam intensity. The $^6Li(n,\alpha)$ reaction cross section is known within 5 % uncertainty up to 1 MeV and about 10 % uncertainty above 1 MeV [23], so it can be used as a reference reaction in order to cancel out the neutron intensity uncertainty and detection efficiency uncertainty, as long as target stoichiometry is well characterized and does not change during the measurement.

For the in-beam background estimate, we plan to measure neutron-induced reactions on a blank tantalum under the same condition as the $^{16}O(n,\alpha)$ reaction study. A first step is to investigate the realistic background on this reaction due to a backing material in solid target, then we can optimized our setup or different choice of backing materials. We plan to investigate the stoichiometry and the stability of the target before and after neutron irradiation.

As discussed in Section 2, the previous (n,α) measurements [2, 9] discussed the difficulty from the poor neutron energy resolution, since the neutrons were produced from $D(d,n)^3He$ or $^3H(p,n)$ reactions. The combination of the production gas cell and entrance and exit windows could add large energy straggling and broadening. At LANSCE, the neutron energy resolution is deduced from the detection timing resolution of time of flight, therefore the cathode signal is optimized to provide the best timing resolution.

5 Outlook

With the development of this multi-purpose LENZ chamber, we plan to study (n,p) reaction cross sections. Various evaluations used Hauser-Feshbach formalism to calculate neutron-induced charged particle cross sections up to 20 MeV, since experimental data sets lack in this entire energy range. Due to the limited available neutron energies, which were generated from charged-particle accelerators, data sets exist only at specific energies. However, these evaluations show significant discrepancy in the overall shape, but good agreement at around 14 MeV, which is the only energy range for existing data sets on the reaction of $^{77}Se(n,p)$ [21]. This similar behavior has been found in the comparison of different evaluations on the cross section of $^{75}As(n,p)$ reaction, which has great interest for nuclear applications. LANSCE is an ideal place to measure the (n,p) cross sections in the broad energy range of 1-20 MeV, therefore we plan to measure the reactions of $^{77}Se(n,p)$. This effort also can improve understanding of nucleosynthesis in astrophysics via using better predicted reaction rates on neutron-induced charged particle reactions. The survey on the nuclear input sensitivity in p-process nucleosynthesis [24] identified several (n,p) and (n,α) reactions as potential branching points to affect the final p-process abundance. The LENZ instrument at LANSCE can study on these nuclear reactions to impact on understanding of heavy element productions in various stellar environments.

Acknowledgment

This work benefits from the LANSCE accelerator facility and is supported by the U.S. Department of Energy under contracts DE-AC52-06NA25396.

References

[1] G. Giorginis *et al.*, *Proceeding of Int. Conf. on Nuclear Data for Science and Technology*, DOI:**10.1051**/ndata:07481 (2007), OECD discussion, CIELO project

[2] V. Khryachkov, I. Bondarenko, V. Pronyaev, P. Prusachenko, A. Sergachev, N. Semenova, and T. Khromyleva, *A workshop of the Collaborative International Evaluated Library Organisation* (2013, Geel, Belgium) https://www.oecd-nea.org/science/wpec/nemea7/presentations.html

[3] M.B. Chadwick, *et al.*, Nucl. Data Sheets **107** 2931 (2006)

[4] S. Harissopulos, H.W. Becker, J.W. Hammer, A. Lagoyannis, C. Rolfs and F. Strieder, Phys. Rev. C **72**, 062801(R) (2005)

[5] CSEWG-Collaboration, Evaluated Nuclear Data File ENDF/B-VI.8, *http://www.nndc.bnl.gov/endf* (2001)

[6] G.M. Hale and M.W. Paris, *Proceeding of Nuclear Measurements, Evaluations and Applications (NEMEA-7) Collaborative International Evaluated Library Organisation*, **NEA/NSC/DOC(2014)13**, 13 (2013)

[7] R.B. Walton, J.D. Clement, and F. Boreli, Phys. Rev. **107**, 1065 (1957)

[8] E.A. Davis, T.W. Bonner, D.W. Worley, and R. Bass, Nucl. Phys. **48**, 169 (1963)

[9] G. Giorginis, V. Khryachkov, V. Corcalciuc, and M. Kievets, *International Conference on Nuclear Data for Science and Technology*, DOI:**10.105**/ndata:07481 (2007)

[10] A. Plompen, *A workshop of the Collaborative International Evaluated Library Organisation* (2013, Geel, Belgium) https://www.oecd-nea.org/science/wpec/nemea7/presentations.html

[11] K.K. Sekharan, A.S. Divatia, M.K. Mehta, S.S. Kerekatte, and K.B. Nambiar, Phys. Rev. **156**, 1187 (1967)

[12] J.K. Bair and F.X. Hass, Phys. Rev. C **7**, 1356 (1973)

[13] P.W. Lisowski, C.D. Bowman, G.J. Russell, and S.A. Wender, Nucl. Instr. Meth. A **106**, 208 (1990)

[14] R.C. Haight, *Proceeding of Int. Conf. on Nuclear Data for Science and Technology*, DOI:**10.1051**/ndata:07518 (2007)

[15] A. Goepfert *et al.*, Nucl. Instr. Meth. A **441**, 438 (2000)

[16] N. Ito, M. Baba, S. Matsuyama, I. Matsuyama and N. Hirakawa, Nucl. Instr. Meth. A **337**, 474 (1994)

[17] CAEN Electronic Instrumentation, *http://www.caen.it*

[18] G.M. Hale, *Private communication* (2015)

[19] Micronsemiconductor, *http://www.micronsemiconducntor.co.uk*

[20] S. Kunieda, R.C. Haight, T.Kawano, M.B. Chadwick, *et al.*, Phys. Rev. C **85**, 054602 (2012)

[21] D.A. Vermilyea, Acta Metallurgica **1**, 282 (1953)

[22] M. Devlin, T.N. Taddeucci, G.M. Hale and R.C. Haight, *Proceeding of Int. Conf. on Nuclear Data for Science and Technology*, DOI:**10.1051**/ndata:07563 (2007)

[23] S. Kamada, K. Shibata, A. Ichihara, and S. Kunieda, Jour. Nucl. Sci. Tech. **47**, 329 (2010)

[24] W. Rapp, J. Goerres, M. Wiescher, H. Schatz, and F. Kaeppeler, Astrophys. J. **653**, 474 (2006)

 DOI: 10.1051/epjconf/201612206001

A study on the rearrangement corrections to the folding model applied to nucleon inelastic scattering

M. Dupuis[1,a] and E. Bauge[1]

[1] *CEA, DAM, DIF, 91297 Arpajon Cedex*

Abstract. A study of direct neutron induced reactions off spherical nuclei is given in terms of a folding model. Matter densities stemming from the Random Phase Approximation (RPA), implemented with the Gogny D1S force are used to provide the relevant coupling potentials. The impact of rearrangement corrections on cross sections is studied for direct inelastic nucleon scattering off ^{16}O and ^{208}Pb. The rearrangement correction to calculated cross sections is shown to strongly depend on the incident energy and on the excitation multipolarity.

1 Introduction

Energy application and astrophysical models require precise nuclear data for a large range of nuclear masses. The development of microscopic models is necessary to fulfill this need, since such models are able to predict cross sections when nuclear measurement are hard or impossible to achieve, and consequently, when phenomenological models are not well constrained.

We aim at reducing phenomenology in the description of nucleon induced reaction, focusing on the direct inelastic scattering process using a microscopic description. Works along this line can be performed in the framework of the Jeukenne-Lejeune-Mahaux (JLM) folding model [1] to describe nucleon elastic and inelastic scattering off both spherical and deformed targets using densities stemming from mean-field and beyond mean field calculations [2–4]. Besides, Cheon *et al.* [5, 6] pointed out that a simple identification of the g-matrix to the effective interaction is invalid when applied to nucleon inelastic scattering process calculations, and that an additional density-dependent term, the rearrangement potential, appears. This correction accounts for the fluctuation of the density during the transition which thus modifies the interaction. For this reason, this approach is also often called consistent or dynamic treatment of inelastic scattering. An expression of the rearrangement correction to the folding model transition potential was first obtained in [5], using the vibrational model approach that identifies the transition potential shape to the derivative of the optical potential. The rearrangement expression was latter justified from a more fundamental point of view in [6] in the framework of the Watson multiple scattering theory, in the zero momentum transfer limit, for a simple particle-hole excitation. These early works showed that the rearrangement correction has a non-negligible impact on calculated nucleon differential inelastic cross sections. This correction has then been used in some [7–9] yet not all applications of folding models based on various nuclear matter calculations, such as

[a] e-mail: marc.dupuis@cea.fr

the density dependent M3Y interactions [10], the Melbourne g-matrix [11, 12] or the JLM approach [4]. However, when the rearrangement corrections were used, their impact on direct inelastic cross section has been rarely studied. We only account for the work of El-Azab Farid and Satchler [7], performed with a density dependent M3Y interaction, that reports a reduction of 28% of the 65 MeV proton inelastic scattering off ^{208}Pb to the 3_1^- states due to the rearrangement. They showed some important change of α scattering cross sections related to the consistency requirement suggested by Cheon *et al.* Recently, a study of direct inelastic neutron scattering off spherical and axially deformed targets was performed within the JLM folding model [13] with nuclear matter densities obtained from the Quasi-Particle Random Phase Approximation (QRPA) with the Gogny D1S force [14, 15]. In this work, rearrangement corrections to the inelastic cross section were explicitly shown only for the 2_1^+ excitation in ^{208}Pb for medium energy neutron scattering ($E \simeq$7-20 MeV). For this transition at these incident energies, rearrangement was found to reduce the magnitudes of inelastic scattering differential cross sections by 50%. In the present paper, we focus on the characterization of the rearrangement correction to nucleon direct inelastic scattering off spherical nuclei in the context of the JLM folding model. We extend the impact analysis of rearrangement on direct inelastic cross sections of ref. [13] to a wider incident energy range, for both neutron and proton scattering, and for various excitation multipolarities.

The first section provides the main features of the model. A characterization of the rearrangement corrections for nucleon scattering off spherical nuclei is given in the second section. Conclusions are drawn in the last section.

2 JLM folding model for shape elastic and direct inelastic scattering

All details about the JLM folding model for elastic and direct inelastic nucleon scattering can be found in [3, 13] (and references therein), so we only remind here its main features. It is based on a Brueckner Hartree-Fock derivation of the optical potential for nucleon scattering in nuclear matter (NM). For a nucleon incident energy E and neutron and proton density ρ_n, ρ_p, Jeukenne et al. expressed their NM optical potential as:

$$V_\tau^{\rm JLMB}(\rho_n,\rho_p,E) = \lambda_V(E)\Big[V_0(\rho,E) + \tau\lambda_{V_1}(E)\alpha(\rho_n,\rho_p)V_1(\rho,E)\Big] + i\lambda_W(E)\Big[W_0(\rho,E) + \tau\lambda_{W_1}(E)\alpha(\rho_n,\rho_p)W_1(\rho,E)\Big]\,. \quad (1)$$

where $\alpha(\rho_n,\rho_p) = \dfrac{\rho_n-\rho_p}{\rho_n+\rho_p}$, $\tau = +1(-1)$ for neutron (proton) scattering. The energy dependent parameters $\lambda_V, \lambda_{V_1}, \lambda_W, \lambda_{W_1}$ are provided in [3, 16].

An optical potential for nucleon scattering off a finite nucleus can be determined from this NM optical potential using the so-called improved local density approximation (LDA) [1].

Following the work of Cheon *et al.* [5, 6], the transition potential for direct nucleon inelastic scattering can be identified to the derivative of the optical potential with respect to the isoscalar and isovector matter densities. In the context of the JLM folding model, the transition potentials that include rearrangement were specified in [13]. We only remind here their main expressions, where dependencies other than densities are not explicitly written. The transition potential corresponding to a nucleon scattering off a target in its ground state and leading to a discrete excitation $|N\rangle$ reads

$$U = U^{(0)} + U^{(R)}\,, \quad (2)$$

where $U^{(0)}$ is the usual folding transition potential and $U^{(R)}$ corresponds to the rearrangement correction. These potentials read

$$U^{(0)} = \rho^{(0,N)}F_0(\rho) + \tau\rho_1^{(0,N)}F_1(\rho) \quad , \tag{3}$$

$$U^{(R)} = \rho^{(0,N)}\left(\rho\frac{\partial F_0(\rho)}{\partial\rho} + \tau\rho_1\frac{\partial F_1(\rho)}{\partial\rho}\right) , \tag{4}$$

where $\rho^{(0,N)} = \rho_n^{(0,N)} + \rho_p^{(0,N)}$ and $\rho_1^{(0,N)} = \rho_n^{(0,N)} - \rho_p^{(0,N)}$. The $\rho_{n/p}^{(0,N)}$ are the radial transition densities $\rho_{n/p}^{(0,N)}(\mathbf{r}) = \sum_\sigma \langle N|\Psi^\dagger(\mathbf{r},\sigma,t)\Psi(\mathbf{r},\sigma,t)|\tilde{0}\rangle$ with $t = \frac{1}{2}(-\frac{1}{2})$ for neutron (proton). The factors F_0 and F_1 reads:

$$F_0(\rho) = \lambda_V V_0(\rho) + i\lambda_W W_0(\rho) \quad , \quad F_1(\rho) = \frac{1}{\rho}\left[\lambda_V\lambda_{V_1}V_1(\rho) + i\lambda_W\lambda_{W_1}W_1(\rho)\right] . \tag{5}$$

3 Rearrangement corrections in spherical nuclei

In [5], Cheon *et al.* displayed the impact of the rearrangement term on the differential cross section for the reaction 135 MeV ^{16}O(p,p') to the 3_1^- excitation located at $E_x = 6.13$ MeV. In this study, the von Geramb's g-matrix [5] and phenomenological transition densities were used. Here, we perform an equivalent analysis within the JLM folding model described in the previous section and RPA transition densities [17] calculated with the Gogny D1S interaction. Note that, for 3_1^- excitation, the RPA/D1S model provides the reduced electric transition probability $B^{RPA}(E3\uparrow) = 0.125\,(10^4e^2\text{fm}^6)$ (calculation performed in an harmonic oscillator basis with 11 major shells) as compared to the adopted value $B^{exp.}(E3\uparrow) = 0.148\,5\,(10^4e^2\text{fm}^6)$ [18]. The theoretical transition density used in the folding has thus been multiplied by the factor $\sqrt{\frac{B^{exp.}}{B^{RPA}}} = 1.088$ to account for this discrepancy. Cross sections were calculated, within the distorted wave Born approximation, using either the transition potential $U = U^{(0)} + U^{(R)}$, eq. (2), that includes rearrangement corrections, or the transition potential $U^{(0)}$, eq. (3), without these corrections. The results displayed on Fig.1 (left panel) follow those reported by Cheon *et al.*: for this transition at this incident energy, the rearrangement correction reduces the differential cross section at forward angles by an amount of 15%. We extend this study to determine the magnitude of the rearrangement correction for neutron/proton scattering in the incident energy range 10-200 MeV. We display on Fig. 1 (right panel) the ratio of the angular integrated inelastic cross section σ, calculated with the transition potential $U = U^{(0)} + U^{(R)}$, to the one calculated with the transition potential $U^{(0)}$, labeled $\sigma_{(0)}$. For proton scattering (full black curve), the rearrangement correction impacts the most on inelastic cross section at the lowest and highest incident energies. It reduces the inelastic cross section magnitude by 40% at $E = 10$ MeV and 25% at $E = 200$ MeV. The smallest correction is found at $E \simeq 135$ MeV. We note that the proton ratio is shifted to the right with respect to the neutron ratio by 5 to 10 MeV. This is only the consequence of the Coulomb barrier undergone by protons.

We now extend the previous study to neutron and proton scattering off ^{208}Pb using ground state and transition densities predicted by the RPA/D1S nuclear structure model (calculation performed in an harmonic oscillator basis with 15 major shells). Figure 2 (left panel) illustrates the rearrangement renormalization effect on angular distributions for proton inelastic scattering to the 2_1^+ states that lies at 4.085 MeV. The RPA/D1S predicted value is $B^{RPA}(E2\uparrow) = 0.293\,(10^4e^2\text{fm}^4)$, in agreement with the adopted value $B^{exp.}(E2\uparrow) = 0.300\,30\,(10^4e^2\text{fm}^4)$ [26]. The ratios $\frac{\sigma}{\sigma_{(0)}}$ in the incident energy range 7-200 MeV for natural parity Yrast excitations with spin $J = 2, 4, 6$ and $8\,\hbar$ are displayed in Figure 2

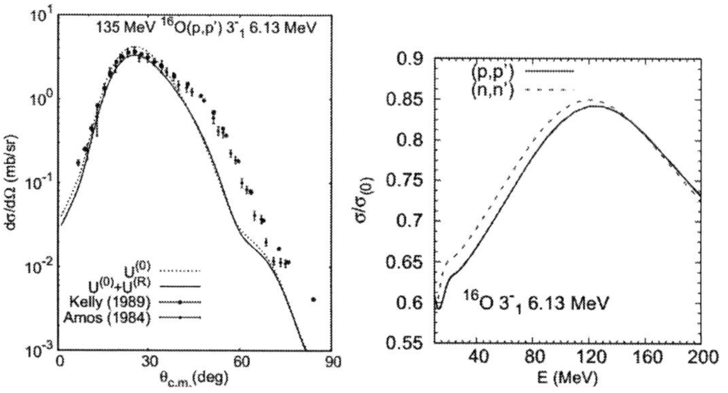

Figure 1. Left panel: differential inelastic cross section for p+^{16}O at the incident energy 135 MeV for the 3_1^- excitation at 6.13 MeV (data from [19, 20]). Full (dashed) black curve: cross section calculated from the transition potential $U = U^{(0)} + U^{(R)}$, ($U^{(0)}$), defined in eqs. (2) to (4). Right panel: direct inelastic cross section ratios $\frac{\sigma}{\sigma_{(0)}}$ for the 3_1^- excitation (see details in the text) for p/n+^{16}O (full black / red dashed curve) in the incident energy range 10-200 MeV.

(right panel). At low energy, the ratio for proton scattering (full black curve in the upper panel) moves rapidly towards unity since the inelastic scattering process is essentially due to Coulomb excitation below the Coulomb barrier. The energy dependence of the cross section ratio is qualitatively the same as the one observed for ^{16}O, but the renormalization effect is stronger, and its minimum is found at 110 MeV, as compared to 135 MeV in the ^{16}O case. This minimum is shifted towards higher energies when the spin of the transition increases, and the magnitude of the rearrangement correction strongly decreases when the spin of the transition increases. As medium energy nucleons mainly probe the nucleus surface, the strong decrease of the inelastic cross section can be related to the reduction of the transition potential at the nuclear surface, caused by the rearrangement correction [13]. To interpret the evolution of the rearrangement impact on the inelastic cross section in the whole energy range and for the various transitions displayed in Figs.1 and 2, we would need to consider two features. First, both the real and imaginary potentials $U^{(0)}$ and $U^{(R)}$ strongly depend on the incident energy, as illustrated in Fig. 3, and on the multipolarity of the excitation. Second, different radial extents of the potentials are probed by the projectile according to the incident energy in such a way that various contributions can produce constructive or destructive interferences. In [7] an analysis of 65 MeV proton inelastic scattering off ^{208}Pb to the 3_1^- states and to the giant monopole resonance revealed that strong interferences occur between the contribution to the inelastic cross section of the inner and outer parts of the transition potentials. This analysis will be extended to the energy range and transitions displayed in the present study. This will be reported in a forthcoming paper.

4 Conclusion

We report on microscopic calculations of direct inelastic scattering cross sections for nucleon induced reactions off ^{16}O, ^{208}Pb in the framework of the JLM folding model. Rearrangement impact on direct inelastic cross sections is shown to strongly depend on the incident energy and the transition multipolarities. An analysis of the relevant transition potentials will be performed to interpret this behavior

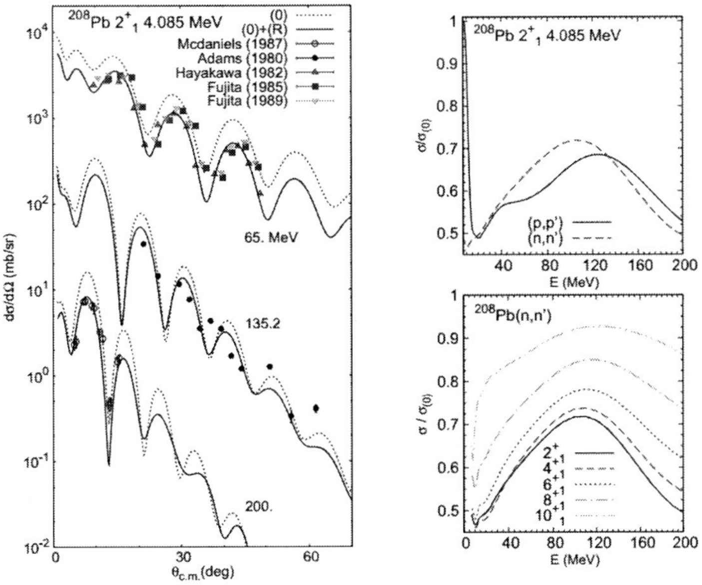

Figure 2. Left panel: differential inelastic cross sections for p+^{208}Pb at incident energies 65 (data from [21–23], 135 (data from [24]) and 200 MeV (data from [25]) for the 2_1^+ excitation lying at 4.065 MeV. Full (dashed) black curves: cross sections calculated from the transition potential $U = U^{(0)} + U^{(R)}$, ($U^{(0)}$), defined in eqs. (2) to (4). Top right panel: inelastic cross section ratios $\frac{\sigma}{\sigma_{(0)}}$ for the 2_1^+ excitation (see details in the text) for p/n+^{208}Pb (full black / red dashed curve) in the incident energy range 7-200 MeV. Bottom right panel: same ratios for neutron inelastic scattering to the 2_1^+, 4_1^+, 6_1^+, 8_1^+ and 10_1^+ states.

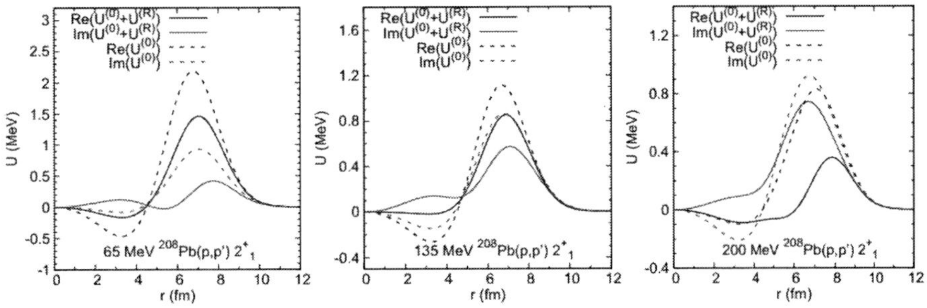

Figure 3. Transition potentials for proton inelastic scattering off ^{208}Pb to the 2_1^+ state at incident energies 65, 135.2 and 200 MeV. Real and imaginary parts of the transition potential $U = U^{(0)} + U^{(R)}$, and $U^{(0)}$, defined in eqs. (2) to (4), are plotted.

for transitions in both spherical and axially deformed nuclei using QRPA [14, 15] nuclear structure input. It appears, at least for the reactions studied here and those reported in [13], that following Cheon *et al.*'s prescription for rearrangement [5] improves the agreement between calculated direct inelastic cross sections and measurements. Note that care should be taken to avoid any double counting of diagrams, such as the second order diagram in the hole line (low density) expansion which could be identified to rearrangement diagrams which are exemplified in [6], and that may be already included in the nuclear matter calculation. The JLM folding model detailed in [1] is based on an optical potential calculated in nuclear matter within the Brueckner Hartree-Fock approximation. Thus second order diagrams in the hole line expansion, such as those described in [27], are not included and the rearrangement corrections of [5, 6] seem to be appropriate in applications with the JLM folding model for vibrational, spin independent, collective states. However, more fundamental work is still needed to clearly establish the expressions of the rearrangement term in the most general case. This will be most certainly helpful in determining which correction needs to be introduced in folding models used in direct nucleon or ion inelastic scattering studies for excitations of various nature, and effective interactions based on various nuclear matter calculations.

References

[1] J.P. Jeukenne, A. Lejeune, C. Mahaux, Phys. Rev. C **16** (1977)
[2] E. Bauge, J. Delaroche, M. Girod, Nucl. Phys. A **654** (1999)
[3] E. Bauge, J.P. Delaroche, M. Girod, Phys. Rev. C **63** (2001)
[4] E. Khan, *et al.*, Nucl. Phys. A **694**, 103 (2001)
[5] T. Cheon, K. Takayanagi, K. Yazaki, Nucl. Phys. A **437**, 301 (1985)
[6] T. Cheon, K. Takayanagi, K. Yazaki, Nucl. Phys. A **445**, 227 (1985)
[7] M. El-Azab Farid, G. Satchler, Nucl. Phys. A **481**, 542 (1988)
[8] D.T. Khoa, G. Satchler, Nucl. Phys. A **668** (2000)
[9] D.T. Khoa, E. Khan, G. Colò, N.V. Giai, Nucl. Phys. A **706** (2002)
[10] M. Dupuis, T. Kawano, J.P. Delaroche, E. Bauge, Phys. Rev. C **83** (2011)
[11] M. Dupuis, S. Karataglidis, E. Bauge, J.P. Delaroche, D. Gogny, Phys. Lett. B **665** (2008)
[12] K. Amos, *et al.*, Adv. in Nucl. Phys. **25** (2000)
[13] M. Dupuis, *et al.*, Eur. Phys. J. A **51** (2015)
[14] S. Péru, H. Goutte, Phys. Rev. C **77** (2008)
[15] S. Péru, G. Gosselin, M. Martini, M. Dupuis, S. Hilaire, J.C. Devaux, Phys. Rev. C **83** (2011)
[16] S. Goriely, J.P. Delaroche, Phys. Lett. B **653** (2007)
[17] J. Blaizot, D. Gogny, Nuc. Phys. A **284**, 429 (1977)
[18] T. Kibédi, R. Spear, Atomic Data and Nuclear Data Tables **80**, 35 (2002)
[19] K. Amos, *et al.*, Nucl. Phys. A **413** (1984)
[20] J. J. Kelly, *et al.*, Phys. Rev. C **39** (1989)
[21] S. I. Hayakawa, *et al.*, Phys. Rev. Let. **49** (1982)
[22] Y. Fujita, *et al.*, Phys. Rev. C **32** (1985)
[23] Y. Fujita, *et al.*, Phys. Rev. C **40** (1989)
[24] G. S. Adams, *et al.*, Phys. Lett. B **91** (1980)
[25] D. K. Mcdaniels, *et al.*, Nuc. Phys. A **467** (1987)
[26] S. Raman, C. Nestor Jr., P. Tikkanen, Atomic Data and Nuclear Data Tables **78**, 1 (2001)
[27] J.P. Jeukenne, A. Lejeune, C. Mahaux, Phys. Rep. **25** (1976)

Which mechanism(s) for the (d,p) reactions? Or can the CDCC approach be used to compute transfer cross sections?

Pierre **Chau Huu-Tai**[1,a]

[1] *CEA,DAM,DIF, F-91297 Arpajon, France*

Abstract. We propose to use the Continuum Discretized Coupled Channels approach to describe both the differential cross sections and the excitation fonctions for the (d,p) reactions.

1 Introduction

Just after the discovery of radioactivity, it was realized that nuclear reactions can be used to explore the structure of nuclei. In particular, since 1932 [1], the deuteron has been widely used as a projectile both to investigate the spectroscopic features of the compound nucleus and to produce isotopes relevant for technological or medical purposes. Indeed as a weakly bound two-nucleon system, this hydrogen isotope can easily transfer one of its component to the target nucleus ${}^{A}_{Z}\mathrm{X}$ to form the compound nucleus ${}^{A+1}_{Z}\mathrm{Y}_{N+1}$ or ${}^{A+1}_{Z+1}\mathrm{Y}_{N}$ and as a charged particle, the incident energy can be accurately chosen allowing a selective study of the reaction. But, in spite of all these advantages and if many differential cross sections are quite well reproduced, the (d,p) or (d,n) excitation functions are often poorly described, meaning that our understanding of the reaction mechanism remains to be improved. It is due to the fact that, for these reactions, many processes (such as direct and multistep transfer, elastic and inelastic breakup...) can occur and will compete and interfere to produce the same compound nucleus. The cross section prediction thus requires to solve at least a quantum three-body problem which includes the continuum effects properly. In 1958, S. Watanabe [2] proposed to build the deuteron optical potential by folding the projectile wave function with the nucleon-target optical potentials in order to reproduce the elastic cross section. Then in the seventies, an improvement of this approach was developped: an approximated way to compute the deuteron elastic and elastic breakup cross sections was proposed by R.C. Johnson and P.J.R. Soper [3] (who used a simplified approach involving an adiabatic approximation) and by G.H. Rawitscher [4]. Within this model (now known as the Continuum Discretized Coupled Channels (CDCC) approach), the total wave function of the deuteron+target system is expanded into a sum of states describing the elastic channel and breakup ones and the cross sections for each channel are derived by solving a set of coupled differential equations. The CDCC has been widely studied and quite sucessfully applied, mainly to predict differential cross sections (see e.g. [5–10, 12–14]). Starting from some CDCC calculations, we will propose a semi-phenomenological method to compute the (d,p) excitation functions.

[a] e-mail: huu-tai.chau@cea.fr

2 A semi-phenomenological model to compute the excitation function using the CDCC approach

2.1 A brief preview of the CDCC philosophy

The CDCC for deuteron induced reactions has been extended in order to also include the target excitations [11]. This extension will be denoted by CDCC* and will be briefly summarized. Within this framework, the total wave function of the d+target system depending on **R**, the deuteron center of mass coordinates and on ρ, the proton-neutron relative coordinates, is written as:

$$|\Psi_{J_T M_T}(\mathbf{R},\rho)\rangle = \sum_{i\,l\,S\,I_p\,L\,J\,I_t} |(i\,l\,S)\,I_p\,L\,J I_t\,;J_T\,M_T\rangle$$

where the channels of the system for a given J_T are labeled by the bin number i (i=0 for the deuteron ground state and $i > 0$ for breakup ones), the deuteron spin $S = 1$, the relative orbital angular momentum l, the angular momentum J, the orbital angular momentum L associated to **R** and the spin of the target I_t where $\vec{S}+\vec{l}=\vec{I_p}$, $\vec{L}+\vec{I_p}=\vec{J}$ and $\vec{I_t}+\vec{J}=\vec{J_T}$.

It satisfies the Schrödinger equation:

$$\hat{H}|\Psi_{J_T M_T}(\mathbf{R},\rho)\rangle = E|\Psi_{J_T M_T}(\mathbf{R},\rho)\rangle$$

After performing a multipole expansion and after left-multiplying the Schrödinger equation by $\langle c|$ and integrating over ρ and the angular variables, the radial parts $u_c(R)$ of the wave function are solutions of a set of coupled differential equations:

$$\left(-\frac{\hbar^2}{2\mu_R}\frac{d^2}{dR^2}+\frac{\hbar^2 L(L+1)}{2\mu_R R^2}+V_p^{(Coul)}-E_i\right)u_c(R) = -\sum_{c'} V_{cc'}(R)u_{c'}(R) \tag{1}$$

where c and c' denote the channels $|(i\,l\,S)\,I_p\,L\,JI_t\,;J_T\,M_T\rangle$ and $|(i'\,l'\,S')\,I'_p\,L'\,J'I'_t\,;J_T\,M_T\rangle$ respectively. The expressions of the form factors $V_{cc'}(R) = \langle c|V(\mathbf{r}_i)|c'\rangle$ can be derived by using the Wigner-Eckart theorem: these form factors depend on some angular momentum couplings, on the target excitation, and on a folding of the nucleon-target optical potentials with the p-n wave functions.

The solutions of these equations are then used to compute the source term of the (d,p) reaction. The equation of this transfer reaction is then solved within the DWBA method *i.e.* it is assumed that the effect of the (d,p) channel on the elastic and breakup ones is weak.

2.2 Computation of the excitation functions

For a given incident energy E_d, the cross section of any reaction involves three contributions

$$\sigma(E_d) = \sigma_{\text{Direct}}(E_d) + \sigma_{\text{Pre-eq.}}(E_d) + \sigma_{\text{CN}}(E_d) \tag{2}$$

where σ_{Direct}, $\sigma_{\text{Pre-eq.}}$ and σ_{CN} are the direct component, the preequilibrium and the Compound Nucleus one, respectively. For the (d,p) cross section, the direct component reads

$$\sigma^{(d,\text{nucleon})}_{\text{Direct}}(E_d) = \sum_{i\leq i_{\max}, l_n, j_n} \sigma_{l_n,j_n}(E_{x,i}, E_d) \tag{3}$$

where $\sigma_{l_n,j_n}(E_{x,i}, E_d)$ denotes the cross section when the nucleon is captured by the target to form the A+1 nucleus into a state whose excitation energy, orbital angular momentum and spin are $E_{x,i}$, l_n, j_n, respectively and where $i_{\max}$ depends on the Q-value of the reaction and on E_d. Therefore to compute the direct component, it is firstly necessary to know all the states that will be populated by the transfer and for each level, the associated cross section $\sigma_{l_n,j_n}(E_{x,i}, E_d)$ has to be computed which can be done by integrating the differential cross section:

$$\sigma_{l_n,j_n}(E_{x,i}, E_d) = \int \frac{d\sigma_{l_n,j_n}}{d\Omega}(E_{x,i}, E_d)\, d\Omega\,.$$

This differential cross section involves the wave function of the captured neutron which is usually derived by assuming that the level with excitation energy $E_{x,i}$ is a pure single-particle state and by using a Woods-Saxon potential whose depth is adjusted to get the correct excitation energy. To take into account that the state is not in fact a pure single-particle one, it is then necessary to introduce a spectroscopic factor, $S_{l_n,j_n,i}$, to renormalize the calculated cross section, which means that, for each level, the experimental cross section must be known at least for one incident energy:

$$\sigma_{l_n,j_n}(E_{x,i}, E_d) = S_{l_n,j_n,i} \int \frac{d\sigma^{\text{calc.}}_{l_n,j_n}}{d\Omega}(E_{x,i}, E_d) d\Omega\,.$$

Thus the total direct cross section reads:

$$\sigma^{(d,\text{nucleon})}_{\text{Direct}}(E_d) = \sum_{i \le i_{\max}, l_n, j_n} S_{l_n,j_n,i} \int \frac{d\sigma^{\text{calc.}}_{l_n,j_n}}{d\Omega}(E_{x,i}, E_d) d\Omega\,.$$

However to do such calculation, one should overcome some difficulties. First, one should infer the levels that will be populated by the transfer reaction from the knowledge of the compound nucleus spectrum. Then, for each of these levels, one should be able to determine the spectroscopic factor which means that, at least, one experimental cross section has been measured. Unfortunetaly these assumptions are not always satisfied.

Therefore the direct component will be computed by using a simplified approach. Indeed, as it has been illustrated in [11], it has been observed that the shape of the excitation function of the integrated cross sections does not depend strongly on the level excitation energy but mainly on the orbital angular momentum, l_n. Thus it will be assumed that:

$$\sigma^{\text{calc.}}_{l_n,j_n}(E_{x,i}, E_d) \propto \tilde{\sigma}_{l_n,j_n}(E_d)\,, \tag{4}$$

which means that the integrated cross sections can be written as the product of an amplitude depending on the excitation energy of the level, $A(E_{x,i})$, times a function depending only on l_n and on the incident energy:

$$\sigma^{\text{calc.}}_{l_n,j_n}(E_{x,i}, E_d) \approx A(E_{x,i})\tilde{\sigma}_{l_n,j_n}(E_d)\,.$$

For each l_n, one can choose a level with a given excitation energy E_{x,i_0} and the associated cross section $\sigma^{\text{calc.}}_{l_n,j_n,i_0}(E_d)$ is then computed:

$$\sigma_{l_n,j_n,i_0}(E_d) \equiv \sigma^{\text{calc.}}_{l_n,j_n}(E_{x,i_0}, E_d) = A(E_{x,i_0})\tilde{\sigma}_{l_n,j_n}(E_d)$$

meaning that

$$\tilde{\sigma}_{l_n,j_n}(E_d) = \frac{1}{A(E_{x,i_0})}\sigma_{l_n,j_n,i_0}(E_d)\,,$$

and that, for all excitation energies with the same l_n:

$$\sigma^{\text{calc.}}_{l_n,j_n}(E_{x,i}, E_d) \approx \frac{A(E_{x,i})}{A(E_{x,i_0})}\sigma_{l_n,j_n,i_0}(E_d)\,.$$

After some transformations, it can be shown that the direct component of excitation function reads:

$$\sigma_{\text{Direct}}^{(d,\text{nucleon})}(E_d) \approx \sum_{l_n,j_n}(\sum_{i\leq i_{\max}} S_{l_n,j_n,i}\frac{A(E_{x,i})}{A(E_{x,i_0})})\sigma_{l_n,j_n,i_0}(E_d). \tag{5}$$

Denoting $\sum_{i\leq i_{\max}} S_{l_n,j_n,i}\frac{A(E_{x,i})}{A(E_{x,i_0})} = W_{l_n,j_n}(E_d)$, the direct contribution is therefore rewritten as:

$$\sigma_{\text{Direct}}^{(d,\text{nucleon})}(E_d) \approx \sum_{l_n,j_n} W_{l_n,j_n}(E_d)\sigma_{l_n,j_n,i_0}(E_d) \tag{6}$$

where the W_{l_n,j_n}'s are non-decreasing bounded functions of E_d thus one deduces that

$$\exists\omega_{l_n,j_n}, \lim_{E_d\to\infty} W_{l_n,j_n}(E_d) = \omega_{l_n,j_n},$$

and that:

$$\sigma_{\text{Direct}}^{(d,\text{nucleon})}(E_d) \underset{\infty}{\tilde{=}} \sum_{l_n,j_n} \omega_{l_n,j_n}\sigma_{l_n,j_n,i_0}(E_d), \tag{7}$$

which means that, at least for large incident energies, the excitation function can be approximated by a linear combination of the individual cross sections, σ_{l_n,j_n,i_0}.

In our approach, the computation of the direct component of the excitation function are performed within three steps:

- First, the individual cross sections σ_{l_n,j_n,i_0} for a few given levels of the A+1 nucleus are calculated within the CDCC*-DWBA framework.
- Then the weights, ω_{l_n,j_n} are deduced by reproducing the experimental cross sections at some given incident energies.
- Using the expression (7), the direct part of the excitation function is obtained overall the incident energy range.

This semi-phenomenological approach will be illustrated in some details for ^{75}As in the subsection 2.3 and it will be compared with some experimental excitation functions for a large set of targets in the subsection 2.4

2.3 Excitation function for ^{75}As

Using the CDCC*-DWBA approach , the individual (d,p) cross sections have been computed for three levels of ^{76}As: the ground state with $(l_n, j_n^\pi) = (4, 7/2^+)$ and two excited levels at 0.044 MeV and 0.366 MeV and with $(l_n, j_n^\pi) = (1, 3/2^-)$, $(l_n, j_n^\pi) = (3, 5/2^-)$, respectively.

Some angular distributions of the ^{75}As(d,p)^{76}As reaction have been measured for deuteron incident at 12 MeV by J.F. Lemming *et al.* [15] and by S. Mordechai *et al.* [16]. Since the neutron wave functions have not been computed for all the excited levels, when the excitation energy of the experimental level is different from the one used in the calculations, the comparison has be done with a "theoretical level" with the same quantum numbers. Some comparisons are plotted in Fig. 1. In these figures, the calculated differential cross sections and the experimental data are represented by the red curves and the black dots, respectively. An overall agreement between calculations and measurements can be observed. The shape of the excitation function computed for the three levels previously mentionned are plotted in arbitrary units in Fig. 2.

This excitation function has been measured by N. Baron *et al.* [17] at E_d = 19.0 MeV and by H.F. Röhm *et al.* [18] for deuteron incident energies ranging from 18.7 to 51.3 MeV. These two sets of data are very different since the cross section measured by H.F. Röhm *et al.* at 18.7 MeV is equal to 105 ± 7.4 mb while the one published by N. Baron *et al.* is equal to 76.9 ± 11.5 mb. >From the shape of the data of H.F. Röhm *et al.*, we have deduced that the calculated cross section with $l_n = 1$ is the main contribution and we have extracted the corresponding weights. The comparison between the calculation and the data is represented in the right panel of the Fig. 2.

2.4 Comparison with some experimental data

Similar calcutions have been performed for nuclei ranging from Al to Hf. A more detailled description of the ingredients used for each nucleus is given in [11]. The comparisons with the experimental data [19–36] are plotted in Fig. 3–14. Three conclusions can be drawn from these comparisons:

- A quite good agreement is obtained for all these nuclei.
- In most of the cases, the contribution of only one excitation function with a given orbital angular momentum l_n is needed to reproduce the experimental data, the value of l_n depending on the neutron number of the target.
- It is amazing to get such an agreeemnt within the whole energy range since one would expect that the compound nucleus and the pre-equilibrium processes would play a role in the reaction mechanism.

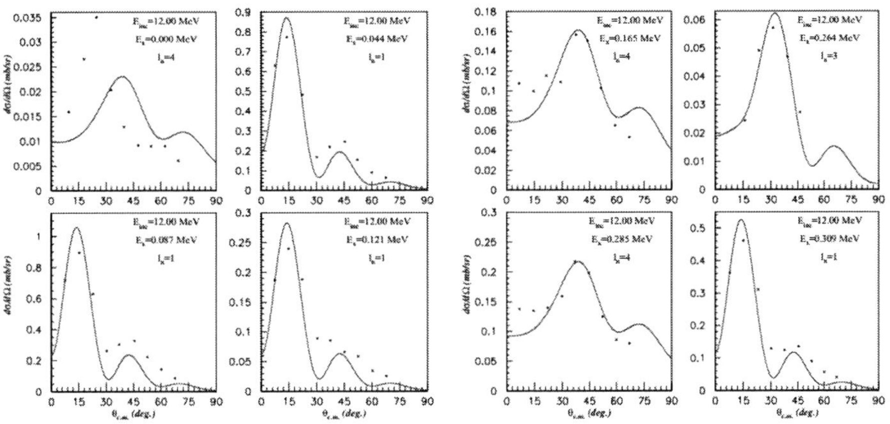

Figure 1. Differential cross sections for the ^{75}As(d,p) reaction. The experimental data (dots) are from [15].

3 Conclusion

In order to get a better description of differential and integral cross sections for the (d,p) reaction, a semi-phenomenological approach has been built. Its starting point are some CDCC*-DWBA calculations of the individual cross sections for a restricted number of states of the compound nucleus since it has been numerically observed that the shape of the integrated cross section as function of the incident energy is almost independent of the excitation energy of the compound nuleus state. It is then shown

 DOI: 10.1051/epjconf/201612206002

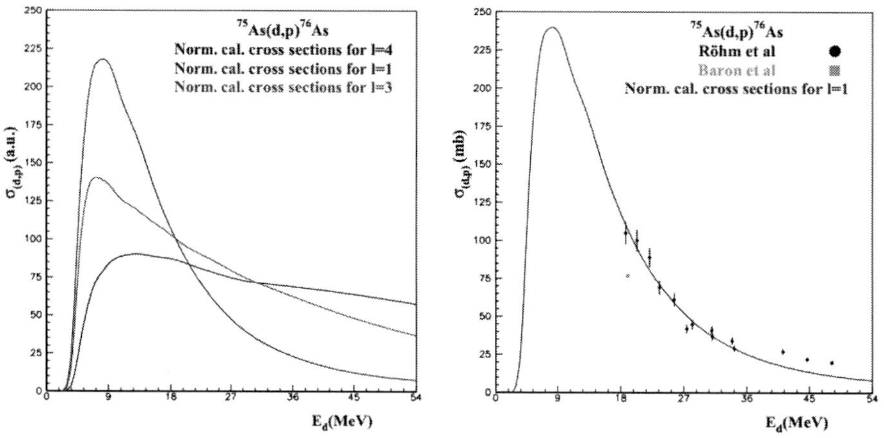

Figure 2. Left panel: integrated cross sections in arbitrary unit for the three levels of ^{76}As defined in the text. Right panel: comparison between the ^{75}As(d,p)^{76}As experimental excitation function [17, 18] and a linear combination of the theoretical excitation functions as defined by Eq. (7).

that the direct part of the excitation function can be written as a linear combination of these individual excitation functions. A quite good agreement between the experimental excitation functions and the calculated ones has been obtained for a large number of nuclei. But despite this agreement, many open questions remain to be answered, in particular it is not clear why there is no room left for other reaction mechanisms such as compound nucleus or pre-equilibrium processes.

References

[1] H.C. Urey, F.G. Brickwedde, G.M. Murphy, Phys. Rev. **39**, 164 (1932).
[2] S. Watanabe, Nucl. Phys. **8**, 484 (1958).
[3] R.C. Johnson, P.J.R. Soper, Phys. Rev. C **1**, 976 (1970).
[4] G.H. Rawitscher, Phys. Rev. C **9**, 2210 (1974).
[5] M. Yahiro, M. Kamimura, Prog. Theor. Phys. **65**, 2046 (1981).
[6] M. Yahiro, M. Kamimura, Prog. Theor. Phys. **65**, 2051 (1981).
[7] M. Yahiro *et al.*, Prog. Theor. Phys. **67**, 1467 (1982).
[8] M. Yahiro *et al.*, Prog. Theor. Phys. Suppl. **89**, 1 (1986).
[9] N. Austern *et al.*, Phys. Rep. **154**, 125 (1987).
[10] P. Chau-Huu Tai, Nucl. Phys. A **773**, 56 (2006).
[11] P. Chau-Huu Tai, Eur. Phys. J. A **51**, 166 (2015).
[12] M. Masaki *et al.*, Nucl. Phys. A **573**, 1 (1994).
[13] K. Hirota *et al.*, Nucl. Phys. A **628**, 547 (1998).
[14] M. Yamagushi *et al.*, Nucl. Phys. A **747**, 3 (2005).
[15] J.F. Lemming *et al.*, Nucl. Phys. A **239** 412 (1975).

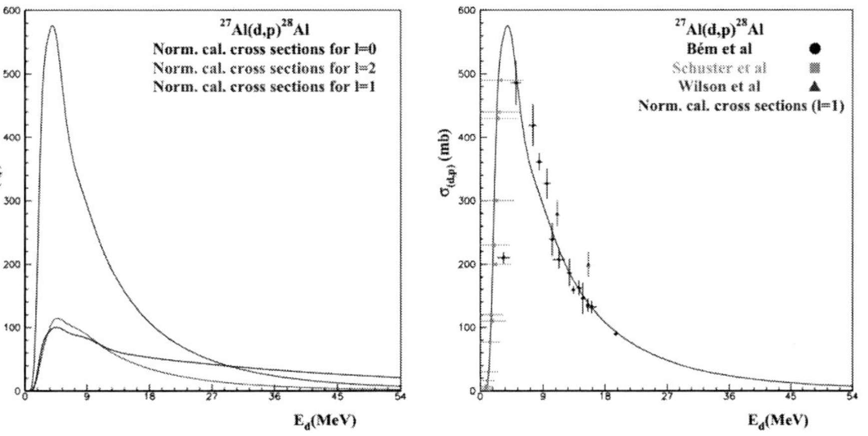

Figure 3. Left panel: integrated cross sections in arbitrary unit for three levels of ^{28}Al. Right panel: comparison between the ^{27}As(d,p)^{28}Al experimental excitation function [19–21] and a linear combination of the theoretical excitation functions as defined by Eq. (7).

[16] S. Mordechai *et al.*, Nucl. Phys. A **272** 82 (1976).

[17] N. Baron, B.L. Cohen, Phys. Rev. **129**, 2636 (1963) doi: 10.1103/PhysRev.129.2636.

[18] H.F. Röhm, H. Münzel, J. Inorg. Nucl. Chem. **34**,1773 (1972) doi: 10.1016/0022-1902(72)80523-2.

[19] E. Schuster, K. Wohlleben, Int. J. Appl. Radiat. Isot. **19** 471 (1968) doi: 10.1016/0020-708X(68)90100-2.

[20] R.L. Wilson *et al.*, Phys. Rev. C **13**, 976 (1976) doi: 10.1103/PhysRevC.13.976.

[21] P. Bém *et al.*, Phys. Rev. C **79**, 044610 (2009) doi: 10.1103/PhysRevC.79.044610.

[22] L. Závorka *et al.*, J. Korean Phys. Soc. **59**, 1961 (2011) doi: 10.3938/jkps.59.1961.

[23] M.U. Khandaker *et al.*, Nucl. Instrum. Methods B **316**, 33 (2013) doi: 10.1016/j.nimb.2013.08.032.

[24] B. Kiraly *et al.*, Nucl. Instrum. Methods B **267**, 15 (2009) doi: 10.1016/j.nimb.2008.11.005.

[25] F. Ditrói *et al.*, Nucl. Instrum. Methods B **270**, 61 (2012) doi: 10.1016/j.nimb.2011.10.010.

[26] R. Michel, M. Galas, Int. J. Appl. Radiat. Isot. **34**, 1325 (1983) doi: 10.1016/0020-708X(83)90023-6.

[27] F. Ditrói *et al.*, Nucl. Instrum. Methods B **268**, 2571 (2010) doi: 10.1016/j.nimb.2010.06.033.

[28] H.I. West Jr *et al.*, Phys. Rev. C **47**, 248 (1993) doi: 10.1103/PhysRevC.47.248.

[29] F. Ditrói *et al.*, Appl. Radiat. Isot. **70**, 574 (2012) doi: 10.1016/j.apradiso.2011.12.034.

[30] F. Simonelli *et al.*, Radiochim. Acta **98**, 187 (2010) doi: 10.1524/ract.2010.1709.

[31] F. Tárkányi *et al.*, Nucl. Instrum. Methods B **316**, 22 (2013) doi: 10.1016/j.nimb.2013.08.031.

[32] J. Lange, H. Münzel, Radiochim. Acta **9**, 66 (1968) doi: 10.1524/ract.1968.9.23.66.

[33] A. Hermanne *et al.*, Nucl. Instrum. Methods B **267**, 727 (2009) doi: 10.1016/j.nimb.2008.12.017.

[34] F. Tárkányi *et al.*, Nucl. Instrum. Methods B **266**, 3529 (2008) doi: 10.1016/j.nimb.2008.05.123.

 DOI: 10.1051/epjconf/201612206002

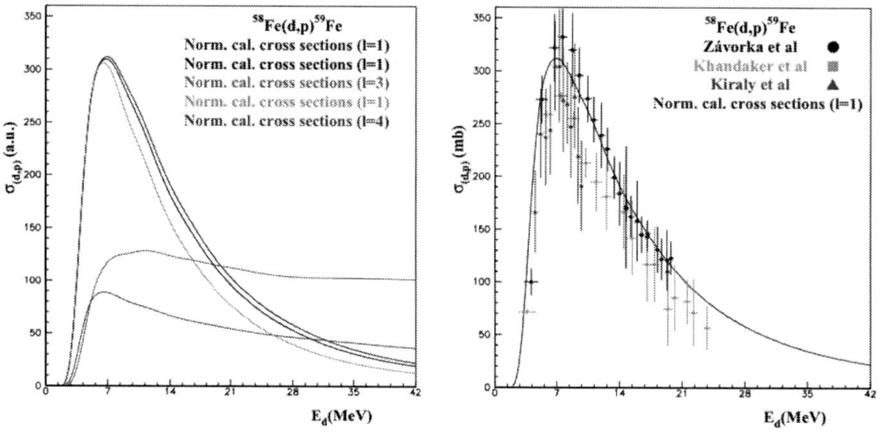

Figure 4. Left panel: integrated cross sections in arbitrary unit for five levels of ^{58}Fe. Right panel: comparison between the ^{58}Fe(d,p)^{59}Fe experimental excitation function [22–25] and a linear combination of the theoretical excitation functions as defined by Eq. (7).

[35] A. Hermanne *et al.*, Nucl. Instrum. Methods B **311**, 102 (2013) doi: 10.1016/j.nimb.2013.06.014.
[36] S. Takács *et al.*, Nucl. Instrum. Methods B **268**, 3443 (2010) doi: 10.1016/j.nimb.2010.08.016.

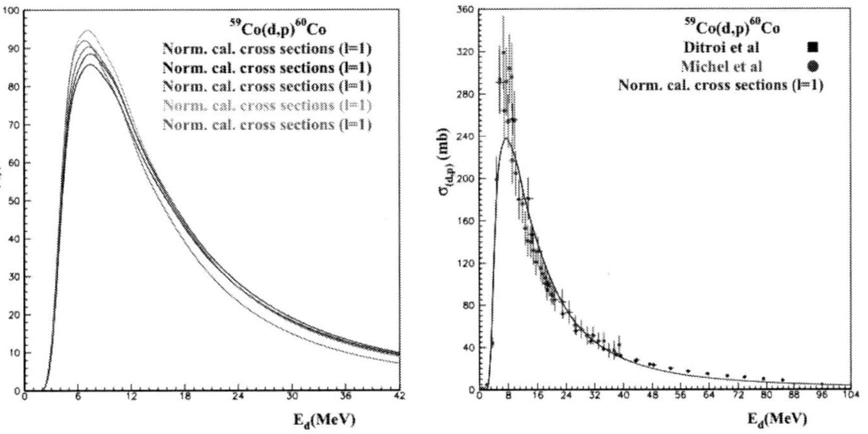

Figure 5. Left panel: integrated cross sections in arbitrary unit for five levels of ^{59}Co. Right panel: comparison between the ^{59}Co(d,p)^{60}Co experimental excitation function [26, 27] and a linear combination of the theoretical excitation functions as defined by Eq. (7).

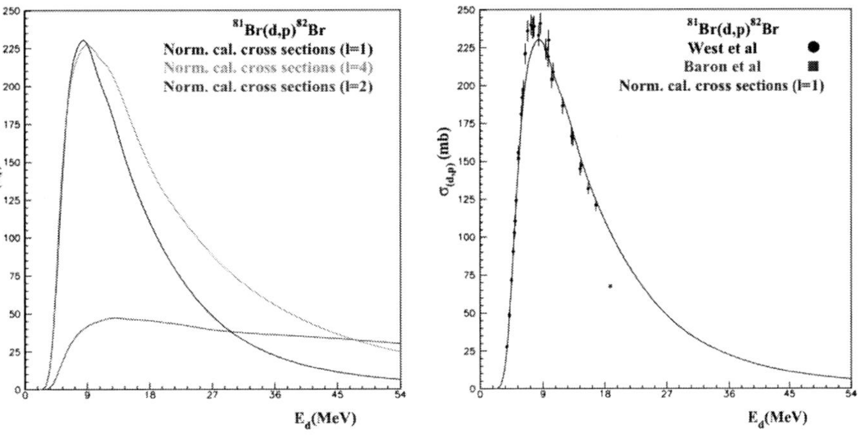

Figure 6. Left panel: integrated cross sections in arbitrary unit for three levels of ^{81}Br. Right panel: comparison between the ^{81}Br(d,p)^{82}Br experimental excitation function [17, 28] and a linear combination of the theoretical excitation functions as defined by Eq. (7).

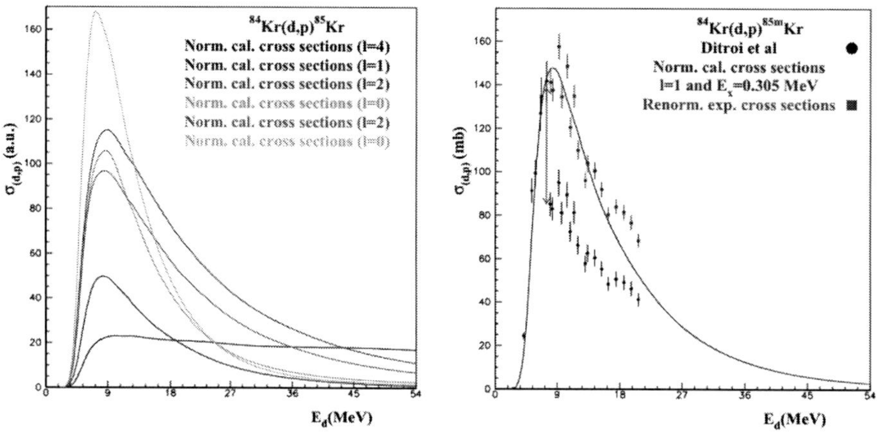

Figure 7. Left panel: integrated cross sections in arbitrary unit for six levels of ^{84}Kr. Right panel: comparison between the ^{84}Kr(d,p)^{85m}Kr experimental excitation function [29] and a linear combination of the theoretical excitation functions as defined by Eq. (7).

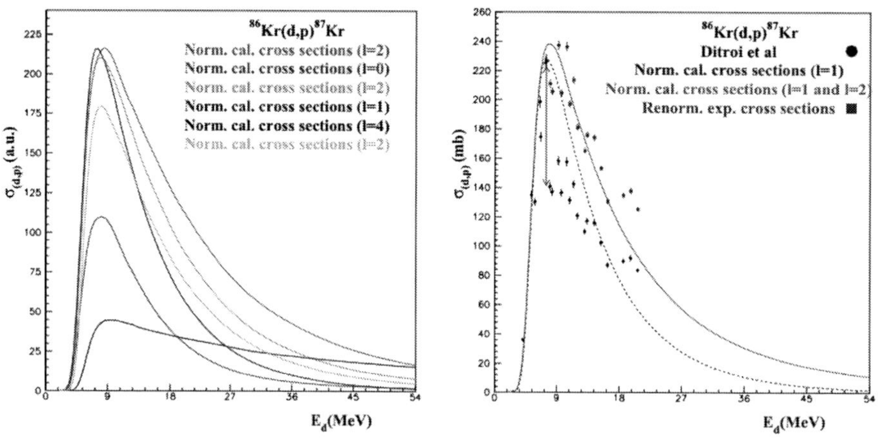

Figure 8. Left panel: integrated cross sections in arbitrary unit for six levels of ^{86}Kr. Right panel: comparison between the ^{86}Kr(d,p)^{87}Kr experimental excitation function [29] and a linear combination of the theoretical excitation functions as defined by Eq. (7).

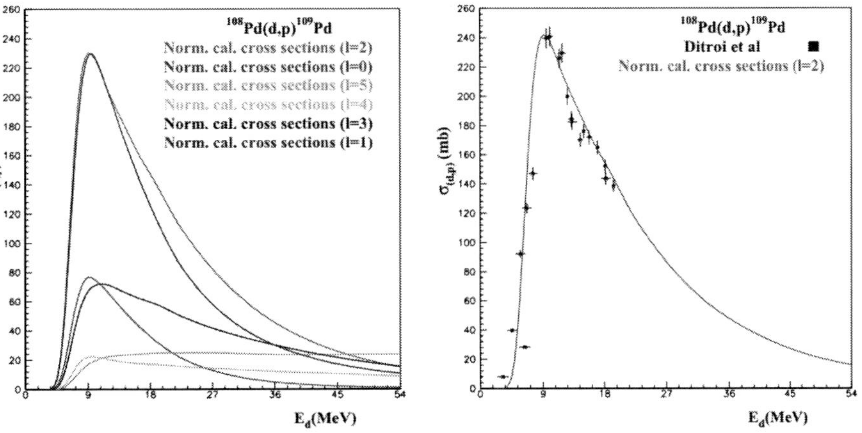

Figure 9. Left panel: integrated cross sections in arbitrary unit for six levels of ^{180}Pd. Right panel: comparison between the ^{108}Pd(d,p)^{109}Pd experimental excitation function [25] and a linear combination of the theoretical excitation functions as defined by Eq. (7).

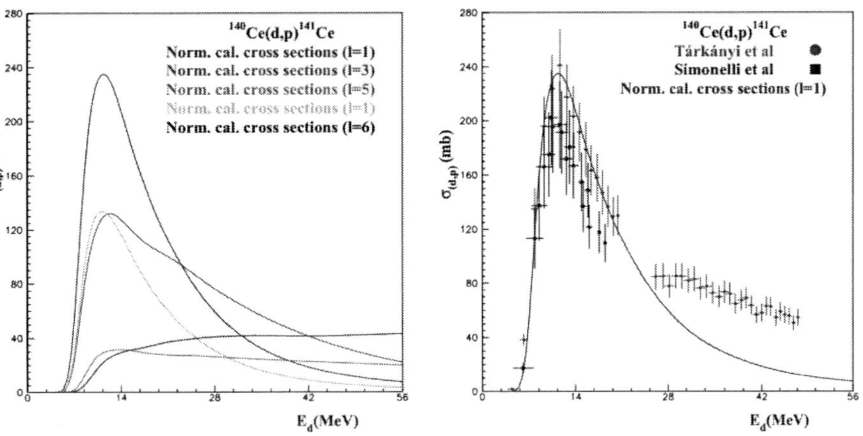

Figure 10. Left panel: integrated cross sections in arbitrary unit for five levels of ^{140}Ce. Right panel: comparison between the ^{140}Ce(d,p)^{141}Ce experimental excitation function [30, 31] and a linear combination of the theoretical excitation functions as defined by Eq. (7).

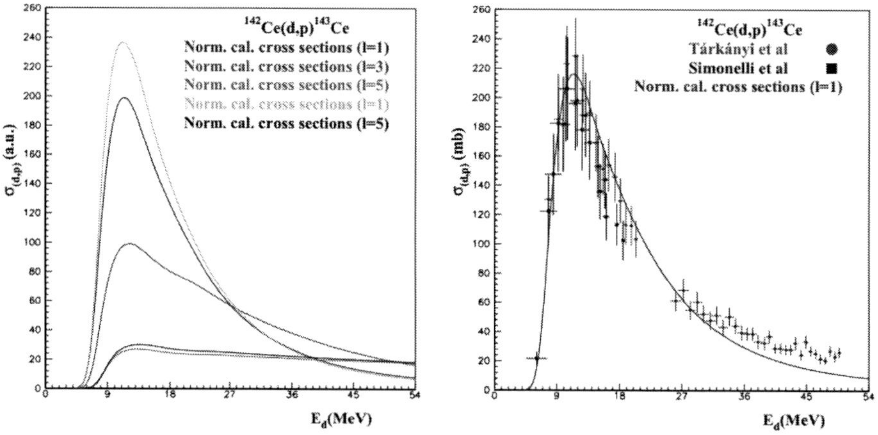

Figure 11. Left panel: integrated cross sections in arbitrary unit for five levels of ^{142}Ce. Right panel: comparison between the ^{142}Ce(d,p)^{143}Ce experimental excitation function [30, 31] and a linear combination of the theoretical excitation functions as defined by Eq. (7).

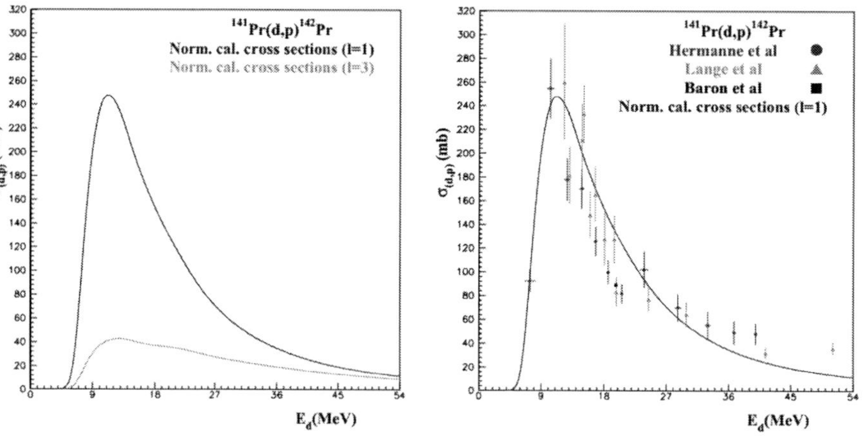

Figure 12. Left panel: integrated cross sections in arbitrary unit for two levels of ^{141}Pr. Right panel: comparison between the ^{141}Pr(d,p)^{142}Pr experimental excitation function [17, 32, 33] and a linear combination of the theoretical excitation functions as defined by Eq. (7).

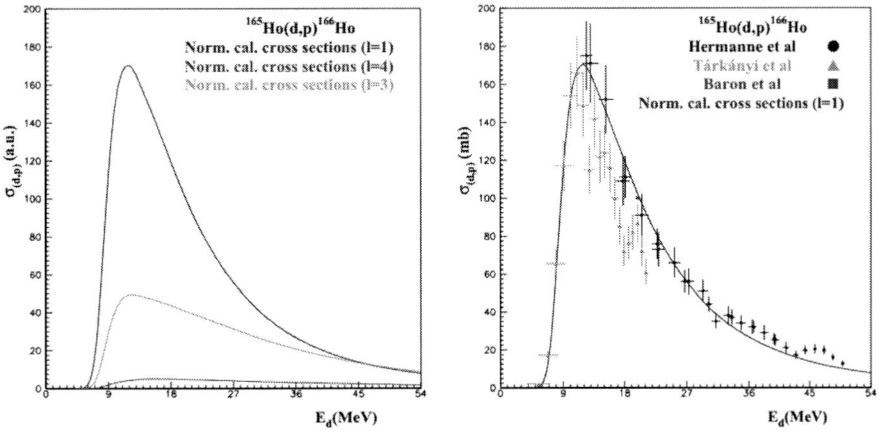

Figure 13. Left panel: integrated cross sections in arbitrary unit for three levels of ^{165}Ho. Right panel: comparison between the ^{165}Ho(d,p)^{166}Ho experimental excitation function [17, 34, 35] and a linear combination of the theoretical excitation functions as defined by Eq. (7).

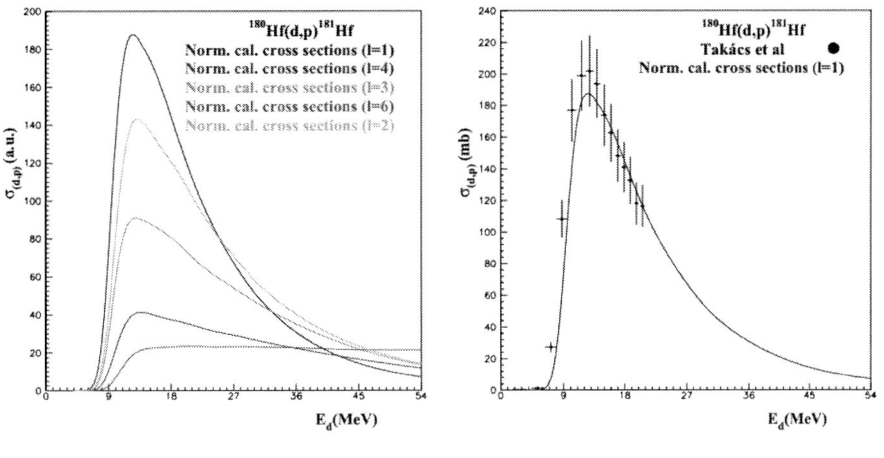

Figure 14. Left panel: integrated cross sections in arbitrary unit for five levels of ^{181}Hf. Right panel: comparison between the ^{180}Hf(d,p)^{181}Hf experimental excitation function [36] and a linear combination of the theoretical excitation functions as defined by Eq. (7).

DOI: 10.1051/epjconf/201612206003

Microscopic effective reaction theory for direct nuclear reactions

Kazuyuki Ogata[1,a], Kosho Minomo[1], Masakazu Toyokawa[2], Michio Kohno[1], Takuma Matsumoto[2], Masanobu Yahiro[2], Yuma Kikuchi[1,3], Tokuro Fukui[1,b], Kazuki Yoshida[1], and Kazuhito Mizuyama[1,c]

[1]*Research Center for Nuclear Physics, Osaka University, Ibaraki, Osaka 567-0047, Japan*
[2]*Department of Physics, Kyushu University, Fukuoka 812-8581, Japan*
[3]*RIKEN Nishina Center, RIKEN, Wako 351-0198, Japan*

Abstract. Some recent activities with the microscopic effective reaction theory (MERT) on elastic, inelastic, breakup, transfer, and knockout processes are reviewed briefly. As a possible alternative to MERT, a description of elastic and inelastic scattering with the continuum particle-vibration coupling (cPVC) method is also discussed.

1 Introduction

Nowadays, a microscopic description of various types of direct nuclear reactions based on the multiple scattering theory (MST) for nucleus-nucleus scattering [1] is feasible. A key prescription of this approach is, as adopted in the traditional direct reaction studies [2], to select essential degrees of freedom in describing a direct reaction process of interest with a desired accuracy. For example, if a breakup reaction of a projectile P into two fragments c and v with a target nucleus A is considered, it will be natural to adopt an A + c + v three-body model. The model Hamiltonian contains A-c and A-v one-body potentials. For them, we use the nucleus-nucleus MST. Then the three-body scattering problem can accurately be solved by sophisticated direct reaction models such as the continuum-discretized coupled-channels method (CDCC) [5–7], the dynamical eikonal approach (DEA) [8, 9], and Faddeev–Alt-Grassberger-Sandhas (Faddeev-AGS) theory [10, 11]. This will be one of the most efficient microscopic approaches to direct nuclear reactions of various kind. We call it *the microscopic effective reaction theory (MERT)*. Extension of MERT to four-body breakup reactions [12–14] and a reaction process involving an excitation of a constituent of P [15, 16] will be straightforward.

Obviously, the concept of MERT is different from that of the ab-initio calculation of nuclear reactions [3, 4], i.e., MERT is a theoretical framework *in a model space*. This means, at least at this moment, MERT is not designed to describe all the phenomena triggered by an incident particle colliding with a target nucleus. However, once a model space is appropriately set, MERT can be regarded as a reaction theory having a predictive power for scattering observables of various projectiles and target nuclei in a wide range of incident energies.

[a]e-mail: kazuyuki@rcnp.osaka-u.ac.jp
[b]Present address: Nuclear Data Center, Japan Atomic Energy Agency, Tokai, Ibaraki 319-1195, Japan
[c]Present address: Nuclear Data Center, Japan Atomic Energy Agency, Tokai, Ibaraki 319-1195, Japan

In Sect. 2, we briefly review some recent activities with MERT. The application of the continuum particle-vibration coupling (cPVC) method to elastic and inelastic scattering processes, as a possible alternative to MERT, is also discussed. Section 3 is devoted to a summary and future perspective.

2 Direct reaction studies with MERT

2.1 Three nucleon force effect on elastic and inelastic scattering

One of the hot topics in nuclear physics is the three nucleon force (3NF) effect on nuclear reaction observables. We adopt the chiral effective field theory (Ch-EFT) [17, 18] to treat the nucleon-nucleon (*NN*) two-body and three-body interactions on the same footing. We then construct an *NN* g-matrix interaction in the infinite nuclear matter taking into account the 3NF effect [19, 20]. As discussed in Ref. [7], the g-matrix interaction is assumed to be a good approximation to the *NN* effective interaction appearing in the nucleus-nucleus MST [1], except at very low energies.

We show in Fig. 1 the cross sections and analyzing powers in proton elastic scattering on several target nuclei at 65 MeV [21]. It should be noted that the calculation contains no free adjustable parameters. The results in Fig. 1 therefore show the predictability of the g-matrix folding calculation to proton elastic scattering for not only the cross section but also the spin observable. We do not find a significant 3NF effect on these observables because the proton elastic scattering probes mainly the surface region of the target nucleus in which the density is low and the 3NF is expected to be small. On the other hand, we see a somewhat large 3NF effect on nucleus-nucleus elastic and inelastic scattering cross sections [24, 25]. For details, see the references above as well as the proceedings paper for CNR*15 by K. Minomo and his collaborators.

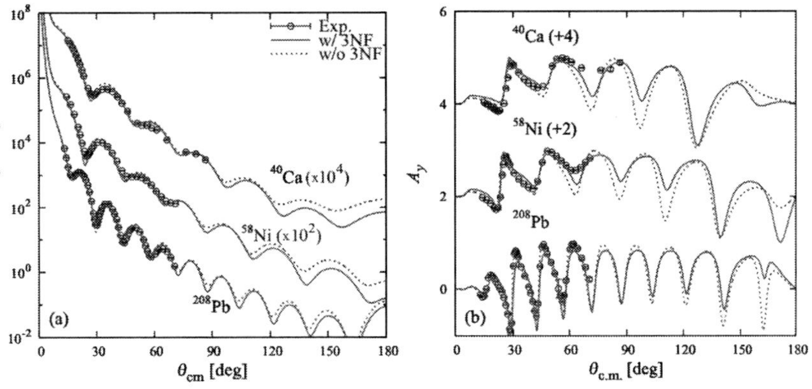

Figure 1. (a) Cross sections and (b) analyzing powers for proton elastic scattering at 65 MeV compared with the experimental data [22, 23]. The solid (dashed) lines represent the results with (without) the 3NF effect. Each cross section is multiplied by the number beside the line, whereas each analyzing power is shifted up by the number. This figure is taken from Ref. [21].

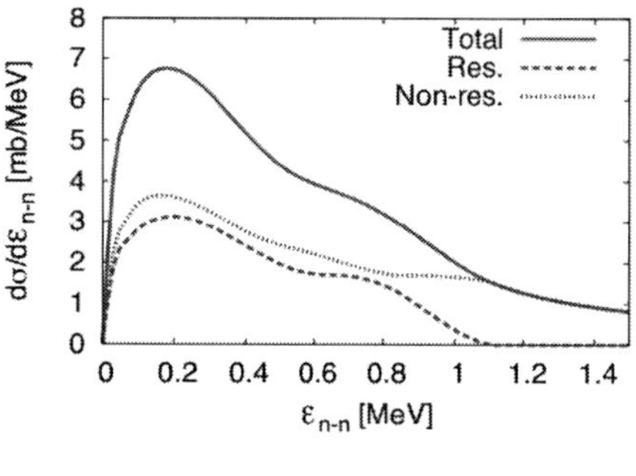

Figure 2. n-n invariant mass spectra [26] (solid line). The decay cross sections for the 2_1^+ resonant state and the nonresonant three-body continuum states are shown by the dashed and dotted lines, respectively.

2.2 Decay mode of the 2_1^+ sate of ^{6}He

Another important subject in the physics of unstable nuclei is the decay mode of a three-body system that is not stable against the particle emission. In Ref. [26], we studied the $\alpha + n + n$ three-body decay of the 2_1^+ sate of ^{6}He populated by the ^{6}He breakup process on the ^{12}C nucleus at 250 MeV/nucleon. The essential feature of this study is both the formation and decay of the particle-unbound state of ^{6}He are described on the same footing. For this purpose we have developed a new framework that is an extended version of the method of complexed-scaled solutions of the Lippmann-Schwinger equation (CSLS) [27] utilizing the transition matrix for the four-body breakup evaluated by CDCC [12, 13]. The n-^{12}C and α-^{12}C distorting potentials are evaluated by the g-matrix folding model. The CDCC-CSLS calculation can thus be regarded as an application of MERT to a four-body breakup/decay process.

The solid line in Fig. 2 shows the n-n invariant mass spectra. One sees a peak at around 0.2 MeV of the n-n relative energy (ϵ_{nn}) and a shoulder structure at $\epsilon_{nn} \sim 0.8$ MeV. The former is due to the n-n final state interaction (FSI) in the decaying process, which has already been found in the Coulomb breakup of ^{6}He [28]. On the other hand, the latter appears only in the nuclear breakup of ^{6}He, in which the 2_1^+ resonant state is expected to be well populated. We then put a gate on the three-body breakup energy ϵ corresponding to the 2_1^+ state, i.e., $\epsilon = 0.98 \pm 0.14$ MeV. The result is shown by the dashed line that is regarded as the decay cross section from the 2_1^+ resonance. The dotted line is obtained by subtracting the dashed line from the solid line, which corresponds to the decay of the nonresonant continuum states. One sees that the low-energy peak appears in both lines, whereas the shoulder structure is found only in the dashed line. It should be noted that since ϵ is restricted to around 1 MeV in the resonant decay (dashed line), at the shoulder the three-body energy is almost exhausted by the n-n relative energy.

One can therefore conclude that there are two decay modes of the 2_1^+ state of ^{6}He. One is the decay mode in which the two neutrons are emitted in correlation with each other. The other is the mode in which the two neutrons are emitted to the opposite directions. The latter, the back-to-back decay mode, will be important in particular, because it is free from the FSI and may suggest a di-neutron structure in the 2_1^+ state of ^{6}He. For more detailed discussion, the reader is referred to Ref. [26].

2.3 Three-body dynamics in the ^{8}B(d, n)^{9}C reaction

Direct nuclear reactions are sometimes utilized as an alternative way of extracting a cross section that is difficult to measure directly. In Ref. [29] the cross section of the ^{8}B(d, n)^{9}C reaction was measured and thereby that of the ^{8}B(p, γ)^{9}C process was extracted; the so-called asymptotic normalization

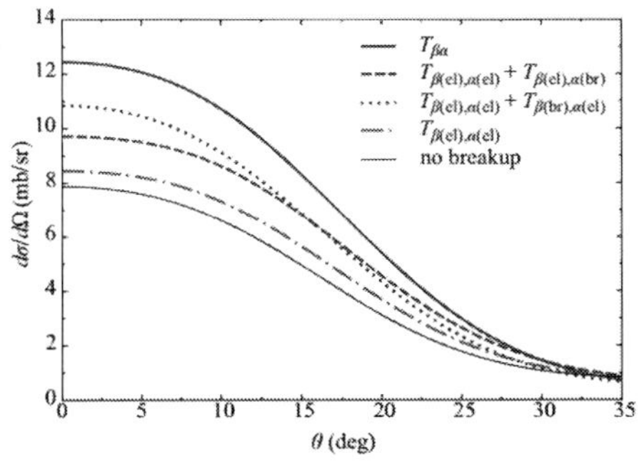

Figure 3. Cross section of the ^{8}B(d,n)^{9}C reaction at 14.4 MeV/nucleon [31]. The meaning of each line is given in the text.

coefficient (ANC) method [30] was adopted to relate the two cross sections. However, the reaction model used to analyze the ^{8}B(d,n)^{9}C reaction was DWBA, in which the breakup channels of d and ^{9}C were not included explicitly.

Recently, we have re-analyzed the ^{8}B(d,n)^{9}C reaction at 14.4 MeV/nucleon by means of the CDCC-BA framework [31]. The key feature is that the breakup channels of both d and ^{9}C are taken into account by CDCC. Figure 3 shows the resulting cross sections as a function of the center-of-mass scattering angle. The thick solid line is the result of CDCC-BA, the dashed (dotted) line corresponds to the calculation with including the breakup channels of d (^{9}C) only, and the dash-dotted line shows the result with neglecting all the breakup channels. Note that in these four calculations breakup channels are taken into account by CDCC in evaluating the three-body wave functions in the initial and final channels; the selection of channels are done in the evaluation of the transition matrix for the transfer process. On the other hand, the thin solid line is the result obtained with neglecting the breakup channels in the calculation of the three-body wave functions.

The small difference between the thin solid line and the dash-dotted line suggests that the three-body wave function in the elastic channel is not affected by the coupling to the breakup states, in each of the initial and final channels. Therefore the difference between the thick solid line and the dash-dotted line is due to the contribution of the transfer process through the breakup states of d or ^{9}C. It is found that the inclusion of the breakup states of d (^{9}C) increases the transfer cross section by about 23% (38%). This finding will strongly suggest the importance of the breakup states of both d and ^{9}C in the analysis of the ^{8}B(d,n)^{9}C reaction at 14.4 MeV/nucleon, to reliably determine the cross section of the ^{8}B(p,γ)^{9}C process. Consequently, however, the ^{8}B(p,γ)^{9}C cross section decreased by about 51% and the result is significantly different from the value extracted by ^{9}C breakup reactions [32]. Further investigation will be necessary to draw a conclusion on the ^{8}B(p,γ)^{9}C cross section.

Another notice is that in Ref. [31] a phenomenological nucleon-nucleus optical potential was adopted. This is because that the reliability of the g-matrix folding model will be questionable at around 10 MeV. A microscopic framework that provides a reliable optical potential at lower energies will be very important; the cPVC method [33] (see Sect. 2.5) and the dispersive optical model (DOM) [34, 35] will be utilized for this purpose.

2.4 Asymmetry of parallel momentum distribution of (p,pN) reaction residues

Elucidation of the single particle (s.p.) properties of nuclei and the underlying many-body correlations is one of the most important goals of nuclear physics. Nucleon knockout reactions have widely been used to study the s.p. nature of nuclei. In Refs. [36, 37], a systematics of the so-called reduction factor R_S for knockout reactions was discussed; R_S is defined by the ratio of the spectroscopic factor S experimentally determined to that obtained by a shell model (SM) calculation. If R_S is unity, it implies that the SM perfectly describes the s.p. property of the nucleus. If R_S is lower than unity, a missing correlation in the SM calculation is suggested. Through $(e, e'p)$ studies, R_S was found to be around 0.6–0.7 for stable nuclei. This is mainly because that the short-range correlation is not taken into account in the standard SM calculations. However, it was shown in Refs. [36, 37] that R_S is around 0.2–0.4 for the proton (neutron) s.p. state in neutron (proton) rich nuclei. This suggests a surprisingly large amount of missing correlations in such cases. Their finding can be summarized as a strong ΔS dependence of R_S, where ΔS is the difference between the proton and neutron separation energies.

On the other hand, systematic studies of R_S with nucleon transfer reactions [38, 39] found that its ΔS dependence is much weaker than in Refs. [36, 37]. Furthermore, it was shown in Ref. [40] that if the intranuclear cascade (INC) model is adopted, R_S extracted from one neutron knockout process from ^{14}O completely deviates from the systematics shown in Refs. [36, 37]. These counterarguments suggest that R_S depends on reaction processes and/or reaction models adopted.

It should be noted that in the analysis of nucleon knockout reactions discussed in Refs. [36, 37], the Glauber model [41, 42] was adopted. The Glauber model is based on the eikonal and adiabatic (sudden) approximations, and the latter is valid only when the energy and momentum transfer (ω-q) is small. Since $R_S \sim 0.2$–0.4 is found when a tightly-bound nucleon is knocked out, the reliability of the Glauber model is not very clear. In fact, in such cases the parallel momentum distribution (PMD) of the reaction residue shows an asymmetric shape, whereas the result of the Glauber model is always symmetric because of the adiabatic approximation [40]. The asymmetry of the PMD can thus be a clue about the reliability of the Glauber model, and that to the understanding of the reaction (-model) dependence of R_S. However, the mechanism that develops the asymmetry of the PMD has not been understood well.

In Ref. [43], we have clarified how the asymmetric PMD is developed in the ^{14}O$(p, pn)^{13}$O at 100 MeV/nucleon. The distorted wave impulse approximation (DWIA) with the eikonal approximation to the distorted waves is adopted as a reaction model; the DWIA does not rely on the adiabatic approximation and safely applied to the knockout process for the tightly-bound neutron in ^{14}O having the separation energy of 23.2 MeV.

First, we show the role of the phase-volume (PV) that guarantees the energy and momentum conservation. For clear discussion, we work in the plane-wave limit, i.e., the PWIA is adopted. The solid line in the left panel of Fig. 4 shows the PWIA result without the PV, which can be regarded as

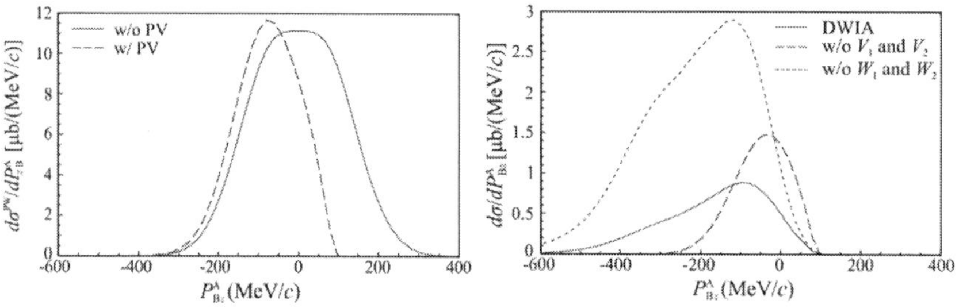

Figure 4. PMD of ^{14}O$(p, pn)^{13}$O at 100 MeV/nucleon [43]. See the text for detail.

a Glauber-like calculation. As expected, the PMD is purely symmetric. The inclusion of the PV gives a quite sharp cut of the PMD on the high momentum side, which results in a narrowing of the width of the PMD, and a reduction of the integrated cross section.

We then discuss the distortion effect. There are two types of distortion: one is the absorption caused by the imaginary part (W) of the optical potential (U) and the other is the attraction by the real part (V) of U. As one sees from the two dashed lines in the left and right panels, the absorption reduces the absolute value of the PMD but hardly changes its shape. On top of that, the attraction develops the low momentum tail of the PMD (the solid line in the right panel). This can be understood as follows. In the projectile-rest frame, the "target" proton comes from the $+z$ direction and hit the neutron in ^{14}O. After the collision, the proton and the struck neutron are emitted to the $-z$ direction, and the reaction residue (^{13}O) feels an attraction from the two nucleons. Thus, ^{13}O prefers to have a momentum to the $-z$ direction, which results in the well-developed low momentum tail of the PMD. It should be remarked that the attraction effect does not change the integrated value of the PMD but significantly reduces its peak height. In other words, if a reaction model without the attraction effect is adopted, the peak height of the PMD will be overestimated. Consequently, a smaller value of the "experimental" S, hence R_S, will be extracted.

Thus, the PV and attraction effects are found to generate the asymmetry of the PMD knockout reactions with a proton target; in fact, the asymmetric PMD reflects the asymmetry in the three-body kinematics. These two effects, which cannot be taken into account in the Glauber model calculation, are expected to exist also in knockout reactions with nucleus targets. Inclusion of them in the analysis of these reactions will be interesting and important. Alternatively, systematic studies of R_S using the (p, pN) reactions will be highly desirable; DWIA based on MERT will play an important role for this purpose. For more detailed discussion, see Ref. [43].

2.5 cPVC description of scattering observables

As an alternative to MERT, one may apply a nuclear many-body model to scattering problems. In Ref. [33], the continuum particle-vibration coupling (cPVC) method was developed; the cPVC method treats the collective mode and the s.p. motion of individual nucleons simultaneously. We have applied the cPVC method to the neutron scattering on ^{16}O below 30 MeV [44]. A distinguished feature of the calculation is that the NN effective interaction adopted (SkM*) has no imaginary part. In other words, the absorption in the n-^{16}O scattering is described by a coupling to the channels that are explicitly taken into account. This means there is no flux "loss" out of the model space adopted in the cPVC calculation.

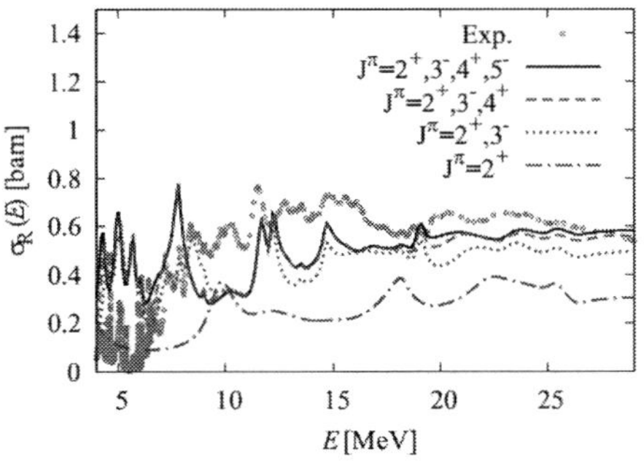

Figure 5. Total reaction cross section of neutron on ^{16}O [44]. The solid, dashed, dotted, and dash-dotted lines correspond to the calculation with the maximum multipolarity of 5, 4, 3, and 2 of the vibration mode, respectively. Experimental data are taken from Ref. [45].

Figure 5 shows the total reaction cross section σ_R. It should be remarked that the peak structure of σ_R is reproduced quite well, which suggests the cPVC calculation can describe the doorway states of the n + ^{16}O system. It is obvious that the g-matrix folding model cannot reproduce the energy dependence of σ_R in this energy region. The cPVC method has been successfully applied to also the proton inelastic scattering on ^{24}O at 62 MeV/nucleon [46].

3 Summary and future perspective

We have briefly reviewed some direct nuclear reaction studies by means of the microscopic effective reaction theory (MERT). The application of the continuum particle-vibration coupling (cPVC) method to low-energy scattering was also introduced. The microscopic description of various reaction processes is an interesting and challenging subject of theoretical nuclear physics. It is also important for an application to nuclear technology. Recently, a nuclear transmutation study for reduction and resource recycle of high level radioactive wastes has launched as an impulsing paradigm change through disruptive technologies (ImPACT) program. Development of MERT for multi-nucleon knockout and transfer reactions will be indispensable in understanding the measured cross section data of long-lived fission products (LLFPs) in inverse kinematics, and to generate neutron reaction data of LLFPs theoretically. In future, as the next step toward the description of compound nuclear reactions, it will also be important to develop MERT for pre-equilibrium processes; the semiclassical distorted wave model (SCDW) [47–51] will be suitable for this purpose.

Acknowledgment

This work has been supported in part by Grant-in-Aid of the Japan Society for the Promotion of Science (JSPS) and by the ImPACT Program of the Council for Science, Technology and Innovation (Cabinet Office, Government of Japan).

References

[1] M. Yahiro, K. Minomo, K. Ogata, and M. Kawai, Prog. Theor. Phys. **120**, 767 (2008).
[2] G. R. Satchler, *Direct Nuclear Reactions* (Oxfrod University Press, 1983).
[3] P. Navrátil, S. Quaglioni, I. Stetcu, and B. R. Barrett, J. Phys. G **36**, 083101 (2009).
[4] P. Navrátil and S. Quaglioni, Phys. Rev. Lett. **108**, 042503 (2012).
[5] M. Kamimura, M. Yahiro, Y. Iseri, Y. Sakuragi, H. Kameyama, and M. Kawai, Prog. Theor. Phys. Suppl. **89**, 1 (1986).
[6] N. Austern, Y. Iseri, M. Kamimura, M. Kawai, G. Rawitscher, and M. Yahiro, Phys. Rep. **154**, 125 (1987).
[7] M. Yahiro, K. Ogata, T. Matsumoto, and K. Minomo, Prog. Theor. Exp. Phys. **2012**, 01A206 (2012).
[8] D. Baye, P. Capel, and G. Goldstein, Phys. Rev. Lett. **95**, 082502 (2005).
[9] G. Goldstein, D. Baye, and P. Capel, Phys. Rev. C **73**, 024602 (2006).
[10] L. D. Faddeev, Zh. Eksp. Theor. Fiz. **39**, 1459 (1960) [Sov. Phys. JETP **12**, 1014 (1961)].
[11] E. O. Alt, P. Grassberger, and W. Sandhas, Nucl. Phys. B **2**, 167 (1967).
[12] T. Matsumoto, E. Hiyama, K. Ogata, Y. Iseri, M. Kamimura, S. Chiba, and M. Yahiro, Phys. Rev. C **70**, 061601(R) (2004).

[13] T. Matsumoto, T. Egami, K. Ogata, Y. Iseri, M. Kamimura, and M. Yahiro, Phys. Rev. C **73**, 051602(R) (2006).
[14] M. Rodríguez-Gallardo, J. M. Arias, J. Gómez-Camacho, R. C. Johnson, A. M. Moro, I. J. Thompson, and J. A. Tostevin, Phys. Rev. C **77**, 064609 (2008).
[15] A. M. Moro and J. A. Lay, Phys. Rev. Lett. **109**, 232502 (2012).
[16] J. A. Lay, A. M. Moro, J. M. Arias, and Y. Kanada-En'yo, Phys. Rev. C **89**, 014333 (2014).
[17] E. Epelbaum, W. Glöckle, and Ulf-G. Meißner, Nucl. Phys. A **747**, 362 (2005).
[18] E. Epelbaum, H.-W. Hammer, and Ulf-G. Meißner, Rev. Mod. Phys. **81**, 1773 (2009).
[19] M. Kohno, Phys. Rev. C **88**, 064005 (2013).
[20] M. Kohno, Prog. Theor. Exp. Phys. **2015**, 123D02 (2015).
[21] M. Toyokawa, M. Yahiro, T. Matsumoto, K. Minomo, K. Ogata, and M. Kohno, Phys. Rev. C **92**, 024618 (2015).
[22] H. Sakaguchi *et al.*, Phys. Lett. B **89**, 40 (1979).
[23] H. Sakaguchi *et al.*, Phys. Lett. B **99**, 92 (1981).
[24] K. Minomo, M. Toyokawa, M. Kohno, and M. Yahiro, Phys. Rev. C **90**, 051601(R) (2014).
[25] K. Minomo, M. Kohno, and K. Ogata, Phys. Rev. C, in press [arXiv1509.04459].
[26] Y. Kikuchi, T. Matsumoto, K. Minomo, and K. Ogata, Phys. Rev. C **88**, 021602(R) (2013).
[27] Y. Kikuchi, T. Myo, M. Takashina, K. Katō, and K. Ikeda: Prog. Theor. Phys. **122**, 499 (2009).
[28] Y. Kikuchi, K. Katō, T. Myo, M. Takashina, and K. Ikeda: Phys. Rev. C **81**, 044308 (2010).
[29] D. Beaumel *et al.*, Phys. Lett. B **514**, 226 (2001).
[30] A. M. Mukhamedzhanov and N. K. Timofeyuk, Yad. Fiz. **51**, 679 (1990) [Sov. J. Nucl. Phys. **51**, 431 (1990)].
[31] T. Fukui, K. Ogata, and M. Yahiro, Phys. Rev. C **91**, 014604 (2014).
[32] T. Fukui, K. Ogata, K. Minomo, and M. Yahiro, Phys. Rev. C **86**, 022801(R) (2012).
[33] K. Mizuyama, G. Colò, and E. Vigezzi, Phys. Rev. C **86**, 034318 (2012).
[34] J. M. Mueller *et al.*, Phys. Rev. C **83**, 064605 (2011).
[35] M. H. Mahzoon, R. J. Charity, W. H. Dickhoff, H. Dussan, and S. J. Waldecker, Phys. Rev. Lett. **112**, 162503 (2014).
[36] A. Gade *et al.*, Phys. Rev. C **77**, 044306 (2008).
[37] J. A. Tostevin and A. Gade, Phys. Rev. C **90**, 057602 (2014).
[38] J. Lee *et al.*, Phys. Rev. C **83**, 057602 (2014).
[39] F. Flavigny *et al.*, Phys. Rev. Lett. **110**, 122503 (2013).
[40] F. Flavigny *et al.*, Phys. Rev. Lett. **108**, 252501 (2012).
[41] R. J. Glauber, *in Lectures in Theoretical Physics* (Interscience, New York, 1959), Vol. 1, p.315.
[42] K. Hencken, G. Bertsch, and H. Esbensen, Phys. Rev. C **54**, 3043 (1996).
[43] K. Ogata, K. Yoshida, and K. Minomo, Phys. Rev. C **92**, 034615 (2015).
[44] K. Mizuyama and K. Ogata, Phys. Rev. C **86**, 041603(R) (2012).
[45] G. M. Hale *et al.*, ENDF/B-VI, 10, 825 (2001).
[46] K. Mizuyama and K. Ogata, Phys. Rev. C **89**, 034620 (2014).
[47] Y. L. Luo and M. Kawai, Phys. Rev. C **43**, 2367 (1991).
[48] Y. Watanabe, R. Kuwata, Sun Weili, M. Higashi, H. Shinohara, M. Kohno, K. Ogata, and M. Kawai, Phys. Rev. C **59**, 2136 (1999).
[49] Sun Weili, Y. Watanabe, M. Kohno, K. Ogata, and M. Kawai, Phys. Rev. C **60**, 064605 (1999).
[50] K. Ogata, Y. Watanabe, Sun Weili, M. kohno, and M. Kawai, Nucl. Phys. A **703**, 152 (2002).

DOI: 10.1051/epjconf/201612206003

[51] K. Ogata, Y. Watanabe, Sun Weili, M. Kohno, and M. Kawai, *Proc. of the Kyudai-RCNP international mini symposium on nuclear many-body and medium effects in nuclear interations and reactions (MEDIUM02)* (Kyushu University Press, 2003) p. 231.

Spatial measure of reaction size in proton scattering

Makoto Ito[1,2,a], Masataka Iwasaki[1], Reiji Otani[1], and Masashi Tomita[1]

[1] *Department of Pure and Applied Physics, Kansai University, Yamatecho, 3-3-35, Suita, Japan*
[2] *Research center for Nuclear Phtsics (RCNP), Osaka University, Mihogaoka 10-1, Suita 567-0047, Japan*

Abstract. The microscopic coupled-channel (MCC) calculations for proton + ^{12}C inelastic scattering are performed in the energy range of E_p = 29.95 MeV to 65 MeV. The nuclear interactions for the proton $-^{12}$C system are constructed from the folding model, which employs the internal wave function of ^{12}C, obtained from the 3α resonating group method (3α RGM), and an effective nucleon-nucleon interaction of the density-dependent Michigan three-range Yukawa (DDM3Y). The MCC calculation with the 3α RGM + DDM3Y nicely reproduces all of the differential cross sections for elastic and inelastic scattering in the angular range of $\theta_{c.m.}$ = 30° to 120°. We introduce a scattering radius, which characterizes a spatial size of the scattering area, from partial wave decompositions of an angle-integrated cross section. The scattering radii for the elastic scattering and the various inelastic channels, which involve the rotational or vibrational excitations and the 3α excitations in ^{12}C, are derived. We found that the scattering radii for the inelastic channels with a well developed 3α structure are strongly enhanced in comparison to the scattering radii for the elastic and collective channels. This enhancement of the scattering radius for the 3α channel strongly suggests that the scattering radius is sensitive to a size of the intrinsic structure of the finally excited state in the scattering process.

1 Introduction

Clustering phenomena, in which a nucleus is decomposed into some subunits, are well known to appear in the excited states of the lighter mass system [1]. In ^{12}C, for instance, three α particles are weakly coupled in the excited $0_2{}^+$ state, called Hoyle state, at E_x = 7.65 MeV [2, 3], while they are strongly overlapped to each other in its ground $0_1{}^+$ state. One of characteristic features in cluster states is an extension of a matter radius. Previous theoretical calculations suggest that a radius of a cluster state is larger by about 50% than a radius of a ground state [1–3].

Unfortunately, a size of a cluster state is difficult to measure directly by experiments because a cluster state has a very short life-time. In a naive expectation, however, such a spatial extension may be reflected in observables of scattering phenomena. In recent studies, there are several attempts to get the sign of a spatial extension of a cluster structure, especially in the reactions exciting the 3α state in ^{12}C [4, 5]. In Ref. [5], for instance, a diffraction radius of the Hoyle state in ^{12}C is speculated from the diffraction pattern observed in the differential cross section in the high energy proton scattering. In fact, a sign of an enhanced radius is confirmed for the Hoyle state of ^{12}C.

[a]e-mail: itomk@kansai-u.ac.jp

Recently, we have proposed the method of the scattering radius on the basis of a partial wave decomposition of a coupled-channel problem [6]. The scattering radius is considered to characterize a spatial size of the reaction area for exclusive reactions. We can always derive the scattering radius just by doing the partial wave analysis of a scattering cross section. For the first time this method has been applied to a simple two channel problem in the proton (p) + ^{12}C scattering going to the Hoyle state [6], where the prominent enhancement of the scattering radius in the Hoyle channel has been confirmed.

In this report, the method of the scattering radius is extended to the multi-channel problem in the p + ^{12}C scattering, going to the various inelastic channels that contain the collective excitations of the 2_1^+ and 3_1^- states and the 3α excitations of the 0_2^+ and 2_2^+ states [7]. In the recent studies on ^{12}C, the revised structure calculation has been performed, and the matter radius of the 3α states has been re-investigated [3]. According to the results, the matter radius of the $2_2{}^+$ state is predicted to be strongly enhanced because the $2_2{}^+$ state corresponds to the rotational excited states of the Hoyle 0_2^+ state [3]. Thus, the scattering radius for the $2_2{}^+$ state is especially interesting in connection to the enhanced matter radius of this state. We analyze the scattering radii for various exit channels in connection to the respective excited states of the ^{12}C nucleus.

There is another reason why we focus on the p + ^{12}C scattering. Nucleon scattering from carbon is very important in the fields of nuclear engineering [8] and medical technology [9]. There are several coupled-channel (CC) calculations of the p + ^{12}C inelastic scattering at the energy of $E_p \leq 65$ MeV [10–12]. However, in these CC calculations, the nuclear potentials are treated in a phenomenological manner, and there is no approach of the full microscopic coupled-channels calculation, in which the nuclear interactions are calculated from a microscopic wave function of ^{12}C based on the nucleon degrees of freedom and a reliable nucleon-nucleon (NN) interaction.

In our coupled-channel calculations, which is called the microscopic coupled-channel (MCC) calculation [6, 7, 13, 14], the nuclear interactions of p + ^{12}C are constructed in the full microscopic manner, the folding model. In MCC, we use the internal wave function of ^{12}C calculated from the 3α resonating group method (RGM) [2], while the density-dependent version of the Michigan 3 range Yukawa (DDM3Y) interaction [15–17] are employed for the effective NN interaction. The DDM3Y interaction was derived not in a phenomenological manner but from the microscopic calculation of infinite nuclear matter [17]. The microscopic interactions with 3α RGM + DDM3Y constitute the MCC calculation and hence, the MCC improves the shortcomings of the previous CC calculations, in which the phenomenological interactions are used. We apply the MCC calculation to the p + ^{12}C inelastic scattering in the range of $E_p = 29.95$ MeV to 65 MeV and check the reproducibility of the differential cross section. From the MCC calculation, we derive the scattering radii for the various exit channels and analyze their property in connection to the intrinsic structure of the ^{12}C nucleus.

2 Theoretical framework

2.1 Coupled-channel equations

We solve a set of the coupled channel (CC) equations for the proton-^{12}C system, which is given in the symbolic form

$$\left[T_f(\mathbf{R}) + V_{f,f}(\mathbf{R}) - E_f\right]\chi_f(\mathbf{R}) = -\sum_{i\neq f} V_{f,i}(\mathbf{R})\chi_i(\mathbf{R})\,. \tag{1}$$

Here, the subscripts f and i design a channel. $T_f(\mathbf{R})$ represents the kinetic energy of the relative motion of the p + ^{12}C system with a relative coordinate $\mathbf{R}$, while $V_{f,i}(\mathbf{R})$ denotes the coupling potential for the transition from channel i to channel f. E_f represents the total energy in channel f, which is given by the relation of $E_f = E - \epsilon_f$ with a proton's incident energy E and an internal energy of ^{12}C, ϵ_f. $\chi_f(\mathbf{R})$ is the proton-^{12}C relative wave function for channel f, which should be solved in the CC equation. In the present calculation, we include all discrete-excited states of ^{12}C in addition to its ground 0_1^+ state: the rotational state of 2_1^+ (4.44 MeV), the vibrational 3_1^- state (9.64 MeV), the 3α cluster states of 0_2^+ (7.65 MeV), and 2_2^+ (10.3 MeV).

In the pragmatic CC calculation, partial wave expansion with the channel-spin representation is applied to the total wave function, and the radial part of the relative wave function, such as $\chi^J_{(sS_a)IL}(R)$, is solved [13]. Here, the radial part of the relative wave function is classified in terms of the proton spin $\mathbf{s}$, the internal spin of ^{12}C, $\mathbf{S}_a$, the channel spin $\mathbf{I}$ (= $\mathbf{s} + \mathbf{S}_a$), the relative spin $\mathbf{L}$, and the total spin $\mathbf{J}$ (= $\mathbf{I} + \mathbf{L}$). Here the subscript of a represents a suffix used to distinguish the internal state of ^{12}C with the same spin ($S_a = 0_1^+, 0_2^+$, and so on). A set of the quantum numbers of $(sS_a)I$ is a channel, while that of $(sS_a)ILJ$ is usually called the sub-channel. In the present calculation, the proton spin is treated as the spin-less particle and hence, the sub-channel representation is reduced to $(sS_a)ILJ$ ($s = 0$ and $S_a = I$) $\rightarrow S_aLJ$. The CC equation is solved on the basis of the S_aLJ representation.

2.2 Nuclear coupling potential

The coupling potential ($V_{f,i}(\mathbf{R})$) has a component of the nuclear (N) and Coulomb (C) parts, such as

$$V_{f,i}(\mathbf{R}) = V^{(N)}_{f,i}(\mathbf{R}) + V^{(C)}_{f,i}(\mathbf{R}) \ . \tag{2}$$

The nuclear potential is composed of the real ($V_{f,i}(\mathbf{R})$) and phenomenological imaginary potentials ($-iU_f(R)$),

$$V^{(N)}_{f,i}(\mathbf{R}) = \tilde{V}_{f,i}(\mathbf{R}) - iU_f(R)\cdot\delta_{f,i} \ . \tag{3}$$

The real part of the nuclear potential ($\tilde{V}_{f,i}(\mathbf{R})$) is calculated by the folding model [13, 15, 16] which is expressed symbolically as

$$\tilde{V}_{f,i}(\mathbf{R}) = \int \rho_{f,i}(\mathbf{r}) v^{DDM3Y}_{NN}(|\mathbf{r}-\mathbf{R}|)\, d\mathbf{r} \ , \tag{4}$$

where $\mathbf{r}$ denotes the coordinate measured from the center of the mass in the ^{12}C nucleus. Here, $\rho_{f,i}(\mathbf{r})$ represents the diagonal ($f = i$) or transition ($f \neq i$) densities of ^{12}C, which is calculated by the resonating group method (RGM) [2] of the 3α cluster model. In Eq. (4), v_{NN} represents the effective nucleon-nucleon (NN) interaction, which acts between a nucleon contained in ^{12}C and an incident proton. The Coulomb interaction in Eq. (2) is calculated by assuming the uniform charge potential with a radius of $R_C = 1.2A^{1/3}$ (fm), and no Coulomb coupling is included in the off-diagonal transition.

In the present work, the density-dependent Michigan three-range Yukawa (DDM3Y) is used for the effective NN interaction [15]. The explicit form of the DDM3Y interaction is given by

$$v^{DDM3Y}_{NN}(s) = \gamma\left(1 + \alpha e^{-\beta\rho(s)}\right) v^{M3Y}_{NN}(s) \ , \tag{5}$$

with the M3Y interaction of

$$v^{M3Y}_{NN}(s) = \frac{7999}{4}\frac{e^{-4s}}{4s} - \frac{2134}{2.5}\frac{e^{-2.5s}}{2.5s} - J_0\delta(\mathbf{s}) \ . \tag{6}$$

In Eq. (5), the parameters of α, β and γ depend on incident energy per nucleon [15]. Eq. (6) corresponds to the NN interaction for the spin (S) and isospin (T) singlet channel ($S = T = 0$). Thus, there is no contribution from the long-range attractive potential generated by the one-pion exchanges.

The third term simulates the single-nucleon knock-on exchange with the energy-independent strength of $J_0 = -262$ MeV fm^3 [16]. In the present CC calculation, no modification is done for the folding potentials although a normalization factor is often introduced in the folding potential for the nucleus-nucleus scattering [18, 19].

The second term in Eq. (3) represents the imaginary potential, which is included only in the diagonal transition ($f = i$). The imaginary potential is composed of two parts, such as

$$U_f(R) = W_{Vf}(R) + W_{Sf}(R) . \tag{7}$$

Here, $W_{Vf}(R)$ and $W_{Sf}(R)$ are the Woods-Saxon (WS) potentials with volume-type and surface-type form factors, respectively. Three parameters, such as strength (W_{0f}), radius (R_f), and diffuseness (a_f), are included in each of the WS potentials. These six parameters are channel dependent, and they are optimized so as to reproduce the observed differential cross sections as much as possible. In the present calculation, we neglect the imaginary potentials in the off-diagonal transition for simplicity. All the parameters used in the present MCC calculation are listed in Ref. [7]

2.3 Definition of the scattering radius

In Eq. (1), the coupling problem among the sub-channels, such as $S_aLJ \to S'_{a'}L'J$ is solved. The angle-integrated cross section can be decomposed into the partial wave (or sub-channel) component, $\sigma(LL'S_a)$, which is designed by the initial orbital spin L, the final spin of a target S_a, and the final orbital spin L' for the scattering two particles. Since the initial state of the ^{12}C target is in the spin-less state ($S_a = 0^+_1$), the initial orbital spin L is equal to the total spin J ($L = J$).

In order to characterize a size of scattering area to produce the final channel, we simply define the effective orbital spin $\bar{L}$ [6] for the final spin state of (S_a) from the partial cross section $\sigma(LL'S_a)$ like

$$\bar{L}(S_a) = \sqrt{\frac{\sum_{LL'} \tilde{L}^4 \sigma(LL'S_a)}{\sum_{LL'} \tilde{L}^2 \sigma(LL'S_a)}} \tag{8}$$

with $\tilde{L} = \sqrt{L(L+1)}$ which is calculated from the initial orbital spin L. The effective orbital spin is obtained by the fourth and second powers of the magnitude of the orbital spin, $\tilde{L} = \sqrt{L(L+1)}$. In Eq.(8), we simply assume the same functional form as the root-mean-squared radius for the matter density distribution [6]. The scattering radius R_{SC} corresponding to $\bar{L}$ can be simply obtained from the relation of $\bar{L} = kR_{SC}$, where k denotes the wave number of the incident channel, measured in the laboratory system.

The magnitude of R_{SC} depends on the definition of $\bar{L}$ in Eq. (8), and there are other possibilities to define the effective orbital spin. The variation of $\bar{L}$ has been discussed in Ref. [6], and we have pointed out that the definition of Eq. (8) gives a reasonable size for the elastic scattering, which is comparable to the matter radius of the ground state of ^{12}C. Thus, in the present analysis, we also use Eq. (8) for the definition of the effective orbital spin $\bar{L}$ although other possibilities of the functional form should be considered for $\bar{L}$. Other possibility in the definition of $\bar{L}$ will be discussed in a forthcoming article.

3 Results

3.1 Differential cross sections

We show the results of the full MCC calculation with the nuclear interaction in Eqs. (3) and (4). The left and right panels in figure 1 show the results of the differential cross sections of the elastic (0^+_1)

and 0_2^+ channels, respectively [7]. The comparison of the CC calculations (solid curves) with the experiments (asterisks) is done in the energy range of E_p = 29.95 MeV to 65 MeV.

The CC calculation for the elastic (0_1^+) channels nicely reproduces the overall features of the observed differential cross sections except for the backward angles of $\theta_{c.m.} \geq 120°$ at the energies of $E_p \leq 39.95$ MeV. In the MCC calculation, a prominent valley appears at the angle of $\theta_{c.m.} = 130°$, and the calculated cross section quickly increased beyond 130° in comparison to the experimental data. We have also calculated the differential cross section of the inelastic scattering to the 2_1^+ and 3_1^-

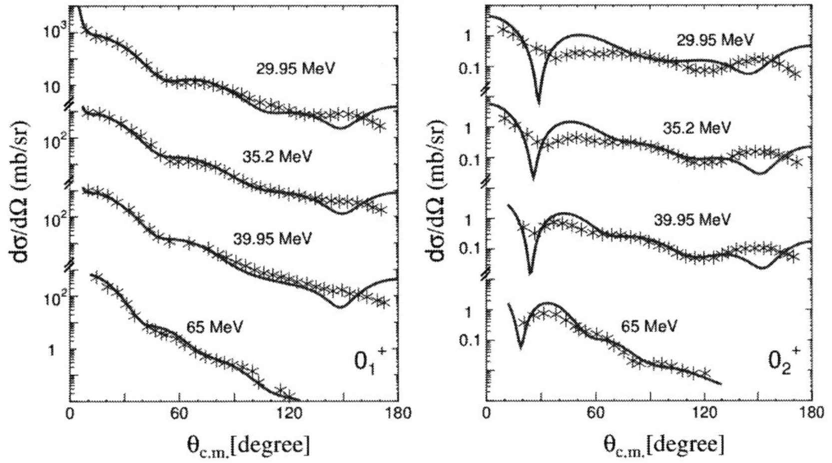

Figure 1. Left panel: Differential cross section of the elastic (0_1^+) scattering in the range of E_{lab} = 29.95 MeV to 65 MeV. The asterisks and the curves represent the experimental data and the theoretical calculations, respectively. Right panel: Same as the left panel except for the inelastic 0_2^+ channel. These figures are taken from Ref. [7].

states, which are the member of the rotational and vibrational excited states from the ground state, respectively, in the same energy range [7]. In these two channels, the angular distributions are nicely reproduced up to $\theta_{c.m.} \leq 130°$ although the calculation for the 3_1^- channel considerably underestimates the experimental cross section at the forward angular range of $\theta_{c.m.} \leq 30°$. The valley-peak structures confirmed in the CC calculation for the 0_1^+ channel are also observed in the these two channels, and the CC calculation is out of phase around $\theta_{c.m.} = 130°$. Therefore, a failure at the backward scattering beyond $\theta_{c.m.} = 130°$, which is the out of phase in the valley-peak structure, is common in all of the collective channels, such as 0_1^+, 2_1^+, and 3_1^-.

The valley-peak structures at the backward angles of $\theta_{c.m.}$ =130° were also observed in the previous CC calculation based on the soft-rotator model [10]. The theoretical picture of ^{12}C in our calculation, which is based on the composite system of the 3α particles, is very different from the approach in Ref. [10], which employs the one-body-like collective motion of ^{12}C. However, the shortcomings in calculating the angular distribution seem to be common in these two models.

The results of the CC calculation for the 0_2^+ state, which has the well-developed 3α structures, are shown in the right panel in Fig. 1 [7]. In all the energy ranges, the theoretical calculations reasonably reproduce the global features of the observed cross sections. However, the deviations of the CC

DOI: 10.1051/epjconf/201612206004

calculation from the experimental data are more prominent in comparison to the 0_1^+ channels. At the backward angles of $\theta_{c.m.} \sim 150°$ the CC calculation is out of phase in comparison to the experimental data. This out of phase is the same feature as the collective channels. Moreover, a sharp valley appears in the CC calculation at the angle of $\theta_{c.m.} = 30°$, and this valley is deviated from the observed data except for the incident energy of $E_p = 65$ MeV.

The channel coupling effect on the 0_2^+ channel is analyzed [7]. The direct coupling of $0_1^+ \to 0_2^+$ is quite weak because the intrinsic structures are very different with each other. However, the coupling of the 0_2^+ and 2_2^+ channels is very strong because the the 2_2^+ state corresponds to the rotational state of the 0_2^+ state [2, 3], which is a weakly coupled state of the 3α particles. Therefore, the angular distribution of the 0_2^+ channel is strongly affected by the coupling to the 2_2^+ channel [7, 14].

The present CC calculations are quite successful in reproducing the differential cross sections of all the final channels in the angular range of $\theta_{c.m.} = 30° \sim 120°$ although the fit to the experimental data depends considerably on the exit channels. There is a common feature in the deviation between the theory and the experiment: the out of phase structure at the backward angle of $\theta_{c.m.} \geq 130°$. However, the overall feature of the angular distributions is reasonably reproduced in the whole angular range.

3.2 Partial cross sections and scattering radii at E_p = 65 MeV

From the CC calculation in Fig. 2, we can obtain the partial cross section, which is explained in Sec. 2.3. In Fig.2, we show the distributions of the partial cross section $\sigma(L)$ for the spin-less channels of the 0_1^+ (asterisks) and 0_2^+ (squares) at $E_p = 65$ MeV. In the case of the spin-less final state, the partial cross section in Eq. (8) is labeled by the incident orbital spin L (or total spin J) because of $S_a = 0$ and $L = L' = J$. In Fig. 2, only the nuclear part of the partial cross sections is plotted, and the Coulomb part is subtracted in the elastic 0_1^+ channel. The magnitude of the partial cross sections is normalized by their total summation of $\sigma_{tot.} = \sum_L \sigma(L)$, such as $\sigma(L)/\sigma_{tot.}$.

In the partial cross sections of these two exit channels, we can observe the different shape: the single peak structure in the elastic channel (asterisks) and the double peaks structure in the 3α channel (solid squares). This double peak structure is originated from the feature of the monopole coupling potential of $0_1^+ \to 0_2^+$. The monopole coupling potential changes its sign at a certain $p-{}^{12}C$ distance and hence, the coupling potential has the one node in the spatial distribution. This nodal behavior is because of the condition of the volume (or nucleon number) conservation. Thus, the double peak structure is considered to mainly reflect the distribution of the coupling potential, while the distribution of the elastic scattering is determined from the diagonal potential in the elastic $p + {}^{12}C_{g.s.}$ channel.

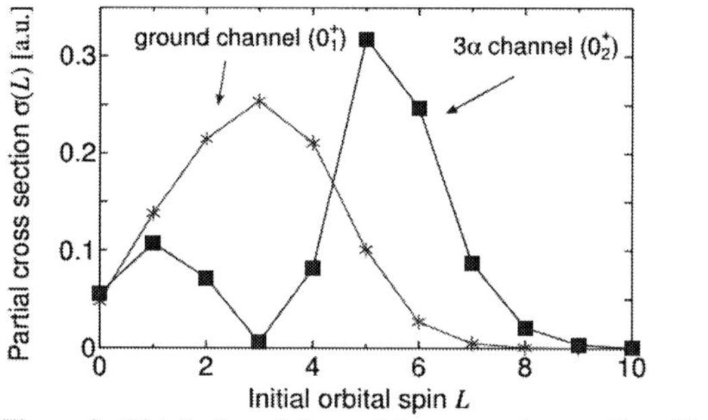

Figure 2. Distribution of the partial cross sections at E_{lab}=65 MeV. The abscissa and the ordinate show the initial orbital spin L and the respective partial cross section $\sigma(L)$, respectively. Both of the partial cross sections are normalized by the total cross section, $\sigma_{tot.} = \sum_L \sigma(L)$.

In this figure, moreover, we can clearly observe the partial cross section of the 0_2^+ channel has more extended distribution than that of the elastic channel. The peak position in the elastic channel exists around $L = 3$, while the higher peak in the 0_2^+ channel locates at $L = 5$. This extension strongly

DOI: 10.1051/epjconf/201612206004

suggests that a size of the scattering area to produce the 0_2^+ channel is larger than that for the spatial area where the elastic scattering occurs.

A characteristic distributions of the partial cross sections can be observed in other inelastic channels. The normalized partial cross sections for the 2_1^+ and 2_2^+ channels, which correspond to the rotational excited channels of the 0_1^+ and 0_2^+ channels, respectively, are shown in Fig. 3. Here the definition of $\sigma(L)$ is $\sigma(L) \equiv \sum_{L'} \sigma(LL'S_a)/\sigma_{tot.}$ and $\sigma_{tot.} = \sum_{LL'} \sigma(LL'S_a)$. Both of the distributions have a single peak structure, and the peak position of the 2_2^+ channel locates at $L = 6$, which is shifted to the high L side in comparison to the peak position of the 2_1^+ channel ($L = 4$). The distribution of the partial cross sections for the 3_1^- channel also spreads in the higher L region with a peak at $L = 5$. although its distribution is not shown in the figure.

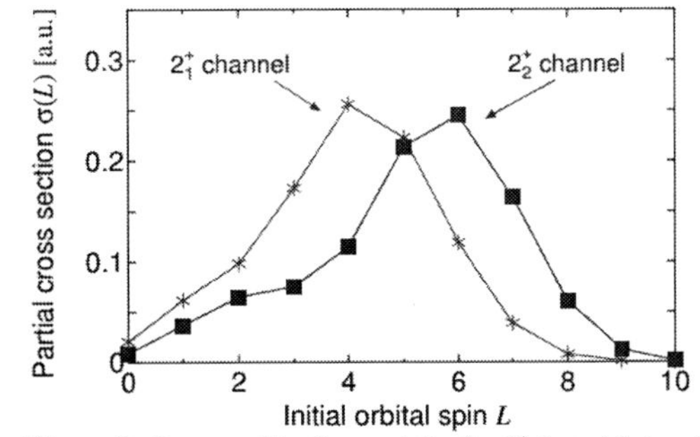

Figure 3. Same as Fig. 3 except for the 2_1^+ (asterisks) and 2_2^+ (squares) channels.

The partial cross sections in Figs. 2 and 3 suggest that the spatial size of the reaction area depends on the final channel. In order to characterize the spatial size of each reaction channel more clearly, the effective orbital spins $\bar{L}$ and the scattering radii, R_{SC} are calculated according to the definition of Eq. (8). The obtained effective orbital spins $\bar{L}$ and the scattering radii R_{SC}, are summarized in table 1. In this table, the matter radii (root mean squared radii, $\bar{r}$) of all the states [2] are also shown in the bottom row for comparison.

	0_1^+	0_2^+	2_1^+	2_2^+	3_1^-
$\bar{L}$	4.6	6.4	5.7	6.8	6.0
R_{sc} [fm]	2.6	3.6	3.2	3.9	3.4
$\bar{r}$ [fm]	2.4	3.5	2.4	4.0	2.8

Table 1. Effective orbital spin $\bar{L}$ and the scattering radius R_{sc} at E_{lab}=65 MeV. The matter radius $\bar{r}$ of the 0_1^+, 0_2^+, 2_1^+, 2_2^+ and 3_1^- states are also shown in the bottom column [2]. R_{sc} and $\bar{r}$ are shown in units of fm.

From table 1, we can see the enhancement of R_{SC} for the inelastic scattering going to the 0_2^+, 2_2^+ and $3_1{}^-$ states, which are the resonance states just above the 3α threshold, in comparison to the scattering radius of the 0_1^+ and 2_1^+ states. The enhancement of the scattering radii in the 0_2^+ and 2_2^+ channels can be attributed to the spatial extension of the 3α structures of these states.

On the contrary, there is a considerable enhancement of the scattering radius in the 3_1^- channel (R_{SC} = 3.4 fm), which is comparable to the scattering radius of the 0_2^+ state (R_{SC} = 3.5 fm), although this state does not have the 3α-cluster structure but the spatially compact structure [1, 2]. The 3_1^- state is a resonant state existing above the 3α threshold of $E_{3\alpha}$=7.27 MeV (E_x=9.6 MeV). Therefore, the 3_1^- state has an extended tail in its radial wave function. This extension is considered to be one of the origins for the enhanced scattering radius of the 3_1^- state.

In particular, the R_{SC} of the 2_2^+ is strongly enhanced, and this enhancement is originated from the peak shift to the higher L region in the respective partial cross section, as shown in Fig. 3. The 2_2^+ state is the 3α rotational state, which has much more extended matter radius, and this large radius has recently been pointed out by Funaki et al. [3]. The scattering radius does not necessarily mean the matter radius of a target nucleus itself but our scattering radius is consistent with the matter radius of the 2_2^+ state. Thus, the scattering radius derived from the partial cross section can be used for the measure of the spatial size, which reflects the intrinsic structure of the final channel.

3.3 Energy systematics of the scattering radii

Finally, we discuss the incident energy dependence of the scattering radii R_{SC}, which is shown in Fig. 4. The results in Fig. 4 represent that the scattering radius does not depend so much on the incident energy. Therefore, the qualitative features of the scattering radius are almost invariant with respective to the variation of the incident energy. In all the energy range considered here, the scattering radius of the 2_2^+ state (triangles) is the largest of all the channels, while the enhancements are commonly observed in the 0_2^+ (squares) and 3_1^- (open circles) channels. In contrast to these enhanced channels, the scattering radii of the ground 0_1^+ (asterisk) and its rotational state of 2_1^+ (solid circles) are clearly suppressed. The magnitude correlation in the scattering radii observed in Fig. 4 is consistent with the matter radii predicted in the 3α cluster model shown in table 1 although the order of $0_2{}^+$ and 3_1^- is inversed only at E_{lab}=29.95 MeV .

The scattering radii shown in Fig. 4 contain the errors, which arise from the theoretical ambiguity of the potential parameters. As shown in Fig. 1, the reproduction of the angular distributions is restricted in the angular range of 30 $^\circ \leq \theta_{c.m.} \leq 120^\circ$, and a simultaneous reproduction over all the channels is a little difficult in a common parameter set of the imaginary potential and the strength of the folding potential. The scattering radii basically depend on the potential parameters which are determined from fitting the theoretical calculation to the experimental data.

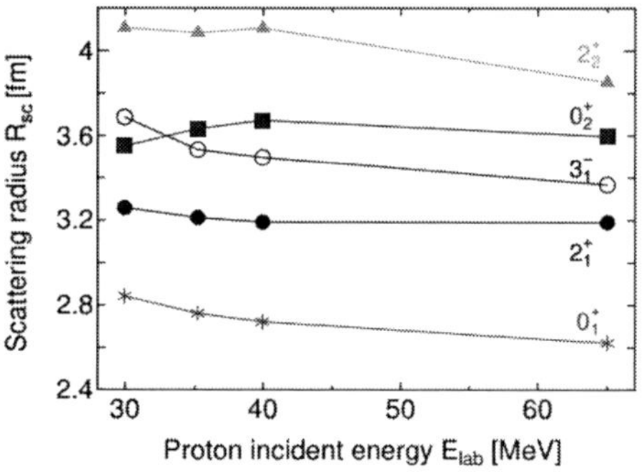

Figure 4. The incident dependence of the scattering radius. Asterisks, solid circles, open circles, sollid squares and solid triangles are the results of the 0_1^+, 2_1^+, 3_1^-, 0_2^+ and 2_2^+ channels, respectively.

In order to estimate the ambiguity of the scattering radii, which is caused by the several choice of the parameter sets (or the reproducibility of the experimental angular distributions), we have searched for the several parameter sets of the potential, which can reproduce each of exit channels separately. For example, one parameter set is optimized for the reproduction of the elastic channel, while the other parameter sets can reproduce only the Hoyle 0_2^+ channel, and so on.

From the searched potential parameter sets, we can determine the various scattering radii, and the standard deviations from the averaged radii are evaluated as the errors contained in the theoretical calculation. The resultant magnitude of the error reaches about ±10%, at most, in all the exit channels although this error bars may be slightly changed in the future parameter search. Thus, the qualitative features of the scattering radius are invariant even if much more tuned potentials in comparison to the present calculation are employed in the CC calculation.

4 Summery and discussion

We have performed the microscopic coupled-channel (MCC) calculation in the proton + ^{12}C scattering in the energy range of E_p = 29.95 MeV to 65 MeV. In the framework of MCC, the real part of the nuclear potentials is derived from the precise wave function of ^{12}C, which is calculated by the 3α resonating group method (3α RGM), and the reliable nucleon-nucleon (NN) interaction, such as the Density-Dependent Michigan 3-range Yukawa (DDM3Y). In addition to the real potential, we have introduced an imaginary potential with the form factor of the Woods-Saxon potential for the diagonal coupling, and the parameters in the imaginary potentials are optimized to reproduce the differential cross sections.

The MCC calculation with 3α RGM + DDM3Y can nicely reproduce the observed angular distributions in the angular region of $\theta_{c.m.}$ = 30° to 120°. The fit to the experimental data in the present MCC calculation is superior to the previous coupled-channel calculations that were based on the phenomenological treatment in the coupling potential [10–12]. Reproduction of the experiments by the microscopically derived nuclear interactions is an important development in the theoretical calculations. However, the reproduction of the detailed structures is still insufficient especially in the backward ($\theta_{c.m.} \geq 120°$) and forward scattering ($\theta_{c.m.} \leq 30°$). In the backward scattering, the theoretical calculations for all the exit channels commonly generate the valley-peak structures, which are out of phase in the oscillating pattern of the angular distribution by comparing with the experimental data. Since the proton scattered to the backward angle strongly feels the interior density of the ^{12}C target, the modification to the interior part of the transition density may be important for the improvement of the fit to the backward scattering.

We have also defined the effective orbital spin $\bar{L}$ from the partial wave components of the angle-integrated differential cross sections. The scattering radius, which is derived from $\bar{L}$, is considered to correspond to the spatial size of the reaction area, which produces the final state in the exclusive reaction. In principle, the scattering radius can always be defined when the partial wave analysis is possible for the differential cross section. In a recent study, effective radii in the deuteron induced reaction are derived from the continuum-discretized coupled-channel, and radii for the inclusive reactions are clearly defined [20]. In contrast to this approach, the present method is useful for the definition of the radius in an exclusive reaction, going to a specific channel.

The method of the scattering radius has been applied to the present MCC calculation for the p + ^{12}C scattering, and the scattering radii of the elastic and inelastic scattering to 0_2^+, 2_1^+, 2_2^+ and 3_1^- states have been derived. We have found a clear enhancement of the scattering radius for the final inelastic channels in comparison to the elastic channel. In particular, we confirmed the enhanced scattering radius in the 3α cluster channels (0_2^+ and 2_2^+). This enhanced scattering radius for the cluster channels is attributed to the spatial extension of the 3α structure.

It is very important to clarify the relation between the scattering radii and the respective matter radii, especially in the inelastic scattering to the Hoyle 0_2^+ state. In Refs. [21, 22], the sensitivity of the differential cross section to the size of the final state is discussed in the inelastic scattering of α + ^{12}C $\rightarrow \alpha$ + ^{12}C(0_2^+). According to the analysis, the oscillating pattern of the differential cross section is insensitive to the size of the 0_2^+ state [21], and the oscillation is mainly determined from the size of the transition potential of $0_1^+ \rightarrow 0_2^+$ [22]. The authors concluded that the oscillation pattern in the differential cross section does not reflect a size of the final 3α structure itself.

The partial cross section of the $0_1^+ \rightarrow 0_2^+$ in the present analysis reveals the double peak structure, and these double peaks are originated from the nodal behavior of the monopole transition [6], in which the transition potential changes the sign at a certain $p - {}^{12}C$ distance. Thus, the global shape of the partial cross section is dominated by the transition potential, and this result is consistent with the results of Refs. [21, 22]. However, there is the possibility that the higher partial wave part in the partial cross section beyond the second peak is sensitive to the size of the final state. The sensitivity of the partial cross section to the size of the diagonal potential of the final 0_2^+ state must be carefully checked in future studies.

In the present coupled-channel calculation, we have treated only the $p + {}^{12}C$ scattering system. The systematic studies are also important to establish the validity of the method of the scattering radius. For example, the analysis of the α and ^{3}He scatterings by ${}^{12}C$ is important because the previous studies have discussed the evolution of the Airy structures in the angular distribution in connection to the enhanced radius of the Hoyle state [18, 19]. Furthermore, the ${}^{16}O$ nucleus is another typical example of the cluster structure. In this nucleus, the $\alpha + {}^{12}C$ [23, 24] and 4α [24] structures appear in the excited states. In particular, a recent study in Ref. [24] suggests that the $0_6{}^+$ state corresponds to the 4α dilute state, and it is expected to have the extended matter radius. Thus, the analysis of the inelastic scattering of proton and α particle by the ${}^{16}O$ nucleus is very interesting. Systematic studies are now underway.

References

[1] K. Ikeda et al., Prog. Theor. Phys. **68**, 1 (1980), and references therein.
[2] Y. Fukushima and M. Kamimura, Proceedings of the International Conference Nuclear Structure, J. Phys. Soc. Jpn. **44**, 225 (1977); M. Kamimura, Nucl. Phys. A **351**, 456 (1981).
[3] Y. Funaki et al., Phys. Rev. C **67**, 051306(R) (2003), and references therein.
[4] A. N. Danilov et al., Phys. Rev. C **80**, 054603 (2009).
[5] K. Iida, S. Koide, A. Kohama, K. Oyamatsu, Mod. Phys. Lett. A **27**, 1250020 (2012).
[6] M. Tomita, M. Iwasaki, R. Otani, and M. Ito, Phys. Rev. C **89**,034619 (2014).
[7] M. Tomita, M. Iwasaki, R. Otani, K. Horio, and M. Ito, Phys. Rev. C **92**, 024609 (2015).
[8] M. B. Chadwick, P. G. Yound et al. Ncul. Sci. and Eng. **131**, 293 (1999).
[9] M. B. Chadwick, P. M. DeLcua, Jr and R. C. Haight, Rad. Protect. Dos. Vol.**70**, 1 (1997).
[10] Efrem Sh. Sukhovitskii et al., Phys. A **640** 147 (1998).
[11] S. Kato et al., Phys. Rev. C **31** 1616 (1985).
[12] T. Nojiri and M. Kamimura and M. Sano, Prog. Theor. Phys. **66**, 1906 (1981).
[13] M Ito, Y. Hirabayashi, and Y. Sakuragi, Phys. Rev. C **66**,034307 (2002); M. Ito, Y. Sakuragi, and Y. Hirabayashi, Phys. Rev. C **63** 064303(2001).
[14] M. Ito, Y. Sakuragi, and Y. Hirabayashi, Eur. Phys. Jour. A **5**, 373 (1999).
[15] M. E. Brandan, G. R. Sarchler, Phys. Rep. **285**, 143 (1997).
[16] G. R. Satchler and W. G. Love, Phys. Rep. **55**, 183 (1979).
[17] G. Bertsch, J. Borysowicz, H. Macmanus and W. G. Love, Nucl. Phys. A **284**, 399 (1977).
[18] S. Ohkubo and Y. Hirabayashi, Phys. Rev. C **70**,041602 (2004).
[19] Sh. Hamada et al., Phys. Rev. C **87**, 024311 (2013).
[20] S. Hashimoto et al., Phys. Rev. C **83**, 054617 (2011).
[21] M. Takashina and Y. Sakuragi, Phys. Rev. C**74**, 054606 (2006).
[22] M. Takashina, Phys. Rev. C**78**, 014602 (2008).
[23] Y. Suzuki, Prog. Theor. Phys. **55**, 1751 (1976); ibid, Prog. Theor. Phys. **56**, 111 (1976).
[24] Y. Funaki et al., Phys. Rev. Lett. **101**, 082502 (2008).

Coupled-rearrangement-channels calculation of the three-body system under the absorbing boundary condition

M. Iwasaki[1], R. Otani[1], M. Ito[1,a], and M. Kamimura[2]

[1] *Department of Pure and Applied Physics, Kansai University, 3-3-35 Yamatecho Suita, Osaka, Japan*
[2] *RIKEN Nishina Center for Accelerator-based Science, RIKEN, Wako, Saitama, Japan*

Abstract. We formulate the absorbing boundary condition (ABC) in the coupled-rearrangement-channels variational method (CRCVM) for the three-body problem. The absorbing potential is introduced in the system of the identical three-bosons, on which the boson symmetry is explicitly imposed by considering the rearrangement channels. The resonance parameters and the strength of the monopole breakup are calculated by the CRCVM + ABC method, and the results are compared with the complex scaling method (CSM). We have found that the results of the ABC method are consistent with the CSM results. The effect of the boson symmetry, which is often neglected in the calculation of the triple α reactions, is also discussed.

1 Introduction

One of the current issues in nuclear physics is the structure of the borromean three-body systems, which mainly appear around Neutron- or Proton-drip lines and its dynamics of continuum states above the particle decay thresholds [1]. The borromean systems around the drip line are weakly binding systems. Since the particle thresholds of the borromean nuclei open at the lower excitation energy than the normal nuclei, a large part of the excited states is embedded in continuum region, which is observed through the breakup reaction, for instance. Therefore, it is very important to describe the structural changes of the borromean systems as well as its breakup dynamics in a consistent manner.

In a naive treatment of the unbound states, a scattering boundary condition should be imposed among the interacting particles, but imposing the scattering boundary condition for three particles is complicated in general. In particular, the exact treatment of the three-body (or few-body) scattering state requires special techniques for the long range interaction, such as the Coulomb force [2]. In order to avoid the complexity of the three-body scattering boundary condition, the scattering problem is often reduced to the bound-state-like problem by transforming the Hamiltonian of a total system into a non-Hermite form. There are two representative methods of the non-Hermitian transformations: the complex scaling method (CSM) [3–5] and the method of the absorbing boundary condition (ABC), or the method of the complex absorbing potential (CAP) [6–11].

In CSM, the radial coordinates and the respective momenta in the Hamiltonian of the total system is simply rotated with the angle of θ in the complex plane, such as $r \to re^{i\theta}$ and $k \to ke^{i\theta}$, while, in

[a] e-mail: itomk@kansai-u.ac.jp

ABC, the negative imaginary potential is placed at the outer region of the system. These two procedures commonly give rise to the damping of the wave function at the asymptotic region. Therefore, the three-body scattering problem can be transformed into the bound-state-like problem, in which the variational calculation using the basis expansion technique is available.

There are many sophisticated applications of CSM to the reaction problem of the three-body continuum states [3, 4]. For example, the CSM calculations are greatly successful in describing the Coulomb breakup into the three-body continuum, such as $^6\mathrm{He} \to \alpha + N + N$ [3]. In the CSM calculations, the rearrangement of the Jacobi coordinates among the three particles, is completely taken into account. The variational calculation with the rearrangement channels is called the coupled-rearrangement-channels variational method (CRCVM), and the CRCVM treatment is essential to obtain the rapid convergence of the binding energy in the three-body system [12].

The application of the ABC method to the three-body problem is still limited although the CSM + CRCVM calculation has already been established in the three-body problem. For example, there are the calculations of the ABC method in the schematic three-body system [7] and the nuclear breakup induced by the deuteron scattering [8], but no coordinate rearrangement is considered in these calculations. This shortcoming in the ABC calculation is because the formulation of the ABC method in CRCVM still remains unclear in contrast to the CSM treatment, in which all the coordinates and momenta are uniformly rotated by the common scaling angle.

In the ABC method, the negative imaginary potentials should included, which are the function of the relative coordinate of the decaying particles. In the single coordinate systems, such as two-body systems, the absorbing potential is uniquely determined, but, in the three-body system with the rearrangement channels, the way to introduce the absorber is not so definite. In order to establish the ABC formalism in CRCVM, we should carefully test the possible ways of the absorbing potential.

Since the application of CSM is difficult in some special problems [10], it is important to develop the ABC method in the CRCVM treatment. In the present report, we present the formulation of the CRCVM + ABC method in the identical three-boson system and demonstrate that the present formulation nicely works in the three-body problem with the rearrangement channels. Here we focus on the calculation of the resonance parameters and the breakup strength induced by the monopole external field. The results of the ABC method are compared with the CSM calculation, and the validity of the ABC method is discussed. Moreover, we investigate the effect of the boson symmetry on the strength function because the boson symmetry has often been neglected in the calculation of the reaction rate of the triple α reaction [13, 14]. In the precise evaluation of the reaction rate, the symmetrization effect may be crucial, and we check the validity of the neglect of the boson symmetry.

2 Framework

2.1 Three-body system with the absorbing boundary condition

In the present calculation, we handle the three-boson system, which has the following total Hamiltonian,

$$\hat{H} = -\frac{\hbar^2}{2\mu_{\mathbf{r}}}\nabla_{\mathbf{r}}^2 - \frac{\hbar^2}{2\mu_{\mathbf{R}}}\nabla_{\mathbf{R}}^2 + \sum_{i>j} V(s_{ij}), \tag{1}$$

where, the first two terms represent the kinetic energies with the Jacobi coordinate of $\mathbf{r}$ and $\mathbf{R}$, which denote the relative coordinate of the first and second particles (1–2) and the coordinate between the center of mass of 1+2 particles and the third particle ((1+2)–3), respectively. $\mu_{\mathbf{r}}$ and $\mu_{\mathbf{R}}$ represent the reduced masses respective to the Jacobi coordinate. In the present calculations, a kinematic factor is set to be $\hbar^2/m = 1/2$ (MeV·fm^2) for simplicity.

The last term denotes the two-body potential of the i–j particles with relative distance s_{ij}, which is assumed to be the Gaussian potential [7], such as

$$V(s) = -2e^{-0.16s^2} + e^{-0.04s^2}. \tag{2}$$

In the ABC method, the total Hamiltonian is modified to include the negative imaginary potential $-i\hat{W}$ with a strength η, such as

$$\hat{H} \rightarrow \hat{H}^\eta = \hat{H} - i\eta\hat{W}. \tag{3}$$

Here we should set the functional form of the absorbing potential $\hat{W}$. In a naive treatment, the absorber should be included in accordance to the Jacobi coordinate in each of the rearrangement channel, such as

$$\hat{W} = \sum_{c=1,2,3} \left(f(r_c - r_a) + f(R_c - R_a)\right). \tag{4}$$

with the form factor of the shifted polynomial function [6, 10],

$$f(r_c - r_a) = (r_c - r_a)^p \theta(r_c - r_a). \tag{5}$$

r_c and R_c represent the absolute values of the Jacobi vectors in the rearrangement channel of c, while r_a show the starting point of the absorber. In Eq. (4), the absorbing potentials are introduced along the Jacobi coordinate in the c-th rearrangement channels, and they are symmetrized on the channel suffix c. Another possible way in introducing the absorber is the imaginary potentials among the pair of two particles,

$$\hat{W} = \sum_{i>j} f(s_{ij} - r_a), \tag{6}$$

where s_{ij} denotes the magnitude of the relative coordinate of the i-th and j-th particles. In Eqs. (4) and (6), we set $p = 4$ and $r_a = R_a$ =14 fm, which are found to be an appropriate setting in the previous study of the two-body system [9–11].

We solve the Schrödinger equation for $\hat{H}^\eta$ in Eq. (3)

$$\hat{H}^\eta \Psi_n^\eta(r, R) = E_n(\eta)\Psi_n^\eta(r, R). \tag{7}$$

The wave function $\Psi_n^\eta(r, R)$ and energy eigenvalue $E_n(\eta)$ for the n-th eigenstate depend on the strength of the absorber η. In the present calculation, the S-wave relative motion is assumed in $\Psi_n^\eta(r, R)$ and hence, it depends on the magnitudes of the Jacobi coordinates, r and R. Eq. (7) is solved by the basis expansion method, in which the oscillating Gaussian basis is employed [5, 12]. In the wave function of $\Psi_n^\eta(r, R)$, the boson symmetry is exactly considered by introducing the rearrangement channels.

2.2 Strength function

We also calculate the strength function of the three-body breakup into the continuum. A general definition of the strength function is given by

$$S(E) = \sum_\nu \left|\langle \Psi_\nu | \hat{O}_\lambda | \Psi_i \rangle\right|^2 \delta(E - E_\nu), \tag{8}$$

where Ψ_i and Ψ_ν are the initial bound state and the ν-th final state having the energy eigenvalue E_ν in continuum, respectively. In Eq. (8), $\hat{O}_\lambda$ denotes the operator of an external field with the rank λ, and the monopole field ($\lambda = 0$) is considered, which are the squared particle coordinate measured from the center of mass for the total system, such as

$$\hat{O}_0 = \sum_{i=1,2,3} \xi_i^2, \tag{9}$$

where the vector ξ_i represents the coordinate of the i-th particle measured from the center of mass for the total system.

The summation of the final state Ψ_ν can be eliminated from Eq. (8) by assuming that the final states of Ψ_ν forms a complete set of the Hamiltonian H. If the Hamiltonian is transformed into H^η by switching on the absorber, we can get the following expression in the coordinate integration;

$$\mathcal{S}_\lambda^\eta(E) = -\frac{1}{\pi}\Im\left[\mathcal{R}^\eta{}_\lambda(E)\right] \tag{10}$$

$$\mathcal{R}_\lambda^\eta(E) = \langle\Psi_i|\hat{O}_\lambda^\dagger \mathcal{G}^\eta(E)\hat{O}_\lambda|\Psi_i\rangle = \int\int D\mathbf{X}D\mathbf{X}'\Psi_i^*(\mathbf{X}')\hat{O}_\lambda^\dagger \mathcal{G}^\eta(E,\mathbf{X}',\mathbf{X})\hat{O}_\lambda\Psi_i(\mathbf{X}) \ . \tag{11}$$

Here $\mathcal{R}_\lambda(E)$ is called the response function, which is represented by a set of the Jacobi coordinate of $\mathbf{X} = (\mathbf{r}, \mathbf{R})$ and $D\mathbf{X} = d\mathbf{r}d\mathbf{R}$. In Eq. (11), $\mathcal{G}^\eta$ denotes the Green function,

$$\mathcal{G}^\eta(E,\mathbf{X}',\mathbf{X}) = \left\langle \mathbf{X}'\left|\frac{1}{E-\hat{H}^\eta}\right|\mathbf{X}\right\rangle, \quad \mathcal{G}^\eta(E) = \frac{1}{E-\hat{H}^\eta} \ , \tag{12}$$

which is defined by the total Hamiltonian in Eq. (3), which includes the absorber with the strength η. The initial wave function of Ψ_i is invariant with respect to the introduction of the absorber and hence, the superscript of η is dropped in Ψ_i.

The calculation of the response function in Eq. (11) can be achieved by assuming the extended completeness relation (ECR) for the discrete solutions under the absorbing boundary,

$$\hat{1} = \sum_n^N |\Psi_n^\eta><\tilde{\Psi}_n^\eta| = \sum_{n\in B}|\Psi_n^\eta><\tilde{\Psi}_n^\eta| + \sum_{n\in R}|\Psi_n^\eta><\tilde{\Psi}_n^\eta| + \sum_{n\in C}|\Psi_n^\eta><\tilde{\Psi}_n^\eta| \ . \tag{13}$$

Here Ψ_n^η denotes the n-th solution of Eq. (7), which is obtained by the basis expansion method. In the last equation, the summation is decomposed into three parts: the bound states (B), the resonant states (R), and the non-resonant continuum states (C). The final expression of the response function, on which the ECR is implemented, is given by

$$\mathcal{R}_\lambda^\eta(E) = \sum_n^N \frac{<\Psi_i|\hat{O}_\lambda^\dagger|\Psi_n^\eta><\tilde{\Psi}_n^\eta|\hat{O}_\lambda|\Psi_i>}{E-E_n(\eta)} = \mathcal{R}_{\lambda B}^\eta(E) + \mathcal{R}_{\lambda R}^\eta(E) + \mathcal{R}_{\lambda C}^\eta(E) \ , \tag{14}$$

where the response function is decomposed into three parts in accordance with Eq. (13). This decomposition is very useful in the analysis of the reaction mechanism of the three-body breakup reaction [4, 11]. For example, we can understand the dominant breakup process, such as the resonance decay or the direct breakup, in which the ground state directly goes into the continuum states. In this response function, the denominator contains a complex number of $E_n(\eta)$, which is the eigenvalue in the ABC solution to Eq. (7). Thus, this function is converged to a finite value even if the excitation energy E is close to the real part of $E_n(\eta)$.

3 Results

3.1 Energy distribution and the resonance parameters

Two panels in figure 1 show the distribution of the energy eigenvalues obtained by solving Eq. (7) for the identical three-bosons system under the absorbing boundary condition. The absorber is introduced in terms of the function of the Jacobi coordinate, which is shown in Eq. (4). The left panel is the result of the weak absorber (η =2.0×10^{-7}), while the right panel is that of the strong absorber (η =2.0×10^{-4}).

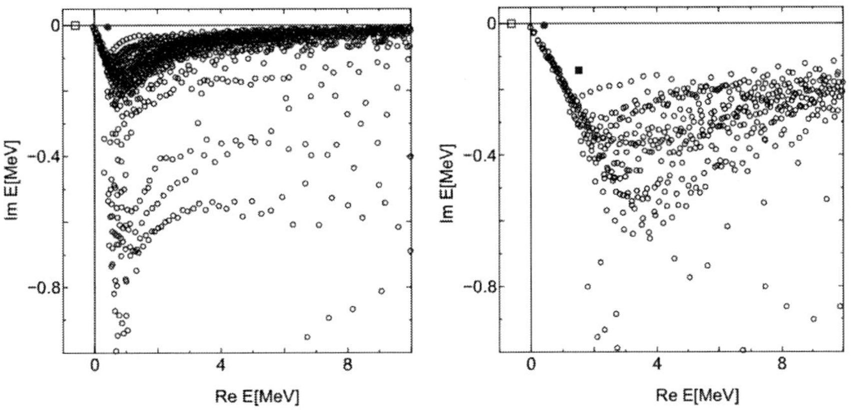

Figure 1. Distribution of the energy eigenvalues in the complex energy plane. The open square and the open circles show the bound state and the non-resonant continuum states, respectively. The solid circle and the solid square show the first and second resonances, respectively. The left panel shows the case of $\eta = 2 \times 10^{-7}$ and the right panel is the case of $\eta = 2 \times 10^{-4}$

If the absorbing potential is switched off, the eigenvalues are located along the real (horizontal) axis. The energy eigenvalues are distributed in the complex energy plane when the absorbing potential is activated. In the negative real energy region, there is a bound state (open square), which is invariant under the absorbing boundary. In the positive energy region, the resonant states are separated from the continuum spectra (open circles), and the separation depends on the strength of the absorber.

In the weak absorber case (left panel), the imaginary part of the continuum energy is not so large, and the continuum states are concentrated around the real axis. Therefore, in the weak absorber case, only one resonance (solid circle) is separated from the continuum states (open circles). On the contrary, the imaginary energy of the continuum states becomes larger when the strength of the absorber is strong, as shown in the right panel. This increased imaginary energy leads to the separation of the second resonance (solid square), as can be confirmed in the right panel.

When the absorber strength η increases continuously, the imaginary energy of the continuum states monotonically increases, while the energies of the resonant states $E_{res}(\eta)$ are almost invariant after the resonances are separated from the continuum. Therefore, the stronger absorber is needed in order to obtain the broad resonant states with a large imaginary energy.

We have also checked the energy distribution calculated from the other type of the absorbing boundary, which is imposed on the particle pair coordinate, shown in Eq. (6). We have confirmed the resultant energy distribution is almost the same as the distribution shown in Fig. 1.

Table 1. Resonance parameters for the first resonance calculated at the optimal strength η_o. The resonance energy (E_R) and the resonance width (Γ_R) are shown in unit of MeV. $< W(\eta_o) >$ represent the expectation value of the absorber at η_o. In the second row, the result of the ABC method is shown, while the results of CSM is shown in the lowest row.

Method	E_R	$\Gamma_R/2$	$< W(\eta_o) >$	η_o
ABC (Jacobi)	0.424	5.239×10^{-3}	6.942×10^{-7}	1.78×10^{-6}
ABC (pair)	0.424	5.239×10^{-3}	7.118×10^{-7}	8.0×10^{-6}
CSM	0.424	5.260×10^{-3}	X	X

Table 2. Same as the table. 1 except for the second resonance.

Method	E_R	$\Gamma_R/2$	$< W(\eta_o) >$	η_o
ABC (Jacobi)	1.500	0.141	4.475×10^{-5}	1.7×10^{-4}
ABC (pair)	1.500	0.144	7.015×10^{-5}	6.3×10^{-4}
CSM	1.498	0.142	X	X

In Fig. 1, the strength of the absorber, η, is fixed to a constant value, but the energy eigenvalues of the resonance basically depend on the strength of η. Therefore, the strength parameter of η should be fixed to be an optimal value [6, 9, 10]. Here we denote the resonance eigenvalue as a function of η by $E_{res}(\eta)$. If the basis set is complete, the correct resonance parameters can be obtained as a limit of $\eta \to 0$. In the finite basis set calculations, however, a finite η gives approximate resonance parameters. The condition for an optimal strength η_o is

$$\left| \eta \frac{dE_{res}(\eta)}{d\eta} \right|_{\eta=\eta_o} = min \ . \tag{15}$$

Here the derivatives $\eta \cdot dE_{res}(\eta)/d\eta$ are obtained simply by employing the generalized Hellmann-Feynman theorem [15] like

$$\left| \eta \frac{dE_{res}(\eta)}{d\eta} \right| = \eta \left| < \tilde{\Psi}^{\eta}_{res}(r,R) \mid \hat{W} \mid \Psi^{\eta}_{res}(r,R) > \right| \equiv < W(\eta) > \ , \tag{16}$$

where $\Psi^{\eta}_{res}(r,R)$ represents the resonance wave function in Eq. (7) calculated with an absorber of the strength η. The tilde in the bra-state means that the complex conjugate is not taken in the radial integration.

An argument to justify the condition of Eq. (15) was presented on the basis of the asymptotic expansion of the function $E(\eta)$. [6, 7]. Let us discuss the optimization of η in a simple argument. If η is very small and the range of the basis function is limited to the shorter region, absorption occurs in region of large distance, where no basis functions exist. Then the resonant eigenvalue becomes almost real. As η increases, the absorption can be efficiently described by a given basis set, and the imaginary part of $E_{res}(\eta)$ increases. When the spatial region of the dominant absorption sufficiently overlaps the region of the basis functions, the imaginary part of $E_{res}(\eta)$ will be close to the true width. If one further increases η, the reflection due to the absorbing potential becomes substantial, and this will cause the imaginary part of $E_{res}(\eta)$ to decrease. Therefore, there is an optimal strength, η_o, for a resonance, in which the reflection from the absorber and the incompleteness of the basis expansion balance each other. The stationary behavior of the resonance energy around $\eta = \eta_o$ have been deeply analyzed in Ref. [10].

The resonance parameters of E_R and Γ_R are calculated at the optimal strength of η_o according to the relation of $E_{res}(\eta_o) = E_R - i\Gamma_R/2$. The resonance parameters for the first resonance are summarized in table 1. In the second row from the top, the resonance parameters calculated from the Jacobi absorber are shown, while the parameters obtained from the pair absorber are shown in the third row. The results of the Jacobi absorber is same as those of the pair absorber. In both of the ABC results, the optimal strength η_o is quite weak ($\sim 10^{-6}$ in Jacobi and $\sim 10^{-5}$ in pair) and hence, the expectation value of $< W(\eta_o) >$ is also small ($\sim 10^{-7}$) in comparison to the magnitude of the resonance parameters. These two results of ABC is consistent to the CSM calculation, shown in the bottom row.

We have done the similar optimization of η for the second resonance, and the results are summarized in table 2. The resonance parameters of the Jacobi ABC method (second row) is almost same as the method of the pair coordinate (third row). The absorber strength for the second resonance ($\eta_o \sim 10^{-4}$) is stronger than that for the first resonance ($\eta_o \sim 10^{-6}$) because the strong absorber is needed to separate the broad resonance from the continuum. Therefore, the reflection effect from the absorber is much more prominent in the second resonance. Due to this reflection, the resonance parameters of ABC considerably deviates from the CSM calculation, shown at the bottom row.

3.2 Strength function of the monopole breakup

We discuss the application of the ABC method to the strength function of the three-body breakup. The strength function can be calculated accroding to the response function in Eq. (14), in which the ABC solutions in Fig. 1 are substituted into Ψ_n^η. Left panel in Fig. 2 shows the strength function from the ground state to the three-body continua, which is induced by the monopole external field. The solid curve shows the result of the ABC method with the Jacobi absorber, while the dashed curve shows the CSM result. The ABC calculation follows the curve obtained by the CSM calculation over all energy region although a small oscillation arise in the higher energy region of $E \geq 3$ MeV, where the magnitude of the strength is smaller by about 10^{-6} than the peak magnitude at $E \sim 0.5$ MeV. We have checked that the pair absorber gives the similar strength function to the result of the Jacobi absorber.

The sharp peak at $E \sim 0.5$ MeV is generated by the contribution from the first resonance, while the broad shoulder at $E \sim 1$ MeV is due to the existence of the second resonance. In order to see the intrinsic contributions in the strength function, the strength decomposition according to Eqs. (13) and (14) is shown in the right panel of Fig. 2. The contribution from the first resonance, which is plotted by the dashed curve, is predominant in the total strength, shown by the solid curve. The dotted curve represents the total contribution of the second resonance plus all of the non-resonant continua. The broad peak of the dotted curve correspond to the bump in the total strength, appearing just below $E \sim 1$ MeV.

In the low energy region of $E \leq 0.3$ MeV, the total strength is almost zero, but the individual contributions from the first resonance (dashed curve) and the other contribution (dotted curve) are finite value with the opposite signs. The total strength is constructed from the cancellation of these contributions. This feature of the cancellation is also seen in the strength function of the two-body breakup, which is studied in the CSM and ABC calculations [3, 4, 11].

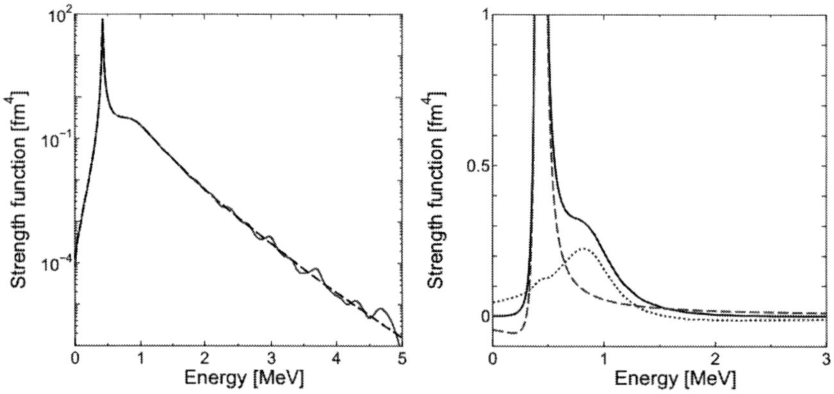

Figure 2. Left panel: Monopole strength functions calculated from the ABC and CSM results. The solid and dashed curves show the ABC and CSM results, respectively. Right panel: Decomposed strength function. The solid curve shows the total strength, while the dashed and dotted curves represent the contributions from the first resonance and the other resonance plus continuum. In these panels, the Jacobi absorber is used for the ABC calculation.

We have also done the decomposition of the strength function on the basis of the CSM treatment [3, 4], in which the application of ECR has already been established. The individual contributions, such as the resonance and the non-resonant continuum, in the strength function is common between ABC and CSM and hence, the ABC calculation is consistent to the CSM calculation.

3.3 Effect of the boson symmetry

Finally, we discuss the effect of the boson symmetry on the strength function, which is taken into account by the rearrangement channels. The effect of the rearrangement channels is investigated for the total binding energy in the three-body system [3, 12]. The inclusion of the rearrangement channel is found to be essential to obtain the rapid convergence of the binding energy, but there is no detailed analysis about the effect of the rearrangement channel on the strength of the three-body breakup reaction.

In the left panel of Fig. 3, the strength functions with and without the boson symmetry are compared. The former and latter results are plotted by the solid and dashed curves, respectively. These strengths are calculated by CSM. The overall shape of the strength is common in these two types of the calculations, but there is a clear difference in the peak structures and the magnitude.

In the result without the boson symmetry (dashed curve), a sharp spike appears just above $E \sim 0.5$ MeV. This spike is generated from the resonant states with a sharp width, which should be excluded by the boson symmetry. Since the boson symmetry restricts the possible phase space among the three particles, the number of the resonances are enhanced in the case without the boson symmetry. Therefore, the neglect of the boson symmetry leads to the formation of the spurious peak structure in the strength function.

As for the magnitude of the strength, the boson symmetry gives a minor effect on the strength at the higher energy side of $E \geq 1$ MeV, where a smooth energy dependence can be seen. However,

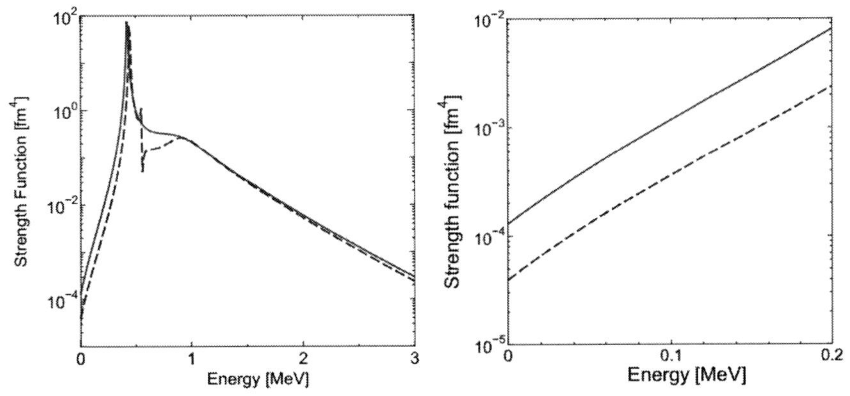

Figure 3. Left panel: Monopole strength functions calculated with and without the boson symmetry. These strengths are calculated by CSM. The solid curve shows the result with boson symmetry, while the dashed curve represent the strength without the symmetry. Right panel: Enlargement of the left panel in the low energy region.

the yield at the low energy side is sensitive to the treatment of the boson symmetry. The low energy strength is enhanced in the calculation with the boson symmetry, as shown by the solid curve. The enlargement of the strength in the low energy region is shown in the right panel of Fig. 3. We can clearly confirm that the strength with the symmetry is enhanced by a factor of three in comparison to the strength without the symmetry.

In recent studies, the reaction rate of the triple α reaction has been investigated by the sophisticated reaction theory [13, 14]. In these calculations, the boson symmetry is neglected, and the model dependence of the reaction rate, such as the number of the channels and the spatial range of the calculation, is discussed in the order of magnitude of about 10^{20}. The effect of the boson symmetry seems to be minor in comparison to the model dependence discussed in Refs. [13, 14]. However, the consideration of the boson symmetry will be important in the precise evaluation of the triple α reaction rate if the order of the magnitude is fixed appropriately.

4 Summary

In sum, we have formulated the method of the absorbing boundary condition (ABC) in the three-body problem with the coupled-rearrangement-channels variational method (CRCVM). The CRCVM + ABC has been applied to the schematic three-boson system, in which the boson symmetry is exactly considered by performing the rearrangement channels. We have introduced the absorbing boundary in two ways: the absorbing potentials in the Jacobi coordinate and the absorber in the particle pair coordinate. The resonance parameters calculated by these two methods are compared, and we have confirmed that the resonance parameters are the same value in these two methods.

The ABC results are also compared with the complex scaling method (CSM), which has already been successful in the application to the CRCVM calculation. The resonance parameters of CSM are consistent to the result of the ABC method although the decay width of the broad resonance in ABC is considerably different from that in the CSM calculation. This deviation is considered to be the reflection effect from the absorber. In the ABC method, the effect of the reflection cannot completely be eliminated even if the strength is optimized to minimize the numerical error. However, the deviation of the decay width is just the order of 10^{-3} MeV, in which the ambiguity in the theoretical model itself will be essential.

The strength function of the three-body breakup is also calculated by the ABC method plus the extended completeness relation (ECR). The strength of the ABC method nicely reproduces the CSM result. The decomposition of the strength functions is also investigated according to ECR. The individual contributions to the total strength are consistent to the CSM strength.

From the present comparison of ABC with CSM, we can conclude that the CRCVM + ABC method, formulated in the present article, nicely works in the three-body problem. The different ways of the absorber, such as the absorbers in the Jacobi coordinate and those in the pair coordinate, give the equivalent results in both the resonance parameters and the strength function if the strength of the absorbers are appropriately fixed. The computational number of the matrix elements is six in the Jacobi absorber (= 2 per channel × 3 channels), while the number is three in the pair absorber (= pair number of three particles). Thus, the number of the matrix elements is reduced in the pair-coordinate absorber, which leads to save the computational time.

In the future subject, we should apply the ABC method to the more realistic system, such as the microscopic three cluster system with the full anti-symmetrization. The calculations of the microscopic cluster model can be easily done in the framework of the generator coordinate method (GCM) [16], but the application of CSM to GCM is still difficult [9]. On the contrary, the application of ABC to GCM has been successful in the two-body system [9]. Therefore, the application of ABC + GCM should be extended to the three-body system, such as the $^{12}C = 3\alpha$ system [17]. The application of ABC to the three-body GCM is now under progress.

References

[1] I. Tanihata, H. Savajols, and R. Kanungo, Prog. Part. Nucl. Phys. **68**, 215 (2013).
[2] A. Deltuva, A. C. Fonseca, and P. U. Sauer, Ann. Rev. Nucl. Part. Sci. **58**, 27 (2008).
[3] S. Aoyama, T. Myo, K. Katō, and K. Ikeda, Prog. Theor. Phys. **116**, 1-35 (2006), and references therein.
[4] T. Myo, K. Katō, S. Aoyama, and K. Ikeda, Phys. Rev. C **63**, 054313 (2001).
[5] S. Ohtsubo, Y. Fukushima, M. Kamimura, E. Hiyama, Prog. Theor. Exp. Phys. **2013**, 073D02 (2013).
[6] U. V. Riss and H-D. Meyer, Jour. Phys. **B**26,4503 (1993).
[7] H. Masui and Y. K. Ho, Phys. Rev. C **65**, 054305 (2002).
[8] M. Ueda et al., Phys. Rev. C **67**, 014606 (2003).
[9] M. Ito and K. Yabana, Prog. Theor. Phys. **113**, 1047 (2005).
[10] Y. Takenaka, et al., Prog. Exp. Theor. Phys. **2014**, 113D04 (2014), and references therein.
[11] M. Iwasaki, et al., Prog. Exp. Theor. Phys. **2015** 023D01, (2015).
[12] E. Hiyama et al., Prog. Part. Nucl. Phys. **51**, 223 (2003).
[13] K. Ogata, M. Kan, and M. Kamimura, Prog. Theor. Phys. Vol. **122**, 1055 (2009).
[14] T. Akahori, Y. Funaki, and K. Yabana, Phys. Rev. C**92**, 022801(R) (2015).
[15] G. Jolicard, C. Leforestier and E. J. Austin, Jour. Chem. Phys. **88**, 1026, (1988).
[16] H. Horiuchi, Suppl. Prog. Theor. Phys. **62**, 90 (1977).
[17] E. Uegaki, S. Okabe, H. Tanaka, Y. Abe, Prog. Theor. Phys. **57**, 1262 (1977).

EPJ Web of Conferences **122**, 07001 (2016) DOI: 10.1051/epjconf/201612207001
*CNR*15*

Fully microscopic description of elastic and inelastic scattering at intermediate incident energies

Kosho Minomo[1,a], Michio Kohno[1], Masakazu Toyokawa[2], Masanobu Yahiro[2], and Kazuyuki Ogata[1]

[1]*Research Center for Nuclear Physics, Osaka University, Ibaraki, Osaka 567-0047, Japan*
[2]*Department of Physics, Kyushu University, Fukuoka 812-8581, Japan*

Abstract. We aim for fully microscopic understanding of many-body nuclear reactions starting from two- and three-nucleon forces based on chiral effective field theory (Ch-EFT). We first construct a g-matrix with the nuclear forces based on Ch-EFT using Brueckner-Hartree-Fock theory, in which the three-nucleon force effects are represented through the density dependence of the g-matrix. Then, the folding model and microscopic coupled-channels method with the g-matrix are applied to nucleon-nucleus and nucleus-nucleus scattering at intermediate incident energies. This new microscopic framework well describes the elastic and inelastic cross sections with no ad-hoc parameters. In addition, the three-nucleon force and coupled-channels effects on many-body nuclear reactions are clarified.

1 Introduction

The microscopic description of many-body nuclear reactions is a fundamental subject in nuclear physics. For this purpose, construction of microscopic optical potentials for nucleon-nucleus and nucleus-nucleus scattering is important. Based on multiple scattering theory [1–3], elastic scattering is described by multistep processes with an effective nucleon-nucleon interaction from the microscopic point of view. So far many types of density-dependent complex interactions (g-matrix interactions) have been proposed as effective interactions based on Brueckner theory.

Recently, the Melbourne group achieved a great success by describing proton scattering with no adjustable parameter. They applied a g-matrix interaction constructed with the Bonn potential [4] to the folding model calculation, and reproduced the measured elastic cross sections and vector analyzing powers in a wide incident-energy range [5]. In addition, the Melbourne interaction was also applied to neutron scattering [6] and scattering of weakly-bound nucleus [7, 8]. Furthermore, total reaction cross sections of scattering of stable and unstable nuclei were well reproduced [9, 10]. In spite of the remarkable progress of microscopic reaction theory in recent years, however, nucleus-nucleus elastic cross sections cannot be well reproduced in some cases [11]. One of the reasons for this failure may be the lack of three-nucleon force (3NF) effects in nucleus-nucleus scattering; the Melbourne interaction does not include the effects explicitly. In fact, the 3NF effects on scattering observables are discussed in nuclear physics [12].

[a]e-mail: minomo@rcnp.osaka-u.ac.jp

In this paper, we introduce a new microscopic reaction theory starting from two- and three-nucleon forces based on chiral effective field theory (Ch-EFT) [13–15]. We first construct a g-matrix with the nuclear forces based on Ch-EFT using Brueckner-Hartree-Fock theory, in which the three-nucleon force effects are represented through the density dependence of the g-matrix. Then, the folding model and microscopic coupled-channels method with the g-matrix are applied to nucleon-nucleus and nucleus-nucleus scattering at intermediate incident energies. One of the advantages of using the chiral interactions is that the two-, three-, and many-body forces are treated systematically, so that the uncertainty of interactions is minimized. At present, the chiral interactions are most reliable for investigating the 3NF effects, although they cannot be applied to high-energy reactions beyond the cutoff energy.

We will show the results of microscopic scattering calculations, without any adjustable parameters, for nucleon-nucelus scattering in Sec. 2 and nucleus-nucleus scattering in Sec. 3. We also clarify the coupled-channels effects together with the 3NF effects on scattering observables. Section 4 is devoted to a summary.

2 Nucleon-nucelus elastic scattering

We first construct a g-matrix with the nuclear forces based on Ch-EFT [13–15] using Brueckner-Hartree-Fock theory; see Ref. [16, 17] for detail. Then, we apply the g-matrix folding model, which is one of the practical methods to construct microscopic optical potentials, with the complex g-matrix interaction for scattering calculations. The detail of the present calculation is written in Ref. [17].

Fig. 1 shows the differential cross sections and analyzing powers of proton elastic scattering at 65 MeV from ^{40}Ca, ^{58}Ni, and ^{208}Pb targets. The solid (dashed) lines show the results with (without) the 3NF effects. We can reproduce well not only the cross sections but also spin observables with the present framework. It should be noted that the present result does not include any adjustable parameters. Hence, the present framework has a predictive power for nucleon-nucleus scattering. We find that the chiral 3NF effects is small for nucleon-nucleus elastic scattering since the scattering occurs mainly in the peripheral region of the target nucleus, in which the density is low.

3 Nucleus-nucleus elastic and inelastic scattering

We now consider nucleus-nucleus elastic and inelastic scattering. In this case, since the coupling effects due to the projectile and target excitations are large, we perform microscopic coupled-channels calculations of ^{16}O-^{16}O and ^{12}C-^{12}C scattering and clarify the coupled-channels effects together with the 3NF effects on scattering observables. Note that we use the Melbourne g-matrix interaction modified with the chiral 3NF effects as suggested in Ref. [11]. The detail of the present calculation is written in Ref. [18].

Figure 2 shows the differential cross sections for the ^{16}O-^{16}O scattering at 70 MeV/nucleon and 44 MeV/nucleon as a function of the scattering angle θ in the center-of-mass system. In each panel, three cross sections corresponding to the 0_1^+ (top), 3_1^- (middle), and 2_1^+ (bottom) states of the ejectile are shown; the other particle is in the ground state in the final state. For the 0_1^+ state, i.e., the elastic scattering, the ratio to the Rutherford cross section is plotted. The solid (dashed) line is the result of the coupled-channels calculation with (without) the 3NF effects, and the dotted (dot-dashed) line corresponds to the result of the one-step calculation with (without) the 3NF effects.

From Fig. 2(a), we see that the calculation including both the 3NF and coupled-channels effects well reproduces the elastic scattering data up to $\theta = 15°$. The 3NF and coupled-channels effects are found to be comparable at forward angles $\theta < 12°$, whereas the 3NF effect is more important than

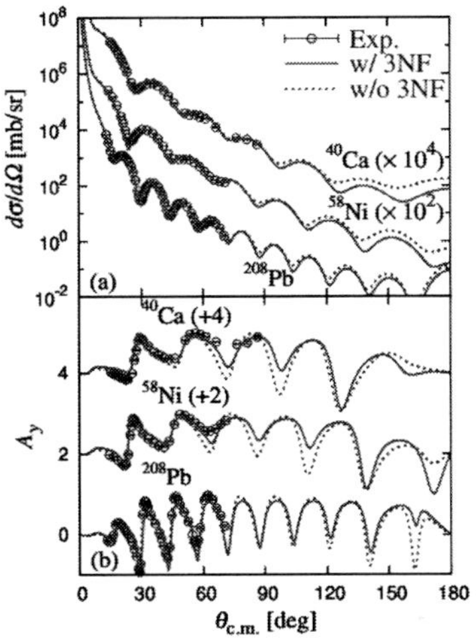

Figure 1. (a) Differential cross sections and (b) analyzing powers of proton elastic scattering at 65 MeV from ^{40}Ca, ^{58}Ni, and ^{208}Pb targets. The solid (dashed) lines show the results with (without) the 3NF effects. Each cross section is multiplied by the number shown in the figure, while each vector analyzing power is shifted up by the number. Experimental data are taken from Ref. [19, 20]. This figure is taken from Ref. [17]

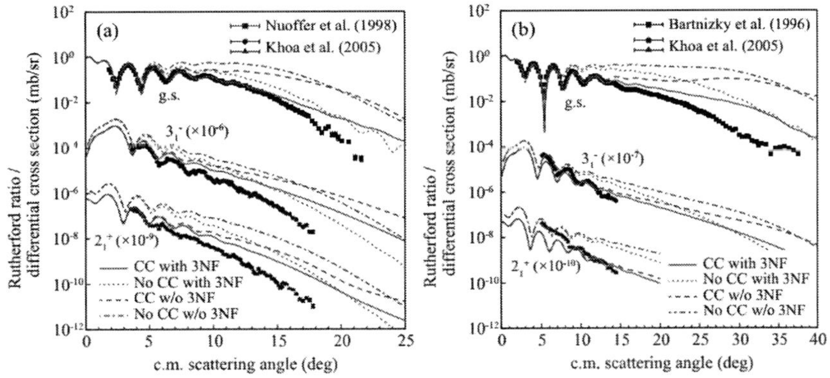

Figure 2. Differential cross sections of elastic and inelastic ^{16}O-^{16}O scattering at (a) 70 MeV/nucleon and (b) 44 MeV/nucleon. The solid (dashed) lines correspond to the result with (without) the 3NF effects, and the dotted (dot-dashed) lines correspond to the result of the one-step calculation with (without) the 3NF effects. The experimental data are taken from Refs. [21–23]. The inelastic cross sections are scaled by the number shown inside the panel for clarity. This figure is taken from Ref. [18].

the coupled-channels effect at larger angles. The 3NF effects change mainly the interior part of the nucleus-nucleus potential, as shown in Ref. [11], which is influential to the scattering at larger angles. On the other hand, it is well known that in general the coupled-channels effects generate the so-called dynamical polarization potential in the nuclear surface region. The separation of the regions sensitive to the 3NF and coupled-channels effects makes the two effects rather independent. In fact, one sees

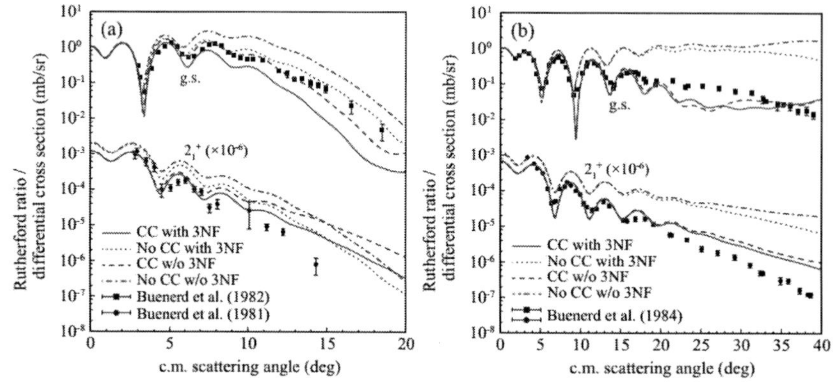

Figure 3. Differential cross sections of elastic and inelastic ^{12}C-^{12}C scattering at (a) 85 MeV/nucleon and (b) 30 MeV/nucleon. The meaning of the lines is the same as in Fig. 2. The experimental data are taken from Refs. [25–27]. The inelastic cross sections are scaled by the number shown inside the panel for clarity. This figure is taken from Ref. [18].

that the 3NF (coupled-channels) effects do not change essentially whether the coupled-channels (3NF) effects are included or not.

For the 3_1^- and 2_1^+ inelastic cross sections, the solid line reproduces the data for $\theta < 5°$ but slightly overshoots at larger angles. One sees that the 3NF effects improve the agreement with the inelastic cross section data. The 3NF effects on the inelastic cross sections appear at large angles as in the elastic cross sections, although the effects are relatively small. In general, inelastic cross sections at very forward angles are mainly determined by the strength of the coupling potential. The negligibly small difference between the solid and dashed lines at forward angles suggests the small 3NF effects on the coupling potentials. It is found that the 3NF effects on the inelastic cross sections come from the changes in the diagonal part of the ^{16}O-^{16}O potential for each channel. It was pointed out in Ref. [24] that the inelastic cross sections may be varied by couplings with even higher excited states. This can be a reason for the discrepancy of the 3_1^- and 2_1^+ cross sections at larger angles and should be investigated in the future.

At 44 MeV/nucleon, the coupled-channels effects are larger and the 3NF effects are smaller than those at 70 MeV/nucleon. This is because at lower energies the reaction takes place in the peripheral low-density region. Nevertheless, the 3NF effects are still non-negligible at large scattering angles. To be precise, the correspondence between the calculation and the data for the elastic cross sections is worsened from that at 70 MeV/nucleon. One of the reasons for this will be the fact that at lower incident energies the higher excited states, which are not taken into account, may play a more significant role.

Figure 3 shows the differential cross sections for the ^{12}C-^{12}C scattering at 85 MeV/nucleon and 30 MeV/nucleon. The meaning of the lines is the same as in Fig. 2. The 0_1^+, 2_1^+, 0_2^+, and 2_2^+ states of 12C are taken into account in the calculation.

At 85 MeV/nucleon, we find larger coupled-channels effects than those in the ^{16}O-^{16}O scattering. As a result, the solid line decreases significantly and undershoots the data. It is found that the coupling between the 0_1^+ and 2_1^+ states plays a dominant role in the ^{12}C-^{12}C scattering. For the inelastic cross section the solid line well agrees with the data. At 30 MeV/nucleon, as shown in Fig. 3(b), the 3NF effects do not affect both the elastic and inelastic cross sections. The measured elastic and inelastic cross sections are well reproduced by the coupled-channels calculations, except for the small difference at large scattering angles.

4 Summary

We have reported the present status of describing elastic and inelastic scattering at the intermediate incident energies. The most essential part of our approach is to use a new g-matrix interaction constructed with the two- and three-nucleon forces based on Ch-EFT. One of the advantages of using the chiral interactions is that the two-, three-, and many-body forces are treated systematically, so that the uncertainty of interactions is minimized. At present, the chiral interactions are most reliable for investigating the 3NF effects. We applied the folding model and the microscopic coupled-channels method with the g-matrix interaction to nucleon- and nucleus-nucleus scattering. For nucleon-nucleus scattering, the present framework well reproduced the measured cross sections and analyzing powers. For nucleus-nucleus scattering, the elastic and inelastic cross sections were also reproduced by considering both 3NF and coupled-channels effects. Furthermore, the roles of 3NF and coupled-channels effects are clarified.

Our goal is to construct a microscopic reaction theory having a predictive power and it will become feasible. We now obtain reliable microscopic optical potentials. These microscopic optical potentials will be applied to description of various reaction processes.

Acknowledgement

This work is supported in part by Grant-in-Aid for Scientific Research from the Japan Society for the Promotion of Science (JSPS) and by the ImPACT Program of Council for Science, Technology and Innovation (Cabinet Office, Government of Japan). The numerical calculations in this work were performed at RCNP.

References

[1] K. M. Watson, Rev. Mod. Phys. **30**, 565 (1958).
[2] A. K. Kerman, H. McManus, and R. M. Thaler, Ann. Phys. **8**, 551 (1959).
[3] M. Yahiro, K. Minomo, K. Ogata, and M. Kawai, Prog. Theor. Phys. **120**, 767 (2008).
[4] R. Machleidt, K. Holinde, and Ch. Elster, Phys. Rep. **149**, 1 (1987).
[5] K. Amos, P. J. Dortmans, H. V. von Geramb, S. Karataglidis, and J. Raynal, *Advanced Nuclear Physics, edited by J. W. Negele and E. Vogt* (Plenum, New York, 2000) Vol. 25, p. 275.
[6] K. Amos and S. Karataglidis, Phys. Rev. C **65**, 057603 (2002).
[7] S. Karataglidis and K. Amos, Phys. Rev. C **87**, 054623 (2013).
[8] M. Toyokawa, K. Minomo, and M. Yahiro, Phys. Rev. C **88**, 054602 (2013).
[9] K. Minomo, T. Sumi, M. Kimura, K. Ogata, Y. R. Shimizu, and M. Yahiro, Phys. Rev. Lett. **108**, 052503 (2012).
[10] S. Watanabe *et al.*, Phys. Rev. C **89**, 044610 (2014).
[11] K. Minomo, M. Toyokawa, M. Kohno, and M. Yahiro, Phys. Rev. C **90**, 051601 (2014).
[12] T. Furumoto, Y. Sakuragi, and Y. Yamamoto, Phys. Rev. C **80**, 044614 (2009).
[13] E. Epelbaum, W. Glöckle, and Ulf-G. Meißner, Nucl. Phys. A**747**, 362 (2005).

[14] E. Epelbaum, H.-W. Hammer, and Ulf-G. Meißner, Rev. Mod. Phys. **81**, 1773 (2009).
[15] K. Hebeler, S. K. Bogner, R. J. Furnstahl, A. Nogga, and A. Schwenk, Phys. Rev. C **83**, 031301(R) (2011).
[16] M. Kohno, Phys. Rev. C **88**, 064005 (2013).
[17] M. Toyokawa, M. Yahiro, T. Matsumoto, K. Minomo, K. Ogata, and M. Kohno, Phys. Rev. C **92**, 024618 (2015).
[18] K. Minomo, M. Kohno, and K. Ogata, Phys. Rev. C **93**, 014607 (2016).
[19] H. Sakaguchi *et al.*, Phys. Lett. B **89**, 40 (1979).
[20] H. Sakaguchi *et al.*, Phys. Lett. B **99**, 92 (1981).
[21] F. Nuoffer *et al.*, Nuovo Cimento A **111**, 971 (1998).
[22] D. T. Khoa *et al.*, Nucl. Phys. A **759**, 3 (2005).
[23] G. Bartnitzky *et al.*, Phys. Lett. B **365**, 23 (1996).
[24] M. Takashina and Y. Sakuragi, Phys. Rev. C **80**, 014605 (2009).
[25] M. Buenerd *et al.*, Phys. Rev. C **26**, 1299 (1982).
[26] M. Buenerd *et al.*, Phys. Lett. B **102**, 242 (1981).
[27] M. Buenerd *et al.*, Nucl. Phys. A **424**, 313 (1984).

Recent developments in heavy-ion fusion reactions around the Coulomb barrier

K. Hagino[1,2,3], N. Rowley[4], and J.M. Yao[1,5,6]

[1] *Department of Physics, Tohoku University, Sendai 980-8578, Japan*
[2] *Research Center for Electron Photon Science, Tohoku University, 1-2-1 Mikamine, Sendai 982-0826, Japan*
[3] *National Astronomical Observatory of Japan, 2-21-1 Osawa, Mitaka, Tokyo 181-8588, Japan*
[4] *Institut de Physique Nucléaire, UMR 8608, CNRS-IN2P3 et Université de Paris Sud, 91406 Orsay Cedex, France*
[5] *School of Physical Science and Technology, Southwest University, Chongqing 400715, China*
[6] *Department of Physics and Astronomy, University of North Carolina, Chapel Hill, North Carolina 27516-3255, USA*

Abstract. The nuclear fusion is a reaction to form a compound nucleus. It plays an important role in several circumstances in nuclear physics as well as in nuclear astrophysics, such as synthesis of superheavy elements and nucleosynthesis in stars. Here we discuss two recent theoretical developments in heavy-ion fusion reactions at energies around the Coulomb barrier. The first topic is a generalization of the Wong formula for fusion cross sections in a single-channel problem. By introducing an energy dependence to the barrier parameters, we show that the generalized formula leads to results practically indistinguishable from a full quantal calculation, even for light symmetric systems such as $^{12}C+^{12}C$, for which fusion cross sections show an oscillatory behavior. We then discuss a semi-microscopic modeling of heavy-ion fusion reactions, which combine the coupled-channels approach to the state-of-the-art nuclear structure calculations for low-lying collective motions. We apply this method to subbarrier fusion reactions of $^{58}Ni+^{58}Ni$ and $^{40}Ca+^{58}Ni$ systems, and discuss the role of anharmonicity of the low-lying vibrational motions.

1 Introduction

Fusion is defined as a reaction in which two separate nuclei combine together to form a (hot) compound nucleus. It is indispensable to understand its reaction dynamics in order to understand synthesis of superheavy elements and nucleosynthesis in stars. Moreover, fusion reactions inherently contain rich physics, as there is a strong interplay in it, especially at energies around the Coulomb barrier, between the nuclear reaction and the nuclear structure. That is, it has been known well that fusion cross sections at subbarrier energies are largely enhanced relative to a prediction of a one-dimensional potential model, due to the couplings of the relative motion between the colliding nuclei to several nuclear intrinsic motions [1–5]. This is in marked contrast to high energy nuclear reactions, in which the reaction dynamics is much simpler and the distorted wave Born approximation (DWBA) often suffices its treatment. Fusion also offers a unique opportunity to study quantum tunneling with many degrees

 DOI: 10.1051/epjconf/201612207002

of freedom. That is, fusion takes place only by quantum tunneling at energies below the Coulomb barrier, and the subbarrier enhancement of fusion cross sections can be viewed as a consequence of a coupling assisted tunneling. Heavy-ion fusion reactions are unique in this respect because a variety of intrinsic degrees of freedom are involved, such as a surface vibration with several multipolarities, various sort of nuclear deformations and the associated rotational motion, and several types of particle transfer processes. This is in contrast to atomic and molecular collisions, in which only a limited types of intrinsic motion are involved. Also, in heavy-ion fusion reactions, the incident energy can be varied in order to study the energy dependence of the tunneling probability, whereas the energy is fixed in many other tunneling phenomena in nuclear physics, such as alpha decays.

In this contribution, we shall discuss two recent theoretical developments in heavy-ion fusion reactions. The first topic is a generalization of the Wong formula for fusion cross sections [6]. Although this celebrated formula yields almost exact results for single-channel calculations for relatively heavy systems such as ^{16}O+^{144}Sm, it tends to overestimate the cross section for light systems such as ^{12}C+^{12}C. We generalize the formula to take account of the energy dependence of the barrier parameters and show that the energy-dependent version reproduces almost perfectly results of a full quantal calculation. We also discuss the deviations arising from the discrete nature of the intervening angular momenta, whose effect can lead to an oscillatory contribution to the excitation function.

The second topic is a semi-microscopic modeling of heavy-ion fusion reactions [7]. For this purpose, we first describe microscopically low-lying collective excitations of atomic nuclei with the multi-reference covariant density functional theory, and combine them with coupled-channels calculations. We use the calculated transition strengths among several collective states as inputs to the coupled-channels calculations. This approach provides a natural way to describe anharmonic multi-phonon excitations as well as a deviation of rotational excitations from a simple rigid rotor. We apply this method to subbarrier fusion reactions of ^{58}Ni+^{58}Ni and ^{40}Ca+^{58}Ni systems, and discuss the role of anharmonicity in subbarrier fusion of these systems.

2 Generalization of the Wong formula for fusion cross sections

Let us first discuss the well-known Wong formula for fusion cross sections. The simplest approach to heavy-ion fusion reactions is to use the potential model, assuming that both the projectile and the target nuclei are inert. Fusion cross sections are then obtained by calculating the S-matrix, S_l, for each partial wave. If all the flux crossing the Coulomb barrier fuses, the fusion cross sections $\sigma_{\rm fus}$ read

$$\sigma_{\rm fus}(E) = \frac{\pi}{k^2}\sum_{l=0}^{\infty}(2l+1)P_l(E), \tag{1}$$

where $k = \sqrt{2\mu E/\hbar^2}$ is the wave number associated with the energy E (μ being the reduced mass) and $P_l(E) = 1 - |S_l|^2$ is the penetrability of the Coulomb barrier for the l-th partial wave.

Wong has derived a compact approximate expression for Eq. (1) [8], which has by now been known as the Wong formula. To this end, Wong introduced the following three approximations:

- *the parabolic approximation.* The total (that is, the nuclear + the Coulomb) inter-nucleus potential $V_0(r)$ for the s-wave is approximated by a parabolic function, that is,

$$V_0(r) \sim V_b - \frac{1}{2}\mu\Omega^2(r - R_b)^2, \tag{2}$$

where V_b, R_b, and $\hbar\Omega$ are the barrier height, the barrier position, and the "barrier curvature", respectively.

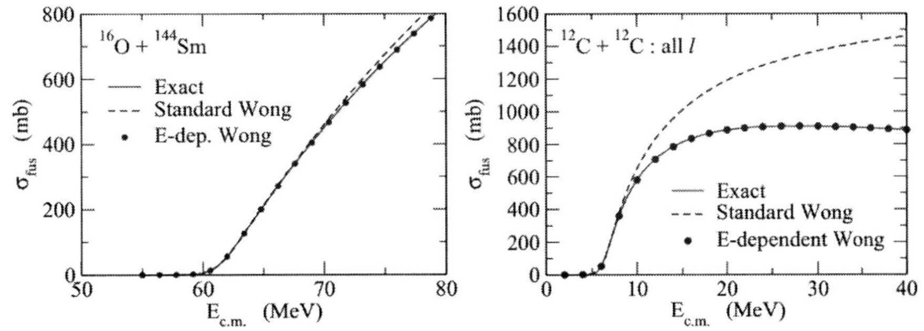

Figure 1. Fusion cross sections for the ^{16}O+^{144}Sm (the left panel) and the ^{12}C+^{12}C (the right panel) systems. The solid lines show the results of a full quantal calculation, while the dashed line shows the fusion cross sections evaluated with the Wong formula, Eq. (6). The filled circles, on the other hand, denote the results of the generalized (the energy dependent) Wong formula given by Eq. (9). The Bose symmetry of the identical spin-0 system is ignored for the ^{12}C+^{12}C system, and both even and odd partial waves are summed up in the cross sections.

- *the l-independent barrier position and curvature.* The barrier position, R_b, and the barrier curvature, $\hbar\Omega$, are assumed to be independent of l. In this case, the effective potential for the l-th partial wave reads,

$$V_0(r) + \frac{l(l+1)\hbar^2}{2\mu r^2} \sim V_b + \frac{l(l+1)\hbar^2}{2\mu R_b^2} - \frac{1}{2}\mu\Omega^2(r-R_b)^2. \tag{3}$$

The penetrability can then be calculated analytically with the Hill-Wheeler formula [9] as,

$$P_l(E) = \frac{1}{1+\exp\left[\frac{2\pi}{\hbar\Omega}\left(V_b + \frac{l(l+1)\hbar^2}{2\mu R_b^2} - E\right)\right]}. \tag{4}$$

- *the continuum approximation for l.* The angular momentum l is treated as a continuous variable and the sum in Eq. (1) is replaced by the integral,

$$\sigma_{\rm fus}(E) = \frac{\pi}{k^2}\sum_{l=0}^{\infty}(2l+1)P_l(E) \to \frac{\pi}{k^2}\int_0^\infty dl\,(2l+1)P_l(E). \tag{5}$$

With these approximations, the fusion cross sections are obtained as [1, 3, 8],

$$\sigma_{\rm fus}(E) = \frac{\hbar\Omega}{2E}R_b^2\ln\left[1+\exp\left(\frac{2\pi}{\hbar\Omega}(E-V_b)\right)\right]. \tag{6}$$

Notice that this formula yields the classical fusion cross sections,

$$\sigma_{\rm fus}(E) \sim \pi R_b^2\left(1-\frac{V_b}{E}\right), \tag{7}$$

at energies well above the Coulomb barrier, $E - V_b \gg \hbar\Omega/2\pi$.

The left panel of Fig. 1 shows the performance of the Wong formula for the $^{16}O+^{144}Sm$ system. The solid line is obtained quantum mechanically by solving the Schrödinger equation for a given inter-nucleus potential, while the dashed line is obtained with the Wong formula, Eq. (6). Even though the Wong formula slightly overestimates the fusion cross sections at energies well above the Coulomb barrier, reflecting the fact that the barrier position gradually decreases as a function of l, the overall agreement is satisfactory. In contrast, the situation drastically changes for light systems. The right panel of Fig. 1 shows a comparison between the quantal and the approximate fusion cross sections for the $^{12}C+^{12}C$ system. Here, in order to simplify the discussion, for the moment we ignore the Bose symmetry of this identical spin-0 system and sum over all even and odd partial waves. One can see that the Wong formula largely overestimates fusion cross sections, and the agreement is much worse than the heavier system, $^{16}O+^{144}Sm$. This is because the centrifugal potential plays a more important role in light systems, and the Coulomb barrier is not rigid against a variation of angular momentum [6].

One can improve the Wong formula by introducing the energy dependence to the barrier parameters $[V_b, R_b, \hbar\Omega]$. To this end, we first introduce the grazing angular momentum l_g, at which the barrier height of the effective potential becomes identical to the incident energy, E. If we denote the barrier position for l_g as R_E, l_g and R_E are related to each other as,

$$\frac{l_g(l_g+1)\hbar^2}{2\mu R_E^2} = E - V_E, \tag{8}$$

where $V_E = V_0(R_E)$ is the sum of the Coulomb and the nuclear potentials at R_E. An idea of the generalized Wong formula is to replace the barrier parameters $[V_b, R_b, \hbar\Omega]$ with the energy dependent ones, $[V_E, R_E, \hbar\Omega_E]$, where $\hbar\Omega_E$ is the barrier curvature for the grazing angular momentum l_g. That is, the generalized version of the Wong formula now reads [6],

$$\sigma_{\rm fus}(E) = \frac{\hbar\Omega_E}{2E} R_E^2 \ln\left[1 + \exp\left(\frac{2\pi}{\hbar\Omega_E}(E - V_E)\right)\right]. \tag{9}$$

Notice that we still assume an l-independent barrier position and curvature in integrating Eq. (5). However, the generalization is such that these parameters are evaluated at the grazing angular momentum for each energy E, rather than using the values for the s-wave. The filled circles in Fig. 1 show the fusion cross sections obtained with the generalized Wong formula, Eq. (9). One can see that the quantal results are well reproduced by introducing the energy dependence to the barrier parameters, both for the $^{16}O+^{144}Sm$ and the $^{12}C+^{12}C$ systems.

One can also go beyond the continuum approximation for the angular momentum sum. Using the Poisson sum formula [10], the next order correction to the Wong formula reads [6, 11],

$$\sigma_{\rm fus}(E) \sim \sigma_{\rm Wong} + \sigma_{\rm osc}, \tag{10}$$

where $\sigma_{\rm Wong}$ is the smooth part of fusion cross sections given by the Wong formula, Eq. (9), while the oscillatory contribution $\sigma_{\rm osc}$ is given by [6, 11],

$$\sigma_{\rm osc} = 2\pi R_E^2 \frac{\hbar\Omega_E}{E} e^{-2\xi} \sin(2\pi l_g), \tag{11}$$

with

$$\xi = \frac{\hbar\Omega_E}{2l_g+1} \cdot \frac{\pi\mu R_E^2}{\hbar^2}. \tag{12}$$

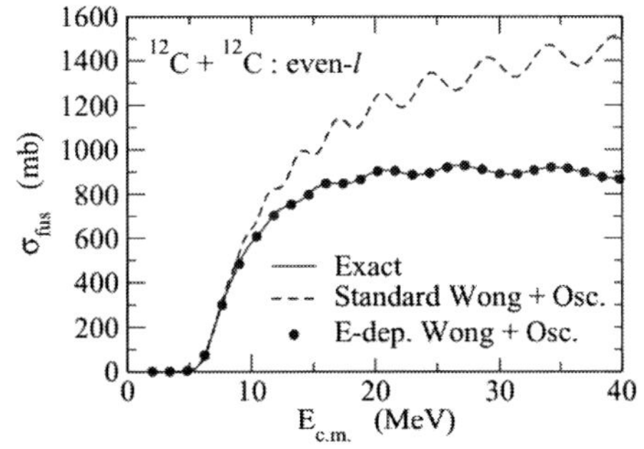

Figure 2. Fusion cross sections for the $^{12}C+^{12}C$ system obtained with even partial waves only. The solid line shows the exact quantal result. The filled circles and the dashed line denote the results of the Wong formula supplemented with the oscillatory contribution, Eq. (13), with and without the energy dependence of the barrier parameters, respectively.

Usually, the oscillatory contribution, σ_{osc}, is much smaller than the smooth part given by σ_{Wong}. There are, however, certain circumstances where the oscillatory contribution becomes significantly large. This is for a light symmetric system of two identical spin-0 bosons, in which all odd partial waves disappear due to the symmetrization of the wave function. In this case, the fusion cross section is given by twice the sum over the even partial waves. The resultant formula now becomes [6, 11],

$$\sigma_{osc} = 2\pi R_E^2 \frac{\hbar\Omega_E}{E} e^{-\xi} \sin(\pi l_g), \tag{13}$$

whereas the smooth part is still given by the Wong formula, Eq. (9). The oscillatory part is now significantly larger than before, since the negative exponent has been reduced by a factor of 2, that is $e^{-2\xi} \to e^{-\xi}$.

Figure 2 shows fusion cross sections for the $^{12}C+^{12}C$ system, obtained by taking only the even partial waves. The meaning of each line is the same as in Fig. 1. One can see that the fusion cross sections significantly oscillate as a function of E. The generalized Wong formula, supplemented with the oscillatory correction, gives an excellent approximation to the exact results.

3 Semi-microscopic modeling of heavy-ion fusion reactions with a beyond-mean field method

The potential model presented in the previous section works for light systems, such as $^{14}N+^{12}C$, but it largely underestimates fusion cross sections for heavier systems, such as $^{16}O+^{154}Sm$, at energies below the Coulomb barrier [1]. It has been well recognized by now that this large enhancement of subbarrier fusion cross sections is caused by the couplings of the relative motion between the colliding nuclei to several nuclear intrinsic degrees of freedom, such as low-lying collective excitations in the colliding nuclei as well as several nucleon transfer processes [1–4]. A natural framework for heavy-ion subbarrier fusion has thus been the coupled-channels method with relevant degrees of freedom [1, 12]. This approach has not only successfully accounted for the subbarrier enhancement of fusion cross sections for many systems but has also provided a natural interpretation of the so called fusion barrier distributions [1, 2, 13, 14].

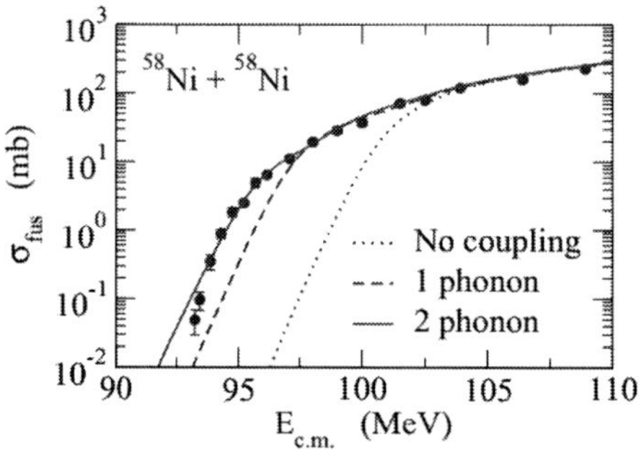

Figure 3. The results of coupled-channels calculations for the fusion cross sections for the ^{58}Ni+^{58}Ni system. The vibrational coupling to multi-quadrupole-phonon states are considered in the harmonic oscillator approximation. The experimental data are taken from Ref. [15].

In the coupled-channels approach, if the projectile nucleus is assumed to be inert, one expands the total wave function of the system in terms of the eigen-functions of the collective states in the target nucleus, $|\varphi_{I0}\rangle$, as

$$\Psi_{LM_L}(\boldsymbol{r}) = \sum_I \frac{u_I(r)}{r} Y_{LM_L}(\hat{\boldsymbol{r}})|\varphi_{I0}\rangle, \tag{14}$$

where I and L are the angular momenta for the target state and for the relative motion, respectively. Here, we have introduced the isocentrifugal approximation [1], and have assumed that L and M_L (thus also M_I) do not change by the excitation of the target nucleus. Substituting Eq. (14) to the projected Schrödinger equation, $\langle\varphi_{I0}|H - E|\Psi_{LM_L}\rangle = 0$, the coupled-channels equations for the radial wave functions $u_I(r)$ read [1],

$$\left[-\frac{\hbar^2}{2\mu}\frac{d^2}{dr^2} + \frac{L(L+1)\hbar^2}{2\mu r^2} + V_0(r) - E + \epsilon_I\right]u_I(r) + \sum_{I'} V_{II'}(r)u_{I'}(r) = 0, \tag{15}$$

where ϵ_I is the energy of the target state I. $V_{II'}(r)$ are the coupling matrix elements given by

$$V_{II'}(r) = \langle\varphi_{I0}|V_{\rm coup}(r, \alpha_{\lambda 0})|\varphi_{I'0}\rangle, \tag{16}$$

where $V_{\rm coup}$ is the coupling potential and $\alpha_{\lambda 0}$ is the excitation operator with a multipolarity λ.

In heavy-ion fusion reactions at energies around the Coulomb barrier, multiple excitations to higher members of collective modes, such as multi-phonon states and high-spin states in the ground state rotational band, often play an important role [2, 14, 16]. As an example, Fig. 3 shows fusion cross sections for the ^{58}Ni+^{58}Ni system obtained with the coupled-channels calculations which take into account the vibrational coupling to quadrupole phonon states. One can see that the coupling to the first excited state is insufficient in order to account for the subbarrier enhancement of fusion cross sections for this system. In order to take into account the multiple excitations in coupled-channels calculations, one usually uses the simple harmonic oscillator model for vibrational nuclei [1]. In this approximation, the n-phonon multiplet states are combined together to form a single effective channel as [1]

$$|n\rangle = \frac{1}{\sqrt{n!}}\left(a^\dagger_{\lambda 0}\right)^n|0\rangle, \tag{17}$$

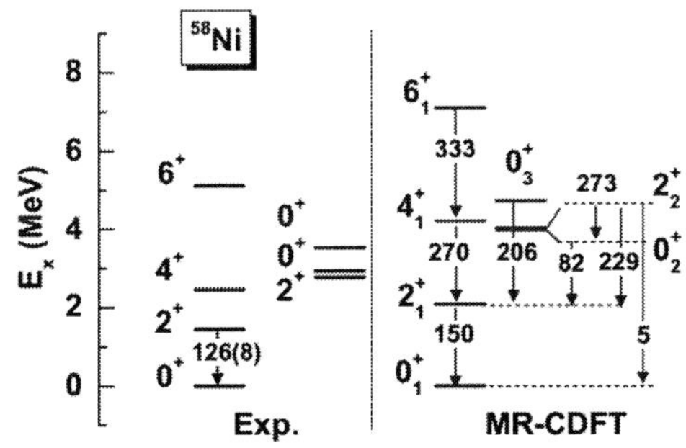

Figure 4. The low-lying energy spectra of ^{58}Ni obtained with the multi-reference covariant density functional theory (MR-CDFT) method with the PC-PK1 force. The arrows indicate the $E2$ transition strengths, given in units of e^2fm^4. The experimental data are taken from Refs. [20, 21].

where $a^\dagger_{\lambda 0}$ is the phonon creation operator and $|0\rangle$ is the vacuum state. The energy of this channel is given by $\epsilon_n = n\hbar\omega$, where $\hbar\omega$ is the oscillator quantum, and the matrix elements of $\alpha_{\lambda 0}$ are related to the phonon numbers as $\langle n|\alpha_{\lambda 0}|n'\rangle \propto \sqrt{n'}\delta_{n,n'-1} + \sqrt{n'+1}\,\delta_{n,n'+1}$. The solid line in Fig. 3 shows the result of the two phonon couplings. One can see that the subbarrier fusion enhancement is now well reproduced.

Even though the harmonic oscillator approximation appears to work well for the ^{58}Ni+^{58}Ni system, most nuclei, including ^{58}Ni, do not have a pure harmonic oscillator spectrum in reality. Concerning the ^{58}Ni nucleus, the degeneracy of the experimentally known two-phonon triplet is considerably broken (see Fig. 4), and also a recent theoretical calculation indicates that the $B(E2)$ strengths among the collective levels in ^{58}Ni deviate largely from the harmonic oscillator limit [17]. Such anharmonic vibrations can be described, *e.g.*, a multi-reference density-functional theory (MR-DFT), which has been rapidly developed for the past decade [18, 19]. This method is based on the so called beyond-mean-field approximation, which incorporates the angular momentum and particle number projections for a mean-field wave function. The quantum fluctuation of the mean-field wave function is also taken into account with the generator coordinate method (GCM).

Figure 4 shows the spectrum of the ^{58}Ni nucleus [7] constructed with the MR-DFT calculation with the covariant density functional with the PC-PK1 interaction [22]. One can see that the excitation energies of the 0^+_2, 2^+_2, and 4^+_1 states are about twice the energy of the 2^+_1 state. On the other hand, it is interesting to notice that the overall pattern of $B(E2)$ values is quite different from what would be expected for a harmonic vibrator, in which the $B(E2)$ value from any of the two-phonon triplet states to the 2^+_1 state is exactly twice the $B(E2)$ value from the 2^+_1 state to the ground state. In particular, the $E2$ transition from the 0^+_2 to the 2^+_1 states is much smaller than that for the $4^+_1 \to 2^+_1$ and the $2^+_2 \to 2^+_1$ transitions. Instead, the 0^+_2 state has a strong transition from the 2^+_2 state, which clearly indicates that the 0^+_2 state is not a member of the two-phonon triplet. Compared to the 0^+_2 state, the $E2$ transition strength for the $0^+_3 \to 2^+_1$ transition is much larger and is comparable to that for the $4^+_1 \to 2^+_1$ and the $2^+_2 \to 2^+_1$ transitions. This suggests that the 0^+_3 state is a better candidate for a member of the two-phonon triplets than the 0^+_2 state, even though the excitation energy is a little large.

In order to take into account such anharmonic nature of the vibrational excitations of ^{58}Ni in the coupled-channels calculations, we first replace the operator $\alpha_{\lambda 0}$ in Eq. (16) with the corresponding

 DOI: 10.1051/epjconf/201612207002

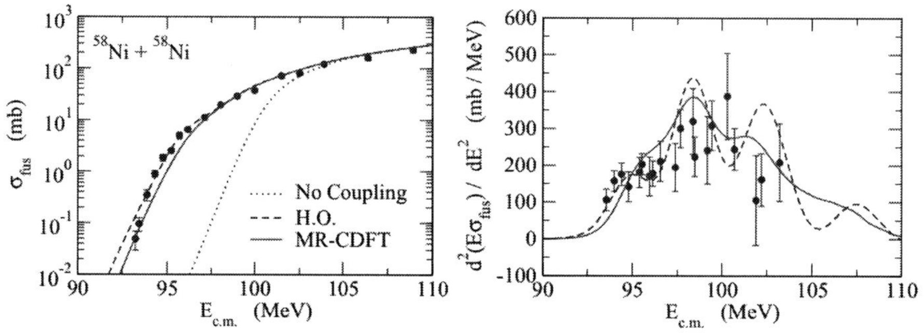

Figure 5. The fusion cross sections (the left panel) and the fusion barrier distributions (the right panel) for the ^{58}Ni+^{58}Ni system. The dashed line is the result of the coupled-channels calculations including the double quadrupole phonon excitations in each ^{58}Ni nucleus in the harmonic oscillator limit. The solid line, on the other hand, is obtained with the multi-reference covariant density functional theory (MR-CDFT) method by including the couplings to the 0_1^+, 2_1^+, 0_3^+, 2_2^+, and 4_1^+ states in ^{58}Ni. The dotted line in the left panel denotes the result in the absence of the channel couplings. The experimental data are taken from Ref. [15] for the fusion cross sections and from Ref. [16] for the fusion barrier distribution.

microscopic multipole operator. That is, we replace the matrix elements of $\alpha_{\lambda 0}$ as,

$$\langle \varphi_{I0}|\alpha_{\lambda 0}|\varphi_{I'0}\rangle \to \frac{4\pi}{3Z_T e R_T^\lambda}\langle \varphi_{I0}|Q_{\lambda 0}|\varphi_{I'0}\rangle, \tag{18}$$

where $Q_{\lambda\mu} = \sum_i r_i^\lambda Y_{\lambda\mu}(\hat{r}_i)$ is the microscopic multipole operator, and R_T and Z_T are the radius and the charge number of the target nucleus, respectively. $|\varphi_{I'0}\rangle$ on the right hand side is a many-body wave function evaluated with the MR-CDFT method. We still use a phenomenological potential for $V_0(r)$ and $V_{\rm coup}(r, \alpha_{\lambda 0})$, and we call our method a semi-microscopic approach.

In heavy-ion fusion reactions, the coupling of the ground state to the lowest-lying collective state is most important. The strength for such coupling can often be estimated from an experimental transition probability. Since it is too much to expect that a MR-CDFT calculation agrees perfectly with experimental data, we introduce an overall scaling factor to the matrix elements, Eq. (18), so that the transition from the lowest-lying collective state to the ground state is consistent with experimental data. The MR-CDFT calculation then provides the relative strengths among collective levels, which are often not available experimentally. The excitation energy, on the other hand, is often known well for many levels, and we simply take them in the calculations whenever they are available experimentally.

Figures 5 shows the fusion cross section $\sigma_{\rm fus}(E)$ and the fusion barrier distribution $D_{\rm fus}(E) = d^2(E\sigma_{\rm fus})/dE^2$ [2, 13] for the ^{58}Ni+^{58}Ni reaction so obtained. The dashed line shows the result of the coupled-channels calculations including up to the double phonon states in the harmonic oscillator limit. All the mutual excitations between the projectile and the target nuclei are included. On the other hand, the solid line in the figure is obtained with the coupling strengths evaluated with the MR-CDFT method. To this end, we include the 0_1^+, 2_1^+, 0_3^+, 2_2^+, and 4_1^+ states in ^{58}Ni in the coupled-channels calculations. Again, all the mutual excitation channels are taken into account. For comparison, the figure also shows the result of no-coupling limit by the dotted line. One can see that the calculations in the harmonic limit overestimate fusion cross sections at the two lowest energies, while the MR-

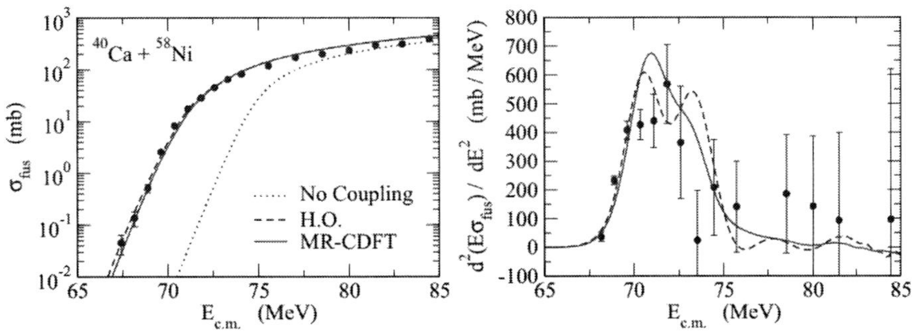

Figure 6. Same as Fig. 5, but for the ^{40}Ca+^{58}Ni system. The experimental data are taken from Ref. [23].

CDFT calculations underpredict fusion cross sections around 95 MeV. The energy dependence of fusion cross sections can be investigated with the fusion barrier distribution, which is plotted in the right panel of the figure. Even though the effect of anharmonicity is not large, one observes that the MR-CDFT calculation leads to a minor improvement by considerably smearing each peak.

We carry out similar calculations also for the ^{40}Ca + ^{58}Ni system [7] (see Fig. 6). The effect of channel coupling is much smaller than in the ^{58}Ni + ^{58}Ni system due to a smaller charge product, $Z_P Z_T$. Nevertheless, we again observe that the anharmonicity effect in ^{58}Ni smears the fusion barrier distribution, leading to a somewhat better agreement with the experimental fusion barrier distribution as compared to the results in the harmonic oscillator limit.

4 Summary

The research field of heavy-ion subbarrier fusion reactions started in the late '70s, when a large enhancement of fusion cross sections was experimentally discovered with respect to the prediction of a simple potential model. The Wong formula has been widely used to estimate fusion cross sections for a given single-channel potential, providing reference cross sections in the absence of channel couplings to discuss the subbarrier enhancement of cross sections. In this contribution, we have first extended the Wong formula by including the energy dependence of the parameters entering the formula, that is, the barrier height, the barrier position, and the barrier curvature. Evaluating these parameters for the grazing angular momentum at each energy, rather than at $l = 0$, we have shown that the energy-dependent version of Wong's formula reproduces the exact result well, even for light systems. We have also derived a compact analytic formula for the oscillatory part of fusion cross sections, which originate from the discrete nature of angular momentum. These oscillatory parts become important in light symmetric systems, for which the symmetrization of the system amplifies the oscillations.

We have then proposed a semi-microscopic approach to heavy-ion subbarrier fusion reactions. The basic idea of this approach is to combine the state-of-the-art nuclear structure calculations with coupled-channels calculations. For this purpose, we have used a multi-reference density functional theory (MR-DFT) based on the beyond-mean-field approach. The MR-DFT provides transition strengths among collective states without resorting to the harmonic oscillator model. We have applied this approach to the ^{58}Ni+^{58}Ni and ^{40}Ca+^{58}Ni fusion reactions, and have found that the anharmonicities smear the fusion barrier distributions, somewhat improving the agreement with the experimental data. It would be an interesting future problem to extend this treatment to heavy-ion elastic and inelastic scattering, for which the double-folding approach is applicable and thus a fully microscopic approach can be developed using the multi-reference density functional theory. That approach would also be useful in applications to nuclear data as well as the problem of nuclear transmutation [24].

Acknowledgements

This work was partially supported by the National Natural Science Foundation of China under Grant Nos. 11305134, 11105111, and the Fundamental Research Funds for the Central University (XDJK2013C028).

References

[1] K. Hagino and N. Takigawa, Prog. Theor. Phys. **128**, 1001 (2012); http://www.nucl.phys.tohoku.ac.jp/ hagino/fusion-corr.pdf
[2] M. Dasgupta, D.J. Hinde, N. Rowley and A.M. Stefanini, Annu. Rev. Nucl. Part. Sci. **48**, 401 (1998).
[3] A.B. Balantekin and N. Takigawa, Rev. Mod. Phys. **70**, 77(1998).
[4] B.B. Back, H. Esbensen, C.L. Jiang and K.E. Rehm, Rev. Mod. Phys. **86**, 317 (2014).
[5] L.F. Canto, P.R.S. Gomes, R. Donangelo, J. Lubian and M.S. Hussein, Phys. Rep. **596**, 1 (2015).
[6] N. Rowley and K. Hagino, Phys. Rev. **C91**, 044617 (2015).
[7] K. Hagino and J.M. Yao, Phys. Rev. **C91**, 064606 (2015).
[8] C.Y. Wong, Phys. Rev. Lett. **31**, 766 (1973).
[9] D.L. Hill and J.A. Wheeler, Phys. Rev. **89**, 1102 (1953).
[10] D.M. Brink, *Semi-Classical Methods for Nucleus-Nucleus Scattering*, (University Press, Cambridge, 1985).
[11] N. Poffé, N. Rowley and R. Lindsay, Nucl. Phys. **A410**, 498 (1983).
[12] K. Hagino, N. Rowley, and A.T. Kruppa, Comp. Phys. Comm. **123**, 143 (1999).
[13] N. Rowley, G.R. Satchler, and P.H. Stelson, Phys. Lett. **B254**, 25 (1991).
[14] J.R. Leigh *et al.*, Phys. Rev. **C52**, 3151 (1995).
[15] M. Beckerman *et al.*, Phys. Rev. **C23**, 1581 (1981).
[16] A.M. Stefanini *et al.*, Phys. Rev. Lett. **74**, 864 (1995).
[17] J.M. Yao, M. Bender, and P.-H. Heenen, Phys. Rev. **C91**, 024301 (2015).
[18] M. Bender, P.-H. Heenen, and P.-G. Reinhard, Rev. Mod. Phys. **75**, 121 (2003).
[19] J.M. Yao, K. Hagino, Z.P. Li, J. Meng, and P. Ring, Phys. Rev. **C89**, 054306 (2014).
[20] National Nuclear Data Center, http://www.nndc.bnl.gov/.
[21] J.M. Allmond *et al.*, Phys. Rev. **C90**, 034309 (2014).
[22] P.W. Zhao, Z.P. Li, J.M. Yao, and J. Meng, Phys. Rev. **C82**, 054319 (2010).
[23] D. Bourgin *et al.*, Phys. Rev. **C90**, 044610 (2014).
[24] E.Sh. Sukhovitskii, Y.-O. Lee, J. Chang, S. Chiba, and O. Iwamoto, Phys. Rev. **C62**, 044605 (2000).

Simulation for thick-target yields of transmutation reactions on radioactive targets, based on inverse kinematics

Shuichiro Ebata[1,a], Masayuki Aikawa[1], and Shotaro Imai[2]

[1]*Nuclear Reaction Data Centre, Faculty of Science, Hokkaido University, Sapporo 060-0810, Japan*
[2]*Institute for the Advancement of Higher Education, Hokkaido University, Sapporo 060-0817, Japan*

Abstract. To dispose of long-lived fission products (LLFP) ejected from nuclear reactor plants is one of the most important tasks on nuclear physics and engineering. The experiments with the radiative target are limited, due to the high radioactivity and chemical property of the target. In consequence, the nuclear reaction data for LLFP are insufficient. In this work, we propose a feasible method to obtain the data for radiative targets using inverse kinematics and simulate specific systems to evaluate the thick-target yields of the nuclear transmutation reactions for LLFP.

1 Introduction

To reduce radioactive waste ejected from nuclear power plants is a significant topic in globe. Nuclear transmutation is a promising technology to dispose of the waste with high radioactivity. The problematic high-level radioactive waste is composed mainly of the minor actinide (MA) and long lived fission products (LLFP). For the reduction of MA, the accelerator-driven system (ADS) has been studied in global, e.g. J-PARC plans to establish the ADS test facility [1]. However, an effective procedure to reduce the LLFP has not been found yet. An essential problem for the development of LLFP transmutation is the insufficiency of the reaction data with a charged particle. Although to accumulate the data is strongly demanded, the experiments with the LLFP target are very few due to the several difficulties which are the high radioactivity and the chemical instability.

Due to the recent progress of the radioactive isotope beam technique, we can utilize LLFP nuclei as the beam. The experiments to access the data of LLFP reactions by the inverse kinematics can be namely performed, while avoiding the radioactive target [2]. When the lump of radioactive waste is assumed, it needs huge effort to obtain many data in a wide energy region. Hence we consider the integration quantity composed by microscopic values and a target matter property: a thick-target yield (TTY). The TTY is a macroscopic value which is defined with corresponding to the reaction, and is composed of the target properties such as the atomic weight A (g mol^{-1}), the density ρ (g cm^{-3}) and also the cross section σ (cm^2). If the TTY is obtained microscopically, the excitation function of the reaction cross section is needed. Therefore, we suggest the method to estimate the TTY of the transmutation reaction from the inverse kinematics, without measurements of microscopic values.

In this paper, we introduce briefly our method and its applications to show the availability of the method. Finally, we summarize the contents and mention the feasibility for the LLFP reaction data.

[a]e-mail: ebata@nucl.sci.hokudai.ac.jp

2 Method

Most fundamental definition of reaction probability Y can be written as a ratio between the numbers of reacted and un-reacted particle. The TTY is equivalent to this Y in the case of a thick target. The differential TTY dY within an infinitesimal thickness dx (cm) can be described as

$$dY = \sigma \frac{N_A \rho}{A_T} dx, \tag{1}$$

where N_A (mol^{-1}) is the Avogadro constant, A_T (g mol^{-1}) and ρ (g cm^{-3}) are the atomic weight and density of the target. Here we introduce the stopping power $S(\varepsilon) = -\frac{A_P d\varepsilon}{d(\rho x)}$ (MeV g^{-1} cm^2) where A_P is the projectile atomic weight and energy $\varepsilon = E/A_P$ is represented in the energy per nucleon. The TTY at the incident energy ε_{in} (MeV/nucleon), can be obtained by integrating dY for the projectile energy ε in the target:

$$Y(\varepsilon_{in}) = \frac{N_A A_P}{A_T} \int_0^{\varepsilon_{in}} \frac{\sigma(\varepsilon)}{S(\varepsilon)} d\varepsilon. \tag{2}$$

The lower limit of energy integration of Eq. (2) corresponds to zero, i.e., the projectile is stopped inside the target. The form of the TTY does not change in the inverse kinematics although A_P and A_T are exchanged, and the $S(\varepsilon)$ is changed to that of the inverse kinematics system.

Here we consider the conversion between the TTYs of the original reaction and of its inverse kinematics. The variable defined in the original reaction ("*forward*") is denoted as Y_{for}, and for its "*inverse*" kinematics as Y_{inv}. The notations P and T for the projectile and target are defined as those of the original reaction. In these notations, a ratio $R(\varepsilon)$ between the differential yields dY_{for} and dY_{inv} can be expressed as

$$R(\varepsilon) = \frac{dY_{for}}{dY_{inv}} = \frac{A_P^2}{A_T^2} \frac{S_{inv}(\varepsilon)}{S_{for}(\varepsilon)}, \tag{3}$$

because the cross section is common among the "*forward*" and "*inverse*" reactions. If we can obtain the Y_{inv} and calculate the stopping powers at each projectile energy ε, the Y_{for} can be estimated without measuring the cross sections [3]. Indeed the stopping power can be calculated accurately by the present simulation code e.g. SRIM2008 (The Stopping power and Range of Ions in Matter) [4].

3 Result

In this section, we show two applications for the availability of the method with the SRIM code, and for a more specific system to access the LLFP data.

3.1 ^{nat}Cu + ^{12}C reaction system

The reaction ^{nat}Cu (^{12}C, X)^{24}Na is considered here because there are some cumulative cross section data with respect to the energy [5]. More detail data in the low energy region is necessary, thus we prepare the excitation function of the cross section, $\sigma(\varepsilon)$, from the spline interpolation of the data, shown as Fig.1 (i). The lowest energy corresponds to the roughly estimated Coulomb barrier U_C of this system: $U_C = e^2 Z_T Z_P / 1.2(A_T^{1/3} + A_P^{1/3})$.

We estimate the TTYs using the $\sigma(\varepsilon)$ and the stopping powers $S_{for}(\varepsilon)$, $S_{inv}(\varepsilon)$ which are calculated by SRIM 2008. The ratio $R(\varepsilon)$ is shown in Fig.1 (ii). The Y_{for} and Y_{inv} at ε_{in}=40 and 100 MeV/nucleon are shown in the Table 1. We can see a plateau behaviour of $R(\varepsilon)$ over ε_{in}=50 MeV/nucleon. The plateau of R appears generally in the high energy region although the conversion energy point depends on the proton numbers of the reaction system, because the behaviour of a stopping power is governed by the Coulomb force. The rise of $R(\varepsilon)$ has a strong dependence on the difference between Z_T and Z_P [3]. Therefore we propose a more convenient conversion relation:

$$Y_{\rm for}(\varepsilon_{\rm in}) \simeq \tilde{R} Y_{\rm inv}(\varepsilon_{\rm in}), \tag{4}$$

where $\tilde{R}$ is constant value. In this example $\tilde{R}$=1.1 is chosen and the TTY $Y^{\rm R}$ estimated by Eq. (4) is shown also in Table 1. The convenient linear relation Eq. (4) well works at both 40 and 100 MeV/nucleon. If we can obtain the energy dependence of TTY in the inverse kinematics system, we can completely calculate the TTY of the original reaction theoretically. Even when we could get the data $Y_{\rm inv}$ at only one energy point, we will estimate an approximate $Y_{\rm for}$ if we use the Eq. (4).

Table 1. The TTYs of $^{\rm nat}$Cu + ^{12}C reaction with $\varepsilon_{\rm in}$=40 and 100 (MeV/nucleon) are listed. $Y^{\rm D}$ and $Y^{\rm R=1.1}$ denote the values calculated by Eq. (2) and evaluated by Eq. (4) with $\tilde{R}$=1.1, respectively.

$\varepsilon_{\rm in}$ (MeV/nucleon)	$Y^{\rm D}_{\rm for}(\varepsilon_{\rm in})$	$Y^{\rm D}_{\rm inv}(\varepsilon_{\rm in})$	$Y^{\rm R=1.1}_{\rm for}(\varepsilon_{\rm in})$
40	0.91×10^{-5}	0.86×10^{-5}	0.94×10^{-5}
100	0.114×10^{-3}	0.103×10^{-3}	0.113×10^{-3}

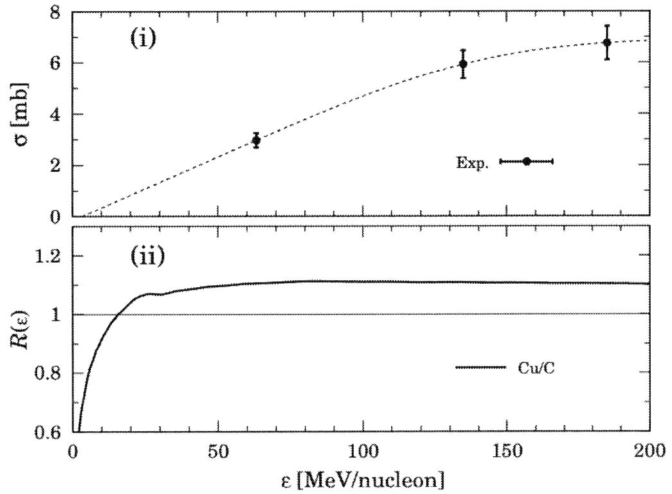

Figure 1. (i) Cross section with respect to ε of the $^{\rm nat}$Cu(^{12}C,X)^{24}Na with experimental data [5], and (ii) the evaluated ratio $R(\varepsilon)$ by SRIM 2008 [4].

DOI: 10.1051/epjconf/201612207003

3.2 Application to the LLFP ^{135}Cs

^{135}Cs is one of the most problematic LLFP, which has the half-life over two million years. We simulate the transmutation reaction of ^{135}Cs bombarded by ^{12}C, through their inverse kinematics. For the simulation, we utilize the PHITS (Particle and Heavy Ion Transport code System) which is a general purpose Monte Carlo particle transport simulator developed under the collaboration among JAEA, RIST, KEK and several other institutes [6]. These simulations for TTYs of ^{12}C(^{135}Cs, X)$^{A'}$Cs at several energies as the inverse kinematics of ^{135}Cs(^{12}C, X)$^{A'}$Cs reaction, are performed by the PHITS which includes an empirical formula to generate reaction cross sections.

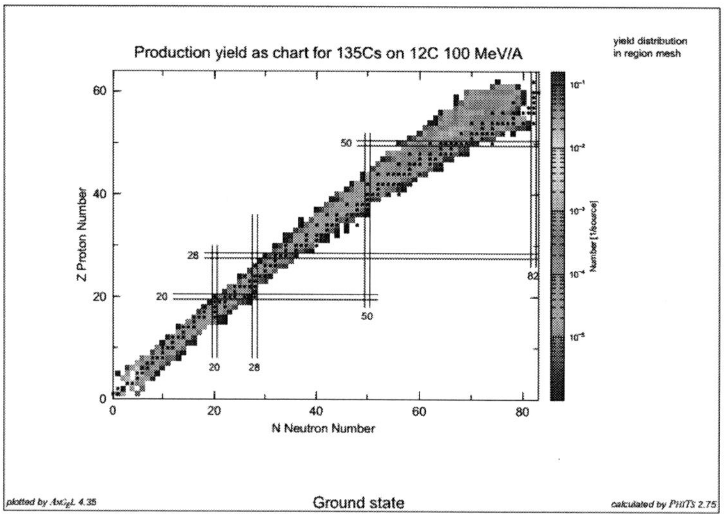

Figure 2. The distribution of fragments for ^{135}Cs + ^{12}C reaction simulated by PHITS 2.75.

Table 2. The PHITS condition for ^{12}C(^{135}Cs, X)$^{A'}$Cs

Beam intensity	1000 pps
Irradiation time	1000 s
^{12}C target thickness	2.0 cm
Beam energy region	20 – 100 MeV/nucleon

The reaction induced ^{135}Cs is equivalent to the objective transmutation reaction. Figure 2 shows the distribution of fragments in the PHITS simulation at 100 MeV/nucleon under the conditions listed in Table 2. To evaluate the TTY of ^{135}Cs(^{12}C, X)$^{A'}$Cs reaction with $A'(\neq 135)$, we sum up the production yields of the reaction in the inverse kinematics: ^{12}C(^{135}Cs, X)$^{A'}$Cs. Although the evaluated TTY does not reach that for all reaction yields, this TTY might be the lower limit of the transmutation reaction. These TTY at each energy from 20 to 100 MeV/nucleon are displayed in Fig.3 as filled circle symbols. The stopping power ratio of the system is calculated by SRIM 2008, which is displayed in Fig.3 with star symbols. Furthermore the TTY of the original reaction is evaluated using Eq. (4) with the constant stopping power ratio $\tilde{R}$=1.06, which is plotted as open circle.

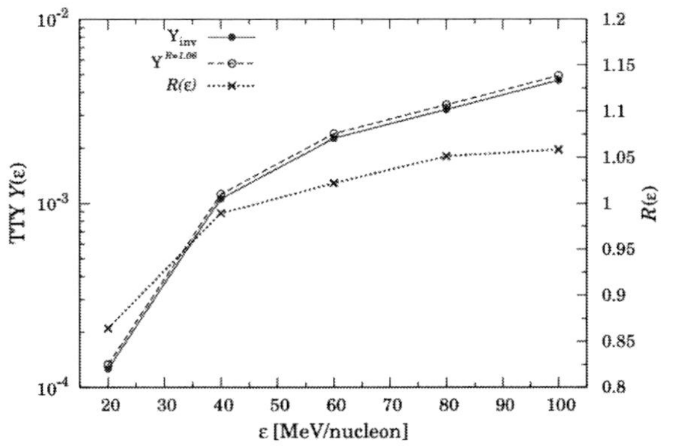

Figure 3. The TTY of ^{12}C(^{135}Cs, X)$^{A'}$Cs (solid circles), the evaluated TTY Y^{R} by Eq. (4) with $\tilde{R}$=1.06 (open circles), and the $R(\varepsilon)$ (stars), are shown.

4 Conclusion

To accumulate the data for the transmutation of LLFP, the recent progress of radioactive isotope beam facility is essential. We reported the conversion method assuming the inverse kinematics using a radioactive beam, to obtain the TTY of the reaction with a radioactive target.

When we apply the method to ^{nat}Cu + ^{12}C, the more convenient conversion relation is found, which is to replace the energy dependent ratio with the constant value. The ratio between differential TTYs (dY_{for}, dY_{inv}) which is governed by the stopping powers decided with the combination of target and projectile, has the saturated behaviour in a high energy region. The linear relation between Y_{for} and Y_{inv} is justified when the reaction cross section in the low energy has only minor contribution to the TTY. We simulate also the ^{135}Cs bombarding on ^{12}C reaction with helps of PHITS and SRIM to estimate the TTY of the LLFP transmutation reaction. The TTYs of breaking up ^{135}Cs to $^{A'}$Cs by ^{12}C are estimated. Although the stopping power ratio converge to about 1.06 in this system, the tendency of the ratio is gentler up to the convergence than the previous example.

In future work, we will apply our method to more specific LLFP subjects and analyse the second particle effects for yields in the Monte Carlo simulation. Although for some LLFP data our method is effective, it has not been finished for other LLFP, yet. We should develop the extended procedure to gather the more accurate data for the transmutation technology.

Acknowledge

This works was funded by ImPACT Program of Council for Science,Technology and Innovation (Cabinet Office, Government of Japan).

References

[1] http://j-parc.jp/Transmutation/en/index.html
[2] H. Wang et al., http://dx.doi.org/10.1016/j.physletb.2015.12.078
[3] M. Aikawa, S. Ebata and S. Imai, Nucl. Instr. Meth. B **353**, 1 (2015)
[4] J. F. Ziegler, J. P. Biersack, M. D. Ziegler, SRIM: the Stopping and Range of Ions in Matter, http://www.srim.org, Nucl. Instr. Meth. B **268**, 1818 (2010)
[5] H. Yashima et al., Phys. Rev. C**66**, 044607 (2002)
[6] T. Saito, et al., J. Nucl. Sci. Technol. 50:9, 913-923 (2013).

EPJ Web of Conferences **122**, 08001 (2016) DOI: 10.1051/epjconf/201612208001
*CNR*15*

Few-Body Problem in Nuclear Reactions

Beyond the horizon of the three-body Faddeev equations

Shinsho Oryu[1,a], Yasuhisa Hiratsuka[1,b], and Takashi Watanabe[1,c]

[1] *Department of Physics, Faculty of Science and Technology, Tokyo University of Science, 2641 Yamazaki, Noda-city, Chiba 278-8510, Japan*

Abstract. Recent progress in few-body problem physics based on the three-body Faddeev equations is reviewed for three-related fields. The first field involves the description of light nuclear reactions in terms of multi-channel three-body Faddeev equations. The second field is the investigation of the two- and three-body threshold behaviors for the NNπ system using the three-body Faddeev equations, where the πD and the NN′ or N-(Nπ) scattering lengths are calculated, and also we show that the NN′ potential has a long range term of $1/r^2$ form. The third is a new Coulomb treatment in terms of a generalized screening range to describe on-shell Coulomb amplitudes which is useful in the three-body Faddeev equations. This procedure reproduces both of the Coulomb phase shift and the wave function from the electron-electron to the heavy-ion-heavy-ion systems.

1 Introduction

1.1 Nuclear reactions

In the numerical approach to the light nuclear reactions, the exchange dynamics of a few-nucleons or the light nuclear-clusters are dominant for large momentum transfer and the details of the nuclear force frequently appear. The present approach is different from the method used for heavy nuclei. Especially, the molecular structure with several alpha-clusters is well developed in light nuclei [1]. Beside the alpha-particle, realistic nuclei are constructed by the proton, neutron, d, t, ^{3}He, ^{12}C, ^{16}O, ^{20}Ne, ^{24}Mg, ^{28}Si, ^{32}S, and so on.

Those structures can be seen by cooling the A-nucleons with the anti-symmetrized molecular dynamics (AMD)-method [2], or by the Jacobi-coordinate based AMD (JAMD) [3]. However, the simplest three-nucleon system has been extensively investigated by the three-body Faddeev equations [4].

The three-body Faddeev equations can theoretically represent all the spectra in the three-body problem, Figure 1. Therefore, one can say that the A-body Faddeev equations are the most powerful method for light nuclear reactions, although the increase in numerical cost and the advances in hardware are always put in the balance. At this stage, the three-cluster Faddeev equations can represent A-nucleon reaction systems by using the coupled-channel method with the three-cluster potentials which will be presented in the section 2.

[a] e-mail:oryu@rs.noda.tus.ac.jp
[b] e-mail:yasuhisa1059@gmail.com
[c] e-mail:watanabe@ph.noda.tus.ac.jp

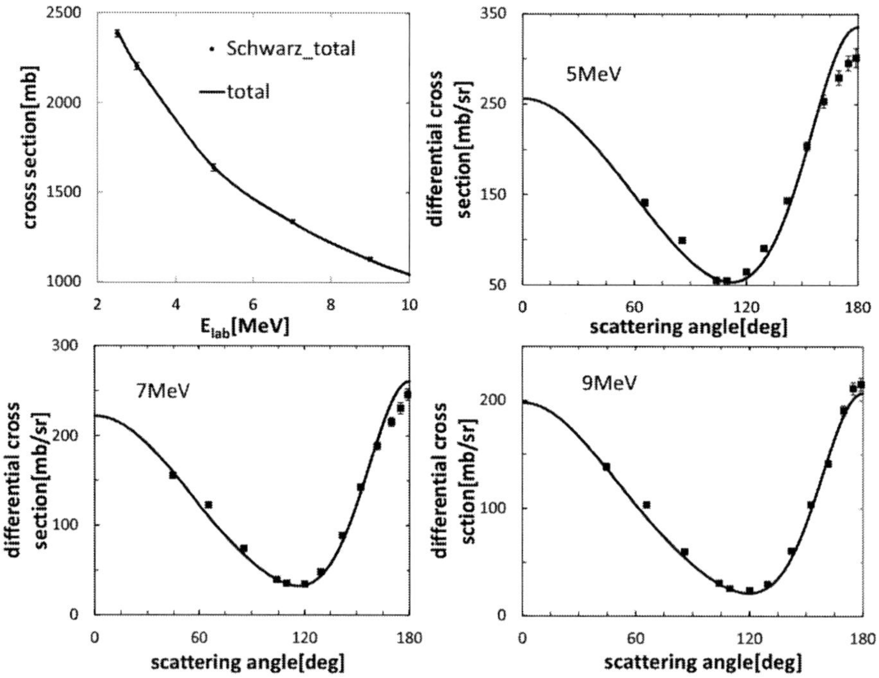

Figure 1. The *nd* total cross section and the differential cross sections for 5MeV, 7MeV and 9MeV. The calculations (solid-line) are done by Eq.(5) in our three-body program. The experimental data are taken from Ref.[5].

1.2 The threshold behaviors

In Lovelace's idea in the early 1960s, he pointed out that the three-body Faddeev equations become "a set of coupled Lippmann-Schwinger equations for the scattering of bound states and unstable particles. The potentials in these scattering equations for composite particles are nonlocal and energy-dependent". In other words, the Hamiltonian below the three-body threshold is different from that above the threshold. Therefore, the eigenvalues are also different. However, we require that the coupled Lippmann-Schwinger equations are analytically continued from the three-body Faddeev equations at the three-body break up threshold. This idea suggests that a composite two-body Hamiltonian is analogous to the three-body Hamiltonian at the threshold between the two-body and the three-body systems. We would like to show our recent progress in section 3.

1.3 The generalized screening range band in the Coulomb problem

There is a long history involving the Coulomb treatment in the few-body problems. Unfortunately, because we had no analytic solution of the three-body Faddeev equations, numerical calculations have been performed using a screened Coulomb potential. Such a screened Coulomb potential becomes a

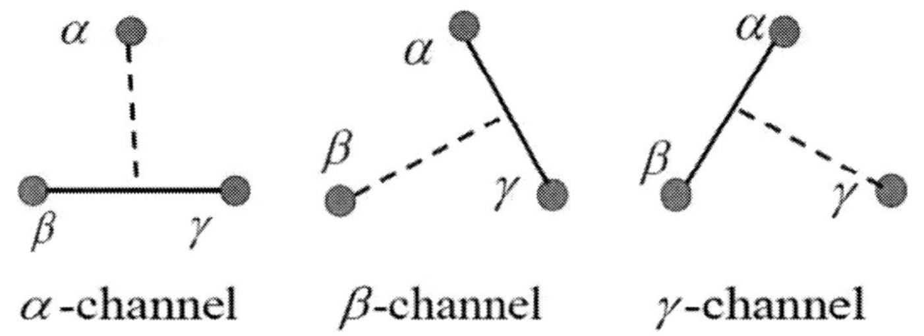

Figure 2. Three particle channels, and Jacobi coordinate.

genuine Coulomb potential with increasing a range, however, the phase shift and the wave function don't converge. It was claimed that in the proton-deuteron scattering calculation using the three-body Faddeev equation with a screened Coulomb potential, convergence of the procedure is found at moderate screening radii [16]; where the screening and a phase shift renormalization approach was employed for including the Coulomb interaction by an on-shell approximation. In this paper, we will show that the screening range is given by a unique band which is an energy-dependent function. Using this range, we obtain the two-body on- and off-shell Coulomb amplitudes for the three-body Faddeev equations in section 4.

2 Multi-channel Faddeev equations

2.1 Single-channel three-body Faddeev equations

Let us define the three *particle*-channels: the α-channel or $\alpha(\beta,\gamma)$-channel indicates (β,γ)-interacting-pair with a spectator α-particle, and the β-channel denotes $\beta(\gamma,\alpha)$, and the γ-channel is $\gamma(\alpha,\beta)$, respectively (see Figure 2). Then we have the three-body Faddeev equations,

$$T^{\alpha} = T_{\alpha} + \sum_{\gamma\neq\alpha} T_{\alpha} G_0 T^{\gamma}, \tag{1}$$

where T^{α} is the three-body T-matrix for the α-initial channel, and T_{α} denotes the two-body T-matrix in the three-body Hilbert space, and G_0 is the three-body free Green's function, respectively.

The equation can be reformulated for the three-body rearrangement amplitudes by putting

$$T^{\alpha} \equiv \sum_{\beta=1,2,3} T_{\alpha\beta}, \tag{2}$$

$$U_{\alpha\beta} \equiv T_{\alpha\beta} - T_{\alpha}\delta_{\alpha\beta} = T_{\alpha} G_0 u_{\alpha\beta} G_0 T_{\beta}, \tag{3}$$

$$X_{\alpha n\beta m} \equiv \psi_{\alpha n} u_{\alpha\beta} \psi_{\beta m} = g_{\alpha n} G_0 u_{\alpha\beta} G_0 g_{\beta m}, \tag{4}$$

where $\psi_{\alpha n}$ and $\psi_{\beta m}$ are the two-body wave functions in the three-body Hilbert space, and $g_{\alpha n}$ and $g_{\beta m}$ are the corresponding two-body form factors. Therefore, $X_{\alpha n\beta m}$ is the three-body rearrangement

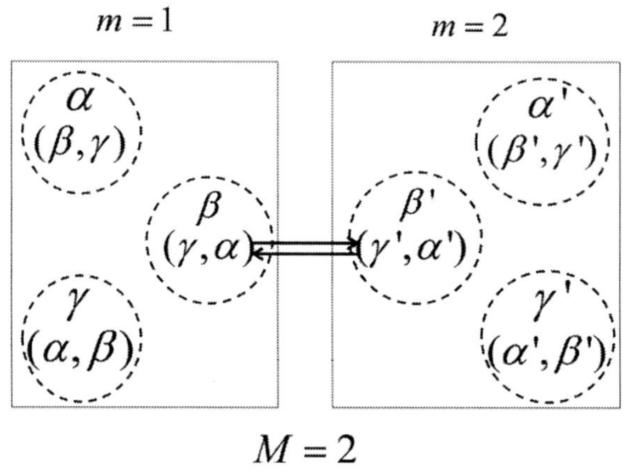

Figure 3. The mixture between two-different three-cluster systems: (α, β, γ) and $(\alpha', \beta', \gamma')$ is performed by the two-body inter-cluster amplitude: $(\gamma, \alpha) \Leftrightarrow (\gamma', \alpha')$ in the case $\beta = \beta'$. The figure shows the multiplicity number: $M = 2$.

amplitudes for the *physical*-states $n(\in \alpha)$, $m(\in \beta)$ which satisfies the following equations,

$$X_{\alpha n,\beta m} = Z_{\alpha n,\beta m} + \sum_{\gamma} \sum_{s,t} Z_{\alpha n,\gamma s} \tau^{\gamma}_{s,t} X_{\gamma t,\beta m}, \quad (5)$$

$$Z_{\alpha n,\beta m} = g_{\alpha n} G_0 g_{\beta m} (1 - \delta_{\alpha\beta}), \quad (6)$$

where $Z_{\alpha n,\beta m}$ gives the Born term of the so called AGS equations [6],[7]. The function $\tau^{\gamma}_{s,t}$ represents the two-body propagators.

2.2 Multi-channel three-body Faddeev [MC3F] equations

The three-cluster systems which consist of A-nucleons, are characterized by the *multiplicity*: M. Each three-cluster dynamics is described by the three-body Faddeev equations; however for $M \geq 2$, the equations are mixed via the two-cluster subsystem with a common spectator particle. Fig.3 illustrates the mixing process for the $M = 2$ case. Let us call the equations the "multi-channel three-body Faddeev" (MC3F) equations where only the two-body sub-amplitudes are input [8],[9]. The two-body sub-amplitudes are obtained by the two-cluster equation with the resonating group method (RGM) [10], which is constructed based on the NN potential where the spurious solutions are extracted by using the so called Orthogonality Condition Model(OCM) [11][12] or the Fish-bone optical potential method (FBOM)[13]. Therefore, our MC3F equations are given by [9],

$$X^{a,b}_{\alpha n,\beta m} = Z^{a,b}_{\alpha n,\beta m} + \sum_{c,d}^{M} \sum_{\gamma}^{3} \sum_{s,t}^{N} Z^{a,c}_{\alpha n,\gamma s} \tau^{c,d}_{\gamma,st} X^{d,b}_{\gamma' t,\beta m}, \quad (7)$$

$$Z^{a,b}_{\alpha n,\beta m} = g^{a}_{\alpha n} G_0 g^{b}_{\beta m} (1 - \delta_{\alpha\beta}) \delta_{ab}, \qquad (\alpha \in a,\ \beta \in b) \quad (8)$$

where the particle channels belong to the multiplicity numbers; i.e. $\alpha \in a$, $\beta \in b$, $\gamma \in c$ & d. Therefore, the equation becomes $3 \times N \times M$ multiple-coupled-integral equations. The dimension of the multiplicity is very large but the equations are manageable using modern computers.

Let us consider an example of Eq.(7) the scattering: ^{4}He+$p \rightarrow {}^4$He+p in the three-cluster systems for $M = 2$ case where the ^{3}He+n+p, and ^{3}H+p+p systems appear for a certain energy region. Since p $(\gamma = \gamma')$ is common in both systems, then the two-body subsystems (^{3}He+$n \leftrightarrow {}^3$H+p) can be mixed by $M = 2$, and then a, b, c, d=1, 2, and α-channel: ^{3}He(n, p), β-channel: n(p,^{3}He) and γ-channel: p(^{3}He,n), while α'-channel: ^{3}H(p, p), β'-channel: p(p,^{3}H) and γ'-channel: p(^{3}H,p). Therefore, the Born term is described only for the same multiplicity number, the mixture of the multiplicity numbers occurs by the two-body propagator: $\tau^{c,d}_{\gamma,st}$. Since, $\gamma \in c$, $\gamma' \in d$ and $\gamma = \gamma'$, then $\tau^{c,d}_{\gamma,st} \equiv \tau^{c,d}_{\gamma\gamma',st}$ [9]. In this method, the most important aspect is the investigation of the cluster-cluster interactions. If we obtain sufficient quality for the cluster-cluster amplitude, the MC3F equation is automatically solved. In alternative approach for such a system, the four-, and five-cluster Faddeev equations are available.

3 Two-body reduction of the three-body Faddeev equations

Let us consider the NNπ three-body system which is the simplest system related to the nuclear potential. The system was extensively studied for πd scattering by Afnan and Thomas [17]. Below the N$'$ $\rightarrow$Nπ break up threshold, the three-body NNπ system is reduced to the quasi two-body NN$'$ system which can be represented by the NN$'$ multi-channel Lippmann-Schwinger (MLS) equations, where N$'$ (or Nπ bound states) is described by the p_{11} state.

3.1 The two-body reduction in the three-body system

Following the Lovelace idea, the MLS equations are analytically continued from the three-body Faddeev equations at the three-body break up threshold $E^{(3)} \equiv E = 0$. The requirement is satisfied by adopting a "quasi two-body free Hamiltonian" $\overline{H}_0^{(3)} \equiv H_0^{(3)} + \epsilon_B$ instead of the "two-body free Hamiltonian" $H_0^{(2)}$. In other words, the zero limit of the binding energy: $\epsilon_B(> 0) \rightarrow 0$ shows that a quasi two-body set: $\{\overline{H}_0^{(3)} \equiv H_0^{(3)} + \epsilon_B, \overline{E} \equiv E + \epsilon_B\}$ becomes a three-body set: $\{H_0^{(3)}, E\}$, they are illustrated in Figure 4. In general, the analytic continuation is performed at every threshold between the A-body and $(A-1)$-body systems. Figure 4 shows the thresholds between the three-body and the two-body systems, and between the two-body and the one-body systems which are generally caused by decreasing the A-body system energy or cooling the A-body system's temperature down.
1) The analytic continuation between 3-body and 2-body is not given directly by the path "a": from the three-body free Hamiltonian and the system energy: $\{H_0^{(3)}, E^{(3)}\}$ to the two-body: $\{H_0^{(2)}, E^{(2)}\}$ with the two-body free Hamiltonian and the system energy, but the path "b" with the quasi two-body Hamiltonian and the system energy $\{\overline{H}_0^{(3)}, E_{\rm cm}^{(3)}\}$.
The Born term of the relativistic three-body Faddeev equations is given by N/D. The denominator is written in terms of the invariant energy S and the individual energies $\omega_j(q_j) \equiv \sqrt{q_j^2 + m_j^2}$ (with j=1, 2, 3),

$$\begin{aligned}
D_{\rm Fadd} &\equiv \sqrt{S} - (\omega_1(q_1) + \omega_2(q_2) + \omega_3(q_3)) && (9)\\
&= (\sqrt{S} - m_1 - m_2 - m_3) - (\omega_1(q_1) + \omega_2(q_2) + \omega_3(q_3) - m_1 - m_2 - m_3) && (10)\\
&= E^{(3)} - H_0^{(3)} && (11)\\
&= (E^{(3)} + \epsilon_B) - (H_0^{(3)} + \epsilon_B) && (12)\\
&= \overline{E}^{(3)} - \overline{H}_0^{(3)} \equiv E_{\rm cm}^{(3)} - \overline{H}_0^{(3)} \equiv D_{\rm E2Q} && (13)
\end{aligned}$$

where the $D_{\rm Fadd}$ function is defined for the original Faddeev equations, and $D_{\rm E2Q}$ is the denominator of the energy dependent two-body quasi (E2Q) potential. The nonrelativistic approximation for the $D_{\rm Fadd}$ function is given by using $\left(\sqrt{S} - m_1 - m_2 - m_3\right) + \epsilon_B \equiv E^{(3)} + \epsilon_B \equiv E + \epsilon_B \equiv E^{(3)}_{\rm cm}$, and from Eqs.(12), (13), we obtain

$$D_{\rm E2Q} \approx E^{(3)}_{\rm cm} - \left(\frac{q_1^2}{2\mu_1} + \frac{p_1^2}{2\nu_1} + \epsilon_B\right) \tag{14}$$

$$= E^{(3)}_{\rm cm} - \left(\frac{\overline{q}_1^2}{2\mu_1} + \frac{\overline{p}_1^2}{2\nu_1}\right) \tag{15}$$

with the reduced masses $\mu_1 = m_1(m_2+m_3)/(m_1+m_2+m_3)$, $\nu_1 = m_2m_3/(m_2+m_3)$. If we define the relations between $\overline{q}_1$ and q_1 by,

$$\frac{\overline{q}_1^2}{2\mu_1} = \frac{q_1^2}{2\mu_1} + \epsilon_B, \tag{16}$$

and substitute Eq.(16) into Eq.(14), and compare with Eq.(15), we obtain

$$\frac{\overline{p}_1^2}{2\nu_1} = \frac{p_1^2}{2\nu_1} \quad \text{with} \quad \overline{\mathbf{p}}_1 = \frac{m_3\overline{\mathbf{q}}_2 - m_2\overline{\mathbf{q}}_3}{m_2+m_3}, \quad \text{and} \quad \mathbf{p}_1 = \frac{m_3\mathbf{q}_2 - m_2\mathbf{q}_3}{m_2+m_3}, \tag{17}$$

where the regions of the MLS equations and the Faddeev equations are defined by

$$0 \le \overline{q}_j < \infty, \quad \text{and} \quad 0 \le q_j < \infty. \tag{18}$$

The transformation from the Faddeev equations to the MLS equations is accomplished with the normalization path "c" by manipulating the Hamiltonian by taking $\overline{p}_1^2/2\nu_1 \to 0$ (or $\overline{\mathbf{p}}_1 \equiv m_3\overline{\mathbf{q}}_2 - m_2\overline{\mathbf{q}}_3 = 0$), and by putting $E^{(3)}_{\rm cm} \to E^{(2)}$ in Eq.(15). Therefore, we obtain

$$D_{\rm E2Q} = E^{(3)}_{\rm cm} - \left(\frac{\overline{q}_1^2}{2\mu_1} + \frac{\overline{p}_1^2}{2\nu_1}\right) \to E^{(2)} - \frac{\overline{q}_1^2}{2\mu_1} \equiv E^{(2)} - H_0^{(2)} \equiv D_{\rm MLS}. \tag{19}$$

2) The path a′: from the two-body free Hamiltonian and the system energy: $\{H_0^{(2)}, E^{(2)}\}$ to the one-body set: $\{H_0^{(1)}, E^{(1)}\}$ is replaced with the path b′ to ensure the analytic continuation to the set $\{\overline{H}_0^{(2)}, E^{(2)}_{\rm cm}\}$, where $H_0^{(1)}$ is the one-body kinetic operator for the center-of-mass motion, and finally $\{\overline{H}_0^{(2)}, E^{(2)}_{\rm cm}\}$ arrives at $\{H_0^{(1)}, E^{(1)}\}$ via the path c′. These paths are illustrated by,

$$D_{\rm MLS} = E^{(2)} - H_0^{(2)} = (E^{(2)} + \zeta) - (H_0^{(2)} + \zeta) \equiv \overline{E}^{(2)}_{\rm cm} - \overline{H}_0^{(2)} \to E^{(1)} - H_0^{(1)} \tag{20}$$

with $E^{(1)} = E^{(2)} + \zeta = E + \epsilon_B + \zeta$. Because the one-body Hamiltonian $H_0^{(1)} (\to H_0^{(1)})$ will vanish due to the zero relative energy and no "one-body potential", then the "eigenenergy" $E^{(1)}$ should be zero. Therefore, we obtain the three-body eigenenergy $E = -\epsilon_B - \zeta$, where ζ is the separation energy from the three-body bound state (see Figure 4).
This means that if the total wave function is written in separable form: $\Phi(q_j, p_j) = \psi_j(q_j)\phi_j(p_j)$ with the two-body wave function: $\phi_j(p_j)$, and the wave function between a particle j and a pair (ki): $\psi_j(q_j)$. However, such a separable form is not always correct. The Coulomb three-body problem, for example, cannot be written in such an expansion. Generally, the eigenequation with single argument is well defined. Hence the variables of the NNπ three-body Faddeev equations are $\{q, p\}$, the NNπ binding energy $E = -\epsilon_d - \epsilon_\pi$ will be obtained by our two-step method via the E2Q potential with single argument, where the integration region changes to $0 \le \overline{q} < \infty$ which starts from the NN′ branch point not from the NNπ threshold.

 DOI: 10.1051/epjconf/201612208001

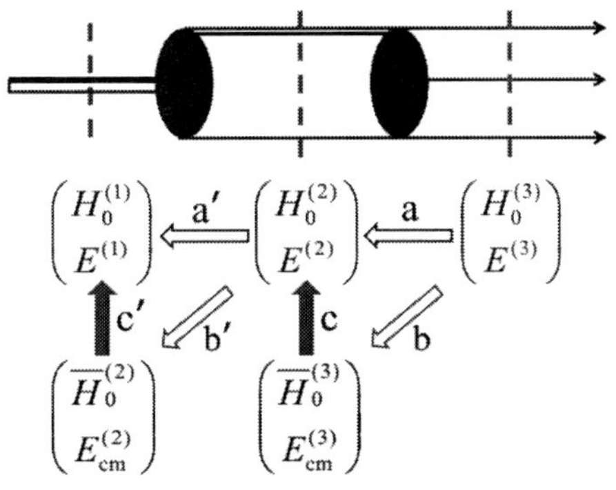

Figure 4. The threshold between the three-body system and the two-body system, and sequentially the threshold between the two-body system and the one-body system appear by cooling the temperature of the system. 1) The path "a": from the three-body free Hamiltonian and the system energy: $\{H_0^{(3)}, E^{(3)}\}$ to the two-body set: $\{H_0^{(2)}, E^{(2)}\}$ is replaced by the path "b" to ensure the analytic continuation with the quasi two-body Hamiltonian of the kinetic part and the system energy: $\{\overline{H}_0^{(3)}, E_{\rm cm}^{(3)}\}$. Finally, the usual procedure from the Faddeev equations to the coupled Lippmann-Schwinger equations is accomplished by the normalization path "c" by putting the two-body relative kinetic energy of the subsystem to zero. 2) The path a′: from the two-body free Hamiltonian and the system energy: $\{H_0^{(2)}, E^{(2)}\}$ to the one-body set: $\{H_0^{(1)}, E^{(1)}\}$ is replaced with the path b′ to ensure the analytic continuation with the set $\{\overline{H}_0^{(2)}, E_{\rm cm}^{(2)}\}$ and finally arrives at $\{H_0^{(1)}, E^{(1)}\}$ via the path c′ with zero kinetic energy.

3.2 The NNπ system below the three-body break up threshold

In the NNπ system, the denominator of the E2Q is given by $\omega_j(k_j) = \sqrt{k_j^2 + m_j^2}$ $(j = 1, 2, 3)$ with the pion mass $m_3 = m_\pi \equiv m$ and the nucleon mass $m_1 = m_2 \equiv M$, and the three-body total energy $\sqrt{S}$. Eqs.(11), (12) become

$$
\begin{aligned}
D_{\rm Fadd} &= E^{(3)} - H_0^{(3)} && (21)\\
&= (E^{(3)} + m) - (H_0^{(3)} + m) && (22)\\
&= \overline{E}^{(3)} - \overline{H}_0^{(3)} \equiv E_{\rm cm}^{(3)} - \overline{H}_0^{(3)} \equiv D_{\rm E2Q}. && (23)
\end{aligned}
$$

The non relativistic approximation of Eq.(23) gives for the process: $(N_2\pi)N_1 \rightarrow N_2(\pi N_1)$ [18],

$$D_{\rm E2Q} \approx (E + m) - \overline{q}_1^2/2m_1 - \overline{q}_2^2/2m_2 - (\overline{\mathbf{q}}_1 - \overline{\mathbf{q}}_2)^2/2m_3 \tag{24}$$

with the energy relation $E^{(3)} \equiv E = \sqrt{S} - 2M - m$, the three-body free energy. The scaling of Eq.(16) is given by,

$$\frac{\overline{q}_{1,2}^2}{2\mu_{1,2}} = \frac{q_{1,2}^2}{2\mu_{1,2}} + m, \tag{25}$$

Therefore, the non relativistic approximation of the Hamiltonian $H_0^{(3)}$ is given by,

$$\begin{aligned} H_0^{(3)} &\approx \sum_j \frac{\overline{q}_j^{\,2}}{2m_j} = \sum_j \frac{q_j^2}{2m_j} + m \\ &= \frac{\overline{q}_{1,2}^2}{\mu_{1,2}} + \frac{\overline{p}_{1,2}^2}{2\nu_{1,2}} = \frac{q_{1,2}^2}{2\mu_{1,2}} + \frac{p_{1,2}^2}{2\nu_{1,2}} + m = \frac{q_3^2}{2\mu_3} + \frac{p_3^2}{2\nu_3} + \alpha, \end{aligned} \tag{26}$$

Eq.(16) gives with the reduced masses for $\mu_{1,2}^{-1} = M^{-1} + (M+m)^{-1}$ and $\mu_3^{-1} = m^{-1} + (2M)^{-1}$

$$\frac{\overline{q}_{1,2}^2}{2m_{1,2}} = \frac{q_{1,2}^2}{2m_{1,2}} + \frac{\mu_{1,2}}{m_{1,2}} m = \frac{q_{1,2}^2}{2m_{1,2}} + \frac{(M+m)}{(2M+m)} m. \tag{27}$$

Using Eqs.(26) and (27), the total kinetic energy becomes

$$\sum_j \frac{\overline{q}_j^2}{2m_j} = \sum_j \frac{q_j^2}{2m_j} + \left(\frac{\mu_1}{m_1} m + \frac{\mu_2}{m_2} m + \frac{\mu_3}{m_3}\alpha\right) = \sum_j \frac{q_j^2}{2m_j} + m. \tag{28}$$

Therefore, we obtain α,

$$\alpha = \frac{-m_3}{(m_1+m_2)} m = -\frac{m^2}{2M}, \tag{29}$$

and also we have a relation for the pion momentum for Eq.(25),

$$\frac{\overline{q}_3^2}{2\mu_3} = \frac{q_3^2}{2\mu_3} - \frac{m^2}{2M}. \tag{30}$$

Finally, from Eq.(30) and the relation

$$\frac{\overline{q}_3^{\,2}}{2\mu_3} + \frac{\overline{p}_3^2}{2\nu_3} = \frac{q_3^2}{2\mu_3} + \frac{p_3^2}{2\nu_3} - \frac{m^2}{2M}, \tag{31}$$

we obtain

$$\frac{\overline{p}_3^2}{2\nu_3} = \frac{p_3^2}{2\nu_3}. \tag{32}$$

Eq.(25) shows that the nucleon momentum $\overline{q}_{1,2}$ is increased and the integral with respect to the $\overline{q}_{1,2}$ compensates for the two-body Nπ-information which is missed in the region $-2\mu_{1,2}m \leq q_{1,2}^2 \leq 0$ by the old treatment, while, Eq.(30) suggests that the pion momentum $\overline{q}_3$ is decreased to become the p_{11} bound state. Therefore, the integration region for $0 \leq q_3^2 \leq \mu_3 m^2/M$ in the old treatment is forbidden in utilizing the two-body NN-information. Now, let us call the new treatment the E2Q method [18].

Consequently, the MLS equations are described by choosing the same particle channels as the Faddeev equations: α=(Nπ)N, β=(πN)N, and γ=(NN)π, but omitting α, β

$$X_{n,m} = V_{n,m} + \sum_{s,t}^{3} V_{n,s} G_0^{st} X_{t,m}, \tag{33}$$

with $\tau_{s,t}^{\alpha} \equiv G_0^{st}$ and the effective potential is given by

$$V_{n,m} = Z_{n,m} + 2\sum_{s,t} Z_{n,\gamma s} \tau_{s,t}^{\gamma} Z_{\gamma t,m}. \tag{34}$$

where the subscripts $n(\in \alpha)$, $m(\in \beta)$, $s,t(\in \gamma)$ (=1, 2, 3) in Eq.(5) denote the physical states with respect of the each particle channel. Therefore, Eq.(33) is the N+N′ → N+N′ scattering equations with the effective NN′ potential of Eq.(34).

3.3 Long range NN′ potential from the NNπ system

Recently, one of us (SO) found that the AGS Born term [7] or the potential of the multi-channel two-body LS equation assumes the well known Yukawa form below the Nπ break up threshold [18].

The leading term of the AGS Born term is given by $\chi = (\sigma^2 + q^2 + q'^2)/2qq'$ with $\sigma^2 = -2m_\pi(E + \epsilon_B)/\Lambda$, and $\Lambda = 1 + m_\pi/m_N \equiv 1 + \Delta$,

$$Z_{N\pi,N\pi}(-\vec{q},\vec{q}\,';E) = \frac{-2g_{N\pi}(\vec{p})m_\pi g_{N\pi}(\vec{p}\,')}{2qq'(\chi\Lambda - x)} \equiv \frac{C^\pi_{N\pi,N\pi}(\mathbf{q},\mathbf{q}')}{2qq'(\chi\Lambda - x)}, \tag{35}$$

where m_π, m_N, E and ϵ_B are the pion mass, the nucleon mass, the three-body energy, and the πN binding energy, respectively. Especially for our purpose, the $N\pi$-binding energy $\epsilon_B \equiv \epsilon_{N\pi}$. The form factor $C^\pi_{N\pi,N\pi}(\mathbf{q},\mathbf{q}') \equiv -2g_{N\pi}(\vec{p})m_\pi g_{N\pi}(\vec{p}\,')$ is defined for simplicity.
Here we expand the Green's function $\{2qq'(\chi\Lambda - x)\}^{-1}$ with respect to the mass factor $\Delta \equiv m_\pi/m_N$,

$$Z_{N\pi,N\pi}(-\mathbf{q},\mathbf{q}';E) = \frac{C^\pi_{N\pi,N\pi}(\mathbf{q},\mathbf{q}')}{2qq'}\sum_{j=0}^{\infty}\frac{(-\Delta)^j\chi^j}{(\chi - x)^{j+1}} = C^\pi_{N\pi,N\pi}(\mathbf{q},\mathbf{q}')\sum_{j=0}^{\infty}\frac{(-\Delta)^j(\sigma^2 + q^2 + q'^2)^j}{[\sigma^2 + (\mathbf{q}-\mathbf{q}')^2]^{j+1}}. \tag{36}$$

Putting $\mathbf{q}' - \mathbf{q} = 2\mathbf{K}$ and $\mathbf{q}' + \mathbf{q} = \mathbf{Q}$, and also $\sigma^2 + 2K^2 + Q^2/2 = [(\sigma^2 + 4K^2) + (\sigma^2 + Q^2)]/2$, gives, with $C^\pi_{N\pi,N\pi}(\mathbf{q},\mathbf{q}') \equiv C^\pi_{N\pi,N\pi}(\mathbf{K},\mathbf{Q})$,

$$\begin{aligned} Z_{N\pi,N\pi}(-\mathbf{q},\mathbf{q}';E) &= C^\pi_{N\pi,N\pi}(\mathbf{K},\mathbf{Q})\frac{1}{(\sigma^2 + 4K^2)}\sum_{j=0}^{\infty}(-\Delta/2)^j\Big[1 + \frac{\sigma^2 + Q^2}{\sigma^2 + 4K^2}\Big]^j \\ &= C^\pi_{N\pi,N\pi}(\mathbf{K},\mathbf{Q})\sum_{k=0}^{\infty}\sum_{j=0}^{\infty}{}_{j+k}C_k\Big(\frac{-\Delta}{2}\Big)^{j+k}\frac{(\sigma^2 + Q^2)^k}{(\sigma^2 + 4K^2)^{k+1}}. \end{aligned} \tag{37}$$

Taking the sum with respect of j, it becomes

$$Z_{N\pi,N\pi}(-\mathbf{q},\mathbf{q}';E) = \frac{2C^\pi_{N\pi,N\pi}(\mathbf{K},\mathbf{Q})}{2 + \Delta}\sum_{k=0}^{\infty}\Big(\frac{-\Delta}{2 + \Delta}\Big)^k\frac{(\sigma^2 + Q^2)^k}{(\sigma^2 + 4K^2)^{k+1}}. \tag{38}$$

One can find that the last equation includes the Yukawa type potential for $k = 0$, and $\mathbf{Q} = \mathbf{0}$, i.e.,

$$Z_{N\pi,N\pi}(-\mathbf{q},\mathbf{q}';E) \approx \frac{2C^\pi_{N\pi,N\pi}(\mathbf{K},\mathbf{0})}{2 + \Delta}\frac{\sigma^2}{(\sigma^2 + 4K^2)}. \tag{39}$$

The form factor $C^\pi_{N\pi,N\pi}(\mathbf{q},\mathbf{q}')$ is a monotonic function; compared to the value of the Green's function at very low energy, one could make it a constant: $C^\pi_{N\pi,N\pi}(\mathbf{q},\mathbf{q}') \approx C_{N\pi,N\pi}$. Therefore, we obtain the Fourier transform of the AGS Born term as,

$$\mathcal{F}\{Z_{N\pi,N\pi}(-\mathbf{q},\mathbf{q}';E)\} \approx \frac{C_{N\pi,N\pi}}{8\pi(2 + \Delta)}\Big(\frac{e^{-\sigma r/2}}{r}\Big) \tag{40}$$

Let us examine the range σ,

$$\sigma^2 = -\frac{2m_\pi}{\Lambda}(E + \epsilon_B) = -2\frac{m_N m_\pi}{m_N + m_\pi}(E + \epsilon_B) > 0 \tag{41}$$

$$\Lambda = \frac{m_N + m_\pi}{m_N} \equiv 1 + \Delta, \tag{42}$$

Table 1. Deuteron binding energy sequence and corresponding rms radii for the $1/r^2$-type n-p triplet potential.

n	E_n [MeV]	E_n/E_{n+1}	$\sqrt{<r_n^2>}$ [m]	$\sqrt{<r_{n+1}^2>}/\sqrt{<r_n^2>}$
1	2.226	171.0	1.97×10^{-15}	13.1
2	1.30×10^{-2}	171.0	2.58×10^{-14}	13.1
3	7.61×10^{-5}	171.0	3.37×10^{-13}	13.1
4	4.45×10^{-7}	171.0	4.40×10^{-12}	13.1
5	2.61×10^{-9}	171.0	5.76×10^{-11}	13.1
6	1.52×10^{-11}	171.0	7.53×10^{-10}	13.1
7	8.91×10^{-14}	171.0	9.85×10^{-9}	13.1

where $E + \epsilon_B \le 0$ should be satisfied, and $E = -\epsilon_B$ generates the long range Coulomb-like potential. Here a special energy average gives the $1/r^2$ type potential [18].
Therefore, we can find a new phenomena at the two-body threshold in the three-body system which is only seen using the Faddeev approach. On the other hand, the three-body threshold satisfies $E = -\epsilon_B \to 0$ which leads to the Efimov effect [14], where $\epsilon_B = 0$ also suggests that the scattering length becomes $\pm\infty$. This means that our theory contains the Efimov effect automatically. Therefore, the theory involves not only the Efimov theory, but also a wider field near the different thresholds which has not been touched by the traditional three-body Faddeev treatment.
It is well known that the $1/r^2$ potential has the energy sequence $E_{n+1} = e^{-2\pi/\mu}E_n$ and the rms radii $r_{n+1} = e^{\pi/\mu}r_n$. If we fit the parameters $C_{N\pi,N\pi}$ and a_0 to the deuteron binding energy 2.226 MeV, then possible quantized binding energies are available only for $l = 0$, in which the quantum number for the modified Bessel function $K_{i\mu}(\kappa r)$ is $\mu = 1.2221$. Therefore, $e^{2\pi/\mu} = 170.98$ and $e^{\pi/\mu} = 13.076$ are obtained. As a consequence, we obtain the energy sequence and rms radii which are seen in the Table 1 [18].

Finally, the first excited state E_2 is around 13keV, however such an energy region has not been measured although the region with more than 50keV was measured. Here, we don't insist that the excited states of the deuteron exist, because the tensor force is not included in fitting the potential depth to obtain the deuteron binding energy. Another question is whether the energy average is unique, however, we confirmed that the calculation without the energy average has also a bound state.

3.4 NN′ and πD scattering lengths

The NN′ and πD scattering lengths are investigated in the NNπ system. Both are measured above the N-N′ and the π-D thresholds but below the NNπ three-body threshold. The N-N′ threshold energy is $E = -m_\pi$ in the original (or traditional) Faddeev calculation which has no singularity at that energy; however, in the E2Q theory $E_{\rm cm} = 0$ is the branch point of the scattering cut and singular. The π-D threshold energy is $E = -\epsilon_d$ of the deuteron binding energy which has no singularity in the kernel of the original Faddeev equations, $D_{\rm Fadd}$ of Eq.(21); however, in the E2Q, equation $E_{\rm cm} = 0$, which is the singularity in $D_{\rm E2Q}$ of Eq.(23). We show the results of the scattering lengths in Table 2. One can see that the discrepancy between the original Faddeev treatments and the E2Q method is significant, although the Nπ potential differences and the corresponding states are also sensitive [19][20].

Finally, we conclude that the three-body calculation below the break up threshold should be carried out using the E2Q method. One should note that the treatment *below* the break up threshold in the Faddeev method is completely different from that *above* the break up threshold, although three-

Table 2. The NN′ and πD scattering lengths are calculated using the original three-body Faddeev equation (Fadd-Org) and the E2Q Faddeev calculation (Fadd-E2Q). Type A is the result for the potential which is given by Thomas [17] and type B is the result for the potential by Fuda [21]. Fadd-E2Q is the present calculated result. Fadd-E2Q results show good agreement with the experimental data [22, 23].

Method	Scattering length [fm]	System/State
Fadd-Org (type A[17])	0.280	NN′ $^3S_1$
Fadd-Org (type B[21])	2.85	NN′ $^3S_1$
Fadd-E2Q (type B[21])	4.66	NN′ $^3S_1$
EXP [22]	5.424±0.004	NN′ $^3S_1$
Fadd-Org (type A[17])	0.033	πD
Fadd-Org (type B[21])	$-0.019 + 0.019i$	πD
Fadd-E2Q (type B[21])	$-0.023 + 0.019i$	πD
EXP [23]	$-0.038 + 0.009i$	πD

body systems, such as the three-nucleon system, with the negligibly small two-body binding energies compared to the pion mass, have been historically investigated without this kind of discussion.

4 Screened Coulomb Potential Method in Universal Variables

4.1 The Coulomb equivalent screening range

Recently we showed that the Lippmann-Schwinger (LS) equation with a screened Coulomb potential in momentum space cannot give the solution. If we assume that the actual solution in momentum space exists, then the on- and half-on-shell auxiliary amplitudes should be zero: $t_l^{\phi}(k,k;z) = t_l^{\phi}(p,k;z) = t_l^{\phi}(k,p';z) = 0$ which is satisfied for the auxiliary potential: $V_l^{\phi}(p,p') = V_l^{C}(p,p') - V_l^{R}(p,p')$ [24][25]. This is the *Lemma* which was introduced in ref. [24]. Therefore, the corresponding phase shift of the on-shell $t_l^{\phi}(k,k;z)$ suggests $\phi_l(k) \equiv \sigma_l(k) - \delta_l^R(k) = 0$.

In the following, we will define the screening range which can reproduce the Coulomb phase shift $\sigma_l(k) = \arg\Gamma(l+1+i\eta(k))$. Using this range, the auxiliary potential can satisfy the *Lemma*.

Using the two-potential theory, we obtain the off-shell Coulomb amplitude, i.e.,

$$t_l^C(p,p';z) = \int_0^\infty\int_0^\infty \overline{\omega}^{\phi}(p,p'';z)t_l^{R\phi}(p'',p''',z)\omega_l^{\phi}(p''',k)\frac{p''^2p'''^2dpdp'}{(2\pi^2)^2} + t_l^{\phi}(p,p';z) \quad (43)$$

with

$$\begin{aligned} t_l^{R\phi}(p,p';z) &= V_l^R(p,p') \\ &+ \int_0^\infty\int_0^\infty V_l^R(p,p'')G_l^{\phi}(p'',p''';z)t_l^{R\phi}(p''',p';z)\frac{p''^2p'''^2dp''dp'''}{(2\pi^2)^2} && (44)\\ \overline{\omega}_l^{\phi}(p,p'';z) &= \delta(p-p'') + t_l^{\phi}(p,p'';z)G_0(p'';z) && (45)\\ \omega_l^{\phi}(p''',p';z) &= \delta(p'''-p') + G_0(p''';z)t_l^{\phi}(p''',p';z) && (46)\\ G_l^{\phi}(p'',p''';z) &= \delta(p''-p''')G_0(p'';z) + G_0(p'';z)t_l^{\phi}(p'',p''';z)G_0(p''';z) && (47)\\ t_l^{\phi}(p,p';z) &= V_l^{\phi}(p,p') + \frac{1}{2\pi^2}\int_0^\infty V_l^{\phi}(p,p'')G_0(p'';z)t_l^{\phi}(p'',p';z)p''^2dp'' && (48) \end{aligned}$$

where a nonsingular $\overline{V}_l^{\phi}$ potential is used instead of the original auxiliary potential V_l^{ϕ} which is equivalent to the pure $\overline{V}_l^{\phi}$ by the Fourier transformation, because $\overline{V}_l^{\phi}$ is well converged to the V_l^{ϕ} potential in the integral as an integrand that is seen in the form of Eq.(43). The modified auxiliary potential is defined by [25]

$$\overline{V}_l^{\phi}(p,p') \equiv \begin{cases} 0 & (p = p') \\ V_l^C(p,p') - V_l^R(p,p') & (p \neq p'). \end{cases} \tag{49}$$

Therefore, the two-potential theory with the Lemma will represent the fully off-shell Coulomb amplitude in momentum space.

4.2 Numerical result of the screening range

We search for the proper screening range parameter using the r-space method such as the Schrödinger equation and the phase shift differential equation with more than three digit accuracy [26], [27]. One may imagine that the Coulomb phase shift is obtained by increasing the distance of the screened Coulomb potential because the screened Coulomb potential converges to the Coulomb potential by increasing the distance. However, this idea is wrong, because the calculated phase shift, which is derived from the screened Coulomb potential not only with the Schrödinger equation but also with the LS equation, does not converge to the Coulomb phase shift by increasing the distance [16][25].

We found that five discrete screening range bands are necessary to reproduce the Coulomb phase shift for $\sigma_0 \leq 4\pi$.

1. The first (lowest) band (black circle in Figure 5) can represent the higher energy region better than the energy which satisfies $\sigma_0(k) = 0$, and the range should be $R = R_0(k) = 0$ fm at $\sigma_0(k) = 0$.
2. The second lower band (black nabla symbol) can be applied to the phase shift region: $0 \leq \sigma_0(k) \leq \pi$ and the range must be $R = R_1(k) = 0$ fm at $\sigma_0(k) = \pi$,
3. and the next band (black square symbol) gives the phase shift region: $\pi \leq \sigma_0(k) \leq 2\pi$, and the range $R = R_2(k) = 0$ fm at $\sigma_0(k) = 2\pi$,
4. and the next band (white circle symbol) applies to: $2\pi \leq \sigma_0(k) \leq 3\pi$, and the range $R = R_3(k) = 0$ fm at $\sigma_0(k) = 3\pi$, and
5. the highest band (black delta symbol) applies to: $3\pi \leq \sigma_0(k) \leq 4\pi$ and the range $R = R_4(k) = 0$ fm at $\sigma_0(k) = 4\pi$, respectively.

We find that $\sigma_0(k) = n\pi$ $(n = 0, 1, 2, \cdots)$ leads to $R_n(k) = 0$ which denotes $V^R(r) = 0$. However, the zero range screened Coulomb property seems to be unusual; then we adopt the finite range of the upper band instead (see Figure 5).

These bands are fitted by the N/D form

$$\mathcal{R} = \frac{a_0 + a_1\eta + a_2\eta^2 + a_3\eta^3 + \cdots}{b_0 + b_1\eta + b_2\eta^2 + b_3\eta^3 + \cdots} \tag{50}$$

where $\mathcal{R} \equiv kR$ is the so-called universal range, and $\eta(k)$ is the Sommerfeld parameter. The parameter $a_0, a_1, \cdots$ and $b_0, b_1, \cdots$ are shown in the Table 3. Therefore, the Coulomb phase shifts of the each system are obtained by taking the individual range $R(k) = \mathcal{R}(k)/k$ and incorporating the individual values of the Sommerfeld parameters.

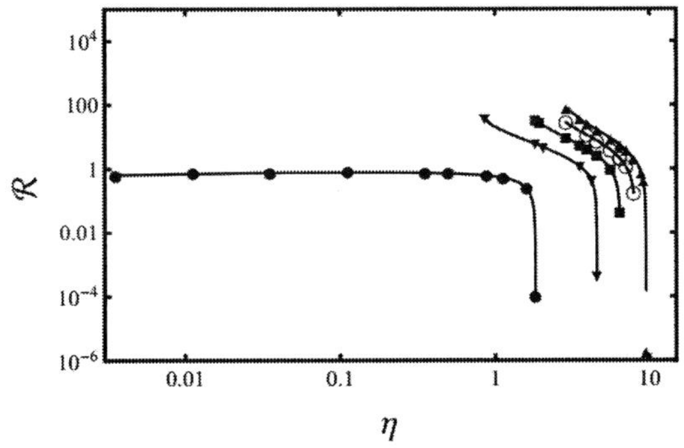

Figure 5. The universal ranges $\mathcal{R}(k)$ are denoted by five different symbols: black circle, black nabla, black square, white circle, and black delta. The universal ranges $\mathcal{R} = \mathcal{R}(k) = \mathcal{R}_n(k)$ $(n = 0, 1, 2, 3, 4)$ are fitted as functions of $\eta = \eta(k)$. 1) $\mathcal{R}_0 = 0$ at $\sigma_0(k) = 0$, 2) $\mathcal{R}_1 = 0$ at $\sigma_1(k) = \pi$, 3) $\mathcal{R}_2 = 0$ at $\sigma_2(k) = 2\pi$, 4) $\mathcal{R}_3 = 0$ at $\sigma_3(k) = 3\pi$, 5) $\mathcal{R}_4 = 0$ at $\sigma_0(k) = 4\pi$. All bands become zero at $\sigma_0(k) = n\pi$ $(n = 0, 1, 2, \cdots)$, however one could replace them by the finite range of the upper band. Each band makes it possible to derive the Coulomb phase shifts for the corresponding energy region.

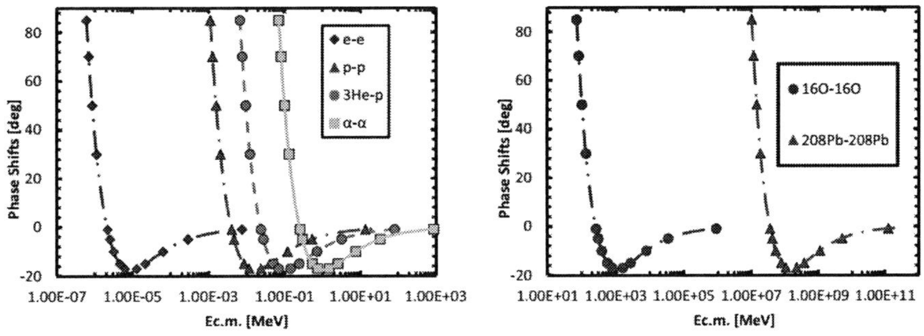

Figure 6. The Coulomb phase shifts in the S-wave are illustrated for several systems from e^--e^- to ^{208}Pb-^{208}Pb. These are calculated using the universal ranges which are given in Figure 5. The present screening range is only available for the Yukawa-type screened Coulomb potential which can reproduce the Coulomb phase shift not only for a wide energy range but also for any system.

Table 3. The parameters of $\mathcal{R} = \mathcal{R}(\eta)$. The parameter setting is defined by Eq. (50).

$\mathcal{R}_n$	k	0	1	2	3	4
$\mathcal{R}_0$	a_k	−0.122527	67.0542	33.4284	−33.6583	
$0.00353 \le \eta \le 1.58$	b_k	−0.14194	90.8834	15.8711	12.7285	
$\mathcal{R}_1$	a_k	5.57504	9.02951	12.9823	−3.20748	
$1.58 \le \eta \le 4.22$	b_k	−0.731152	−0.925015	1.87304	0.68854	
$\mathcal{R}_2$	a_k	1.04150	2.62713	5.58478	−0.64538	
$3.95 \le \eta \le 5.60$	b_k	2.11619	3.39619	3.91962	−2.57886	0.397671
$\mathcal{R}_3$	a_k	3.26102	12.2893	43.3765	−5.55534	
$5.60 \le \eta \le 7.90$	b_k	0.419669	−0.514822	−1.45874	0.924641	0.00556185
$\mathcal{R}_4$	a_k	1.1429	1.8394	3.89062	−0.245209	−0.0191509
$7.07 \le \eta \le 9.13$	b_k	2.12168	3.59564	−1.4752	0.186987	

In order to obtain the on-shell phase shift in the first step of this paper, a screened Coulomb potential is used instead of V^{ϕ}. Therefore, the same phase shifts and range bands are also obtained by the LS equation. However, the long range behavior should be treated by V^{ϕ} in the two potential theory from Eq.(43) to Eq.(48). The range parameters will be obtained only using the auxiliary potential. Therefore, the parameters in Table 3 are tentative.

5 Conclusion

We reviewed our recent works involving the Faddeev equations and related topics. First is the A-body nuclear system. If we imagine that the hardware is powerful enough to calculate the equations, what is the next problems? One of the most fundamental problems is whether the nuclear force is good or not. The second problem is the meson degrees of freedom. An effective many-body force can be written, but the overlaps with the nuclear force and the double counting problem should be carefully checked. Therefore, MC3F is the second best way to treat a realistic nuclear systems. In the analogy of the data analysis, MC3F is an interpolation method with some fitting of data for the inter-cluster amplitudes rather than an extrapolation method with a nuclear force from the A-body Faddeev equations. The inter-cluster amplitudes are given by the RGM, for example, in which the cluster-cluster interaction is constructed analytically from a simple nuclear potential and checked by comparison with the on-shell data. In other words, one can investigate the multi three-cluster system with the same accuracy for the three-nucleon system.

The MC3F method is an extension of the Faddeev equations to many-body systems. On the other hand, the reduction method for the three-body system to the two-body and the one-body systems is discussed in this paper. We found that the integral range is critically changed above and below the break up threshold. The momentum of residual particles after a pion absorption could be increased by releasing the binding energy. A careful investigation of the A-body Faddeev equations may bring new physics by crossing through A-number of thresholds. One example at the three-body break up threshold is the Efimov effect. Fortunately, the Efimov effect [14] was observed in the atomic system [15], although the effect has not been clearly observed in nuclear physics. We mentioned that several phenomena below the three-body break up threshold such as the scattering length in the two-body quasi system, and the bound state in the two-body quasi system as the three-body bound state, may be linked with the three-body force choice. Another interesting phenomenon near the quasi two-body break up threshold is that an energy average of the E2Q could produce the long range NN′ interaction where N′ stands for the pion absorbed nuclear state [18]. Furthermore, we showed that the NN′ and πD calculations by the E2Q method seem to give promising results for the scattering lengths, rather than the original three-body Faddeev calculations (see Table 2). Finally, it should be stressed that the new theory doesn't change the Faddeev theory, but reinterprets it.

Our Coulomb method is a completely new one. The results are useful for many applications.

References

[1] K. Ikeda, H. Horiuchi and S. Saito, Suppl. Prog. Theor. Phys. No.68 (1980) 1.
[2] H. Horiuchi, Nucl. Phys. A**522** 257- (1991).
[3] T. Watanabe and S. Oryu, Prog. Theor. Phys. Lett.**116** 429-434 (2006).
[4] L.D. Faddeev, Zh. Eksperim. i Theor. Fiz. **39**, 1459 (1960) [English transl.:Soviet Phys.-JETP **12**, 1014 (1961)]
[5] P. Schwarz *et. al.*, Nucl. Phys. **A398** 1-18 (1983).
[6] C. Lovelace, Phys. Rev. **135**, B1225 (1964).
[7] E. O. Alt, P.Grassberger, and W. Sandhas, Nucl. Phys. **B2**, 167 (1967).
[8] K. Miyagawa, T. Ueda, T. Sawada and S. Takagi, Nucl. Phys. A459 (1986) 93.
[9] S. Oryu, S. Nemoto and P. U. Sauer, Innovative Computational Methods in Nuclear Many-Body Problems, edited by H. Horiuchi, M. Kamimura, H. Toki, Y. Fujiwara, M. Matsuo and Y. Sakuragi, World Scientific, (1998), 38.
[10] K. Wildermuth and Y.C.Tang, *A Unified Theory of the Nucleus, Clustering Phenomena in Nuclei* , Vol. 1, Vieweg & Sohn, Germany 1977.
[11] S. Saito,Prog. Theor. Phys., **41-3**, 705-722 (1969)
[12] Horiuchi, Prog. Theor. Phys. **51** 1266-1268 (1974)
[13] E. W. Schmid, Z. Phyik A-Atom and Nuclei **297**, 105-114 (1980).
[14] V. Efimov, Phys. Lett. B**33** 563-564 (1970); *ibid.*, Nucl. Phys. **A210**, 157-188 (1973).
[15] T. Kraemer et al.,Nature Letter **400**, 315-318 (2006).
[16] A. Deltuva, A. C. Fonseca, and P. U. Sauer, Phys. Rev. C71, 054005 (2005).
[17] A. W. Thomas, Nucl. Phys. **A258**, 417-446 (1976); *ibid.*, I. R. Afnan and A. W. Thomas, Phys. Rev. **C10**, 109 (1974).
[18] S. Oryu, Phys. Rev. **C86**, 044001 (2012); *ibid.*, Few-Body Syst, **54**, 283-286 (2013).
[19] Y. Hiratsuka, S. Oryu and T. Watanabe, Proc. of the 6th APFB Conf. (Adelaide 2014).
[20] S. Oryu, Y.Hiratsuka, T. Watanabe, Proc of the FB21 (Chicago, 2015).
[21] M. G. Fuda, Phys. Rev. **C52**, 2875-2884 (1995).
[22] O. Dumbrajs *et al.*, Nucl. Phys. **B216**, 277-335 (1983).
[23] P. Hauser *et al.*, Phys. Rev. **C58**, R1869-R1872 (1998).
[24] S. Oryu, Phys. Rev. C**73** 054001-1-16, (2006).
[25] S. Oryu, S. Nishinohara, N. Shiiki, and S. Chiba, Phys. Rev. C75, 021001-1-5 (2007).
[26] Bin B. Liu and H. Crater, Phys. Rev. C**67** 024001-1-36 (2003).
[27] F. Calogero, *Variable Phase Approach to Potential Scattering* Academic, New York, 1967.

Spectra for the $A = 6$ reactions calculated from a three-body resonance model

Mark W. Paris[1,a] and Gerald M. Hale[1,b]

[1] *Theoretical Division (T-2), Los Alamos National Laboratory, Los Alamos NM 87545, USA*

Abstract. We develop a resonance model of the transition matrix for three-body breakup reactions of the $A = 6$ system and present calculations for the nucleon observed spectra, which are important for inertial confinement fusion and Big Bang nucleosynthesis (BBN). The model is motivated by the Faddeev approach where the form of the T matrix is written as a sum of the distinct Jacobi coordinate systems corresponding to particle configurations (α, n-n) and (n, n-α) to describe the final state. The structure in the spectra comes from the resonances of the two-body subsystems of the three-body final state, namely the singlet ($T = 1$) nucleon-nucleon (NN) anti-bound resonance, and the $N\alpha$ resonances designated the ground state ($J^\pi = \frac{3}{2}^-$) and first excited state ($J^\pi = \frac{1}{2}^-$) of the $A = 5$ systems ^{5}He and ^{5}Li. These resonances are described in terms of single-level, single-channel R-matrix parameters that are taken from analyses of NN and $N\alpha$ scattering data. While the resonance parameters are approximately charge symmetric, external charge-dependent effects are included in the penetrabilities, shifts, and hard-sphere phases, and in the level energies to account for internal Coulomb differences. The shapes of the resonance contributions to the spectrum are fixed by other, two-body data and the only adjustable parameters in the model are the combinatorial amplitudes for the compound system. These are adjusted to reproduce the observed nucleon spectra from measurements at the Omega and NIF facilities. We perform a simultaneous, least-squares fit of the tt neutron spectra and the ^{3}He^{3}He proton spectra. Using these amplitudes we make a prediction of the α spectra for both reactions at low energies. Significant differences in the tt and ^{3}He^{3}He spectra are due to Coulomb effects.

1 Introduction

Significant effort over many decades has been invested in measuring and analyzing the reactions of light nuclei ($A \leq 20$). These reactions are important for an array of applications in astrophysical (big bang nucleosynthesis and stellar evolution, for example) and terrestrial (inertial confinement fusion, criticality safety, and measurements standards) settings. Many systems however, particularly those involving unstable projectiles and targets remain unmeasured. There has been much interest and experimental activity lately at ICF facilities, such as OMEGA at the U. of Rochester, and the NIF at Livermore, to measure the particle spectra from three-body breakup reactions initiated by the laser implosion of capsules containing tritium ($t = ^3$H), ^{3}He, or a mixture of the two. Here we describe

[a] e-mail: mparis@lanl.gov
[b] e-mail: ghale@lanl.gov

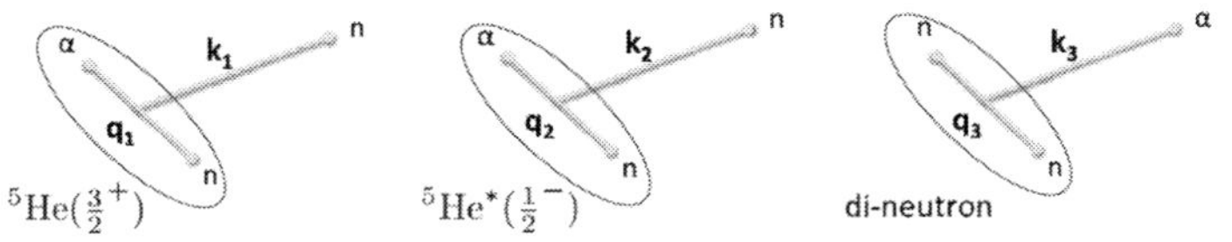

Figure 1: Configurations included in the three-body final state wave function showing the momentum Jacobi coordinates, $\mathbf{q}_i, \mathbf{k}_i$.

an R-matrix-based model for least-squares fitting these spectra that maintains some aspects of the Faddeev method for three-particle scattering. We discuss the simultaneous fit of the $n(p)$-spectra for the reaction $t(t,n)n\alpha$ ($^3\text{He}(^3\text{He},p)p\alpha$) and give predictions for the α spectra.

The next section gives a brief introduction to the concepts of R-matrix theory and summarizes the resonance model, in which the T matrix is expressed as the sum of three Faddeev-like components. When specialized to the case of single-level terms, these expressions form the basis of our present model, which is similar to that of Brune *et al.* [1], but differs from it in detail. The application of these formulas to calculate the spectra of nucleons from the 3-body breakup of the reactions induced by $t+t$ and $^3\text{He}+^3\text{He}$ is described in Section 3. These calculations use charge-symmetric (CS) parameters for the quantities dominated by nuclear forces, but take into account the Coulomb differences in the resonant sub-systems. The conclusions from this study thus far and the outlook for future work are given in Section 4.

2 R-matrix based resonance model

R-matrix theory asserts that, due to the short range of nuclear forces, there exists a finite region of coordinate space beyond which the scattering wave function for an A-nucleon system assumes its known asymptotic form in terms of T-matrix amplitudes. Through the device of the Bloch operator, $\mathscr{L}_B$, acting at the surface enclosing that region, a complete set of eigenfunctions of the operator $H+\mathscr{L}_B$ can be used to expand the scattering wave function inside it as

$$\phi^+_{\mathbf{q}}(\mathbf{x}) = \sum_\lambda c_\lambda(\mathbf{q})\phi_\lambda(\mathbf{x}), \tag{1}$$

$$c_\lambda(\mathbf{q}) = \frac{1}{\sqrt{\pi}}\sum_{\lambda'} A_{\lambda\lambda'}\gamma_{\lambda'}O^{-1}(q)\mathscr{Y}^{j\pi}_{sl}(\hat{\mathbf{q}}). \tag{2}$$

In Eq. (2), $A_{\lambda\lambda'}$ is the level matrix, $\gamma_{\lambda'}$ is the reduced-width amplitude, O is the outgoing-wave solution, and $\mathscr{Y}^{j\pi}_{sl}$ is the spin-angle eigenfunction of total angular momentum and parity.

In the $A = 6$ reactions we consider here, a two-body initial state reacts to give a three-body final state. The assumed configurations and their three-momentum Jacobi coordinates in the final state are shown in Fig.1 for the case of the ^{6}He compound system relevant for the $T(t,n)n\alpha$ reaction. Analogous configurations obtain for the ^{6}Be compound system relevant for the $^3\text{He}(^3\text{He},p)p\alpha$ reaction. These three configurations take into account the the ^{5}He $J^\pi = \frac{3}{2}^+$ ground state, the $^5\text{He}^*$ $\frac{1}{2}^-$ excited state and the di-neutron (nn) correlation. The pair-wise interactions in the final state are treated as in the Faddeev approach to three-particle scattering, but in these applications to the $A = 6$ systems, the initial state cannot be formed from a bound state of any pair of particles in the final state, as is required by the Faddeev theory. Hence, we call it "Faddeev-like."

DOI: 10.1051/epjconf/201612208002

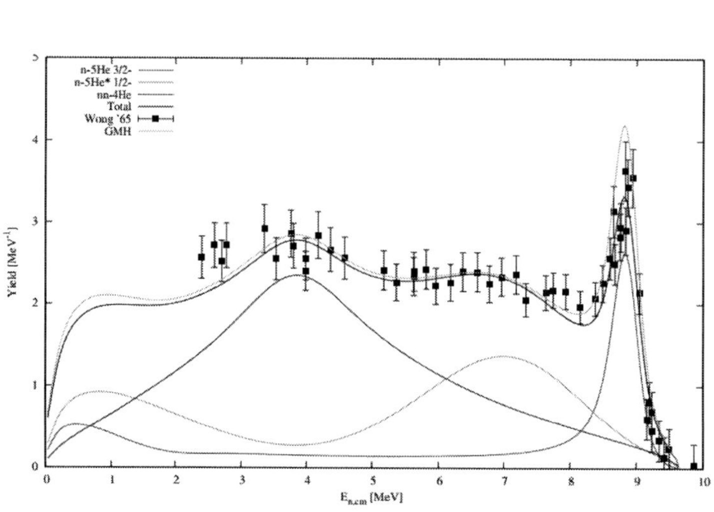

Figure 2: Neutron spectrum from the tt reaction at $E_t = 0.5$ MeV. The data are from Ref. [2], and the colored lines give the 3 Jacobi components of the spectrum.

The three-body breakup transition amplitude can be written as a sum of Faddeev-like components

$$T^{(3)} = \sum_{i=1}^{3} T^{(3)}_{\mathbf{q}_i\mathbf{k}_i\mathbf{k}_0}, \tag{3}$$

$$T^{(3)}_{\mathbf{q}_i\mathbf{k}_i\mathbf{k}_0} = -\pi\langle\chi_{\mathbf{q}_i}\chi_{\mathbf{k}_i}|V_i|\psi^+_{\mathbf{k}_0}\rangle, \tag{4}$$

where $\psi^+_{\mathbf{k}_0}$ is the scattering solution in the initial two-body state, and $\chi_{\mathbf{q}_i}$ and $\chi_{\mathbf{k}_i}$ are the plane-wave solutions for the momenta $\mathbf{q}_i$ and $\mathbf{k}_i$ conjugate to $\mathbf{x}_i$ and $\mathbf{r}_i$, respectively. The transition amplitude is related to the differential cross section as

$$\frac{d^3\sigma}{d\mathbf{k}'_1} = \frac{(2\pi)^4}{v_{\rm rel}} \int d^3k'_2 d^3k'_3 \left|T^{(3)}\right|^2 \delta(\epsilon_f - (\epsilon_i + Q))\delta^{(3)}(\mathbf{k}'_1 + \mathbf{k}'_2 + \mathbf{k}'_3), \tag{5}$$

where $\mathbf{k}'_i$, $i = 1, 2, 3$ are the particle three-momenta in the final state and $v_{\rm rel}$ is the relative velocity of the inital state particles. Relating the plane-wave solution $\chi_{\mathbf{q}_i}$ on the left side of the matrix element [Eq.(4)] to the two-body scattering solution $\phi^-_{\mathbf{q}_i}$ through the Lippmann-Schwinger equation gives Eq. (4) as the distorted-wave T-matrix element

$$T^{(3)}_{\mathbf{q}_i\mathbf{k}_i\mathbf{k}_0} = -\pi\langle\phi^-_{\mathbf{q}_i}\chi_{\mathbf{k}_i}|V_i + V_i G^+_{0i} V_i|\psi^+_{\mathbf{k}_0}\rangle. \tag{6}$$

The R-matrix phenomenology is introduced by making in each of the Jacobi arrangements the expansion of Eqs. (1) and (2) for $\phi^-_{\mathbf{q}_i}$, giving for the 3-body breakup T matrix,

$$T^{(3)} = \sum_{i=1}^{3} \sum_{\lambda_i} c^{(i)}_{\lambda_i}(\mathbf{q}_i)\tilde{T}^{(2)}_{\lambda_i\mathbf{k}_i\mathbf{k}_0}, \tag{7}$$

DOI: 10.1051/epjconf/201612208002

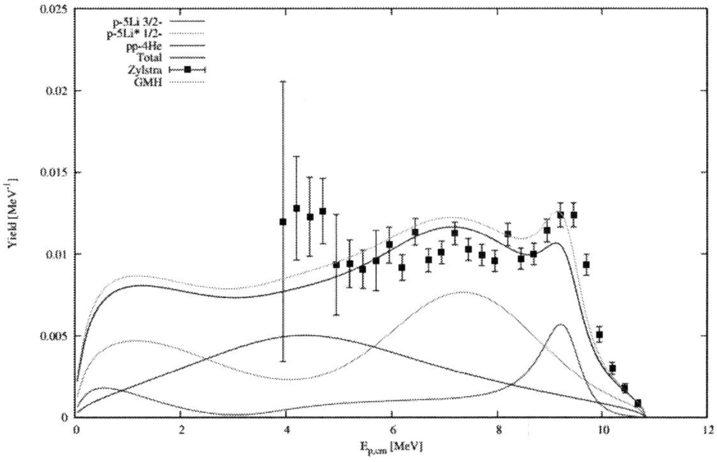

Figure 3: Proton spectrum from the ^{3}He^{3}He reaction at $E_{^3\mathrm{He}} = 0.33$ MeV. The data are from Ref.[3], and the colored lines give the 3 Jacobi components of the spectrum.

with

$$\tilde{T}^{(2)}_{\lambda_i \mathbf{k}_i \mathbf{k}_0} = -\pi\langle \phi^{(i)}_{\lambda_i} \chi_{\mathbf{k}_i} | V_i + V_i G^+_{0i} V_i | \psi^+_{\mathbf{k}_0}\rangle \tag{8}$$

$$= \sum_{Js'l'sl} \mathscr{Y}_{\lambda Js'l'}(\hat{\mathbf{k}}) O^{-1}_{l'}(k) R^L_{\lambda s'l'sl}(\varepsilon) O^{-1}_{l}(k_0) \mathscr{Y}^*_{Jsl}(\hat{\mathbf{k}}_0), \tag{9}$$

a pseudo two-body off-energy-shell T-matrix element expressed in terms of the outgoing-wave R-matrix amplitudes $R^L_{\lambda s'l'sl}(\varepsilon)$. In the applications described below, all the expansions are single-level ($\lambda = \lambda' = 1$), the resonance parameters in the c_λ are determined by outside data, and only the magnitudes of the R-matrix amplitudes $R^L_{\lambda s'l'sl}(\varepsilon)$ are adjusted to fit the experimental spectra. The form of c_λ in the single-level approximation is

$$c_\lambda(\mathbf{q}) \sim \frac{1}{E_R - \frac{i}{2}\Gamma - \epsilon_q} e^{-i\phi_\ell} Y_\ell(\hat{\mathbf{q}}). \tag{10}$$

Here, (E_R, Γ) are the resonance parameters derived from the outside data and ϕ_ℓ is the hard-sphere phase shift.

3 Applications to A = 6 spectra

The single-level resonance parameters for the ground and first-excited states of ^{5}He and ^{5}Li were taken from analyses of $n\alpha$ and $p\alpha$ scattering; these are shown in the first three columns of Tables 4 and 5. Similarly, those for the singlet nn and pp resonances were taken from a charge-independent analysis of NN scattering.

The data fitting has been simplified by making several assumptions for this exploratory calculation. We neglect configuration mixing by considering just the magnitude-squared of each configuration

Channel (resonance)	J^{π}	E_R (MeV)	Γ(MeV)	$\|R^L\|$
^{5}He+n	3/2$^-$	0.99	0.65	0.157
^{5}He*+n	1/2$^-$	6.66	20.6	0.547
^{4}He+(nn)	0$^+$	-0.07	0.	0.338

Figure 4: Parameter values for $T(t, n)n\alpha$.

Channel (resonance)	J^{π}	E_R (MeV)	Γ(MeV)	$\|R^L\|$
^{5}Li+p	3/2$^-$	2.08	2.11	0.157
^{5}Li*+p	1/2$^-$	8.26	19.8	0.547
^{4}He+(pp)	0$^+$	1.56	0.	0.234

Figure 5: Parameter values for ^{3}He(^{3}He, p)$p\alpha$.

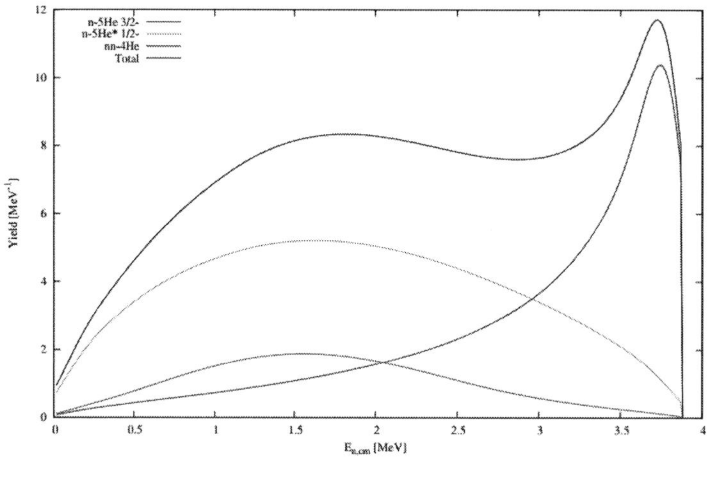

Figure 6: The α-spectrum for $T(t, n)n\alpha$.

amplitude in Eq.(7) and discarding interference terms. This permits the minimization by a least-squares approach. We have also neglected angular dependence of the di-neutron (nn) component of the wave function. The data have been fitted by taking into account the resolution broadening due to the instrument response functions. For the tt n-spectrum from Ref.[2] we use 0.250 MeV width in a Gaussian distribution; the p spectra was Gaussian broadened by 0.230 MeV.

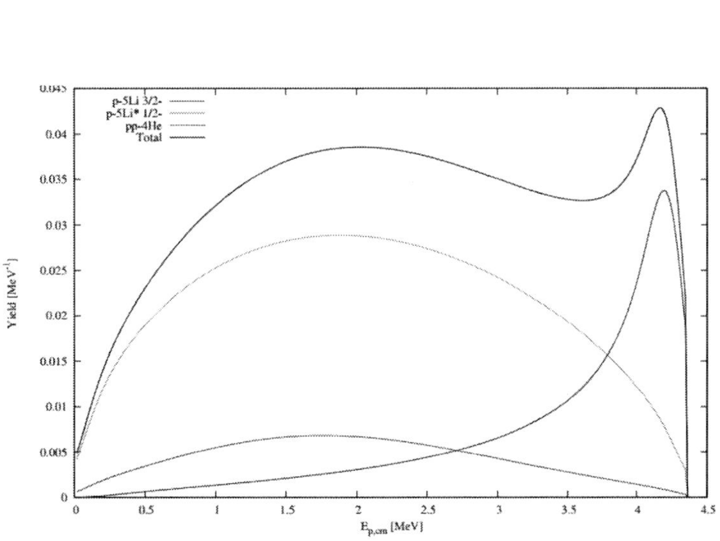

Figure 7: The α-spectrum for ^{3}He(^{3}He, p)$p\alpha$.

The R-matrix amplitudes R^L, in the right-most columns of the tables, were adjusted to simultaneously fit the tt n-spectra of Wong *et al.* [2] and the p-spectra of Zysltra[3], as shown in Figs.2 and 3. Charge symmetry has been used to constrain the R^L parameters in the first two rows of the tables to be the same values. The di-neutron and di-proton configuration parameters, in the third rows of the tables, are varied independently. The amplitudes determined in the simultaneous fit to the n and p spectra were used to predict the α-spectra for the tt and ^{3}He^{3}He reactions, as shown in Fig.6 and Fig.7.

4 Conclusions and outlook

The agreement with experimental $A = 6$ spectra indicates that the main features of the reactions are being captured within the present resonance model. Coulomb differences in the resonances are important, and lead to different branching ratios for the components of the spectra even with CS amplitudes. However, the peaks at the ends of the spectra coming from the ground states of ^{5}He and ^{5}Li are under-predicted, and the details of the interference terms could be improved.

Planned improvements include incorporation of data from ^{3}He t reactions from OMEGA[3] and the recent tt data determined at the National Accelerator Facility[5]. The fitting will be generalized from least-squares to allow the incorporation of configuration mixing. Improvements to the resonance model will be pursued to take into account angular dependence in the nn component of the wave function and dependence on the incident energy.

Acknowledgements

We are indebted to our colleagues Andrew Bacher, Carl Brune, Johan Frenje, Hans Herrmann and particularly Alex Zylstra for valuable discussions about theory and ICF experiments, and for folding our calculations with instrumental response functions in order to compare with their experimental data. This work was supported under the auspices of the US Department of Energy.

References

[1] C.R. Brune *et al.*, Phys. Rev. C **92** 014003 (2015)
[2] C. Wong, J.D. Anderson and J.W. McClure, Nucl. Phys. **71**, 106-112 (1965)
[3] A. B. Zylstra, *private communication* (2015)
[4] J.A. Frenje *et al.*, plenary talk, this conference (2015)
[5] Sayre, D. B. *et al.* Phys. Rev. Lett. 111, 052501 (2013)

R-matrix Analysis for the ^{8}Be System and Features in the p+^{7}Li Reaction Over Resonance Region

Satoshi Kunieda[1,a]

[1] *Japan Atomic Energy Agency, Tokai-mura Naka-gun, Ibaraki 319-1195, Japan*

Abstract. A comprehensive R-matrix analysis is currently underway for the ^{8}Be compound system toward a consistent evaluation of the p+^{7}Li reaction cross-sections over the resonance energy region. In this analysis, the energy eigenvalues are fixed to the level energies given in ENSDF, and we searched for values of the boundary condition parameters as well as the reduced-width amplitudes. It is found that some additional levels are necessary to fit the resonant shape of experimental cross-sections. Besides, the channel radii obtained may be consistent with well-known physical pictures of the atomic nucleus.

1 Introduction

Resonant peaks are observed in the nuclear reaction cross-sections in the lower-energy range where the level structure of compound nucleus shows its very unique character. It is not easy to theoretically predict such resonant cross-sections explicitly as it is still at the stage of scientific challenge. Hence, at this moment, experimental researches are essential to come close to the true values. R-matrix [1] is rigorous and straightforward to the quantum mechanical theory, which relates the nature of the compound nucleus to the S-matrix (hence the cross-sections) under a certain boundary condition at the channel surface. That is, once we measured cross-sections, a consistent data evaluation could be made through the R-matrix analysis. Indeed, a number of analyses have been reported especially for light systems in the world to evaluate cross-sections and/or to understand the nature of the compound nuclei.

In this study, we focus on the analysis of ^{8}Be compound system toward a complete evaluation of the p+^{7}Li cross-sections. It is essentially because the ^{7}Li(p,n)^{7}Be reaction is a promising neutron source to be used in the nuclear science and engineering. Indeed, this reaction is drawing attention also in the medical field as it could be a candidate of neutron-generator in Boron Neutron Capture Therapy [2]. Also, we are interested in the astrophysics, as the ^{7}Li(p,α)^{4}He reaction has an important role in determining the amount of ^{7}Li-depletion in a pre-main sequence phase [3]. Therefore, those reaction cross-sections should be quantitatively known over the relevant energy range.

Let's look at the present status of the p+^{7}Li data evaluation in the world. Although the cross-sections have been measured over the years for those reactions, in-depth evaluation works have not necessarily been reported. The worst situation is found, e.g., in JENDL-4 [4] (or JENDL/HE-2007 [5]) where the p+^{7}Li cross-section is not given at all in spite of its importance. The TENDL library [6]

[a] e-mail: kunieda.satoshi@jaea.go.jp

contains cross-sections virtually for any reactions including p+^{7}Li. However, the p+^{7}Li cross-sections are estimated with the optical and the statistical model, which are beyond the applicable limit for resonant region and also for the few-body system itself. ENDF/B-VII.1 [7] certainly has p+^{7}Li cross-section over a resonance range ($E_p \leq 10$ MeV), in which the data were evaluated though a R-matrix analysis. It is consistent with a number of measurements at around the first peak of ^{7}Li(p,n$_0$)^{7}Be cross-sections (around $E_p = 2$ MeV), however, there are serious discrepancies above $E_p = 3$ MeV. Indeed, it is not easy to achieve a consistent R-matrix analysis over a wide energy range due to a significant number of possible channels and unknown background contributions from bound/distant levels and/or the direct reactions.

A comprehensive R-matrix analysis is currently underway for the ^{8}Be compound system to evaluate the p+^{7}Li cross-sections over the resonance energy region. In this analysis, the energy eigenvalues are fixed to the level energies given in Evaluated Nuclear Structure Data File (ENSDF) [8, 9], instead we search for values of the boundary condition parameters and the reduced-width amplitudes where the unknown background contributions are assumed to be supplemented by putting pseudo levels far above the relevant energy range. We preliminary achieved reasonable fits to measured cross-sections of different reactions simultaneously. Investigating the values of the R-matrix parameters and sources of the resonances, we may obtain a better understanding of natures in the p+^{7}Li reactions.

2 Overview of R-matrix formalism and our computational code

2.1 Short summary of the R-matrix formalism and parameters

The R-matrix approach is a quantum mechanical formalism which connects the nature of the compound system to the scattering observables such as the cross-sections. To be more precise, the internal and external wave-functions are matched at the channel surface under a certain boundary condition, then S-matrix/collision-matrix is calculated as follows.

$$U_{cc'} = e^{-i(\varphi_c+\varphi_{c'})}\left[\delta_{cc'} + 2i\sqrt{P_c P_{c'}}\left(\frac{\mathbf{R}}{\mathbf{I}-\mathbf{RL}}\right)_{cc'}\right], \tag{1}$$

with

$$L_{cc'} = (S_c + iP_c - B_c)\delta_{cc'} . \tag{2}$$

Where the symbol φ_c denotes the hard-sphere scattering phase-shift, while P_c and S_c stand for the penetration and shift factors at the channel surface, respectively. The boundary condition number B_c is virtually the logarithmic derivative of the wave function. It should be noted that the channel is defined explicitly, viz., $c \equiv (\alpha, l, s; J)$ where the symbol α is the partition (the pair of particle and nucleus), l for the angular momentum of the pair, s for the channel spin, and J is the spin of the eigenstates. The matrix $\mathbf{R}$ includes structural information of the compound system, which is necessary to calculate resonant cross-sections. The matrix element is given as follows,

$$R_{cc'} = \sum_\lambda \frac{\gamma_{\lambda c}\gamma_{\lambda c'}}{E_\lambda - E}, \tag{3}$$

where the symbol E_λ denotes the energy eigenvalue for each eigenstate λ and $\gamma_{\lambda c}$ is the reduced-width amplitude for each explicit channel. They are normally treated as the parameters to be deduced from the experimental cross-sections under a certain boundary condition.

DOI: 10.1051/epjconf/201612208003

2.2 Computational code AMUR

AMUR [10, 11] is a multichannel and multilevel R-matrix code to be used for the analysis of the experimental cross-sections, hence for the nuclear data evaluation. Substantial features of the code are listed as follows.

- The code is based on the Wigner-Eisenbud's formalism [12] which is an exact expression of the R-matrix theory. One exception is for the γ-ray channels which we optionally treated by the Reich-Moore approximation [13].
- Both the neutron and charged-particle cross-sections are calculated. It is true also for the differential cross-sections including those for the charged-particle elastic-scattering where the total amplitude is calculated as the sum of the T-matrix element and the Coulomb scattering amplitude.
- If necessary, experimental data can be corrected in the analysis. For examples, the re-normalization and some sources of the resolution are considered as the experimental parameters (those options are not used in this analysis).
- The R-matrix (and the experimental) parameters are deduced through the simultaneous fit to measured cross-sections for the different reactions. We use the generalized least-square method which is equivalent with the approach in the KALMAN/SOK code [14].
- An object-oriented programming technique is used so as to make it easier for us to continue further development/management of the code.

3 R-matrix analysis for the ^{8}Be system

3.1 Energy eigenvalues for the compound system

Through the R-matrix analysis, in principle, it is possible to deduce the energy eigenvalue E_λ and the corresponding J^π for the compound system from experimental cross-sections. However, we found it is not so easy to do this essentially because of the reasons as follows.

- A number of broad resonances overlap/interfere in the relevant energy range, which makes it difficult to identify the each position explicitly.
- Since experimental cross-sections are limited, we may fall into mis-assignments of J^π even for the major levels.
- The analysis is very time consuming since number of channels is significant due to different combination of the partitions, angular momentum and channel spins. For instance, more than 20 channels could be involved only in just the $J^\pi = 2^+$ state in the relevant energy range.

Therefore, in this analysis, we take the energy eigenvalues and the corresponding J^π of ^{8}Be from the latest version of ENSDF [8, 9]. Indeed, the level information given in such a database had been evaluated based on a number of measurements for the nuclear structure. In this sense, present analysis inherits the knowledge from a long-standing accumulation of such studies.

3.2 Channels considered

In the present analysis, we tried to include all the explicit channels as much as possible except for the γ-ray channels. For examples, the channels considered are listed in Table 1 for $J^\pi = 2^-$ and 2^+ where the channels with high angular momentums ($l = 3, 4$) were preliminary neglected as we found they were less important in the relevant energy range.

Table 1. Possible and considered channels for $J^\pi = 2^\pm$.

J^π	Partition	l	s	considered ?
2^-	n+^{7}Be$_{g.s.}$	0	2	yes
		2	1, 2	yes
		4	2	no
	n+^{7}Be$^*_{1st}$	2	0, 1	yes
	p+^{7}Li$_{g.s.}$	0	2	yes
		2	1, 2	yes
		4	2	no
	p+^{7}Li$^*_{1st}$	2	0, 1	yes
	d+^{6}Li$_{g.s.}$	1	1, 2	yes
		3	1, 2	no
2^+	n+^{7}Be$_{g.s.}$	1	1, 2	yes
		3	1, 2	no
	n+^{7}Be$^*_{1st}$	1	1	yes
		3	1	no
	p+^{7}Li$_{g.s.}$	1	1, 2	yes
		3	1, 2	no
	p+^{7}Li$^*_{1st}$	1	1	yes
		3	1	no
	d+^{6}Li$_{g.s.}$	0	2	yes
		2	0, 1, 2	no
		4	2	no
	α+^{4}He$_{g.s.}$	2	0	yes

Figure 1 illustrates the excited levels of ^{8}Be together with the final states in the residual nuclei for the p+^{7}Li reaction. We include both the ground and the first excited states for the residuals ^{7}Be and ^{7}Li whereas only the ground state is considered for ^{6}Li and ^{4}He. The other levels, such as the second excited level in ^{7}Be and ^{7}Li ($J^\pi = 7/2^-$) could also be populated in the relevant energy range. Unfortunately, we did not include such states in the present analysis, since corresponding measured cross-sections, such as for ^{7}Li$(p, p_2)^7$Li* and ^{7}Li$(p, n_2)^7$Be*, have not been reported. Therefore, we need to admit the present analysis could have some incompletenesses due to the lack of the coupling effects from those neglected channels.

3.3 Parameters to be searched for

In most cases, the boundary condition parameters are fixed to some plausible values (e.g., $R_c \sim 1.2A^{1/3} + 0.8$ (fm) and $B_c \sim l$ for the neutron reaction), then all the other R-matrix parameters E_λ and $\gamma_{\lambda c}$ are searched for. Such an approach could be reasonable for the neutron reaction as a number of successful analyses had been reported. However, it is not clear to be reasonable also for the charged-particle reactions. In this analysis, we decided to fix the E_λ value to the level energy taken from ENSDF, instead the boundary condition parameters are treated as free parameters to be searched for as well as the $\gamma_{\lambda c}$ values.

Since the energy range of the analysis is limited while there should be interferences from a number of levels which exist below the reaction threshold and above the upper-energy end of the analysis. Furthermore, there must be the direct-reactions which may go through the collective excitation of the

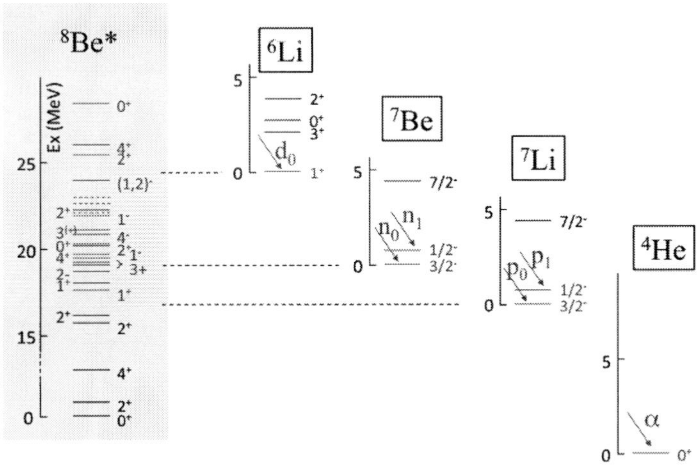

Figure 1. Excited levels of ^{8}Be and final states of the residual nuclei considered in this analysis. Note the level energies and corresponding J^π are taken from ENSDF.

target, the isobaric analog state, the breakup of the residual nuclei and so on. In this study (although we are still looking for more reasonable solutions), we tried to supplement such contributions by putting pseudo levels far above the relevant energy range, in which the reduced-width amplitudes for those distant levels are fitting parameters.

3.4 Experimental cross-sections to be fitted

Experimental data preliminary used are listed in Table 2 and 3 for the cross-sections and differential cross-sections, respectively. Those data were collected from the EXFOR database [15]. We fitted cross-sections for the different reactions which share the same compound nucleus ^{8}Be* simultaneously, viz., for the n+^{7}Be, p+^{7}Li, d+^{6}Li and α+^{4}He reactions.

4 Results and discussion

4.1 Fitted results

Figure 2 shows example results of the cross-sections obtained by the R-matrix fits together with typical experimental data up to E_p = 10 MeV. Although the present results are still preliminary, reasonable fits had been achieved not only for the major reaction ^{7}Li$(p, n_0)^7$Be but also for the other reactions. It is also true for the differential cross-sections as example results are illustrated in Figs 3 and 4 for the ^{7}Li$(p, n_0)^7$Be, ^{7}Li$(p, p_0)^7$Li and ^{7}Li$(p, p_1)^7$Li* reactions. However, more reasonable fits are expected

Table 2. Experimental cross-sections preliminary used in this work

Reaction	Author	(Year)	$E_{inc.}$ (MeV)
^{7}Li$(p, p_1)^7$Li	Presser *et al.* [16]	(1972)	1.7 – 6.0
	Mateus *et al.* [17]	(2002)	0.7 – 2.4
^{7}Li$(p, n_0)^7$Be	Burke *et al.* [18]	(1974)	2.1 – 3.8
	Sekharan *et al.* [19]	(1976)	1.9 – 4.2
	Poppe *et al.* [20]	(1976)	4.0 – 26.0
^{7}Li$(p, n_1)^7$Be	Presser *et al.* [16]	(1972)	2.4 – 6.0
	Poppe *et al.* [20]	(1976)	4.5 – 26.0
^{7}Li$(p, \alpha)^4$He	Cassagnou *et al.* [21]	(1962)	1.4 – 4.8
	Mani *et al.* [22]	(1964)	3.0 – 11.9
	Spraker *et al.* [23]	(2000)	0.04 – 0.1
^{6}Li$(d, p_0)^7$Li	Mcclenahan *et al.* [24]	(1975)	0.5 – 3.5
	Dunjiu *et al.* [25]	(1985)	0.5 – 2.5
^{6}Li$(d, p_1)^7$Li	Mcclenahan *et al.* [24]	(1975)	0.4 – 3.4
	Dunjiu *et al.* [25]	(1985)	1.0 – 2.5
^{6}Li$(d, \alpha)^4$He	Mcclenahan *et al.* [24]	(1975)	0.5 – 3.4
	Gould *et al.* [26]	(1985)	2.2 – 5.9
	Dunjiu *et al.* [25]	(1985)	0.5 – 2.5
^{4}He$(\alpha, p_0)^7$Li	King *et al.* [27]	(1977)	38 – 50
^{4}He$(\alpha, p_1)^7$Li	King *et al.* [27]	(1977)	38 – 50
^{4}He$(\alpha, n)^7$Li	King *et al.* [27]	(1977)	38 – 50
^{7}Be$(n, p)^7$Li	Gledenov *et al.* [28]	(1987)	thermal – 416 eV
^{7}Be$(n, p_0)^7$Li	Koehler *et al.* [29]	(1988)	thermal – 13.5 keV

to be obtained since we have not yet investigated the sign of the reduced-width amplitude for all the levels and channels.

4.2 Features in the ^{7}Li$(p, n_0)^7$Be reaction

Figure 5 plots the calculated ^{7}Li$(p, n_0)^7$Be cross-sections and those from $J^\pi = 1^\pm$, $2^\pm$ and $3^\pm$ which are found to be major contributors on this reaction. The source of the largest peak observed around $E_p = 2$ MeV is quite simple as we confirmed it comes from the 3^+ level at $E_x = 19.235$ MeV of ^{8}Be*. On the other hand, the situation seems to be rather complicated for the second largest peak around $E_p = 5$ MeV where a number of different J^π contribute to the resonant cross-section. Looking at ENSDF, $J^\pi = 3^{(+)}$ at $E_x = 21.5$ MeV is suggested for the main source of the resonance where the positive parity is assigned preliminary. In this analysis, let us suggest that both the $J^\pi = 3^+$ (p-wave) and 3^- (d-wave) levels are necessary to reproduce the experimental cross-sections. This is highly possible since the results obtained are also consistent with the measured differential cross-sections as illustrated in Fig 6. It is noticeable that the background from $J^\pi = 2^-$ is so huge that extends over a very wide energy range. This is a tail from the $J^\pi = 2^-$ (s-wave) at $E_x = 18.91$ MeV exists below the reaction threshold, which suggests it is important to analyze cross-section in a wide energy range as much as possible to determine the background contributions with a reasonable accuracy.

Table 3. Experimental differential cross-sections preliminary used in this work

Reaction	Author	(Year)	$E_{inc.}$ (MeV)	θ (degree)
^{7}Li$(p, p_0)^7$Li	Bashkin *et al.* [30]	(1951)	0.9 – 3.7	164 [*l*]
	Warters *et al.* [31]	(1953)	0.4 – 1.4	50, 70, 89,2, 110, 130, 143.4, 160 [*c*]
	Fasoli *et al.* [32]	(1964)	3.0 – 5.5	65.8, 70.1, 90.0, 97.4, 109.9, 125.3, 140.7, 149.5, 168.3 [*c*]
	Gleyvod *et al.* [33]	(1965)	2.5 – 12.1	70, 90, 130, 150, 166 [*c*]
	Kilian *et al.* [34]	(1969)	2.7 – 10.6	60, 90, 120,140 [*l*]
	Peneta *et al.* [35]	(2012)	1.5 – 7.0	140, 150, 160, 170 [*l*]
^{7}Li$(p, p_1)^7$Li*	Bashkin *et al.* [30]	(1951)	0.9 – 3.7	164 [*l*]
	Fasoli *et al.* [32]	(1964)	3.0 – 5.5	65.8, 70.1, 90.0, 97.4, 109.9, 125.3, 140.7, 149.5, 168.3 [*c*]
	Gleyvod *et al.* [33]	(1965)	2.5 – 12.1	70, 90, 130, 150, 166 [*c*]
	Kilian *et al.* [34]	(1969)	2.7 – 10.6	60, 90, 120,140 [*l*]
	Guzhovskij *et al.* [36]	(1984)	3.0 – 8.0	55 [*l*]
	Peneta *et al.* [35]	(2012)	1.5 – 7.0	140, 150, 160, 170 [*l*]
^{7}Li$(p, n_0)^7$Be	Austin *et al.* [37]	(1962)	2.0 – 3.5	0 [*l*]
	Burke *et al.* [18]	(1974)	1.9 – 3.8	0 – 160 [*l*]
	Borchers *et al.* [38]	(1963)	3.2 – 10.0	0 [*l*]
	Elbakr *et al.* [39]	(1972)	2.2 – 5.5	0 [*l*]
$(p, n_0 + n_1)$	Meadows [40]	(1972)	3.4 – 7.0	3, 15, 50 [*l*]
^{7}Li$(p, n_1)^7$Be*	Guzhovskij *et al.* [36]	(1984)	3.0 – 8.0	55 [*l*]
	Elbakr *et al.* [39]	(1972)	2.5 – 5.5	0 [*l*]
^{7}Li$(p, \alpha)^4$He	Peneta *et al.* [35]	(2012)	1.5 – 7.0	140, 150, 160, 170 [*l*]
^{6}Li$(d, p_0)^7$Li	Elywn *et al.* [41]	(1977)	1.0 – 1.0	15 – 165 [*l*]
^{6}Li$(d, p_1)^7$Li	Elywn *et al.* [41]	(1977)	1.0 – 1.0	15 – 165 [*l*]
^{6}Li$(d, n_0)^7$Be	Elywn *et al.* [41]	(1977)	0.2 – 0.9	0 – 148.5 [*l*]
^{6}Li$(d, n_1)^7$Be	Elywn *et al.* [41]	(1977)	0.2 – 0.9	0 – 148.5 [*l*]
^{6}Li$(d, \alpha)^4$He	Elywn *et al.* [41]	(1977)	0.1 – 1.0	15 – 165 [*l*]

Table 4. Channel radii preliminary obtained for each partition

particle	r_{exp} (fm)	nuc.	r_{exp} (fm)	r_c (fm)
n	-	^{7}Be	-	3.32
p	0.80	^{7}Li	2.43	3.68
d	2.00	^{6}Li	2.54	6.08
α	1.68	^{4}He	1.68	2.84

4.3 Channel radii obtained

Since we fixed the energy eigenvalues of the compound to the level energies of ^{8}Be* given in ENSDF, the boundary condition parameters could be searched for in the fitting procedure. Table 4 lists the channel radii obtained, in which we preliminary assumed the value is different for each particle-nucleus pair in this analysis. We found that the values obtained are somewhat larger than the experi-

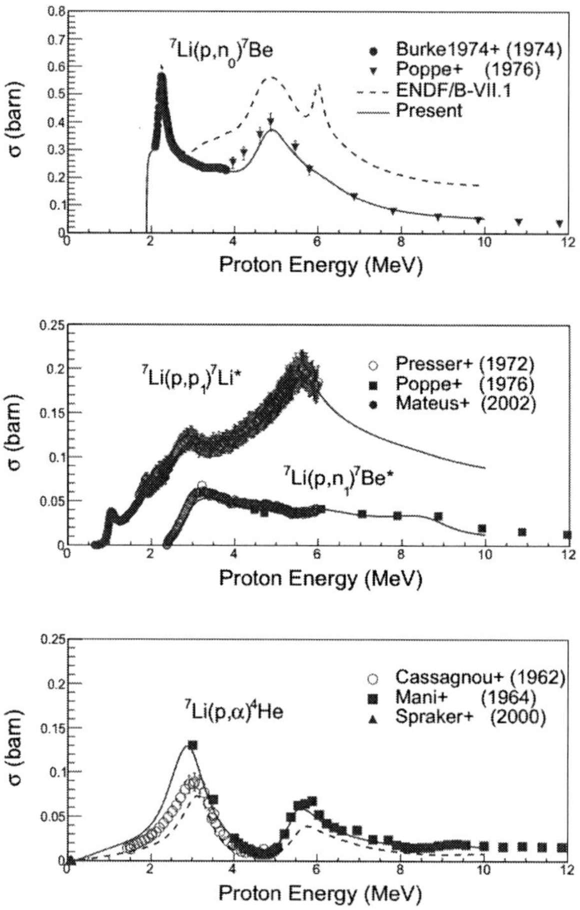

Figure 2. Example results of the cross-sections obtained by the R-matrix fits together with typical experimental data up to E_p = 10 MeV.

mental r.m.s charge radii [42] for the nuclei. Imagine that the nucleus has a certain diffuseness at the surface, and the particle itself is not the point-particle. Therefore, the present values may be consistent with well-known physical pictures of the atomic nucleus.

4.4 Correlation matrices of the cross-sections

Figure 7 illustrates the correlation matrices obtained in the present R-matrix fit for ^{7}Li$(p, n_0)^7$Be and ^{7}Li$(p, \alpha)^4$He cross-sections. Since the R-matrix theory is straightforward to the quantum mechanics,

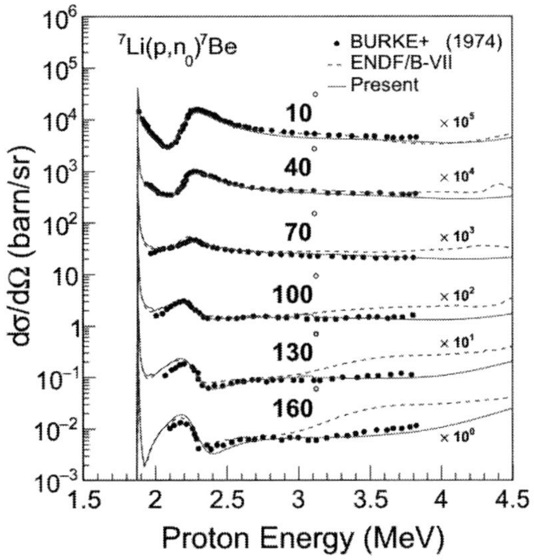

Figure 3. Example results of the R-matrix fit for $^7\mathrm{Li}(p,n_0)^7\mathrm{Be}$ differential cross-sections together with the experimental data of Burke *et al.* [18]

the correlation matrices obtained visualize some features in the reactions. For example, correlation becomes local at around the resonance peaks both in the reactions. This is a consequence of the physical constraint from the theory in which the collision matrix (S-matrix) is always unitary. Another noticeable feature is a very strong correlation observed above ~7 MeV both in the reactions. The main source of such a strong correlation is the background for which we gave the pseudo levels to supplement the contributions from the bound/distant levels and direct reactions.

5 Summary

A comprehensive R-matrix analysis is currently underway for the ^{8}Be compound system toward a consistent evaluation of the p+^{7}Li cross-sections over the resonance energy region. In this analysis, the energy eigenvalues were fixed to the level energies given in ENSDF, instead we tried to search for values of the boundary condition parameters as well as the reduced-width amplitudes for each channel. Reasonable fits were obtained simultaneously for all the measured cross-sections we analyzed, where the agreements appeared to be better than the existing evaluations up to E_p =10 MeV. It should be noted that we included first excited state of ^{7}Li and ^{7}Be for the final states while such a detaied analysis have not beed reported.

In the course of the analysis, it was found that some additional levels were necessary to achieve a reasonable fit to the resonant shape of experimental cross-sections. Besides, the channel radii obtained

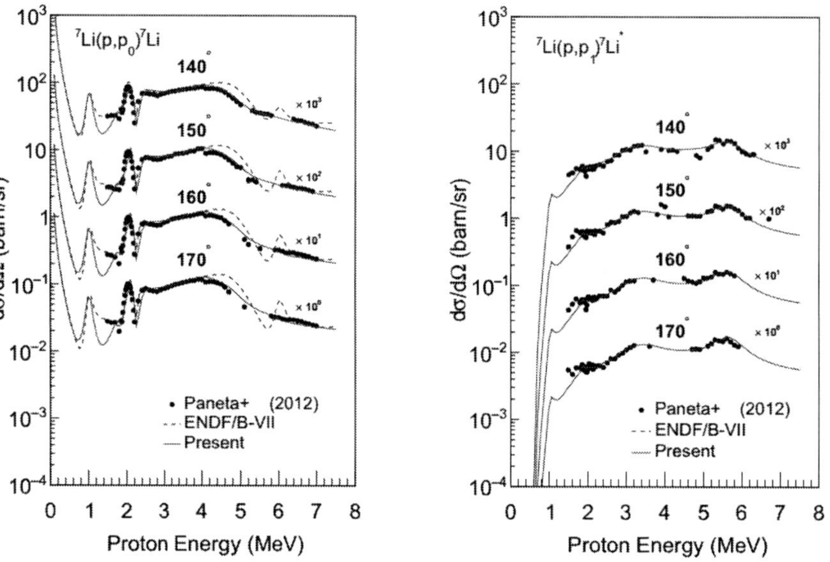

Figure 4. Example results of the R-matrix fit for ^{7}Li$(p, p_0)^7$Li and ^{7}Li$(p, p_1)^7$Li differential cross-sections together with the experimental data of Peneta *et al.* [35]

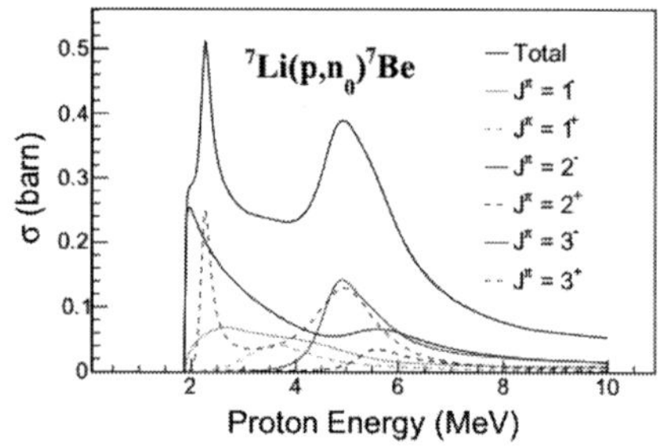

Figure 5. Contributions from the different J^π for the ^{7}Li$(p, n_0)^7$Be cross-section.

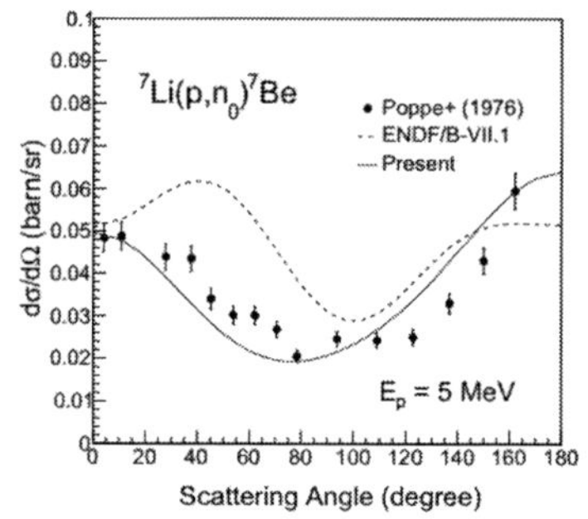

Figure 6. Calculated and experimental [20] differential cross-sections for the ^{7}Li$(p, n_0)^7$Be reaction at E_p =5 MeV.

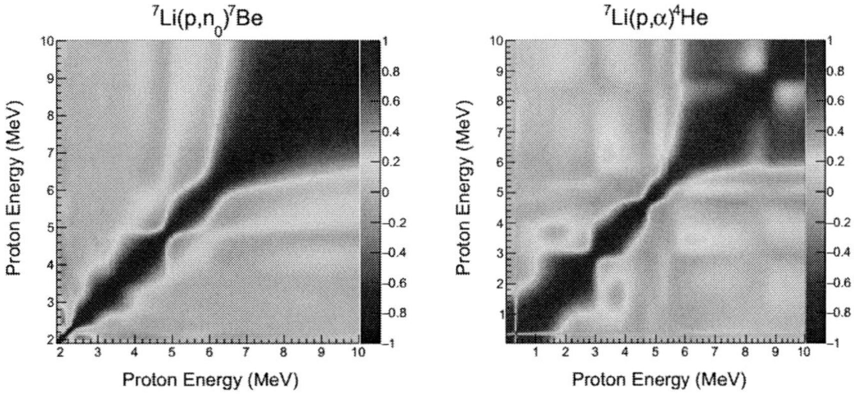

Figure 7. Correlation matrices obtained in the R-matrix fit for ^{7}Li$(p, n_0)^7$Be (left) and ^{7}Li$(p, \alpha)^4$He (right) cross-sections.

may be consistent with well-known physical pictures of the atomic nucleus. However, at the same time, the present analysis left worrisome points,

- We neglected channels for the reactions such as ^{7}Li$(p, p_2)^7$Li* and ^{7}Li$(p, n_2)^7$Be*.
- The direct (including breakup) reactions were supposed to be supplemented by the tails from the distant levels.
- The hard-sphere well was assumed to calculate the potential scattering phase-shifts.

Although we have not yet understood if such approximations are acceptable, present knowledge obtained could be an useful guide toward a consistent evaluation of the p+^{7}Li cross-section in the resonance energy range. To understand more detailed natures in the reactions, measured differential cross-sections are needed for the second excitation state of ^{7}Li and ^{7}Be. Also, the optical model may give some interpretations for those approximations.

References

[1] A.M. Lane, R.G. Thomas, Rev. Mod. Phys. **30**, 257 (1958)
[2] B. Bayanov, V. Belov, S. Taskaev, Journal of Physics: Conference Series **41**, 460 (2006)
[3] P. Ventura, A. Zeppieri, I. Mazzitelli, F. D'Antona, Astronomy and Astrophysics **331**, 1011 (1988)
[4] K. Shibata, O. Iwamoto, T. Nakagawa, N. Iwamoto, A. Ichihara, S. Kunieda, S. Chiba, K. Furutaka, N. Otuka, T. Ohsawa et al., J. Nucl. Sci. Technol. (Tokyo) **48**, 1 (2011)
[5] Y. Watanabe, K. Kosako, S. Kunieda, S. Chiba, R. Fujimoto, H. Harada, M. Kawai, F. Maekawa, T. Murata, H. Nakashima et al., J. Korean Phys. Soc. **59**, 1040s (2011)
[6] A. Koning, D. Rochman, Nuclear Data Sheets **113**, 2841 (2012)
[7] M. Chadwick, M. Herman, P. Obložinský, M. Dunn, Y. Danon, A. Kahler, D. Smith, B. Pritychenko, G. Arbanas, R. Arcilla et al., Nuclear Data Sheets **112**, 2887 (2011), special Issue on ENDF/B-VII.1 Library

[8] *Evaluated nuclear structure data file*, http://www.nndc.bnl.gov/ensdf/
[9] D. Tilley, J. Kelley, J. Godwin, D. Millener, J. Purcell, C. Sheu, H. Weller, Nuclear Physics A **745**, 155 (2004)
[10] S. Kunieda, T. Kawano, M. Paris, G. Hale, K. Shibata, T. Fukahori, Nucl. Data Sheets **118**, 250 (2014)
[11] S. Kunieda, T. Kawano, M. Paris, G.M. Hale, K. Shibata, T. Fukahori, Nucl. Data Sheets **123**, 159 (2015)
[12] E.P. Wigner, L. Eisenbud, Phys. Rev. **72**, 29 (1947)
[13] C.W. Reich, M.S. Moore, Phys. Rev. **111**, 929 (1958)
[14] T. Kawano, H. Matsunobu, T. Murata, A. Zukeran, Y. Nakajima, M. Kawai, O. Iwamoto, K. Shibata, T. Nakagawa, T. Ohsawa et al., J. Nucl. Sci. Technol. (Tokyo) **37**, 327 (2000)
[15] N. Otuka, E. Dupont, V. Semkova, B. Pritychenko, A. Blokhin, M. Aikawa, S. Babykina, M. Bossant, G. Chen, S. Dunaeva et al., Nuclear Data Sheets **120**, 272 (2014)
[16] G. Presser, R. Bass, Nucl. Phys., A **182**, 321 (1972)
[17] R. Mateus, A.P. Jesus, B. Braizinha, J. Cruz, J.V. Pinto, J.P. Ribeiro, Nucl. Instrum. Methods in Physics Res., Sect.B **190**, 117 (2002)
[18] C.A. Burke, M.T. Lunnon, H.W. Lefevre, Phys. Rev. C **10**, 1299 (1974)
[19] K.K. Sekharan, H. Laumer, B.D. Kern, F. Gabbard, Nuclear Instrum.and Methods in Physics Res. **133**, 253 (1976)
[20] C.H. Poppe, J.D. Anderson, J.C. Davis, S.M. Grimes, C. Wong, Phys. Rev. C **14**, 438 (1976)
[21] Y. Cassagnou, J.M.F. Jeronymo, G.S. Mani, A. Sadeghi, P.D. Forsyth, Nucl. Phys. **33**, 449 (1962)
[22] G.S. Mani, R. Freeman, F. Picard, A. Sadeghi, D. Redon, Nucl. Phys. **60**, 588 (1964)
[23] M. Spraker, R.M. Prior, M.A. Godwin, B.J. Rice, E.A. Wulf, J.H. Kelley, D.R. Tilley, H.R. Weller, Phys. Rev. C **61**, 015802 (2000)
[24] C.R. McClenahan, R.E. Segel, Phys. Rev. C **11**, 370 (1975)
[25] C. Dunjiu, Z. Enchen, J. Chenglie (1985), unpublished
[26] C.R. Gould, J.M. Joyce, J.R. Boyce, *Cross section measurements for charged particle induced reactions on Li-6** (1975), Vol. 2, p. 697, conf.on Nucl.Cross-Sect.and Techn.,Washington 1975
[27] C.H. King, S.M. Austin, H.H. Rossner, W.S. Chien, Phys. Rev. C **16**, 1712 (1977)
[28] Y.M. Gledenov, T.S. Zvarova, I. Kvitek, M.P. Mitrikov, R.S. Mitrikova, Y.P. Popov, V.I. Salatsky, F.V. Zuan, *Investigation of the Decay of ^{8}Be Excited up to Neutron Binding Energy* (1987), p. 298, 37.Conf.Nucl.Spectrosc.and Nucl.Struct.,Jurmala 1987
[29] P.E. Koehler, C.D. Bowman, F.J. Steinkruger, D.C. Moody, G.M. Hale, J.W. Starner, S.A. Wender, R.C. Haight, P.W. Lisowski, W.L. Talbert, Phys. Rev. C **37**, 917 (1988)
[30] S. Bashkin, H.T. Richards, Phys. Rev. **84**, 1124 (1951)
[31] W.D. Warters, W.A. Fowler, C.C. Lauritsen, Phys. Rev. **91**, 917 (1953)
[32] U. Fasoli, D. Toniolo, G. Zago, Nuovo Cimento **34**, 542 (1964)
[33] R. Gleyvod, N.P. Heydenburg, I.M. Naqib, Nucl. Phys. **63**, 650 (1965)
[34] K. Kilian, G. Clausnitzer, W. Durr, D. Fick, R. Fleischmann, H.M. Hofmann, Nucl. Phys. A **126**, 529 (1969)
[35] V. Paneta, A. Kafkarkou, M. Kokkoris, A. Lagoyannis, Nucl. Instrum. Methods in Physics Res., Sect.B **288**, 53 (2012)
[36] B.J. Guzhovskij, S.N. Abramovich, V.H. Protopopov, Izvestiya Akademii Nauk KazSSSR,Ser.Fiz.-Mat. **4**, 24 (1984)
[37] S.M. Austin, Bulletin of the American Physical Society **7**, 269 (1962)

[38] R.R. Borchers, C.H. Poppe, Phys. Rev. **129**, 2679 (1963)
[39] S.A. Elbakr, I.J. Van Heerden, W.J. McDonald, G.C. Neilson, Nuclear Instrum.and Methods in Physics Res. **105**, 519 (1972)
[40] J.W. Meadows, Argonne National Laboratory report series, ANL-7910, 17 (1972)
[41] A.J. Elwyn, R.E. Holland, C.N. Davids, L. Meyer-Schutzmeister, J.E. Monahan, F.P. Mooring, W. Ray, J. , Phys. Rev. C **16**, 1744 (1977)
[42] I. Angeli, Atomic Data and Nuclear Data Tables **87**, 185 (2004)

EPJ Web of Conferences **122**, 08004 (2016) DOI: 10.1051/epjconf/201612208004
*CNR*15*

Nucleon and triton production from nucleon-induced reactions on ^{7}Li

Yukinobu Watanabe[1,a], Hairui Guo[1,4], Kohei Nagaoka[1], Takuma Matsumoto[2], Kazuyuki Ogata[3], and Masanobu Yahiro[2]

[1]*Department of Advanced Energy Engineering Science, Kyushu University, Kasuga, Fukuoka 816-8580, Japan*
[2]*Department of Physics, Kyushu University, Fukuoka 812-8581, Japan*
[3]*Research Center for Nuclear Physics, Osaka University, Ibaraki, Osaka 567-0047, Japan*
[4]*Institute of Applied Physics and Computational Mathematics, Beijing, 100094, China*

Abstract. Nucleon (N) and triton production from nucleon-induced reactions on ^{7}Li at an incident energy of 14 MeV are analyzed by using three-body continuum discretized coupled channels method (CDCC), final state interaction (FSI) model, and sequential decay (SD) model. The CDCC is used to describe nucleon and triton production via breakup continuum channels, ^{7}Li(N,N$'$)^{7}Li*$\rightarrow t + \alpha$. Triton production from $p(n)$ + ^{7}Li $\rightarrow t$ + ^{5}Li(^{5}He) channel and nucleon production from sequential decay of the ground-state ^{5}Li(^{5}He) are calculated by the FSI model and the SD model, respectively. The calculated double differential cross sections for both nucleon and triton production are in good agreement with experimental ones except at relatively low nucleon emission energies.

1 Introduction

In fusion technology, lithium is an important element for tritium breeding in D-T fusion reactors [1] as well as a candidate material for neutron converter in the intense neutron source of International Fusion Materials Irradiation Facility (IFMIF) [2]. From the point of view of the safety in IFMIF, it is also important to predict tritium production from nuclear reactions between lithium and nucleon emitted via deuteron stripping reaction. Regardless of the great significance of nucleon induced reactions on lithium, the experimental data are not necessarily sufficient. Therefore, theoretical study of these nuclear reactions is needed to provide accurate and reliable nuclear data, especially double differential cross sections (DDXs) of nucleon and triton production from ^{7}Li.

Since ^{7}Li is a weakly bound nucleus which can easily break up into two fragments, namely, ^{7}Li $\rightarrow t + \alpha$, the breakup channels are expected to be important reaction channels in nucleon induced reactions on ^{7}Li [3]. Thus, understanding of the breakup channels is essential to analyses of N + ^{7}Li reactions. However, conventional theoretical models such as DWBA method and statistical models cannot describe them well, because the breakup channels have to be described in the three-body system consisting of N, α and t, the breakup states of ^{7}Li are in continuum, and the breakup channel has

[a]e-mail: watanabe@aees.kyushu-u.ac.jp

a strong coupling with the elastic scattering channel, Therefore, systematic study of the ^{7}Li breakup mechanism is an interesting and meaningful subject not only in fusion applications, but also in nuclear physics itself.

Recently, Ichinkhorloo *et al.* [4] have applied three-body continuum discretized coupled channels method (CDCC)[5, 6] with the JLM N-N interaction [7] to n +^{7}Li scattering for incident energies below 24 MeV. Their result reproduces well experimental angular distributions for neutron elastic and inelastic scattering from ^{7}Li and double differential (n, xn) cross sections at relatively high emission energies for incident energies between 11.5 and 24.0 MeV. However, they did not analyze triton production from the breakup process of ^{7}Li.

In our early work [3], both neutron and proton scatterings from ^{7}Li were analyzed over a wide incident energy range up to 150 MeV using three-body CDCC with the JLM N-N interaction considering only P and F internal states of ^{7}Li. The energy dependence of the normalization factors of the JLM N-N interaction was investigated and its empirical expression was obtained from the analysis. The CDCC calculation reproduced the experimental data of neutron total, proton reaction cross sections, and elastic and inelastic scattering angular distributions well. The importance of the breakup effect was demonstrated by analyzing the effect of the coupling between breakup and elastic channels on differential elastic scattering cross section. However, we did not analyze simultaneously nucleon and triton production processes via the breakup channels.

In the present work, we calculate DDXs for nucleon and triton production from the following two reaction processes:

$$p(n) +^7 \mathrm{Li} \rightarrow p'(n') +^7 \mathrm{Li}^* \rightarrow p'(n') + \alpha + t, \qquad (a)$$

$$p(n) +^7 \mathrm{Li} \rightarrow t +^5 \mathrm{Li}(^5\mathrm{He}) \rightarrow t + p'(n') + \alpha, \qquad (b)$$

where ^{7}Li* denotes the ^{7}Li in the breakup continuum states excited by nucleon inelastic scattering. Both nucleon and triton emissions via the breakup channel (a) in proton and neutron induced reactions at 14 MeV are analyzed using the full CDCC. The nucleon and triton production DDXs from the breakup channel are obtained consistently, which is a more straightforward way to demonstrate the applicability of CDCC to the analysis of the breakup channels of nucleon induced reactions on ^{7}Li. In addition, the triton and nucleon emissions from the reaction process (b) are also calculated by a final state interaction (FSI) model[8] and a sequential decay (SD) model[9], respectively.

Section 2 describes a method of calculating nucleon and triton production DDXs in terms of CDCC, FSI model, and SD model. In Sect. 3, calculated results are compared with experimental data and discussed. Finally, Section 4 gives a summary and conclusion.

2 Theoretical Model

Nucleon and triton production processes from the breakup channel (a) and the reaction channel (b) of nucleon induced reactions on ^{7}Li are analyzed by using the following hybrid approach consisting of CDCC, FSI, and SD models.

2.1 CDCC for analysis of the breakup channel

CDCC is an extension of coupled channels methods, in which the breakup continuum states of the weakly bound nucleus are truncated and discretized to the finite number of discrete states in terms of the relative linear and angular momentum of its fragments, and the discrete states are treated similarly to the usual bound inelastic states in the coupled channels method. The detailed formulation of CDCC for analyzing nucleon induced reactions on ^{7}Li is given in our previous paper [3]. The reactions are analyzed using the N + t + α three-body model. The breakup continuum states of ^{7}Li are discretized by the pseudostate method [10, 11], and the diagonal and coupling potentials are obtained by folding the JLM effective N-N interaction with transition densities. In the work, the relative angular momentum l was truncated as $l \leq 3$, and even l states (S and D states) were ignored because their contributions are expected to be smaller than those of odd states. In the present work, the even states are considered as well.

Double differential (n, n') and (p, p') cross sections from the breakup channels (a) in the laboratory system are given by

$$\frac{d^2\sigma}{dE_{\rm N}^L d\Omega_{\rm N}^L} = \sum_{lI} \frac{1}{2I+1} \frac{\pi\mu_r}{\hbar^2 k P_0^2} \frac{m_{\rm N} + M_{\rm Li}}{M_{\rm Li}} \left(\frac{E_{\rm N}^L}{E_N^C}\right)^{1/2} \times \sum_{m_{I_0} m_I} \left| \sum_{L_0 L J} \sqrt{2L+1} (I_0 m_{I_0} L_0 0 \mid J m_{I_0})(I m_I L m_{I_0} - m_I \mid J m_{I_0}) Y_{L m_{I_0} - m_I}(\Omega_{\rm N}^C) S_{lIL}^J(k) \right|^2, \quad (1)$$

where $E_{\rm N}^L$ and $\Omega_{\rm N}^L$ are the energy and emission direction of the emitted nucleon in the laboratory system, and $E_{\rm N}^C$ and $\Omega_{\rm N}^C$ are those in the center-of-mass system. $m_{\rm N}$ and $M_{\rm Li}$ are the masses of incident nucleon and ^{7}Li, respectively. μ_r and $\hbar k$ are the reduced mass and relative momentum of the $t - \alpha$ system, respectively. $\hbar P_0$ is the incident momentum. l, I and m are the orbital angular momentum, the total spin and its projection on the z axis in the t-α system of ^{7}Li, respectively. L and J are the orbital and total angular momenta in the N - ^{7}Li system. The subscript 0 means the ground state of ^{7}Li. $S_{lIL}^J(k)$ is the S-matrix element.

According to Refs. [13–15], the triple differential cross section of the breakup fragments in the laboratory system is expressed as

$$\frac{d^3\sigma}{d\Omega_t dE_t d\Omega_\alpha} = \frac{2\pi}{\hbar} \frac{\mu}{P_0} \left|T_{fi}\right|^2 \rho(E_t), \quad (2)$$

where Ω_t and Ω_α represent the emission direction of t and α, respectively. E_t is the emission energy of t and μ is the reduced mass of the N - ^{7}Li system. $\rho(E_t)$ is the phase space factor [13, 14, 16] expressed as

$$\rho(E_t) = \frac{m_t m_\alpha m_{\rm N} P_t P_\alpha}{(2\pi\hbar)^6 \left[m_\alpha + m_{\rm N} + \dfrac{m_\alpha (\boldsymbol{P}_t - \boldsymbol{P}_{tot}) \cdot \boldsymbol{P}_\alpha}{P_\alpha^2} \right]}, \quad (3)$$

where m_t and m_α are the masses of t and α, respectively. $\hbar \boldsymbol{P}_t$ and $\hbar \boldsymbol{P}_\alpha$ are the momenta of the emitted t and α, respectively. $\boldsymbol{P}_{tot}$ is the total momentum of the three-body system. The transition matrix element, T_{fi}, is given by [13, 15]

$$T_{fi} = i \frac{(2\pi\hbar)^3}{\sqrt{2\mu\mu_r}} \frac{1}{k\sqrt{PP_0}} \sum_{lILJM} \sqrt{2J+1} [Y_l(\hat{\boldsymbol{k}}) \otimes Y_L(\hat{\boldsymbol{P}})]_{JM} e^{i(\delta_{lIk} + \sigma_{lIk})} S_{lIL}^J(k), \quad (4)$$

where $\hbar \boldsymbol{P}$ is the momentum of N relative to ^{7}Li. $\hat{\boldsymbol{k}}$ and $\hat{\boldsymbol{P}}$ are the direction of $\hbar \boldsymbol{k}$ and $\hbar \boldsymbol{P}$, respectively. δ_{lIk} and σ_{lIk} are the nuclear and Coulomb phase shifts between t and α, respectively.

The triton production DDX from the breakup channel (a) is then obtained by

$$\frac{d^2\sigma}{d\Omega_t dE_t} = \int d\Omega_\alpha \frac{d^3\sigma}{d\Omega_t dE_t d\Omega_\alpha}. \quad (5)$$

2.2 Final state interaction model

The FSI model [8] is used for the calculation of triton production from the reaction channel (b). ^{5}He and ^{5}Li are considered as α - n and α - p systems, respectively. The formula is expressed as

$$\left(\frac{d^2\sigma}{d\Omega_t dE_t}\right)_{\mathrm{FSI}} = N_F \sin^2\beta_l \frac{F_l^2(k_{\mathrm{N}\alpha}a) + G_l^2(k_{\mathrm{N}\alpha}a)}{(k_{\mathrm{N}\alpha}a)^2}\rho(E_t), \tag{6}$$

where N_F is the normalization factor determined by fitting to experimental data, F_l and G_l are the first-order spherical Bessel functions for the α - n system and the Coulomb wave functions for the α - p system, k denotes the wave number of $p(n)$ in α - $p(n)$ system, a denotes the channel radius, and ρ is the phase space factor. β_l means for the α - $p(n)$ phase shift, which is calculated by

$$\beta_l = \tan^{-1}[\frac{1}{2}\Gamma_\lambda/(E_\lambda + \Delta_\lambda - E_{\mathrm{N}\alpha})], \tag{7}$$

with

$$\Delta_\lambda = \frac{1}{2}\Gamma_\lambda(F_l F_l' + G_l G_l'), \tag{8}$$

$$\frac{1}{2}\Gamma_\lambda = \frac{\gamma^2 k_{\mathrm{N}\alpha} a}{(F_l^2 + G_l^2)}. \tag{9}$$

The parameters necessary to obtain β_l are taken from Ref. [17] for the α - p system and Ref. [18] for the α - n system: $E_\lambda = -3.3$ MeV, $a = 3.0$ fm and $\gamma^2 = 8.23$ MeV for ^{5}Li, and $E_\lambda = -4.93$ MeV, $a = 3.0$ fm and $\gamma^2 = 7.89$ MeV for ^{5}He.

2.3 Sequential decay model

Nucleon emission DDXs from the decay of ground-state ^{5}Li(^{5}He) in the reaction channel (b) can be calculated with the SD model [9]. The formula is expressed as

$$\frac{d^2\sigma}{d\Omega_{\mathrm{N}'} dE_{\mathrm{N}'}} = \frac{\sigma_t^{FSI}}{2\pi}\int L_1(E_{\mathrm{N}} \to E_{^5\mathrm{Li}(^5\mathrm{He})})L_2(E_{^5\mathrm{Li}(^5\mathrm{He})} \to E_{\mathrm{N}'})\Lambda(\eta_{\mathrm{NN}'})dE_{^5\mathrm{Li}(^5\mathrm{He})}, \tag{10}$$

where σ_t^{FSI} is the triton production cross section obtained with FSI model. L_1 denotes the probability that an incident nucleon with energy E_{N} produces an intermediate ^{5}Li(^{5}He) with energy $E_{^5\mathrm{Li}(^5\mathrm{He})}$. L_2 denotes the probability that the intermediate ^{5}Li(^{5}He) with energy $E_{^5\mathrm{Li}(^5\mathrm{He})}$ produces a nucleon with energy $E_{\mathrm{N}'}$. $\Lambda(\eta_{\mathrm{NN}'})$ means the probability of the cosine of the angle between the direction of incident nucleon and emitted nucleon being $\eta_{\mathrm{NN}'}$. L_1 and L_2 are expressed as

$$L_1 = \pi\Theta_1(m_{\mathrm{N}} + M_{\mathrm{Li}})\left[\frac{m_{\mathrm{N}} m_t M_{^5\mathrm{Li}(^5\mathrm{He})}}{m_t + M_{^5\mathrm{Li}(^5\mathrm{He})}} E_{\mathrm{N}}\left(\frac{M_{\mathrm{Li}}}{m_{\mathrm{N}} + M_{\mathrm{Li}}} E_{\mathrm{N}} + Q_1\right)\right]^{-1/2}, \tag{11}$$

$$L_2 = \pi\Theta_2\left[\frac{m_{\mathrm{N}} m_\alpha E_{^5\mathrm{Li}(^5\mathrm{He})} Q_2}{M_{^5\mathrm{Li}(^5\mathrm{He})}(m_{\mathrm{N}} + m_\alpha)}\right]^{-1/2}, \tag{12}$$

where Q_1 and Q_2 are the reaction Q values for ^{5}Li(^{5}He) production and ^{5}Li(^{5}He) decay, respectively. Θ_1 is the angular distribution of the intermediate ^{5}Li(^{5}He) which is obtained from the triton production DDX calculated with FSI. The nucleon emission is assumed to be isotropic, then the angular distribution of the nucleon emission in the N-α system is expressed as

$$\Theta_2 = \frac{1}{4\pi}. \tag{13}$$

3 Results and discussion

3.1 Proton induced reaction on ^{7}Li at 14 MeV

The proton and triton emission DDXs from the breakup channel (*a*) in proton induced reactions on ^{7}Li are calculated within the framework of CDCC. Fig. 1 shows comparisons of the CDCC result and the experimental data [19] of inclusive proton emission DDXs at an incident energy of 14 MeV. The experimental proton emission data contain the influence of contamination in the target, such as hydrogen(H), carbon(C), and oxygen(O), which is denoted in the figures. The contributions corresponding to unbound $^{3/2}$P, $^{1/2}$P, $^{1/2}$S, $^{3/2}$D, $^{5/2}$D, $^{7/2}$F and $^{5/2}$F states of ^{7}Li are denoted by dashed, dotted, dash-dotted, dash-dot-dotted, short-dashed, short-dotted and short-dash-dotted lines, respectively. The total CDCC result including (*p*,*p*') scattering from the bound states of ^{7}Li is shown by the thick solid lines. Four peaks seen in the total CDCC result represent the elastic, inelastic scattering to the first excited state, $7/2^-$ and $5/2^-$ resonance states, respectively, from the high emission energy end. The CDCC calculation reproduces the experimental data reasonably well in the relatively high emission energy region, whereas the calculated results underestimate the experimental data at low emission energies remarkably. The underestimation is caused by neglecting the other reaction channels including the reaction channel (*b*) which cannot be calculated with the present CDCC as discussed later.

The CDCC result of the triton production DDX from the breakup channel (*a*) at 14 MeV is shown by the thick solid lines in Fig. 2. The calculated results overestimate slightly the experimental data [20] at low emission energies. It may be due to overestimation of the contribution from the unbound $^{3/2}$P state of ^{7}Li. In addition, the calculated results underestimate considerably the experimental data at relatively high emission energies. The discrepancy at high emission energies is due to the contribution from the reaction channel (*b*), as mentioned later.

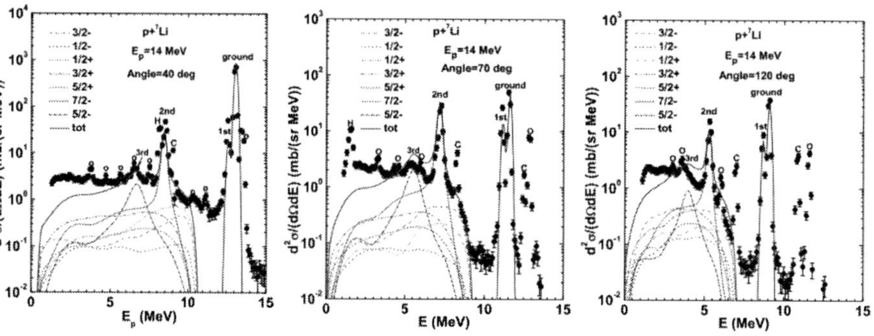

Figure 1. CDCC result of the proton emission DDX for p + ^{7}Li reaction at 14 MeV compared with the experimental data [19]. The dashed, dotted, dash-dotted, dash-dot-dotted, short-dashed, short-dotted and short-dash-dotted lines denote the contributions corresponding to unbound $^{3/2}$P, $^{1/2}$P, $^{1/2}$S, $^{3/2}$D, $^{5/2}$D, $^{7/2}$F and $^{5/2}$F states of ^{7}Li, respectively.

Since it is expected that the discrepancy between the CDCC calculations and experimental DDXs seen in Figs. 1 and 2 is due to neglecting the contribution from the reaction channel (*b*), the triton production DDXs from the reaction channel (*b*) are predicted by the FSI model, and the nucleon emission from the sequential decay of the intermediate ^{5}Li(^{5}He) in (*b*) is calculated with the SD model. In Fig. 3, the triton production DDXs calculated with the FSI model are shown by the dashed

DOI: 10.1051/epjconf/201612208004

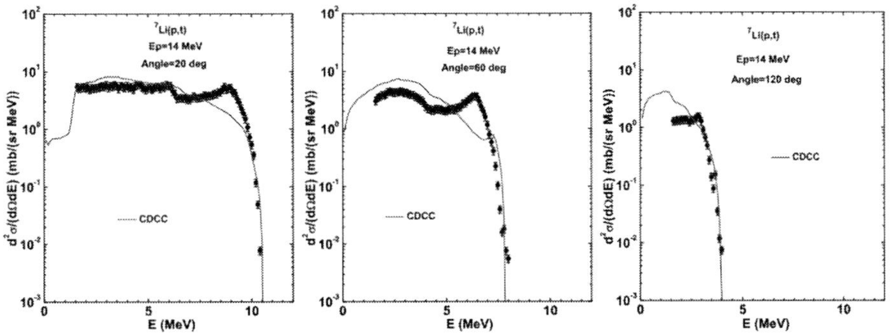

Figure 2. Comparison of triton production DDXs calculated by CDCC with experimental data for p + ^{7}Li reaction at 14 MeV [20].

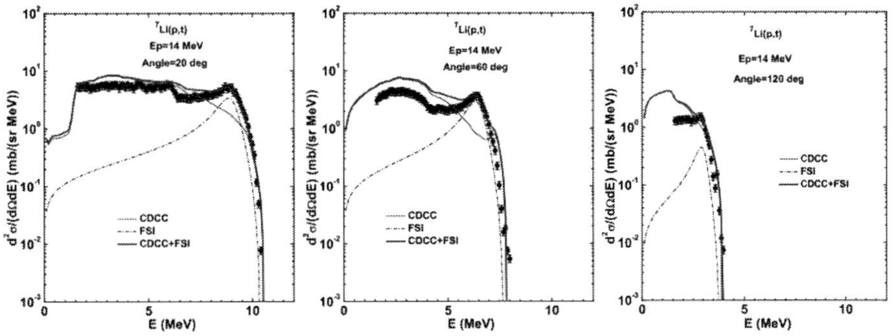

Figure 3. The calculated triton production DDX compared with experimental data for p + ^{7}Li reaction at 14 MeV. The explanation of each line is given in the text.

lines. The normalization factor N_F in Eq. (6) for each angle is determined so as to reproduce the experimental broad peak seen in the high emission energy region. The summation of FSI and CDCC calculations reproduce the experimental data reasonably well over a wide range of triton emission energy, although the calculation overestimates slightly the experimental data in the low emission energy region. On the other hand, the proton emission from the reaction channel (b) for the p + ^{7}Li reaction at 14 MeV shown by the dashed lines in Fig. 4 contributes mainly to the low emission energy region. The summation of the contribution from the reaction channel (b) and the CDCC result is shown by the thick solid lines. The result improves the agreement with the experimental data considerably well. For the proton emission DDX, however, it still underestimates the experimental data for small angles in the low emission energy region because of the other reaction processes contributing to this energy region still exist, such as (p, np) and $(p, 2p)$ reactions.

Figure 5 shows the calculated angular distribution of 14 MeV (p, t) reaction on ^{7}Li. The dashed line denotes the component calculated by the FSI model, where the cross section at 0° is assumed

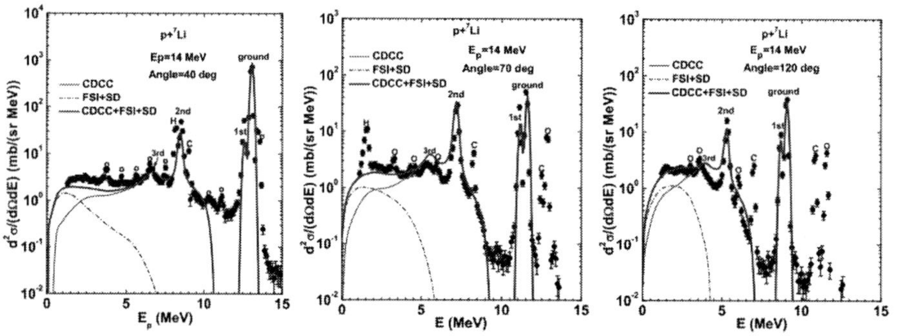

Figure 4. The calculated proton emission DDX from both reaction channels (*a*) and (*b*) compared with experimental data for p + ^{7}Li reaction at 14 MeV. The explanation of each line is given in the text.

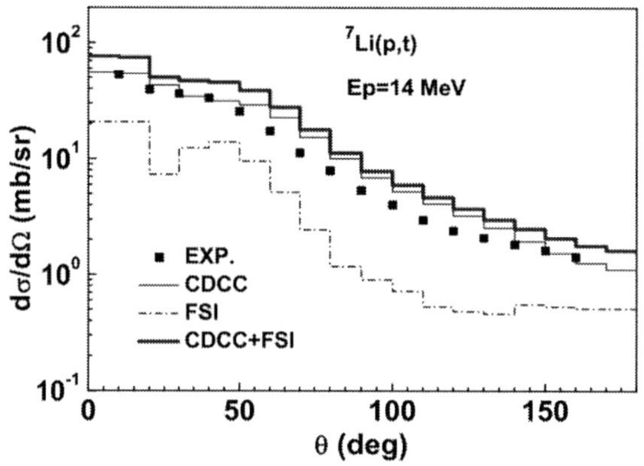

Figure 5. Comparison of the angular distribution of the ^{7}Li (*p*,*t*) reaction at 14 MeV with the experimental data (solid circles). The thin solid line, the dashed line and the thick solid line denote the CDCC result, the FSI result and the summation of CDCC and FSI results, respectively.

to be the same as that at 10° because of no experimental data. The thick solid line represents the summation of the CDCC component (thin solid line) and the FSI one, the shape of which is consistent with the experimental data [20], although the magnitude is a little larger.

3.2 Neutron induced reaction on ^{7}Li at 14 MeV

The neutron emission DDXs from the breakup channel (*a*) in neutron induced reactions on ^{7}Li are calculated with CDCC. In Fig. 6, comparisons of the CDCC results and the experimental data [21] of the inclusive neutron emission DDXs are shown at an incident energy of 14.2 MeV. The first peak of the experimental data from high emission energy end denotes the summation of elastic and first inelastic scattering components because of the poor energy resolution, and the second and third peaks correspond to the inelastic scattering to the $7/2^-$ and $5/2^-$ resonance states, respectively. The CDCC calculation results are in good agreement with the experimental data in the relatively high neutron emission energy region, while the calculated results underestimate the experimental data in the low emission energy region. As shown later in Fig. 7, the underestimation is due to the other missing reaction channels, such as the reaction channel (*b*), which cannot be calculated within the framework of the present CDCC.

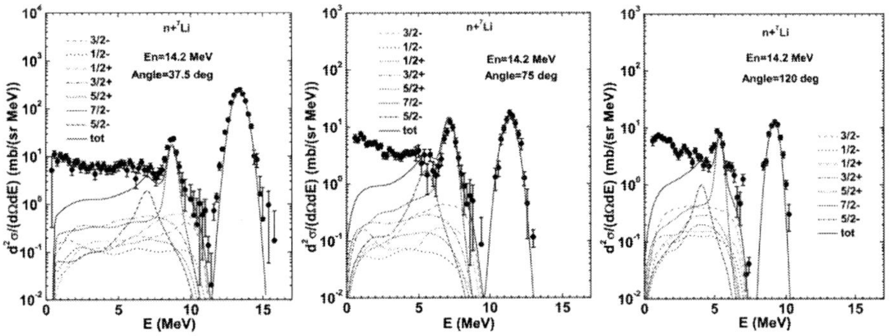

Figure 6. CDCC result (solid lines) of the neutron emission DDX for n + ^{7}Li reaction at 14.2 MeV compared with the experimental data[21]. The dashed, dotted, dash-dotted, dash-dot-dotted, short-dashed, short-dotted and short-dash-dotted lines denote the contributions corresponding to unbound $^{3/2}$P, $^{1/2}$P, $^{1/2}$S, $^{3/2}$D, $^{5/2}$D, $^{7/2}$F and $^{5/2}$F states of ^{7}Li, respectively.

Next, DDXs of inclusive neutron and triton emissions from both the reaction channels (*a*) and (*b*) at incident energies of 14.2 and 14 MeV are calculated by using the hybrid approach of CDCC+FSI+SD, and shown in Figs. 7 and 8, respectively. The FSI calculation for the reaction channel (*b*) requires the experimental data of triton production DDXs. However, only the experimental data [22] are available for the emission angle of 0° at 14 MeV. We assume that the shape of (n, t) angular distributions for the reaction channel (*b*) is same as that of the (p, t) ones given in Fig. 5. Then, the FSI result of the (n, t) angular distribution is obtained by normalizing that of the (p, t) angular distribution at 0°. From Figs. 7 and 8, it can be seen that neutron and triton production from the reaction channel (b) contributes mainly to the low neutron and high triton emission energy regions. The summation of both the contributions from the reaction channels (a) and (b) is denoted by the thick solid line. The summation results are in good agreement with the experimental data [21, 22] except for the neutron emission DDXs in the low emission energy region where it is expected that the other reaction processes contributing to this energy region still exist, such as (n, $2n$) reaction.

4 Summary and conclusion

The nucleon and triton production processes in nucleon induced reactions on ^{7}Li at 14 MeV are analyzed by using the three-body CDCC, final state the interaction (FSI) model, and the sequential decay (SD) model. The double differential cross sections (DDXs) for nucleon and triton production from the reaction process via breakup channels of ^{7}Li,

$$p(n) +^7 \mathrm{Li} \rightarrow p'(n') +^7 \mathrm{Li}^* \rightarrow p'(n') + \alpha + t,$$

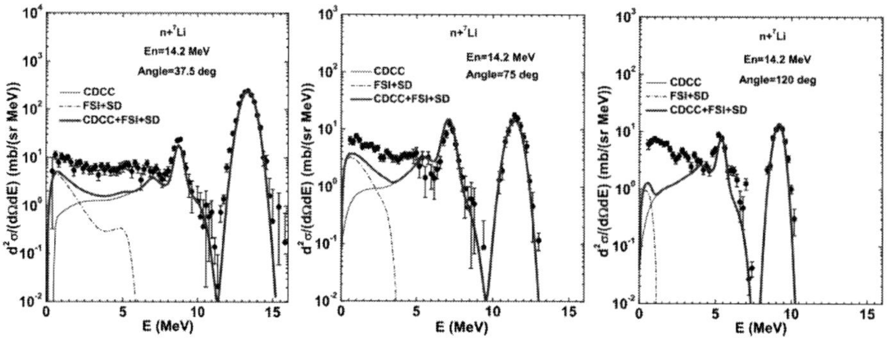

Figure 7. Same as Fig. 4 but for the n + ^{7}Li reaction at 14.2 MeV.

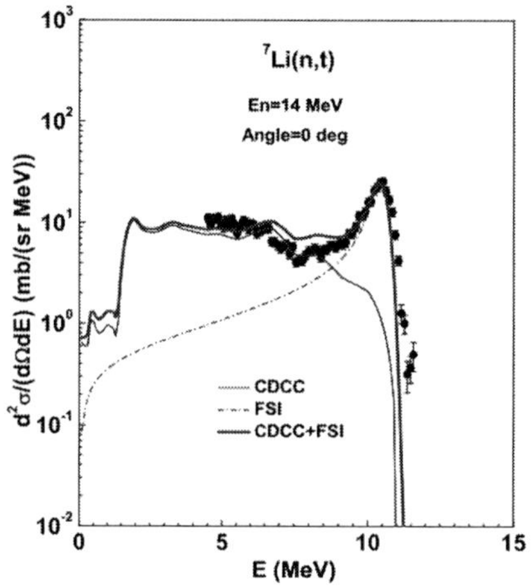

Figure 8. Same as Fig. 3 but for the n + ^{7}Li reaction at 14 MeV.

are calculated using the CDCC. The calculated result for nucleon emission reproduces the experimental DDXs fairly well at relatively high emission energies. The CDCC results of triton production DDXs overestimate slightly the experimental data at low triton emission energies. This may be because the contribution from the unbound $^{3/2}$P state of ^{7}Li is overestimated.

Next, the triton and nucleon emission DDXs from the reaction channel,

$$p(n) +^7 \text{Li} \rightarrow t +^5 \text{Li}(^5\text{He}) \rightarrow t + p'(n') + \alpha,$$

are calculated with the FSI and SD models, respectively. The result supplements the remarkable underestimation seen in the CDCC results at high triton emission energies and low nucleon emission energies.

Through the present analysis, it has been clarified that the breakup continuum channels play an essential role in triton production from nucleon induced reactions on ^{7}Li at 14 MeV. Since the present FSI model needs the experimental triton emission DDXs to determine the magnitude of cross sections, a theoretical model which can describe transfer or knock-on (p,t) reaction leading to three body final states will be required for further study of the incident energy dependence of the FSI component. In addition, a statistical decay model calculation will be necessary to compensate the underestimation seen in the low nucleon emission energy region by considering nucleon emissions from highly excited states of ^{7}Li*, such as $(n,2n)$ or $(p,2p)$ reactions.

Acknowledgments

We would like to thank Satoshi Chiba for helpful discussion and comments. We are also grateful to Norihiko Koori for providing the experimental data of ^{7}Li(p,xp) and (p,xt) reactions at 14 MeV and useful comments on the measurement. This work was supported by Grant-in-Aid for Scientific Research of the Japan Society for the Promotion of Science (No. 22560820).

References

[1] ITER - the way to new energy, http://www.iter.org.
[2] P. Garin, M. Sugimoto, J. Nucl. Mater. **417**, 1262 (2011), and references therein.
[3] H. Guo, Y. Watanabe, T. Matsumoto, K. Ogata, M. Yahiro, Phys. Rev. C **87**, 024610 (2013).
[4] D. Ichinkhorloo, Y. Hirabayashi, K. Katō, M. Aikawa, T. Matsumoto, S. Chiba, Phys. Rev. C **86**, 064604 (2012).
[5] M. Kamimura, M. Yahiro, Y. Iseri, Y. Sakuragi, H. Kameyama, M. Kawai, Prog. Theor. Phys. Suppl. **89**, 1 (1986).
[6] M. Yahiro, K. Ogata, T. Matsumoto, K. Minomo, Prog. Theor. Exp. Phys. **2012**, 01A206 (2012), and references therein.
[7] J.P. Jeukenne, A. Lejeune, C. Mahaux, Phys. Rev. C **16**, 80 (1977).
[8] T. Rausch, H. Zell, D. Wallenwein, W-Von. Witsch, Nucl. Phys. A **222**, 429 (1974).
[9] T. Beynon, B. Sim, Ann. Nucl. Energy **15**, 27 (1988).
[10] A.M. Moro, J.M. Arias, J. Gomez-Camacho, I. Martel, F. Perez-Bernal, R. Crespo, F. Nunes, Phys. Rev. C **65**, 011602 (2001).
[11] T. Matsumoto, T. Kamizato, K. Ogata, Y. Iseri, E. Hiyama, M. Kamimura, M. Yahiro, Phys. Rev. C **68**, 064607 (2003).
[12] R.J. Spiger, and T.A. Tombrello, Phys. Rev. **163**, 964 (1967).
[13] Y. Iseri, M. Yahiro, M. Kamimura, Prog. Theor. Phys. Suppl. **89**, 84 (1986).
[14] G.G. Ohlsen, Nucl. Instr. Meth. **37**, 240 (1965).
[15] M. Yahiro, Y. Iseri, M. Kamimura, M. Nakano, Phys. Lett. B **141**, 19 (1984).
[16] H. Fuchs, Nucl. Instr. Meth. **200**, 361 (1982).
[17] A. Barnard, C. Jones, J. Weil, Nucl. Phys. **50**, 604 (1964).
[18] G.L. Morgan, R.L. Walter, Phys. Rev. **168**, 1114 (1968).
[19] N. Koori, I. Kumabe, M. Hyakutake, K. Orito, K. Akagi, A. ida, Y. Watanabe, K. Sagara, H. Nakamura, K. Maeda, T. Nakashima, M. Kamimura, Y. Sakuragi, JEARI-M 89-167 (1989).
[20] N. Koori, I. Kumabe, M. Hyakutake, Y. Watanabe, K. Orito, K. Akagi, A. Iida, M. Eriguchi, Y. Wakuta, K. Sagara, H. Nakamura, K. Maeda, T. Nakashima, JEARI-M 91-009 (1991).

[21] S. Chiba, M. Baba, H. Nakashima, M. Ono, N. Yabuta, S. Yukinori, N. Hirakawa, J. Nucl. Sci. Technol. **22**, 771 (1985).
[22] V. Valković, Nucl. Phys. **60**, 581 (1964).

EPJ Web of Conferences **122**, 08005 (2016) DOI: 10.1051/epjconf/201612208005
*CNR*15*

The scattering cross sections for ^{6,7}Li + *n* reactions

D.Ichinkhorloo [1], M. Aikawa [1], S. Chiba [2], Y.Hirabayashi [3], and K. Katō [1]

[1]*Nuclear Reaction Data Centre, Faculty of Science, Hokkaido University, Sapporo 060-0810, Japan*
[2]*Research Laboratory for Nuclear Reactors, Tokyo Institute of Technology, Tokyo 152-8550, Japan*
[3]*Information Initiative Center, Hokkaido University, Sapporo 060-0811, Japan*

Abstract. We investigate the continuum-discretized coupled-channel analysis to the integrated elastic and inelastic scattering cross sections for ^{6,7}Li + *n* reactions. We used the Jeukenne-Lejeune-Mahaux effective nucleon-nucleon interaction and optical model potential at incident neutron energies below and above 10 MeV, respectively. The calculated elastic and inelastic scattering cross sections with observed incident energies are almost in good agreement with experimental and evaluated data.

1 Introduction

The Li + *n* reactions are important not only from the basic interest but also from the application point of view. Lithium isotopes will be used as a tritium-breeding material in the *d-t* fusion reactors. Therefore accurate nuclear data are required for n- and p-induced reactions.

In the previous works [1-3], we have successfully studied cross sections for the ^{6,7}Li + *n* elastic and inelastic scattering angular distributions and neutron spectra applying the continuum-discretized coupled-channel method (CDCC) method [4] with (α - d) + n and (α - t) + n models. It was found that the observed cross section data for incident energies from 7.47 to 24 MeV can be reproduced by the present cluster model with one normalization parameter for the imaginary part of the Jeukenne-Lejeune-Mahaux effective nucleon-nucleon (JLM) [5] interaction. More recently, H. Guo et al. [6] have analyzed both neutron and proton scatterings from ^{6,7}Li in wide incident energies up to 150 MeV, and demonstrated the applicability of CDCC to nucleon scatterings from ^{6,7}Li. They analyzed neutron total cross sections, proton reaction cross sections and differential cross sections for nucleon elastic and inelastic scatterings. However, it is still difficult to reproduce low energy data below 14.1 MeV in these frameworks.

In this work, we extended the CDCC analysis to the integrated elastic and inelastic scattering cross sections for ^{6,7}Li at incident neutron energies below 10 MeV by using optical model potential (OMP) [7,8] and above 10 MeV by using JLM. We adjust the normalization constants for the OMP, because the agreement of the calculated cross sections with the data in very low incident energies of the neutron is insufficient without any adjustments. The energy dependent normalization constants, real part λ_V and imaginary part λ_W, of the OMP and JLM are determined explicitly from the integrated elastic cross section data, respectively. Comparing the results of calculations with experimental data, we discuss that the present CDCC calculations, which reproduce the experimental data observed in incident energies higher than 10 MeV with the single folding potential of the JLM and in lower energies with introducing the normalization factors for the cluster folding potential of the OMP.

2 Method and Model

We prepare the wave functions of the bound and α-d,t scattering states of ^{6,7}Li in the similar way as previous work [1-3] in the CDCC method.

In ^{6}Li case, the binding energy of the 1^+ ground state is observed as 1.47 MeV with respect to the ^{6}Li→ α + d threshold, and the low-energy part of the α-d scattering phase shifts in the S-wave (l=0) and D-wave (l=2) have been obtained experimentally. The excited 3^+, 2^+ and 1^+ states of ^{6}Li are observed at 2.18, 4.31, and 5.68 MeV, respectively, which are considered to be the triplet resonance state in the α-d, D-wave. According to the cluster model, the wave functions for the ground state (1^+) and the exited states are written as

$$\phi_{Il}(^6Li;k) = A\{\varphi(\alpha)[\varphi_I(d)\otimes u_l(k,r)]_I\} \tag{1}$$

where $\varphi(\alpha)$ and $\varphi_I(d)$ stand for the internal wave functions of the alpha and deuteron clusters, respectively.

The interaction between α and d has central and spin-orbit parts, which are parametrized by a two-range Gaussian form and by a two-range Gaussian-derivative form, respectively;

$$\begin{aligned} V^{CE}{}_l(r) &= v_{1,l}e^{-(r/r_{1,l})^2} + v_{2,l}e^{-(r/r_{2,l})^2} + V_{CL}(r) \\ V_l^{SO}(r) &= v_{1,l}^{SO}re^{-(r/r_{1,l})^2} + v_{2,l}^{SO}re^{-(r/r_{2,l})^2} \\ V_{CL}(r) &= \begin{cases} Z_1Z_2e^2/r, & r \geq R_{CL} \\ (Z_1Z_2e^2/(2R_{CL}))(3-r^2/R_{CL}{}^2), & r < R_{CL}. \end{cases} \end{aligned} \tag{2}$$

They are chosen l -dependently so as to reproduce well the energies of the ground and excited states and the α-d scattering phase shifts. The parameter values are listed in Table 1.

Table 1. The parameters of the effective central and spin orbit potentials between α and d for l = 0 and 2

Parameters	$r_{1,l}$ (fm)	$r_{2,l}$ (fm)	$v_{1,l}$ (MeV)	$v_{2,l}$ (MeV)	$v_{1,l}^{(SO)}$ (MeV)	$v_{2,l}^{(SO)}$ (MeV)	R_{CL} (fm)
l = 0	2.191	1.607	-105.85	46.22	--	--	3.00
l = 2	2.377	1.852	-82.00	26.00	-2.31	1.42	3.00

The Schrödinger equation of the ^{6}Li + n scattering system, which is described by using the n + α + d three body model, is written as

$$\left[K_R + K_r + V_{d\alpha}(r) + U_{dn}(r_{dn}) + U_{\alpha n}(r_{\alpha n}) - E\right]\Psi_{JM}^{CDCC} = 0, \tag{3}$$

where E is the energy of the total system, vector $\boldsymbol{r}$ is the relative coordinate between α and d, $\boldsymbol{R}$ the one between the center of mass of the d- α pair and n, and $\boldsymbol{r}_{dn}(\boldsymbol{r}_{\alpha n})$ denotes the relative coordinate between two particles $d(\alpha)$ and n. Operators K_r and K_R are kinetic energies associated with $\boldsymbol{r}$ and $\boldsymbol{R}$, respectively, and $V_{d\alpha}(r)$ is the interaction between d and α. The total wave function Ψ_{JM} with the total angular momentum J and its projection M on z-axis is expanded in terms of the orthonormal set of eigenstates ϕ_{Il} of $H(^6\text{Li})$ for the α-d system;

$$\begin{aligned} \Psi_{JM}^{CDCC} &= \sum_L y_{JM}^{l_0 I_0 L}\phi_0(r)\hat{\chi}_{\gamma_0}(P_0,R)/R \\ &+ \sum_{i=1}^{N}\sum_{l=0}^{l_{max}}\sum_I\sum_L y_{JM}^{lIL}\hat{\phi}_{iIl}(r)\hat{\chi}_\gamma(\hat{P},R)/R, \end{aligned} \tag{4}$$

where the spin and angular parts are described as

$$y_{JM}^{lIL} = \left[[i^l Y_l(\Omega_r)\otimes\eta_d]\otimes i^l Y_L(\Omega_R)\right]_{JM}\varphi(\alpha)\varphi(d), \tag{5}$$

with

$$\hat{\chi}_{\gamma_0}(P_0,R) = \chi_{\gamma_0}(P_0/R), \quad \gamma_0 = (0,l_0,I_0,L,J) \tag{6}$$

$$\hat{\chi}_\gamma(\hat{P}_\gamma, R) = W_\gamma \chi_\gamma(\hat{P}_\gamma / R) \quad \gamma = (i, l, I, L, J)$$

On the right hand side of Eq.(4), the first term presents the elastic channel denoted by γ_0 and the second one corresponds to the discretized breakup channels, each denoted by γ. The expansion-coefficient χ_γ in Eq.(6) represents the relative motion between n and ^{6}Li, and L is the orbital angular momentum regarding $\boldsymbol{R}$.

Table 2. Parameters of the optical potentials for d+n and α+n at half the neutron incident energy.

System	V_0(MeV)	r_0(fm)	a_0(fm)	W_D(MeV)	r_{WD}(fm)	a_{WD}(fm)
$d+n$	65.8	1.574	0.501	4.59	1.511	0.517
$\alpha+n$	47.0	2.098	0.660	9.52	2.009	0.280

The interaction U_{dn} ($U_{\alpha n}$) between d (α) and n is taken to be the optical potential for d+n (α+n) scattering. In Table 2, the parameters are shown explicitly. For simplicity, the spin dependence of the interaction is neglected. In this study, we adopt $p + \alpha$ scattering at 31MeV [7] instead of n-α interactions. For $n + d$ scattering, we used a neutron potential parameter set presented by Wilmore et al. [8] at lower incident energies. However, these optical potentials cannot reproduce the calculated cross section of the ^{6}Li + n scattering. We introduce normalization parameters for real and imaginary parts of the cluster folding potential. We also use the JLM interaction based on a single folding model in the incident neutron energy region above 10 MeV. In the previous works [1-3], we reported the CDCC calculation with JLM interaction.

Table 3. The parameters of the effective central and spin orbit potentials between α and t for l=1 and 3

Parameters	$r_{1,l}$ (fm)	$r_{2,l}$ (fm)	$v_{1,l}$ (MeV)	$v_{2,l}$ (MeV)	$v^{(SO)}_{1,l}$ (MeV)	$v^{(SO)}_{2,l}$ (MeV)	$r^{(SO)}_{1,l}$ (fm)	$r^{(SO)}_{2,l}$ (fm)
$l = 1$ (3/2$^-$)	2.447	--	-84.70	--	-0.99	-0.67	4.900	2.447
$l = 1$ (1/2$^-$)	2.447	--	-89.50	--	-0.30	-0.11	4.900	2.447
$l = 3$	2.608	--	-75.65	--	-1.05	--	2.466	2.447

In ^{7}Li case, we prepare the wave functions of the bound and α-t scattering states of ^{7}Li in the similar way as the ^{6}Li + n reaction analysis in the CDCC method. The binding energy of the 3/2$^-$ and 1/2$^-$ bound states are observed as -2.47 MeV and -1.99 MeV, respectively, with respect to the ^{7}Li→ α + t threshold, and the low-energy part of the α-t scattering phase shifts in the P-wave (l =1) and F-wave (l = 3) have been obtained experimentally. The excited 7/2$^-$ and 5/2$^-$ states of ^{7}Li are observed at 4.65 MeV and 6.60 MeV, which are considered to be the triplet resonance state in the α-t relative motion with the F-wave.

The potential between α and t clusters is chosen l-dependently so as to reproduce well the energies of the ground state, and the excited states, and the α-t scattering phase shifts. The parameters are listed in Table 3.

Table 4. Parameters of the optical potentials for t+n and α+n at half the neutron incident energy.

System	V_0(MeV)	r$_0$(fm)	a_0(fm)	W_D(MeV)	r_{WD}(fm)	a_{WD}(fm)
$t+n$	50.35	2.146	0.144	0.520	2.164	0.378
$\alpha+n$	47.00	2.098	0.660	9.520	2.009	0.280

The interaction U_{tn} ($U_{\alpha n}$) between $t(\alpha)$ and n is taken to be the optical potential for the t+n (α+n) scattering (Table 4). For simplicity, the spin dependence of the interaction is neglected. In this study, we adopt $p + \alpha$ scattering at 31 MeV [7] instead of n-α interactions. For $n + t$ scattering, we used a neutron potential parameter set presented by Wilmore et al. [8] at lower incident energies. However, these optical potentials cannot reproduce the experimental cross sections. Therefore, we introduce normalization parameters for real and imaginary parts. We also use the JLM interaction based on a single folding model in the incident neutron energy region above 14.1 MeV.

3 Result and discussion for ^{6}Li

In this calculation, we analyze the integrated elastic and inelastic scattering cross sections of the ^{6}Li+n collision at incident neutron energies below 10 MeV by using optical model potential (OMP) and above 10 MeV by using JLM for the ^{6}Li-n folding potential. For the JLM single-folding potential, we take the normalization factors λ_v=1.0 and λ_w =0.2 for the real and imaginary parts, respectively. These values indicate that the small imaginary potential is needed while the real part has no any adjustment parameter. This is consistent with the previous studies [1-3]. However, because the JLM potential has not been constructed for incident neutron energies lower than about 10 MeV [5], we cannot use the JLM single-folding potential for these energies. Instead of the microscopic JLM single-folding potential, we employ a more phenomenological cluster folding potential in lower energies.

We determine normalization factors, real part λ_v and imaginary part λ_w, of the cluster folding ^{6}Li-n potential using α-n and d-n OMPs from measured integrated elastic cross section data. The obtained factors are expressed as the following linearly energy dependent form:

$$\lambda_V(E) = 1.6 - 0.12\mathrm{E_n}$$
$$\lambda_w(E) = 0 \qquad (7)$$

Figure 1 shows the differential cross sections of the ^{6}Li + n elastic scattering with incident energies between 1.5 and 24.0 MeV. One can see that the results of the CDCC calculation represented with solid lines are in good agreement with the experimental data.

For inelastic scattering, Fig.1 shows the angular distributions to the 3^+ resonance state of ^{6}Li for E_n=5.98, 7.5, 8.17, 10.27, 14.1, 18.0 and 24.0 MeV. The calculated differential cross sections are obtained by integrating the breakup cross section to several discretized 3^+ solutions obtained around the resonance energy region. We can see that the CDCC calculation can also reproduce the inelastic observed cross sections together with the elastic ones.

We also calculate the integrated elastic and inelastic cross sections between 1.5 and 24.0 MeV. The integrated elastic cross sections for ^{6}Li agree with the evaluated data (JENDL-3.3) and other measurements within the experimental uncertainties, as shown in Fig. 2. In Fig.3, the integrated inelastic cross-section values for the 2.186 MeV state of ^{6}Li are almost in good agreement with the evaluation data of JENDL-3.3 and the experimental data.

4 Result and discussion for ^{7}Li

In this calculation, we analyze the integrated elastic and inelastic scattering cross sections of the ^{7}Li+n collision at incident neutron energies below 14.1 MeV by using the optical model potential (OMP) and above 14.1 MeV by using the JLM ^{7}Li-n folding potential. For the JLM single-folding potential, we take the normalization factors λ_v = 1.0 and λ_w = 0.2 for the real and imaginary parts, respectively. These values of the normalization factors indicate that the small imaginary potential is needed while the real part has no any adjustment parameter.

We determine normalization factors, real part λ_v and imaginary part λ_w, of the cluster folding ^{7}Li-n potential using α-n and t-n OMPs from the integrated elastic cross section data. For the OMP cluster-folding potential, we assume λ_w = 0 for all incident neutron energies lower than 14.1 MeV. The normalization factor λ_v is determined at each incident energy below 5 MeV and introduced a linear energy-dependent form from 5 MeV to 14.1MeV. The obtained results of λ_v are presented in Table 5.

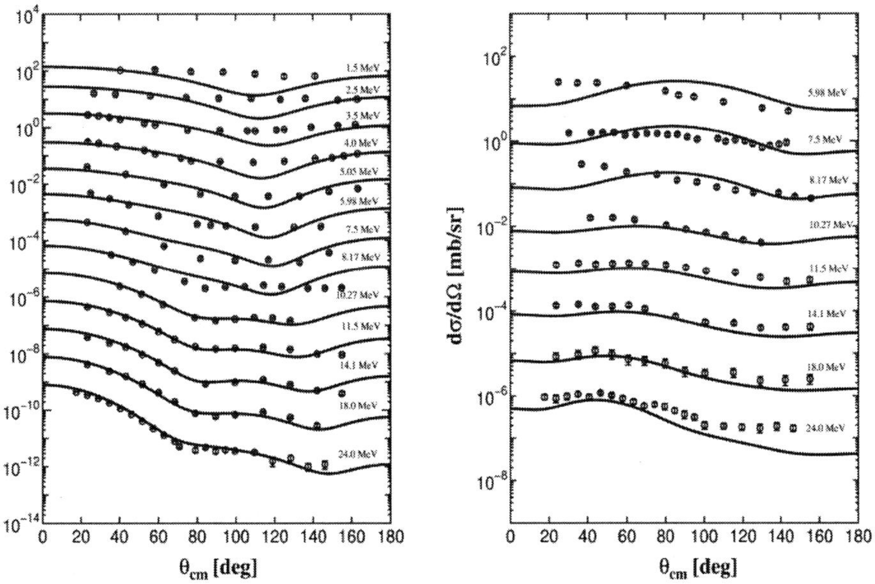

Figure 1. Elastic and inelastic angular distribution of the differential cross sections for the ^{6}Li + *n* scattering for incident energies between 1.5 and 24.0 MeV. The solid lines and open circles correspond to the calculated data and experimental data. The data are subsequently shifted downward by a factor 1/10.

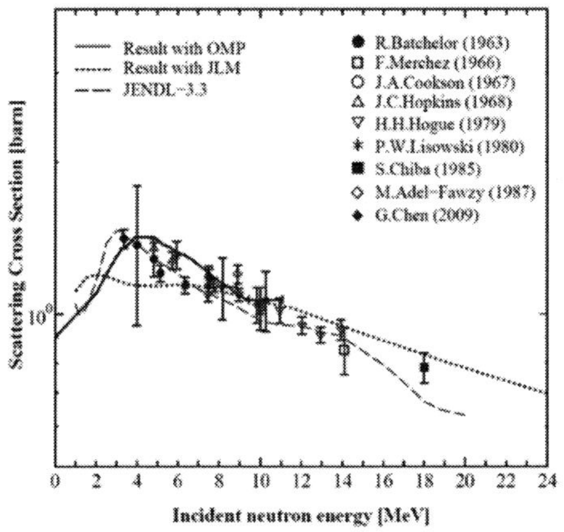

Figure 2. The integrated elastic scattering cross sections of ^{6}Li, in comparison with the evaluated data and experimental data.

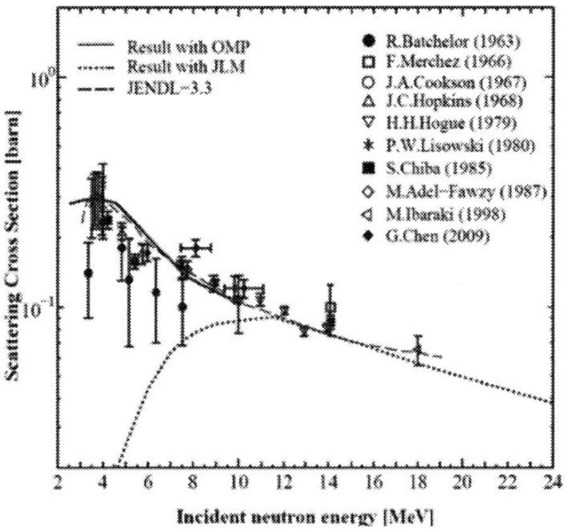

Figure 3. The integrated inelastic scattering cross sections for the 3^{+} state of ^{6}Li, in comparison with the evaluated data and experimental data.

DOI: 10.1051/epjconf/201612208005

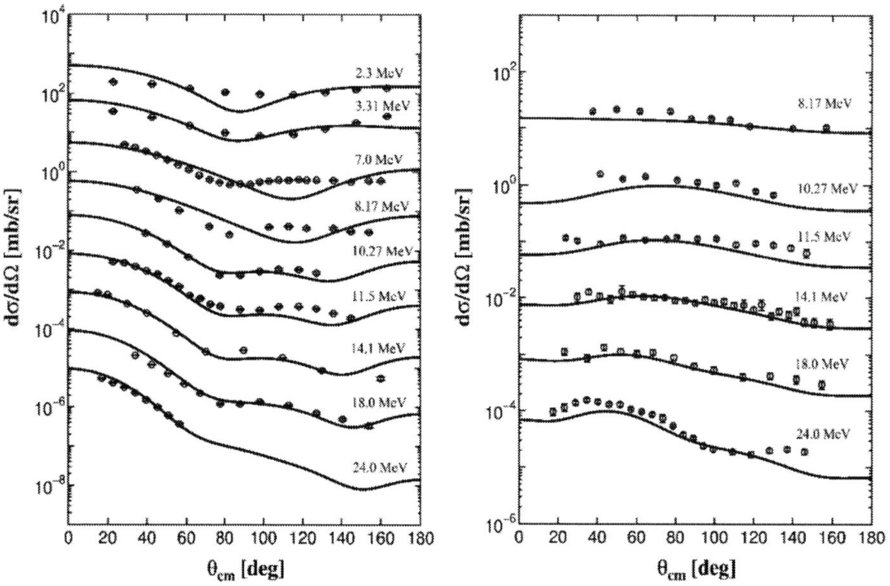

Figure 4. Elastic and inelastic angular distribution of the differential cross sections for the ^{7}Li + n scattering for incident energies between 2.3 and 24.0 MeV. The solid lines and open circles correspond to the calculated data and experimental data. The data are subsequently shifted downward by a factor 1/10.

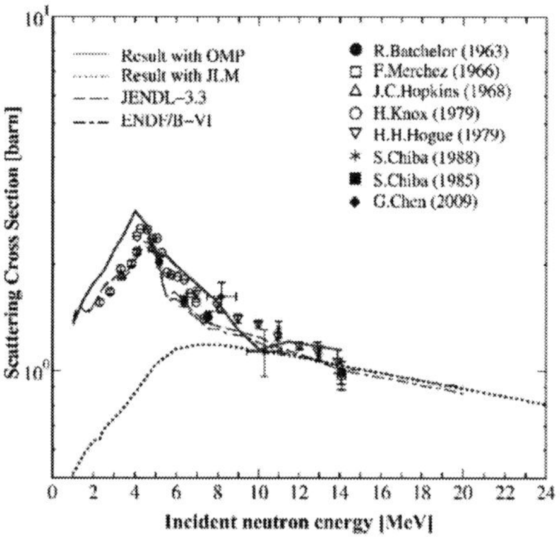

Figure 5. The integrated elastic scattering cross sections of ^{7}Li, in comparison with the evaluated data and experimental data.

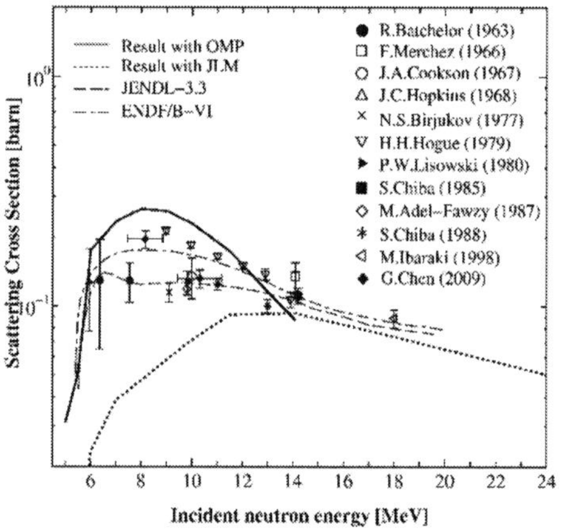

Figure 6. The integrated inelastic scattering cross sections for the 4.65 MeV state of ^{7}Li, in comparison with the evaluated data and experimental data.

Table 5. The normalization factors λ_v

Neutron energy E_n (MeV)	Normalization factors $\lambda_v(E_n)$
1.0	0.760
1.5	0.740
2.0	1.050
2.5	1.117
3.0	1.220
3.5	1.400
4.0	1.360
4.5	1.340
5.0	1.000
$5.5 \leq E_n \leq 14.1$	$1.7-0.07E_n$

The left hand panel of Fig. 4 shows the differential cross sections of the ^{7}Li + n elastic scattering with incident energies between 2.3 and 24.0 MeV. One can see that the results of the CDCC calculation represented with solid lines are in good agreement with the experimental data. The angular distributions show rather flat in low energy cases in comparison with high energy ones. It seems to be due to increase of the s-wave contribution. For inelastic scattering, the right hand panel of Fig.4 also shows the angular distributions to the 7/2$^-$ resonance state of ^{7}Li for E_n = 8.17, 10.27, 11.5, 14.1, 18.0 and 24.0 MeV. The calculated differential cross sections are obtained by integrating the breakup cross section to several discretized 7/2$^-$ solutions obtained around the resonance energy region. We can see that the CDCC calculation can also reproduce the inelastic observed cross sections together with the elastic ones.

We also calculate the integrated elastic and inelastic cross sections between 1.0 and 24.0 MeV. The integrated elastic cross sections for ^{7}Li almost agree with the evaluated data (JENDL-3.3 and ENDF/B-VI) and other measurements within the experimental uncertainties, as shown in Fig. 5. In Fig.6, the integrated inelastic cross-section values for the 4.65 MeV state of ^{7}Li are almost in good agreement with the evaluation data and the experimental data.

5 Summary

Applying the CDCC framework to the ^{6}Li (α-d)+n and ^{7}Li (α-t)+n models, we investigated the integrated neutron elastic and inelastic scattering cross sections for the ^{6}Li and ^{7}Li targets, respectively, at incident neutron energies below 10 MeV using the cluster-folding of the optical model potentials and above 10 MeV using the JLM single-folding potential. Energy dependence of the normalization factors, λ_v and λ_w, of the cluster folding potential is introduced and determined from measured integrated elastic cross sections. The CDCC calculation gives a satisfactorily good agreement with the experimental data.

Although our model calculation is still difficult to reproduce low energy data below 14.1 MeV without adjusting nomarlization factor. This may suggest limitation of applying our model. Therefore it is desirable to extend the CDCC method with other reactions processes such as compound reaction.

Acknowledgements

This work was supported by Japan Society for Promotion of Science (JSPS) for "'R&D' Platform Formation of Nuclear Reaction Data in Asian Countries (2010-2013)", Asia-Africa Science Platform Program and Grant-in-Aid for Publication of Scientific Research Results (No. 257005). Furthermore, this work was supported by Ministry of Education and Science of Republic Kazakhstan for Promotion of Science (the grant No. 3106/GF4).

References

1. T. Matsumoto, D.Ichinkhorloo, Y.Hirabayashi, K.Katō, and S.Chiba, Phys. Rev. C **83**, 064611, (2011)

2. D.Ichinkhorloo, T.Matsumoto, Y.Hirabayashi, K.Katō and S.Chiba, J. Nucl. Sci. Technol. Vol.**48**, No. 11, pp. 1357-1460 , Sep. 2011
3. D.Ichinkhorloo, Y.Hirabayashi, K.Katō, M. Aikawa, T.Matsumoto, and S.Chiba, Phys. Rev. C **83**, 064604, (2012)
4. M.Kamimura, M.Yahiro, Y.Iseri, Y.Sakuragi, H.Kameyama and M.Kawai, Prog. Theor. Phys. Suppl. No. 89 (1986),
5. J.-P. Jeukenne, A. Lejeune, and C. Mahaux, Phys. Rev. C **16**, 80 (1977).
6. H.Guo, Y.Watanabe, T.Matsumoto, K.Ogata and M.Yahir, Phys. Rev. C **87**, 024610 (2013).
7. S. J. Burger and G. Heymann, Nucl. Phys. A **243**, 461 (1975)
8. D.Wilmore and P.E.Hodgson ,Nucl. Phys. 55, 673 (1964)
9. G.Chen, X.Ruan, Z.Zhou, J.Zhang, B.Qi, X.Li, H.Huang, H.Tang, Q.Zhong, J.Jiang, B.Xin, J.Bao, and L.Chen, Nuclear Science and Engineering **163**, 272–284 (2009)
10. A.B.Smith et al, J,NP/A,373,305,8201 (1982)
11. H.H.Hogue et al., J. Nucl. Sci. Eng., **69**, 22 (1979)
12. P.W.Lisowski et al., LA-8342, Los Alamos Scientific Laboratory (1980)
13. J.A.Cookson et al., J. Nucl. Phys. A, **91**, 273 (1967)
14. M.Adel-Fawzy et al., Nucl. Instrum. Methods, **169**, 533 (1980)
15. R.Batchelor et al., J. Nucl. Phys., **47**, 385 (1963)
16. J.C.Hopkins et al., J. Nucl. Phys. A, **107**, 139 (1968)
17. F.Merchez et al., J. Physique Colloque, **27**, 1, 61 (1966)
18. S.Chiba al., J. Nucl. Sci. Technol., **25**, 2, 210 (1985)
19. M.Ibaraki et al., JAERI-Research 98-032, Japan Atomic Energy Research Institute (1998)

Measurement of the ^{6}Li(n,α) neutron standard cross-section at the GELINA facility

Kaj Jansson[1,a], Ali Al-Adili[1], Riccardo Bevilacqua[2], Cecilia Gustavsson[1], Franz-Josef Hambsch[3], Stephan Pomp[1], and Marzio Vidali[3]

[1] *Department of Physics and Astronomy, Uppsala University, Box 516, 751 20 Uppsala, Sweden*
[2] *European Spallation Source, Box 176, 221 00 Lund, Sweden*
[3] *European Commission, Joint Research Centre, Institute of Reference Materials and Measurements (IRMM) Retieseweg 111-2440 Geel, Belgium*

Abstract. The ^{6}Li(n,α) reaction cross-section is commonly used as a reference cross section. However, it is only considered a neutron standard up to 1 MeV. For higher energies, there are discrepancies of several per cents between recent measurements and evaluated data files. In order to extend and establish ^{6}Li(n,α) as a neutron standard above 1 MeV these discrepancies must be resolved.
Our measurement at the GELINA facility at JRC-IRMM in Geel, Belgium is ongoing. We are using a double twin Frisch-grid setup to detect both α-particles from two ^{6}Li targets and fission products from two ^{235}U reference targets. Our targets have thick backings but are employed in pairs, one forward facing and one backward facing. In this way we still cover, in principle, a solid angle of 4π. We present some preliminary results showing that the existing cross-section data is well reproduced around the resonance at 240 keV. The final data taking will start in the beginning of 2016, when the GELINA facility goes online again after a few months of shut down.

1 Introduction

1.1 The cross section

The ^{6}Li(n,α) reaction cross-section is an established standard. One reason for that, is its relatively high cross section, while another reason is its high Q-value, of almost 5 MeV. A large Q-value enables the reaction products to give large clear signals in many detector setups, e.g. ionisation chambers. Depending on the setup, one or both of the product particles may be detected.

Currently, it is only considered a neutron standard up to 1 MeV. In order to extend and establish ^{6}Li(n,α) as a neutron standard above 1 MeV some issues must be resolved.

Up to 1 MeV the cross section from different experiments agree fairly well, but above 1 MeV to about 2.5 MeV discrepancies exist. The first resonance of ^{6}Li(n,α) occurs at a neutron energy of 0.240 MeV in the laboratory frame of reference. It corresponds to the $\frac{5}{2}^-$ excited state at 7.454 MeV in the compound nucleus [1]. Around 2 MeV three other levels in the compound nucleus contribute to

[a] e-mail: kaj.jansson@physics.uu.se

resonances in the ^{6}Li(n,α) reaction. However, their widths are much larger (4.71, 2.75 and 4.37 MeV compared to 0.080 MeV for the 7.454 MeV level) [1] and, therefore, the resonance structure is more smeared out.

In the 1-2.5 MeV region at least two independent experiments have reported discrepancies. One experiment was performed using an ionization chamber similar to the one used at JRC-IRMM (see Sect. 2.2) [2], while the second one employed solid state detectors [3]. Both experiments see a several percent higher cross section than, e.g., the ENDF/B-VII evaluation [4] in this energy region.

It is speculated [5] that one of the reasons the evaluation based on previous measurements shows a smaller cross section is the *Particle Leaking* effect. Due to the incoming neutron momentum there is a possibility of both products (t and α) being emitted in the forward direction (in the laboratory frame of reference), but with large polar angles ($\theta \lesssim 90°$) for moderate neutron energies. In this case the detector response will look like as if only one particle, with the combined energy of both the triton and the α-particle, has entered the chamber. If this effect is overlooked it can lead to an underestimation of the cross section since the pulse height is much higher than for a normal α-particle event and might be discarded because of that. This is why the effect is called Particle Leaking.

The aim of this measurement is to allow for an extension of the standard energy range from 1 MeV to a few MeV. Therefore, any discrepancies in this region must be resolved before the cross section can be trusted as a standard.

2 Experimental setup

2.1 GELINA

The measurement is conducted at the *GEel LINear Accelerator* (GELINA) [6, 7] neutron source which resides at the *Joint Research Centre, Institute for Reference Materials and Measurements* (JRC-IRMM) in Geel, Belgium. It provides a pulsed continuous neutron energy spectrum that is peaked around 1-2 MeV but extends up to 20 MeV. To increase the neutron flux in the keV range a moderator is available.

The measurement of ^{6}Li(n,α) at GELINA is one out of three related experiments conducted at the JRC-IRMM. Detector setups similar to the one presented here have been used to measure ^{10}B(n,α)^{7}Li [8], also at the GELINA facility, and a measurement of the ^{6}Li(n,α)t cross section has been conducted at the VdG facility [9].

The experimental setup was viewing the neutron production target from a distance of 60 m. In the beginning of 2015 the setup was moved to a closer position (10 m) and was then also equipped with a moderated beam, in order to collect more data for lower neutron energies. The increased neutron flux at this distance allowed us to test modifications to the setup much quicker.

2.2 Gridded ionisation chamber

The detector setup consists of a twin Frisch-grid ionisation chamber which will measure the α-particles from the ^{6}Li(n,α)t reaction as well as the fission products from the ^{235}U(n,f) reference reaction. A schematic view of the *Gridded Ionisation Chamber* (GIC) is depicted in Fig. 1. The cylindrical chamber has one shared gas volume normally kept at a pressure slightly above 1 bar (the gas used was P-10), but the uranium measurement is confined to the front compartment of the chamber and the lithium measurement to the back. Targets with thick (0.5 mm) backings were used and two targets of ^{6}Li and ^{235}U respectively are employed in a back-to-back position in order to form a combined doubled sided target. The combined target in each compartment act as the cathode.

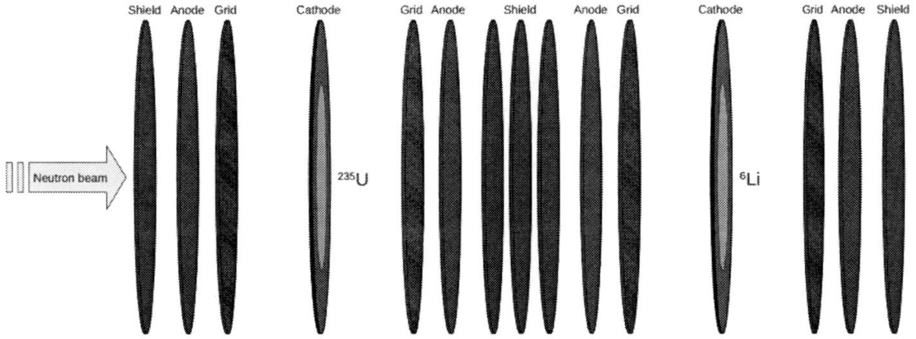

Figure 1. Schematic of the ionisation chamber. The chamber axis coincides with the neutron beam axis. Each cathode consists of two targets facing forward respectively backward. In total the chamber has 10 independent electrodes, five in each compartment separated by shields. The shields are interconnected and kept at the same potential.

Both lithium targets consisted of LiF, which was vacuum deposited on a aluminium backing and the deposited layers had similar masses, 2014 ± 23 and 2610 ± 40 µg, respectively. The 0.5 mm thick backings made it impossible for any reaction products to traverse the target to the other side of the cathode (the range of a 5 MeV triton in aluminium is only about 100 µm [10]). The uranium targets were of similar areal densities, 193 and 222 µg/cm^2, but due to different diameters they had different masses, 3071 ± 46 and 8520 ± 40 µg ^{235}U, respectively. The isotopic mass ratio of ^{235}U in the uranium target material were 97.663(3)% and 99.9336(14)% for the two targets, respectively.

On each side of each cathode, the chamber was equipped with an anode and a grid. The induced signals on each electrode were recorded by a 100 MHz digitiser. In theory one can calculate the third knowing two of the signals, but in practise it is better to record all of them in order to reduce uncertainties due to electronic drift [11]. The purpose of the grids is to shield the anodes and in this experiment they consist of conductive meshes (rather than parallel wires, which is another option).

What distinguish the gridded chambers from normal ionisation chambers is that the grids enable measurements of both particle energy and emission angle. The grid shields the anode from the ionisation electrons until they pass through it. The charge inducted on the anode will be independent on angle and proportional to the total number of electrons. On the grids however, the induced charge will be proportional to the cosine of the emission angle, θ. By treating the whole cloud of N electrons as a single charge, $-Ne$, and applying the Shockley-Ramo theorem [12] the total induced charges on the cathodes, anodes and grids can be derived as

$$Q_{\text{Cathode}} = -Ne\left(1 - \frac{\overline{X}}{D}\cos\theta\right) \tag{1}$$

$$Q_{\text{Anode}} = -Ne \tag{2}$$

$$Q_{\text{Grid}} = Ne\frac{\overline{X}}{D}\cos\theta, \tag{3}$$

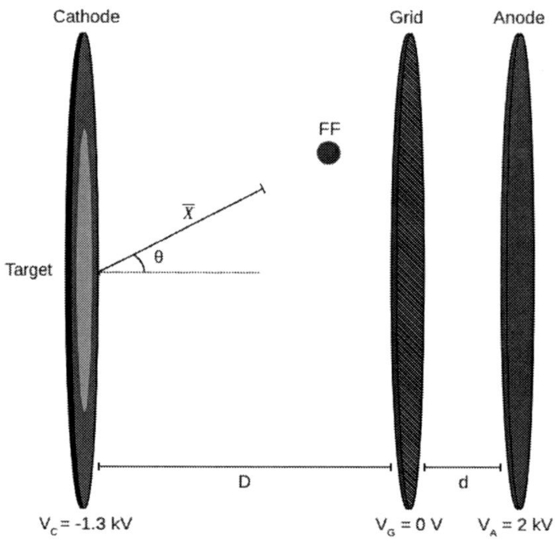

Figure 2. Schematic of the chamber electrodes. The emitted particle will leave a track of liberated electrons. The mean distance from the particle origin of these electrons is denoted $\overline{X}$. The grid-cathode distance, D, is 40 mm and the grid-anode distance, d, is 7 mm.

where $\overline{X}$ is the distance of the electron cloud's centre-of-gravity from the cathode along the particle track and D is the distance between the cathode and the grid. The notation is further illustrated in Fig. 2. For all types of signals we define the pulse height of the signal (PH) to be the digitised voltage level a short time after the event minus the voltage level right before the event (the baseline). The PH is therefore proportional to the total induced charge in Eq. 1-3.

These derivations were based on a perfect grid. In reality the grid will allow part of the electric field to leak through and therefore the anode is affected by the electrons even before they pass through the grid. The effect is in the order of a few percent and can be corrected for. A parametrisation of the so called grid inefficiency has been made by Göök et al., derived for different types of grids [13].

Since the last region the electrons will pass through is the one between the grid and the anode (the region where positive charge is induced on the grid), any loss of electrons will manifest itself as a more negative grid PH. In order for the electrons to pass through the grid without losses, the field strength between the grid and the anode must be several times higher than the field strength between the cathode and the grid. This has been investigated by Bevilacqua et al. [14] for mesh grids, which behave differently to parallel-wire grids.

3 Analysis

3.1 Signals

The digital sampling of the waveforms enables corrections in the post-analysis and has shown to be superior to analogue equipment [11]. However, the collection of stored waveforms needs to be iterated in order to extract the relevant information. The parameters that are of interest, are in this case the time and the signal PH. To extract this information each signal was numerically fitted to a function consisting of line segments. Three segments, whereof two horizontal (four parameters in total), were used for cathode and anode signals while the grid signals required four segments (six parameters). The anode and grid signals from an α-particle event are depicted in Fig. 3 together with the corresponding fitted functions. The choice of fit function represents a compromise between the number of needed parameters, computation complexity and how well the function captures the signal features.

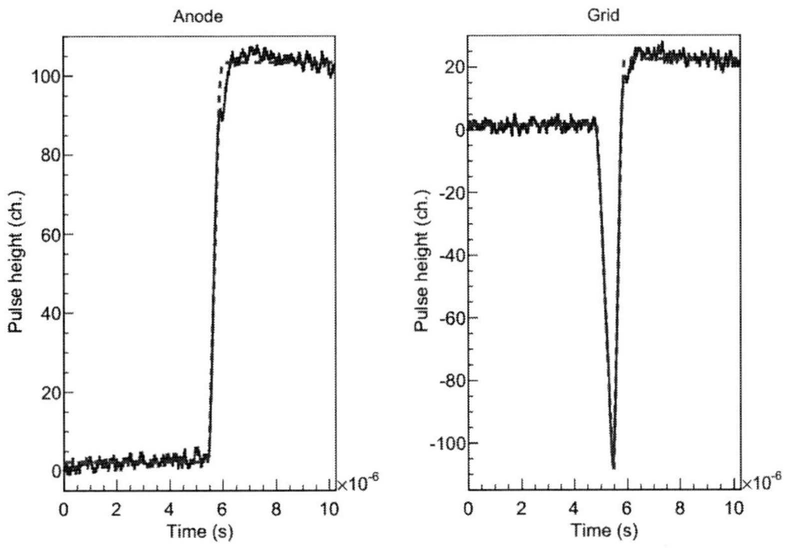

Figure 3. An example of anode and grid signals from an α-event. The anode signal in the left panel is easily distinguished from the grid signal in the right panel. The dashed lines represent the fits made to the signals.

By fitting the signals rather than just extracting the PH, more information become available, e.g. the signals' rise time. The deduced PHs from the fitted signals were compared to the PHs obtained by applying a noise reducing $CR\text{-}RC^4$ filter and then extracting the PH. No preference for either method could be seen. The χ^2-distribution from the fit results corresponded to the expected distribution considering the number of degrees of freedom as well as the uncertainty (noise) of the baseline.

3.2 Time-of-flight

The trigger was constructed by combining both cathode signals after they were fed through a CFD module. The recorded timestamp, t_{TS}, can be turned into the ToF, t_{ToF}, if a reference time, t_0, is known. The reference time can be found by observing the γ-flash coming from the neutron production target or by identifying resonances. The ToF is then given by,

$$t_{\mathrm{ToF}} = t_{\mathrm{TS}} - t_0 + L/c, \quad (4)$$

where L is the distance between the neutron production target and the relevant cathode. Once the neutron ToF is deduced the kinetic energy of the neutron, T_n, is given by:

$$T_n = m_n(\gamma - 1) = \frac{m_n}{\sqrt{1 - \frac{L^2}{t_{\mathrm{ToF}}^2 c^2}}} - m_n, \tag{5}$$

where $m_n = 939.565379\,\mathrm{MeV}$ is the neutron mass.

We observed a correlation between the anode PH and the ToF when t_{TS} was used without any correction. Since all signals were digitised, we could correct the timestamp based on an extrapolation of the slope of the cathode signal to the baseline. This procedure removed the time walk completely.

Bismuth and lead plates were placed in the neutron beam in front of the chamber during the measurements, in order for resonances to be visible in the ToF spectrum. Unfortunately, the statistics has not been enough in order to resolve and identify most of them. The resonance structure of ^{235}U can serve the same purpose as the beam filters. In this case many peaks have been identified and were used to calibrate the ToF in the eV to keV region.

Even though piled up events are rare in our data, we have implemented a routine that searches for such events. This is done by analysing the differentiated signal. If the piled up signals can be resolved the routine is also able to correct the digitised signal by removing one of the signals. The removal is performed by setting the derivative to zero in the region where the piled up event resides and integrating.

3.3 Fission events

Due to the much larger energy deposited in the gas by FFs, a PH threshold separated the FFs unambiguously from the α-particles. Due to the backing thickness, only one of the two FFs is detected for each fission event. The mass of the FF is, for each single event, unknown and can only be estimated based on its energy. The neutron energy does not reach more than a few MeV, so its influence on the fragment kinematics will be small.

The analysis starts by plotting the grid PH versus the anode PH as in Fig. 4. For each anode channel the fission events are projected onto the grid axis. The projection is used to determine for which grid values the distribution drops to half its maximum. These grid values, G_0 and G_1, correspond to $\cos\theta = 0$ and $\cos\theta = 1$ (or $\cos\theta = -1$ for particles emerging from the backwards facing target). From Eq. 3 it is known that the grid signal is proportional to $|\cos\theta|$ which allows us to derive the following formula:

$$|\cos\theta| = \frac{G(Q_G) - G_0}{G_1 - G_0}. \tag{6}$$

The value of G_0 is always very close to zero but the value of G_1 will change with the fragment energy since it incorporates $\overline{X}$ which in turn depends on the stopping range of the heavy ion in question. It turns out that G_1 can be well parametrised by two line segments as shown in Fig. 4.

The fragments emitted in an angle close to 90° will suffer from larger energy losses in the target since the effective thickness of the target is proportional to $\frac{1}{\cos\theta}$. Corrections for the energy losses in the target can be made by observing the mean energy of all events for a certain cosine value. The inverse cosine is plotted against the anode PH, A, and a straight line, $A = \frac{k}{|\cos\theta|} + m$, is fitted to the data where the losses are not too big. We can estimate the anode PH, in the absence of energy loss in the target, with

$$A_{\mathrm{no\ loss}} = A(\theta) - \frac{k}{|\cos\theta|}, \tag{7}$$

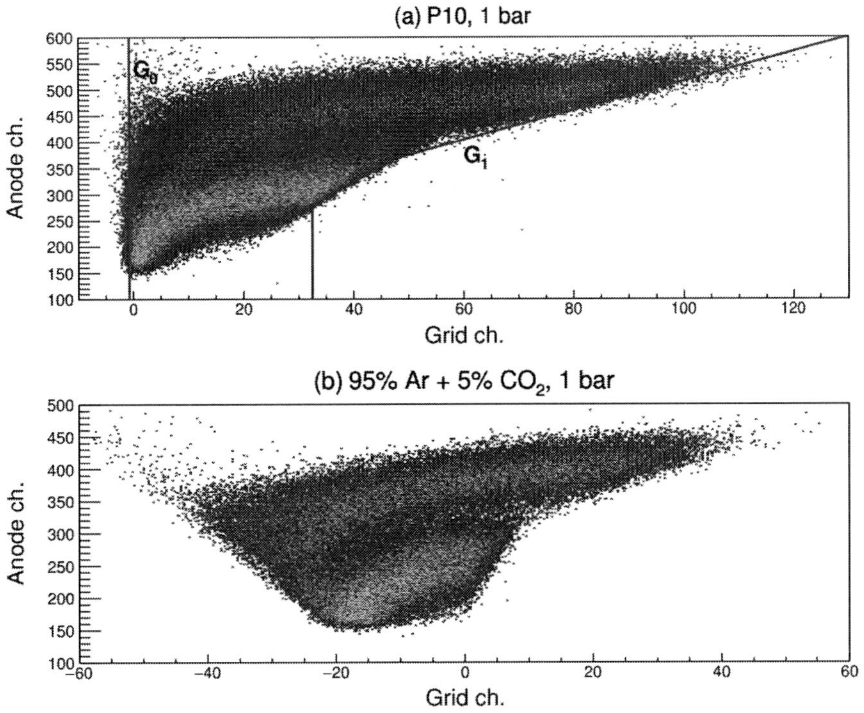

Figure 4. The top panel (a) shows the grid signal PH plotted against the anode signal PH, for all fission events in the forward direction when using P10 as the ionisation gas. The lines represent the limits G_0 and G_1 (see Eq. 6), where the distribution height drops to half its maximum and defines $\cos\theta = 0$ and $\cos\theta = \pm 1$. The bottom panel (b) shows the same plot but with 95% Ar + 5%CO_2 as the ionisation gas. The extensive negative grid signals in this case shows that the two gases have quite different properties when it comes to electron losses, if the same electric potentials are used.

through extrapolation of the fitted line by letting $\frac{1}{\cos\theta} \rightarrow 0$. For gracing angles the energy losses are too big to be well corrected for and some particle losses are unavoidable, for the final results only the events with $|\cos\theta| > 0.3$ are used currently.

As previously mentioned the reduced field strengths, both between cathode and grid as well as between grid and anode, are important parameters. The kind of gas and the type of grid determines how, e.g., drift velocity, recombination and electron captures are affected by the pressure and voltage settings. Negative grid PHs can be a sign of a bad configuration and often indicate electron losses. More investigations like the one in ref. [14] are needed.

3.4 Lithium events

The two reaction products (^{4}He and ^{3}H) have comparable masses to the neutron's, which means that the kinematics are strongly dependent on both the neutron energy and the emission angles in

 DOI: 10.1051/epjconf/201612208006

the laboratory frame of reference. For all, except very low, neutron energies the two products will not be emitted in opposite directions (in the laboratory frame of reference) due to the momentum transfer of the incoming neutron. One effect of this is the *Particle Leaking* which was discussed in the Introduction.

The anodes can be calibrated by installing an α-particle source but this has not yet been done. Instead the only way of determining the deposited energy is by kinematical calculations based on the neutron energy. The α-particle emission angle is determined analogously to how it was done for the fission fragments. To be able to do kinematic calculations and transform the system into the centre-of-momentum (CoM) frame, the anodes must be calibrated. This was done by calculating the theoretical α-particle energy using the neutron energy determined by the ToF technique at $\cos\theta = 0$ and $\cos\theta = \pm 1$. The two end points had to be independently calibrated since the energy loss in the target does not scale linearly with particle energy. The laboratory energy were determined for each event by interpolation between the two end point calibrations.

Once the laboratory properties of the α-particle are known they can be transformed into the CoM frame by Lorentz transformation and energy and momentum conservation. One finds that

$$T_\alpha^* = \gamma\left(E_\alpha - \beta\cos\theta\sqrt{E_\alpha^2 - m_\alpha^2}\right) - m_\alpha \quad (8)$$

$$\cos\theta^* = \pm\sqrt{1 - \frac{E_\alpha}{E_\alpha^*}(1 - \cos^2\theta)}, \quad (9)$$

where E denotes total relativistic energy, T denotes kinetic energies, m denotes mass and all CoM properties are marked by *. The negative solution in Eq. 9 is applied for the backward facing side while the positive is applied for the forward facing side. In addition $\cos\theta^*$ is multiplied by -1 if $\cos\theta_{\text{limit}} > \cos\theta$ where θ_{limit} is given by

$$\theta_{\text{limit}} = \arctan\sqrt{\frac{(m_n + m_{\text{Li}})^2}{m_n m_\alpha}\frac{m_t}{m_\alpha + m_t}\frac{E_n + Q}{E_n}}, \quad (10)$$

in order to take into account that the forward laboratory angle might have reached into the back hemisphere of the CoM frame.

4 Preliminary results

4.1 Fission spectra

The result of changing the P10 gas to 95% Ar+5% CO_2 was lots of events with negative grid PHs (see Fig. 4). Properties that can affect the electron losses, like the electron drift velocity and temperature as well as recombination and capture rates, are not the same for the two gases. The new argon mixture needed higher voltage settings. However, our high-voltage supplies were already operated close to their maximum setting so a switch back to P10 was made, since proper settings for this gas could be obtained.

For FFs of gracing angles, the effective thickness of the target becomes large. Consequently, the energy losses will be large too, making accurate corrections more difficult. Sometimes the fragments will not even escape the target. One can see that some losses of events are occurring in the cosine spectra in Fig. 5.

To avoid being affected by particle losses, currently only the fission events where $|\cos\theta| > 0.3$ are used. The ratio of event counts in the forward and backward direction was compared to the ratio of ^{235}U atoms in the corresponding targets and the difference to the expected value was 1.2%.

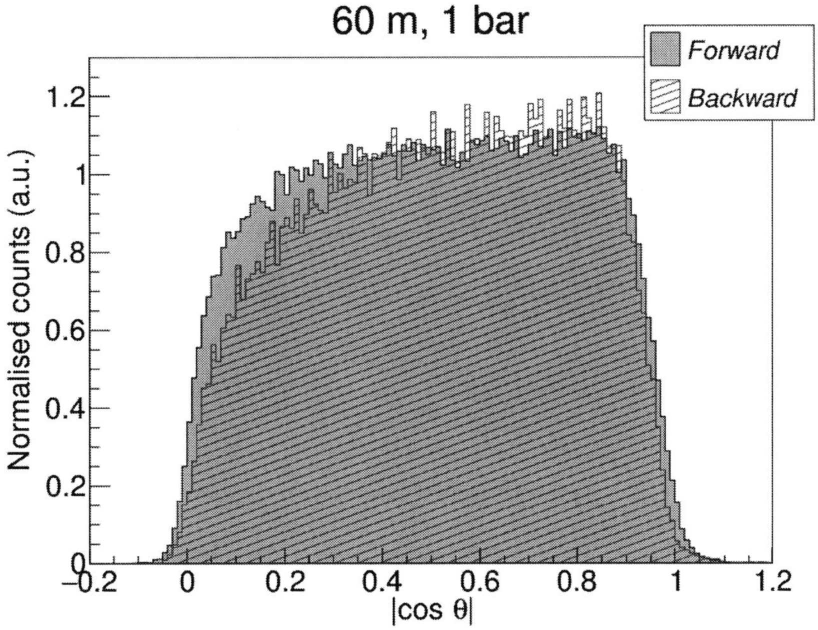

Figure 5. Cosine distributions of the events emerging from the forward and backward facing fission targets respectively. The data comes from the measurement at 60 m distance and 1 bar pressure. When $|\cos\theta| \lesssim 0.3$ the energy losses in the target soon become large and in the backward direction one sees a larger loss of events due to this. We see systematically broader distributions from the older target (here put in the backward position) which has smaller diameter and is slightly thinner. The spectra have been normalised based on the events where $|\cos\theta| > 0.3$. Apart from losses due to the target, the distributions are quite well aligned and somewhat isotropic.

Depicted in Fig. 6 are the two energy spectra from the forward and backward directions. They match each other well, but some small discrepancies can be seen at lower energies.

4.2 Preliminary cross section

A preliminary cross section has been calculated and compared to the ENDF/B-VII.1 evaluation [4] (see Fig. 7). The experimental cross section shows a tendency of being systematically too low except for low energies, but around the resonance region the agreement is fairly good. The measured cross section peaks at 243 keV, close to the expected energy. With better statistics the binning can be made finer and the neutron-energy is expected to improve when more resonances are utilized in the ToF calibration (see Sect. 3.2).

For energies below ~150 keV or above 1 MeV large systematic differences to the evaluation are seen. In the low energy region the measurement overestimates the cross section due to a troublesome background. The origin of this background is, by the time of writing, unknown. Also for the higher energies a cluster of background events partly overlaps with the α-particle distributions making it difficult to correctly select the appropriate events. However, it is less severe than for the lower energies.

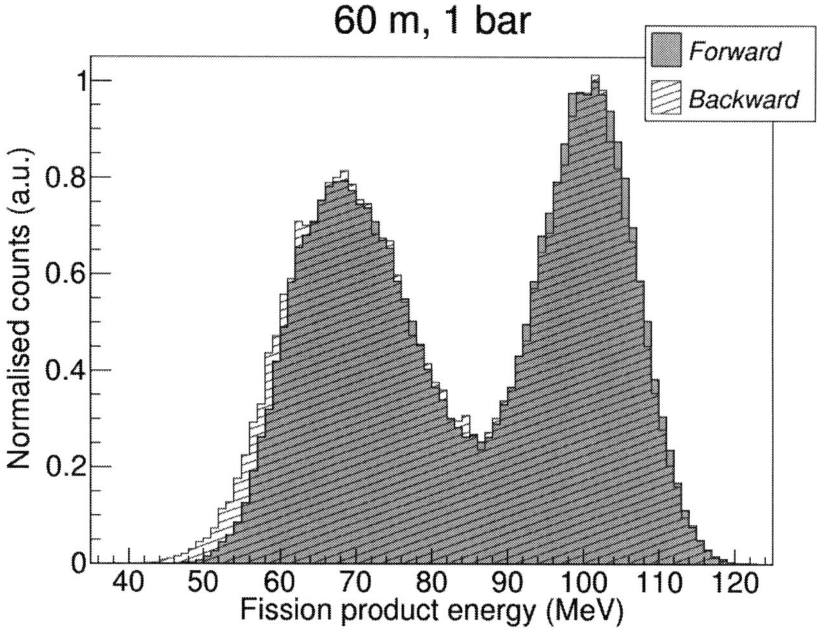

Figure 6. Anode energy spectra for events with $|\cos\theta| > 0.3$ from the data taken at 60 m distance and 1 bar pressure. The two peaks in each spectrum were used to roughly calibrate the x-axis and were positioned at 68 and 101 MeV, respectively, based on GEF calculations [15] of the thermal fission of ^{235}U. Both spectra were normalised with respect to the number of ^{235}U atoms in their respective target. We see a low energy tail from the older target (here put in the backward position) which has smaller diameter and is slightly thinner.

Part of the underestimation of the cross section at higher energies can be explained by the Particle Leaking effect, discussed earlier in Sect. 1.1, which becomes more important the higher the neutron energy is. It has not yet been corrected for.

4.3 Improving the experimental conditions

During 2015 the whole setup was moved to a shorter flightpath of about 10 m. This allowed us to gather statistics and identify problems much faster since the neutron flux is more than an order of magnitude larger at the shorter distance. At the new position the background situation, for both low and high energies, seems to have changed and does not disturb the α-distributions as much as at the 60 m position. A comparison of the 10 m data to the ENDF cross section was not possible due to a misalignment of the chamber that was not discovered in time to correct it before the accelerator stopped, in June 2015.

If a 1 bar pressure is used with the setup it will not fully stop all the tritons while a 2 bar pressure will do the trick. The events that was caused by neutrons of similar energies (similar ToF) were analysed together which allows for identifying the different ion species as band shaped clusters in a

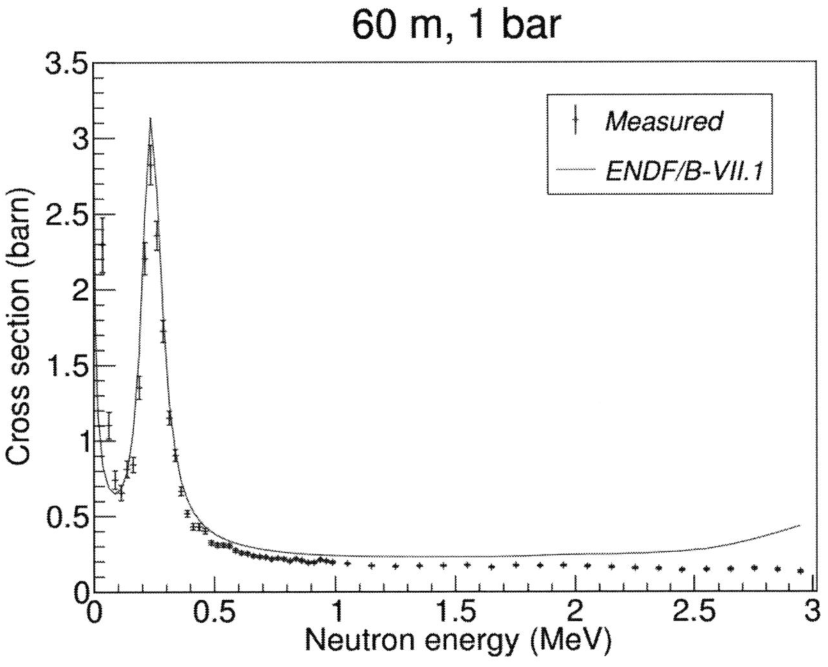

Figure 7. Preliminary cross section of the ^{6}Li(n,α) reaction, the error bars represent statistical uncertainties only. To obtain absolute values the ^{235}U cross section from the ENDF/B-VII.1 evaluation [4] has been used. The measured fission counts have been extrapolated in the $|\cos\theta| \in [0, 0.3)$ region assuming an isotropic emission. Agreement in the low and high energy region is not so good but around the resonance region our measured data agrees well with the evaluation. Only data from the 60 m measurement was used due to a misalignment of the chamber at 10 m.

grid-versus-anode plot (see Fig. 8). The partially stopped tritons can sometimes interfere with the α-particle distribution and this is why the pressure was increased to 2 bar. Fully stopping the tritons also leaves more available information for the analysis. Unfortunately, we also seem to get more negative grid signals for the α-particles and the fission fragment distribution became distorted due to the higher energy losses in the ionisation gas.

5 Conclusion and outlook

Although some problems remain, the analysis of the 60 m data has been able to replicate the evaluated ^{6}Li(n,α) cross section from ENDF/B-VII.1 [4] in the energy region from hundreds of keV to about 1 MeV. The prominent issues were mainly background related although upon moving the whole setup to a different location, closer to the neutron production, the background became less problematic.

The linear accelerator was shut down in July (2015) and will start again in beginning of 2016. Then we plan to have two separate chambers that we will be placed at 30 m distance from the source.

 DOI: 10.1051/epjconf/201612208006

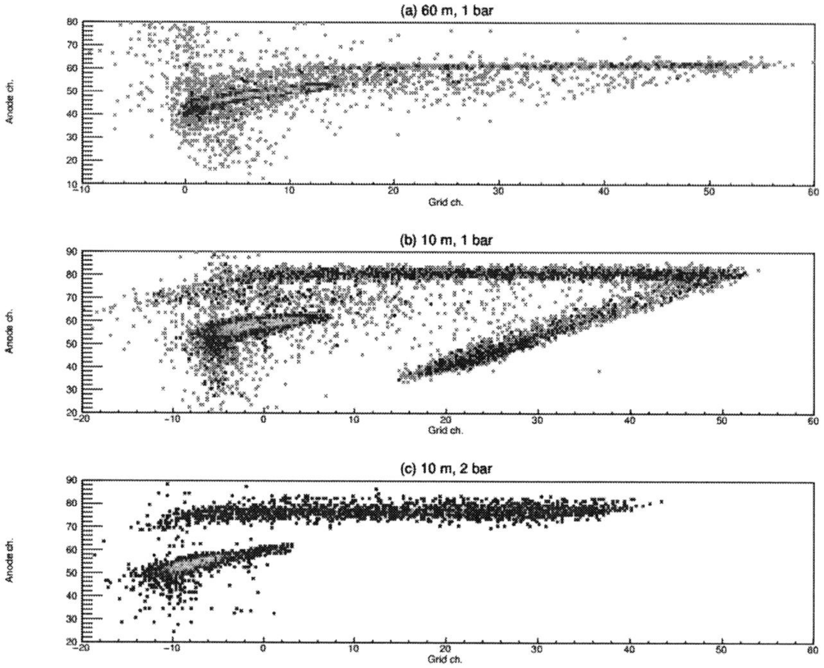

Figure 8. Forward directed events from the Li target. The top panel, (a), shows the neutron energy interval 200-250 keV from the measurement at 60 m distance and 1 bar pressure. The lower panels, (b) and (c), show similarly the neutron energy interval 6-7 keV from the measurement at 10 m distance at 1 bar and 2 bar pressure, respectively. The lower cluster of events in each panel, respectively, are α-particles. The top cluster are tritons which are only partly stopped when the chamber pressure is 1 bar. The α-particle distribution is bent due to both the frame boost and to energy losses in the target. In the middle panel, (b), some CFD thresholds had been adjusted and therefore the punch-through kink is clearly seen in (b) but not in (a). By fully stopping the tritons much less events fall between the two clusters.

The Li chamber will have a higher pressure in order to also stop tritons while the U chamber will keep the current pressure of about 1 bar.

References

[1] D. Tilley, C. Cheves, J. Godwin, G. Hale, H. Hofmann, J. Kelley, C. Sheu, H. Weller, Nuclear Physics A **708**, 3 (2002)

[2] G. Zhang, J. Chen, G. Tang, Y. Gledenov, M. Sedysheva, G. Khuukhenkhuu, Nucl. Instrum. Meth. A **566**, 615 (2006)

[3] M. Devlin, T.N. Taddeucci, G.M. Hale, R.C. Haight, J.M. O'Donnell, AIP Conference Proceedings **1090**, 215 (2009)

[4] M. Chadwick, M. Herman, P. Obložinský, M. Dunn, Y. Danon, A. Kahler, D. Smith, B. Pritychenko, G. Arbanas, R. Arcilla et al., Nuclear Data Sheets **112**, 2887 (2011), special Issue on ENDF/B-VII.1 Library

[5] R. Bencardino, R. Bevilacqua, G. Giorginis, F.-J. Hambsch, Nucl. Sci. Eng. **177**, 68 (2014)

[6] Institute for Reference Materials and Measurements (Geel), Unit Standards for Nuclear Safety, Security and Safeguards, *EUFRAT NUCLEAR FACILITIES AT JRC-IRMM*, Accessed: 2015-03-10, `https://ec.europa.eu/jrc/sites/default/files/eufrat-facilities.pdf`

[7] European Commission, Joint Research Centre, *Linear electron accelerator facility*, Accessed: 2015-03-10, `https://ec.europa.eu/jrc/en/research-facility/linear-electron-accelerator-facilit`

[8] F.J. Hambsch, I. Ruskov, Nucl. Sci. Eng. **163**, 1 (2009)

[9] R. Bevilacqua, F.-J. Hambsch, R. Bencardino, G. Giorginis, M. Vidali, L. Lamia, I. Ruskov, Nuclear Data Sheets **119**, 104 (2014)

[10] J.F. Ziegler, M. Ziegler, J. Biersack, Nucl. Instrum. Meth. B **268**, 1818 (2010), 19th International Conference on Ion Beam Analysis

[11] A. Al-Adili, F.-J. Hambsch, S. Oberstedt, S. Pomp, Sh. Zeynalov, Nucl. Instrum. Meth. A **624**, 684 (2010)

[12] W. Shockley, Journal of Applied Physics **9**, 635 (1938)

[13] A. Göök, F.-J. Hambsch, A. Oberstedt, S. Oberstedt, Nucl. Instrum. Meth. A **664**, 289 (2012)

[14] R. Bevilacqua, A. Göök, F.-J. Hambsch, N. Jovančević, M. Vidali, Nucl. Instrum. Meth. A **770**, 64 (2015)

[15] K.H. Schmidt, B. Jurado, C. Amouroux, Tech. Rep. 24, Nuclear Energy Agency (2014)

EPJ Web of Conferences **122**, 09001 (2016) DOI: 10.1051/epjconf/201612209001
*CNR*15*

Towards many-body based nuclear reaction modelling

Stéphane HILAIRE[1a] and Stéphane Goriely[2]
[1]*CEA, DAM, DIF, F-91297 Arpajon, France*
[2]*Institut d'Astronomie et d'Astrophysique, Université Libre de Bruxelles, CP-226, 1050 Brussels, Belgium*

Abstract. The increasing need for cross sections far from the valley of stability poses a challenge for nuclear reaction models. So far, predictions of cross sections have relied on more or less phenomenological approaches, depending on parameters adjusted to available experimental data or deduced from systematic expressions. While such predictions are expected to be reliable for nuclei not too far from the experimentally known regions, it is clearly preferable to use more fundamental approaches, based on sound physical principles, when dealing with very exotic nuclei. Thanks to the high computer power available today, all the ingredients required to model a nuclear reaction can now be (and have been) microscopically (or semi-microscopically) determined starting from the information provided by a nucleon-nucleon effective interaction. This concerns nuclear masses, optical model potential, nuclear level densities, photon strength functions, as well as fission barriers. All these nuclear model ingredients, traditionally given by phenomenological expressions, now have a microscopic counterpart implemented in the TALYS nuclear reaction code. We are thus now able to perform fully microscopic cross section calculations. The quality of these ingredients and the impact of using them instead of the usually adopted phenomenological parameters will be discussed. Perspectives for the coming years will be drawn on the improvements one can expect.

1 Introduction

To model a nuclear reaction, two complementary paths can be considered. On the one hand, one can use empirical approaches, and on the other hand, many-body based approaches can be chosen. In the first case, nuclear reaction inputs are usually obtained from phenomenological expressions whose parameters are fine-tuned to fit experimental data. This enables generally to produce accurate cross sections which are however only reliable close to the mass or energy range in which the adjustments have been performed. On the contrary, more fundamental approaches, based on sound physical principles, do not provide the same fitting flexibility but presumably have a higher predictive power when dealing with exotic nuclei far from the experimentally known regions. With the development of high performance computers, systematic studies and the calculation of all nuclear reaction inputs required to model a nuclear reaction have become more and more feasible. During the last two decades, significant efforts have been engaged towards this aim, and all the traditional

[a] Corresponding author: stephane.hilaire@cea.fr

phenomenological ingredients now have a microscopic counterpart which can be used to perform nuclear reaction predictions.

We first discuss in Section 2, the iterative process required to improve the quality of these ingredients and illustrate some of the evolutions the various iterations have made possible. We then review in Section 3, the main inputs currently available online. We finally conclude on what is planned in a close future, in particular in reducing the still present sources of uncertainties.

2 A step by step process

Except for light nuclei, there is nowadays no direct link between the nucleon-nucleon interaction and the prediction of a nuclear reaction without any severe restriction on the target mass. Therefore, several intermediate steps are required, as illustrated in Figure 1, to establish a connection between a nucleon-nucleon interaction and a nuclear reaction code.

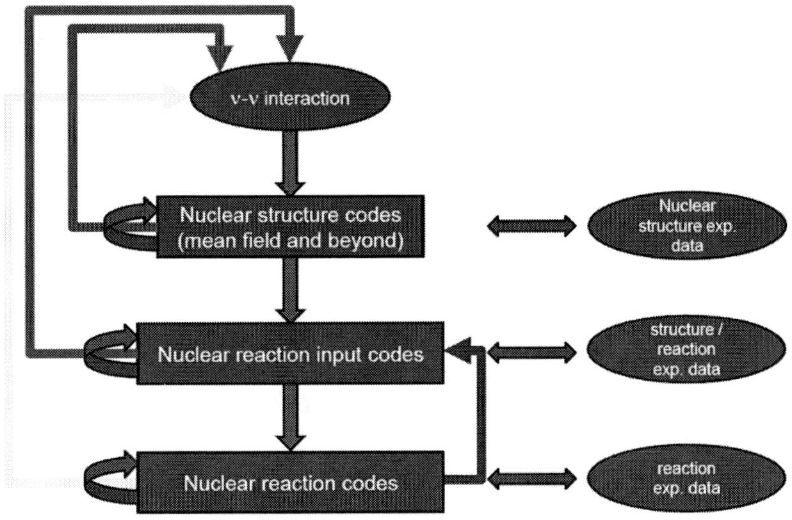

Figure 1. Schematic view of the connections required to link a nucleon-nucleon (ν-ν) effective interaction with a nuclear reaction code. The green arrows indicate a transmission of input to modelling codes whose results are then compared with relevant experimental data (double red arrow). At each step, one can either loop back to fine-tune the parameters of the nucleon-nucleon effective interaction (red solid arrows) or improve the modelling under consideration (curved arrow) in order to improve the agreement with experiment. At each step of this process, different experimental data are considered as constraints. Up to now, the results produced by nuclear reaction codes are not used to fine-tune the ν-ν effective interaction parameters (light grey arrow) but have led to modification in the nuclear reaction input codes as illustrated by the blue arrow.

The first step has been under development for the last 15 years and concerns the prediction of nuclear structure properties. In our case, since we are interested in predicting observables for all target masses, we use the mean-field Hartree-Fock-Bogoliubov (HFB) approach which is currently the only method capable of covering the whole nuclear chart starting from a unique nucleon-nucleon effective interaction. Such a mean-field approach has proven its capacity to reproduce relevant experimental data (nuclear masses, radii, deformations, densities as well as spectroscopic information) with an accuracy comparable to the best phenomenological models. Within this framework, Skyrme-type

DOI: 10.1051/epjconf/201612209001

effective forces have been extensively used [1-16] reproducing a wide set of observables, either experimentally determined or based on realistic ab-initio-type calculations.

The second step of this iterative sequential process consists in using the nuclear structure results to estimate nuclear reaction ingredients, such as nuclear level densities [17-18], optical potentials [19-21], gamma-ray strength functions [22-23] and fission paths [8,24]. This second step also requires the use of specific approaches implemented in what we call "nuclear reaction input codes" (NRIC). When the result obtained with one of this NRIC does not agree sufficiently with experimental data, as in the first step, one can either modify the nucleon-nucleon effective interaction or add new features to the modelling in the nuclear reaction input codes to correct approximations. A typical example of the first type of situation, where the interaction turns out to be responsible for a deviation of the theory from experiment is shown in Figure 2. In this figure, one can observe the impact of the effective interaction on the nuclear level densities. The BSk13 interaction has been in this case parameterized to suppress the drift observed with the BSk9 parametrization when comparing the theoretical level densities prediction of s-wave neutron mean spacing at the neutron binding energy with experimental data. Another example of such an adjustment of the interaction parameters has also been performed using fission barriers as constraints [8]. Fission barriers are not strictly speaking experimental data but rather "reaction input data" (see Figure 1) derived from nuclear reaction analysis. Extracting them from nuclear structure predictions is not at all trivial and requires specific methods [8,24]. Of course, when the interaction is modified, one has to try to keep, at the same time, all the properties already obtained with previous parameterizations at the same level of quality.

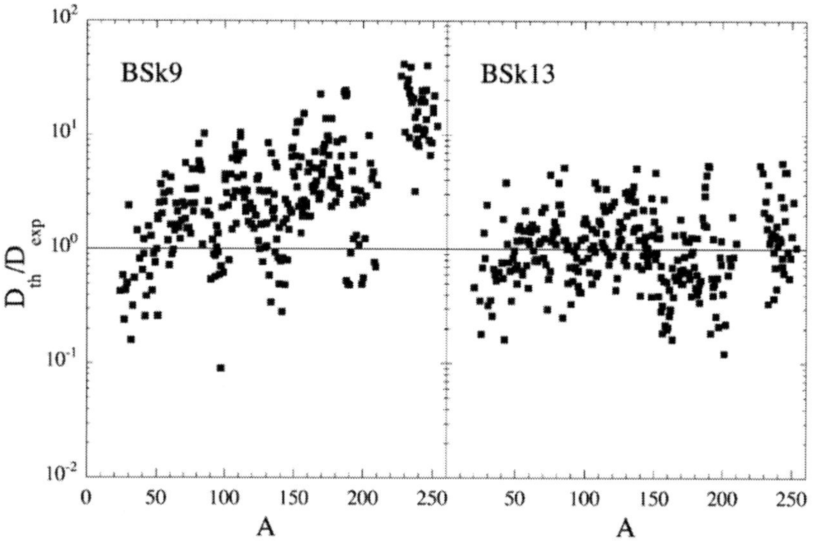

Figure 2. Ratio of the theoretical s-wave neutron resonances mean spacing (D_{th}) to the experimental data compiled in [25] using the combinatorial level density method [17] with the BSk9 [6] and BSk13 [7] interactions.

Once the quality reached at this second step is found to be reasonable, all the nuclear reaction inputs are then introduced in a nuclear reaction code to estimate nuclear cross sections and study the quality of the obtained predictions with respect to experimental reaction data, which in practice mainly means reaction cross sections even if recently fission yields have also been considered [26]. Again, when the agreement with experiment is not satisfactory, several actions are possible, as suppressing approximations in the nuclear reaction codes or modifying the nuclear reaction input codes. However, the number of approximations, the model uncertainties and the computing time are still too large to link directly the nucleon-nucleon effective interaction and the reaction codes outputs, so that it remains impossible to adjust the interaction parameters directly on reaction cross sections, even if recent developments might change this restriction in the future [27]. An example of a result obtained at this third step which has implied a modification of the nuclear reaction code producing nuclear level densities is illustrated in Figure 3 and discussed in details in [28].

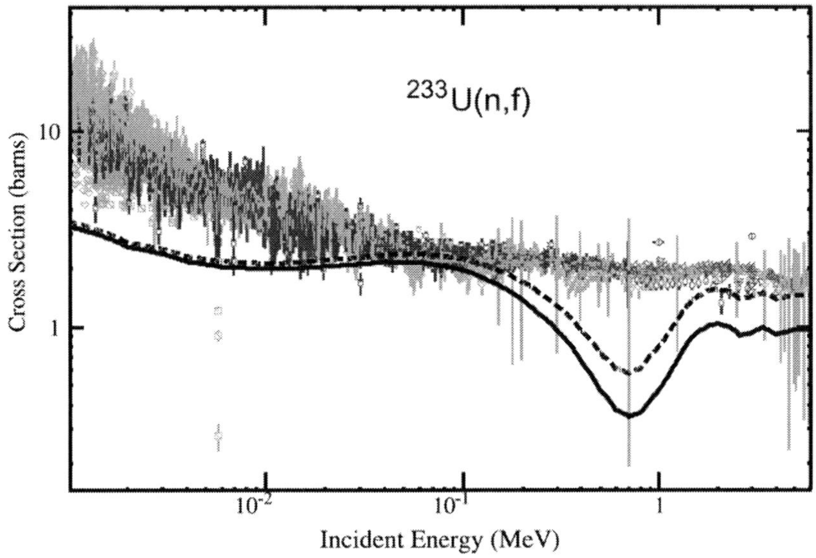

Figure 3. ^{233}U(n,f) cross sections computed using microscopic nuclear level densities [17] and fission paths [8]. The dashed and full curves correspond to the predictions obtained with the EMPIRE nuclear reaction code [29] considering or not left-right asymmetry of the outer barrier.

This figure shows that the combinatorial nuclear level density performed in [17] used too crude an approximation to account for the impact of collective vibrational levels, which turned out to have a significant impact when applied to determined fission transition states [28]. This deficiency was then solved by improving the method used to perform combinatorial level density calculation with a more rigorous treatment of vibrational levels [18].

3 Current status within the Skyrme-HFB approach

As already mentioned in Section 2, several Skyrme-type effective interactions have been derived during the last fifteen years or so to try to reproduce at best nuclear structure observables. The first attempt, in 2001 [1] aimed at proving, that it was possible to reach a low root-mean-square (rms) deviation with respect to all experimental masses available at that time. Since then, most of the subsequent models [1-16] were developed to further explore the parameter space widely or to take into account additional constraints. These include in particular a sensitivity study of the mass model accuracy and extrapolation to major changes in the description of the pairing interaction [3,7,10], the spin-orbit force [16] or the nuclear matter properties, such as the effective mass [4], the symmetry energy [6,13,14] and the stability of the equation of state [10,13]. All nuclear reaction inputs required for a nuclear reaction code have been determined coherently using the BSk14 interaction [8] obtained in 2007. The more recent Skyrme interactions have not yet been used for such an extensive project.

Tables of nuclear level densities, E1 gamma-ray strengths, nuclear matter densities, fission paths have all been determined from the sole BSk14 interaction and are available either in the RIPL-3 [25] library, in the TALYS code [29] or the BRUSLIB nuclear data library [30]. In the latest TALYS version, however, the theoretical nuclear masses are those obtained with BSk17 [11], the first interaction providing a root mean square deviation with experimental masses lower than 600 keV.
Using such ingredients has already shown interesting features which can clearly not be addressed with traditional phenomenological approaches. An example, illustrated in Figure 4, concerns the possibility offered to study continuum spin and parity distribution deviating from the statistical hypothesis with a significant impact on isomer production [30].

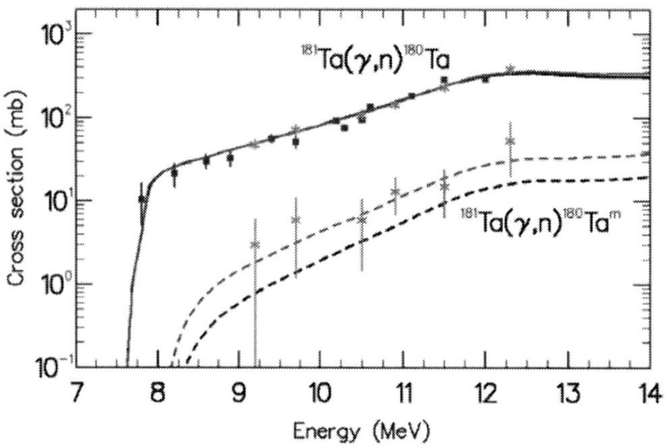

Figure 4. Comparison between experimental and theoretical cross section. Solid lines and squares [31,32] correspond to the total photoreaction cross section. Dashed lines and stars [31] correspond to the isomeric state production. Red (resp. black) curves have been obtained using the combinatorial (resp. statistical) model for nuclear level densities.

As can be observed, a significantly larger isomer production is obtained when non-statistical nuclear level densities are used in the theoretical calculation. This is directly related to the spin distribution with spin higher than 5 ħ in ^{180}Ta which is bigger with a combinatorial model than with the traditional Fermi Gas type formula assuming a Wigner law for the spin distribution [31]. The predictive power of microscopic nuclear reaction inputs has also been investigated in fission cross section calculation either to provide one with the possibility to perform systematic fission cross section predictions [33] or following a consistent methodology [34] used to produce high quality evaluated files [35] for applications. In the first case [33], it has been shown that the quality of the prediction is not at the level required for practical applications, mainly because the microscopic fission barrier heights still suffer from a lack of accuracy, but provided a global deformation-independent normalization is applied, a satisfactory estimate of the fission cross section for non-energy applications can be reached.

4 Conclusions and perspectives

A project to derive all nuclear reaction input from a unique nucleon-nucleon effective interaction has been conducted for more than fifteen years. The level of quality reached using a Skyrme interaction is encouraging but the approach still suffers from approximations or lack of accuracy. A way to remove part of them consists in improving the coherence throughout the whole code systems, by using, for instance, finite range Gogny-type interactions [35,36,37] known to be also very accurate to reproduce nuclear structure observables [38]. Work along this path has already been engaged starting by improving the accuracy of the Gogny force with respect to nuclear masses [37] as well as computing level density tables [39] and nuclear matter densities required for optical model calculation. The next step, presently underway, consists in computing gamma-ray strength functions and fission barriers. We believe that such a project is a way to progress and be able in the future to obtain satisfactory nuclear reaction predictions on the bases of reliable and accurate microscopic inputs only.

References

1. M. Samyn, S. Goriely, P.-H Heenen, J.M. Pearson, F. Tondeur, Nucl. Phys. A **700**, 142 (2002)
2. S. Goriely, M. Samyn, P.-H Heenen, J.M. Pearson, F. Tondeur, Phys. Rev. C **66**, 024326 (2002)
3. M. Samyn, S. Goriely, J.M. Pearson, Nucl. Phys. A **725**, 69 (2003)
4. S. Goriely, M. Samyn, M. Bender, J. M. Pearson, Phys. Rev. C **68**, 054325 (2003)
5. M. Samyn, S. Goriely, M. Bender, J.M. Pearson, Phys. Rev. C **70**, 044309 (2004)
6. S. Goriely, M. Samyn, J.M. Pearson, M. Onsi, Nucl. Phys. A **750**, 425 (2005)
7. S. Goriely, M. Samyn, J.M. Pearson, Nucl. Phys. A **773**, 279 (2006)
8. S. Goriely, M. Samyn, J.M. Pearson, Phys. Rev. C **75**, 064312 (2007)
9. S. Goriely, J.M. Pearson, Phys. Rev. C **77**, 031301 (2008)
10. N. Chamel, S. Goriely, J.M. Pearson, Nucl. Phys. A **812**, 72 (2008)
11. S. Goriely, N. Chamel, J.M. Pearson, Phys. Rev. Lett. **102**, 152503 (2009)
12. S. Goriely, J.M. Pearson, Phys. Rev. C **80**, 065804 (2009)
13. S. Goriely, N. Chamel, J. M. Pearson, Phys. Rev. C **82**, 035804 (2010)
14. S. Goriely , N. Chamel, J. M. Pearson, Phys. Rev. C **88**, 024308 (2013)
15. S. Goriely , N. Chamel, J. M. Pearson, Phys. Rev. C **88**, 061302 (2013)
16. S. Goriely, Nucl. Phys. A **933**, 68 (2015)
17. S. Hilaire, S. Goriely, Nucl. Phys. A **779**, 63 (2006)
18. S. Goriely, S. Hilaire, Phys. Rev. C **78**, 064307 (2008)
19. E. Bauge, J.P. Delaroche, M. Girod, Phys. Rev. C **58**, 1118 (1998)
20. E. Bauge, J.P. Delaroche, M. Girod, Phys. Rev. C **63**, 024607 (2001)
21. S. Goriely, J.P. Delaroche, Phys. Lett. B **653**, 178 (2007)
22. S. Goriely, E. Khan, Nucl. Phys. A **706**, 217 (2002)
23. S. Goriely, E. Khan, M. Samyn, Nucl. Phys. A **739**, 331 (2004)
24. M. Samyn, S. Goriely, J.M. Pearson, Phys. Rev. C **72**, 044316 (2005)
25. R. Capote et al, Nucl. Data Sheets **110**, 3107 (2009)
26. J.F. Lemaitre, S. Panebianco, J.L. Sida, S. Hilaire, Phys. Rev. C **92**, 034617 (2015)
27. G. Blanchon, M. Dupuis, H.F. Arellano, N. Vinh Mau, Phys. Rev. C **91**, 014612 (2015)
28. M. Sin et al., *Proceedings if the International Conference on Nuclear Data for Science and Technology- ND 2007,* (EDP Science Volume 1, eds. O. Bersillon, F. Gunsing, E. Bauge, R. Jacqmin, S. Leray, Apr. 22-27, 2007, Nice, France, 2008) p. 313
29. A.J. Koning, S. Hilaire, M. Duijvestijn, *Proceedings if the International Conference on Nuclear Data for Science and Technology- ND 2007,* (EDP Science Volume 1, eds. O. Bersillon, F. Gunsing, E. Bauge, R. Jacqmin, S. Leray, Apr. 22-27, 2007, Nice, France, 2008) p. 211, see also http://www.talys.eu
30. Y. Xu, S. Goriely, A. Jorissen, G.L. Chen, M. Arnould, Astronomy & Astrophysics **549**, A106 (2013), see also http://www.astro.ulb.ac.be/bruslib
31. S. Goko et al., Phys. Rev. Lett. **96**, 192501 (2006)
32. H. Utsunomiya et al., Phys. Rev. C **67**, 015807 (2003)
33. S. Goriely et al. Phys. Rev. C 79, 024612 (2009)
34. S. Goriely et al., Phys. Rev. C 83, 034601 (2011)
35. P. Romain, B. Morillon, H. Duarte, Nucl. Data Sheets **131**, 222 (2016)
36. J.F. Berger, M. Girod, D. Gogny, Comp. Phys. Commun. **63**, 365 (1991)
37. F. Chappert. M. Girod, S. Hilaire, Phys. Lett. B **668**, 420 (2008)

38. S. Goriely, S. Hilaire, M. Girod, S. Peru, Phys. Rev. Lett. **102**, 242501 (2009)
39. J.P. Delaroche et al., Phys. Rev. C **81**, 014303 (2010)
40. S. Hilaire, M. Girod, S. Goriely, A.J. Koning, Phys. Rev. C **86**, 064317 (2012)

Unified studies of structure and reactions in light unstable nuclei

Makoto Ito[1,a]

[1] *Department of Pure and Applied Physics, Kansai University, 3-3-35 Yamate-cho, Suita, 564-8680, Japan*

Abstract. The generalized two-center cluster model (GTCM), which can treat covalent, ionic and atomic configurations in general systems with two inert cores plus valence nucleons, is formulated in the basis of the microscopic cluster model. In this model, the covalent configurations constructed by the molecular orbital (MO) method and the atomic (or ionic) configuration obtained by the valence bonding (VB) method can be described in a consistent manner. GTCM is applied to the light neutron-rich system, ^{10,12}Be = α + α + XN (X = 2,4), and the unified studies of the structural changes and the reaction problem are performed. In the structure study, the calculated energy levels are characterized in terms of the chemical bonding like structures, such as the covalent MO or ionic VB structures. The chemical bonding structures changes from level to level within a small energy interval. In the unbound region, the structure problem with the total system of α + α + XN and the reaction problem, induced by the collision of an asymptotic VB state of α+^{6,8}He, are combined by GTCM. The properties of unbound resonant states are discussed in a close connection to the reaction mechanism, and some enhancement factors originated from the properties of the intrinsic states are predicted in the reaction observables. The unified calculation of the structures and the reactions is applied to the Coulomb shift problem in the mirror system, such the ^{10}Be and ^{10}C nuclei. The Coulomb displacement energy of the mirror systems are discussed.

1 Introduction

In light neutron-excess systems, many kinds of molecular structures are discussed from the viewpoint of the clustering phenomena. In particular, much attention has been concentrated on Be isotopes from theoretical and experimental approaches. The Be isotopes can be considered as typical examples of two-center superdeformed systems which build on an α + α rotor of ^{8}Be. Various cluster structures are generated by the coupling the two α degrees of freedom and the single particle motion of the excess neutrons. The coupled structures of the two α cores and the valence neutrons are analogous to the so-called chemical-bonding structure in molecular systems.

The low-lying states can be nicely understood by the molecular orbital (MO) picture, in which the excess neutrons perform the single-particle motion in the covalent orbitals, such as π^- or σ^+, around two α cores [1–3]. The typical application of the MO model is demonstrated in ^{10}Be = $\alpha + \alpha + N + N$ [2]. In the MO model, the covalent orbitals are constructed by the technique of the linear combination of the atomic orbital (LCAO) [1, 2]. Since the nuclear interaction of two α cores is weak, the $\alpha - \alpha$ relative motion strongly couples to the configuration of the LCAO covalent orbitals. For example, the

[a] e-mail: itomk@kansai-u.ac.jp

σ^+ orbit, which has the enlarged distribution along the $\alpha - \alpha$ axis, enhances the α–α clustering. The σ^+ orbital, which contains a large S-wave component, plays an important role for the formation of the low-lying intruder 0^+ state [3].

On the contrary, in the highly-excited region of $E_x \geq 10$ MeV, recent experiments revealed the existence of the interesting resonant states, which dominantly decay into the ^{4,6,8}He fragments [4–6]. The data has been accumulated especially in ^{12}Be, and these resonances are interpreted in terms of the ionic or atomic configurations of ^{x}He + ^{y}He, in which valence neutrons are trapped in one of an α cores [7, 8]. The localization of the valence neutrons around a core is very different from the MO picture. In the quantum chemistry, such the localized orbitals are usually treated by the so-called the valence bonding (VB) method. Therefore, the observation of the highly-excited resonances means that the structural changes occurs from the low-lying MO structures to the highly-excited VB structures.

The structural change from MO to VB is an interesting research subject. In order to handle these two configurations in a consistent manner, the traditional MO model based on the LCAO technique should be improved to cover the VB configuration. Recently, we have formulated the generalized two-center cluster model (GTCM), which is available to treat both of the MO and VB configurations [7, 8]. The GTCM can also describe not only the structure change of MO and VB but also the reaction process induced by the collision of the VB configuration in a unified manner. We have applied GTCM to the ^{10,12}Be nuclei, and the unified studies of the structure change and the respective reaction problem, such as the slow scattering of α + ^{6,8}He, have been performed [7, 8].

The studies on the mirror system of ^{10,12}Be is the naive extension of GTCM, and the analysis of the Coulomb shift is another interesting research subject. In the previous studies, for example, a characteristic Coulomb effect is known as the so-called Thomas-Ehrman shift (TES) in ^{17}O and ^{17}F [9]. In the ^{16}O core plus one nucleon model, the Coulomb shift in ^{17}F = ^{16}O + P is exceptionally suppressed for the S-wave orbit. Such the suppression of the Coulomb repulsion is expected in the α cluster structure because the development of the cluster structure usually involves the large mixture of the S-wave component. In the covalent MO picture, for example, the σ^+ orbit with a large mixture of the S-wave component enhance the $\alpha-\alpha$ clustering, while two nuclei dominantly perform the S-wave relative motion in the ionic VB states. Therefore, there is a possibility that the Coulomb suppression strongly reflects the development of the α cluster structure in the MO or VB configuration.

In this article, we demonstrate the unified studies of the structure changes in ^{10,12}Be and the reactions induced by the slow α + ^{6,8}He scattering. Moreover, we extend GTCM to the mirror systems of ^{10}Be (= $\alpha + \alpha + 2N$) and ^{10}C (= $\alpha + \alpha + 2P$), and the the Coulomb shift of these mirror systems is discussed in connection to the cluster structures.

2 Theoretical framework

2.1 Generalized Two-center Cluster Model

The framework of the generalized two-center cluster model (GTCM) corresponds to the extension of the microscopic cluster model, which is based on the generator coordinate method (GCM) [10]. The detailed explanation of GTCM has already been published in Refs. [7, 8], and we briefly explain the formulation of GTCM by showing the example of two α plus two valence neutrons, such as ^{10}Be = α + α + $2N$ and ^{10}C = $\alpha + \alpha + 2P$. In this model, the basis function is given by

$$\Phi^{J^\pi K}_{\mathbf{m}}(S) = \hat{P}^{J^\pi}_K \mathcal{A}\left\{\psi_L(\alpha)\psi_R(\alpha)\prod_{j=1}^{2}\varphi_j\left(m_j\right)\right\}_S . \quad (1)$$

The α-clusters at the left (L) and right (R) side center are shown by $\psi_i(\alpha)$ ($i = L, R$), and they are placed with a relative distance parameter of S, which is the generator coordinate. The internal structure of the α cluster is expressed by the $(0s)^4$ configurations in the harmonic oscillator (HO) potential. The width parameter of the HO potential is set to $b = 1.46$ fm. The single-particle wave function for the valence neutrons localized around the α cluster is given by an atomic orbital (AO) $\varphi(m_j)$, which is labeled by the AO indices of $m_j = (p_k, i, \tau)$. Here, p_k ($k = x, y, z$) means the $0p$-orbitals around i-th core ($i = L$ or R) with the spin τ ($= \uparrow$ or $\downarrow$). In Eq. (1), $\mathbf{m}$ represents a set of AOs for the valence neutrons, $\mathbf{m} = (m_1, m_2)$. The intrinsic basis functions with the full anti-symmetrization $\mathcal{A}$ are projected to the eigenstate of the total spin J, its intrinsic angular projection K, and the total parity π by the projection operator $\hat{P}^{J^\pi}_K$.

The total wave function is finally given by taking the superposition over S, $\mathbf{m}$ and K as

$$\hat{\Psi}^{J^\pi}_\nu = \int dS \sum_{\mathbf{m},K} C^{J^\pi\nu}_{\mathbf{m}K}(S)\, \Phi^{J^\pi K}_{\mathbf{m}}(S) \ . \tag{2}$$

In the present calculation, we consider all the possible $0p$ configurations for the AO basis with $\mathbf{m}$, which leads to the complete formation of the covalent, ionic, and atomic configuration around two α cores. The coefficients $C^{J^\pi\nu}_{\mathbf{m}}(S)$ in Eq. (2) for the ν-th eigenstate are determined by solving a coupled channel GCM (Generator Coordinate Method) equation [10]

$$\left\langle \Phi^{J^\pi}_{\mathbf{m}}(S) \,|\, \left(\hat{H} - E^{J^\pi}_\nu\right) | \Psi^{J^\pi}_\nu \right\rangle = 0 \ . \tag{3}$$

The bound states below the α decay threshold are calculated by a diagonalization procedure, while the scattering boundary condition is explicitly imposed above the threshold, which will be explained in the later section. The solution of GCM equation can automatically reproduce the MO or VB structure, the formation of which depends on the situation. For example. the MO configurations are generated at the ground and low-lying bound states, while the VB configuration predominant in the continuum region above the α threshold.

In the present studies, the intrinsic configuration is restricted to the axial symmetric case ($K = 0$) because we mainly consider the structure change in the $J^\pi = K^\pi = 0^+$ states. The restriction to the axial symmetric configurations can also be justified for the low-energy reaction phenomena in the finite spin states ($J^\pi \neq 0^+$). The validity of the axial symmetric configuration is deeply discussed in the studies of the sub-barrier fusion reaction based on the time-dependent wave packet approach [11].

2.2 Hamiltonian and effective nucleon-nucleon interaction

The total Hamiltonian $\hat{H}$ in the present calculation is composed of the nucleon degrees of freedom and its explicit form is

$$\hat{H} = \sum_i^{12} \hat{t}_i - \hat{T}_{c.m.} + \sum_{i<j}^{12} \hat{v}^{(C)}_{ij} + \sum_{i<j}^{12} \hat{v}^{(LS)}_{ij} + \sum_{i<j}^{12} \hat{v}^{(Coul)}_{ij} \ , \tag{4}$$

where i and j denote a number of nucleons. $\hat{t}$ represents a single nucleon's kinetic energy, while $\hat{v}^{(C)}_{ij}$ and $\hat{v}^{(LS)}_{ij}$ are the two body nucleon-nucleon (NN) interactions for the central and spin-orbit parts, respectively. $\hat{v}^{(Coul)}_{ij}$ represents the Coulomb interaction, which includes the projection operator to the proton state. The kinetic energy of the center of mass ($c.m.$) motion, shown by $\hat{T}_{c.m.}$, is subtracted from the summation of the nucleon's kinetic energy. In the present treatment, the $c.m.$ motion of the total system can be exactly eliminated [10].

As for the NN interaction, we use the Volkov No.2 [12] and the G3RS [13] for the central $\hat{v}_{ij}^{(C)}$ and spin-orbit $\hat{v}_{ij}^{(LS)}$ parts, respectively. The functional form of the Volkov No.2 interaction is

$$\hat{v}_{ij}^{(C)} = v^{(C)}(r_{ij})(W - MP_\sigma P_\tau + BP_\sigma - HP_\tau) \quad . \tag{5}$$

Here $v^{(C)}(r)$ is given by the superposition of Gaussian, while W, $MP_\sigma P_\tau$, BP_σ and HP_τ denote the Wigner, Majorana, Bartlett and Heisenberg exchanges. The G3RS spin-orbit interaction has the functional form of

$$\hat{v}_{ij}^{(LS)} = v^{(LS)}(r_{ij})P(^3O)\mathbf{L}_{ij} \cdot \mathbf{S}_{ij} \ , \tag{6}$$

where $P(^3O)$ represents the projection operator on the triplet-odd state for the two nucleon channel. $\mathbf{L}_{ij} = \mathbf{r}_{ij} \times \mathbf{p}_{ij}$ and $\mathbf{S}_{ij} = \mathbf{s}_i + \mathbf{s}_j$ are the relative and total nucleons' spins for interacting i-th and j-th nucleons, respectively. The form factor of $v^{(LS)}(r)$ is also constructed by the superposition of Gaussian.

A combination of Volkov. No. 2 and G3RS has already been applied to the previous MO models [2, 3]. In the MO model, these interactions are successful in reproducing various properties observed in the low-lying states of ^{10}Be, especially the formation of the intruder 0_2^+ state [3]. The details of the interaction parameters adopted in the present GTCM calculation are shown in Ref. [7]. These parameters are obtained by modifying the parameters used in the MO model [2, 3], and they can nicely describe the properties of the low-lying states in ^{10,12}Be [7, 8]. Furthermore, the adopted parameters can also reproduce the threshold energies for the various ^{x}He + ^{y}He channels, as shown in Ref. [7]. In ^{12}Be, the energy difference of α+^{8}He$_{g.s.}$ and ^{6}He$_{g.s.}$+^{6}He$_{g.s.}$ in the calculation is a little larger than that in the experimental observation (about 1MeV), but the relative differences among the thresholds are reasonably reproduced by the present NN interaction.

Since the threshold energy corresponds to the order of magnitude of the interaction strength to dissociate a compound system into a pair of He clusters, the reproduction of the threshold energy is crucial in discussing the appearance and disappearance of the cluster degrees of freedom. This means that the correct treatment of the threshold is indispensable for the unified studies of the structures and reactions, where dynamical changes of the cluster structure prominently occurs. Therefore, the NN interactions tuned in the present study, which reproduce the thresholds for various He decays as well as the low-lying states, can be considered quite reliable for the unified calculation of the structure and the reactions.

2.3 Scattering boundary condition

The superposition of the basis function in Eq. (1) over the possible **m** can reproduce the binary ^{x}He + ^{y}He dimer states in the asymptotic region of the distance parameter S [7]. This binary states are the same as the asymptotic channels in the nuclear reaction problem. Thus, we can naturally impose the scattering boundary condition on the total wave function in Eq. (2), and the eigenvalue equation of Eq. (3) can be solved under the scattering boundary condition. There are two methods in solving the scattering problem: Kohn-Hulthén-Kato (KHK) method [8] and the closed state method [7]. These two methods are completely equivalent if the converged results are obtained within a given model space. Here we mainly explain the closed state method because this method is very convenient in the analysis of the formation and decays of the resonant levels excited in a reaction process. In the closed state method, a coupling of the closed compound states under the bound state approximation and the open channels with the scattering boundary is directly solved.

Let us explain the basic idea of the closed state method by showing an example of the ^{12}Be nuculeus. In this system, we consider two reactions induced by the scattering of ^{8}He$_{g.s.}$ by an α

target: the elastic scattering of α+^{8}He$_{g.s.}$ $\rightarrow$ α+^{8}He$_{g.s.}$ and two neutron transfer reaction of α+^{8}He$_{g.s.}$ $\rightarrow$ ^{6}He$_{g.s.}$+^{6}He$_{g.s.}$. In a simple transition between these two open channels, the scattering (S) matrix reveals just a smooth energy dependence as will be shown in the later section. Thus, it is insufficient to describe resonance phenomena, in which there is a complicated energy dependence in S matrix. In order to handle resonance phenomena, we introduce the intermediate quasi-bound states, which have the more complicated structures than that of the open channels. The quasi-bound states are obtained by solving the eigenvalue problem in Eq. (3) in a finite-basis expansion method, in which the $\alpha - \alpha$ distance parameter S is limited inside of the nuclear interaction range. This quasi-bound state is called the closed states because they are confined inside of the interaction region and never decay into asymptotic region. Reaction processes can be described in terms of the coupling of the open channels and the closed compound states: An incident wave in α+^{8}He$_{g.s.}$ can couple to the closed states as well as final channels, and the incident wave is scattered into a final channel through the transitions among the closed states.

According to this simple image, the total wave function, which takes into account a coupling of open channels and closed states, can be constructed by a linear combination of these two parts like

$$\Psi^{J^\pi(+)} = \sum_{\beta L} \varphi_{\beta L}^{J^\pi(+)}(\mathbf{R}_\beta) + \sum_{\nu} b_\nu \hat{\Psi}_\nu^{J^\pi} , \tag{7}$$

where the first term $\varphi_{\beta L}^{J^\pi(+)}(\mathbf{R}_\beta)$ stands for the open channels labeled by a channel subscript βL. Here β means a set of the quantum number of a combination of the scattering particles (excitation energy, internal spin, and so on). $\varphi_{\beta L}^{J^\pi(+)}(\mathbf{R}_\beta)$ is given by the superposition of the channel basis, such as

$$\varphi_{\beta L}^{J^\pi(+)}(\mathbf{R}_\beta) = \sum_{S} B_{\beta L}(S) \chi_{\beta L}^{J^\pi(+)}(\mathbf{R}_\beta, S) . \tag{8}$$

In Eq. (8), $\chi_{\beta L}^{J^\pi(+)}(\mathbf{R}_\beta, S)$ shows the channel basis [7, 8], which is constructed from a specific linear combination of the AO basis in Eq. (1), while $B_{\beta L}(S)$ means the amplitude for the respective channel basis. The superscript (+) of $\chi_{\beta L}^{J^\pi(+)}$ means that the scattering boundary condition for the incident and final open channels is explicitly imposed, according to the prescription of the KHK method [8].

The second term in Eq. (7) represents the closed compound states confined within the interaction region. The closed states are constructed by a linear combination of the solutions $\hat{\Psi}_\nu^{J^\pi}$ for Eq. (2) with an amplitude of b_ν. $\hat{\Psi}_\nu^{J^\pi}$ is calculated by the standard diagonalization procedure, in which a range of the distance parameter S is restricted inside of the interaction region.

We solve the Schrödinger equation $(H - E)\Psi^{J^\pi(+)} = 0$ with the wave function assumed in Eq. (7). The equation can be reduced by eliminating the coefficients of b_ν in Eq. (7). When we substitute the total wave function in Eq. (7) into $(H - E)\Psi^{J^\pi(+)} = 0$, the mixing coefficient b_ν for the closed state $\Psi_\nu^{J^\pi}$ can be written in the form of

$$b_\nu = \frac{1}{E - E_\nu} < \hat{\Psi}_\nu^{J^\pi} |(H - E)| \sum_{\beta L} \varphi_{\beta L}^{J^\pi(+)} > . \tag{9}$$

Here, E_ν is an energy eigenvalue of the intrinsic states, which satisfy the eigenvalue equation of Eq. (3) with the abbreviation of $E_\nu^{J^\pi} \rightarrow E_\nu$. Using Eq. (9), we can eliminate b_ν from the original Schrödinger equation and derive the effective equation for the open channels as

$$(H - \mathcal{V}(E) - E) \sum_{\beta L} \varphi_{\beta L}^{J^\pi(+)} = 0 \tag{10}$$

with the energy-dependent non-local interaction of

$$\mathcal{V}(E) = \sum_{\nu}(H - E)|\Psi_{\nu}^{J^{\pi}} > \frac{1}{E - E_{\nu}} < \Psi_{\nu}^{J^{\pi}}|(H - E) . \quad (11)$$

This non-local interaction corresponds to the so-called dynamic polarization potential, which represents the coupling of the open channels and the intrinsic states, which have the more complicated structure than the incident and exit channels [14]. Equation (10) for the open channels can be solved by the standard technique of the KHK method [8], and the S matrix elements for the transition from the incident and exit channels are obtained.

3 Results

3.1 Chemical bonding structure in ^{10,12}Be

The energy levels from the bound region to the unbound region are shown in the left (^{10}Be) and right (^{12}Be) panels of Fig. 1 ($J^{\pi} = 0^{+}$). In this figure, the ordinates represents the total energy with respect to the α threshold, and the energy levels become the bound and resonant states below and above the α threshold energy, respectively. The energy levels are shown up to the excitation energy of $E_x \leq 15$ MeV. In this calculation, bound states are obtained by the diagonalization procedure, while resonant states are identified from the energy dependence of the scattering matrix in the complex plane.

In this figure, the total reaction probability, which is defined by $1-\Re S_{el}$ with the elastic S matrix of S_{el} is also shown. The S matrix is calculated from the closed state method, explained at the previous section. In the reaction probability of ^{10}Be, broad structure can be seen, while, in ^{12}Be, there is complicated behavior in the reaction probability, which implies the existence of many resonant states with narrow width. At the resonance energy position, the various chemical bonding-like structures are generated, as shown by the schematic images attached on these two panels.

The obtained 0^{+} states are classified into three categories according to analysis of the wave function as follows:

(i) Compact MO states: In ^{10}Be, the MO state of $(\pi^{-}_{3/2})^2$ and $(\pi^{-}_{1/2})^2$, which form the parallel and anti-parallel spin-orbit pair, are generated at the ground 0^{+}_{1} and the highest 0^{+}_{4} state, respectively. In ^{12}Be, the four neutrons are in MO motion around the two α cores in two bound states ($(\pi^{-}_{3/2})^2(\sigma^{+}_{1/2})^2$ in 0^{+}_{1} and $(\pi^{-}_{3/2})^2(\pi^{-}_{1/2})^2$ in 0^{+}_{2}), appearing below the α+^{8}He$_{g.s.}$ threshold. The $\pi^{-}_{3/2}$ and $\pi^{-}_{1/2}$ orbits, corresponding to the $0p_{3/2}$ and $0p_{1/2}$ orbits in the shell model limit, respectively, gain the binding energy in the case of the compact $\alpha - \alpha$ distance. Therefore, the MO states containing the $\pi^{-}_{3/2}$ and $\pi^{-}_{1/2}$ orbits have the spatially compact structures.

(ii) Extended VB states with ^{x}He + ^{y}He: Above the α decay threshold, ionic He-dimer states, which have the VB structure of excess neutrons, are generated in both ^{10,12}Be. In ^{10}Be, for instance, the α+^{6}He(2^{+}_{1}) structure appears in the 0^{+}_{3} state, which is a resonant state around $E \sim 4$ MeV. The box in the left panel corresponds to the 0^{+} state having the dominant α + ^{6}He$_{g.s.}$ structure, but this 0^{+} state is not prominent resonance but continuum like state. The hight of the box shows the width of the α + ^{6}He$_{g.s.}$ continuum state. As shown in the reaction probability, this continuum state is observed as a broad peak in the α + ^{6}He$_{g.s.}$ scattering. The resonant states, 0^{+}_{3}, 0^{+}_{4}, and 0^{+}_{6} in ^{12}Be have the He-dimer configuration of the corresponding threshold, α + ^{8}He$_{g.s}$, ^{6}He$_{g.s.}$ + ^{6}He$_{g.s.}$, and ^{5}He$_{g.s.}$ + ^{7}He$_{g.s.}$, respectively. In these He-dimer resonances, the relative wave function between two He nuclei is largely extended and hence, these resonant states can be interpreted in terms of the molecular resonances, composed of two He fragments.

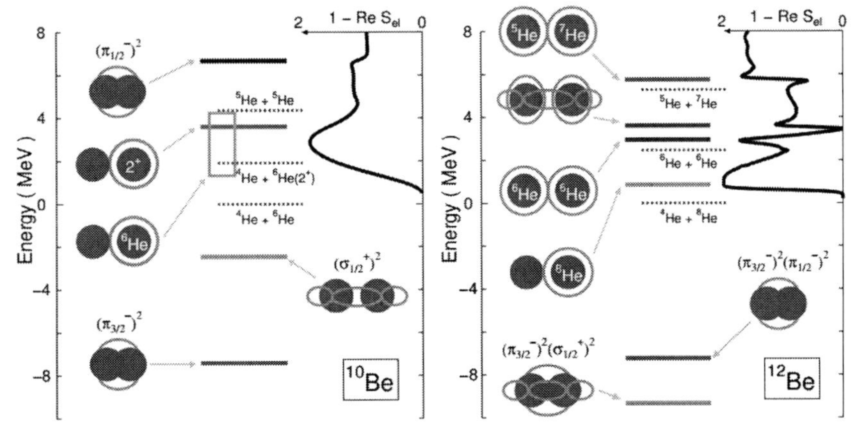

Figure 1. (Color online) Left panel: The energy spectra of ^{10}Be ($J^\pi = 0^+$). The solid curve on the right sides shows the total reaction probability of the α+^{6}He$_{g.s.}$ scattering with a respective scale at the top most axis. The dotted lines represent the threshold energy of the open channels considered in the calculation. The box represents the broad continuum state. Right panel: The same as left panel except for ^{12}Be. These figures are taken from Ref. [7]

(iii) Covalent SD state: The 0_2^+ state of ^{10}Be has the $(\sigma_{1/2}^+)^2$ configuration. The structure of 0_5^+ state in ^{12}Be is based on ^{10}Be(0_2^+). The main component of this state is the product of $(\sigma_{1/2}^+)^2 \otimes$ (^{5}He + ^{5}He). Due to the $(\sigma_{1/2}^+)^2$ formation, these states have a large clustering (S ~5 fm) in comparison with the touching distance of two α particles (~ 2.8 fm).

In both the systems, the compact MO states appear in the bound region below the α threshold, while the compact MOs and the extended VB states coexist in continuum region; specifically, the spatially compact and extended states coexist in the same energy region. This is in marked contrast to the naive expectation that the spatial extension of two α cores monotonically increase as a function of an excitation energy. In ^{10}Be, we can observe a wide variety in the structure change, which are induced by the coupling of the two excess neutrons. In ^{12}Be with four excess neutrons, furthermore, more levels are generated in the same energy region of $E_x \leq 15$ MeV, and hence the structure changes becomes much more prominent in comparison to ^{10}Be.

Let us compare the level scheme in Fig. 1 with the recent experiments. In ^{10}Be, shown in the left panel of Fig. 1, the theoretical energy of the intruder 0_2^+ state nicely reproduces the observed excitation energy. Furthermore, the broad 0^+ strength is identified by the experiment by Rogachev et al. [16], and the energy position of the observed strength corresponds to the calculated continuum 0^+ state just above the α + ^{6}He$_{g.s.}$ threshold. The calculation of ^{12}Be also reproduces the recent observations nicely. In the bound region, the excitation energy of the intruder 0_2^+ state in ^{12}Be is much lower than that in ^{10}Be, and the energy of 0_2^+ is almost same as the observed energy. In continuum region, theoretical calculation generates four levels with a close energy spacing, and such a degenerate feature of the energy levels are observed in the recent experiment of the inelastic scattering of ^{12}Be [5, 6]. In the inelastic scattering in Ref. [5], the observed 0^+ strength reveals the overlapping resonances with an energy distance of ~ 1 MeV. Furthermore, there are at least two observed levels consistent with the calculated levels; the calculated 0_3^+ and 0_5^+ states has a dominant decay width into the α+^{8}He$_{g.s.}$ and ^{6}He$_{g.s.}$+^{6}He$_{g.s.}$ channels, respectively, and the two levels having similar decay schemes are observed around the same energy regions [5, 6]. The decay scheme will be discussed in the later sections.

3.2 Threshold energies and the resonant levels

In the $N = Z$ nuclear system, the α cluster structures are known to appear according to the threshold rule, in which the cluster structures appear around the excitation energy close to the threshold energy dissociating into the corresponding cluster configuration [15]. It is interesting to focus on the relation of the continuum states in ^{10,12}Be and threshold energies from the viewpoint of the threshold rule. A comparison of the levels in continuum region with the respective He-decay thresholds is summarized in two panels of Fig. 2. In these panels, the four-body (^{10}Be $=\alpha + \alpha + 2N$) and six-body (^{12}Be$=\alpha + \alpha + 4N$) channels are also plotted although the degree of freedom for these many-body decay is not considered in the present calculation. Since, in ^{10}Be, the calculated threshold difference, $\Delta E = E_{th}(^5\mathrm{He}_{g.s.}+^5\mathrm{He}_{g.s.}) - E_{th}(\alpha+^6\mathrm{He}(2_1^+))$ is larger by about 1.4 MeV than the experimental value [7], the thresholds and the energy levels and thresholds above $E_{th}(\alpha+^6\mathrm{He}(2_1^+))$ are shifted by 1.4 MeV to the lower energy side. Due to a similar reason, in ^{12}Be, the thresholds and the energy levels above $E_{th}(^6\mathrm{He}_{g.s.}+^6\mathrm{He}_{g.s.})$ are shifted by 1.3 MeV.

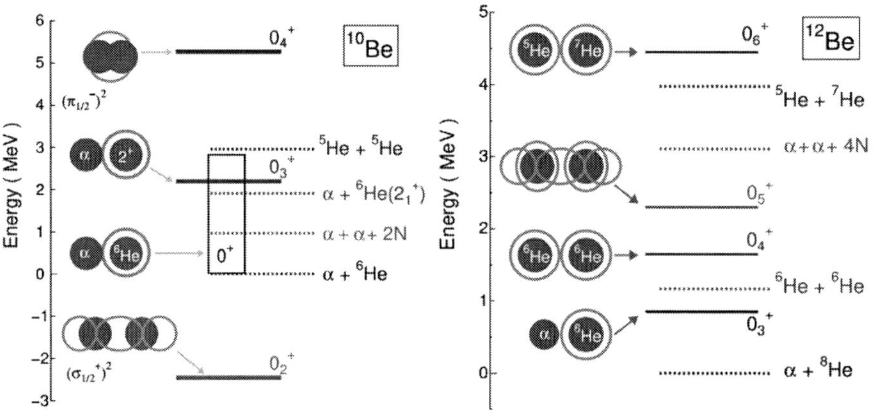

Figure 2. (Color online) Left panel: Energy levels in the higher energy region in ^{10}Be. The threshold for the binary He channels (calculation) and the $\alpha + \alpha + 2N$ (experiment) are shown by the dotted lines, while the energy levels are plotted by the solid lines. Right panel: The same as left panel but for ^{12}Be. In ^{10}Be, the threshold and the energy levels above the α+^{6}He(2_1^+) threshold is shifted to the lower energy side by about 1.4 MeV, while, in ^{12}Be, those above the ^{6}He$_{g.s.}$+^{6}He$_{g.s.}$ threshold is shifted to the lower energy side by about 1.3 MeV. These figures are taken from Ref. [7]

In the left panel (^{10}Be), there is one to one correspondence between calculated levels and the threshold energies although the 0^+ state in a box is a broad continuum-like state. The discrete bound or resonant levels are distributed in an energy range of about 8 MeV, which corresponds to about 3 MeV energy interval on average. In marked contrast to the level scheme in ^{10}Be, all the energy levels in ^{12}Be shown in the right panel are confined within the range of $\sim$ 4 MeV, and the energy interval between neighboring levels is just $\sim$1 MeV. Although the energy interval in ^{12}Be is much smaller than that in ^{10}Be, we can still confirm the one to one correspondence between the energy levels and thresholds.

Because of the small spacings of the threshold energies in ^{12}Be, the mixing of the He-dimer components is strong, although the dominant component can be identified in each of the energy levels. The dominant populations in the $0_1^+ \sim 0_6^+$ levels are investigated in Ref. [7], and they have been found to be about 50 % on average. Thus, each level contains impurities besides a main component. Such a mixture of the impurities is important for reducing the decay width of the resonant levels, leading to the formation of the sharp resonance [7].

3.3 Analysis of the low-energy α scattering

3.3.1 α + ^{8}He resonant scattering

By solving exactly the effective equation for the open channels shown in Eq. (10), we calculate the S matrices for the elastic scattering and neutron transfer reactions, which are shown in Fig. 3. In this figure, S matrices $S_{f,i}$ for the transitions from the incident α + ^{8}He$_{g.s.}$ channel (i) to the final channels (f), ^{6}He$_{g.s.}$ + ^{6}He$_{g.s.}$ or ^{5}He$_{g.s.}$ + ^{7}He$_{g.s.}$, are plotted as a function of the center-of-mass energy of the incident channel. In this figure, we can clearly understand the important role of the intrinsic states $\Psi_\nu^{J^\pi}$ in forming the sharp resonances.

The dotted curves represent the S matrices calculated only from the coupling of the open channels $\varphi_{\beta L}^{J^\pi(+)}$, which is included in the first term of Eq. (7). No resonant structures with sharp widths exist in the results of the open channels. In the full solutions shown by the solid curves, the sharp structures are generated by the coupling with the intrinsic states (double squares labeled by $0_3^+ \sim 0_6^+$). Each peak is generated by a linear combination of all the intrinsic states, but the dominant contribution in the individual peak is a single intrinsic state, which is indicated by the arrows. Since the intrinsic states include the component of not only the open channels, but also of many kinds of closed channels, which have a higher excitation energy than the resonance energy, the coupling with the intrinsic states plays an important role in reducing the resonance width.

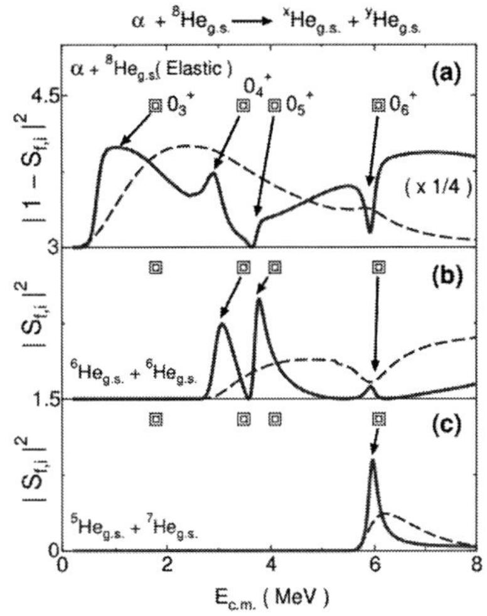

Figure 3. (Color online) S matrices for the central collision of α+^{8}He$_{g.s.}$ (J^π=0$^+$). Panel (a) shows the S matrix of the elastic scattering, while (b) and (c) show those for the final channels of ^{6}He$_{g.s.}$+^{6}He$_{g.s.}$ and ^{5}He$_{g.s.}$+^{7}He$_{g.s.}$, respectively. The dashed and solid curves represent the results calculated with the three open channels and the full solutions of Eq. (8), respectively. The double squares represent the energy levels obtained by the diagonalization of the total Hamiltonian. The S matrices of the elastic scattering are divided by a factor of four to compare with the magnitude of the non-elastic S matrices. This figure is taken from Refs. [7, 8].

It is naively expected that the coupled-channel calculation between the incident and exit channels is sufficient in describing the two-neutron transfer reaction, α + ^{8}He$_{g.s}$ → ^{6}He$_{g.s.}$ + ^{6}He$_{g.s.}$. In the present calculation of the open channels, the $^{5}He_{g.s.}$ + $^{7}He_{g.s.}$ channel, which is never considered in the naive two neutron transfer approach, is included. However, no sharp resonances appear even if the one-neutron transfer channel is coupled to the incident and final channels. The inclusion of the intrinsic states, which have a more complicated structure than those of the incident and exit channels, is essential in describing the formation of resonances with sharp widths.

The present calculation predicts the sharp resonances corresponding to the four levels, $0_3^+ \sim 0_6^+$, shown in Fig. 2, in all the exit channels. The ratio of S matrices for different exit channels gives important information on the intrinsic structures of the compound system, ^{12}Be. In particular, the resonant behavior around $E_{c.m.} \sim 4$ MeV ($E \sim 1.3$ MeV with respect to the $^{6}He_{g.s}$ + $^{6}He_{g.s.}$ threshold) gives an interesting result. At the resonance energy corresponding to 0_5^+, the magnitude of the respective S matrix element increases up to ~ 1. Since the unitary condition of $\sum_f |S_{f,i}|^2 = 1$ holds in the present calculation without any imaginary potentials, this enhancement means that the incident flux flows mainly into the $^{6}He_{g.s.}$ + $^{6}He_{g.s.}$ channel. Around this enhancement region, S_{el} is quite small due to the unitary condition, and $|1 - S_{el}|^2$ becomes ~ 1. (Fig. 3(a)). This situation corresponds to the so-called shadow scattering in the elastic channel, in which the incident flux is strongly absorbed inside the interaction region and scattered as edge waves around the boundary of the strong absorber.

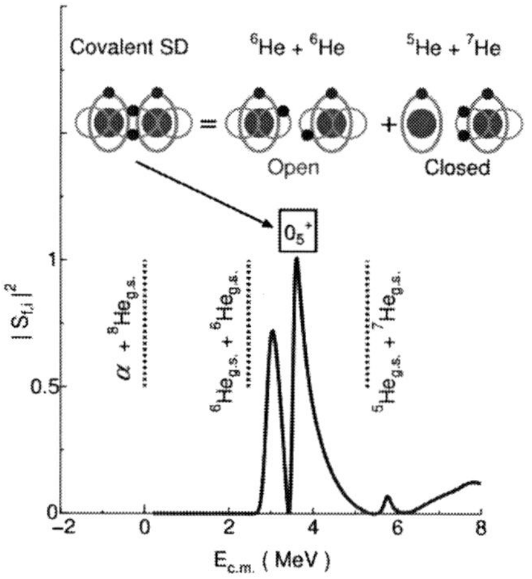

Figure 4. (Color) S matrix for the two neutron-transfer reactions, α + $^{8}He_{g.s.} \rightarrow$ $^{6}He_{g.s.}$ + $^{6}He_{g.s.}$ (J^{π}=0^{+}). The dotted lines represent the threshold energy of the open channels considered in the calculation. In the upper part, the schematic picture of the covalent SD state is shown. This figure is taken from Refs. [7, 8]

The result of $|S_{f,i}|^2 \sim 1$ means that the 0_5^+ state mainly decays into not α + $^{8}He_{g.s.}$ but $^{6}He_{g.s.}$ + $^{6}He_{g.s.}$. Such a characteristic decay scheme is due to the intrinsic structure of 0_5^+, which has the covalent SD structure as explained in the previous section. The S matrix of two neutron transfer and the threshold energies of open channels are shown in Fig. 4. Since the σ^+ orbital is constructed by $0p_z(L) - 0p_z(R)$ where $0p_z(C)$ represents the $0p_z$-wave with the center C, the covalent SD is given by a linear combination of the (^{6}He + ^{6}He) + (^{5}He + ^{7}He) configurations, as shown in the schematic picture in Fig. 4. Here, ^{6}He + ^{6}He and ^{5}He + ^{7}He are constructed by a coherent mixing of the asymptotic channels of the respective configurations. A component of α + $^{8}He_{g.s.}$, which is an open channel at the resonance energy, is suppressed in this configuration. Furthermore, the resonance energy is lower than the threshold of $^{5}He_{g.s.}$ + $^{7}He_{g.s.}$, which is the lowest open channel in the ^{5}He + ^{7}He configuration. Therefore, the decay channel of the covalent SD is mainly restricted to $^{6}He_{g.s.}$ + $^{6}He_{g.s.}$. This is the reason why the S matrix corresponding to the covalent SD reaches the unitary limit of $|S_{f,i}|^2 \sim 1$. Due to this limitation of the decay channel, the covalent SD becomes an isomeric state with a width of $\Gamma \sim$400 keV.

The enhancement of the decay into $^{6}He_{g.s.}$ + $^{6}He_{g.s.}$ can be supported by a recent measurement of an inelastic α scattering on ^{12}Be [5]. In this study, the excitation energy spectrum of ^{12}Be was obtained from the invariant mass of the decay products, $^{6}He_{g.s.}$ + $^{6}He_{g.s.}$ and α + $^{8}He_{g.s.}$. The contributions of 0^+ were extracted by the multipole decomposition analysis (MDA) [5]. A sharp resonance was suggested in the $^6He_{g.s.}$ + $^6He_{g.s.}$ decay channel about 1.2 MeV above the decay threshold. In contrast, the corresponding resonance was not identified in the α+$^8He_{g.s.}$ channel, suggesting a smaller partial decay width compared to the $^6He_{g.s.}$ + $^6He_{g.s.}$ channel [5]. These observations are consistent with the predicted nature of the covalent isomeric state in terms of excitation energy and decay scheme. Furthermore, the low-energy ^{8}He scattering by an α target has been measured at GANIL [17]. Thus, the direct comparison of the theoretical calculation of the α scattering with the respective experiment will be possible in future studies.

3.3.2 α + ^{6}He resonant scattering

The S matrices in the α+$^6He_{g.s.}$ reaction with J^π = 0^+ are calculated by applying the standard technique of the Kohn-Hulthén-Kato method [7, 8]. The calculated S matrices are shown in Fig. 5. In the panels of (a), (b) and (c), the S matrices going to the elastic α + $^6He_{g.s.}$, α + $^6He(2_1^+)$, $^5He_{g.s.}$ + $^5He_{g.s.}$ (I=0) channels are plotted, respectively. We can confirm the broad resonant structures in the individual panels. There are two dents in the elastic channel (a) at $E_{c.m.}$ ~ 4 MeV and 7 MeV, while two peaks appear at the respective dent-energies in the non-elastic channels of (b) and (c). This anti-correlation of the S matrices means that the incident flux from the elastic channel mainly absorbed in the specific exit channels, α + $^6He(2_1^+)$ and $^5He_{g.s.}$ + $^5He_{g.s.}$ (I=0), and resonances are generated in these exit channels. These broad structures are in marked contrast to the sharp structures appearing in the α + $^8He_{g.s.}$ reaction shown in Fig. 3. Since the number of the excess neutrons in ^{10}Be (N=2) is smaller than that in ^{12}Be (N=4), the attraction from excess neutrons is not so strong in ^{10}Be in a comparison to ^{12}Be. Due to this reduced attraction, the resonance width becomes much broader in ^{10}Be.

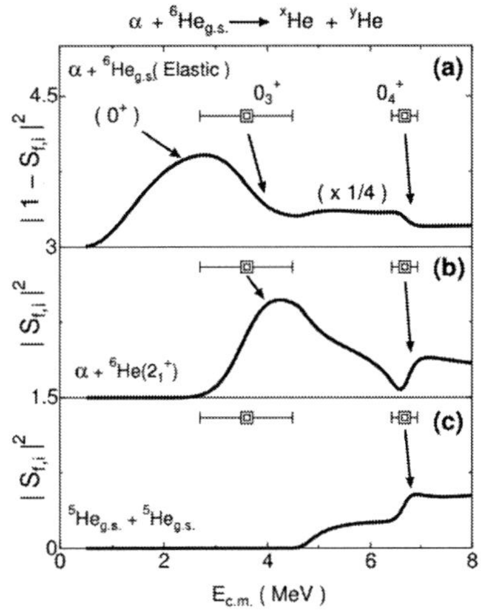

Figure 5. S matrices for the central collision of α+$^6He_{g.s.}$ (J^π=0^+). Panel (a) shows the S matrix of the elastic scattering, while (b) and (c) show those for the final channels of α+$^6He(2_1^+)$ and $^5He_{g.s.}$+$^5He_{g.s.}$ (channel spin I=0), respectively. The double squares represent the resonance energy obtained by the absorbing boundary condition, while the error bars, attached to the double squares, show the decay width of the resonance. The S matrices of the elastic scattering are divided by a factor of four. This figure is taken from Ref. [7]

We also apply the absorbing boundary condition (ABC) in order to identify the resonance energy and width more clearly [18, 19]. In the ABC method, a negative imaginary potential is placed outside of the interaction region of the total system, and the continuum component contained in the wave function is absorbed. Due to this absorption, the resonant states are identified in the complex energy plane, and the resonance energy and the decay width can be calculated. The basic formulation of the ABC method is given in Refs. [18, 19]. The identified resonant levels are shown by the double squares with the error bars in all the panels of Fig. 5. The lower resonance at $E_{c.m.}$ ~ 3.8 MeV and higher resonance to $E_{c.m.}$ ~ 6.8 MeV correspond to the 0_3^+ and 0_4^+ states, obtained in the structure analysis shown in the left panel of Fig. 1. The 0_3^+ state, having a dominant component of α + ^{6}He(2_1^+) $L = 2$, corresponds to the extended VB state in Fig. 1, which is realized by the higher oscillation of the $\alpha - \alpha$ relative wave function. The 0_4^+ state is the compact MO states in Fig. 1, having the dominant configuration of $(\pi_{1/2}^-)^2$.

In the S matrix for the inelastic scattering of α + ^{6}He$_{g.s.}$ $\rightarrow \alpha$ + ^{6}He(2_1^+) (Panel (b) of Fig. 5), there is a strong peak at $E_{c.m.} \sim 4$ MeV, which nicely corresponds to the 0_3^+ state in resonance energy and width. As indicated by the arrow, the enhancement of the S matrix at $E_{c.m.} \sim 4$ MeV is generated by the existence of the resonant 0_3^+ state, which are identified under the absorbing boundary condition. Since the 0_3^+ state has a well developed α + ^{6}He(2_1^+) structure, this state is strongly excited in the final channel of the inelastic scattering. On the contrary, in one neutron transfer channel of ^{5}He$_{g.s.}$ + ^{5}He$_{g.s.}$, shown in the panel (c), we can also confirm a small peak at $E_{c.m.} \sim 7$ MeV near the resonance of the 0_4^+ state. As shown by the arrow, this small peak is generated by the resonance of the 0_4^+ state, which has the MO configuration of $(\pi_{1/2}^-)^2$ with a spatially compact structure. Because of a large difference in the structure between α+^{6}He$_{g.s.}$ and $(\pi_{1/2}^-)^2$, it is difficult to excite the 0_4^+ resonance from the incident channel of α+^{6}He$_{g.s.}$. This is the reason why the S matrix is not so strongly enhanced at the resonance energy of the 0_4^+ state. In addition to these two peaks, a broad structure exists at $E_{c.m.} \sim 3$ MeV in the elastic channel (Panel (a)). This broad peak is not a resonance but a continuum-like state because we cannot find any respective resonance in the calculation of the absorbing boundary condition.

The reaction mechanism of the resonance formation in the S matrices shown in Fig. 5 is basically same as that discussed in ^{12}Be; specifically, the resonant structures are generated by the coupling of the intrinsic resonant states shown by the double circles and the scattering of the open channels, such as α + ^{6}He$_{g.s.}$, α + ^{6}He(2_1^+), and ^{5}He$_{g.s.}$ + ^{5}He$_{g.s.}$. The restricted calculation, which contains only these three open channels, cannot generate the sharp resonant structure in the scattering probability shown in Fig. 5.

The calculation of the S matrix is consistent to the recent observations. The broad 0^+ continuum in the calculation correspond to the broad 0^+ strength observed by Rogachev et al. [16], while a resonance-like 0^+ state is observed in the α scattering at the energy position close to the theoretical 0_3^+ resonance [20]. In Ref. [20], the low-energy scattering of α+^{6}He$_{g.s.}$ $\rightarrow \alpha$+^{6}He(2_1^+) has been performed, and the respective angular distributions has been analyzed. A candidate of $L = 2$ resonance is suggested at the energy of $E_{c.m.} = 4.4$ MeV ($E_x \sim 11.8$ MeV), which is almost the same as the resonance energy of the theoretical 0_3^+ state with the α+^{6}He(2_1^+) $L = 2$ structure. In a deep comparison of the theoretical calculation with the experiments, more precise experimental data and the extended calculation on the differential cross section are indispensable.

3.4 Coulomb shift in ^{10}Be and ^{10}C

We apply the GTCM to ^{10}C = $\alpha + \alpha + P + P$, which is the mirror system of ^{10}Be, and the Coulomb shift is investigated. The GTCM calculation for ^{10}C is performed in the similar manner by adding the Coulomb interaction of $\alpha - P$. We calculate the bound and continuum spectra under the bound and scattering boundary conditions, respectively. Figure 6 shows the energy levels of ^{10}Be (left levels) and ^{10}C (right levels). In the individual spectra, the lowest four 0^+ levels are plotted except for the broad continuum bump at $E_{c.m.} \sim 3$ MeV shown in Figs. 1 and 5. As we have already confirmed in Fig. 1, the internal structures of the 0_1^+ and 0_2^+ states are the MO configurations of $(\pi_{3/2}^-)^2$ and $(\sigma_{1/2}^+)^2$, respectively, in which two valence nucleons perform the single particle motion around two α cores. On the contrary, the unbound 0_3^+ and 0_4^+ states have the α + ^{6}He(2^+) and $(\pi_{1/2}^-)^2$, respectively.

In the ^{10}C spectra, all of the energy levels shift to the higher energy region because of the Coulomb repulsion of the $\alpha - P$ system. The magnitudes of the shifts are not common among the levels but they depend on the energy levels. The shift of the 0_1^+ state is 2.3 MeV, which is the largest of all the energy levels. This prominent shift is due to the spatially compact structure of the $(\pi_{3/2}^-)^2$ configuration. The $\pi_{3/2}^-$ orbit, which has the dominant component of the $0p_{3/2}$ orbit, strongly reduce the $\alpha - \alpha$ distance, and the 0_1^+ state has a spatially compact structure similar to the shell model-like structure. Therefore, the effect of the Coulomb repulsion is prominent for the compact 0_1^+ state if the valence neutrons are replaced by the protons.

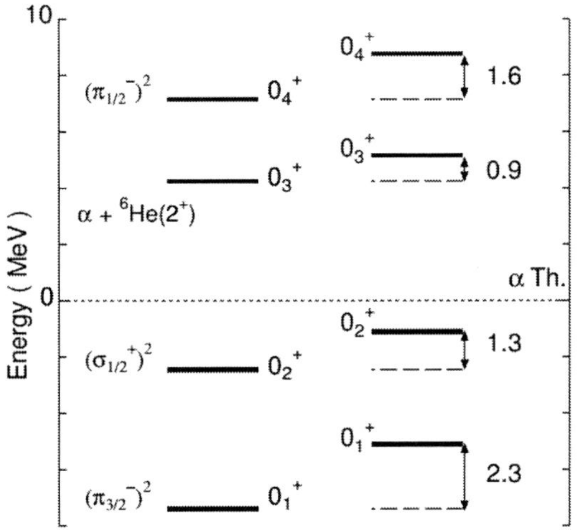

Figure 6. Energy levels of the ^{10}Be (left levels) and ^{10}C (right levels) nuclei. The α threshold is set to be zero energy. At the ^{10}C spectra, the corresponding levels to ^{10}Be are plotted by the dashed line. The numbers attached to the levels show the Coulomb shifts. The shifts are shown in units of MeV.

On the contrary, the Coulomb shift of the 0_2^+ state is 1.3 MeV, which is smaller than the shift of the ground 0_1^+ state, 2.3 MeV. The reduction of the shift for 0_2^+ is due to the spatially extended structure of the $\sigma_{1/2}^+$ orbit, which has a large mixture of the S-wave component. The large mixing of the S-wave component in the valence orbit leads to the enhancement of the $\alpha - \alpha$ clustering. Since the repulsive effect of the Coulomb interaction is suppressed in the spatially extended configuration, the Coulomb shift becomes small in the 0_2^+ state. Due to the difference of the Coulomb repulsion, the excitation energy of the 0_2^+ in ^{10}C is smaller by about 1 MeV than that of 0_2^+ in ^{10}Be. This shift of the excitation energy is consistent to the recent compilation of the energy levels of the ^{10}C nucleus [21].

In the continuum energy region above the α threshold, the difference of the Coulomb shift is also seen in the 0_3^+ and 0_4^+ states; 0.9 MeV for the former state and 1.6 MeV for the latter state. Since the difference of the Coulomb barrier is about 1.5 MeV between α + ^{6}He and α + ^{6}Be, the Coulomb shift of the 0_3^+ state is suppressed in comparison to the shift of the Coulomb barrier. The 0_3^+ state has a dominant component of resonant α + ^{6}He(2$^+$) state with the D-wave relative motion, but there is a considerable mixture of α + $^6\mathrm{He}_{g.s.}$ with the S-wave component. This S-wave mixture is due to the existence the broad α + $^6\mathrm{He}_{g.s.}$ strength just below the 0_3^+ state, as shown in Fig. 5. Therefore, the 0_3^+ state has the extended and dilute binary structure, which leads to the suppression of the Coulomb shift. On the contrary, the 0_4^+ state has the dominant component of the compact MO configuration of

$(\pi^-_{1/2})^2$, which has a large amplitude of the wave function inside of the interaction region. Thus, the Coulomb shift is enhanced in the compact 0^+_4 state.

In all the spectra, the Coulomb shift is reduced for the spatially extended state (0^+_2 and 0^+_3), which has a large mixture of the S-wave component, while the shift is enhanced for the compact non-S-wave states (0^+_1 and 0^+_4). Therefore, these suppressed Coulomb shift confirmed in the present calculation are basically same as the Thomas-Ehrman Shift (TES), which was originally discussed in the ^{17}O and ^{17}F systems [9]. The Coulomb suppression in ^{10}Be and ^{10}C, which are shown in Fig. 6, correspond to the extension of the TES to the cluster degrees of freedom of $\alpha + \alpha + 2N$ and hence, the present shift should be called the "Cluster Thomas-Ehrman shift (Cluster TES)". The shift of the excitation energy in the Cluster TES is about 1 MeV for the 0^+_2 state. This shift is much larger than the shift discussed in the traditional TES, which is about 0.4 MeV for the $1/2^+$ state with respect to the ground $5/2^+$ state.

Although we have discussed the energy levels in the mirror systems of ^{10}Be and ^{10}C, the present calculation also predicts the asymmetry in the monopole transition from 0^+_1 to 0^+_2; specifically, the $0^+_1 \rightarrow 0^+_2$ strength in ^{10}C is about twice as large as the respective strength in ^{10}Be. Due to the suppression of the Coulomb shift for the 0^+_2 state, the excitation energy of 0^+_2 in ^{10}C is reduced in comparison to the respective excitation energy in ^{10}Be. This reduction ^{10}C. A detailed analysis of the monopole transitions will be shown in forthcoming papers.

4 Summary and discussion

We have formulated the generalized two-center cluster model (GTCM), which can handle the formation of the compact MO structure and the extended VB structure in a consistent manner [7, 8]. The basic structure of these two configurations, MOs and VBs, is almost same as the electrons' bonds employed in quantum chemistry, but GTCM can describe their dynamical transition induced by the Hamiltonian of the total many-body systems; specifically, this model can generate the compact MO structure at a small $\alpha - \alpha$ distance, while, in the asymptotic region, the extended VB structure are naturally formed.

First, GTCM has been applied to the structure problem in ^{10,12}Be, and the energy spectra have been calculated under the appropriate boundary conditions. In ^{10,12}Be, there is one to one correspondence between the energy levels and the respective threshold. This correspondence means that the excited levels appear according to the threshold rule. The structure of the neutrons' orbit prominently changes from level to level as a function of the excitation energy. In the bound region, the covalent MO structures are dominant, and the ionic VB structures mainly appears in the unbound region. The bound and resonant levels are compared with all the recent observations. The results of the GTCM calculation is completely consistent to the level scheme observed over a wide energy region, which contains the low-lying intruder 0^+_2 below the α decay threshold and the continuum states above the α threshold [4–6]. In particular, the results of ^{12}Be is nicely reproduce the degenerate feature of the levels in the unbound retion [5].

Secondly, GTCM is applied to the resonant scattering of ^{8,6}He by an α target. In $\alpha + {}^8\text{He}_{g.s.}$, we have formulated the closed state method to handle the scattering problem in a close relationship with the structure information on the compound system of ^{12}Be. In the closed state method, the coupling of the closed state, which are obtained from the bound-state-like boundary condition, and the open channels is explicitly solved. The formation of the sharp resonance can be naturally described by the mixing of the open channels and the closed states, which contains all the structure correlations inside of the interaction region. We have focused on two neutron transfer, $\alpha + {}^8\text{He}_{g.s.} \rightarrow {}^6\text{He}_{g.s.} + {}^6\text{He}_{g.s.}$, and pointed out a characteristic decay scheme in the central $J^\pi = 0^+$ collision. Similar reaction analysis has been performed for the $\alpha + {}^6\text{He}_{g.s.}$, and we have predicted a prominent resonance in the inelastic scattering of α+^{6}He$_{g.s.}$ $\rightarrow$ α+^{6}He(2^+). In both of the systems, the obtained reaction scheme is consistent to the recent observations on the unbound states above the α threshold [5, 6, 16, 17, 20].

Finally, we have applied GTCM to ^{10}C = $\alpha + \alpha + 2P$, which is the mirror partner of ^{10}Be = $\alpha + \alpha + 2N$, and the Coulomb shift in the lowest four 0^+ states is analyzed. The present calculation predicts the suppression of the Coulomb shift for the spatially extended states of $0^+_{2,3}$, which are induced by the development of the α clustering. Since the spatially-extended cluster structure involves the large mixture of the S-wave component, the Coulomb suppression is attributed to the anomalous feature of the S-wave orbit. The suppressed Coulomb shift presented here is basically same as the Thomas-Ehrman Shift (TES), which is discussed in the S-wave orbit for the single nucleon degree of freedom [9]. Therefore, we call the shift of the excitation energy of 0^+_2 found in ^{10}Be and ^{10}C "Cluster TES".

In stable $N = Z$ systems, the energy positions and their decay width into the α particle were the signatures in identifying an excited 0^+ level as an α cluster state. Although the Coulomb shift is meaningless in the $N = Z$ system, the Coulomb shift originated from the Cluster TES will be a new sign to identify the cluster structure in $N \neq Z$ systems. In the analysis of the Coulomb shift, the unified treatment of the structure and reaction is very useful because a large part of the excited energy levels in $N \neq Z$ sytems appears as the resonant state above the particle decay threshold. The systematic studies of various mirror systems are now under progress.

References

[1] H. Furutani et al., Suppl. Prog. Theor. Phys. **68**, 193 (1980); M. Seya, M. Kohn, and S. Nagata, Prog. Theor. Phys. **65**, 204 (1981).
[2] N. Itagaki and S. Okabe, Phys. Rev. C **61**, 044306 (2000).
[3] N. Itagaki, S. Okabe and K. Ikeda, Phys. Rev. C **62**, 034301 (2000), and references therein.
[4] M. Freer et al., Phys. Rev. Lett.82, 1383 (1999) ; Phys. Rev. C63, 034301 (2001).
[5] A. Saito *et al.*, Mod. Phys. Lett. A **25** 1858 (2010), and references therein.
[6] Z. H. Yang, *et al.*, Phys. Rev. Lett. **112**, 162501 (2014), and references therein.
[7] M. Ito, Phys. Rev. C **85**, 044308 (2012); M. Ito and K. Ikeda, Rep. Prog. Phys. **77**, 096301 (2014), and references therein.
[8] M. and N. Itagaki, Phys. Rev. C **78** 011602(R) (2008); M. Ito and D. Suzuki, Phy. Rev. C **84** 014608 (2011), and references therein.
[9] Bohr A and Mottelson B R 1969 Nuclear Structure vol I (New York: Benjamin) p 320.
[10] H. Horiuchi, Suppl. Prog. Theor. Phys. **62**, 90 (1977).
[11] M. Ito, K. Yabana, T. Nakatsukasa and M. Ueda, Phys. Lett. B **637**, 53 (2006).
[12] A. B. Volkov, Nucl. Phys. **74**, 33 (1965).
[13] N. Yamaguchi, T. Kasahara, S. Nagata, and Y. Akaishi, Prog. Theor. Phys. **62**, 1018 (1979).
[14] M. Ito, Y. Hirabayashi, and Y. Sakuragi, Eur. Phys. Jour. **5**, 373 (1999).
[15] K. Ikeda, H. Horiuchi, and S. Saito, Prog. Theor. Phys. **68**, 1 (1980), and references therein.
[16] G. V. Rogachev et al., Prog. Theor. Phys. Suppl. **196**, 184,(2012), and references therein.
[17] M. Freer, Private Communication (2014).
[18] Y. Takenaka, et al., Prog. Exp. Theor. Phys. **2014**, 113D04 (2014), and references therein.
[19] M. Iwasaki, R. Otani, Y. Takenaka, and M. Ito, Prog. Exp. Theor. Phys. **2015** 023D01, (2015).
[20] D. Suzuki et al., Phys. Rev. C **87**, 054301 (2013).
[21] V. Z. Goldberg and G. V. Rogachev, Phys. Rev. C**86**, 044314 (2012).

EPJ Web of Conferences **122**, 09003 (2016) DOI: 10.1051/epjconf/201612209003
*CNR*15*

Forging the link between nuclear reactions and nuclear structure

W. H. Dickhoff[1,a]

[1] *Department of Physics, Washington University, St. Louis, Missouri 63130, USA*

Abstract. A review of the recent applications of the dispersive optical model (DOM) is presented. Emphasis is on the nonlocal implementation of the DOM that is capable of describing ground-state properties accurately when data like the nuclear charge density are available. The present understanding of the role of short- and long-range physics in determining proton properties near the Fermi energy for stable closed-shell nuclei has relied mostly on data from the $(e, e'p)$ reaction. Hadronic tools to extract such spectroscopic information have been hampered by the lack of a consistent reaction description that provides unambiguous and *undisputed* results. The DOM, conceived by Claude Mahaux, provides a unified description of both elastic nucleon scattering and structure information related to single-particle properties below the Fermi energy. We have recently introduced a nonlocal dispersive optical potential for both the real and imaginary part. Nonlocal absorptive potentials yield equivalent elastic differential cross sections for ^{40}Ca as compared to local ones but change the ℓ-dependent absorption profile suggesting important consequences for the analysis of nuclear reactions. Below the Fermi energy, nonlocality is essential for an accurate representation of particle number and the nuclear charge density. Spectral properties implied by $(e, e'p)$ and $(p, 2p)$ reactions are correctly described, including the energy distribution of about 10% high-momentum protons obtained at Jefferson Lab. The nonlocal DOM allows a complete description of experimental data both above (up to 200 MeV) and below the Fermi energy in ^{40}Ca. It is further demonstrated that elastic nucleon-nucleus scattering data constrain the spectral strength in the continuum of orbits that are nominally bound in the independent-particle model. Extension of this analysis to ^{48}Ca allows a prediction of the neutron skin of this nucleus that is larger than most predictions made so far.

1 Introduction

The properties of a nucleon that is strongly influenced by the presence of other nucleons have traditionally been studied in separate energy domains and frameworks. Positive energy nucleons are described by fitted optical potentials mostly in local form [1, 2] which generate the distorted waves employed in the analysis of nuclear reactions. Bound nucleons have been analyzed with static potentials. At first, this lead to an independent-particle model (IPM) which is then modified by the interaction between valence nucleons in a very limited configuration space as in traditional shell-model calculations [3, 4]. No attention is paid in this approach to the larger energy domain of giant resonances or effects associated with fast nucleons. These high-momentum nucleons have been experimentally identified as

[a] e-mail: wimd@wuphys.wustl.edu

playing a role in the properties of nuclei and play a very important role in the success of *ab initio* methods to describe nuclei that rely on interactions with substantial repulsive cores. The link between nuclear reactions and nuclear structure is however possible by considering these potentials as representing different energy domains of one underlying nucleon self-energy provided by the framework of Green's function theory [5]. This idea was implemented in the DOM by Mahaux and Sartor [6]. By employing dispersion relations, the method provides a critical link between the physics above and below the Fermi energy with both sides being influenced by the absorptive potentials on the other side. Such a link is not present in traditional optical potentials that have been constructed by independently fitting the real and imaginary parts.

When a Hamiltonian for the description of nonrelativistic nucleons is employed as a starting point for the description of nuclei, a perturbation expansion can be derived for the nucleon single-particle propagator [5]. A compact formulation for the propagator leads to the Dyson equation which relates the propagator to a noninteracting one used as a starting point and the so-called self-energy. The latter acts as the potential in a Schrödinger-like equation derived from the Dyson equation with solutions that identify the properties of nucleons that are either removed or added to a ground-state nucleus. In the latter case, the solutions describe all aspects of elastic proton or neutron scattering from a target ground state. Removal properties can be employed to generate expectation values of one-body operators in the ground state like densities, as well as a contribution to the ground-state energy when it is dominated by the effects of two-body interactions. The potential (self-energy) is energy dependent, complex in certain energy domains, and its real and imaginary part obey a dispersion relation. Furthermore, at the Hartree-Fock (HF) level the self-energy is already nonlocal and all theoretical calculations of the self-energy confirm that both the real and imaginary part of the self-energy are nonlocal. The traditional use of the optical potential as local and nondispersive is therefore a gross misrepresentation of the actual physics and is in need of updating to a nonlocal and dispersive implementation. The work discussed in this contribution represents a step in this direction.

While the DOM provides an intermediary between theory and experiment, the ultimate goal is to be able to accurately calculate the nucleon self-energy in an *ab initio* manner. Currently we are very far from this goal but it should be in our sights with the help of the experience gained from the DOM. In Sec. 2 a brief outline of the ingredients of the DOM are presented. Results are presented for a fully nonlocal implementation of the DOM for ^{40}Ca in Sec. 3 including results for the particle spectral function that quantitatively clarify the removal of single-particle (sp) strength from mean-field orbits. In Sec. 4 recent results for ^{48}Ca are presented with emphasis on the neutron skin which emerges as a quantity that can be extracted from the fit to the available data when results for ^{40}Ca are employed as a starting point. Some brief conclusions and outlook are provided in Sec. 5.

2 Green's functions and the dispersive optical model

The self-energy $\Sigma_{\ell j}$ provides the critical ingredient to solve the Dyson equation for the nucleon propagator $G_{\ell j}$. Employing an angular momentum basis, it reads

$$G_{\ell j}(k,k';E) = \frac{\delta(k-k')}{k^2} G^{(0)}(k;E) + G^{(0)}(k;E) \int dq\, q^2\, \Sigma_{\ell j}(k,q;E)\, G_{\ell j}(q,k';E). \tag{1}$$

The noninteracting propagators $G^{(0)}$ only contain kinetic energy contributions. The solution of this equation generates $S_{\ell j}(k;E) = \text{Im}\, G_{\ell j}(k,k;E)/\pi$, the hole spectral density, for negative continuum energies. The spectral strength at E, for a given ℓj, is given by

$$S_{\ell j}(E) = \int_0^\infty dk\, k^2\, S_{\ell j}(k;E). \tag{2}$$

For discrete energies one solves the eigenvalue equation for the overlap functions

$$\phi^n_{\ell j}(k) = \left\langle \Psi^{A-1}_n \middle| a_{k\ell j} \middle| \Psi^A_0 \right\rangle, \tag{3}$$

for the removal of a nucleon with momentum k and discrete quantum numbers ℓ and j [7]. The removal energy corresponds to $\varepsilon^-_n = E^A_0 - E^{A-1}_n$ with the normalization for such a solution α_{qh} given by

$$S^n_{\ell j} = \left(1 - \left. \frac{\partial \Sigma_{\ell j}(\alpha_{qh}, \alpha_{qh}; E)}{\partial E} \right|_{\varepsilon^-_n} \right)^{-1}, \tag{4}$$

which is referred to as the spectroscopic factor. We note that from the solution of the Dyson equation below the Fermi energy, one can generate the one-body density matrix by integrating the non-diagonal imaginary part of the propagator up to the Fermi energy and therefore access the expectation values of one-body operators in the ground state including particle number, kinetic energy and charge density [5]. The latter is obtained by folding the point density with the nucleon form factors [8]. For positive energies, it was already realized long ago that the reducible self-energy provides the scattering amplitude for elastic nucleon scattering [9].

The self-energy fulfills the dispersion relation which relates the physics of bound nucleons to those that propagate at positive energy [5]. It contains a static correlated HF term and dynamic parts representing the coupling in the $A \pm 1$ systems that start and end at the Fermi energies for addition ($\varepsilon^+_F = E^{A+1}_0 - E^A_0$) and removal ($\varepsilon^-_F = E^A_0 - E^{A-1}_0$), respectively. The latter feature is particular to a finite system and allows for discrete quasi particle and hole solutions of the Dyson equation where the imaginary part of the self-energy vanishes. It is convenient to introduce the average Fermi energy $\varepsilon_F = \frac{1}{2}\left[\varepsilon^+_F - \varepsilon^-_F\right]$ and employ the subtracted form of the dispersion relation calculated at this energy [6, 7]

$$\begin{aligned} \mathrm{Re}\,\Sigma_{\ell j}(k,k';E) = \Sigma_{\ell j}(k,k';\varepsilon_F) - \mathcal{P}\int_{\varepsilon^+_F}^{\infty} \frac{dE'}{\pi} \mathrm{Im}\,\Sigma_{\ell j}(k,k';E') \left[\frac{1}{E-E'} - \frac{1}{\varepsilon_F - E'} \right] \\ + \mathcal{P}\int_{-\infty}^{\varepsilon^-_F} \frac{dE'}{\pi} \mathrm{Im}\,\Sigma_{\ell j}(k,k';E') \left[\frac{1}{E-E'} - \frac{1}{\varepsilon_F - E'} \right], \end{aligned} \tag{5}$$

where $\mathcal{P}$ represents the principal value. The beauty of this representation was recognized by Mahaux and Sartor [6, 10] since it provides a link with empirical information both for the real part of the nonlocal self-energy at the Fermi energy (probed by a multitude of HF calculations) as well as through empirical knowledge of the imaginary part of the optical potential also constrained by experimental data. Consequently Eq. (5) yields a dynamic contribution to the real part linking both energy domains around the Fermi energy. Empirical information near ε_F is emphasized by Eq. (5) because of the E'^{-2}-weighting in the integrands. The real self-energy at the Fermi energy will be denoted by Σ_{HF}.

Early implementations of the DOM employed a local form of the imaginary part of the potential and transformed the nonlocality of Σ_{HF} to a local energy-dependent potential [6]. Traditional volume and surface absorption terms with Woods-Saxon form factors were employed. Initial revival of the DOM in St. Louis contained the idea that a simultaneous fit to ^{40}Ca and ^{48}Ca allows an extrapolation to more neutron-rich calcium isotopes generating a data-driven extrapolation to the drip line [11, 12]. As will be seen later, a clear motivation to generate more neutron scattering data on ^{48}Ca emerged from this work culminating in a comprehensive DOM article for several relevant parts of the chart of nuclides [13] while still employing local potentials.

The DOM thus provides an ideal strategy to predict properties for exotic nuclei by utilizing extrapolations of the potentials towards the respective drip lines [11, 13, 14]. While it is possible to correct the artificial energy dependence of the HF self-energy for some properties near the Fermi energy like spectroscopic factors, it was necessary to employ the approximate expressions of limited accuracy for the properties of nucleons below the Fermi energy like occupation numbers that were developed by Mahaux and Sartor [6] . By restoring the proper treatment of nonlocality in the HF contribution, it was possible to overcome this problem [7] although the local treatment of the absorptive potentials yielded a poor description of the nuclear charge density and particle number.

Recently a fully nonlocal treatment of these potentials for the nucleus ^{40}Ca was presented [15] with the aim to include *all* available data below the Fermi energy that can be linked to the nucleon single-particle propagator [5] while maintaining a correct description of the elastic-scattering data. The result is a DOM potential that can be interpreted as the nucleon self-energy constrained by all available experimental data up to 200 MeV. Such a self-energy allows for a consistent treatment of nuclear reactions that depend on distorted waves generated by optical potentials as well as overlap functions and their normalization for the addition and removal of nucleons to discrete final states. The re-analysis of such reactions may further improve the consistency of the extracted structure information. Extending this version of the DOM to $N \neq Z$ allows for predictions of properties that require the simultaneous knowledge of both reaction and structure information since at present few weakly-interacting probes are available for exotic nuclei [16].

3 Results for ^{40}Ca

The nonlocal DOM potential generated in Ref. [15] was motivated by theoretical calculations of the self-energy for ^{40}Ca which either emphasized long-range correlations (LRC) [17] or short-range correlations (SRC) [18]. In Ref. [17] the results of the *ab initio* Faddeev random phase approximation (FRPA) which emphasizes the coupling of nucleons to low-lying collective states and to giant resonances, were compared with local DOM results. The nonlocality of the DOM potentials was represented by simple Gaussians which allow an analytic projection onto orbital angular momentum. We can now compare volume integrals for the potentials generated in Ref. [15] with these *ab initio* results. In Fig. 1 we compare these volume integrals for the imaginary part of the potentials as a function of energy in a domain which spans 200 MeV on either side of the Fermi energy for several orbital angular momenta [19]. We observe that the DOM results have been constrained by level information, the experimental charge density,high-momentum removal data from Jefferson Lab [20, 21] extending to very negative energies as well as available elastic scattering data up to 200 MeV. The FRPA calculations were performed in a limited configuration space which is clearly illustrated in Fig. 1. We note that the agreement between DOM and FRPA results is particularly good in the vicinity of the Fermi energy but the quality of the FRPA calculations is limited to about 60 MeV on either side of the Fermi and also deteriorates with increasing angular momentum illustrating the limitations of the configuration space.

The traditional good quality of DOM fits of elastic scattering data is maintained with the nonlocal implementation as illustrated in Fig. 2 for both proton and neutron elastic scattering [15]. Similar statements pertain to the reaction cross section for protons and total and reaction cross section for neutrons [15]. The quality of the description of data below the Fermi energy is however dramatically improved over earlier DOM calculations with a local absorptive potential [7]. In Fig. 3 the experimental charge density of ^{40}Ca [22] is compared with the DOM fit [15]. The accurate description of the charge density can only be obtained by incorporating a nonlocal absorptive potential which decreases the admixture of higher orbital angular momentum components in the ground state sufficiently fast to allow particle number to be described properly. We note that local absorptive potentials exhibit no such orbital angular momentum dependence and overestimate particle number substantially [7].

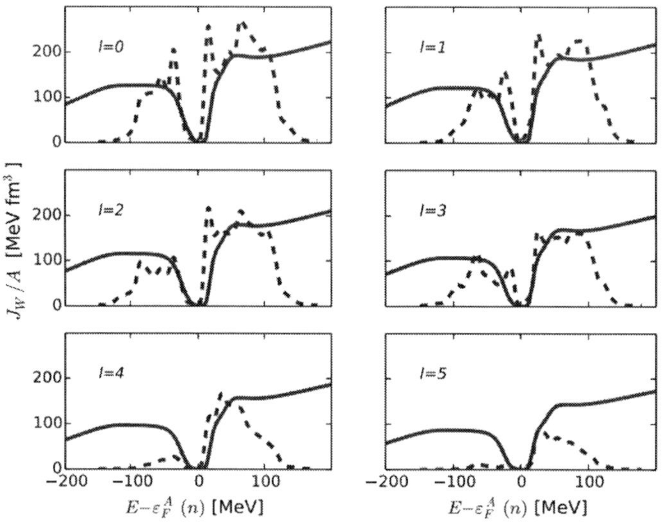

Figure 1. Comparison of volume integrals for different orbital angular momenta for the nonlocal DOM (solid) and FRPA (dashed) calculations in a wide energy domain. The curves are centered at the neutron Fermi energy for ^{40}Ca.

Since elastic scattering data only depend on the asymptotic form of the elastic scattering wave function, it is important to realize that by employing the nonlocal DOM in a wide energy domain, it is possible to assess where the spectral strength of mostly occupied states resides in the continuum [23]. The corresponding particle spectral function for a finite system requires the complete reducible self-energy $\Sigma^{\rm red}$. This reducible self-energy is calculated by employing a momentum-space scattering code discussed in Ref. [18]. In an angular-momentum basis, iterating the irreducible self-energy Σ to all orders, yields

$$\Sigma^{\rm red}_{\ell j}(k,k';E) = \Sigma_{\ell j}(k,k';E) + \int dqq^2\, \Sigma_{\ell j}(k,q;E)\, G^{(0)}(q;E)\, \Sigma^{\rm red}_{\ell j}(q,k';E), \tag{6}$$

where $G^{(0)}(q;E) = (E - \hbar^2 q^2/2m + i\eta)^{-1}$ is again the free propagator. The propagator is then obtained from the Dyson equation in the following form [5]

$$G_{\ell j}(k,k';E) = \frac{\delta(k-k')}{k^2} G^{(0)}(k;E) + G^{(0)}(k;E)\, \Sigma^{\rm red}_{\ell j}(k,k';E)\, G^{(0)}(k';E). \tag{7}$$

The on-shell matrix elements of the reducible self-energy in Eq. (6) are sufficient to describe all aspects of elastic scattering like differential, reaction, and total cross sections as well as polarization data [18]. While only this element is strictly related to data, the DOM analysis in its nonlocal implementation correctly and simultaneously describes all relevant data at all energies (up to 200 MeV) both below and above the Fermi energy [15]. We therefore assume that the DOM reducible self-energy is

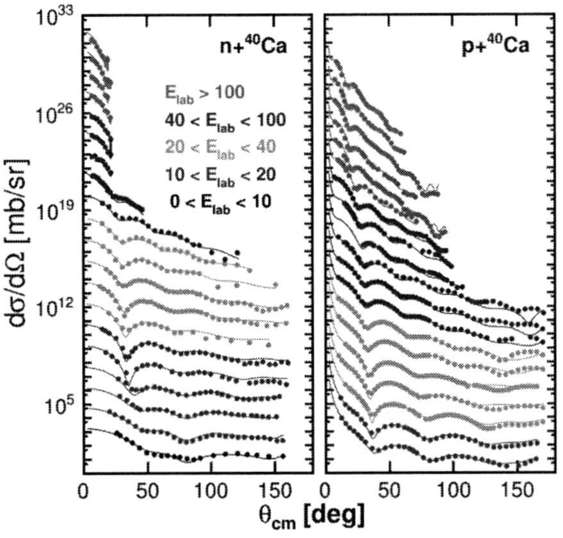

Figure 2. Calculated and experimental elastic-scattering angular distributions of the differential cross section for both$n$+^{40}Ca and p+^{40}Ca. Data for each energy are offset for clarity. References to the data can be found in Ref. [13].

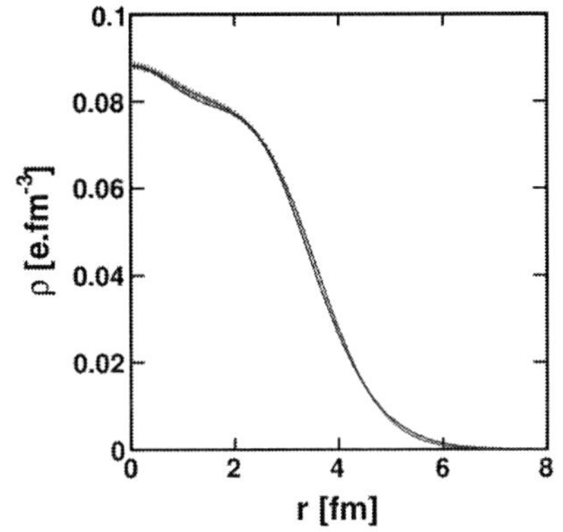

Figure 3. Comparison of experimental charge density of ^{40}Ca [22] (thick hashed line representing a 1% error) with the DOM fit (solid curve).

reasonably unique and may therefore provide insight into the depletion of the Fermi sea. The spectral representation of the particle part of the propagator, referring to the $A + 1$ system, appropriate for a treatment of the continuum and possible open channels is given by [6]

$$G^p_{\ell j}(k, k'; E) = \sum_n \frac{\phi^{n+}_{\ell j}(k) \left[\phi^{n+}_{\ell j}(k')\right]^*}{E - E^{*A+1}_n + i\eta} + \sum_c \int_{T_c}^{\infty} dE' \frac{\chi^{cE'}_{\ell j}(k) \left[\chi^{cE'}_{\ell j}(k')\right]^*}{E - E' + i\eta}. \tag{8}$$

Overlap functions for bound $A + 1$ states are given by

$$\phi^{n+}_{\ell j}(k) = \left\langle \Psi^A_0 \right| a_{k\ell j} \left| \Psi^{A+1}_n \right\rangle, \tag{9}$$

whereas those in the continuum are given by

$$\chi^{cE}_{\ell j}(k) = \left\langle \Psi^A_0 \right| a_{k\ell j} \left| \Psi^{A+1}_{cE} \right\rangle \tag{10}$$

indicating the relevant channel by c and the energy by E. Excitation energies in the $A + 1$ system are with respect to the A-body ground state $E^{*A+1}_n = E^{A+1}_n - E^A_0$. Each channel c has an appropriate threshold indicated by T_c which is the experimental threshold with respect to the ground-state energy of the A-body system. The overlap function for the elastic channel can be explicitly calculated by solving the Dyson equation while it is also possible to obtain the complete spectral density for $E > 0$

$$S^p_{\ell j}(k, k'; E) = \sum_c \chi^{cE}_{\ell j}(k) \left[\chi^{cE}_{\ell j}(k')\right]^*. \tag{11}$$

The overlap functions for bound particle states and their normalization (spectroscopic factor) can be calculated in the same fashion as the corresponding quantities for hole states [18]. The spectral density at continuum energies is obtained from crossing the branch cut on the real axis. In practice, this requires solving the scattering problem twice at each energy so that one may employ

$$S^p_{\ell j}(k, k'; E) = \frac{i}{2\pi} \left[G^p_{\ell j}(k, k'; E^+) - G^p_{\ell j}(k, k'; E^-)\right] \tag{12}$$

with $E^{\pm} = E \pm i\eta$, and only the elastic-channel contribution to Eq. (11) is explicitly known. Equivalent expressions pertain to the hole part of the propagator $G^h_{\ell j}$ [6].

The calculations are performed in momentum space according to Eq. (6) to generate the off-shell reducible self-energy and thus the spectral density by employing Eqs. (7) and (12). Because the momentum-space spectral density contains a delta-function associated with the free propagator, it is convenient for visualization purposes to consider a Fourier transform to coordinate space

$$S^p_{\ell j}(r, r'; E) = \frac{2}{\pi} \int dk k^2 \int dk' k'^2 j_\ell(kr) S^p_{\ell j}(k, k'; E) j_\ell(k'r'), \tag{13}$$

which has the physical interpretation for $r = r'$ as the probability density $S_{\ell j}(r; E)$ for adding a nucleon with energy E at a distance r from the origin for a given ℓj combination. By employing the asymptotic analysis to the propagator in coordinate space following *e.g.* Ref. [5], one may express the elastic-scattering wave function that contributes to Eq. (11) in terms of the half on-shell reducible self-energy obtained according to

$$\chi^{elE}_{\ell j}(r) = \left[\frac{2mk_0}{\pi\hbar^2}\right]^{1/2} \left\{ j_\ell(k_0 r) + \int dk k^2 j_\ell(kr) G^{(0)}(k; E) \Sigma_{\ell j}(k, k_0; E) \right\},$$

where k_0 is related to the scattering energy in the usual way. We subtract this contribution to the spectral function given by its absolute square from $S_{\ell j}(r; E)$ in Fig. 4 for different energies in the

DOI: 10.1051/epjconf/201612209003

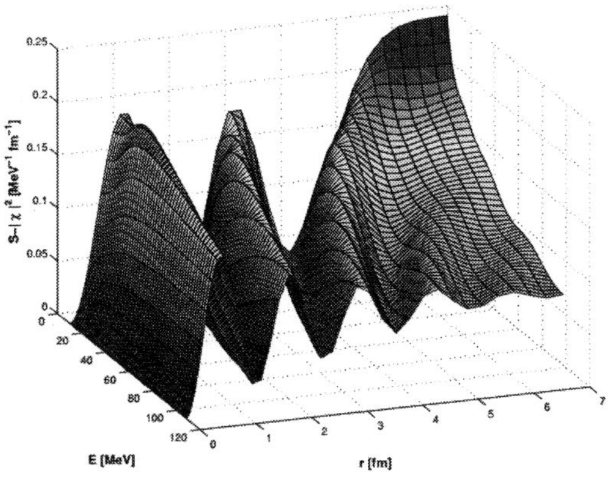

Figure 4. Difference between the particle spectral function for $s_{1/2}$ and the contribution of the elastic-scattering wave function multiplied by r^2, as a function of both energy and position. Asymptotically with r, this difference is constant and determined only by the inelasticity.

channel with $\ell = 0, j = 1/2$. Asymptotically at large distances, the influence of other open channels is represented by an almost constant shift whereas, inside the range of the potential, a pattern related to the absorptive properties of the potential and the orbits that are occupied emerges.

The presence of strength in the continuum associated with mostly-occupied orbits (or mostly empty but $E < 0$ orbits) is then obtained by double folding the spectral density in Eq. (13) in the following way

$$S^{n+}_{\ell j}(E) = \int drr^2 \int dr' r'^2 \phi^{n-}_{\ell j}(r) S^{p}_{\ell j}(r, r'; E) \phi^{n-}_{\ell j}(r'), \tag{14}$$

using an overlap function

$$\sqrt{S^{n}_{\ell j}} \phi^{n-}_{\ell j}(r) = \left\langle \Psi^{A-1}_{n} \middle| a_{r\ell j} \middle| \Psi^{A}_{0} \right\rangle, \tag{15}$$

corresponding to a bound orbit with $S^{n}_{\ell j}$ the relevant spectroscopic factor and $\phi^{n-}_{\ell j}(r)$ normalized to 1 [18]. In the case of an orbit below the Fermi energy, this strength identifies where the depleted strength resides in the continuum. The occupation number of this orbit is given by an integral over a corresponding folding of the hole spectral density

$$S^{n-}_{\ell j}(E) = \int drr^2 \int dr' r'^2 \phi^{n-}_{\ell j}(r) S^{h}_{\ell j}(r, r'; E) \phi^{n-}_{\ell j}(r'), \tag{16}$$

where $S^{h}_{\ell j}(r, r'; E)$ provides equivalent information below the Fermi energy as $S^{p}_{\ell j}(r, r'; E)$ above. An important sum rule is valid for the sum of the occupation number for the orbit $n_{n\ell j}$ and its depletion number $d_{n\ell j}$ [5]

$$1 = n_{n\ell j} + d_{n\ell j} = \int_{-\infty}^{\varepsilon_F} dE \; S^{n-}_{\ell j}(E) + \int_{\varepsilon_F}^{\infty} dE \; S^{n+}_{\ell j}(E), \tag{17}$$

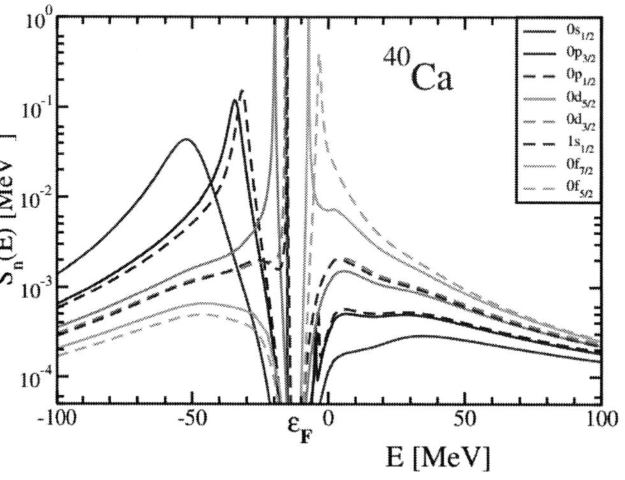

Figure 5. Calculated spectral strength, both below and above the Fermi energy, for bound orbits in ^{40}Ca. The spectral strength is constrained by elastic-scattering data, level structure, charge density, particle number, and the presence of high-momenta below the Fermi energy [15].

equivalent to $a^\dagger_{n\ell j}a_{n\ell j} + a_{n\ell j}a^\dagger_{n\ell j} = 1$. Strength above ε_F, as expressed by Eq. (14), reflects the presence of the imaginary self-energy at positive energies. Without it, the only contribution to the spectral function comes from the elastic channel. The folding in Eq. (14) then involves integrals of orthogonal wave functions and yields zero. Because it is essential to describe elastic scattering with an imaginary potential, it automatically ensures that the elastic channel does not exhaust the spectral density and therefore some spectral strength associated with IPM bound orbits also occurs in the continuum.

We display in Fig. 5 the results of the DOM spectral function for the most relevant bound orbits in ^{40}Ca including the hole spectral function of Ref. [15]. Because the DOM analysis assumes that the imaginary part of the self-energy starts at ε_F, the spectral strength is a continuous function of the energy. The method of solving the Dyson equation for $E < 0$ is very different than that for $E > 0$. The continuity of the curves at $E = 0$ confirms the numerical aspects of both of these calculations. Below the Fermi energy, the spectral strength contains peaks associated with the $0s_{1/2}, 0p_{3/2}, 0p_{1/2}, 0d_{5/2}, 1s_{1/2}$, and $0d_{3/2}$ orbits with narrower peaks for orbits closer to the Fermi energy. Their strength was calculated for the overlap functions associated with the location of the peaks by solving the Dyson equation without the imaginary part but with self-consistency for the energy of the real part [7]. The strength of these orbits above the Fermi energy exhibits systematic features displaying more strength when the IPM energy is closer to the continuum threshold [24]. We make this observation quantitative by listing the integrated strength according to the terms of Eq. (17) in Table 1. For the depletion we integrate from 0 to 200 MeV which corresponds to the energy domain constrained by data in the DOM. We also include the $0f_{7/2}$ and $0f_{5/2}$ spectral functions in Fig. 5 and corresponding results in Table 1 noting that the strength in the continuum from 0 to 200 MeV further rises to 0.202 and 0.320, respectively. From ε_F to 0 the strength for these states is also included in the sum and decreases from 0.722 to 0.591, respectively. This illustrates that there is a dramatic increase of strength into the continuum when the IPM energy approaches this threshold. Such orbits

Table 1. Occupation and depletion numbers for bound orbits in ^{40}Ca. The $d_{nlj}[0, 200]$ depletion numbers have only been integrated from 0 to 200 MeV. The fraction of the sum rule in Eq. (17) is given in the last column.

orbit	$n_{n\ell j}$	$d_{n\ell j}$ [0,200]	$n_{n\ell j} + d_{n\ell j}$ $[\varepsilon_F, 200]$
$0s_{1/2}$	0.926	0.032	0.958
$0p_{3/2}$	0.914	0.047	0.961
$1p_{1/2}$	0.906	0.051	0.957
$0d_{5/2}$	0.883	0.081	0.964
$1s_{1/2}$	0.871	0.091	0.962
$0d_{3/2}$	0.859	0.097	0.966
$0f_{7/2}$	0.046	0.202	0.970
$0f_{5/2}$	0.036	0.320	0.947

correspond to valence states in exotic nuclei [14, 25, 26]. The $1p_{3/2}$ and $1p_{1/2}$ spectral functions are not shown as they mimic the behavior of the $0f_{7/2}$ distribution but their presence causes the wiggles in the $0p_{3/2}$ and $0p_{1/2}$ spectral functions due to slight nonorthogonality. The sensitivity to the separation from the continuum is associated with the pronounced surface absorption necessary to describe the elastic-scattering data in this energy range. The few percent missing strength most likely resides above 200 MeV and is associated with the effect of SRC as discussed in Ref. [18].

4 Results for ^{48}Ca

Recent developments of the DOM have concentrated on the description of ^{48}Ca employing nonlocal potentials [19]. With adequate data for elastic proton scattering and the availability of an accurate charge density for this nucleus, it possible to answer the question how the properties of protons are changed when eight neutrons are added to ^{40}Ca. With recent new data for elastic neutron scattering [13] and the notion that neutrons in ^{40}Ca are as well described as protons due to isospin symmetry, it is also possible to assess the consequences for neutrons. In particular, the rms radius of the neutron distribution of ^{48}Ca may be receiving experimental scrutiny from a parity-violating elastic scattering experiment at Jefferson Lab [27] clarifying important properties of nuclei with neutron excess relevant for the physics of neutron stars. It may well be that available neutron data for ^{48}Ca already indirectly constrain information related to the neutron properties in the ground state. We note that the particle spectral function associated with elastic scattering shown in Fig. 4 for an $\ell = 0$ neutron in ^{40}Ca clearly exhibits evidence of awareness of ground-state properties as it exhibits a two-node structure at low energy as required by the presence of mostly occupied $0s_{1/2}$ and $1s_{1/2}$ orbits. Similar plots for d-waves exhibit a single node, whereas higher ℓ-values display no node [19].

In complete analogy with previous local DOM analyses an asymmetry-dependent potential proportional to $(N - Z)/A$ is included for the analysis of ^{48}Ca. Nonlocality is also implemented for this component while the $N = Z$ potential associated with ^{40}Ca is kept the same except for the radius parameters that are determined by ^{48}Ca data [19]. Whereas proton and neutron spectral functions for ^{40}Ca as illustrated in Fig. 5 are equivalent except for a shift in energy to take into account the Coulomb potential for protons, this is no longer the case for ^{48}Ca. It is therefore necessary to treat proton elastic scattering in momentum space in such a way that the relevant information can be obtained but the full treatment of the Coulomb potential is avoided. The appropriate method has been explored in Refs. [28–30] and we have adopted it here by screening the Coulomb potential in such a way that its contribution is completely included in the domain where the nuclear potential acts [19]. The results

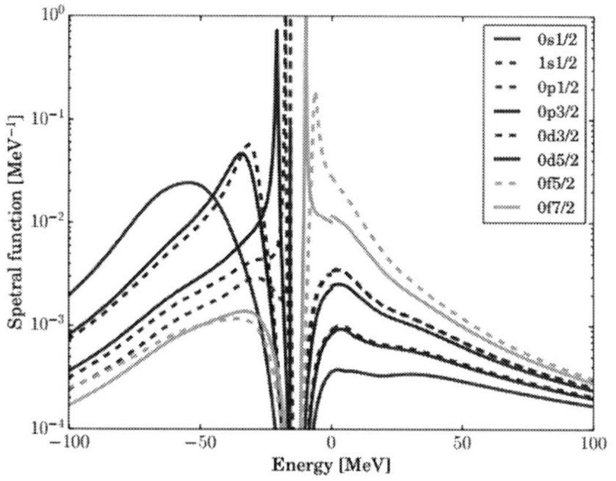

Figure 6. Calculated spectral strength, both below and above the Fermi energy, for proton bound orbits in ^{48}Ca. The spectral strength is constrained by elastic-scattering data, level structure, charge density, and particle number.

for the proton spectral functions for 48Ca are shown in Fig. 6 and exhibit a slight discontinuity at $E = 0$ where the treatment of the Coulomb potential at positive energy is particularly delicate. We note that the $0s_{1/2}$ spectral function in Fig. 6 has a wider peak than the corresponding result for ^{40}Ca as the width of this orbit is not properly constrained in ^{48}Ca. One would not expect a substantial difference between these nuclei however for this quantity. Similarly, there are no experimental high-momentum removal constraints for this nucleus which would be very helpful in clarifying the change from $N = 20$ to $N = 28$ for protons with high momenta.

Near the Fermi energy sufficient experimental data exist and it is possible to make a quantitative comparison of spectroscopic factors between these two nuclei as shown in Table 2. We first note that the spectroscopic factors of the $0d_{3/2}$ and $1s_{1/2}$ orbits in ^{40}Ca are larger by about 10-15% than the numbers extracted in the analysis of the $(e, e'p)$ reaction of Ref. [31]. The relativistic analysis of the $(e, e'p)$ reaction in Ref. [32] clarified that the treatment of nonlocality leads to different distorted proton waves as compared to conventional non-relativistic optical potentials, yielding about 10-15% larger spectroscopic factors. No matter where the absolute values are situated, it is clear from Table 2 that proton spectroscopic factors for these orbits are smaller in ^{48}Ca whereas neutron ones are larger. Since these results are strongly constrained by experimental elastic scattering data, it appears to provide some support for the notion that the spectroscopic factors of the more strongly bound va-

Table 2. Spectroscopic factors for orbits in the immediate vicinity of the Fermi surface in ^{40}Ca and ^{48}Ca. Proton and neutron spectroscopic factors in ^{40}Ca are identical.

orbit	^{40}Ca	$p\ ^{48}$Ca	$n\ ^{48}$Ca
$1s_{1/2}$	0.78	0.71	0.83
$0d_{3/2}$	0.76	0.65	0.80
$0f_{7/2}$	0.73	0.59	0.84

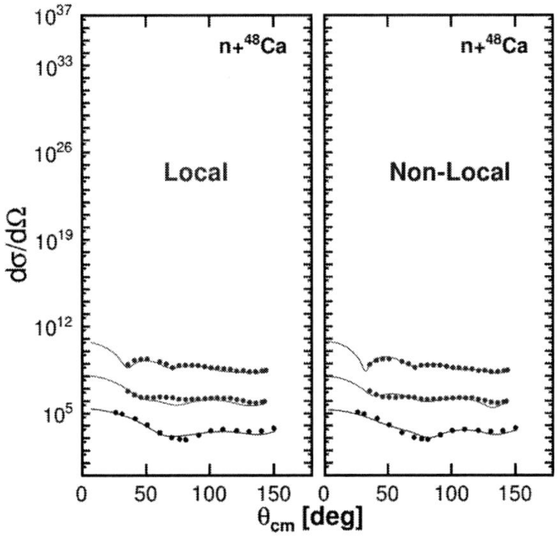

Figure 7. Calculated and experimental elastic-scattering angular distributions of the differential cross section for n+^{48}Ca in the local (left panel) [13] and nonlocal DOM implementation [19]. Data for each energy are offset for clarity. References to the data can be found in Ref. [13].

lence orbits of the minority nucleon decrease whereas the ones for the less bound majority nucleons increases with increasing nucleon asymmetry [25]. For the particle spectroscopic factor of the proton $0f_{7/2}$ orbit a similar observation can be made. We note that for neutrons this orbit switches from being mostly empty to mostly occupied between these two nuclei but also in this case the spectroscopic factor increases from ^{40}Ca to ^{48}Ca. For protons it is quite clear what the explanation of this decrease of the valence spectroscopic factors amounts to. Elastic scattering reaction cross section data clearly show more surface absorption for protons in ^{48}Ca than for ^{48}Ca [12]. The correspondingly larger imaginary part of the potential will therefore remove additional strength in ^{48}Ca from valence proton orbits as compared to ^{40}Ca leading to an unavoidable reduction of the spectroscopic factors.

In a previous local DOM analysis of elastic neutron scattering from ^{48}Ca [13] some difficulties were encountered in describing the three sets of available differential cross section. As shown in Fig. 7 the nonlocal fit is apparently capable of describing these data quite satisfactorily when the fits are compared side by side. The fit of the proton elastic scattering data is of the same quality as achieved with the local DOM [19]. Results for the fits to reaction cross sections for both protons and neutrons and total cross sections for neutrons are also well represented in the nonlocal DOM [19].

Finally, results for the charge density of ^{48}Ca are compared with the experimental results [22] together with results for the neutron distribution and the weak charge in Fig. 8. The quality of the charge density fit is similar as for ^{40}Ca shown in Fig. 3. As the main difference between the proton and neutron distribution is associated with $0f_{7/2}$ neutrons, it is reasonable to expect the interior densities to be comparable as shown in Fig. 8. This observation suggests that at higher energy where elastic scattering for protons is dominated by volume physics the protons interact with a similar com-

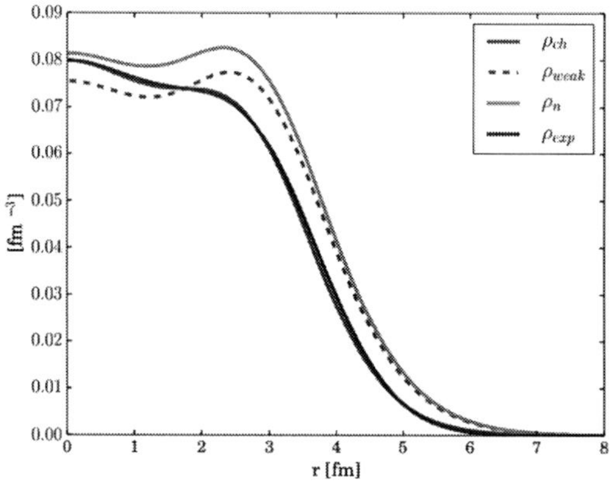

Figure 8. Comparison of calculated and experimental charge density for ^{48}Ca. The calculated weak charge distribution (dashed line as well as the neutron matter distribution are plotted as well [19].

position of nucleons in ^{40}Ca and ^{48}Ca. This notion is confirmed by comparing volume integrals of the imaginary part of the potentials as a function of energy for these nuclei [19].

The result for the neutron skin accordingly is quite large $R_n - R_p = 0.249 \pm 0.023$ fm considerably larger than a prediction of an *ab initio* coupled-cluster calculation [33] and larger than most mean-field results compiled in Ref. [34]. The error quoted for the nonlocal DOM result is obtained by scrambling the available elastic scattering data with the assigned experimental errors for neutron scattering on ^{48}Ca. A clear impetus therefore exists to extend our nonlocal DOM analysis to a description of ^{208}Pb where a measurement for the neutron skin has already been reported [35] but unfortunately with a large error bar.

5 Conclusions and outlook

The overview presented in this contribution clearly demonstrates that it is possible to fruitfully employ the link between nuclear reactions and structure as provided by the DOM. The biggest step forward in recent years has been the implementation of fully nonlocal and dispersive optical potentials for Ca isotopes allowing for an accurate fit of ground-state properties like the nuclear charge density. The analysis of scattering data employing the particle spectral density furthermore documents the depletion of orbits that are mostly occupied in the ground state. Apparently the present DOM implementation is also capable of *predicting* ground-state properties like the neutron skin with suggestive results that clamor for experimental probing. Future work on ^{208}Pb is clearly called for. In addition it is important to point out that the nonlocal analysis leads to distorted waves that are quite different from those obtained from local potentials potentially leading the new results when *e.g.* transfer reactions are analyzed with nonlocal potentials [36]. DOM ingredients based on local potentials already generate more consistent results for the analysis of transfer reactions [37].

Acknowledgement

The material presented here is based upon work supported by the U.S. Department of Energy, Office of Science, Office of Nuclear Physics under Award number DE-FG02-87ER-40316 and by the U.S. National Science Foundation under grant PHY-1304242. At this point I would like to acknowledge the work of Hossein Mahzoon whose thesis research has been instrumental in generating most of the results reported in this contribution. The work of other collaborators is also very much appreciated.

References

[1] R.L. Varner, W.J. Thompson, T.L. McAbee, E.J. Ludwig, T.B. Clegg, Phys. Rep. **201**, 57 (1991)
[2] A.J. Koning, J.P. Delaroche, Nucl. Phys. **A713**, 231 (2003)
[3] B.A. Brown, Prog. Part. Nucl. Phys. **47**, 517 (2001)
[4] E. Caurier, G. Martínez-Pinedo, F. Nowacki, A. Poves, A.P. Zuker, Rev. Mod. Phys. **77**, 427 (2005)
[5] W.H. Dickhoff, D. Van Neck, *Many-Body Theory Exposed!, 2nd edition* (World Scientific, New Jersey, 2008)
[6] C. Mahaux, R. Sartor, Adv. Nucl. Phys. **20**, 1 (1991)
[7] W.H. Dickhoff, J. Phys. G **37**, 064007 (2010)
[8] B.A. Brown, S.E. Massen, P.E. Hodgson, J. Phys. G: Nucl. Phys. **5**, 1655 (1979)
[9] J.S. Bell, E.J. Squires, Phys. Rev. Lett. **3**, 96 (1959)
[10] C. Mahaux, R. Sartor, Phys. Rev. Lett. **57**, 3015 (1986)
[11] R. Charity, L.G. Sobotka, W.H. Dickhoff, Phys. Rev. Lett. **97**, 162503 (2006)
[12] R.J. Charity, J.M. Mueller, L.G. Sobotka, W.H. Dickhoff, Phys. Rev. C **76**, 044314 (2007)
[13] J.M. Mueller, R.J. Charity, R. Shane, L.G. Sobotka, S.J. Waldecker, W.H. Dickhoff, A.S. Crowell, J.H. Esterline, B. Fallin, C.R. Howell et al., Phys. Rev. C **83**, 064605 (2011)
[14] R.J. Charity, W.H. Dickhoff, L.G. Sobotka, S.J. Waldecker, Eur. Phys. J. A **50**, 23 (2014)
[15] M.H. Mahzoon, R.J. Charity, W.H. Dickhoff, H. Dussan, S.J. Waldecker, Phys. Rev. Lett. **112**, 162503 (2014)
[16] W.H. Dickhoff, D. Van Neck, S.J. Waldecker, R.J. Charity, L.G. Sobotka, Phys. Rev. C **82**, 054306 (2010)
[17] S.J. Waldecker, C. Barbieri, W.H. Dickhoff, Phys. Rev. C **84**, 034616 (2011)
[18] H. Dussan, S.J. Waldecker, W.H. Dickhoff, H. Müther, A. Polls, Phys. Rev. C **84**, 044319 (2011)
[19] M.H. Mahzoon, *Ph.D. Thesis* (Washington University in St. Louis, 2015)
[20] D. Rohe et al., Phys. Rev. Lett. **93**, 182501 (2004)
[21] D. Rohe, *Habilitationsschrift* (University of Basel, 2004)
[22] H. de Vries, C.W. de Jager, C. de Vries, At. Data Nucl. Data Tables **36**, 495 (1987)
[23] H. Dussan, M.H. Mahzoon, R.J. Charity, W.H. Dickhoff, A. Polls, Phys. Rev. C **90**, 061603 (2014)
[24] N. Michel, W. Nazarewicz, M. Ploszajczak, Phys. Rev. C **75**, 031301(R) (2007)
[25] A. Gade et al., Phys. Rev. Lett. **93**, 042501 (2004)
[26] O. Jensen, G. Hagen, M. Hjorth-Jensen, B. Alex Brown, A. Gade, Phys. Rev. Lett. **107**, 032501 (2011)
[27] J. Mammei et al., *CREX: Parity-violating measurement of the weak charge distribution of* ^{48}Ca *to 0.02 fm accuracy*, http://hallaweb.jlab.org/parity/prex/ (2013)
[28] A. Deltuva, A.C. Fonseca, P.U. Sauer, Phys. Rev. Lett. **95**, 092301 (2005)

[29] A. Deltuva, A.C. Fonseca, P.U. Sauer, Phys. Rev. C **71**, 054005 (2005)
[30] A. Deltuva, A.C. Fonseca, P.U. Sauer, Annu. Rev. Nucl. Part. Sci. **58**, 27 (2008)
[31] G.J. Kramer et al., Phys. Lett. B **227**, 199 (1989)
[32] J.M. Udias, P. Sarriguren, E. Moya de Guerra, E. Garrido, J.A. Caballero, Phys. Rev. C **51**, 3246 (1995)
[33] G. Hagen, et al, Nature Physics (2015)
[34] F.J. Fattoyev, J. Piekarewicz, Phys. Rev. C **86**, 015802 (2012)
[35] A. Abrahamyan et al., Phys. Rev. Lett. **108**, 112502 (2012)
[36] A. Ross, L.J. Titus, F.M. Nunes, M.H. Mahzoon, W.H. Dickhoff, R.J. Charity, Phys. Rev. C **92**, 044607 (2015)
[37] N.B. Nguyen, S.J. Waldecker, F.M. Nuñes, R.J. Charity, W.H. Dickhoff, Phys. Rev. C **84**, 044611 (2011)

Nuclear effects in protonium formation low-energy three-body reaction: $\bar{p} + (p\mu)_{1s} \to (\bar{p}p)_\alpha + \mu^-$

Strong $\bar{p}$–p interaction in $\bar{p} + (p\mu)_{1s}$

Renat A. Sultanov[1,a] and Dennis Guster[b]

[1] *Department of Information Systems, BCRL & Integrated Science and Engineering Laboratory Facility (ISELF) at St. Cloud State University, St. Cloud, MN 56301-4498, USA*

Abstract. A three-charge-particle system ($\bar{p}$, μ^-, p^+) with an additional matter-antimatter, i.e. $\bar{p}$–p^+, nuclear interaction is the subject of this work. Specifically, we carry out a few-body computation of the following protonium formation reaction: $\bar{p} + (p^+\mu^-)_{1s} \to (\bar{p}p^+)_{1s} + \mu^-$, where p^+ is a proton, $\bar{p}$ is an antiproton, μ^- is a muon, and a bound state of p^+ and its counterpart $\bar{p}$ is a protonium atom: $Pn = (\bar{p}p^+)$. The low-energy cross sections and rates of the Pn formation reaction are computed in the framework of a Faddeev-like equation formalism. The strong $\bar{p}$–p^+ interaction is approximately included in this calculation.

1 Introduction

Obtaining and storing of low-energy antiprotons ($\bar{p}$) is of significant scientific and practical interest and importance in current research in atomic and nuclear physics [1–4]. For example, with the use of slow $\bar{p}$'s it would be possible to make low temperature ground state antihydrogen atoms $\bar{H}_{1s}$ - a bound state of $\bar{p}$ and e^+, i.e. a positron. The two-particle atom represents the simplest and stable anti-matter species. By comparing the properties of the hydrogen atom H and $\bar{H}$ it would be possible to test the fundamentals of physics, such as CPT theorem [5]. In this connection it is important to mention the metastable antiprotonic helium atoms too, i.e. atomcules, such as $\bar{p}^3He^+$ and $\bar{p}^4He^+$ [6]. In the field of the $\bar{p}$ physics these Coulomb three-body systems play a very important role. For example, with the use of high-precision laser spectroscopy atomclues allow us to measure $\bar{p}$'s charge-to-mass ratio and other fundamental constants in the standard model [7].

Together with the atomcules and the $\bar{H}$ atoms there is now a significant interest in protonium (Pn) atom too: a bound state of $\bar{p}$ and p^+ [8–10]. The two-particle system is also named as antiprotonic hydrogen. In the atomic scale it is a heavy and a very small system with strong Coulomb and nuclear interactions. An interplay between these interactions occurs inside the atom. The last circumstance is responsible for interesting resonance and quasi-bound states in Pn [11]. Therefore, Pn represents a very useful tool to study, for example, the anti-nucleon–nucleon ($\bar{N}N$) interaction potential [12, 13] and annihilation processes [14, 15]. Additionally, as we already mentioned, the interplay between Coulomb and nuclear forces plays a significant role in the $\bar{p}$ and p quantum dynamics [16]. The $\bar{p}+p^+$

[a] e-mail: rasultanov@stcloudstate.edu
[b] e-mail: dcguster@stcloudstate.edu

elastic scattering has also been considered in many papers, see for example review [14]. Additionally, *Pn* formation is related to charmonium - a hydrogen-like atom ($\bar{c}c$), i.e. a bound state of a c-antiquark ($\bar{c}$) and c-quark [17]. Because of the fundamental importance of protonium and problems related to its formation, as far as bound and quasi-bound states, resonances and spectroscopy, the two-particle atom have attracted much attention last decades.

There are a number of few-body collisions to make *Pn* atoms at low temperatures. For example, the following three-charge-particle reaction is one of them:

$$\bar{\mathrm{p}} + (\mathrm{p}^+e^-)_{1s} \rightarrow (\bar{\mathrm{p}}\mathrm{p}^+)_{\alpha'} + e^-, \tag{1}$$

First of all, this collision represents a Coulomb three-body system and has been considered in many theoretical works in which different methods and computational techniques have been applied [18–20]. We also would like to point out, that because in the process (1) a heavy particle, i.e. proton, is transferred from one negative "center", e^-, to another, $\bar{\mathrm{p}}$, it would be difficult to apply a computational method based on an adiabatic (Born-Oppenheimer) approach [21]. Additionally, it would be useful to mention here, that experimentalists use another few-body reaction to produce *Pn* atoms, i.e. a collision between a slow $\bar{\mathrm{p}}$ and a positively charged molecular hydrogen ion, i.e. H_2^+:

$$\bar{\mathrm{p}} + \mathrm{H}_2^+ \rightarrow (\bar{\mathrm{p}}\mathrm{p}^+)_{\alpha'} + \mathrm{H}. \tag{2}$$

In the current work, however, we consider another three-body system of the *Pn* formation reaction. Specifically, we compute the cross-sections and rates of a very low energy collision between $\bar{\mathrm{p}}$ and a muonic hydrogen atom H_μ, i.e. a bound state of p^+ and a negative muon:

$$\bar{\mathrm{p}} + (\mathrm{p}\mu)_{1s} \rightarrow (\bar{\mathrm{p}}\mathrm{p}^+)_{\alpha} + \mu^-. \tag{3}$$

Here, α=1s, 2s or 2p is *Pn*'s final quantum atomic state. Because of the μ^- participation in the reaction (3), at very low energy collisions *Pn* can be formed in an extremely small size (compact in the atomic scale) ground and close to ground states α. In these states the hadronic nuclear forces between $\bar{\mathrm{p}}$ and p are much stronger than in the reaction (1) and, probably, would be very effective in order to study them. The size of the *Pn* atom in its ground state is only $a_0(Pn) = \hbar^2/(e_0^2 m_p/2) \sim 50$ fm! At such small distances the Coulomb interaction between $\bar{\mathrm{p}}$ and p^+ becomes extremely strong. The corresponding binding energy in the *Pn* atom without the inclusion of the nuclear $\bar{\mathrm{p}}$–p^+ interaction is $E_n(Pn) = -e_0^4 m_p/2/(2\hbar n^2) \sim -10$ keV. Here, we took $n = 1$, $\hbar$ is the Planck constant, e_0 is the electron charge, and m_p is the proton mass. In connection with this we would like to make a comment. The real $\bar{\mathrm{p}}$–p^+ binding energy, i.e. with inclusion of the strong interaction, can have a large value. This value may be comparable or even larger than m_p. Therefore, it might be necessary to use a relativistic treatment to the reaction (3) in the output channel [22]. The situation with very strong Coulomb interaction inside *Pn* can also be a reason of vacuum polarization forces. The Casimir forces can contribute (it might be quite significant) to the final cross sections and rates of the *Pn* formation reaction (3). It would be interesting to take into account all these physical effects in the framework of the reaction (3) and compute their influence on its rate. Hopefully, soon experimentalists will be able to carry out high quality measurements of the reaction (3). Thereafter one could compare the new results with corresponding theoretical data and fit (adjust) the $\bar{\mathrm{p}}$–p^+ strong interaction in the framework of the theoretical calculations in order to reproduce the laboratory data. This would allow us to better know and understand the strong $\bar{\mathrm{p}}$–p^+ interaction and the annihilation processes.

In the first order approximation reaction (3) can be considered as a three-charged-particle system (123) with arbitrary and comparable masses: m_1, m_2, and m_3. It is shown in Fig. 1. The strong $\bar{\mathrm{p}}$–p^+ interaction can be included approximately after a solution of the Coulomb three-body problem. A few-body method based on a Faddeev-like equation formalism is applied. In this approach

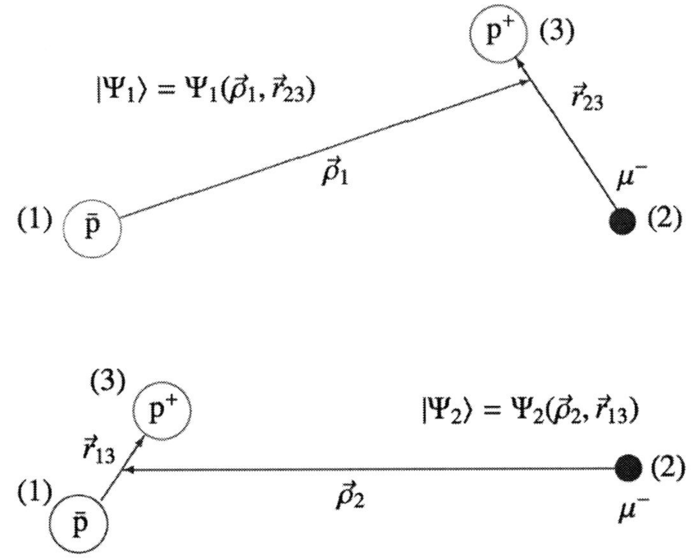

Figure 1. Two asymptotic spacial configurations of the 3-body system (123), or more specifically ($\bar{\mathrm{p}}, \mu^-, \mathrm{p}^+$) which is considered in this work. The few-body Jacobi coordinates $(\vec{\rho}_i, \vec{r}_{jk})$, where $i \neq j \neq k = 1, 2, 3$ are also shown together with the 3-body wave function components Ψ_1 and Ψ_2: $\Psi = \Psi_1 + \Psi_2$ is the total wave function of the 3-body system.

the three-body wave function is decomposed in two independent Faddeev-type components [23, 24]. Each component is determined by its own independent Jacobi coordinates. Since, the reaction (3) is considered at low energies, i.e. well below the three-body break-up threshold, the Faddeev-type components are quadratically integrable over the internal target variables $\vec{r}_{23}$ and $\vec{r}_{13}$. They are shown in Fig. 1.

The next section represents the notation pertinent to the three-charged-particle system (123), the basic few-body equations, boundary conditions, detailed derivation of the set of coupled one-dimensional integral-differential equations suitable for numerical calculations and the numerical computational approach used in the current paper. The muonic units, i.e. $e = \hbar = m_\mu = 1$, are used in this work, where $m_\mu = 206.769\ m_e$ is the mass of the muon. The proton (anti-proton) mass is $m_p = m_{\bar{p}}$=1836.152 m_e.

2 A three-charge-particle system: $\bar{\mathrm{p}}$, μ^- and p^+

As we already mentioned in Introduction, a quantum-mechanical few-body approach is applied in this work. A coordinate space representation is used. This approach is based on a reduction of the total three-body wave function Ψ on two or three Faddeev-type components [24]. When one has two negative and one positive charges, only two asymptotic configurations are possible below the total energy break-up threshold. The situation is explained in Fig. 1 in the case of the title three-body system.

Therefore, one can decompose Ψ only on two components and write down a set of two coupled equations [25, 26]. A modified close coupling method is applied in order to solve these equations [27–29]. This means to carry out an expansion of the Faddeev-type components into eigenfunctions of the subsystem Hamiltonians. This technique provides an infinite set of one-dimensional integral-differential equations [30, 31]. Within this formalism the asymptotic of the full three-body wave function contains two parts corresponding to two open channels [32].

2.1 An infinite set of coupled integral-differential few-body equations

One could use the following system of units: $e = \hbar = m_3 = 1$. We denote antiproton $\bar{\mathrm{p}}$ by 1, a negative muon μ^- by 2, and a proton p^+ by 3. Before the three-body breakup threshold two cluster asymptotic configurations are possible in the three-body system, i.e. (23)−1 and (13)−2. As we mentioned above, by their own Jacobi coordinates $\{\vec{r}_{j3}, \vec{\rho}_k\}$ as shown in Fig. 1:

$$\vec{r}_{j3} = \vec{r}_3 - \vec{r}_j, \quad \vec{\rho}_k = \frac{(\vec{r}_3 + m_j\vec{r}_j)}{(1+m_j)} - \vec{r}_k, \quad (j \neq k = 1,2). \tag{4}$$

Here $\vec{r}_\xi$, m_ξ are the coordinates and the masses of the particles $\xi = 1, 2, 3$ respectively. This suggests a Faddeev formulation which uses only two components. A general procedure to derive such formulations is described in work [25]. In this approach the three-body wave function is represented as follows:

$$|\Psi\rangle = \Psi_1(\vec{r}_{23}, \vec{\rho}_1) + \Psi_2(\vec{r}_{13}, \vec{\rho}_2), \tag{5}$$

where each Faddeev-type component is determined by its own Jacobi coordinates. Moreover, $\Psi_1(\vec{r}_{23}, \vec{\rho}_1)$ is quadratically integrable over the variable $\vec{r}_{23}$, and $\Psi_2(\vec{r}_{13}, \vec{\rho}_2)$ over the variable $\vec{r}_{13}$. To define $|\Psi_l\rangle$, $(l = 1, 2)$ a set of two coupled Faddeev-Hahn-type equations can be written:

$$\left(E - \hat{H}_0 - V_{23}(\vec{r}_{23})\right)\Psi_1(\vec{r}_{23}, \vec{\rho}_1) = \left(V_{23}(\vec{r}_{23}) + V_{12}(\vec{r}_{12})\right)\Psi_2(\vec{r}_{13}, \vec{\rho}_2), \tag{6}$$

$$\left(E - \hat{H}_0 - V_{13}(\vec{r}_{13})\right)\Psi_2(\vec{r}_{13}, \vec{\rho}_2) = \left(V_{13}(\vec{r}_{13}) + V_{12}(\vec{r}_{12})\right)\Psi_1(\vec{r}_{23}, \vec{\rho}_1). \tag{7}$$

Here, $\hat{H}_0$ is the kinetic energy operator of the three-particle system, $V_{ij}(r_{ij})$ are paired interaction potentials $(i \neq j = 1, 2, 3)$, E is the total energy.

Now, let us present the equations (6)-(7) in terms of the adopted notation

$$\left(E + \frac{1}{2M_k}\Delta_{\vec{\rho}_k} + \frac{1}{2\mu_j}\Delta_{\vec{r}_{j3}} - V_{j3}\right)\Psi_i(\vec{r}_{j3}, \vec{\rho}_k) = \left(V_{j3} + V_{jk}\right)\Psi_{i'}(\vec{r}_{k3}, \vec{\rho}_j), \tag{8}$$

here $i \neq i' = 1, 2$, $M_k^{-1} = m_k^{-1} + (1+m_j)^{-1}$ and $\mu_j^{-1} = 1 + m_j^{-1}$. In order to separate angular variables, the wave function components Ψ_i are expanded over bipolar harmonics:

$$\{Y_\lambda(\hat{\rho}) \otimes Y_l(\hat{r})\}_{LM} = \sum_{\mu m} C^{LM}_{\lambda\mu lm} Y_{\lambda\mu}(\hat{\rho}) Y_{lm}(\hat{r}), \tag{9}$$

where $\hat{\rho}$ and $\hat{r}$ are angular coordinates of vectors $\vec{\rho}$ and $\vec{r}$; $C^{LM}_{\lambda\mu lm}$ are Clebsh-Gordon coefficients; Y_{lm} are spherical functions [33]. The configuration triangle of the particles (123) is presented in Fig. 2 together with the Jacobi coordinates $\{\vec{r}_{23}, \vec{\rho}_1\}$ and $\{\vec{r}_{13}, \vec{\rho}_2\}$ and angles between them. The centre-of-mass of the whole three-body system is designated as O. The centre-of-masses of the two-body subsystems (23) and (13) are O_1 and O_2 respectively. Substituting the following expansion:

$$\Psi_i(\vec{r}_{j3}, \vec{\rho}_k) = \sum_{LM\lambda l} \Phi^i_{LM\lambda l}(\rho_k, r_{j3}) \left\{Y_\lambda(\hat{\rho}_k) \otimes Y_l(\hat{r}_{j3})\right\}_{LM} \tag{10}$$

into (8), multiplying this by the appropriate biharmonic functions and integrating over the corresponding angular coordinates of the vectors $\vec{r}_{j3}$ and $\vec{\rho}_k$, we obtain a set of equations which for the case of the central potentials has the form:

$$\Big(E + \frac{1}{2M_k\rho_k^2}\Big\{\frac{\partial}{\partial\rho_k}(\rho_k^2\frac{\partial}{\partial\rho_k}) - \lambda(\lambda+1)\Big\} + \frac{1}{2\mu_j r_{j3}^2}\Big\{\frac{\partial}{\partial r_{j3}}(r_{j3}^2\frac{\partial}{\partial r_{j3}}) - l(l+1)\Big\}$$
$$-V_{j3}\Big)\Phi^i_{LM\lambda l}(\rho_k, r_{j3}) = \int d\hat{\rho}_k \int d\hat{r}_{j3} \sum_{\lambda' l'} W^{(ii')LM}_{\lambda l\lambda' l'} \Phi^{i'}_{LM\lambda' l'}(\rho_j, r_{k3}), \tag{11}$$

where the following notation has been introduced:

$$W^{(ii')LM}_{\lambda l\lambda' l'} = \big\{Y_\lambda(\hat{\rho}_k)\otimes Y_l(\hat{r}_{j3})\big\}^*_{LM}\big(V_{j3}+V_{jk}\big)\big\{Y_{\lambda'}(\hat{\rho}_j)\otimes Y_{l'}(\hat{r}_{k3})\big\}_{LM}. \tag{12}$$

To progress from (11) to one-dimensional equations, we apply a modified close coupling method, which consists of expanding each component of the wave function $\Psi_i(\vec{r}_{j3}, \vec{\rho}_k)$ over the Hamiltonian eigenfunctions of subsystems:

$$\hat{h}_{j3} = -\frac{1}{2\mu_j}\nabla^2_{\vec{r}_{j3}} + V_{j3}(\vec{r}_{j3}). \tag{13}$$

Thus, following expansions can be applied:

$$\Phi^i_{LM\lambda l}(\rho_k, r_{j3}) = \frac{1}{\rho_k}\sum_n f^{(i)LM}_{nl\lambda}(\rho_k) R^{(i)}_{nl}(r_{j3}), \tag{14}$$

where functions $R^i_{nl}(r_{j3})$ are defined by the following equation:

$$\Big(E^i_n + \frac{1}{2\mu_j r_{j3}^2}\Big\{\frac{\partial}{\partial r_{j3}}(r_{j3}^2\frac{\partial}{\partial r_{j3}}) - l(l+1)\Big\} - V_{j3}\Big)R^i_{nl}(r_{j3}) = 0. \tag{15}$$

Substituting Eq. (14) into (11), multiplying by the corresponding functions $R^i_{nl}(r_{j3})$ and integrating over $r_{j3}^2 dr_{j3}$ yields a set of integral-differential equations for the unknown functions $f^i_{nl\lambda}(\rho_k)$:

$$2M_k(E - E^i_n)f^i_\alpha(\rho_k) + \Big(\frac{\partial^2}{\partial\rho_k^2} - \frac{\lambda(\lambda+1)}{\rho_k^2}\Big)f^i_\alpha(\rho_k) = 2M_k\sum_\alpha\int_0^\infty dr_{j3}r_{j3}^2\int d\hat{r}_{j3}\int d\hat{\rho}_k\rho_k\rho_j \times \frac{Q^{ii'}_{\alpha\alpha'} f^{i'}_{\alpha'}(\rho_j)}{\rho_j}, \tag{16}$$

where

$$Q^{ii'}_{\alpha\alpha'} = R^i_{nl}(r_{j3}) W^{(ii')LM}_{\lambda l\lambda' l'} R^{i'}_{n'l'}(r_{k3}). \tag{17}$$

For brevity one can denote $\alpha \equiv nl\lambda$ ($\alpha' \equiv n'l'\lambda'$), and omit LM because all functions have to be the same. The functions $f^i_\alpha(\rho_k)$ depend on the scalar argument, but this set is still not one-dimensional, as formulas in different frames of the Jacobi coordinates:

$$\vec{\rho}_j = \vec{r}_{j3} - \beta_k\vec{r}_{k3},\ \ \vec{r}_{j3} = \frac{1}{\gamma}(\beta_k\vec{\rho}_k + \vec{\rho}_j),\ \ \vec{r}_{jk} = \frac{1}{\gamma}(\sigma_j\vec{\rho}_j - \sigma_k\vec{\rho}_k), \tag{18}$$

with the following mass coefficients:

$$\beta_k = m_k/(1+m_k),\ \ \sigma_k = 1 - \beta_k,\ \ \gamma = 1 - \beta_k\beta_j\ \ (j \neq k = 1, 2), \tag{19}$$

clearly demonstrate that the modulus of $\vec{\rho}_j$ depends on two vectors, over which integration on the right-hand sides is accomplished: $\vec{\rho}_j = \gamma\vec{r}_{j3} - \beta_k\vec{\rho}_k$. Therefore, to obtain one-dimensional integral-differential equations, corresponding to equations (16), we will proceed with the integration over variables $\{\vec{\rho}_j, \hat{\rho}_k\}$, rather than $\{\vec{r}_{j3}, \hat{\rho}_k\}$. The Jacobian of this transformation is γ^{-3}. Thus, we arrive at a set of one-dimensional integral-differential equations:

$$2M_k(E - E_n^i)f_\alpha^i(\rho_k) + \Big(\frac{\partial^2}{\partial\rho_k^2} - \frac{\lambda(\lambda+1)}{\rho_k^2}\Big)f_\alpha^i(\rho_k) = \frac{M_k}{\gamma^{-3}}\sum_{\alpha'}\int_0^\infty d\rho_j S_{\alpha\alpha'}^{ii'}(\rho_j,\rho_k)f_{\alpha'}^{i'}(\rho_j), \quad (20)$$

where functions $S_{\alpha\alpha'}^{ii'}(\rho_j,\rho_k)$ are defined as follows:

$$S_{\alpha\alpha'}^{ii'}(\rho_j,\rho_k) = 2\rho_j\rho_k\int d\hat{\rho}_j\int d\hat{\rho}_k R_{nl}^i(r_{j3})\big\{Y_\lambda(\hat{\rho}_k)\otimes Y_l(\hat{r}_{j3})\big\}_{LM}^*\big(V_{j3}+V_{jk}\big)$$
$$\times\big\{Y_{\lambda'}(\hat{\rho}_j)\otimes Y_{l'}(\hat{r}_{k3})\big\}_{LM}R_{n'l'}^{i'}(r_{k3})\,. \quad (21)$$

In the next section we show that fourfold multiple integration in equations (21) leads to a one-dimensional integral and the expression (21) could be determined for any orbital momentum value L:

$$S_{\alpha\alpha'}^{ii'}(\rho_j,\rho_k) = \frac{4\pi}{2L+1}[(2\lambda+1)(2\lambda'+1)]^{\frac{1}{2}}\rho_j\rho_k\int_0^\pi d\omega\sin\omega R_{nl}^i(r_{j3})\big(V_{j3}(r_{j3})$$
$$+V_{jk}(r_{jk})\big)R_{n'l'}^{i'}(r_{k3})\sum_{mm'}D_{mm'}^L(0,\omega,0)C_{\lambda0lm}^{Lm}C_{\lambda'0l'm'}^{Lm'}Y_{lm}(\nu_j,\pi)Y_{l'm'}^*(\nu_k,\pi)\,, \quad (22)$$

where $D_{mm'}^L(0,\omega,0)$ are Wigner functions, ω is the angle between $\vec{\rho}_j$ and $\vec{\rho}_k$, ν_j is the angle between $\vec{r}_{k3}$ and $\vec{\rho}_j$, ν_k is the angle between $\vec{r}_{j3}$ and $\vec{\rho}_k$ (please see Fig. 2). Finally, we obtain an infinite set of coupled integral-differential equations for the unknown functions $f_\alpha^1(\rho_1)$ and $f_{\alpha'}^2(\rho_2)$ [31], i.e. $f_{\alpha(\alpha')}^{i(i')}(\rho_{i(i')})$ $(i \neq i' = 1,2)$, α and α' belong to two different sets of three-body quantum numbers:

$$\Big((k_n^i)^2 + \frac{\partial^2}{\partial\rho_i^2} - \frac{\lambda(\lambda+1)}{\rho_i^2}\Big)f_\alpha^i(\rho_i) = g\sum_{\alpha'}\sqrt{\frac{(2\lambda+1)(2\lambda'+1)}{(2L+1)}}\int_0^\infty d\rho_{i'}f_{\alpha'}^{i'}(\rho_{i'})\int_0^\pi d\omega\sin\omega$$
$$\times R_{nl}^i(r_{i'3})\big(V_{i'3}(r_{i'3}) + V_{ii'}(r_{ii'})\big)R_{n'l'}^{i'}(r_{i3})\rho_{i'}\rho_i\sum_{mm'}D_{mm'}^L(0,\omega,0)C_{\lambda0lm}^{Lm}C_{\lambda'0l'm'}^{Lm'}Y_{lm}(\nu_i,\pi)Y_{l'm'}^*(\nu_{i'},\pi). \quad (23)$$

The total angular momentum of the three-body system is L. Next in Eq. (23):

$$g = \frac{4\pi M_i}{\gamma^3}, \quad k_n^i = \sqrt{2M_i(E - E_n^{i'})}, \quad (24)$$

where $E_n^{i'}$ is the binding energy of the subsystem ($i'3$). Also:

$$M_1 = \frac{m_1(m_2+m_3)}{(m_1+m_2+m_3)} \text{ and } M_2 = \frac{m_2(m_1+m_3)}{(m_1+m_2+m_3)} \quad (25)$$

are the reduced masses. Further: $D_{mm'}^L(0,\omega,0)$ the Wigner functions, $C_{\lambda0lm}^{Lm}$ the Clebsh-Gordon coefficients, Y_{lm} are the spherical functions, ω is the angle between the Jacobi coordinates $\vec{\rho}_i$ and $\vec{\rho}_{i'}$, ν_i is

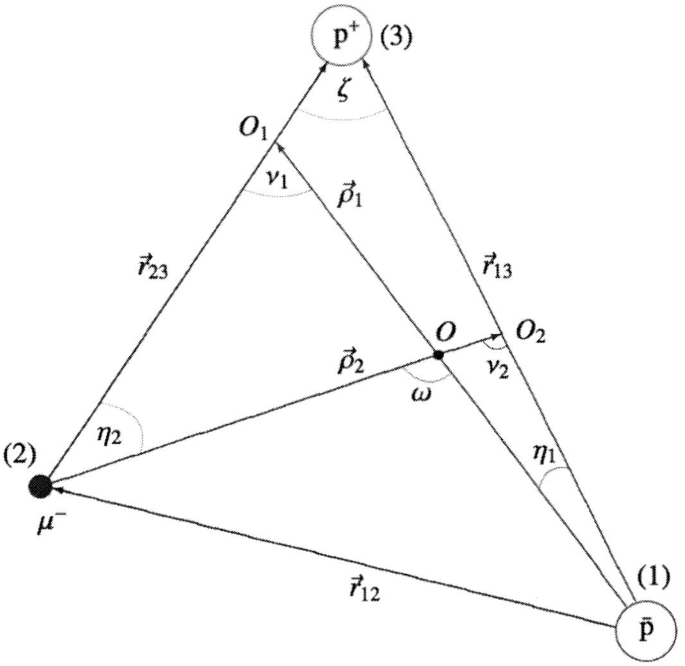

Figure 2. The title three-charge-particle system $\bar{p}, \mu^-$ and p (or p^+ - proton) and system's configurational triangle (123) are shown together with the few-body Jacobi coordinates (vectors): $\{\vec{\rho}_1, \vec{r}_{23}\}$ and $\{\vec{\rho}_2, \vec{r}_{13}\}$. Additionally, $\vec{r}_{12}$ is the vector between two negative particles in the system. The needed for detailed few-body treatment geometrical angles between the vectors such as $\eta_{1(2)}, \nu_{1(2)}, \zeta$ and ω are also presented in this figure.

the angle between $\vec{r}_{i'3}$ and $\vec{\rho}_i$, $\nu_{i'}$ is the angle between $\vec{r}_{i3}$ and $\vec{\rho}_{i'}$. One can show that:

$$\begin{aligned} \sin \nu_i &= \frac{\rho_k}{r_{kj}\gamma} \sin \omega, \\ \cos \nu_i &= \frac{\beta\rho_i + \rho_k \cos \omega}{\gamma r_{kj}}, \\ \gamma &= 1 - \frac{m_i m_{i'}}{(m_i + 1)(m_{i'} + 1)}. \end{aligned} \tag{26}$$

2.1.1 Angular integrals

The details of the derivation of the angular integrals $S^{ii'}_{\alpha\alpha'}(\rho_j, \rho_k)$ (22) are explained below in this section. The configuration triangle, $\Delta(123)$, is determined by the Jacobi vectors $(\vec{r}_{j3}, \vec{\rho}_k)$ and should be considered in an arbitrary coordinate system $OXYZ$. In this initial system the angle variables of the three-body Jacobi vectors $\{\vec{r}_{j3}, \vec{\rho}_k\}$ have the following values: $\hat{r}_{j3} = (\theta_j, \phi_j)$, $\hat{\rho}_k = (\Theta_k, \Phi_k)$, $j \neq k =$

1, 2. Let us adopt a new coordinate system $O'X'Y'Z'$ in which the axis $O'Z'$ is directed over the vector $\vec{\rho}_k$, $\triangle(123)$ belongs to the plain $O'X'Z'$ and the vertex $k = 1$ of $\triangle(123)$ coincides with the origin O' of the new $OX'Y'Z'$. The new angle variables of the Jacobi vectors in the $O'X'Y'Z'$ system have now the following values: $\hat{r}'_{j3} = (\nu_k, \pi)$, $\hat{\rho}'_k = (0,0)$, $\hat{r}'_{k3} = (\eta_k, \pi)$, $\hat{\rho}'_j = (\omega, \pi)$, here $k = 1$ and $j = 2$. The spatial rotational transformation from $OXYZ$ to $O'X'Y'Z'$ has been done with the use of the following Euler angles $(\Phi_k, \Theta_k, \varepsilon)$ [33]. Taking into account the transformation rule for the bipolar harmonics between new and old coordinate systems, one can write down the following relationships [33]:

$$\left\{Y_\lambda(\hat{\rho}_k) \otimes Y_l(\hat{r}_{j3})\right\}^*_{LM} = \sum_m (D^L_{Mm}(\Phi_k, \Theta_k, \varepsilon))^* \left\{Y_\lambda(\hat{\rho}'_k) \otimes Y_l(\hat{r}'_{j3})\right\}^*_{Lm} \tag{27}$$

and

$$\left\{Y_{\lambda'}(\hat{\rho}_j) \otimes Y_{l'}(\hat{r}_{k3})\right\}_{LM} = \sum_{m'} D^L_{Mm'}(\Phi_k, \Theta_k, \varepsilon) \left\{Y_{\lambda'}(\hat{\rho}'_j) \otimes Y_{l'}(\hat{r}'_{k3})\right\}_{Lm'}, \tag{28}$$

where $D^L_{Mm}(\Phi_k, \Theta_k, \varepsilon)$ are the Wigner functions [33]. The fourfold multiple angular integration $\int d\hat{\rho}_j \int d\hat{\rho}_k$ in Eq. (21) can be written in the new variables and be symbolically represented as $\int d\hat{\rho}_j \int d\hat{\rho}_k = \int_0^\pi d\omega \sin\omega \int_0^{2\pi} d\varepsilon \int_0^{2\pi} d\Phi_k \int_0^\pi \sin\Theta_k d\Theta_k$. Next, taking into account the normalizing condition for the Wigner functions [33]:

$$\int_0^{2\pi} d\varepsilon \int_0^{2\pi} d\Phi_k \int_0^\pi \sin\Theta_k d\Theta_k (D^L_{Mm}(\Phi_k, \Theta_k, \varepsilon))^* D^L_{Mm'}(\Phi_k, \Theta_k, \varepsilon)) = \frac{8\pi^2}{2L+1}\delta_{mm'} \tag{29}$$

one can obtain the following formula:

$$S^{ii'}_{\alpha\alpha'}(\rho_j, \rho_k) = 2\rho_j\rho_k \sum_m \frac{8\pi^2}{2L+1} \int_0^\pi d\omega \sin\omega R^i_{nl}(r_{j3}) \left\{Y_\lambda(0,0) \otimes Y_l(\hat{r}'_{j3})\right\}^*_{Lm}$$
$$\times (V_{j3} + V_{jk}) \left\{Y_{\lambda'}(\hat{\rho}'_j) \otimes Y_{l'}(\hat{r}'_{k3})\right\}_{Lm'} R^{i'}_{n'l'}(r_{k3}). \tag{30}$$

Now, let us make the next transformation of $\triangle(123)$ in which the vertex $j = 2$ of $\triangle(123)$ coincides with the centre O' of the $O'X'Y'Z'$ and $O'XYZ$, however the axis $O'Z''$ is directed along $\vec{\rho}_j$ and $\triangle(123)$ belongs to the plain $O'X''Z''$. This transformation, which converts the coordinate frame $O'X'Y'Z'$ into $O'X''Y''Z''$ is characterized by the following Euler angles $(0, \omega, 0)$. Therefore the vectors $(\vec{r}_{k3}, \vec{\rho}_j)$ have the following new variables: $\hat{r}''_{k3} = (\nu_j, \pi)$, $\hat{\rho}''_j = (0,0)$. As a result of this rotation one can write down the following relationship:

$$\left\{Y_{\lambda'}(\hat{\rho}'_j) \otimes Y_{l'}(\hat{r}'_{k3})\right\}_{Lm} = \sum_{m'} D^L_{Mm'}(0, \omega, 0) \left\{Y_{\lambda'}(\hat{\rho}''_j) \otimes Y_{l'}(\hat{r}''_{k3})\right\}_{Lm'} \tag{31}$$

and obtain the following result:

$$S^{ii'}_{\alpha\alpha'}(\rho_j, \rho_k) = 2\rho_j\rho_k \sum_{mm'} \frac{8\pi^2}{2L+1} \int d\omega \sin\omega R^i_{nl}(r_{j3}) \left\{Y_\lambda(0,0) \otimes Y_l(\hat{r}'_{j3})\right\}^*_{Lm} (V_{j3} + V_{jk})$$
$$\times D^L_{mm'}(0, \omega, 0) \left\{Y_{\lambda'}(0,0) \otimes Y_{l'}(\hat{r}''_{k3})\right\}_{Lm'} R^{i'}_{n'l'}(r_{k3}). \tag{32}$$

Now by taking into account that $Y_{lm}(0,0) = \delta_{m,0}\sqrt{(2l+1)/4\pi}$ [33], the bipolar harmonics in (32) are:

$$\{Y_\lambda(0,0) \otimes Y_l(\nu_k, \pi)\}^*_{Lm} = \sqrt{\frac{2\lambda+1}{4\pi}} C^{Lm}_{\lambda 0 lm} Y^*_{lm}(\nu_k, \pi), \tag{33}$$

$$\left\{Y_{\lambda'}(0,0) \otimes Y_{l'}(\nu_j, \pi)\right\}_{Lm'} = \sqrt{\frac{2\lambda'+1}{4\pi}} C^{Lm'}_{\lambda' 0 l' m'} Y_{l'm'}(\nu_j, \pi), \tag{34}$$

with the use of these relationships we finally get the convenient for numerical computations Eq. (22). In conclusion, we would like to note that rotational transformations of a coordinate system $OXYZ$ might also be useful in the theory of molecular collisions. In addition, few useful formulas for the triangle (123) are presented below: $\sin\nu_1 = \rho_2/(\gamma r_{23})\sin\omega$, $\sin\nu_2 = \rho_1/(\gamma r_{13})\sin\omega$ and $\cos\nu_1 = 1/(\gamma r_{23})(\beta\rho_1 + \rho_2\cos\omega))$, $\cos\nu_2 = 1/(\gamma r_{13})(\alpha\rho_2 + \rho_1\cos\omega))$.

2.2 Boundary conditions, numerics, cross sections and the reaction rates

To find a unique solution to Eqs. (23) appropriate boundary conditions depending on the specific physical situation need to be considered. First we impose:

$$f_{nl}^{(i)}(0) \sim 0. \tag{35}$$

Next, for the three-body charge-transfer problems we apply the well known **K**−matrix formalism. This method has already been applied for solution of three-body problems in the framework of the Schrődinger equation [34, 35] and coordinate space Faddeev equation [36]. For the present rearrangement scattering problem with $i+(j3)$ as the initial state, in the asymptotic region, it takes two solutions to Eq.(23) to satisfy the following boundary conditions:

$$\begin{cases} f_{1s}^{(i)}(\rho_i) \underset{\rho_i\to+\infty}{\sim} \sin(k_1^{(i)}\rho_i) + K_{ii}\cos(k_1^{(i)}\rho_i) \\ f_{1s}^{(j)}(\rho_j) \underset{\rho_j\to+\infty}{\sim} \sqrt{v_i/v_j}K_{ij}\cos(k_1^{(j)}\rho_j)\,, \end{cases} \tag{36}$$

where K_{ij} are the appropriate coefficients, and v_i $(i = 1, 2)$ is a velocity in channel i. With the following change of variables in Eq. (23):

$$\mathrm{f}_{1s}^{(i)}(\rho_i) = f_{1s}^{(i)}(\rho_i) - \sin(k_1^{(i)}\rho_i), \tag{37}$$

(i=1, 2) we get two sets of inhomogeneous equations which are solved numerically. The coefficients K_{ij} can be obtained from a numerical solution of the FH-type equations. The cross sections are given by the following expression:

$$\sigma_{ij} = \frac{4\pi}{k_1^{(i)2}}\left|\frac{\mathbf{K}}{1-i\mathbf{K}}\right|^2 = \frac{4\pi}{k_1^{(i)2}}\frac{\delta_{ij}D^2 + K_{ij}^2}{(D-1)^2 + (K_{11}+K_{22})^2}, \tag{38}$$

where $(i, j = 1, 2)$ refer to the two channels and $D = K_{11}K_{22} - K_{12}K_{21}$. Also, from the quantum-mechanical unitarity principle one can derive that the scattering matrix $\mathbf{K} = \begin{pmatrix} K_{11} & K_{12} \\ K_{21} & K_{22}\end{pmatrix}$ has the following important feature:

$$K_{12} = K_{21}. \tag{39}$$

In this work, the relationship (39) is checked for all considered collision energies in the framework of the 1s, 1s+2s and 1s+2s+2p modified close coupling approximation Eqs. (10) and (14).

The solution of the Eqs. (6)-(7) involving both components $\Psi_{1(2)}$ required that we apply the expansions (10) and (14) over the angle and the distance variables respectively. However, to obtain a numerical solution for the set of coupled Eqs. (23) we only include the -s and -p waves in the expansion (10) and limit n up to 2 in the Eq. (14). As a result we arrive at a truncated set of six coupled integral-differential equations, since in $\Psi_{1(2)}$ only 1s, 2s and 2p target two-body atomic wave-functions are included. This method represents a modified version of the close coupling approximation with six expansion functions. The set of truncated integral-differential Eqs. (23) is solved by a discretization procedure, i.e. on the right side of the equations the integrals over ρ_1 and ρ_2 are replaced by sums using the trapezoidal rule [37] and the second order partial derivatives on the left side are discretized using a three-point rule [37]. By this means we obtain a set of linear equations for the unknown coefficients $f_\alpha^{(i)}(k)$ $(k = 1, N_p)$:

$$\left[k_n^{(1)2} + D_{ij}^2 - \frac{\lambda(\lambda+1)}{\rho_{1i}^2}\right] f_\alpha^{(1)}(i) - \frac{M_1}{\gamma^3}\sum_{\alpha'=1}^{N_s}\sum_{j=1}^{N_p} w_j S_{\alpha\alpha'}^{(12)}(\rho_{1i},\rho_{2j}) f_{\alpha'}^{(2)}(j) = 0, \tag{40}$$

$$-\frac{M_2}{\gamma^3}\sum_{\alpha=1}^{N_s}\sum_{j=1}^{N_p} w_j S_{\alpha'\alpha}^{(21)}(\rho_{2i},\rho_{1j}) f_\alpha^{(1)}(j) + \left[k_{n'}^{(2)2} + D_{ij}^2 - \frac{\lambda'(\lambda'+1)}{\rho_{2i}^2}\right] f_{\alpha'}^{(2)}(i) = B_{\alpha'}^{21}(i). \tag{41}$$

Here, coefficients w_j are weights of the integration points ρ_{1i} and ρ_{2i} $(i = 1, N_p)$, N_s is the number of quantum states which are taken into account in the expansion (14). Next, D_{ij}^2 is the three-point numerical approximation for the second order differential operator: $D_{ij}^2 f_\alpha(i) = (f_\alpha(i-1)\delta_{i-1,j} - 2f_\alpha(i)\delta_{i,j} + f_\alpha(i+1)\delta_{i+1,j})/\Delta$, where Δ is a step of the grid $\Delta = \rho_{i+1} - \rho_i$. The vector $B_{\alpha'}^{21}(i)$ is: $B_{\alpha'}^{(21)}(i) = M_2/\gamma^3 \sum_{j=1}^{N_p} w_j S_{\alpha' 1s0}^{(21)}(i,j)\sin(k_1\rho_j)$, and in symbolic-operator notations the set of linear Eqs. (40)-(41) has the following form:

$$\sum_{\alpha'=1}^{2\times N_s}\sum_{j=1}^{N_p} \mathbf{A}_{\alpha\alpha'}(i,j)\vec{f}_{\alpha'}(j) = \vec{b}_\alpha(i). \tag{42}$$

The discretized equations are subsequently solved by the Gauss elimination method [38]. As can be seen from Eqs. (40)-(41) the matrix **A** should have a so-called block-structure: there are four main blocks in the matrix: two of them related to the differential operators and other two to the integral operators. Each of these blocks should have sub-blocks depending on the quantum numbers $\alpha = nl\lambda$ and $\alpha' = n'l'\lambda'$. The second order differential operators produce three-diagonal sub-matrixes [31].

However, there is no need to keep the whole matrix **A** in computer's operating (fast) memory. The following optimization procedure shows that it would be possible to reduce the memory usage by at least four times. Indeed, the numerical equations (40)-(41) can be written in the following way: $D_1 f^1 - M_1\gamma^{-3}S^{12}f^2 = 0$, and $-M_2\gamma^{-3}S^{21}f^1 + D_2 f^2 = b$. Here, D_1, D_2, S^{12} and S^{21} are sub-matrixes of **A**. Now one can determine that: $f^1 = (D_1)^{-1}M_1/\gamma^3 S^{12}f^2$, where $(D_1)^{-1}$ is reverse matrix of D_1. Thereby one can obtain a reduced set of linear equations which are used to perform the calculations: $\left[D_2 - M_1M_2\gamma^{-6}S^{21}(D_1)^{-1}S^{12}\right]f^2 = b$ [31].

To solve the coupled integral-differential equations (23) one needs to first compute the angular integrals Eqs. (22). They are independent of energy E. Therefore, one needs to compute them only once and then store them on a computer's hard drive (or solid state drive) to support future computation of other observables, i.e. the charge-transfer cross-sections at different collision energies.

The sub-integral expressions in (22) have a very strong and complicated dependence on the Jacobi coordinates ρ_i and $\rho_{i'}$. To calculate $S_{\alpha\alpha'}^{(ii')}(\rho_i,\rho_{i'})$ at different values of ρ_i and $\rho_{i'}$ an adaptable algorithm has been applied together with the following mathematical substitution: $\cos\omega = (x^2 - \beta_i^2\rho_i^2 - \rho_{i'}^2)/(2\beta_i\rho_i\rho_{i'})$. The angle dependent part of the equation can be written as the following one-dimensional integral:

$$S_{\alpha\alpha'}^{(ii')}(\rho_i,\rho_{i'}) = \frac{4\pi}{\beta_i}\frac{[(2\lambda+1)(2\lambda'+1)]^{\frac{1}{2}}}{2L+1}\int_{|\beta_i\rho_i-\rho_{i'}|}^{\beta_i\rho_i+\rho_{i'}} dx R_{nl}^{(i)}(x)\left[-1+\frac{x}{r_{ii'}(x)}\right]R_{n'l'}^{(i')}(r_{i3}(x))$$
$$\times\sum_{mm'} D_{mm'}^L(0,\omega(x),0)C_{\lambda 0lm}^{Lm}C_{\lambda'0l'm'}^{Lm'}Y_{lm}(\nu_i(x),\pi)Y_{l'm'}^*(\nu_{i'}(x),\pi). \tag{43}$$

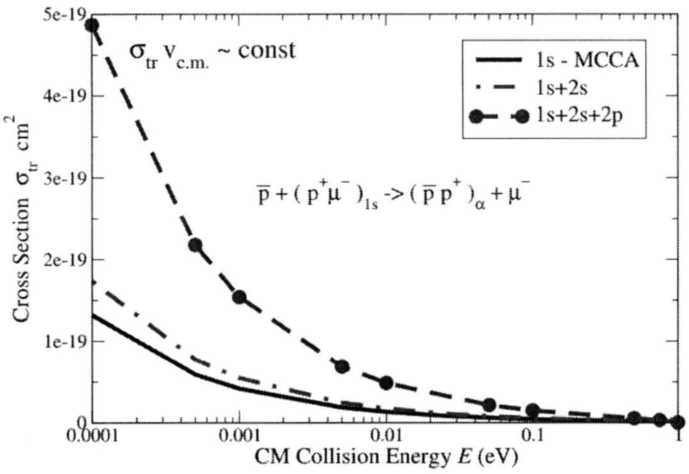

Figure 3. This figure shows our numerical result for the low-energy proton transfer reaction integral cross section σ_{tr} in the three-charge-particle collision $\bar{p} + H_\mu \to (\bar{p}p)_\alpha + \mu^-$, where H_μ is a muonic hydrogen atom: a bound state of a proton and a negative muon. In this work only the reaction final channel with α=1s in considered in the framework of the 1s, 1s+2s and 1s+2s+2p modified close-coupling approximation (MCCA) approach.

An adaptive algorithm which is incorporated in a FORTRAN subroutine from [39] is used in this work in order to carry out the angle integration in (43). This recursive computer program, QUADREC, is a better, modified version of the well known program QUANC8 [38]. QUADREC provides a much higher quality, stable and more precise integration than does QUANC8 [39]. The expression (43) differs from zero only in a narrow strip, i.e. when $\rho_i \approx \rho_{i'}$. This is because in the considered three-body system the coefficient β_i is approximately equal to one. Therefore, in order to obtain numerically reliable converged results it is necessary to adequately distribute a very large number of discretization points (up to 6000) between 0 and ~80 muonic units.

3 Results

In this section we report our computational results. The *Pn* formation three-body reaction is computed. A Faddeev-like equation formalism (6)-(7) has been applied. The few-body approach is presented in previous sections. In order to solve the coupled equations (6)-(7) two different independent sets of target expansion functions have been used (14). This method allows us to avoid the over-completeness problem and the two targets are treated equivalently. The main goal of this work is to carry out a reliable quantum-mechanical computation of the cross sections and corresponding rates of the title *Pn* formation reaction at low and very low collision energies. It would also be interesting to include and estimate a contribution of the strong $\bar{p}$–p interaction. The coupled integral-differential Eqs. (23) have been solved numerically for the case of the total angular momentum $L = 0$ in the framework of the two-level 2×(1s), four-level 2×(1s+2s), and six-level 2×(1s+2s+2p) close coupling approximations in Eq. (14). The sign "2×" indicates that two different sets of expansion functions are applied. The following boundary conditions (35), (36), and (37) have been used. To compute the charge transfer cross

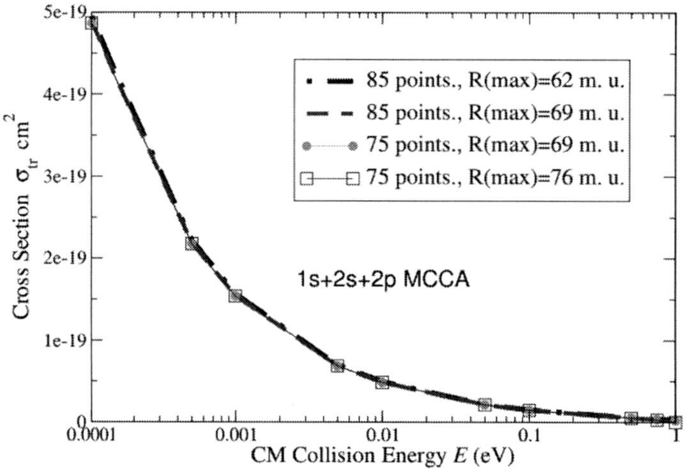

Figure 4. Numerical convergence results for the low-energy proton transfer reaction integral cross section σ_{tr} in $\bar{\text{p}} + \text{H}_\mu \rightarrow (\bar{\text{p}}\text{p})_\alpha + \mu^-$, where H_μ is a muonic hydrogen atom and $\alpha = 1s$. In these calculations only $1s + 2s + 2p$ modified close-coupling approximation (MCCA) approach is used.

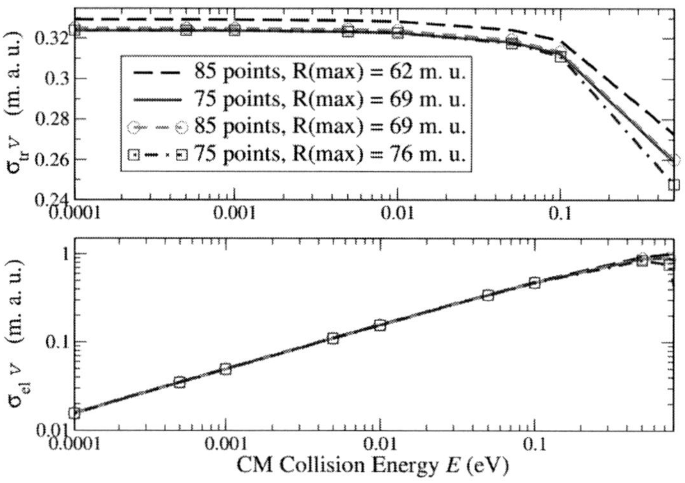

Figure 5. Numerical convergence results for the low-energy proton transfer reaction integral cross section σ_{tr} and elastic scattering cross section σ_{el} multiplied by the collision velocity in $\bar{\text{p}} + \text{H}_\mu \rightarrow (\bar{\text{p}}\text{p})_\alpha + \mu^-$, where H_μ is a muonic hydrogen atom and $\alpha = 1s$. In these calculations only $1s+2s+2p$ modified close-coupling approximation (MCCA) approach is used.

Table 1. The total *Pn* formation cross section σ_{tr} in the three-body reaction (3), when $\alpha = 1$, and a product of this cross section and the corresponding center-of-mass velocities $v_{c.m.}$ between $\bar{p}$ and H_μ. The results are prsented together with the corresponding unitarity ratio between the K_{12} and K_{21} coefficients of the scattering K-matrix.

	2×(1*s*) - MCCA Approach		2×(1*s* + 2*s*) - MCC Approach		
E, eV	σ_{tr}, cm^2	$\sigma_{tr} \times v_{c.m.}$, m.a.u	σ_{tr}, cm^2	$\sigma_{tr} \times v_{c.m.}$, m.a.u	$\chi(E)$, Eq. (39)
0.0001	0.1319E-18	0.8776E-01	0.1733E-18	0.1153	0.9976
0.0005	0.5897E-19	0.8776E-01	0.7749E-19	0.1153	0.9976
0.0010	0.4170E-19	0.8776E-01	0.5479E-19	0.1153	0.9976
0.0050	0.1865E-19	0.8776E-01	0.2450E-19	0.1153	0.9976
0.0100	0.1319E-19	0.8775E-01	0.1733E-19	0.1153	0.9976
0.0500	0.5895E-20	0.8773E-01	0.7748E-20	0.1153	0.9977
0.1000	0.4168E-20	0.8771E-01	0.5478E-20	0.1153	0.9977
0.5000	0.1860E-20	0.8753E-01	0.2450E-20	0.1153	0.9982

sections the expression (38) has been applied. The constructed equations satisfy the Schrődinger equation exactly. For the energies below the three-body break-up threshold they exhibit the same advantages as the Faddeev equations [24], because they are formulated for the wave function components with correct physical asymptotes. To solve the equations, a close-coupling method is applied, which leads to an expansion of the system's wave function components into eigenfunctions of the subsystem (target) Hamiltonians providing with a set of one-dimensional integral-differential equations after the partial-wave projection. A further advantage of the Faddeev-type method is the fact that the Faddeev-components are smoother functions of the coordinates than the total wave function.

Below we report our new computational results. We compare some of our findings with the corresponding data from older work [40]. The *Pn* formation cross section in the reaction (3) are shown in Fig. 3. Here we used only $1s$, $1s + 2s$ and $1s + 2s + 2p$ states within the modified close-coupling approximation (MCCA) approach. One can see that the contribution of the $2s$- and $2p$-states from each target is becoming even more significant while the collision energy becomes smaller. It would be useful to make a comment about the behaviour of $\sigma_{tr}(\varepsilon_{coll})$ at very low collision energies: $\varepsilon_{coll} \sim 0$. From our calculation we found that the p^+ transfer cross sections $\sigma_{tr} \rightarrow \infty$ as $\varepsilon_{coll} \rightarrow 0$. However, the p^+ transfer rates, λ_{tr}, are proportional to the product $\sigma_{tr} \times v_{c.m.}$ and this trends to a finite value as $v_{c.m.} \rightarrow 0$. Here $v_{c.m.} = \sqrt{2\varepsilon_{coll}/M_k}$ is a relative center-of-mass velocity between the particles in the input channel of the three-body reactions, and M_k is the reduced mass. To compute the proton transfer rate the following formula is used:

$$\lambda_{tr} = \sigma_{tr}(\varepsilon_{coll} \rightarrow 0) v_{c.m.} \qquad (44)$$

Therefore, additionally, for the process (1) we compute the numerical value of the following important quantity: $\Lambda(Pn) = \lambda_{tr} = \sigma_{tr}(\varepsilon_{coll} \rightarrow 0) v_{c.m.} \approx$ const, which is proportional to the actual *Pn* formation rate at low collision energies. Table 1 includes our data for this important parameter together with the *Pn* formation total cross section. All these results are obtained in the framework of the 2×(1*s*) and 2×(1*s* + 2*s*) approximations. The sign "2 × " means, as we mentioned, that two sets of the expansion functions from each target are applied. Therefore, for example, in the case of the $1s$ MCCA approach two expansion functions are used in our calculations. However, in the case of the $1s + 2s$ MCCA approach four expansion functions are applied. Additionally, as we mentioned above, the unitarity relationship, i.e. Eq. (39), is checked for all considered collision energies E. It is seen, that $\chi(E)$ posses a fairly constant value close to one. Figs. 4 and 5 show our convergence results. In these cases we used different number of integration points, namely 75 and 85 per one muonic radius length and different values of the up level of integration, namely 62, 69 and 76 m. a. u. Thus, the maximum amount of the integration number used in this work is $N_{max} = 76 \times 85 = 6460$. It is seen that the results are in a good agreement with each other for the transfer and elastic cross sections.

In the framework of the $2\times(1s + 2s + 2p)$ MCCA approach, i.e. when six coupled Faddeev-Hahn-type integral-differential equations are solved, our result for the Pn formation rate has the following value: $\Lambda_{1s2s2p}(Pn) = \sigma_{tr} \times v_{c.m.} \approx 0.32$ m.a.u. The corresponding rate from the work [40] is: $\Lambda'(Pn) \approx 0.2$ m.a.u. Both results are in a fairly good agreement with each other and were multiplied by *5* as in [40]. In conclusion, because of the complexity of the few-body system and the method, in this work only the total orbital momentum $L = 0$ has been taken into account. It was adequate in the case of the slow and ultraslow collisions discussed above. However, inclusion of the strong proton-antiproton interaction in this work by just shifting the ground state energy level of the Pn atom [13] increased our results by $\sim 50\%$. In a future work it would be useful to take into account higher atomic target states like 3s+3p+3d+4s+... as well as the continuum spectrum. This calculation would be an interesting investigation. However, at very low energy collisions, which are considered in this paper, all these channels are closed and located far away from the actual collision energies. At the same time the main contribution from s- and p-states (polarization) is taken into account. A direct inclusion of the strong $\bar{p}$–p interaction [14, 15] would be a very interesting future investigation in the framework of the three-body reaction (3).

References

[1] M. Hori and J. Waltz, Prog. Part. Nucl. Phys. **72**, 206 (2013).
[2] W. A. Bertsche, E. Butler, M. Charlton, and N. Madsen, J. Phys. B: At. Mol. Opt. Phys. **48**, 232001 (2015).
[3] G. Gabrielse et al., (ATRAP Collaboration), Phys. Rev. Lett. **106** 073002 (2011).
[4] G. B. Andresen et al., (ALPHA Collaboration), Phys. Rev. Lett. **105** 013003 (2010).
[5] Y. Yamazaki and S. Ulmer, Ann. Phys. (Berlin), **525**, 493 (2013).
[6] R. S. Hayano, M. Hori, D. Horvath, and E. Widmann, Rep. Prog. Phys. **70**, 1995 (2007).
[7] M. Hori et al., Nature **475**, 484 (2011).
[8] N. Zurlo et al., (ATHENA Collaboration), Phys. Rev. Lett. **97**, 153401 (2006).
[9] L. Venturelli et al., Nucl. Instrum. Methods Phys. Res., Sect. B **261**, 40 (2007).
[10] E. L. Rizzini et al., Europ. Phys. J. Plus, **127**, 1 (2012).
[11] I. S. Shapiro, Phys. Rep. **35**, 129 (1978).
[12] J. Hrtánková and J. Mareš, Nucl. Phys. A **945**, 197 (2015).
[13] J. -M. Richard and M. E. Sainio, Phys. Lett. B **110**, 349 (1982).
[14] E. Klempt, F. Bradamante, A. Martin, J.-M. Richard, Phys. Rep. **368**, 119 (2002).
[15] E. Klempt, C. Batty, J.-M. Richard, Phys. Rep. **413**, 197 (2005).
[16] L. N. Bogdanova, O. D. Dalkarov, and I. S. Shapiro, Ann. Phys. **84**, 261 (1974).
[17] T. Barnes, AIP Conf. Proc. **796**, 53 (2005).
[18] X. M. Tong, K. Hino, N. Toshima, Phys. Rev. Lett. **97**, 243202 (2006).

[19] K. Sakimoto, Phys. Rev. A **88** 012507 (2013).
[20] B. D. Esry and H. R. Sadeghpour, Phys. Rev. A **67**, 012704 (2003).
[21] M. Born and R. Oppenheimer, Ann. Phys., Liepzig, **84**, 457 (1927).
[22] T. Ueda, Prog. Theor. Phys. **62**, 1670 (1979.)
[23] L. D. Faddeev, Zh. Eksp. Teor. Fiz. **39** 1459 (1960) [Sov. Phys. JETP **12** 1014 (1961)].
[24] L. D. Faddeev and S. P. Merkuriev, *Quantum Scattering Theory for Several Particle Systems*, (Kluwers Academic Publishers, Dordrecht, 1993); L. D. Faddeev, *Mathematical Aspects of the Three-Body Problem in the Quantum Scattering Theory*, (Israel, Program for Scientific Translation, Jerusalem, 1965).
[25] Y. Hahn and K. Watson, Phys. Rev. A **5** 1718 (1972).
[26] Y. Hahn, Nucl. Phys. A **389**, 1 (1982).
[27] R. A. Sultanov and S. K. Adhikari, Phys. Rev. A **61**, 227111 (2000).
[28] R. A. Sultanov and D. Guster, J. Phys. B: At. Mol. Opt. Phys. **46** 215204 (2013).
[29] R. A. Sultanov and D. Guster, Hyperfine Interact. **228**, 47 (2014).
[30] R. A. Sultanov, D. Guster, and S. K. Adhikari, Few-Body Syst. **56**, 793 (2015).
[31] R.A. Sultanov and D. Guster, J. Comp. Phys. **192**, 231 (2003).
[32] S. P. Merkuriev, Ann. Phys. **130**, 395 (1980).
[33] D. A. Varshalovich, A. N. Moskalev, and V. L. Khersonskii, *Quantum Theory of Angular Momentum*, (World Scientific, Singapore, 1988).
[34] A. Adamczak, C. Chiccoli, V. I. Korobov, V. S. Melezhik, P. Pasini, L. I. Ponomarev, and J. Wozniak, Phys. Lett. B **285**, 319 (1992).
[35] J.S. Cohen and M.C. Struensee, Physical Review A **43**, 3460 (1991).
[36] A. A. Kvitsinsky, J. Carbonell, C. Gignoux, Physical Review A **51**, 2997 (1995).
[37] M. Abramowitz and I. A. Stegun, *Handbook of Mathematical Functions: with Formulas, Graphs, and Mathematical Tables*, (Dover Publications, New York, 1965).
[38] G. E. Forsythe, M. A. Malcolm, and C. B. Moler, *Computer Methods in Mathematical Computations*, (Prentice-Hall, Inc., Englewood Cliffs, New Jersey 1977).
[39] A. N. Berlizov and A. A. Zhmudsky, arXiv:physics/9905035v2.
[40] A. Igarashi and N. Toshima, Eur. Phys. J. D **46**, 425 (2008).

EPJ Web of Conferences **122**, 09005 (2016) DOI: 10.1051/epjconf/201612209005
*CNR**15*

Level and width statistics of the open many-body systems

Shoujirou Mizutori[1,a] and Hirokazu Aiba[2,b]

[1] *Kansai University of Welfare Sciences*
[2] *Kyoto Koka Women's College*

Abstract. The level and width statistics of the two kinds of the random matrix models coupled to the continuum are analyzed. In the first model, the gaussian orthogonal ensemble with random couplings to the continuum, not only the width statistics deviate from the Porter-Thomas distribution due to the super-radiant mechanism, but also the distribution of the nearest neighbor level spacings shows deviation from the Wigner one simultaneously. In the second model, the two body random ensemble with correlated couplings to the continuum, the correlation between the target and the compound states leads to the global energy dependence of the widths. Within the narrow energy interval where states with widths deviating from the global energy dependence lie, the distributions behave similar way with the case of the random couplings. Namely, the deviation of statistics of the nearest neighbor level spacings from the Wigner distribution and the deviation of the width statistics from the Porter-Thomas distribution take place simultaneously within the models we investigated.

1 Introduction

The description of the isolated neutron resonances(INR) has close connection with the random matrix theory and the phenomenon of the quantum chaos.

To begin with, N. Bohr's compound nuclear picture[1], which tried to explain the narrow widths of the neutron resonances, implies the chaotic motion of the nucleons in the nuclei. In the 50's, Wigner's random matrix theory(RMT) was proposed as the theory of the INR[2]. The nearest neighbor level spacings(NNLS) following the Wigner surmise, and the widths following the Porter-Thomas(PT) distribution, are well explained by the RMT. In the 80's, the idea of the nuclear data ensemble(NDE) was proposed[3], in which the combination of the data of different nuclei is treated as an ensemble, and regarded as a strong evidence for the RMT. At the same time, it has been established that the RMT is a generic model of the quantum chaos. Hence, there has been no question that the INR are well described by the RMT and are typical examples of the quantum chaos.

However in 2010, it was claimed that the widths of the INR of Th isotopes do not follow the PT distribution with 99.997% significance[4]. Further claim was raised that the precise re-examination of the NDE indicates the exclusion of the RMT with 98.17% significance[5]. Although the significance level claimed there might be overestimated[6], these new analysis gave a great doubt on the RMT.

[a] e-mail: mizutori@tamateyama.ac.jp
[b] e-mail: h_aiba@mail.koka.ac.jp

This issue is not restricted to whether a model is correct or not. As the RMT has close relation with the quantum chaos, the problem is concerned with the whole picture of the compound nuclear states. It is natural that the nuclear theory is nudged by the claim[7].

There are several arguments that the distribution of the width may not directly connected with the chaoticity. The expectation that the widths follow the Porter-Thomas distribution arises from the assumption that the randomness of the matrix elements of the intrinsic states reflects directly to the width distribution. There are however several possibilities giving rise to the deviation. The penetration factor may have a modified energy dependence [8], the neutron width may be different from the simple square of the coupling to the continuum due the super-radiant mechanism[9], and/or the randomness of the matrix elements of the intrinsic states may not manifest in the couplings to the continuum because of the correlation between the compound states and the target plus one neutron states[10].

Super-radiance is a phenomenon which occurs in the limit of the strong coupling to the continuum. In such a situation, only a few, the same number with the open channels, states monopolize the coupling to the continuum, and the remaining states have only small widths. If the strength is modestly large, the collectivization is weak and some states have relatively large widths than the others, which may cause the deviation from the PT distribution. Although the super-radiant mechanism together with the correlation between the compound and the target nuclei may explain the deviation of the width statistics from the PT distribution, the effect of the same mechanism on the level statistics has not been investigated so far. Since the level interaction occurs in the complex plane, there must be some effect on the level statistics as is seen in a past result[11]. In this talk, adopting the same model with ref.[9] and ref.[10], we investigate the relation of the width statistics and the level statistics.

2 The model

The energies and the widths of the neutron resonances are given[12] as the eigenvalues

$$\mathcal{E}_f = \varepsilon_f - i\frac{1}{2}\Gamma_f \tag{1}$$

of the effective Hamiltonian

$$\mathcal{H} = H - i\sum_c \mathscr{P}^c W_c^T \cdot W_c, \tag{2}$$

where the real part H is the Hamiltonian of the compound nuclei with bound state approximation, and W_c stands for the coupling between the bound state and the channel vector associate with the channel c. $\mathscr{P}^c$ is the strength of the continuum coupling for the channel c.

We investigate two models. In one model, we adopt the Gaussian Orthogonal Ensemble (GOE) as H and random numbers with normal distribution as each elements of W_c [9]. We call this model GOE + random coupling (GOE+RC) model. In the other model, we adopt the two-body random ensemble (TBRE) as H and the values of the overlaps between the compound states and the target plus one-particle states as the elements of W_c [10]. This we call TBRE + correlated coupling (TBRE+CC) model. In both cases, the discussion is restricted in the case of single channel. So we omit the subscript c hereafter.

The deviations from the GOE limit, namely the Wigner surmise in the case of NNLS and the PT distribution in the case of widths, are characterlized in terms of the parameter β of the Brody distribution and the degree of freedom ν of the χ^2 distribution, respectively. Here,

$$P_B(S) = aS^{\beta}\exp(-bS^{\beta+1}), \quad a = (\beta+1)b, \quad b = \left\{\Gamma\left(\frac{\beta+2}{\beta+1}\right)\right\}^{\beta+1}, \tag{3}$$

is the Brody distribution and

$$f(x;\nu,\langle x\rangle) = \frac{(1/2)^{\nu/2}}{\Gamma(\nu/2)}\left(\frac{x}{\langle x\rangle}\right)^{\nu/2-1}\exp\left(\frac{-\nu x}{2\langle x\rangle}\right) \tag{4}$$

is the χ^2 distribution.

2.1 GOE+random coupling model

We first check the GOE+RC model. Here, H is composed of the random numbers. Namely,

$$H_{ij} \in N(0, 1+\delta_{ij}), \tag{5}$$

where $N(r,s)$ is the normal distribution with average r and variance s. W is also composed of the random numbers,

$$W_i \in N(0,1). \tag{6}$$

The coupling constant $\mathscr{P}$ is scaled as

$$\tilde{\kappa} = \mathscr{P}/\sqrt{\Omega}, \tag{7}$$

where Ω is the dimension of the random matrix.

2.2 TBRE+correlated coupling model

The basic ingredients of the TBRE are the two body interactions,

$$V = \sum_{i>j,k>n} v_{ijkn}c_i^{\dagger}c_j^{\dagger}c_nc_k \tag{8}$$

$$v_{ijkn} \in N(0, 1+\delta_{ik}\delta_{jn}) \tag{9}$$

where $c_i^{\dagger}$ is the creation operator of the single particle. The number of levels is expressed as l.

The basis states with particle number m are constructed as

$$|\alpha; m\rangle = \Pi^{m}_{n(\alpha)=1}c^{\dagger}_{i_{n(\alpha)}}|0\rangle \tag{10}$$

where $\{i_{n(\alpha)}\}$ indicate a set of m single particle labels, and the basis states with particle number $d = (m-1)$ are constructed as

$$|\mu; d\rangle = \Pi^{d}_{n(\mu)=1}c^{\dagger}_{i_{n(\mu)}}|0\rangle. \tag{11}$$

The matrix elements of the real part of the Hamiltonian are

$$H(m)_{\alpha\beta} = \langle\alpha; m|\sum_{i>j,k>n} v_{ij,kn}c_i^{\dagger}c_j^{\dagger}c_nc_k|\beta; m\rangle, \tag{12}$$

while those of the target states are

$$H(d)_{\mu\nu} = \langle\mu; d| \sum_{i>j,k>n} v_{ij,kn} c_i^\dagger c_j^\dagger c_n c_k |\nu; d\rangle. \tag{13}$$

The coupling with continuum is assumed to occur through the channel vector state $|\mathscr{C}; m\rangle$. We choose as this state, in the TBRE+CC model, adding a single particle to the target ground state, $|\sigma_0; d\rangle$. To define the single-particle, we make use of the eigenstates of the density matrix,

$$\rho_{ij} = \delta_{ij} - \langle\sigma_0; d|c_i c_j^\dagger|\sigma_0; d\rangle. \tag{14}$$

Namely we compose a creation operator $c_a^\dagger$ as

$$c_a^\dagger = \sum_i f_{ai} c_i^\dagger$$

$$\sum_j \rho_{ij} f_{aj} = n_a f_{ai}.$$

Then the channel vector is constructed as

$$|\mathscr{C}; m\rangle = |a + \sigma_0; m\rangle \equiv \mathcal{N}\{c_a^\dagger|\sigma_0; d\rangle\} \tag{15}$$

where $\mathcal{N}$ indicates taking normalization. Another assumption is that the compound states with energy lower then the target ground state do not couple with the continuum.

With these two assumptions, the coupling matrix elements to the continuum in the basis of the eigenstate of the compound Hamiltonian are expressed as

$$W_\lambda = \langle a + \sigma_0; m|\lambda; m\rangle \quad \text{for } E_\lambda > 0 \tag{16}$$

$$= 0 \quad \text{for } E_\lambda < 0 \tag{17}$$

where E_λ is the energy of the compound state λ relative to the target ground state.

The coupling strength with the continuum is controlled in terms of the parameter κ, defined as

$$\kappa = \mathscr{P}/\lambda, \tag{18}$$

where $\lambda^2 = \frac{1}{\Omega}\mathrm{Tr}(H^2)$ and Ω is the size of the Hamiltonian $H(m)$.

3 Results

3.1 GOE+RC model

Figure 1 shows the deviation from the GOE limit when the coupling to the continuum increases. The filled squares show the parameter β in eq.(3) while the filled circles are for the degrees of freedom ν of the χ^2 distribution in eq.(4). As is shown, when $\tilde{\kappa}$ is increased, both β and ν decrease, untile $\tilde{\kappa}$ becomes around 1. With increasing $\tilde{\kappa}$ further, both β and ν become close to unity.

To show the reason of this behavior, we plotted the positions of the eigenvalues in the $\mathcal{E} - \Gamma$ plane for $\tilde{\kappa} = 0.1$(panel a), 1.0(panel b), 10.0(panel c) in fig. 2. As is seen in panel a), the widths do not have any energy dependence for small $\tilde{\kappa}$. For $\tilde{\kappa} = 1.0$ we see some widths relatively large but not isolated from others around $\mathcal{E} = 0$ in panel b). In this situation, these states with large widths may have close real energies to other states with small widths, since the energy repulsion takes place in the complex

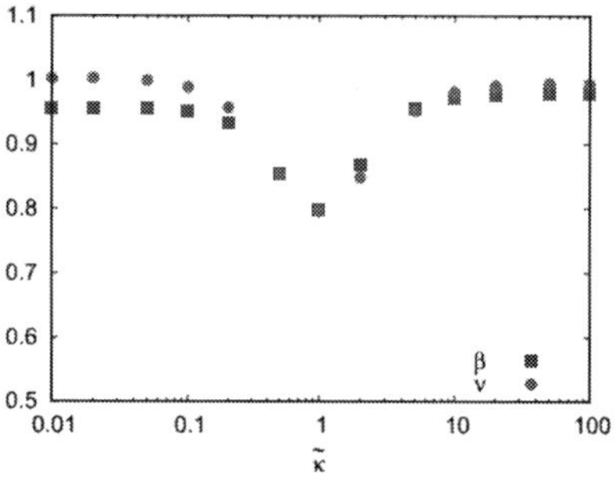

Figure 1. Fitted results of β, the parameter for the Brody distribution, and ν, the degree of the freedom in χ^2-distribution, are shown as functions of $\tilde{\kappa}$, the parameter controlling the coupling to the continuum. Filled squares are for β while filled circles are for ν. The dimension of the Hamiltonian is 924, coincides with the dimension of the Hamiltonian of TBRE for l=12 and m=6. The result is calculated from the accumulation of the 50 realizations of the random matrix.

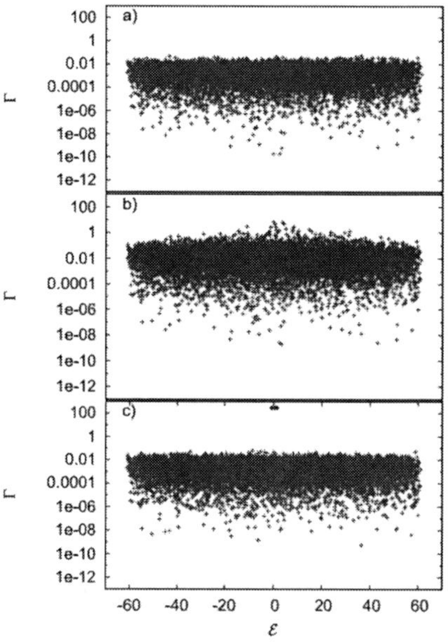

Figure 2. The positions of the complex eigenvalues in the $\mathcal{E} - \Gamma$ plane for $\tilde{\kappa}$ = 0.1 (a), 1.0 (b), 10.0 (c). The dimension of the Hamiltonian = 924 and the plots are accumulated results of 10 realizations of the random matrix.

plane. Therefore, the reason which drive to the deviation of the width distribution from the PT one causes also the deviation of the NNLS distribution from Wigner one, as states with large difference in width(the imaginary axis) may have small level distance(the real axis). With further increase of $\tilde{\kappa}$ to 10.0, as is seen in panel c), states with large widths (one state for each realization) are separated from the other states and have no significance in the statistics, while the widths of the other states follow PT distribution as they are created by small, random real perturbations to a state with large width.

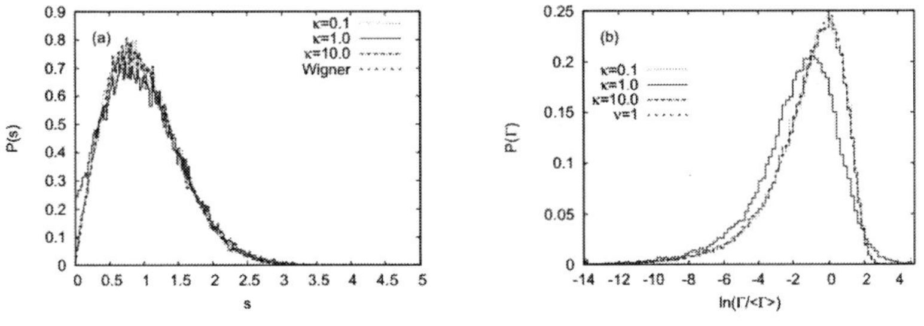

Figure 3. The NNLS distribution (a) and the width distribution(b) obtained from GOE+RC model for $\tilde{\kappa}$ = 0.1(dotted line), $\tilde{\kappa}$ = 1.0 (solid line), $\tilde{\kappa}$ = 10.0 (dashed line) compared to the Wigner one (a) or Porter-Thomas one(b). Ω = 924, and the number of realizations is 50. For $\tilde{\kappa}$ = 0.1 and $\tilde{\kappa}$ = 10.0 the results are close to the Wigner one or PT one while for $\tilde{\kappa}$ = 1.0 the results are different.

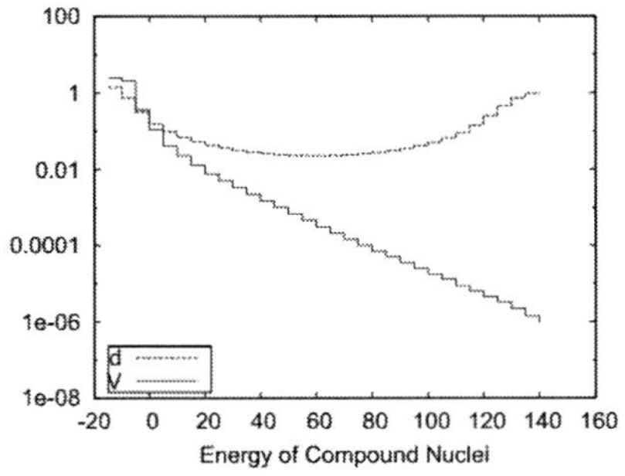

Figure 4. The E_λ dependence of the size of the coupling to the continuum $V_\lambda = W_\lambda^2$, compared with the level spacing d. The values are the results of the averaging over an energy interval $\Delta_E = 5$ and 50 realizations. In this figure, the values for E_λ <, which are not included in the effective Hamiltonian (2), are also plotted.

In fig. 3, the NNLS distribution (a) and the width distribution(b) are compared for $\tilde{\kappa}$ = 0.1, 1.0, 10.0 together with the simple GOE prediction, namely the Wigner one (a) or the PT one(b). They are the combined plots for 10 realizations of the random matrix. In both $\tilde{\kappa}$ = 0.1 case and $\tilde{\kappa}$ = 10.0 case, the distributions are very close to the GOE ones, while for $\tilde{\kappa}$ = 1.0 there are significant differences.

3.2 TBRE + correlated coupling model

The most significant difference of the TBRE+CC model from the GOE+RC model is the strong E_λ dependence of the coupling W_λ. In fig. 4, the average of the coupling strength W_λ^2 over an interval Δ_E and realizations is plotted as a function of E_λ. For small κ, the coupling can be treated perturbatively, hence the widths have the similar dependence on $\mathcal{E}_f$.

In fig. 5, positions of the complex eigenvalues are plotted on the $\mathcal{E} - \Gamma$ plane for the TBRE+CC model with l = 14 and m = 7 and κ = 0.1(panel a), 1.0(panel b) and 10.0(panel c). They are the combined plots for 10 realizations of the random matrix. Contrary to the case of the GOE+RC model, seen in fig 2, the widths have overall energy dependence in all cases.

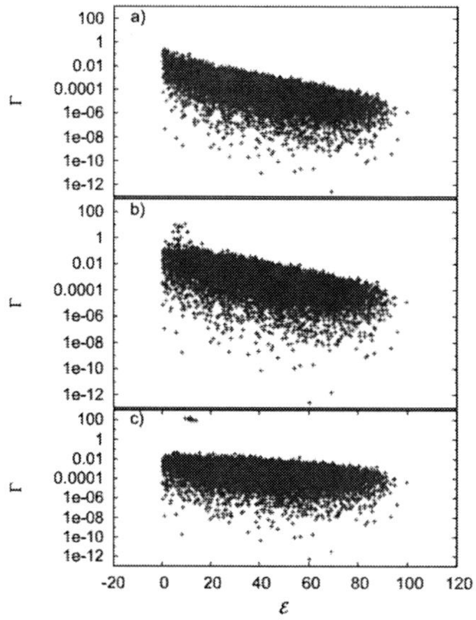

Figure 5. The positions of the complex eigenvalues in the $\mathcal{E} - \Gamma$ plane for $\kappa = 0.1$ (a), 1.0 (b), 10.0 (c). The Hamiltonian is for l=14 and m=7, and the plots are accumulated results of 10 realizations.

Another difference is that the energy interval where the states with large widths deviating from the global energy dependence for $\kappa \simeq 1$ exist is much narrower in the case of the TBRE+CC model. (Compare the panel (b) of figs. 2 and 5.)

This strong dependence of average Γ on $\mathcal{E}_f$ derives a completely different distribution of Γ, as is seen in [10], if the states all over the energy region are considered. However, in the real situation of INR, neutron widths are analyzed within very small intervals compared with the whole energy region of the excited states of nuclei. Therefore, we restricted our analysis within a small energy interval in which the effect of the global energy dependence of the average width is, if not negligible, small.

In fig. 6, the NNLS distribution (a) and the width distribution(b) obtained from the eigenvalues in the interval $5.0 \leq \mathcal{E} \leq 25.0$ are compared for $\kappa = 0.1, 1.0, 10.0$ together with the Wigner one(a) or PT one(b). Both $\kappa = 0.1$ case and $\kappa = 10.0$ case, the distributions are very close to the Wigner one (a) or PT one, while for $\kappa = 1.0$ there are significant difference.

In fig. 7, we plot β and ν obtained from the TBRE+CC model with l=14, m=7. The analysis is done in the energy interval $5 \leq \mathcal{E} \leq 20$. Although less clear than in the result of the GOE+RC model, the deviation from the unity is seen in the intermediate region of coupling strength around $\kappa \simeq 1$.

Such deviation is seen only in the limited energy region. In fig. 8, the dependence of ν on κ in the different energy interval is compared. Unlike the interval $5 \leq \mathcal{E} \leq 20$, the behavior in the interval $20 \leq \mathcal{E} \leq 35$ is less significant, and is invisible in the interval $35 \leq \mathcal{E} \leq 50$.

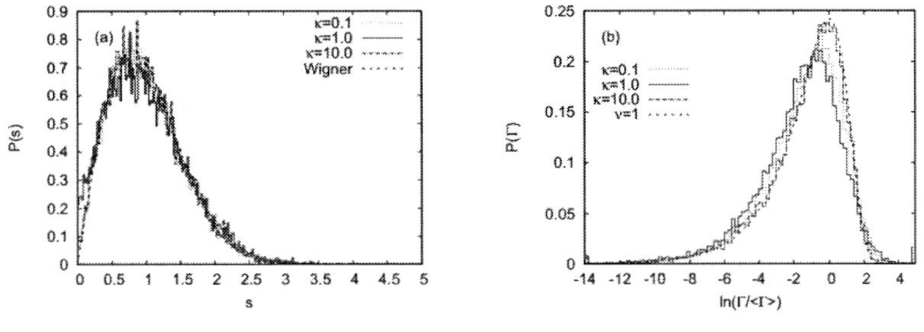

Figure 6. The NNLS distribution (a) and the width distribution(b) obtained from the TBRE+CC model in the energy interval $5.0 \leq \mathcal{E} \leq 25.0$ for $\kappa = 0.1$(dotted line), $\kappa = 1.0$ (solid line), $\kappa = 10.0$ (dot-dashed line) compared to the Wigner one (dashed line) (a) or PT one (b). $l = 14$ and $m = 7$, and the number of realizations is 50. For $\kappa = 0.1$ and $\kappa = 10.0$ the results are close to the Wigner one or the PT one while for $\kappa = 1.0$ the results are different.

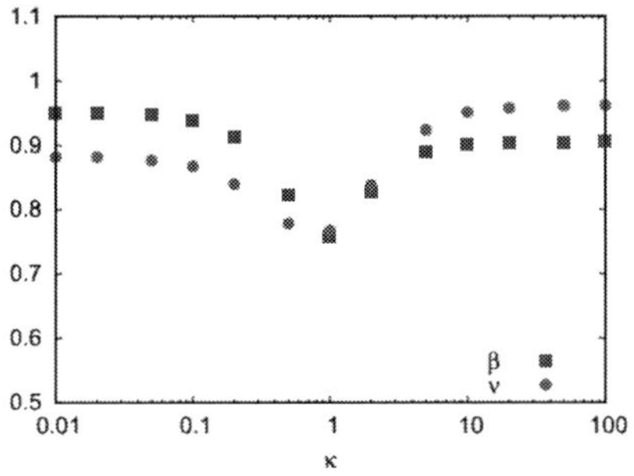

Figure 7. β (square) and ν (circle) obtained from the result of the TBRE+CC model with l=14, m=7. States in the energy interval $5 \leq \mathcal{E} \leq 20$ are considered. The number of realizations is 50.

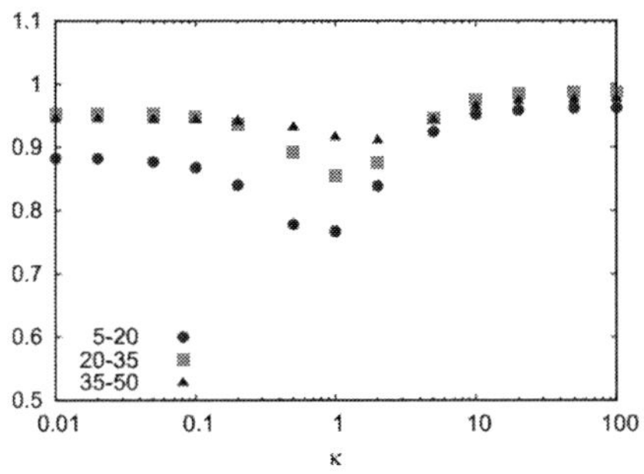

Figure 8. ν obtained from the TBRE+CC model with l=14, m=7. The circles are the result s obtained in the energy interval $5 \leq \mathcal{E} \leq 20$ (same with the ones in fig. 7), while the pluses (the crosses) are the ones in the interval $20 \leq \mathcal{E} \leq 35$ ($35 \leq \mathcal{E} \leq 50$), respectively. The number of realizations is 50.

4 Conclusion

Motivated by the recent claim that the neutron resonance width distribution does not support the random matrix theory, we examine the consequence of the super-radiant mechanism in the strong continuum coupling and the effect of the correlation between the target and the compound nuclei on the statistics of both the levels and the widths simultaneously, by controlling the strength to the continuum.

If the coupling to the continuum is large, several states have relatively large widths, resulting the deviation from the GOE limit both in the nearest neighbor level spacing distribution and the width distribution. However in the limit of strong coupling, the behaviors of both levels and widths are restored to the GOE ones, because only one state monopolizes the coupling to the continuum in this limit, and the rest states have the normal statistical property.

The effect of the correlation between the target and the compound states results in the global excitation energy dependence of the widths. Therefore, the width distribution for whole spectrum is completely different from the GOE one. However, if the width distribution is evaluated within a relatively narrow energy interval, it approximately follows the Porter-Thomas distribution as long as the coupling to the continuum is week.

When the coupling to the continuum becomes large, both the NNLS distribution and the width distribution deviate from the ones of the GOE in the energy interval where the states with large widths crowd. However, in other energy intervals away from "width concentrating region", the distributions are similar to the GOE ones. With further increase of the strength, both the NNLS distribution and the width distribution come again close to the GOE ones.

The deviation of the width distribution from the Porter-Thomas one and the deviation of the NNLS distribution from the Wigner one occur coincidently within the model we adopted.

References

[1] N. Bohr, Science **137**, 344 (1936)
[2] C.E. Porter, *Statistical Theories of Spectra: Fluctuations* (Academic, New York, 1965)
[3] R. U. Haq, A. Pandey, and O. Bohigas, Phys. Rev. Lett. **48**, 1086 (1982)
O. Bohigas, R. U. Haq, and A. Pandey, *Nuclear Data for Science and Technology* (Reidel, Dordrecht, 1983) 809
[4] P. E. Koehler et al, Phys. Rev. Lett. **105**, 072592 (2010)
[5] P. E. Koehler, Phys.Rev. C **84**, 034312 (2011)
[6] J. F. Shriner Jr, H. A. Weidenmüller, and G. E. Mitchell, arXiv:1209.2439v3 (2014)
[7] E. G. Reich, Nature, **466**, 1032 (2010)
[8] H.A. Weidenmüller, Phys. Rev. Lett. **105**, 232501(2010)
[9] G. L. Celardo, N. Auerbach, F. M. Izrailev, and V. G. Zelevinsky, Phys. Rev. Lett. **106**, 042501(2011)
[10] A. Volya, Phys. Rev. C **83**, 044312(2011)
[11] S. Mizutori and V. G. Zelevinsky, Zeit. für Phys. A **346**, 1(1993)
[12] H. Feshbach, Ann. Phys. (NY) **5**, 357(1958)
C. Mahaux and H. A. Weidenmüller, *Shell-model approach to nuclear reactions* (North Holland, Amsterdam, 1969)

Calculation of Beta Decay Half-Lives and Delayed Neutron Branching Ratio of Fission Fragments with Skyrme-QRPA

Futoshi Minato[1a]

[1] *Nuclear Data Center, Japan Atomic Energy Agency, Tokai 319-1195, Japan*

Abstract. Nuclear β-decay and delayed neutron (DN) emission is important for the r-process nucleosynthesis after the freeze-out, and stable and safe operation of nuclear reactors. Even though radioactive beam facilities have enabled us to measure β-decay and branching ratio of neutron-rich nuclei apart from the stability line in the nuclear chart, there are still a lot of nuclei which one cannot investigate experimentally. In particular, information on DN is rather scarce than that of $T_{1/2}$. To predict $T_{1/2}$ and the branching ratios of DN for next JENDL decay data, we have developed a method which comprises the quasiparticle-random-phase-approximation (QRPA) and the Hauser-Feshbach statistical model (HFSM). In this work, we calculate fission fragments with $T_{1/2} \leq 50$ sec. We obtain the rms deviation from experimental half-life of 3.71. Although the result is still worse than GT2 which has been adopted in JENDL decay data, DN spectra are newly calculated. We also discuss further subjects to be done in future for improving the present approach and making next generation of JENDL decay data.

1 Introduction

Nuclear data center (NDC) at Japan Atomic Energy Agency (JAEA) provides various kinds of nuclear data [1]. One of them is the decay data, which includes information concerning nuclear decay properties such as half-life, branching ratio of delayed neutron (DN), β-spectra, and so on. The data is practically used to evaluate the amount of radioactivity of unstable nuclei followed by a nuclear reaction and a fission. Such a work is important in the decomissioning of nuclear reactor and astrophysical applications. NDC has released the JENDL fission product decay data in 2011 (JENDL/FPD-2011) [1], which covers decay data of 1,284 nuclides. It is able to reproduce experimentally measured decay heat generated by fission products successfully by incorporating the total absorption gamma spectroscopy (TAGS) data. Recently, NDC started to distribute a new decay data, JENDL decay data file 2015 (JENDL/DDF-2015) [2], from November 2015. The file contains 3,237 nuclides ranging from the lightest elements including H-1 and n, up to Rf(Z=104). The decay data is also provided in JEFF [3] and ENDF [4] nuclear data groups. Recent development of accelerators which utilize radioactive beams makes it possible to investigate unstable nuclei which one hasn't been able to study, and the renewal of the decay data becomes also active.

Although the JENDL/FPD-2011 has succeeded to reproduce the decay heats, it is pointed out that it cannot reproduce the total amount of DN after a nuclear fission. A part of the problem might be

[a]e-mail: minato.futoshi@jaea.go.jp

[1]available from our website, http://wwwndc.jaea.go.jp/index.html

attributed to incorrect evaluation of the fission yield data as discussed in Ref. [5], and the other might be due to the accuracy of decay data. Besides the problem, JENDL/FPD-2011 didn't contain the DN spectra, which are considered as one of the promising quantities applicable for nuclear nondestructive inspection (the situation remains unchanged in JENDL/DDF-2015). Experimental data of DN spectra is limited strongly. Although new experiments to investigate DN are planned at several experimental facilities in the world, we still need to depend on theoretical methods to derive most of DN information. In case of JENDL/FPD-2011 and JENDL/DDF-2015, the decay data concerning unmeasured β-decay are calculated with Gross Theory 2 (GT2) [6], which has been also widely used to discuss and compare with experimental data. However, the approach uses a statistical treatment in the level density so that it cannot provide a detailed nuclear structure (*i.e.*, spin, parity and energy of excited levels), although it is able to reproduce half-lives systematically well. Besides, recent measurements of neutron-rich nucleus indicate that the GT2 fails to reproduce half-lives of some nuclei close to the neutron drip-line [7]. One of the reasons might be due to that GT2 is based on a phenomenological method whose parameters are adjusted by known experimental data. Therefore, it is required to make an alternative theoretical method to calculate data on β-decay for next JENDL decay data.

To calculate β-decay and DN, we have developed a new approach, which comprises quasiparticle-random-phase-approximation (QRPA) and the Hauser-Feshbach statistical model (HFSM) [8]. This approach is capable to calculate β-decay and DN emission of nuclei on the nuclear chart systematically as well as GT2. In addition, the formalism starts from the two-body effective interaction, and can take into account the nuclear deformation and the pairing effects consistently, so that it can provide a more reliable prediction of half-life for nuclei which have not been measured experimentally. Moreover, it can give a detailed nuclear structure of excited states. The systematical calculations of half-lives and branching ratio of DN using QRPA have been performed not only by non-relativistic but also relativistic approaches [9–11]. Finite amplitude method, which is equivalent to QRPA, was also applied to systematical β-decay calculation [12]. However, they don't discuss DN spectra. The data of DN spectra is only available in ENDF/B-VII.1 decay data [4], in which FRDM+QRPA is used to calculate unmeasured half-lives [13] and CoH code to calculate DN spectra [14]. However, our approach is more sophisticated as compared to FRDM+QRPA with respect to full microscopic and respecting self-consistency of the effective interaction in the ground and excited states.

So far, we have examined the reproduction power of our approach. However, we have found that it gave large deviations from experimental half-lives, especially in odd-A and odd-odd nuclei. The reason is clear; we have assumed that valence nucleon(s) is(are) unperturbatively put on the last orbit outside even-even core nuclei and its β-transition is treated in a manner of the single particle one. In this work, we assess two approaches for better description of odd nuclei, that is, the blocking effect approximation (BEA) and the equal filling approximation (EFA), and re-examined the reproduction power.

2 Theoretical Method

To calculate β-decays and branching ratios of DN, three processes are carried out. At first, the ground state of nucleus is derived with the Skyrme-Hartree-Fock + BCS method. The effective two-body interaction that we used is the SkO' parameter set [15]. The pairing force which accounts for the short-range correlation between identical particles (isospin $T = 1$) is given in the volume-type form,

$$v(\vec{r}, \vec{r}') = P(S = 0)P(T = 1)v_{01}\delta(\vec{r} - \vec{r}'), \quad (1)$$

where $P(S = 0)$ and $P(T = 1)$ are the projection operator for $S = 0$ and $T = 1$, respectively, and v_{01} the parameter representing the pairing strength. Because we adopt the zero-range force as the pairing

DOI: 10.1051/epjconf/201612210001

force, a pairing active space is introduced. We use the same approach as the reference [16]. For proton and neutron, we set $v_{01} = 256.1, 258.1$ MeV fm^3 [15], respectively. The calculation is performed in the cylindrical-coordinate system, where the axial symmetry is assumed. The single-particle levels in the continuum are taken into account by discretizing them in the boundary box. We use 16 fm with the step size $\Delta z = \Delta\rho = 0.8$ fm for the boundary condition.

In the second step, excited states of the daughter nucleus are calculated by QRPA based on the ground state calculated by the Skyrme-Hartree-Fock + BCS. The phonon creation operator in our formalism is given by

$$Q_\nu^\dagger = \sum_{pn} X_{pn}^\nu \alpha_p^\dagger \alpha_n^\dagger - Y_{pn}^\nu \alpha_n \alpha_p, \tag{2}$$

and the excited state can be approximated by

$$|\nu\rangle = Q_\nu^\dagger|\mathrm{QRPA}\rangle \sim \mathrm{Q}_\nu^\dagger|0\rangle, \tag{3}$$

where $|\nu\rangle$ is ν-th excited state, $|\mathrm{QRPA}\rangle$ the QRPA ground state, and $|0\rangle$ the Skyrme-Hartree-Fock + BCS ground state. The coefficients X and Y are derived by solving the QRPA-equation,

$$\begin{pmatrix} A & B \\ -B^* & -A^* \end{pmatrix} \begin{pmatrix} X \\ Y \end{pmatrix} = E_{QRPA} \begin{pmatrix} X \\ Y \end{pmatrix}, \tag{4}$$

where submatrices A and B are expressed by

$$\begin{aligned} A_{pnp'n'} =& (E_p + E_n)\delta_{pp'}\delta_{nn'} + (u_p v_n u_{p'} v_{n'} + v_p u_n v_{p'} u_{n'})\langle p\bar{n}'|v|\bar{n}p'\rangle \\ &+ (u_p u_n u_{p'} u_{n'} + v_p v_n v_{p'} v_{n'})\langle pn|v|p'n'\rangle \\ B_{pnp'n'} =& -(u_p v_n v_{p'} u_{n'} + v_p u_n u_{p'} v_{n'})\langle pn'|v|\bar{n}\bar{p}'\rangle - (u_p u_n v_{p'} v_{n'} + v_p v_n u_{p'} u_{n'})\langle pn|v|\bar{p}'\bar{n}'\rangle \end{aligned} \tag{5}$$

The factors, u_i and v_i, are the BCS coefficients. The matrix elements of the particle-hole residual interaction in Eq. (5) are derived self-consistently from the second derivative of energy with respect to the density matrix, while those of the particle-particle one are calculated using the similar force as Eq. (1), given by

$$v(\vec{r}, \vec{r}') = P(S = 1)P(T = 0)v_{10}\delta(\vec{r} - \vec{r}'). \tag{6}$$

For the excited states, we consider only the Gamow-Teller transition, and the 1st forbidden transitions are not taken into account in this work, so that $T = 1$ pairing force does not take part in the particle-particle channel in Eq. (5). We don't consider $T = 0$ pairing when solving the BCS gap equation because its effect on the ground state is considered to be small for N>Z nuclei. Therefore, v_{10} in Eq. (6) is treated as a free parameter in our formalism. We determine it by reproducing experimental half-lives of even-even nuclei. Then, we take the average values for each isotopes and use it for systematical calculation. In case of odd-Z nucleus with proton number Z, v_{10} of $Z + 1$ nucleus is used assuming that property of even-even core nucleus is not changed much. The parameters v_{10} used in our work are listed in Tab. 1. We introduce the model space of single particle energy up to 20 MeV and two quasi-particle energy up to 60 MeV in solving the QRPA equation.

Once we obtain the excited states of the daughter nucleus, we can calculate the half-life, $T_{1/2}$, which is given by,

$$\frac{1}{T_{1/2}} = \frac{\lambda_\beta}{\ln 2} = \frac{(g_A/g_V)^2}{D}\int_0^{Q_\beta} \sum_\nu \left|\langle\nu|\sigma\tau^-|0\rangle\right|^2 E_e\sqrt{E_e^2 - m_e^2}(E - E_e)^2 F(Z, E_e)dE_e, \tag{7}$$

where $F(Z, E_e)$ is the Fermi function accounting for the final state interaction between nucleus and outgoing electron and $D = 6163.4$ s. The Q_β values are derived in the same way as Ref. [11].

Table 1. Isocaler $T = 0$ pairing strength, v_{10} (in unit of MeV fm^3), used in this work.

Element	Fe	Co	Ni	Cu	Zn	Ga	Ge	As	Se
	0	380	380	208	208	175	175	0	0
Element	Br	Kr	Rb	Sr	Y	Zr	Nb	Mo	Tc
	68	68	0	0	0	0	0	0	0
Element	Ru	Rh	Pd	Ag	Cd	In	Sn	Sb	Te
	0	160	160	303	303	220	220	75	75
Element	I	Xe	Cs	Ba	La	Ce	Pr	Nd	Pm
	0	0	0	0	190	190	0	0	260
Element	Sm	Eu	Gd	Tb	Dy				
	260	380	380	0	0				

In the third step, the neutron emission from the excited states are calculated. We assume that the daughter nucleus is a compound nucleus and apply the Hausher-Feshbach statistical model, which is implemented in the code CCONE [17], to estimate neutron evaporation probabilities. The global neutron optical potential of Ref. [18] and EGLO model are used to calculate neutron and γ transmission coefficients, respectively. For level densities, the formula of Gilbert and Cameron [19] is used, and the parameters are taken from Mengoni-Nakajima's [20]. Then, we obtain the neutron emission probability, P_n^ν, and γ emission probability, P_γ^ν, from the ν-th excited state. The branching ratios of delayed neutron are then calculated by the relation

$$P_{\beta n} = \frac{\sum_\nu \lambda_\nu P_n^\nu}{\sum_\nu \lambda_\nu}, \tag{8}$$

where $\lambda_\nu = \ln 2/T_{1/2,\nu}$ are the β-decay rate from the ground state of parent nucleus to ν-th excited state of daughter nucleus. CCONE code also provide a neutron spectrum from an excited state so that we can obtain a DN spectrum by summing them over the excited states. The spectra is normalized as done in Ref. [14].

2.1 Odd-A and Odd-Odd nuclei

To calculate odd-A and odd-odd nuclei, we tested the blocking effect approximation (BEA) [21] and the equal filling approximation (EFA) [22], for the ground state. For BEA, we assume the time reversal symmetry, namely the state occupied by a valence particle, which corresponds to the blocking state, have $u_i^2 = u_{\bar{i}}^2 = v_i^2 = v_{\bar{i}}^2 = 0.5$. The ground state is sought by blocking single particle levels near the Fermi energy one by one to find the lowest total energy. For EFA, the obtained BCS coefficients, which satisfy the odd particle number, are also used in QRPA. Both for BEA and EFA, we omit the coupling between valence particle and core nucleus in calculating QRPA.

3 Result

We first discuss the result of half-lives calculated by EFA and BEA. Figure 1 shows the half-lives of nickel and cobalt isotopes, comparing with the experimental data taken from [23]. For nickel isotopes, the calculations both for EFA and BEA reproduce the experimental data reasonably. For cobalt isotopes, EFA provides the half-lives in the same order as the experimental data, while BEA underestimates them strongly. The half-life is not only governed by β-transition of core nucleus

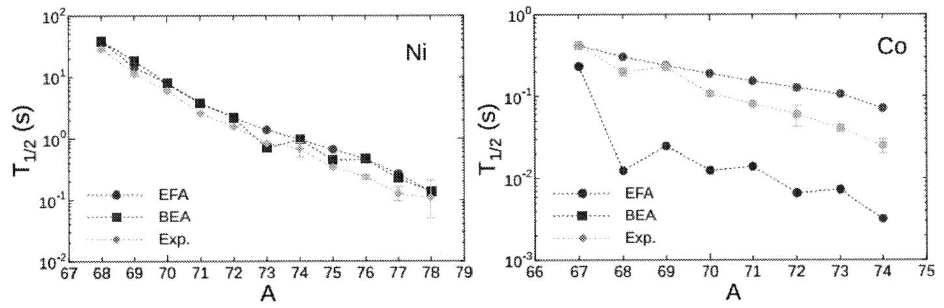

Figure 1. Half-lives of nickel and cobalt isotopes calculated by EFA and BEA, comparing with the experimental data.

but also that of valence nucleon. As a consequence, the half-life changes significantly depending on which orbit the valence nucleon occupies for BEA. On the other hand, since EFA averages the occupation probability of valence nucleon over the orbits near the Fermi energy, contribution of the valence nucleon to the half-life is less sensitive than BEA. In the present approach, the ground state is sought for the lowest total energy, but we omit higher-order effects caused by rotational and vibrational coupling. If we would include those effects, we would obtain a better result. However, it requires a more sophisticated formulation and a much longer computational time. For further improvement of the decay data evaluation, it will be important. In the following, β-decay and DN are calculated with EFA approach, which gives a more consistent result to experimental data than BEA.

Figures 2 and 3 show the results of half-lives of $Z = 26 \sim 33$ and $Z = 34 \sim 41$, respectively, calculated by the present approach. For comparison, KTUY+GT2 [6] (hereafter, GT2) and experimental data taken from [23] are also shown. From $Z = 26$ to 33 shown in Fig. 2, the present approach is comparable to GT2. On the other hand, the present approach shows shorter half-lives than GT2 systematically from $Z = 34$ to 41 shown in Fig. 3 and GT2 is relatively closer to experimental data than the present approach except Rb isotopes. We notice that from neutron number $N = 64$, the present approach shows dumps, which give deviations from experimental data as well as GT2. At the moment, we don't find the physical answer making this dump. This point is the subject to be remedied in future.

Figures 4 and 5 show the results of branching ratio of DN for $Z = 26 \sim 33$ and $Z = 34 \sim 41$, respectively, comparing with GT2. The result of the present work shows a sudden increase and decrease, while GT2 shows a monotonic increase as a function of the mass number A and roughly follows the experimental data. The defect of the present work in the branching ratio is, of course, due to incorrect reproduction of β-strength function. However, this disappointing result might be improved by including the 1st forbidden transition. As discussed in the paper [9], the 1st forbidden transition produces an additional β-strength in low energy region below neutron threshold, and then the DN emission probabilities are reduced to a moderate value.

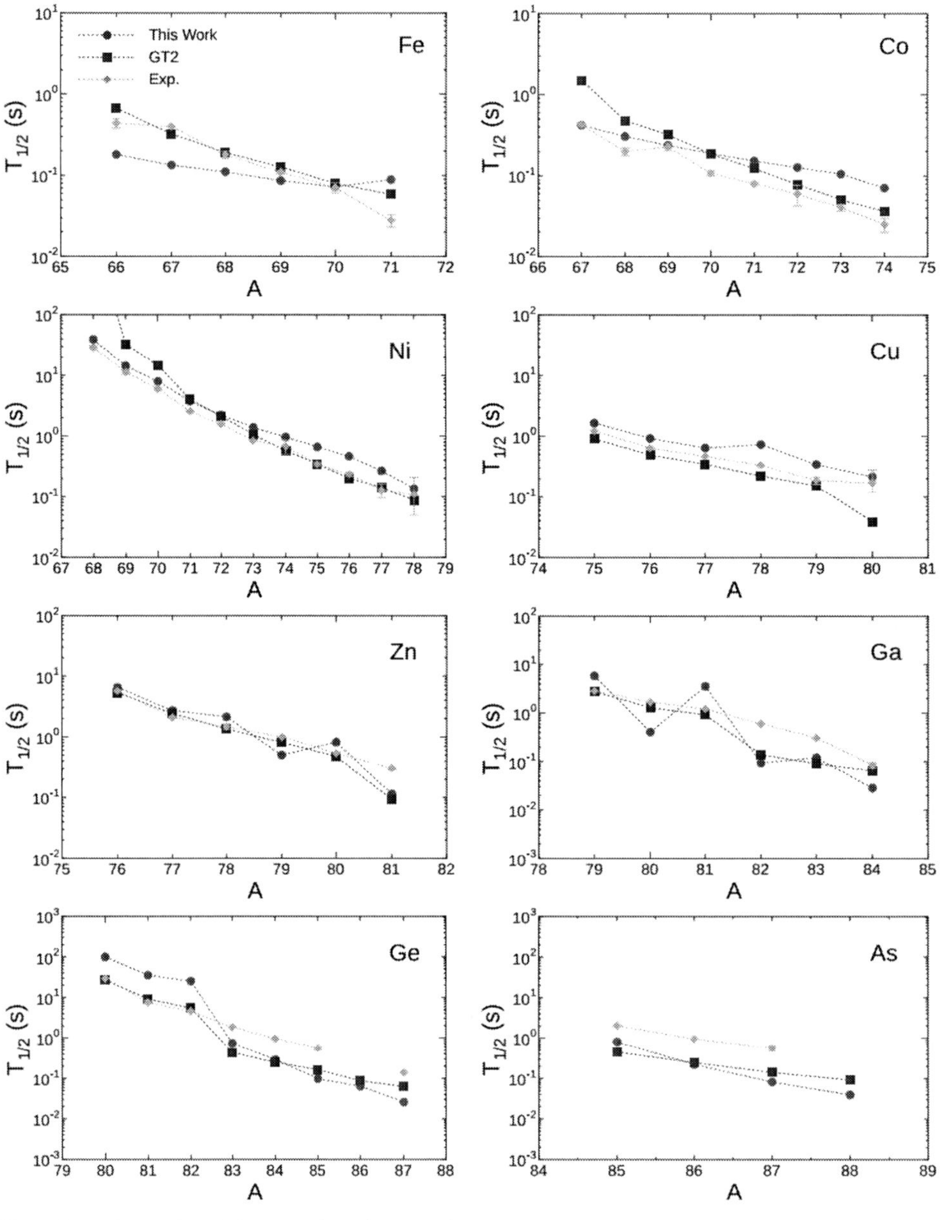

Figure 2. Half-lives of Z = 26 to 33 nuclei. The result of BEA and EFA are shown.

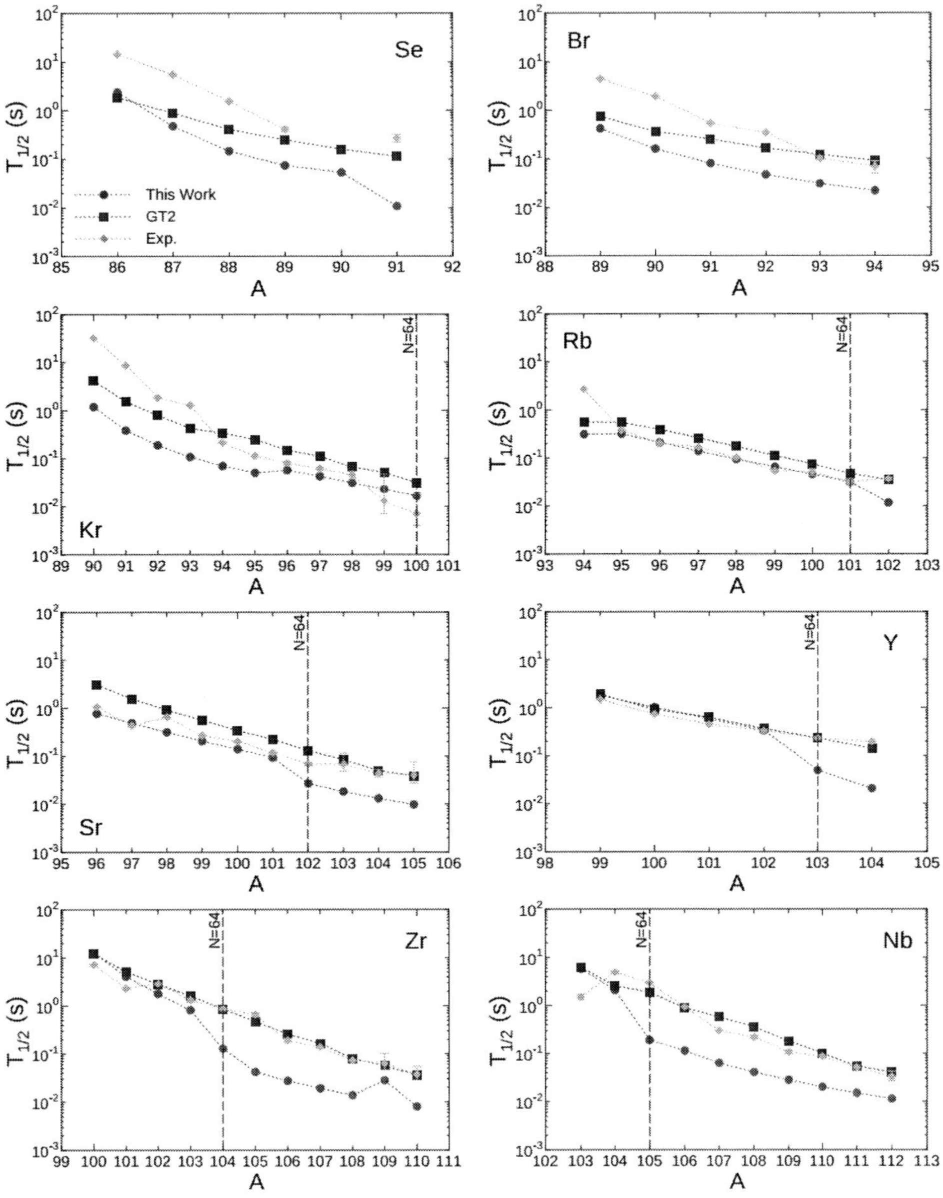

Figure 3. Same as Fig. 2, but for Z = 34 to 41 nuclei. The vertical line in the panels indicates N = 64.

DOI: 10.1051/epjconf/201612210001

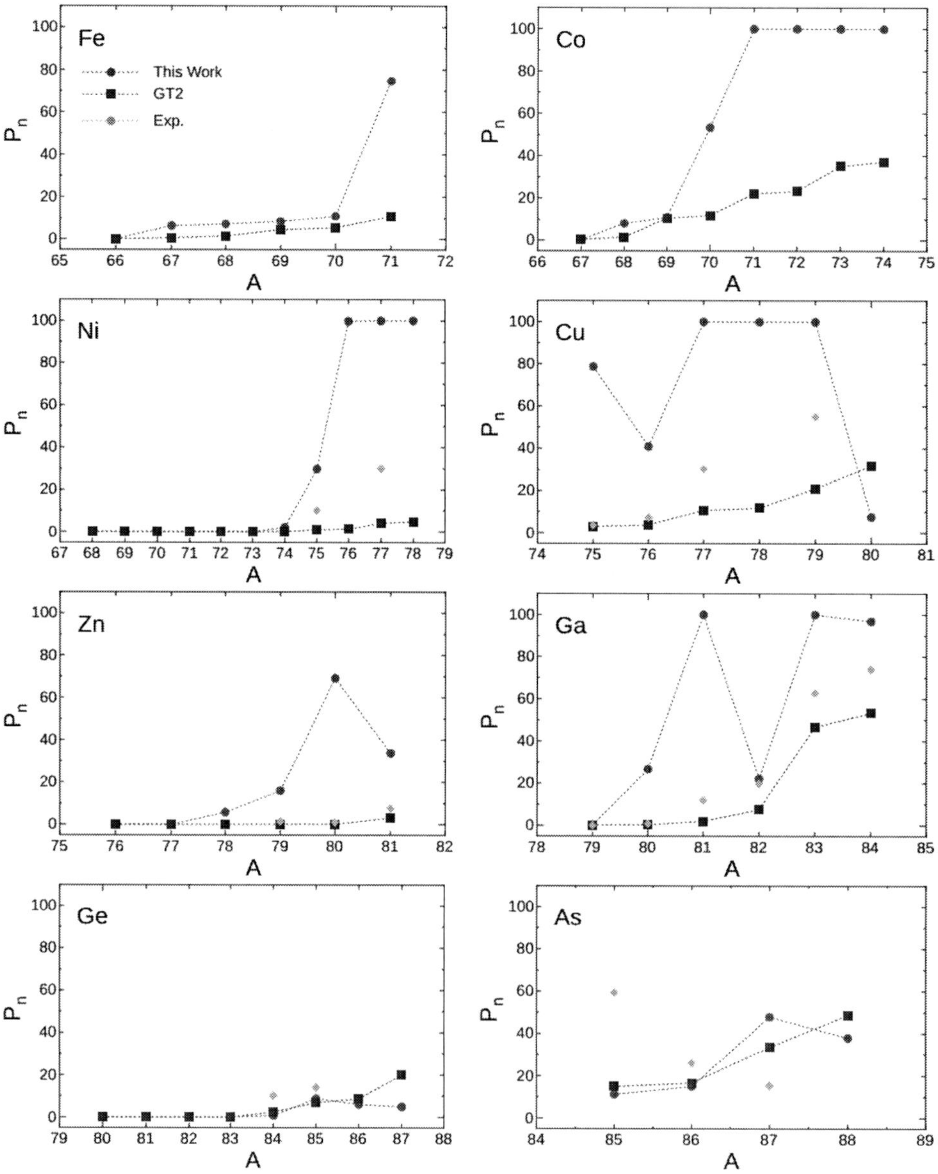

Figure 4. Calculated branching ratios of DN for Z = 26 to 33 nuclei, comparing with the experimental data taken from [23] and GT2.

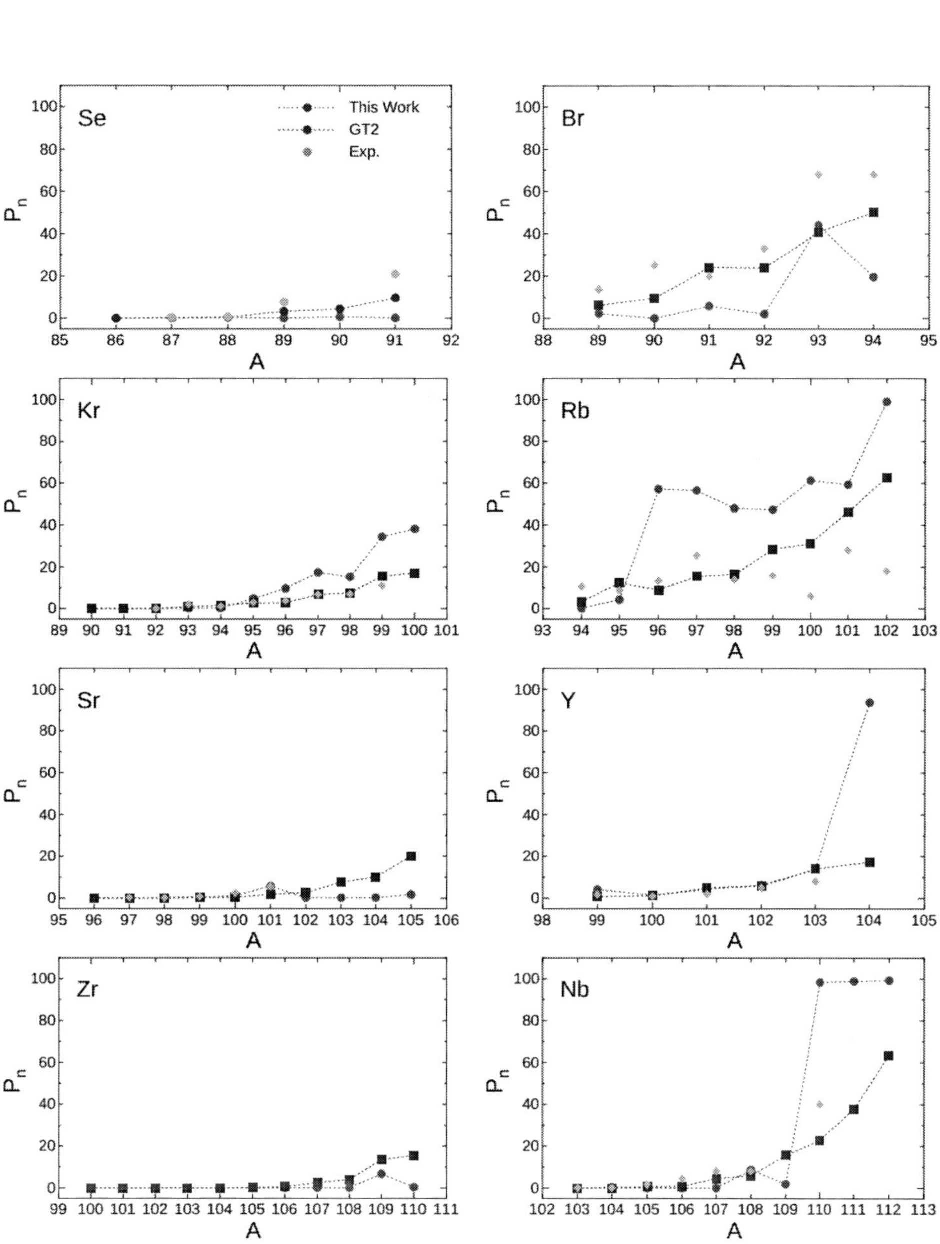

Figure 5. Same as fig. 5, but for Z = 34 to 41 nuclei.

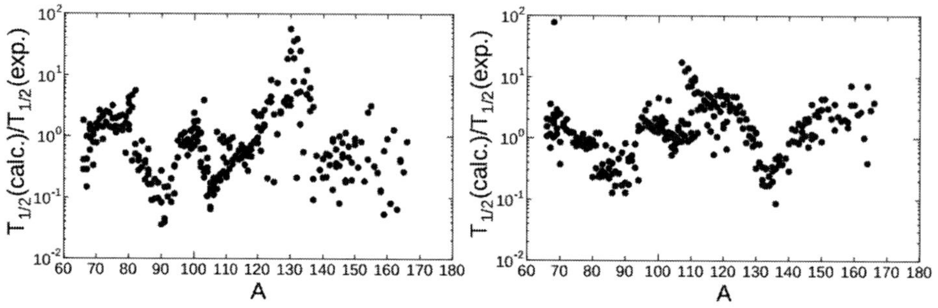

Figure 6. Ratio of the half-lives (the top panels) and the branching ratios (the bottom panels) to the experimental data. The left panels are for the present result, and the right ones are for the GT2.

Figure 6 shows the ratio of the half-lives and the branching ratios defined by

$$\sigma_{rms}(T_{1/2}) = \exp\left(\sqrt{\frac{1}{N}\sum_i^N \log\left(\frac{T_{1/2}(th.)}{T_{1/2}(exp.)}\right)}\right)$$

$$\sigma_{rms}(P_{\beta n}) = \exp\left(\sqrt{\frac{1}{N}\sum_i^N \log\left(\frac{P_{\beta n}(th.)}{P_{\beta n}(exp.)}\right)}\right). \tag{9}$$

For 280 nuclei from $Z = 24$ to 64 with $T_{1/2} \leq 50$ sec, the present work provides $\sigma_{RMS}(T_{1/2}) = 3.71$, which is improved well as compared to 4.90 of the previous work [8]. On the other hand, for 114 nuclei whose experimental data are available, $\sigma_{RMS}(P_{\beta n}) = 9.44$ is not improved so much (in the previous work, 9.85). GT2 gives a better value, $\sigma_{RMS}(T_{1/2}) = 2.76$ and $\sigma_{RMS}(P_{\beta n}) = 3.16$. At the moment, the present approach doesn't take into account the 1st forbidden transition, on the other hand, GT2 does. The better quality of GT2 might be due to this point. Therefore, we plan to include the 1st forbidden transition now.

Finally, Figs. 7 and 8 illustrate the DN spectra of free chosen nuclei, in which the results are compared with those given in the ENDF/B-VII.1 decay data library. Despite the theoretical approaches calculating β-strength function are different, the spectra of ENDF and our approach are similar for most of calculated nuclei. As we mentioned above, if we take into account the 1st forbidden transition, the result might be changed.

4 Conclusion

We have developed the method which combines QRPA and HFSM. To improve the description of odd-A and odd-odd nuclei, we examined the blocking effect approximation and the equal filling approximation in the ground state. We found that the latter method gives a better reproduction in half-lives systematically. However, the reproduction power of the present approach both for half-lives and branching ratios of delayed neutron was still worse than that of GT2. This deviation may be attributed from the omission of 1st forbidden transition in our formalism, which becomes more effective as the neutron and proton shell gaps become larger for neutron-rich side as well as heavier nuclei. We also omitted the coupling between valence particle(s) and core nucleus, i.e., core polarization effect. We

DOI: 10.1051/epjconf/201612210001

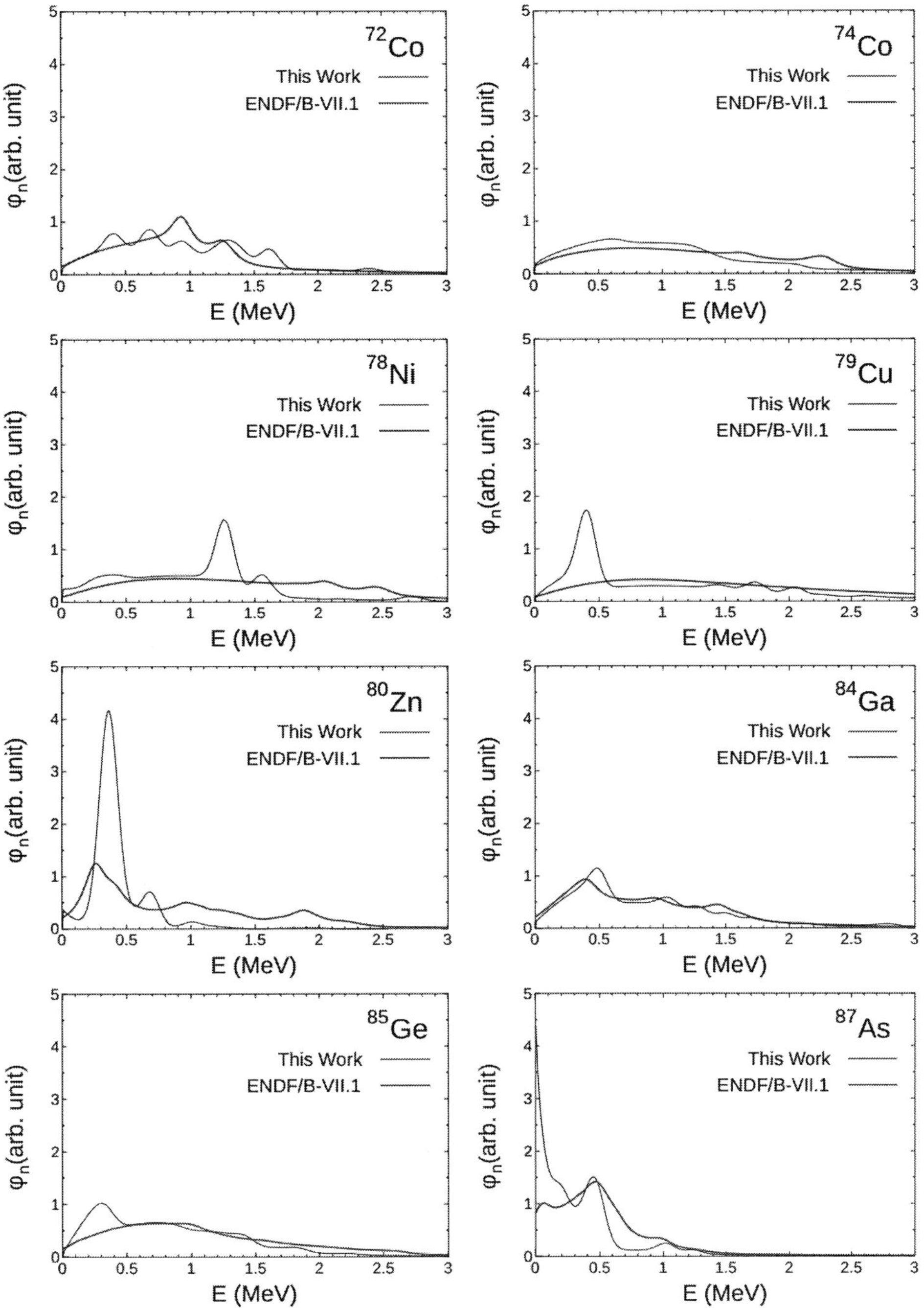

Figure 7. Delayed neutron spectra for free chosen nuclei.

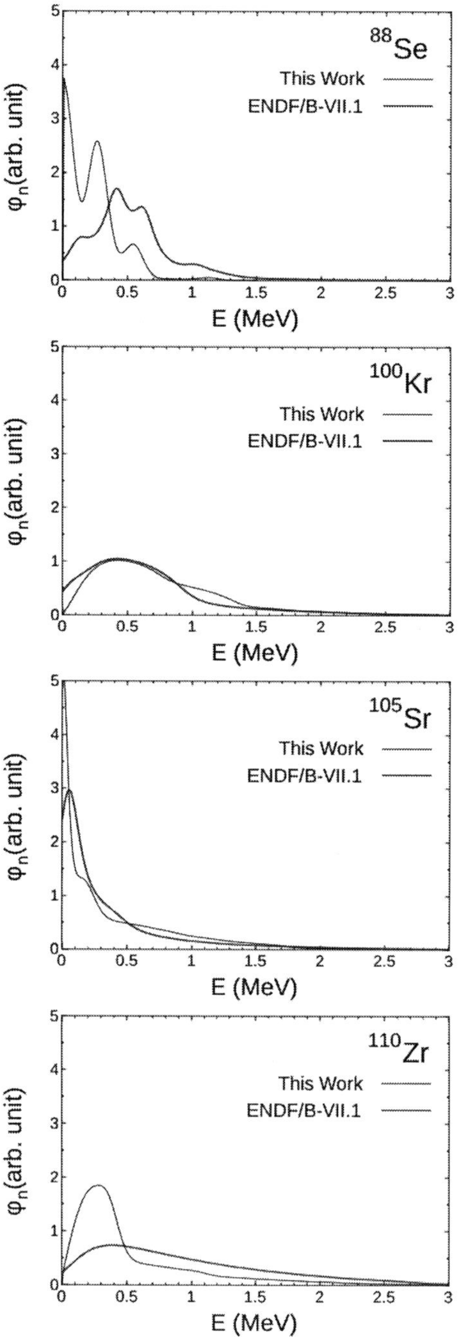

Figure 8. Same as Fig. 7

will need to include it in future for a more realistic and reliable calculation. In addition, we used the same isoscaler $T = 0$ pairing strength v_{10} as the even-even for odd-mass nuclei. It would be more practical to use v_{10} determined for each isotopes individually. To use a mass dependent v_{10} as used in [11] will also improve our result, which is next step of our work.

References

[1] J. Katakura, JAEA-Data/Code 2011-025 (Mar 2012).
[2] J. Katakura, to be submitted.
[3] M.A. Kellett, O. Bersillon, R.W. Mills, JEFF Report 20, OECD/NEA (2009).
[4] M.B. Chadwick et al., Nucl. Data Sheets **112** (2011) 2887.
[5] J. Katakura, F. Minato, K. Ohgama, to be published in EPJ Web of Conferences, Wonder 2015: Fourth International Workshop on Nuclear Data Evaluation for Reactor Applications, Oct. 2015, Aix-en-Provence, France.
[6] T. Tachibana, M. Yamada, Proc. Inc. Conf. on exotic nuclei and atomic masses, Arles, 1995, eds. M. de Saint Simon and O. Sorlin (Editions Frontueres, Gif-sur-Yvette, 1995) p.763; T. Yoshida and T. Tachibana, J. Nucl. Sci. Technol., **37** (2000) 491.
[7] G. Lorusso et al., Phys. Rev. Lett. **114** (2015) 192501.
[8] F. Minato, O. Iwamoto, JPS Conf. Proc. **6** (2015) 030053.
[9] I.N. Borzov, Nucl. Phys. A**777** (2006) 645.
[10] M. Martini, S. Péru, S. Goriely, Phys. Rev. C **89** (2014) 044306.
[11] T. Marketin, L. Huther, G. Martinez-Pinedo, arXiv:1507.07442 (2015).
[12] M.T. Mustonen, J. Engel, arXiv:1510.02136 (2015).
[13] P. Möller, B. Pfeiffer, K.-L. Kratz, Phys. Rev. C **67** (2003) 055802.
[14] T. Kawano, P. Möller, W.B. Wilson, Phys. Rev. C **78** (2008) 054601.
[15] P.-G. Reinhard, D.J. Dean, W. Nazarewicz et al., Phys. Rev. C **60** (1999) 014316.
[16] M. Bender, K. Rutz, P.-G. Reinhard, J.A. Maruhn, Phys. Rev. C **60** (1999) 034304.
[17] O. Iwamoto, J. Nucl. Sci. Tech. **44** (2007) 687.
[18] A.J. Koning, J.P. Delaroche, Nucl. Phys. **713** (2003) 231.
[19] A. Gilbert, A.G.W. Cameron, Can. J. Phys. **43** (1965) 1446.
[20] A. Mengoni, Y. Nakajima, J. Nucl. Sci. Technol. **31** (1994) 151.
[21] P. Ring and P. Schuck, *The Nuclear Many Body Problem*, Springer, Berlin (1980).
[22] Sara Perez-Martin, L.M. Robledo, Phys. Rev. C **78** (2008) 014304.
[23] National Nuclear Data Center, Brookhaven National Laboratory, NuDat 2.6, http://www.nndc.bnl.gov/nudat2/

EPJ Web of Conferences **122**, 10002 (2016) DOI: 10.1051/epjconf/201612210002
*CNR*15*

Analysis of Electron and Antineutrino Energy Spectra from Fissile Samples under Irradiation based on Gross Theory of Beta-decay

T. Yoshida[1,a], T. Tachibana[2], and S. Chiba[1]

[1] *Tokyo Institute of Technology, Ookayama, Meguro-ku, Tokyo*
[2] *Senior High School of Waseda University, Kamishakujii, Nerima-ku, Tokyo*

Abstract. We applied the gross theory of β-decay to calculate the reactor electron and antineutrino ($\bar{\nu}_e$) spectra emitted from ^{235,238}U and ^{239,241}Pu by summing up all the contributions from a large number of decaying fission-products (FPs). We make it clear what kinds of transition types and FP nuclides are important to shape the lepton spectra. After taking the ambiguity in the current data for fission yields and Q_β-values into account, we suggested a possibility that the high-energy part of the widely referred electron-spectra by Schreckenbach *et al.*, almost only one experimental data set available now, might possibly be too low. Arguments on a special role of the odd(Z)-odd(N) nuclides and on the consistency between U-238 and other fissiles in the experimental data lead to the importance of a new and independent measurement of electron energy spectra which could be converted into the reactor $\bar{\nu}_e$ spectra.

1 Introduction

Recently increasing attention is paid to the technology of antineutrino-based monitoring of nuclear-reactor operation from the outside for non-proliferation purposes. Precise determination of the antineutrino energy-spectra is one of the key issues there. Reliable knowledge of the reactor antineutrino spectra is also required for interpreting the neutrino oscillation experiments for basic research of physics. One way to achieve this is to convert the measured electron spectra from fissionable samples into the antineutrino spectra using the energy conservation of leptons in the β-decay process. Another is to compose the spectrum by summing up all the contributions from a large number of decaying fission products (FPs).

2 Calculation Method

In order to describe the aggregate behaviour of FPs in a reactor core, we start with the calculation of the number of each FP nuclide considering its creation and destruction. By summing up the contributions from them, we can calculate the electron and the antineutrino ($\bar{\nu}_e$) spectra as

$$I_e(E_e) = \sum N_i \lambda_i I_i^e(E_e) \tag{1}$$

[a] e-mail: tyoshida@nr.titech.ac.jp

$$I_{\bar{\nu}}(E_{\bar{\nu}}) = \sum N_i \lambda_i I_i^{\bar{\nu}}(E_{\bar{\nu}}), \tag{2}$$

where N_i and λ_i are the number density and the decay constant of the i-th FP. The point is how to obtain the spectra, $I_i^e(E_e)$ and $I_i^{\bar{\nu}}(E_{\bar{\nu}})$, characteristic of each nuclide i. Here E_e and $E_{\bar{\nu}}$ are the kinetic energies of the electron and the antineutrino. In order to calculate $I_i^e(E_e)$ and $I_i^{\bar{\nu}}(E_{\bar{\nu}})$, we applied the Gross Theory of beta decay first developed by Takahashi, Koyama and Yamada[1–3] and then improved by Kondoh and Tachibana *et al.*[4, 5]. The details of these calculations related to eqs.(1) and (2) are given elsewhere[6]. The decay constants λ_i and the decay chains were taken from JENDL FP Decay Data File 2011 along with the fission yields from JENDL Fission Yield Data File 2011 [7]. Hereafter we call these two files together as JENDL simply.

3 Calculation Results

The calculated results for electron spectra are shown in Fig. 1. The experimental data cited here are from a series of extensive measurements by the group of Schreckenbach *et al.* [8–11]. They measured the energy spectra of electron from fissionable-nuclide samples under irradiation using a spectrometer called BILL [12] at the Institute of Laue-Langevin. Hereafter we call this the 'BILL experiment.' The time-lapse from the beginning of irradiation, in accordance to which the number density N_i in eqs.(1) and (2) were calculated here, is indicated in parentheses. In addition to this, another set of experimental data is available [13]. Though this measurement is rather old and covers only the energy range up to 8.5 MeV for ^{235}U, it is also an elaborative work (we call this the 'UI experiment' after University of Illinois).

As is seen in Figs. 2 and 3, the main contributor to the antineutrino spectra is the Gamow-Teller transition. This is followed by the first-forbidden transition which covers 15% (around 2-3 MeV) ~ 20% (above 6 MeV) of the emitted antineutrinos. These two types overwhelm the Fermi-transition. Needless to say, this transition-wise break down of the antineutrino spectra is almost the same as the case of the corresponding electron spectra, which are not shown in Fig.1 so as to save the space. In these figures, three sets of $\bar{\nu}_e$ spectra converted from the measured electron spectra not only by measured themselves but by other people by their own sophisticated method.

4 Discussions

4.1 Below 8 MeV

Table 1 shows the C/E values for the electron spectra at typical energy points. Though the agreement between the calculation and the measurement is not always satisfactory for the practical applications previously mentioned, it should be underlined that the overall trends of the spectra below ~8 MeV are reproduced fairly well in the absolute-value basis without any tuning of the parameters in the theory. There seems to be two noticeable points in Table 1. One is the fact the calculations are persistently larger than the measured values over the whole energy range. This suggests that the overestimation might be related to the calculation of the number of beta-decay events rather than to the individual spectrum shape of each FP nuclide. This exact point stimulates us to investigate further, where we treat the FP decay-heat data including the β-ray spectra at very short cooling times on the same basis as the reactor lepton-spectra dealt with here. Another point is that the overall trend of the C/E for ^{238}U is different from others; that is, the C/E (~1.1) is closer to unity and less dependent on the electron energy. Considering that there had been a 25-years lapse of time after the last ^{241}Pu measurements and that the neutron source had been changed from the ILL-reactor to the FRM-II facility, a possibility cannot be excluded that some unknown bias exists between the new ^{238}U and the old ^{235}U and Pu measurements, or otherwise in both. At least we cannot exclude this possibility one-hundred percent.

4.2 Above 8 MeV

In all the cases from ^{235}U to ^{241}Pu, the same kind of the conspicuous overestimation against the BILL data is seen above ~8 MeV. It should be noted, however, the antineutrino with energy higher than 8 MeV has little importance as far as the current detection technique using the reaction $\bar{\nu}_e + p \rightarrow n + \bar{e}$ is premised. It is another puzzle that the UI curve, which is fairly consistent with the BILL-curve up to 7 MeV, suddenly deviates upward from BILL. It is noteworthy that even the contribution solely from the nuclides with $Q_\beta = 10 \sim 12$MeV exceeds the converted the $\bar{\nu}_e$ spectra above 8 MeV as is seen in Fig.4. Here it must be recognised that the gross theory is an average theory which describes the average property of decaying nuclides specified by their mass-number and even/odd-ness. Therefore, the prediction power of this theory for an aggregate behaviour of a small number of FPs is inherently limited and it must be applied very carefully to the reactor lepton spectrums above 8 MeV, namely, our concerns here.

Table 2 lists the top 10 contributors which constitute the high energy part of the $\bar{\nu}_e$ spectra at 9 MeV. It should be noted that the 8 odd-odd nuclides in Table 2 cover 78% of the spectral value at 9 MeV. As the origin of the large discrepancy in the $\bar{\nu}_e$ spectra above 8 MeV, we first think about the fission yield data. The effect of whole replacement of the fission yields for the summation calculation from JENDL to JEFF-3.1[14] seems to be rather restricted as is seen in Fig.5. When we see in details, however, a big differences between JENDL (or ENDF-B/VII[15]) and JEFF3.1 can not be ignored. Figure 6 shows the mass yield for A=88 to which important contributers Br-88 and As-88 in Table 2 belong. The charge-yield curves in JENDL and ENDF/B-VII.1 has an unexpected shoulder on the lower-*Z* side of the peak in contrast to the smooth behaviour seen in the JEFF-3.1. This is seen also in mass-chains A=86, 96, 100, 102, and 131 among others. This different behaviour gives rise to the spectrum variation between JENDL and JEFF as we saw in Fig.5 as a whole. Anyway, the uncertainty in the fission yields does not seem to be the origin of the big discrepancy above 8 MeV. In addition,as is argued in ref.[6], neither the ambiguity in Q_β does explain the discrepancy.

4.3 Selection Rules for Odd-Odd Nuclide Decay

Here we have to remember fact that the prediction power of the gross theory for an aggregate behaviour of a small number of FPs is less reliable than large number cases because individual properties of few nuclides can play a decisive role beyond the realm of this *average* theory. From Table 2 we see that a limited number of odd(*Z*)-odd(*N*) nuclides dominate the highest-energy part of the $\bar{\nu}_e$ spectra where transitions to the ground-state, if possible, might play the most important role.

Any β-decay of an odd-odd nuclide end up in an even-even nuclide whose low-lying levels are sparse compared to odd-*A* and odd-odd nuclides, and this enhances the importance of its ground-state transition. As the ground-state of any even-even nuclide is determined to be 0^+, we can easily estimate the effect of the selection rules on the β-transition to the low-lying states provided we know the spin-parity of the decaying odd-odd nuclide (see the last column of Table 2 where the latest NDS evaluation is shown). This kind of consideration was successfully applied to the reactor decay heat problem[17], and is expected to mitigate the serious disagreement above 8 MeV in the electron and the $\bar{\nu}_e$ spectra.

5 Concluding Remarks

We applied the gross theory of β-decay to calculate the reactor electron and antineutrino ($\bar{\nu}_e$) spectra emitted from ^{235,238}U and ^{239,241}Pu along with the summation methods. Though the agreement between the calculation and the measurement is not always satisfactory for the practical applications in nuclear technology and basic particle physics, it should be underlined that the overall trends of the spectra below ~8 MeV are reproduced fairly well in the absolute-value basis without any tuning of the parameters in the theory, which leaves promising room for further improvement. One aspect is a possible modification of the treatment of the spin-parity effect of decaying odd-odd nuclides. It would hopefully remedy the overestimation against the BILL-experimental data seen above 8 MeV both in the $\bar{\nu}_e$ and the electron spectra. Throughout the present arguments, however, we could not exclude a possibility that the lepton spectra by Schreckenbach *et al.* (practically the only one extensive series of measurements) are not free from any systematic bias. Generally speaking it might be risky to rely on a single experimental data set in practical applications where very high precision is required, even if their assumed experimental errors are small.

References

[1] K. Takahashi, M. Yamada, Progr. Theor. Phys. **41**, 1470-1503 (1969).
[2] S. I. Koyama, K. Takahashi, M. Yamada, Progr. Theor. Phys. **44**, 663-687 (1970).
[3] K. Takahashi, Progr. Theor. Phys. **45**, 1466-1492 (1971).
[4] T. Kondoh, T. Tachibana, M. Yamada, Progr. Theor. Phys. **74**, 708-724 (1985).
[5] T. Tachibana, M. Yamada, Y. Yoshida, Progr. Theor. Phys. **84**, 641-657 (1990),
[6] T. Yoshida, T. Tachibana, N. Hagura, S. Chiba, Progr. Nucl. Energy, **88**, 175-182 (2016).
[7] J. Katakura, JENDL FP Decay Data File 2011 and Fission Yield Data File 2011, Japan Atomic Energy Agency, Report JAEA-Data/Code 2011-25 (2011).
[8] K. Schreckenbach, G. Colvin, W. Gelletly, F. Von Feilitzsch, Phys. Lett. **160B**, 325-330 (1985).
[9] F. Von Feilitzsch, A. A. Hahn, K. Schreckenbach, Phys. Lett. **118B**, 162-166 (1982).
[10] A. A. Hahn, K. Schreckenbach, W. Gelletly, F. Von Feilitzsch, G. Colvin, B. Krusche, Phys. Lett. **B218**, 365-368 (1989).
[11] N. Haag, A. Gütlein, L. Hofmann, L. Oberauer, W. Potzel, K. Schreckenbach, Phys. Rev. Lett. **112**, 122501-1-5 (2014).
[12] W. Mampe, K. Schreckenbach, P. Jeuch, B. P. K. Maier, F. Braumandl, J. Larysz, T. von Egidy, Nucl. Instr. Meth. **154**, 127-149 (1978).
[13] N. Tsoulfanidis, B. W. Wehring, M. E. Wyman, Nucl. Sci. Eng. **43**, 42-53 (1971).
[14] M. A. Kelett, O. Bersillon, R. W. Mills, The JEFF-3.1/-3.1.1 radioactive decay data and fission data, OECD NEA Report N.6287, Nuclear Energy Agency/OECD (2009).
[15] M. B. Chadwick, et al., Nucl. Data Sheets **112**, 2887-2996 (2011).
[16] M. Wang, G. Audi, A. H. Wapstra, F. G. Kondev, M. MacCormick, X. Xu, B. Pfeifer, Chinese Phys. C **36**, 1603-2014 (2012).
[17] T. Yoshida, Nucl. Sci. Engn. **63**, 376-390 (1977).

Table 1. C/E (calculated/experimental) values of the electron spectra from ^{235,238}U and ^{239,241}Pu samples at typical energy points

Energy (MeV)	U-235	U-238	Pu-239	Pu-241
2.0	1.18	-	1.25	1.28
3.0	1.22	1.11	1.32	1.29
4.0	1.15	1.11	1.29	1.24
5.0	1.08	1.07	1.24	1.20
6.0	1.01	1.11	1.25	1.26
7.0	1.12	-	1.29	1.46
8.0	3.75	-	2.38	2.91
9.0	16.2	-	-	10.5

Table 2. Top 10 nuclides contributing to U235 $\bar{\nu}_e$ spectrum at 9 MeV

Nuclide	Q (MeV)	Fractional Contribution	Spin-Parity
Rb-94	10.29	0.240	3^- (Abriola,2006)
Y-98m	9.23	0.121	4 or 5 (Singh,2003)
As-86	11.54	0.087	not given (Negret, 2015)
Br-90	10.35	0.077	not given (Browne, 1997)
Y-100m	9.46	0.075	3, 4 or 5 (Singh, 2008)
Br-88	8.96	0.068	2^- (McCutchan,2014)
Rb-95	9.28	0.065	-
As-84	10.09	0.042	3^- (Abriola,2009)
As-88	12.60	0.038	not given (McCutchan,2014)
Br-91	9.80	0.035	-
F.R.Sum	-	0.848	-

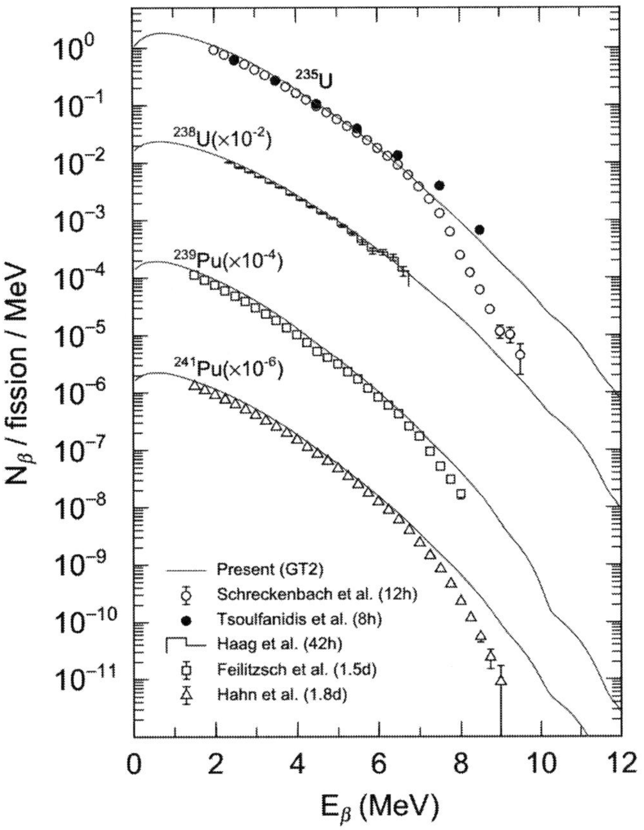

Figure 1. Comparison of the aggregate electron spectra from $^{235,238}U$ and $^{239,241}Pu$ samples between the present calculation and the measurements

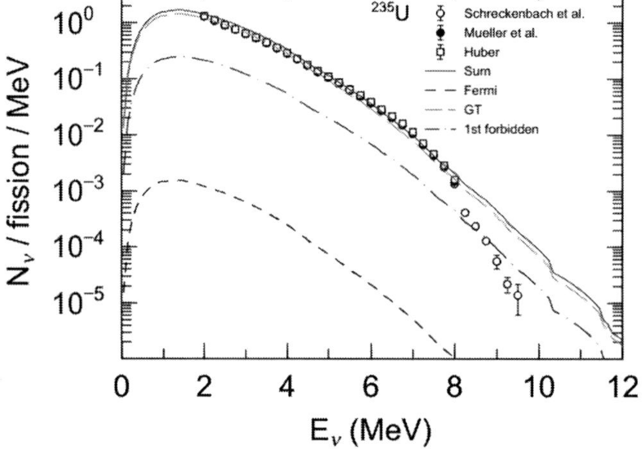

Figure 2. Calculated antineutrino spectrum from ^{235}U sample and its component-wise breakdown. For reference are shown three spectra converted from the electron spectrum measured by Schreckenbach *et al.*

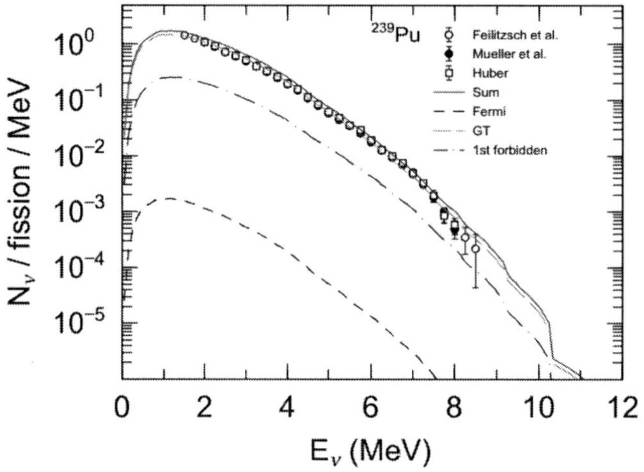

Figure 3. Calculated antineutrino spectrum from ^{239}Pu sample and its component-wise breakdown. For reference are shown three spectra converted from the electron spectrum measured by Feilitzsch *et al.*

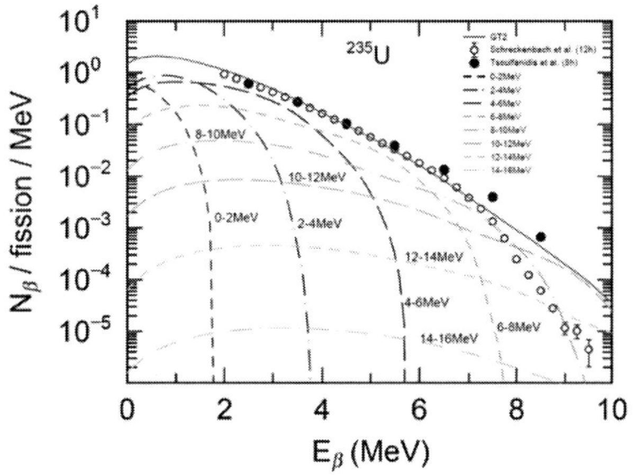

Figure 4. Calculated antineutrino spectrum from ^{235}U sample and its component-wise breakdown. For reference are shown three spectra converted from the electron spectrum measured by Schreckenbach *et al.*

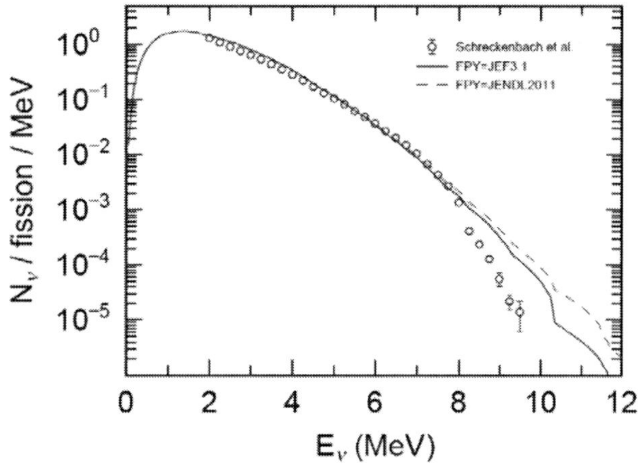

Figure 5. Change induced by replacement of JENDL FP yileds by JEFF yields in calculated antineutrino spectrum from ^{235}U

DOI: 10.1051/epjconf/201612210002

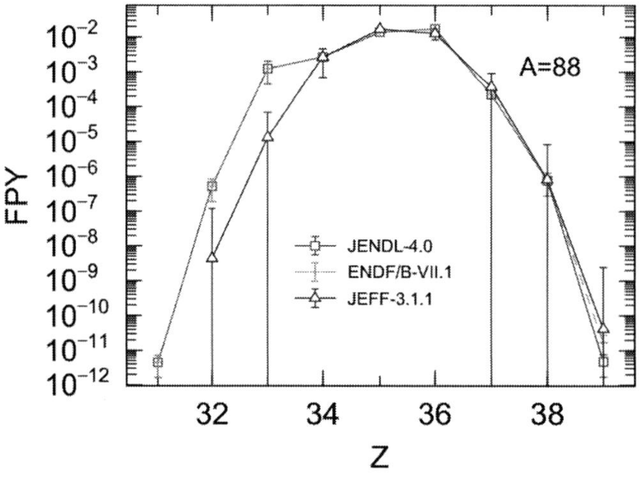

Figure 6. Charge yield for thermal-neutron induced fission of ^{235}U (A = 88)

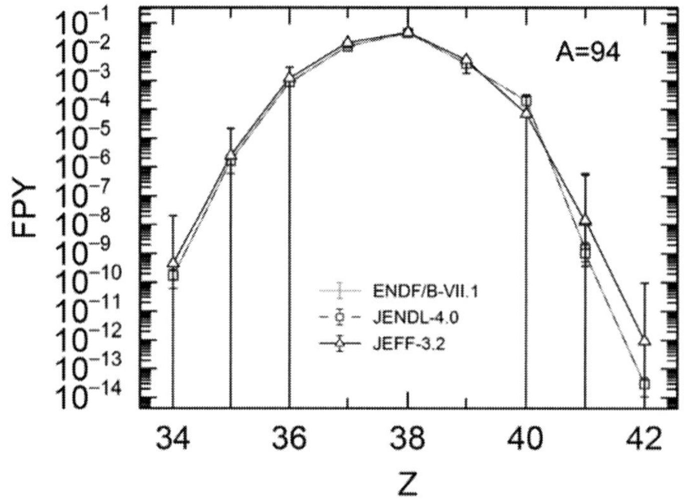

Figure 7. Charge yield for thermal-neutron induced fission of ^{235}U (A = 94)

EPJ Web of Conferences **122**, 10003 (2016) DOI: 10.1051/epjconf/201612210003
*CNR*15*

Benchmark calculation for radioactivity inventory using MAXS library based on JENDL-4.0 and JEFF-3.0/A for decommissioning BWR plants

Ken-ichi Tanaka

The Institute of Applied Energy, 1-14-2 Nishi-shinbashi, minato-ku, Tokyo, 105-003, Japan

Tokyo Institute of Technology, 2-12-N1-9 Ookayama, Meguro-ku, Tokyo 152-8550, Japan

Abstract. We performed benchmark calculation for radioactivity activated in a Primary Containment Vessel (PCV) of a Boiling Water Reactor (BWR) by using MAXS library, which was developed by collapsing with neutron energy spectra in the PCV of the BWR. Radioactivities due to neutron irradiation were measured by using activation foil detector of Gold (Au) and Nickel (Ni) at thirty locations in the PCV. We performed activation calculations of the foils with SCALE5.1/ORIGEN-S code with irradiation conditions of each foil location as the benchmark calculation. We compared calculations and measurements to estimate an effectiveness of MAXS library.

1 Introduction

According to the Japanese law, any operating organization of decommissioning of nuclear facilities such as electric power companies needs to submit an application for decommissioning and to receive the authorization. Tasks for the decommissioning planning start with radiological characterization of the facilities. Reliable information about radioactivity inventory obtained from the task is much important to establish a decommissioning strategy and to perform decommissioning activities safely and rationally.

We performed benchmark calculation of radioactivity of activation foils irradiated by neutrons in a Primary Containment Vessel (PCV) of a Boiling Water Reactor (BWR) by using SCALE5.1/ORIGEN-S code [1] with MAXS library [2]. MAXS library is activation cross section data library for activation calculation in the PCV of the BWR.

Neutron energy spectra in the PCV of the BWR tend to show a peak in epi-thermal energy range. Thus, use of ORIGEN-S cross section library that was developed by collapsing cross sections with a typical neutron energy spectrum of a Light Water Reactor (LWR) is considered to be inappropriate under the condition in the PCV. In particular, applying it to estimations of a radionuclide generated by a reaction that has threshold energy would make a large ambiguity. To give values of cross sections by weighting with the neutron energy spectra in the PCV would be one of solutions.

A new multi-group neutron activation cross section library MAXS [2] was developed to solve the problem mentioned above in JAEA based on JENDL-4.0 and JEFF-3.0/A. Activation cross sections are collapsed by using various neutron energy spectra in the PCV (e.g. core internals, a reactor pressure vessel, PCV internals and a Biological Shielding Wall (BSW)). Furthermore, data of cross section for a number of reactions such as (n,γ), (n,p), (n,α),(n,2n) and so on were considered to generate the library.

We performed activation calculations at activation foils in the PCV of the BWR by using SCALE5.1/ORIGEN-S with MAXS library as a benchmark calculation. We compared calculated radioactivities with measured ones to estimate the effectiveness.

2 Outline of benchmark problem

2.1 A typical spectrum of a LWR and neutron energy spectra in the PCV of the BWR

A typical spectrum of the LWR used as a weighting function in a generation of ORIGEN library is shown in Figure 1 [3].

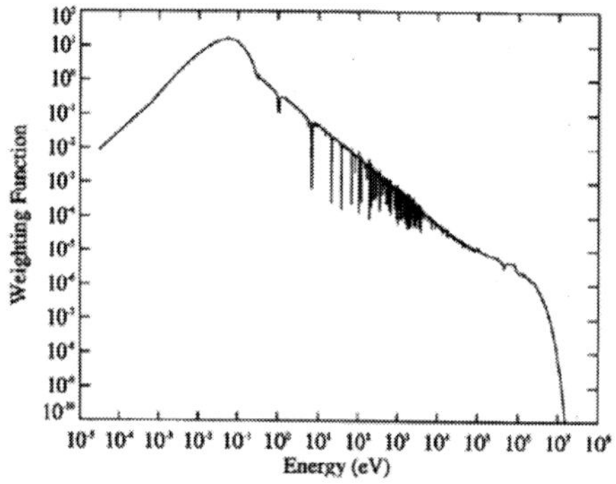

Figure 1 A typical neutron energy spectrum of a Light Water Reactor used as a weighting function in a generation of ORIGEN library [3]

The typical spectrum is comprised of three kinds of spectra; thermal Maxwellian spectrum in thermal energy range, 1/E spectrum in epi-thermal energy range and fission spectrum in fast energy range. The shape of the typical spectrum has been given as results from neutron transport phenomena which are supposed to be observed in the LWR. Emitting neutrons by nuclear fission and moderation by light water are being considered in the phenomena. However, the phenomena would be too simple to apply to a complicated structure like inside the PCV of the BWR.

Neutron energy spectra in the PCV of the BWR tend to show a peak in epi-thermal energy range as seen in Figure 2 [4]. The right side of Figure 2 shows position of neutron energy spectra drawn in the left side. Positions ①, ②, ③ and ④ are in a Reactor Pressure Vessel (RPV). Fission spectrum is dominant in the spectra at these positions. The spectra show a peak in epi-thermal energy range at positions ⑤ and ⑦ which are located inside the PCV. The last one at position ⑥ is in concrete of a Biological Shielding Wall (BSW). A shape of the spectrum in the BSW is almost similar to that of the typical spectrum.

DOI: 10.1051/epjconf/201612210003

Various kinds of neutron transport phenomena occur in the PCV. Spectra around a reactor core such as positions ①, ②, ③ and ④ are almost similar to a fission spectrum. Neutrons leak from the reactor core and are moderated in coolant water flowing through an area between a shroud, which surrounds the reactor core, and a reactor pressure vessel (RPV). Although a spectrum in the area shows a shape which similar to the typical LWR spectrum, thermal neutrons are absorbed during penetrating the RPV. Because of a mechanism of the transport phenomena mentioned here, spectra in the PCV tend to show a peak in epi-thermal range such as positions ⑤ and ⑦.

According to a comparison between the shape of the typical spectrum of a LWR and those of spectra in the PCV of the BWR, it would be inappropriate to apply ORIGEN library, which is generated with the typical spectrum, to activation calculation in the PCV. In the benchmark calculations, we applied MAXS library, which is generated by collapsing with neutron energy spectrum at position ⑦, to activation calculation for radioactiveities measured with activation foils in the PCV.

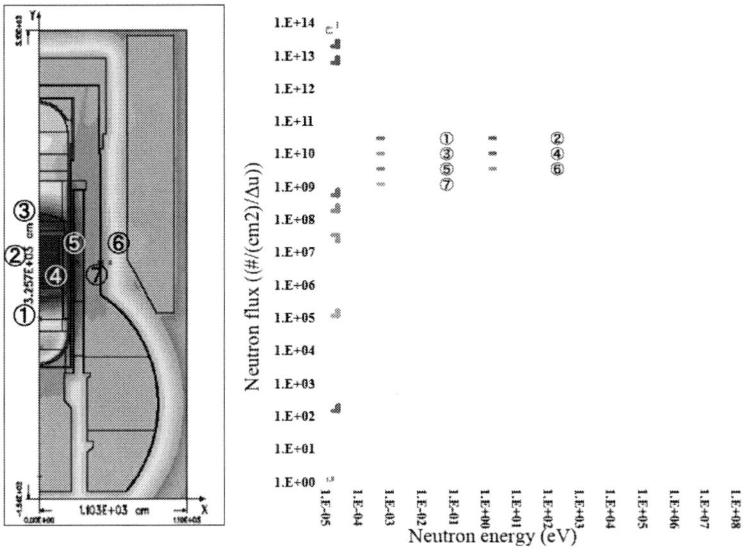

Figure 2 Neutron energy spectra in the PCV of the BWR

2.2 Radioactivity measurements with activation foil in the PCV

We measured radioactivities induced by neutrons in the PCV with three types of activation foils as foil activation detectors; Gold (Au) foils, Gold foils covered by Cadmium (Au+Cd) foils and Nickel (Ni) foils. We can obtain radioactivities induced by neutron-flux of thermal energy range (thermal-flux: energy less than 1.0 eV) from a difference between measurements of Au foils and Au+Cd foils because Cd cover cuts off thermal-flux on the Au+Cd foils, which measures radioactivities induced by neutron-flux of epi-thermal energy range (epi-thermal-flux: energy within 1.0 eV to 1.0 MeV). We used Ni foils as threshold detectors to measure radioactivities induced by neutron-flux of fast energy range (fast-flux: energy higher than 1.0 MeV).

Activation foils at location 1 to 15 were irradiated from Feb. 17th 2001 to Feb.25th 2002; 373 days. Further, we added fifteen locations at 16 to 30 from Oct.26th 2004 to Nov.11th 2005; 381 days.

Locations of activation foil in the PCV are shown in Figure 3. We set foils at thirty locations where characteristics of neutron transport phenomena in the PCV can be observed [4]. To observe neutron streaming caused by a Feed Water (FW) pipe, we set 6 sets of activation foils around the pipe as shown in the upper right corner in the figure.

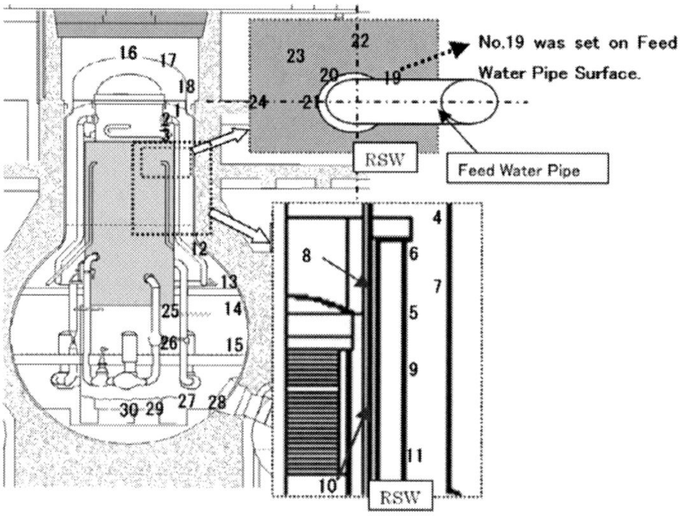

Figure 3 Locations of activation foil in the PCV of the BSW

3 Activation calculations

3.1 Calculation conditions

We performed activation calculations on irradiation conditions of each foil location by using SCALE5.1/ORIGEN-S [1]. We input spectrum indexes, which are input data of ORIGEN-S code, given by three group neutron-fluxes at each location, which were obtained from two dimensional (2D) distribution calculation of neutron-flux in our previous work [4]. Distribution of spectrum indexes such as RES, Fast and Φ_{th} are shown in Figure 4 (Here, Φ_{th} indicates distribution of thermal flux). We extracted values of the spectrum indexes at each foil location from the distributions to input ORIGEN-S code. We also input irradiation histories as mentioned above.

In the calculations of this work, we applied MAXS library collapsing with the spectrum of ⑦ in Figure 2. The spectrum of ⑦ is typical one in the PCV which shows a peak in epi-thermal energy range.

Applying the library with the spectrum of ⑦ to the activation calculations in the PCV would be appropriate.

However, applying the library to the calculation of foil locations around the FW pipe might be inappropriate because effects of neutron-streaming around the FW pipe make distribution of neutron-flux around it complex.

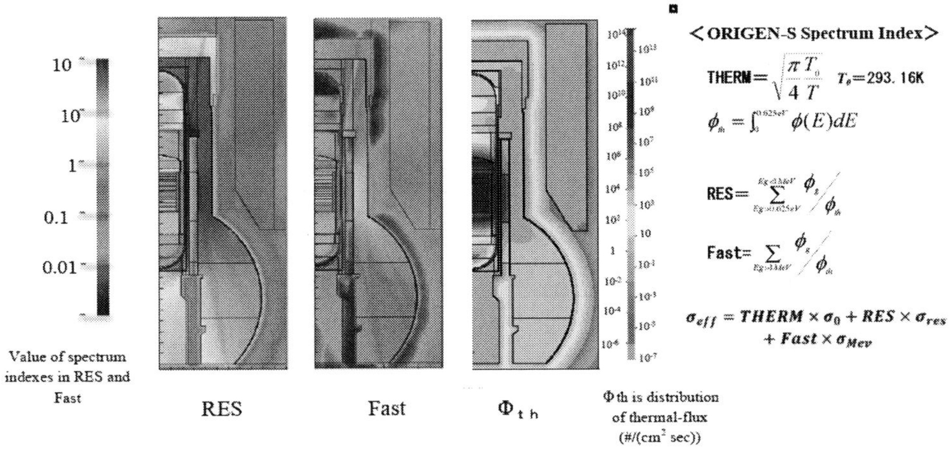

Figure 4 Distribution of ORIGEN spectrum indexes

3.2 Calculation results

In works to be related to estimations of radioactivity for decommissioning, reliability of calculated radioactivity is verified with measured radioactivity. We judge calculation radioactivity distribution "Reliable", when comparison between calculated value and measured value (C/M) is less than one order magnitude ($1.0<C/M<10.0$) [5]. C/Ms of the calculations with ORIGEN library and with MAXS library at each foil location are shown in Figure 5. C/Ms of Au foils and Ni foils, which were calculated both with ORIGEN library and with MAXS library, are shown in the figure. When C/M nears 1.0, it means that calculated value simulates measured value well. C/Ms with MAXS library are smaller than those with ORIGEN library by almost half.

In C/Ms of Au foils, MAXS library is effective to improve the reliability of the activation calculations in the PCV except locations around the PCV as seen in the figure. C/Ms of these locations were less than 1.0. C/M, which is less than 1.0, indicates that calculation is underestimation. We suppose that the underestimation would result from the complex phenomena around FW pipe as mentioned above.

DOI: 10.1051/epjconf/201612210003

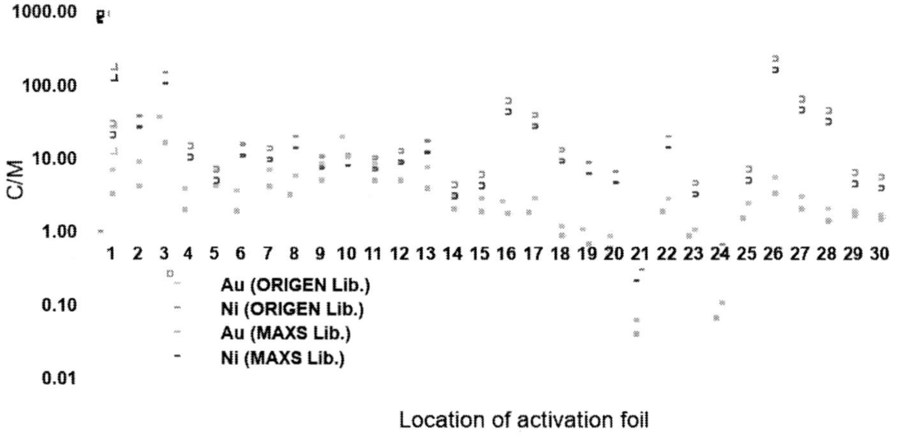

Figure 5 Comparisons between calculated and measured radioactiveities (C/M) with ORIGEN library and with MAXS library at each foil location

In C/Ms of Ni foils, almost all C/Ms with ORIGEN library were larger than 10.0. These means calculations were overestimation. Although applying MAXS library improved C/Ms, Applying it was not so effective as those of Au foils. Activation reaction of Ni foil has threshold energy. ORIGEN-S code performs activation calculation with not fast neutron-flux but with total neutron-flux. Since using total neutron-flux as irradiation neutron would cause the overestimation, the effectiveness of the library could not made clear. C/Ms at foil locations around the FW pipe shows similar tendency as Au foils.

4 Summary and conclusions

We performed activation calculations at the locations in the PCV of the BWR where radioactiveities of both Au and Ni were measured as benchmark calculation for MAXS library. To examine an applicability of the library, we compered calculated radioactivities with measured ones. For Au foils, the calculations provided the applicability of applying MAXS library except around FW pipe where the complex phenomena of neutron transport would occur. We suppose that the library would be effective to the other nuclides of which activation reaction cross section spread through whole energy range that the activation calculations deal with. On the other hand, the applicability of the library for Ni foils were not so clear because of the threshold energy of activation reaction of Ni.

In this work, only one spectrum was used to generate the library. The underestimations shown around FW pipe would be improved by applying libraries which were generated with appropriate spectra in the PCV. Activation calculations of nuclides, which have threshold energy of activation reaction, activation calculations might need not only libraries generated with appropriate spectra but also neutron-flux of irradiation.

References

1. I.C.Gauld et.al, ORNL/TM - 2005/39 Version5.1 Vol.II, Book1, Sect.F7, (2006)
2. K. Okumura et.al, *Proc. of SND 2014*□ 150716 (2014)
3. K. Tanaka et al, PNE,**85**: 254-270 (2015)
4. I.C.Gauld et.al, ORNL/TM - 2005/39 Version5.1 Vol.III, Sect.M6, (2006)
5. IAEA, *Safety Series No.100*, ISBN 9201240899 (1989)

EPJ Web of Conferences **122**, 11001 (2016) DOI: 10.1051/epjconf/201612211001
*CNR*15*

Cluster production within antisymmetrized molecular dynamics

Akira Ono[1,a]

[1] *Department of Physics, Tohoku University, Sendai 980-8578, Japan*

Abstract. Clusters are quite important at various situations in heavy-ion collisions. Antisymmetrized molecular dynamics was improved to take into account the correlations to form light clusters, such as deuterons and α particles, and light nuclei composed of several clusters. The momentum fluctuations of emitted particles are also taken into account by a simple method. Formation of fragments and light clusters in a wide range of heavy-ion collisions was well described with a single set of model parameters. Fragmentation in a proton induced reaction was also well reproduced by introducing cluster correlations. Calculated results demonstrate strong impacts of clusters in various observables including those usually regarded as probes of the density dependence of symmetry energy.

1 Introduction

Fragments and light clusters are copiously produced in various kinds of nuclear reactions such as in heavy-ion collisions. For example, in Xe + Sn central collisions at 50 MeV/nucleon, the INDRA data [1] show that only 10% of total protons in the system are emitted as free protons and all the other protons are bound in light clusters and fragments in the final state. About 20% of protons are bound in α particles, about 10% are bound in clusters of $A = 2$ and 3, and about 60% are bound in heavier fragment nuclei. Even though the incident energy is raised, to 250 MeV/nucleon for example, the free-proton ratio is still about 20% as shown by the FOPI data [2] for the central Au + Au collisions. In these heavy-ion collisions, the system is first compressed and then expands relatively rapidly. The formation of fragments and light clusters should be closely related to the property of such expanding nuclear matter. On the other hand, the target nucleus can also break up into fragments in light-ion induced reactions, as in the experiments [3], even though the excitation of the nucleus does not primarily induce strong collective expansion. Formation of fragments and clusters is one of the essential features of these reactions, since they are produced in most of the events in these reactions with the participation of most of the nucleons in the system.

It has been a longstanding problem to understand fragment formation from microscopic points of view. Transport theories, which solve the time evolution of the many-nucleon system or its one-body distribution, have generally been a powerful approach to understand important aspects of collision dynamics in the Fermi energy domain and in higher energies. Among transport models, the molecular dynamics models [4–7] may be expected to be suitable for fragmentation reactions because many-nucleon correlations are handled at least in the sense of classical dynamics. However, studies have

[a] e-mail: ono@nucl.phys.tohoku.ac.jp

shown that some quantum features are required in order to properly describe fragment formation. The antisymmetrized molecular dynamics (AMD) approach [5] takes into account the full antisymmetrization of many-nucleon wave function, which is certainly an important quantum feature. Emergence of many fragmentation channels is also a crucial idea to consistently understand the fragment formation and the single-particle dynamics [8, 9]. Furthermore, light clusters are highly quantum objects because they have only one (or a few) bound state.

When light clusters are produced in reactions, they are probably still surrounded by other particles. The property of a cluster in idealistic nuclear medium is typically calculated with an inmedium Schrödinger equation as in the calculation by Röpke [11]. The effect of Pauli blocking is taken into account. A cluster can exist as a bound state, depending on the density of the medium and the momentum of the cluster relative to the medium. Then, in actual systems, many clusters may coexist and change the nuclear matter properties and the collision dynamics. During the time evolution, clusters will be repeatedly created and broken by reactions such as $p+n+X \leftrightarrow d+X'$, $p+n+d+X \leftrightarrow \alpha+X'$ and many others.

Recently the heavy-ion data from Texas A&M have been analyzed to extract the cluster composition and other properties of nuclear matter as functions of the evolution of the temperature and the density [12]. Comparison of the extracted equilibrium constants for the cluster composition with various equation-of-state (EOS) models suggests necessity of a cluster suppression mechanism such as medium effects [13]. Nevertheless, in general, the link between heavy-ion collisions and EOS should be carefully checked because of the dynamical nature of collisions.

It is therefore demanded that the transport models should solve the time evolution of collisions by correctly treating cluster correlations. However, transport models have been developed based on the single-nucleon motions, and only few models treat cluster correlations explicitly. In a version of BUU by Danielewicz *et al.* [14], clusters up to ^{3}H and ^{3}He have been introduced as new particle species and the equations for the distribution functions are coupled by the collision terms representing various reactions of clusters and nucleons. A recent calculation by this model demonstrates that collision dynamics is influenced by the clusters [15]. On the other hand, in the antisymmetrized molecular dynamics (AMD) approach, we have introduced cluster correlations by improving the two-nucleon collision procedure [16–18]. The aims of this article are to report the present status of this approach and to demonstrate the importance of clusters in the studies of heavy-ion collisions.

2 AMD approach

AMD uses a single Slater determinant of Gaussian wave packets [5],

$$\langle \mathbf{r}_1 \cdots \mathbf{r}_A | \Phi(Z) \rangle = \det_{ij}\Big[\varphi_{\mathbf{Z}_i}(\mathbf{r}_j)\chi_{\alpha_i}(j)\Big], \tag{1}$$

where the spatial wave functions of nucleons are given by

$$\langle \mathbf{r} | \varphi_{\mathbf{Z}} \rangle = \Big(\frac{2\nu}{\pi}\Big)^{3/4} \exp\Big\{-\nu\Big(\mathbf{r} - \frac{\mathbf{Z}}{\sqrt{\nu}}\Big)^2 + \frac{1}{2}\mathbf{Z}^2\Big\} \tag{2}$$

and χ_α is the spin-isospin wave function with $\chi_\alpha = p\uparrow, p\downarrow, n\uparrow$, or $n\downarrow$. Thus the many-body state $|\Phi(Z)\rangle$ is parametrized by a set of complex variables $Z \equiv \{\mathbf{Z}_i;\ i = 1, \ldots, A\}$, where A is the number of nucleons in the system. The width parameter $\nu = (2.5\ \mathrm{fm})^{-2}$ is treated as a constant parameter common to all the wave packets. The Gaussian wave packet $|\varphi_{\mathbf{Z}}\rangle$ has the minimum uncertainty $\Delta x \Delta p = \frac{1}{2}\hbar$ with $\Delta x^2 = 1/(4\nu)$ and $\Delta p^2 = \hbar^2\nu$.

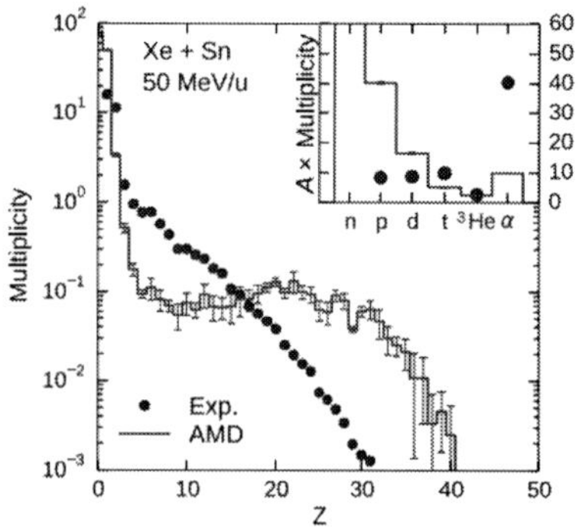

Figure 1. A result of AMD without cluster correlations (red historgram) for the fragment charge distribution in central Xe + Sn collisions at 50 MeV/nucleon. The INDRA data [1] are shown by blue points. The inset shows the number of nucleons in light particles.

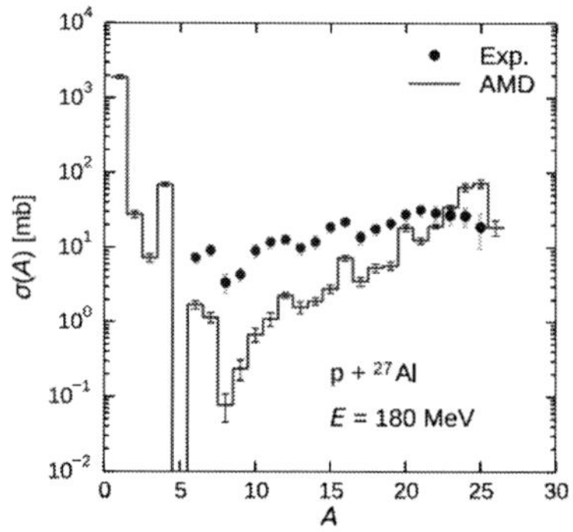

Figure 2. A result of AMD without cluster correlations (red histogram) for the fragment production cross sections in $p+$ Sn collisions at 180 MeV. The experimental data [3] are shown by blue points.

The time evolution of the wave-packet centroids is determined by the motion in the mean field, derived from the time-dependent variational principle, and the stochastic two-nucleon collision process. The former is written as

$$i\hbar \sum_{j\tau} C_{i\sigma,j\tau} \frac{dZ_{j\tau}}{dt} = \frac{\partial \mathcal{H}}{\partial Z^*_{i\sigma}}, \qquad \mathcal{H}(Z) = \frac{\langle \Phi(Z)|H|\Phi(Z)\rangle}{\langle \Phi(Z)|\Phi(Z)\rangle} - \frac{3\hbar^2 \nu}{2M} A + T_0(A - N_{\rm F}(Z)) \tag{3}$$

with a hermitian matrix $C_{i\sigma,j\tau}$ and the expectation value $\mathcal{H}$ of the effective Hamiltonian, such as the Skyrme interaction, with subtraction of spurious kinetic energies of fragments [5]. During the time evolution, any pair of two nucleons will collide with a suitable probability when the distance between them is small. A two-nucleon collision is treated as a quantum mechanical transition from an AMD state $|\Phi_i\rangle$ to another AMD state $|\Phi_f\rangle$ specified by a relative momentum between the scattered two nucleons $(p_{\rm rel}, \Omega)$. The transition rate is expressed as

$$v d\sigma = \frac{2\pi}{\hbar} |\langle \Phi_f|V|\Phi_i\rangle|^2 \delta(E_f - E_i) \frac{p_{\rm rel}^2 dp_{\rm rel} d\Omega}{(2\pi\hbar)^3}, \tag{4}$$

assuming a unit normalization volume. In general, medium modification is introduced for the scattering matrix elements. It should be noted that some medium effect still exists also in the $p_{\rm rel}$-dependence of the final state energy E_f. The Pauli blocking for the scattered nucleons is taken into account [5].

2.1 AMD without clusters

In the usual treatment of two-nucleon collisions, only the states of the scattered two nucleons are changed in the final state $|\Phi_f\rangle$. See Ref. [5] for the precise description of the method which employs 'physical coordinates'.

It has been known for a long time that this very basic version of AMD does not well reproduce the fragmentation in collisions of heavy nuclei. For example, in Xe + Sn central collisions at 50 MeV/nucleon (see Fig. 1), the relatively heavy nuclei are produced too frequently and the yield of lighter intermediate mass fragments is too small. The production of α particles is seriously underestimated and nucleons are overproduced. The result after calculating the statistical decay of primary fragments is shown as well as in all the other figures in this article. A similar problem was observed in the Ca + Ca collisions at 35 MeV/nucleon [19]. The basic version of AMD was applied to the fragmentation in the $p + {}^{27}\mathrm{Al}$ reaction at 180 MeV [20], where the breakup of the target nucleus does not happen sufficiently to reproduce the yield of light fragments. A similar result is shown in Fig. 2 calculated with a new code by turning off extensions. These problems cannot be solved by adjusting the effective interaction or the inmedium nucleon-nucleon cross sections.

In many cases, the reproduction of fragment yield was improved very much by introducing wave-packet splitting as a quantum branching process based on the single-particle motions with decoherence [8, 9, 19, 21]. However, a consistent reproduction of the yields of light clusters and light intermediate mass fragments has not been achieved by a single set of model parameters for all the reaction systems. This may suggest that the correlations to produce light clusters should be treated more explicitly.

It should be noted that the AMD wave function is suitable for describing cluster correlations in the ground state and low-lying excited states of nuclei [9]. When a cluster is placed in a nucleus within AMD, a reasonable value of the cluster-binding energy is obtained depending on the location and the center-of-mass momentum of the cluster [10]. The important question is, however, whether such clusterized states are realized with correct probabilities during the evolution of reactions. In fact, after a two-nucleon collision, one (or both) of the scattered nucleons may accidentally form a cluster with surrounding nucleon(s). The probability is governed by the classical phase space or the classical density of states $\tilde{D}(E)$ of the internal degrees of freedom of a cluster to be formed. The density of states is $\tilde{D}(E) = 0$ for $E < E_{\mathrm{gs}}$ below the ground state energy of the cluster, and it starts at $E = E_{\mathrm{gs}}$ continuously. Since the cluster-binding energy $|E_{\mathrm{gs}}|$ is small due to the cancellation of the potential and kinetic energies, the bound phase-space volume $\int_{E_{\mathrm{gs}}}^{0} \tilde{D}(E)dE$ cannot be large enough to correspond to a single quantum bound state. This is the reason why the cluster correlations do not emerge dynamically with reasonable probabilities. It is necessary to treat cluster correlations more explicitly as in the next subsection.

2.2 AMD with clusters

Light clusters have only one (or a few) bound state so that the density of states of the internal degrees of freedom is $D(E) = \delta(E - E_{\mathrm{gs}}) + D_{\mathrm{continuum}}(E)$. As mentioned above, the contribution of the discrete bound state cannot be suitably taken into account by the classical phase space of the wave packet centroids. To overcome this problem, an extension has been introduced [16] to allow direct formation of light clusters with $A = 2, 3$ and 4 in the final state $|\Phi_f\rangle$ in Eq. (4) for each two-nucleon collision. Namely, in this new method with cluster correlations, when two nucleons N_1 and N_2 collide, we consider the process

$$N_1 + N_2 + B_1 + B_2 \rightarrow C_1 + C_2 \tag{5}$$

in which each of the scattered nucleons N_j ($j = 1, 2$) may form a cluster C_j with a spectator particle B_j. This process includes the collisions without cluster formation as the special case of $C_j = N_j$ with empty B_j. The transition rate of the cluster-forming process is given by Eq. (4) with the suitable choice of the final state $|\Phi_f\rangle$. When a cluster is formed, the corresponding wave packets are placed at the same phase-space point, i.e., the cluster internal state is represented by the harmonic-oscillator $(0s)^n$

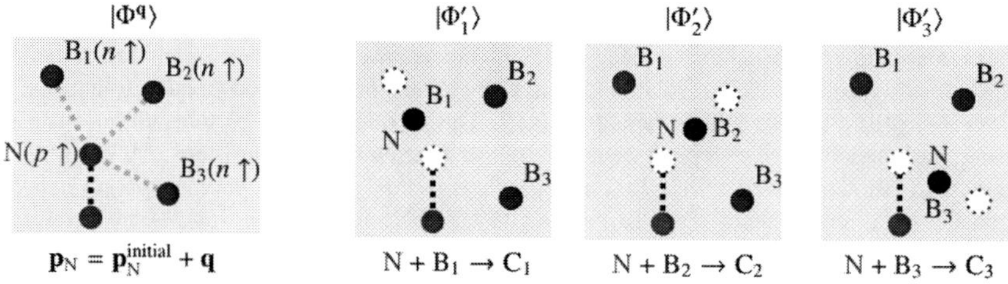

Figure 3. An example of deuteron formation in a two-nucleon collision, in which one of the scattered nucleon N (proton with up-spin) may form a deuteron cluster with any one of B_1, B_2 or B_3 (neutrons with up-spin). The filled circles depict the wave packet centroids. From Ref. [16].

configuration. Denoting the initial and final states of the $N_j + B_j$ system by $|\varphi_j\rangle$ and $|\varphi'_j\rangle$, respectively, we have the transition rate

$$vd\sigma = \frac{2\pi}{\hbar} |\langle\varphi'_1|\varphi_1^{\mathbf{q}}\rangle|^2 \, |\langle\varphi'_2|\varphi_2^{-\mathbf{q}}\rangle|^2 \, |M|^2 \, \delta(E_f - E_i) \frac{p_{\text{rel}}^2 dp_{\text{rel}} d\Omega}{(2\pi\hbar)^3}, \tag{6}$$

where $|\varphi_j^{\pm\mathbf{q}}\rangle = e^{\pm i\mathbf{q}\cdot\mathbf{r}_j}|\varphi_j\rangle$ are the states after the momentum transfer $\pm\mathbf{q}$ to the nucleons N_j $(j = 1,2)$, and (p_{rel}, Ω) is the relative momentum between N_1 and N_2 in these states. The matrix element $|M|^2$ is the same as for the usual two-nucleon collisions. We use an average value of $|M|^2$ evaluated at p_{rel} and that evaluated at the initial relative momentum. The angular momentum conservation is taken into account by adjusting the collective rotation of the part near the colliding nucleons N_1 and N_2.

The actual situation of a two-nucleon collision requires more considerations because there are many possible ways of forming a cluster for each N of the scattered nucleons N_1 and N_2. For a scattered nucleon N, we first consider the possibility that N may form a cluster with one of the nucleons $\{B_k;\ k = 1,2,\ldots\}$ which have the same spin-isospin state (see Fig. 3). This spin-isospin state that is studied first is randomly decided. The cluster-formed state is denoted by $|\Phi'_k\rangle$ which is obtained, by first changing the state to $|\Phi^{\mathbf{q}}\rangle$ by the momentum transfer $\mathbf{q}$ to N, and then moving the two wave packets of N and B_k to the same phase-space point without changing their center of mass. Since the different final states are not orthogonal $\mathcal{N}_{kl} = \langle\Phi'_k|\Phi'_l\rangle \neq \delta_{kl}$, the probability that N forms a cluster with one of $\{B_k\}$ should be calculated as

$$P = \sum_{kl} \langle\Phi^{\mathbf{q}}|\Phi'_k\rangle \mathcal{N}_{kl}^{-1} \langle\Phi'_l|\Phi^{\mathbf{q}}\rangle = \sum_k |v_k|^2, \tag{7}$$

$$v_k = \sum_l \mathcal{N}_{kl}^{-1/2} \langle\Phi'_l|\Phi^{\mathbf{q}}\rangle, \tag{8}$$

This probability is calculated with an approximation that the many-body state is a direct product of wave packets centered at the 'physical coordinates' [5]. With the calculated probability P, a cluster will be formed with one of $\{B_k\}$. It is somewhat arbitrary which one of $\{B_k\}$ should be chosen with what probability. In the present calculations, we choose B_k with the relative weight $|v_k|^{2\gamma}$ with the parameter $\gamma = 2.0$. This partial probability should replace the overlap probability $|\langle\varphi'_1|\varphi_1^{\mathbf{q}}\rangle|^2$ or $|\langle\varphi'_2|\varphi_2^{-\mathbf{q}}\rangle|^2$ in Eq. (6). With the rest of the probability $(1-P)$, the particle N does not form a cluster with a nucleon of this spin-isospin state. The procedure is repeated for other spin-isospin states for B. The particle N in the above description should be regarded as a cluster, instead of a scattered nucleon, if a (sub)cluster has been already formed in previous steps of the repetition. Thus the formation of light clusters is considered up to an α particle. It should be noted that the probability factors for different combinations of the formed clusters (C_1, C_2) decided in this way are functions of $p_{\rm rel}$ (or the momentum transfer $\mathbf{q}$) and the value of $p_{\rm rel}$ that conserves the energy depends on (C_1, C_2). A numerical procedure has been developed to treat this situation consistently.

Even when the cluster formation is introduced, the many-body state is always represented by an AMD wave function which is a Slater determinant of nucleon wave packets. The time evolution of the many-body state is solved just as usual without depending on whether some of the wave packets form clusters due to collisions in the past (except for the cluster-cluster binding process and for the momentum fluctuation in the next subsections). This is in contrast to the case of a version of BUU by Danielewicz *et al.* [14] where clusters are treated as new particle species. In our approach, a cluster may be broken by the mean field or by a two-nucleon collision between a nucleon in it and another nucleon in the system like a process $d + X \to n + p + X'$. It is also possible that, in the final state of this two-nucleon collision, the same cluster is formed again like an elastic process $d + X \to d + X'$. Thus various kinds of cluster reactions are taken into account without introducing many parameters. We only need to assume some inmedium two-nucleon cross sections (or matrix elements). In the calculations in the next section, we have chosen a so-called screened cross section [15]

$$\sigma_{\rm NN} = \sigma_0 \tanh(\sigma_{\rm free}/\sigma_0), \quad \sigma_0 = y\rho^{-2/3}, \quad y = 0.85. \tag{9}$$

In our previous works [10, 16, 17], it was necessary to introduce a reduction factor for the overall cluster-formation probabilities to reproduce multifragmentation data. In the present work, however, such a reduction factor does not seem necessary any longer when the momentum fluctuation is taken into account (see Sec. 2.4).

2.3 Cluster-cluster binding process

Many of light nuclei (Li, Be *etc.*) have only one or a few bound states which may be regarded as bound states of internal clusters. The quantum-mechanical probability of forming such a nucleus is not consistent with the semiclassical phase space with which it can be formed in the standard treatment of AMD. Therefore, for a better description, inter-cluster correlation is introduced as a stochastic process of binding clusters.

The basic idea is to replace the relative momentum between clusters by zero if moderately separated clusters ($1 < R_{\rm rel} < 7$ fm) are moving away from each other with a small relative kinetic energy ($\mathbf{R}_{\rm rel} \cdot \mathbf{V}_{\rm rel} > 0$ and $\frac{1}{2}\mu V_{\rm rel}^2 < 8$ MeV). In addition to these conditions, linking is allowed only if each of the two clusters is one of the three closest clusters of the other when the distance is measured by $[(\mathbf{R}_{\rm rel}/3\ {\rm fm})^2 + (\mathbf{V}_{\rm rel}/0.25c)^2]^{1/2}$, so that linking usually occurs in dilute environment. Non-clustered nucleons are treated here in the same way as clusters but two nucleons are not allowed to be linked. Two clusters also should not be linked if they can form an α or lighter cluster due to the combination of their spins and isospins. It is possible that more than two clusters are linked by this condition. However, only in the case that the mass number of the linked system is in the range $6 \le A_{\rm CC} \le 10$, the binding is performed for the linked system. The binding eliminates the velocities of clusters in the center-of-mass frame of the linked system. This combined system is called CC below.

The energy conservation should be achieved by scaling the relative radial momentum between a third particle and the center-of-mass of CC. The angular momentum is also conserved by adjusting the collective rotation of the part near CC. A reasonable way to choose a third particle may be to find a cluster which has participated in a collision that formed one of the clusters in CC. However, since we do not keep the full history of collisions in our computation, we choose a particle (a cluster or a non-clustered nucleon) that has the minimal value of

$$\frac{(r + 7.5\text{ fm})(1.2 - \cos\theta)}{\min(\varepsilon_{\parallel},\ 5\text{ MeV})} \times \begin{cases} 1/2.7 & \text{if the particle is a non-clustered nucleon} \\ 1/2.1 & \text{if the particle is in an already combined system CC}' \\ 1 & \text{otherwize} \end{cases} \tag{10}$$

as the third particle for energy conservation, where r and $\varepsilon_{\parallel}$ are the distance and the radial component of the kinetic energy for the relative motion between the third particle and CC. The factor with the angle θ between the relative coordinate ($\mathbf{r}$) and velocity ($\mathbf{v}$) is introduced so as to favor the case of $\mathbf{r} \parallel \mathbf{v}$.

Clusters in already-formed combined systems are handled in some exceptional ways. If the third particle selected for the energy conservation belongs to a combined system (CC′) formed at a former time step, then CC′ is treated as the particle for the energy conservation and the internal velocities in CC′ are also eliminated to stabilize it. In the process to search CC, two clusters are not linked if they belong to different already-formed combined systems. If two clusters belongs to the same already-formed combined system, the condition $\mathbf{R}_{\text{rel}} \cdot \mathbf{V}_{\text{rel}} > 0$ is not required. Two-nucleon collisions are forbidden among nucleons within the same already-formed combined system, for a technical reason to keep the identity of the combined system.

2.4 Momentum fluctuation of emitted particles

The momentum fluctuation is also a longstanding problem in molecular dynamics approach for quantum systems. Due to the uncertainty principle, any single-particle state has a certain momentum width Δp as far as the particle is localized in coordinate space like a Gaussian wave packet. This is a reasonable description of nucleons in a nucleus. When a particle is emitted, however, it is reasonable to project the many-body state onto the states in each of which the emitted particle has a definite momentum. This idea has been introduced into AMD by the wave-packet splitting as a quantum branching process based on the single-particle motion with decoherence [8, 9, 19, 21]. However, this method treats only the independent motions of nucleons in the mean field.

In the present work with cluster correlations, we employ the simplest method proposed in Ref. [22] with a straightforward extension to clusters. A particle (a nucleon, a cluster or a combined system of clusters) is regarded as emitted if there is no other particle within the distance of 4 fm. When a particle is emitted, a momentum fluctuation is randomly given to it according to the momentum distribution of the wave packet $f(\mathbf{p}) \propto e^{-(\mathbf{p}-\mathbf{P})^2/2\Delta p^2}$ with $\Delta p^2 = A\hbar^2\nu$, where A is the mass number of the particle and ν is the width parameter of the nucleon wave packets. We search a part of the system that is likely a source which emitted the particle. The relative momentum between the particle and the emitting source is scaled for the consistency with the method of the subtraction of the spurious kinetic energies of fragments (see Ref. [22]). The center-of-mass momentum and the internal momenta of the emitting source are adjusted for the conservation of the total momentum and the total energy, respectively. The angular momentum is also conserved by adjusting the collective rotation of the emitting source.

The momentum fluctuation of emitted particles not only affects the spectra of those emitted particles, but also influences the stability of the rest of the system. Through the momentum and energy conservation, the fluctuation acts as pressure to reduce the expansion of the emitting source and/or it lowers the internal energy of the emitting source on average. Due to these effects, the two-nucleon collision process with clusters can be done without artificial adjustments, and the cluster-cluster binding process can be done with a relatively natural way, compared to the previous studies [10, 16, 17].

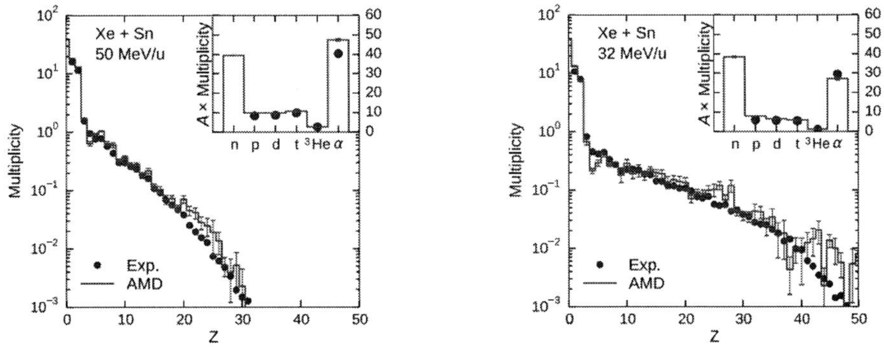

Figure 4. Fragment charge distribution in central Xe + Sn collisions ($0 < b < 2$ fm) at the incident energies of 50 (left) and 35 (right) MeV/nucleon, calculated by AMD with cluster correlations. The inset shows the multiplicities of light particles multiplied by the mass number. The INDRA experimental data are taken from Ref. [1].

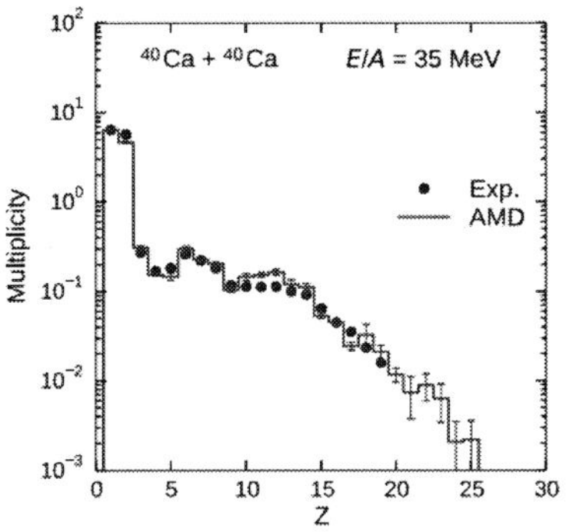

Figure 5. Fragment charge distribution in violent events in $^{40}Ca+^{40}Ca$ collisions at 35 MeV/nucleon. A filter for the experimental setup has been applied to the calculated events, and the violent events are selected by the same condition as for the experimental data [24].

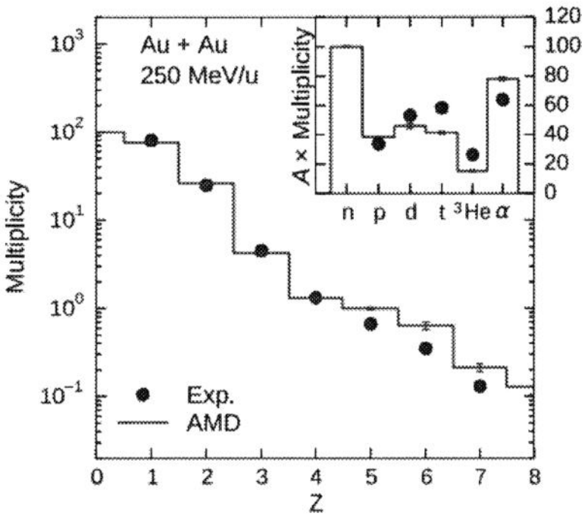

Figure 6. Fragment charge distribution in central Au + Au collisions at 250 MeV/nucleon. The calculated result by AMD with cluster correlations is compared to the FOPI experimental data [2].

3 Basic characters of multifragmentation

In this section, the calculated fragment charge distributions for multifragmentation in some central collisions are shown in comparison with experimental data. All the calculations were performed with the AMD with cluster correlations with the same model parameters as described in the previous section. The Skyrme SLy4 interaction was adopted with the spin-orbit term neglected. For high energy collisions at 250 MeV/nucleon, the momentum dependence of the Skyrme interaction was modified

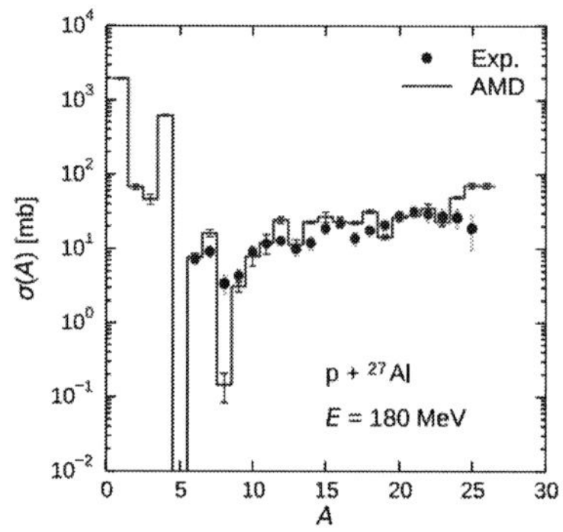

Figure 7. Fragment production cross sections in p + Sn collisions at 180 MeV calculated by AMD with cluster correlations. The experimental data [3] are shown by blue points.

in a similar way to Ref. [23] as described in Ref. [17]. The decays of excited fragments at the end of the AMD calculation were calculated by a statistical decay code.

Similar results were presented in Ref. [18]. The results here, however, have been updated with the improved framework as explained in the previous section.

Figure 4 shows the comparison for Xe + Sn collisions at 50 and 32 MeV/nucleon. The light particle multiplicities (multiplied by the particle mass number) are also shown in the inset. The overall distribution of the fragments ($Z \geq 3$) and the individual multiplicities of light charged particles are all reproduced very well. The deviations in the yields around $Z = 4$ at both energies have not been well understood. As well as the cluster correlations (to form clusters up to α particles), the cluster-cluster correlations are quite important to reproduce the data at both incident energies simultaneously.

The fragment multiplicities are calculated by the same model for violent collisions of ^{40}Ca + ^{40}Ca at 35 MeV/nucleon (Fig. 5) and central collisions of Au + Au at 250 MeV/nucleon (Fig. 6). The reproduction of data is almost satisfactory if we consider the wide difference of the system sizes and the incident energies. The α particles in Ca + Ca at 35 MeV/nucleon are underproduced in contrast to the overproduction in Xe + Sn at 50 MeV/nucleon. However, the reproductions of these two systems are much more consistent than with any other previous versions of AMD. The yield of relatively heavy fragments in the Au + Au collision may depend on the details of the cluster-cluster binding process.

Fragmentation in $p+^{27}$Al at 180 MeV was also studied with cluster correlations. The result in Fig. 7 should be compared with Fig. 2 to find a strong influence of cluster correlations to produce medium mass fragments sufficiently. In the case with cluster correlations, we also find that the fragment yields for $A \lesssim 15$ is sensitive to the assumed inmedium two-nucleon cross sections. To reproduce the data, in the calculation for Fig. 7, it was necessary to choose a weaker reduction of cross sections in medium [$y = 4$ in Eq. (9)] and a weaker condition for Pauli blocking than in other calculations for heavy-ion collisions.

The yields of clusters and fragments in these various reaction systems represent the essential information on how many-body spatial correlations develop in expanding systems with various expansion speeds and/or thermal excitations. A consistent description of this feature is an important milestone to enable precise studies of other features, as in the following section.

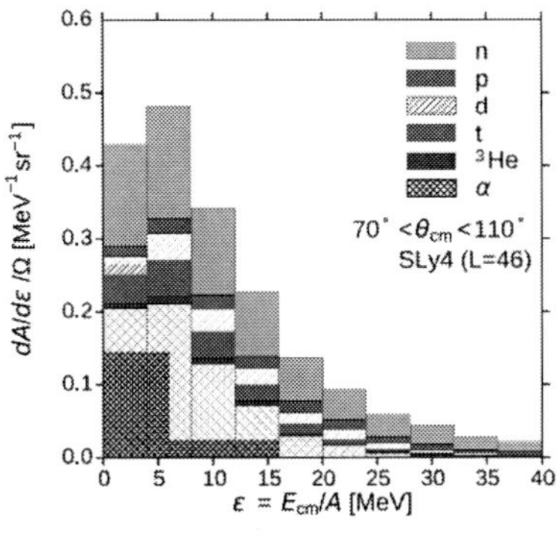

Figure 8. The calculated mass-weighted distributions of the energies $\varepsilon = E/A$ of light particles emitted to the angles $70° < \theta < 110°$ in the center-of-mass system of ^{124}Sn + ^{124}Sn central collisions at 50 MeV/nucleon.

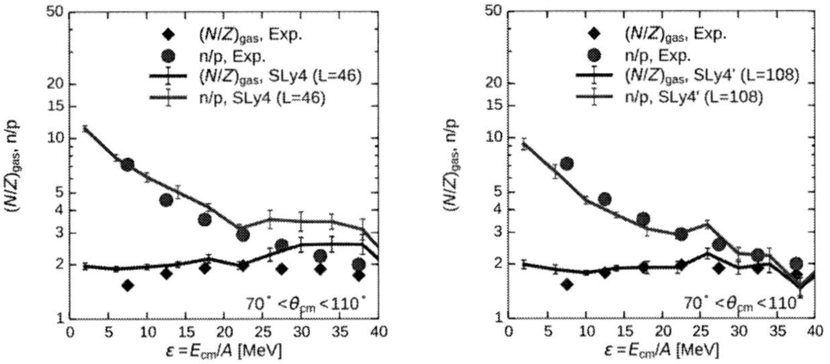

Figure 9. The spectral n/p ratio of free neutrons and protons (red) and the $(N/Z)_{\text{gas}}$ ratio for the sum of light particles (black). The lines show the calculated results by AMD with cluster correlations. The adopted effective interaction for the left panel is the Skyrme SLy4 interaction corresponding to a relatively soft symmetry energy $L = 46$ MeV, and that for the right panel is a modified version of SLy4 which corresponds to a stiffer symmetry energy $L = 108$ MeV. The experimental data from Ref. [26] are shown by points.

4 Clusters and the symmetry energy effects

4.1 Neutron-proton ratio at 50 MeV/nucleon

We choose here a typical system of central ^{124}Sn+^{124}Sn collisions at 50 MeV/nucleon. Such a neutron-rich system is suitable to see the effect of the density dependence of the symmetry energy $S(\rho)$ or the parameter $L = 3\rho_0(dS/d\rho)_{\rho_0}$. In the early stage of collisions, the difference of the calculated neutron and proton densities in the high density region shows a very clear effect of the density dependence of the symmetry energy. In the late stage of the reaction, the calculation clearly shows that the symmetry energy influences the degree of fractionation, i.e., how much the gas part is neutron rich compared with the liquid part of the system. Fragment nuclei with $Z \gtrsim 3$ are usually regarded as belonging to the liquid part, while the gas part consists of emitted nucleons and light clusters. The effects in some final observables studied with a version of AMD were shown in Ref. [25]. Direct information of the liquid part may be obtained from the isotope distribution of fragment nuclei. The width of the isotope distribution has been found to be sensitive to the symmetry energy parameter L, as well as the mean value of the distribution. However, in order to obtain a robust constraint on the L parameter from the comparison with experimental data, it is necessary to better understand the decay of primary fragments. On the other hand, direct information of the gas part should be observed in the yields and spectra of nucleons and light clusters (see Ref. [26] and references therein). It is evidently important to have a good theoretical understanding of emission of clusters.

In order to better understand the composition of the gas part, Fig. 8 shows the distributions of the kinetic energy per nucleon $\varepsilon = E_{\rm cm}/A$ for the nucleons and clusters emitted in transverse directions $70^\circ < \theta_{\rm cm} < 110^\circ$. Each colored area shows the number of the specific cluster multiplied by the cluster mass number A, so that the figure shows the decomposition of nucleons into clusters for each ε. Except for the high energy part, the α cluster formation is quite important since about half of gas nucleons are bound in α particles. The neutron-proton ratio is $(N/Z)_{\rm gas} \approx 2$ for the sum of these particles ($A \le 4$). This is significantly larger than the ratio $(N/Z)_{\rm system} = 1.48$ for the total system, which is understood as the effect of neutron fractionation. Since the same numbers of neutrons and protons are bound in many α particles (and deuterons), only a small number of protons and ^{3}He are left. Consequently the free n/p ratio and the $t/^3$He ratio can be very large, due to the simple relation

$$\frac{N_{\rm gas} - 2n_\alpha}{Z_{\rm gas} - 2n_\alpha} \gg \frac{N_{\rm gas}}{Z_{\rm gas}}, \tag{11}$$

with n_α being the number of α particles. In Fig. 9, the free n/p ratio is compared with the experimental values from Ref. [26]. The n/p ratio can be very large ($\sim$ 10 at low energies) due to the above-mentioned effect of the α-particle formation. The exact value of the n/p ratio is probably quite sensitive to the details of the emission of α particles. On the other hand, the $(N/Z)_{\rm gas}$ ratio for the sum of these light particles takes a moderate value ($\sim$ 2) without a strong dependence on the energy ε.

It seems that the present calculation with the Skyrme SLy4 interaction (corresponding to $L = 46$ MeV) is overestimating the n/p and $(N/Z)_{\rm gas}$ ratios compared to the data as shown in the left part of Fig. 9. In order study the effect of the density dependence of the symmetry energy, we also performed calculations with a force obtained by changing the density dependent term in the SLy4 force

$$v_\rho^{(L=46)} = \tfrac{1}{6}t_3(1 + x_3P_\sigma)\rho(\mathbf{r}_1)^\alpha\delta(\mathbf{r}_1 - \mathbf{r}_2) \tag{12}$$

to

$$v_\rho^{(L=108)} = \tfrac{1}{6}t_3(1 + x_3'P_\sigma)\delta(\mathbf{r}_1 - \mathbf{r}_2)\rho(\mathbf{r}_1)^\alpha + \tfrac{1}{6}t_3(x_3 - x_3')\rho_0^\alpha P_\sigma\delta(\mathbf{r}_1 - \mathbf{r}_2). \tag{13}$$

By choosing $x_3' = -0.5$, we have a force corresponding to $L = 108$ MeV with the same equation of state of symmetric nuclear matter and with the same $S(\rho_0)$ as the original SLy4 force. The results with this stiffer symmetry energy ($L = 108$ MeV) are shown in the right part of Fig. 9. A clear effect of the density dependence of the symmetry energy is seen in the n/p and $(N/Z)_{\rm gas}$ ratios by comparing the left and right parts. This effect is consistent with the stronger fractionation for the lower value of L which corresponds to the larger value of the symmetry energy $S(\rho)$ at subsaturation densities $\rho \lesssim \rho_0$. It might be possible to obtain a constraint on L by the comparison with experimental data. However, one should carefully check the consistency with other observables because these ratios may be quite sensitive to the details of cluster formation.

4.2 Nuetron-proton dynamics at 300 MeV/nucleon

Heavy-ion collisions at several hundred MeV/nucleon provide us an opportunity to explore compressed nuclear matter. In this subsection, we will discuss the impacts of clusters in the dynamics

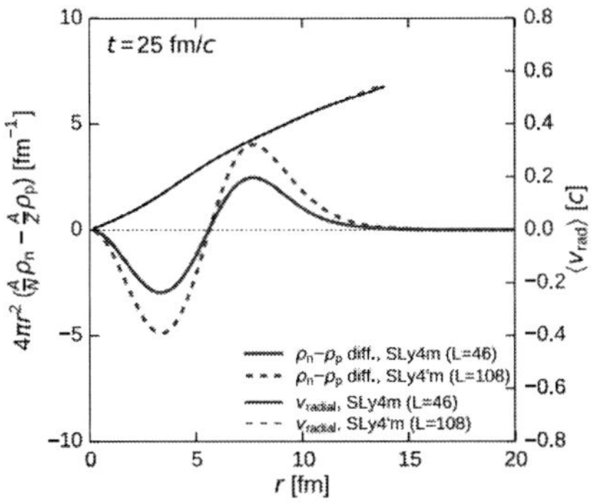

Figure 10. The difference between neutron and protons densities defined as in the text as a function of the distance from the center of the system at the time $t = 25$ fm/c in ^{132}Sn + ^{124}Sn central collisions at 300 MeV/nucleon. The collective radial velocity is also shown. From Ref. [18].

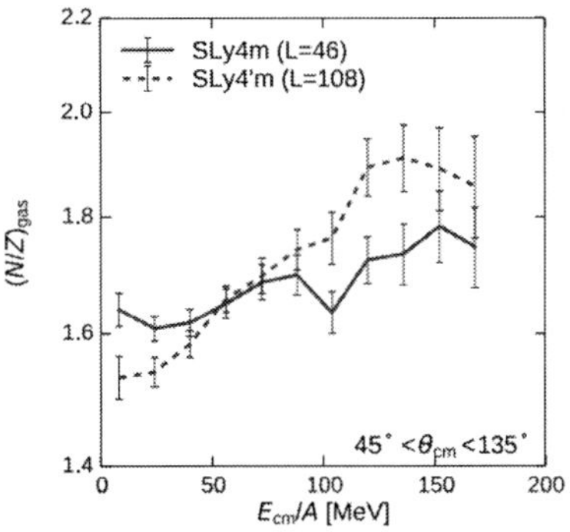

Figure 11. The $(N/Z)_{\rm gas}$ ratio for the sum of emitted particles $1 \leq A \leq 4$ for the final state of the same reaction system. From Ref. [18].

of ^{132}Sn + ^{124}Sn central collisions at 300 MeV/nucleon by employing the result of the AMD calculation with cluster correlations of Ref. [17]. The details of this calculation are different from Sec. 2. The same discussion for the dynamics of neutrons and protons was given in Ref. [18]. The pion observables are discussed in Ref. [17].

In the AMD calculation with clusters, the maximum density slightly higher than $2\rho_0$ is reached at about $t = 20$ fm/c, in ^{132}Sn + ^{124}Sn central collisions at 300 MeV/nucleon. In such neutron rich systems, the neutron-proton ratio in the compressed part of the system is sensitive to the value of the symmetry energy at high densities [27]. Figure 10 shows the neutron-proton density difference at $t = 25$ fm/c defined by

$$\Delta_{np}(r) = 4\pi r^2\Big[\frac{A}{N}\rho_n(r) - \frac{A}{Z}\rho_p(r)\Big] \tag{14}$$

for the angle-averaged densities $\rho_n(r)$ and $\rho_p(r)$. The coefficients are defined by using A, N and Z of the reaction system so that we have $\int_0^\infty \Delta_{np}(r)dr = 0$. The results are shown by two lines for the effective interactions corresponding to $L = 46$ MeV (solid) and $L = 108$ MeV (dashed). As expected, the central high-density part becomes less neutron rich than the exterior low-density part, and the effect is stronger for the stiffer symmetry energy.

An important question is, of course, how this effect at an early stage can be observed in experiments. The charged pion ratio can be a good probe [17, 27]. We here explore another complementary possibility of using spectra of nucleons and clusters. A hint is found in Fig. 10 in which the radial expansion velocity is also shown. The almost linear dependence on the distance from the center suggests that the system is very simply expanding in this high energy collision. Therefore the exterior part, which is neutron rich, will contribute to the high velocity particles in the final state, while the neutron-poor interior part will be observed as low velocity particles. In fact, this very simple picture seems correct in the calculation as in Fig. 11 which shows the $(N/Z)_{\rm gas}$ spectral ratio for the sum of light particles $1 \leq A \leq 4$. The relation between the two lines for the different density dependences of the symmetry energy in the $(N/Z)_{\rm gas}$ spectral ratio is similar to $\Delta_{np}(r)$ in Fig. 10.

It should be noted that in the literature the n/p spectral ratio has been found to be insensitive to the density dependence of the symmetry energy, while it is sensitive to the neutron-proton effective mass difference [28], based on the stochastic mean field (SMF) calculation without cluster correlations. On the other hand, the present AMD calculation takes into account the cluster correlations. In fact, when we turn off the cluster correlations in AMD, we find the sensitivity to the symmetry energy is lost in the $(N/Z)_{\rm gas}$ spectral ratio. This suggests that the expansion dynamics is largely influenced by cluster correlations, which is understandable because the number of particles (namely clusters and nucleons) is smaller in the calculation with cluster correlations than without them. The interaction between particles ceases at an earlier time with cluster correlations, while the interaction between nucleons continues longer in the calculations without clusters so that the final neutron and proton spectra are modified from those at early times. This effect is similar to what we found in a different context in the comparison of AMD and SMF [29].

5 Summary

Clusters are important in heavy-ion collisions, not only simply because they are emitted, but also because formation and existence of light clusters influence very much the global reaction dynamics and the bulk nuclear matter properties. AMD has been extended to include cluster correlations in the final states of two-nucleon collisions. The binding of several clusters to form a nucleus should also be considered. The momentum fluctuations of emitted particles also play an important role. The consistent reproduction of basic character of multifragmentation is much more satisfactory than before. This suggests that the excited nuclear many-body systems realized in heavy-ion collisions have stronger cluster correlations than usually expected from single-particle motions in the mean field picture. The fragmentation in a proton induced reaction was also found to be well described when cluster correlations are taken into account. Some observables, such as the n/p spectral ratio, are sensitive to the α-particle formation. If cluster correlation is strong, the expansion is simple in collsions at 300 MeV/nucleon so that the high-density effect of the symmetry energy is reflected almost directly in the $(N/Z)_{\rm gas}$ spectral ratio.

Acknowledgments

This work was supported by JSPS KAKENHI Grant Numbers 21540253 and 24105008. The numerical calculations were partly carried out on SR16000 at YITP in Kyoto University.

References

[1] S. Hudan *et al.*, Phys. Rev. C **67** (2003) 064613.
[2] W. Reisdorf *et al.*, Nucl. Phys. A **848** (2010) 366.
[3] K. Kwiatkowski *et al.*, Phys. Rev. Lett. **50** (1983) 1648.
[4] J. Aichelin, Phys. Rep. **202**, 233 (1991).
[5] A. Ono, H. Horiuchi, T. Maruyama and A. Ohnishi, Prog. Theor. Phys. **87** (1992) 1185.
[6] T. Maruyama, K. Niita, T. Maruyama, and A. Iwamoto, Prog. Theor. Phys. **98** (1997) 97.
[7] T. Maruyama, K. Niita, K. Oyamatsu, T. Maruyama, S. Chiba, and A. Iwamoto, Phys. Rev. C **57** (1998) 655.
[8] A. Ono and H. Horiuchi, Porg. Part. Nucl. Phys. **53**, 501 (2004).
[9] Y. Kanada-En'yo, M. Kimura and A. Ono, Prog. Theor. Exp. Phys. **2012** (2012) 01A202.

[10] A. Ono, J. Phys.: Conf. Ser. **569** (2014) 012086.
[11] G. Röpke, Nucl. Phys. A **867** (2011) 66.
[12] L. Qin *et al.*,, Phys. Rev. Lett. **108** (2012) 172701.
[13] M. Hempel, K. Hagel, J. Natowitz, G. Röpke and S. Typel, Phys. Rev. C **91** (2015) 045805.
[14] P. Danielewicz and G.F. Beartsch, Nucl. Phys. A **533** (1991) 712.
[15] D.D.S. Coupland, W.G. Lynch, M.B. Tsang, P. Danielewicz and Y. Zhang, Phys. Rev. C **84** (2011) 054603.
[16] A. Ono, J. Phys.: Conf. Ser. **420** (2013) 012103.
[17] N. Ikeno, A. Ono, Y. Nara, and A. Ohnishi, in preparation.
[18] A. Ono, a talk given at the 12th International Conference on Nucleus-Nucleus Collisions, June 21–26, Catania, Italy.
[19] A. Ono and H. Horiuchi, Phys. Rev. C **53** (1996) 2958.
[20] Y. Tosaka, A. Ono and H. Horiuchi, Phys. Rev. C **60** (1999) 064613.
[21] A. Ono, S. Hudan, A. Chbihi, J. D. Frankland, Phys. Rev. C **66** (2002) 014603.
[22] A. Ono and H. Horiuchi, Phys. Rev. C **53** (1996) 845.
[23] C. Gale, G. Bertsch, and S. Das Gupta, Phys. Rev. C **35** (1987) 1666.
[24] K. Hagel *et al.*, Phys. Rev. C **50** (1994) 2017.
[25] A. Ono, J. Phys.: Conf. Ser. **436** (2013) 012068.
[26] D.D.S. Coupland *et al.*, arXiv:1006.4546 [nucl-ex] (2014).
[27] B. A. Li, Phys. Rev. Lett. **88** (2002) 192701.
[28] V. Giordano, M. Colonna, M. Di Toro, V. Greco and J. Rizzo, Phys. rev. C **81** (2010) 044611.
[29] M. Colonna, A. Ono and J. Rizzo, Phys. Rev. C **82** (2010) 054613.

Clustering effects in fusion evaporation reactions with light even-even N=Z nuclei. The ^{24}Mg and ^{28}Si cases

L. Morelli[1,a], M. Bruno[1], M. D'Agostino[1], G. Baiocco[2], F. Gulminelli[3], M. Cinausero[4], M. Degerlier[5], D. Fabris[6], F. Gramegna[4], T. Marchi[4], S. Barlini[7], M. Bini[7], G. Casini[7], N. Gelli[7], G. Pasquali[7], S. Piantelli[7], and S. Valdrè[7].

[1] *Dipartimento di Fisica e Astronomia dell'Università and INFN, Sezione di Bologna, Bologna, Italy.*
[2] *Dipartimento di Fisica dell'Università and INFN, sezione di Pavia, Pavia, Italy.*
[3] *CNRS, UMR6534, LPC, Caen, France and ENSICAEN, UMR6534, LPC, Caen, France.*
[4] *INFN, Laboratori Nazionali di Legnaro, Legnaro (Padova), Italy.*
[5] *University of Nevsehir, Science and Art Faculty, Physics Department, Nevsehir, Turkey.*
[6] *INFN, Sezione di Padova,, Padova, Italy.*
[7] *Dipartimento di Fisica dell'Università and INFN, Sezione di Firenze, Firenze, Italy.*

Abstract.
In the recent years, cluster structures have been evidenced in many ground and excited states of light nuclei [1, 2]. The decay of highly excited states of ^{24}Mg is studied in fusion evaporation events completely detected in charge in the reactions ^{12}C+^{12}C and ^{14}N+^{10}B at 95 and 80 MeV incident energy, respectively, and compared to the results of a pure statistical model [3, 4].
Inclusive variables are in general well reproduced by the model. We found clear deviations from the statistical model if we select emission channels involving multiple α particles which are more probable than expected from a purely statistical behavior. Data from ^{12}C+^{12}C reaction have been analyzed in order to study the decay of the Hoyle state of ^{12}C* with two different selections: peripheral binary collisions and 6α decay channel in central events. To continue the investigation on light systems, we have recently measured the ^{16}O+^{12}C reaction at three different beam energies, namely E_{beam} = 90, 110 and 130 MeV. Preliminary results are presented.

1 INTRODUCTION

The NUCL-EX collaboration has recently started an experimental campaign of exclusive measurements of fusion-evaporation reactions with light nuclei as interacting partners. The aim is to progress in the understanding of statistical properties of the decay of light nuclei at excitation energies above particle emission thresholds and to measure observables linked to the presence of cluster structures in nuclear excited levels.

The statistical decay of hot nuclei and nuclear clustering are very active research topics in nuclear physics [1]. Using the statistical theory of Compound Nucleus (CN) decay, the detailed output of a fusion-evaporation reaction is uniquely predicted under the knowledge of nuclear ground state

[a] e-mail: luca.morelli@bo.infn.it

properties and level densities. The knowledge of level densities is not only important for the understanding of nuclear structure [5], but it is also required for different applications of nuclear physics, from nucleosynthesis calculations to reactor science.

Despite the interest of this issue, few studies, mainly based on inclusive experiments, exist on the evaporation of very light nuclei (A $\sim$ 20 region) at relatively high excitation energy ($\varepsilon^* \approx 3$ A.MeV). The interest on this mass - excitation energy region is easily justified: from the experimental point of view, measuring light nuclei means low multiplicity events, which, together with a high detection coverage and high energy and angular resolution, leads to the possibility of achieving a complete event reconstruction, thus having a full control on the reaction mechanism.

In addition, the knowledge of the decay mechanism could help in the backtracing procedure in multifragmentation events. Indeed in the statistical framework fragments formed in the freeze-out volume are mainly in this energy and mass region.

Nuclear structure signatures are especially evident in light nuclei, even at high excitation energy. By using an exclusive channel selection and a highly constrained statistical code, it is possible to put into evidence deviations from a statistical behavior in the decay of the hot sources formed in the collision. Indeed some excited states of different nuclei in this mass region are known to present pronounced cluster structures. These correlations may persist in the ground state along some selected isotopic chains [6] and, according to the Ikeda diagrams [2], α-clustered excited states are sizeably expected at high excitation energies close to the multi-alpha decay threshold in all even-even N = Z nuclei.

The subject of α-clustering has been a central issue in nuclear physics and has witnessed a gain of interest in recent years. On the theoretical side, highly sophisticated ab-initio calculations have shown pronounced cluster features in the ground state of a large number of light nuclei [6]. Concerning experimental research, rotational bands consistent with α-cluster structures have been identified in different even-even light nuclei and shown to persist even along their isotopic chains. Such effects can be experimentally seen as an excess of cluster production with respect to the prediction of the statistical model, provided that the ingredients of the latter are sufficiently constrained via experimental data. A campaign of exclusive measurements of fusion-evaporation reactions with light nuclei as interacting partners is currently carried on by our collaboration. In the framework of this campaign, the ^{12}C+^{12}C and ^{14}N + ^{10}B reactions have been measured in order to study the decay of the same ^{24}Mg compound nucleus, populated at the same excitation energy but through different entrance channels. The results obtained in our first measurements are briefly reported in the following sections.

A very interesting clustering effect is for instance that of the so-called Hoyle state, i.e. the excited state 0^+ at 7.65 MeV of ^{12}C, which plays a decisive role in stellar nucleosynthesis of ^{12}C.

Recently different experiments have studied the low-lying ^{12}C excited states and in particular the Hoyle state at 7.65 MeV excitation energy [7] and several interesting features of these states have been pointed out [8–14].

In several cases experimental results have interpreted the Hoyle state as mainly sequential through an intermediate ^{8}Be$_{gs}$ decay [8–12]; other experimental data show sizeable amount of instantaneous three α-particle decay [13, 14] where the three α particles are characterized by low kinetic energies and low kinetic energy dispersion. Most of the results indicate a nearly complete agreement with a sequential decay with a very small amount (0.2 % [12]) of the instantaneous one.

In this work we show some preliminary results from the analysis of semiperipheral ^{12}C + ^{12}C reactions where the projectile decays in three α-particles passing through the Hoyle state.

The results compatible with indications for non-sequential decay have been obtained in reactions involving heavy ions [13, 14]. We have therefore investigated the Hoyle state when obtained in the decay chain of ^{24}Mg* formed in central collisions and in particular we have analyzed six α-particles in the final stage of the ^{24}Mg* decay.

We will compare the experimental results both for central and peripheral collisions with the prediction of a model based on Hauser-Feshbach formalism which includes light nuclei ecxcited states. The main features of this model have been extensively described in [4]. The predictions have been filtered in order to be compared to experimental data, taking into account detector characteristics such as thresholds, energy and angular resolutions. Data have been also compared to filtered simple model calculations for simultaneous decay (DDE), i.e. three α-particles simultaneously emitted from the excited Carbon and for a linear chain (DDL), i.e. two α-particles emitted back to back with the third α-particle at rest in the ^{12}C frame.

Similary to the ^{24}Mg*, the compound nucleus ^{28}Si* can be formed through fusion of α-cluster stable nuclei in the entrance channel and in the future of light radioactive beams provided by SPES. It is interesting for such systems that they can be studied at different beam energies, thus extracting information on how (and to what extent) structure effects are still at play in the decay of hot nuclei at different excitation regimes.

2 THE EXPERIMENTS

The experiments were performed at the LNL (Laboratori Nazionali di Legnaro), with ^{12}C, ^{14}N and ^{16}O beams provided by the XTU TANDEM accelerator. We have studied the two systems ^{12}C + ^{12}C and ^{14}N + ^{10}B, leading to the same compound nucleus ^{24}Mg* at the same excitation energy. The decay of the ^{28}Si* compound nucleus has been investigated using ^{16}O +^{12}C reaction at three different energies. For these last reactions the analysis has been focused up to now on the necessary preliminary checks on the quality of collected data. Charge (and mass, where the information is available) identification and energy calibration are in progress.

The apparatus has been described in [15]. Here we recall the main features. The experimental setup is composed of the GARFIELD apparatus and the Ring-Counter (RCo) annular detector, fully equipped with digital electronics [16]. The forward polar angles ($7^o \div 17^o$) are covered by a three stage apparatus (RCo), ionization chamber (IC), silicon strip s (Si) and CsI(Tl) scintillators. The angular resolution is $\sim \pm 0.7^o$ for the polar angle and $\sim \pm 11^o$ for the azimuthal one for particles detected in the scintillators. The performances of the RCo allow to reach energy determination with the accuracy of percent, charge identification of particles and fragment with threshold as low as $0.8 \div 1$ AMeV and mass identification of light isotopes with ~ 6 AMeV threshold through ΔE - E technique in Si-CsI(Tl) and/or pulse shape analysis (PSA)in the CsI [15]. The remaning polar angles ($\theta = 30^o \div 170^o$) are covered by a two stage, drift ionization chamber - CsI(Tl) scintillator apparatus (GARFIELD). Energy is determined with accuracy of the order of few percent and reaction products are identified in charge and mass (for light products) with a threshold around 1 AMeV [15]. The combination of the two devices allows a nearly-4π coverage, which, combined with a high granularity, permits to measure the charge, the energy and the emission angles of nearly all the charged reaction products, with an excellent discrimination of the different reaction mechanisms.

3 RESULTS

3.1 Central ^{12}C + ^{12}C and ^{14}N +^{10}B reaction

The selection of the fusion-evaporation mechanism is based on the coincidence between LCP's and a fragment detected at forward angles (RCo). At the same time we ask the complete charge detection and longitudinal momentum conservation.

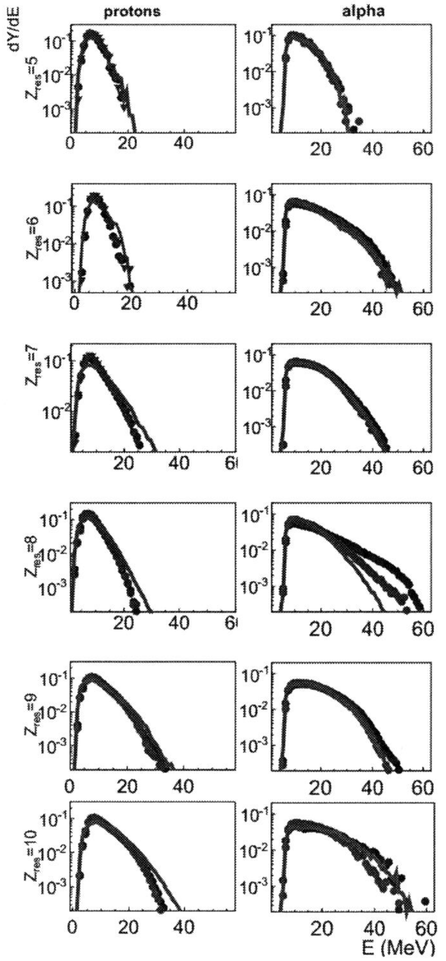

Figure 1. Proton (left part) and α (right part) energy spectra in complete Z_{det}=12 events detected in coincidence with a residue of charge Z_{res}, indicated in each figure column. Data (black and blue for $^{12}C+^{12}C$ and $^{14}N+^{10}B$ respectively) are compared to model calculations (lines) for $^{12}C+^{12}C$ (see text for details). All distributions are normalized to unitary area.

We compare experimental data to the predictions of a dedicated Monte Carlo Hauser-Feshbach (HFℓ) code for the decay of the compound nucleus [3, 4], explicitly including all the experimentally measured particle unstable levels from the archive NUDAT2(http://www.nndc.bnl.gov/nudat2/). For both reactions the best reproduction of the systematics of the fusion cross sections is obtained assuming two different maximum values (J_{0max}) for the angular momentum distribution of the hot fused source [17]. In particular we use J_{0max}=18 $\hbar$ for ^{12}C+^{12}C reaction and J_{0max}=15 $\hbar$ for ^{14}N + ^{10}B reaction and we adopt the same diffusness parameter $\Delta J = 2$ for both reactions. Energy spectra for protons and α particles detected at GARFIELD angles are plotted in figure 1, for residues of different charge, for the ^{14}N reaction, and compared to HFℓ [4] calculations and to data for the ^{12}C+^{12}C reaction. For both reactions, a good reproduction of proton and α energy spectra is achieved in all channels. Some discrepancies are present in events with an Oxygen residue, where the energy tails for α particles are not reproduced by the model, especially for the ^{12}C+^{12}C case. It has been already shown in [18] that this discrepancy is mostly due to the an extra experimental cross section for channels of the type (2α, $^{16}O^{gs/*}$)[1]. Such outgoing channels populated in the ^{12}C+^{12}C collisions can be attributed to an entrance channel effect, given the α-like structure of reaction partners and produced fragments. Indeed, for the ^{14}N +^{10}B case inclusive distributions are better reproduced by calculations. This seems not to exhaust the total discrepancy with HFℓ calculations, as we infer from the energy spectrum of α particles in coincidence with Oxygen and with ^{20}Ne for the ^{14}N reaction.

Thanks to the completeness of event reconstruction, an estimate of the dissipated energy [18] can be extracted to further investigate all the channels involving alpha particles in both reactions: $Q_{kin} = \sum_{i=1}^{2} E_{\alpha_i} + E_{Z_{Res}} - E_{beam}$, where E_{α_i} and $E_{Z_{Res}}$ are, respectively, the laboratory energy of α particles and residues, and E_{beam} is the energy of the incident projectile. Figure 2 displays the obtained Q_{kin} distributions for ^{12}C+^{12}C (black dots) and ^{14}N+^{10}B (blue dots) reactions, taking into account the different initial Q-value for the two reactions. In particular we plot only the channels with the maximum α multiplicity associated to the residue of charge Z.

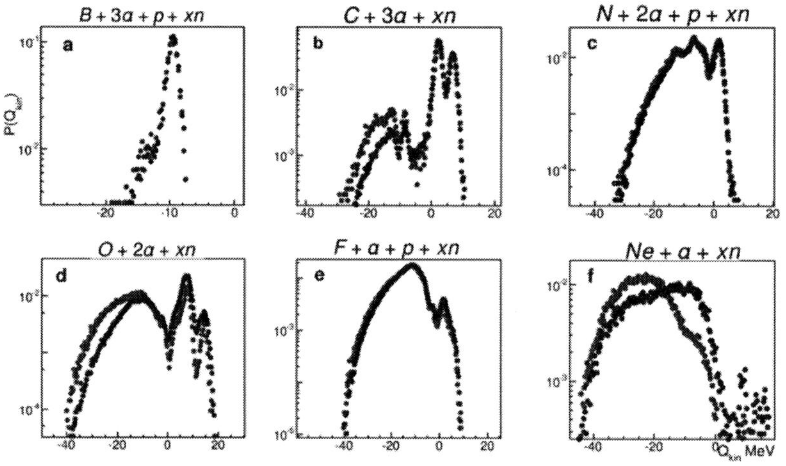

Figure 2. Q_{kin}-values distribution when the residue of charge from 5 to 10, indicated in each panel, is detected in coincidence with the maximum possible number of α-particles. Data are from ^{12}C+^{12}C (black dots) and from ^{14}N+^{10}B (blue dots) reactions.The Q_{kin} is normalized to the Q-value of ^{14}N+^{10}B data. For more details see text.

In the Q_{kin} distribution for the channel (2α, $^{16}O^{gs/*}$ figure 2d), the two peaks correspond to α-decay chains, starting from the ^{24}Mg* compound nucleus and leaving a ^{16}O residue either in its ground state

[1] $O^{gs/*}$ indicate the ^{16}O either in the ground or in an excited state

Table 1. Most probable experimental channel and its branching ratio together with the value predicted by the HFℓ calculations. Errors on the experimental value (about 5%) take into account both statistical error and the possible particle contamination

Z_{res}	Channel	BR_{exp} NB (%)	$BR_{HF\ell}$ NB (%)	BR_{exp} CC (%)	$BR_{HF\ell}$ CC (%)
5	$^{11-xn}$B+3α+p+xn	99	100	99	100
6	$^{12-xn}$C+3α+xn	96	74	98	78
7	$^{15-xn}$N+2α+p+xn	90	95	91	95
8	$^{16-xn}$O+2α+xn	56	15	63	15
9	$^{19-xn}$F+α+p+xn	91	93	92	88
10	$^{22-xn}$Ne+α+nx	47	3	26	3

or in one of its excited bound states. The Q-values $Q_{kin} = -15.78$ MeV for ^{12}C and $Q_{kin} = -0.8$ MeV for ^{14}N reactions correspond to the opening of the 4-body channel ^{15}O+$n + \alpha + \alpha$. Neutrons are not detected in the experiments, and the broader distribution observed for lower Q_{kin} values is due to events in which neutron(s) emission has taken place. A difference in the relative population of less dissipative events is evident between the ^{12}C+^{12}C (black dots) and ^{14}N+^{10}B (blue dots) reactions in figure 2 [2]. In particular, a much higher percentage of (2α,^{16}O) events populates the less dissipative Q-value region in the ^{12}C sample. This larger deviations are evident in the relative population of the different regions for all the even-Z residues, while we observe a very good agreement in Q_{kin} distributions for the odd-Z residue. This difference between the two data-sets confirms a possible larger contribution of direct reactions for the ^{12}C+^{12}C experiment. Since a residual deviation is observed in figure 2 for α particles emitted in coincidence with an Oxygen in the ^{14}N reaction, we turn now to an estimation of the possible α clustering effects for both reactions, both in the entrance channel and in the excited ^{24}Mg.

In order to put into evidence the differences in terms of branching ratio, not only between data samples, but also considering the HFℓ predictions for both reactions we present, in Table 1, the most populated channel in the experimental sample for each residue. We can see that the BR (branching ratio) of the dominant decay channels is reasonably well reproduced by the statistical model for odd Z residues, while discrepancies can be seen for even-Z ones.

The evaporation chains leading to a final Carbon or Oxygen or Neon residue show a preferential α decay in both reactions. A possible interpretation of this α excess could be the presence of residual α correlations in the excited ^{24}Mg or in its daughter nucleus ^{20}Ne, populated irrespective of the entrance channel of the reaction.

3.2 ^{12}C* Hoyle state in ^{24}Mg* decay: the 6-α channel

We now turn to examine events where the whole available mass and charge is found as α particles. This selection has been performed in [4] and the inclusive results show compatibility with a sequential α-particle emission. To improve the statistics we have analyzed here also events where only five α-particles have been detected and extracted the properties of the sixth α-particle from momentum and energy conservation. We have reconstructed the energy dissipated by the quantity Q_{kin} previously defined ($Q_{kin} = \sum_{i=1}^{6} E_i - E_{beam}$). In figure 3 (left panel) the Q_{kin} distribution is shown, together with the cut used for the analysis.

In order to reconstruct the intermediate ^{12}C* disassembling into 6 alphas of the ^{24}Mg* formed in central collisions we have selected the three (out of six) α-particles corresponding to the minimum

[2]with less dissipative we indicate events with Q_{kin} greater than the neutron emission threshold

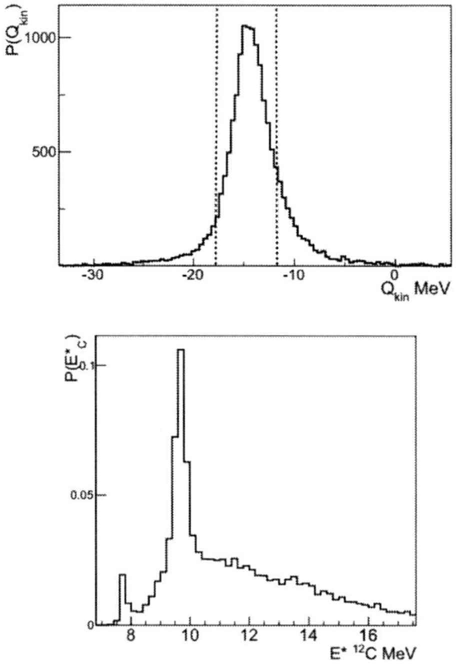

Figure 3. Upper panel: Q_{kin} distribution for the six α-particles events. The vertical lines indicate the events retained for the analysis. Lower panel: Reconstructed $^{12}C^*$ excitation energy from the three α-particles with the minimum total energy in central collisions.

$^{12}C^*$ excitation energy. This is shown in figure 3 (upper panel). The two lowest $^{12}C^*$ excited states are present at 7.65 and 9.64 MeV.

The lowest energy peak in the lower panel of figure 3 corresponds to the very well known Hoyle state, which has been studied in past years by several authors [8–11, 13, 14]. The main debate consists in the interpretation of the decay of this state as sequential, via the $^8Be_{gs}$ formation, or if there is a contribution of instantaneous breakup. Different observables have been proposed to clarify this point and we will show our results for some of these observables.

The first observable is the minimum relative energy (see figure 4) of two out of the three α-particles assigned to an excited $^{12}C^*$ produced along the $^{24}Mg^*$ decay chain. Despite of the low statistics, the results are compatible with the ones obtained for HFℓ, indicating that the decay mainly proceeds through the intermediate formation of a $^8Be_{gs}$. The same is true for the Dalitz plot where data and HFℓ predictions show the same configuration (see figure 5). As usual one defines the two coordinates of the Dalitz plot as $x_d = \sqrt{3}(e_1 - e_2)$ and $y_d = 2e_3 - e_1 - e_2$, where $e_{i,j,k} = E_{i,j,k}/(E_i + E_j + E_k)$ are the α-particle energies in the ^{12}C frame, normalized to the total energy.

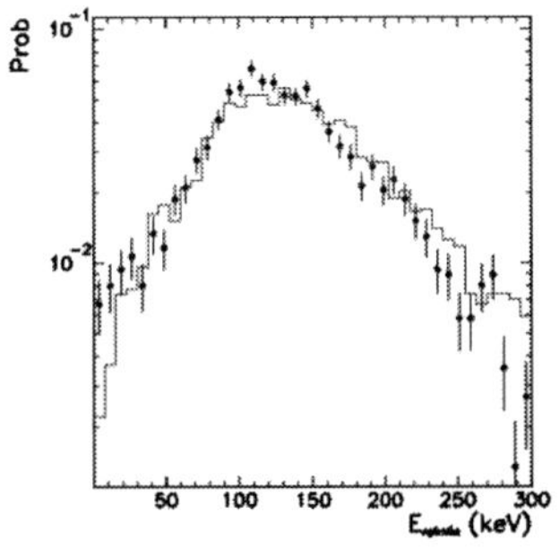

Figure 4. Central collisions: minimum relative α-α energy (MeV).The continuous lines are the predictions of HFℓ predictions.

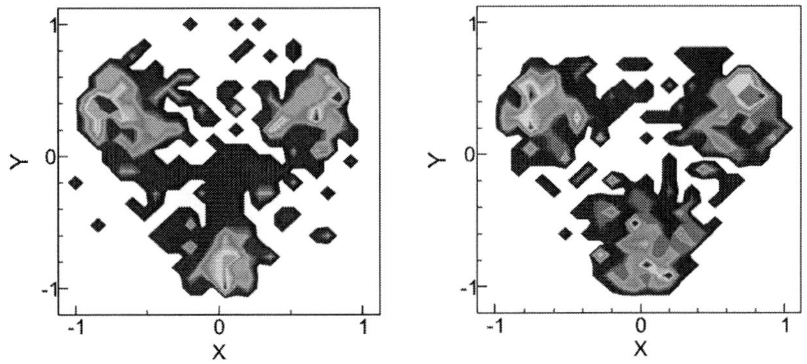

Figure 5. Central collisions. Left panel: experimental energy Dalitz plot. Right panel: HFℓ Dalitz plot.

All the results for central collisions are compared to HFℓ predictions, showing a very good agreement. All the results obtained in inelastic channels in $^{12}C+^{12}C$ reaction [19] and in various reactions at different energies [8–12] indicate a very low limit of this contribution of the order of permil. On the other side reactions with heavier ions at higher energies, involving more complex systems, seem to be compatible with a simultaneous decay of some percent [13, 14]. Our results for central reactions with the formation of $^{12}C^*$ Hoyle state in the decay chain of $^{24}Mg^*$, give no indications of deviations from the sequential decay mechanisms.

3.3 The Hoyle state in peripheral collisions

With a different selection we repeated the same analysis performed for central collisions, with the aim of investigating the decay mechanism of the Hoyle state in different conditions. We have selected events where three α-particles are detected in the forward cone (RCo) and nothing is detected in the rest of the solid angle. If we want to unambiguously identify "true" α-particles we have to exploit the ΔE-E Si-CsI or the pulse shape analysis in CsI(Tl) scintillators [16]. This implies that the threshold for the identification is about 24 MeV and consequently that the kinetic energy of ^{12}C nucleus emitting three α-particles is relatively high. The recoil nucleus ^{12}C is therefore very slow and its energy not sufficient to reach the detector, even with a very thin target. In figure 6(left panel) the energy spectrum of the reconstructed ^{12}C quasi-projectile is shown, as calculated from the sum of the energy of the three α-particles:

$$E_{tot} = \sum_{i=1}^{3} E_i + E_{rec}$$

where E_{rec} is the recoil energy of the ^{12}C quasi-target.

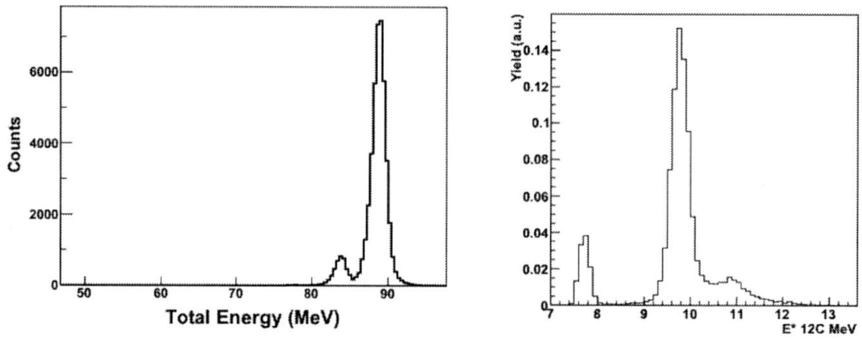

Figure 6. Total energy spectrum for "true" α-particles (left panel). The ^{12}C excitation energy spectrum (right panel) is presented for the "true" α-particles selecting the ^{12}C target in the ground state.

As it can be easily seen most of the kinetic energy is peaked at $\approx$88 MeV and a small peak is present at about 84 MeV, corresponding to the ^{12}C quasi-target at about 4.4 MeV excitation energy, the first ^{12}C excited level. The excitation energy of the ^{12}C quasi-projectile can be calculated as:

$$E^* = \sum_{i=1}^{3} E_i - \frac{p_C^2}{2m_C} + E_{th}$$

E^* is presented in Figure 6 (right panel) (E_{th} = -Q) where Q =-7.272 MeV is the Q-value for the decay of ^{12}C in three α-particles. Three excited ^{12}C quasi-projectile levels are present for "true" α-particles ($\approx$ 7.7, 9.6 and 10.8 MeV). We have analyzed data corresponding to the first 7.65 MeV ^{12}C* excited state, the first observable is the minimum relative energy of two out of the three α-particles (figure 7). The results are compatible with the ones obtained for central collisions, indicating that the decay mainly proceeds through the intermediate formation of a ^{8}Be.

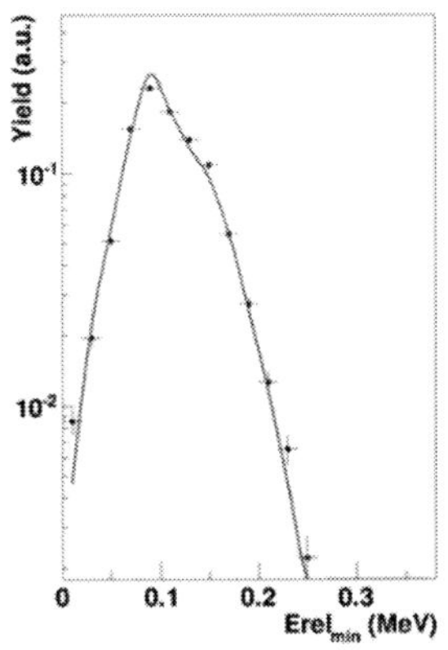

Figure 7. Peripheral collisions: minimum relative α-α energy (MeV).The continuous lines are the predictions of HFℓ predictions.

The other observable proposed in the previous section is the Dalitz-plot calculated for the α-particle energies in the ^{12}C frame, normalized to the total energy. The results from data and HFℓ calculations are shown in figure 8.

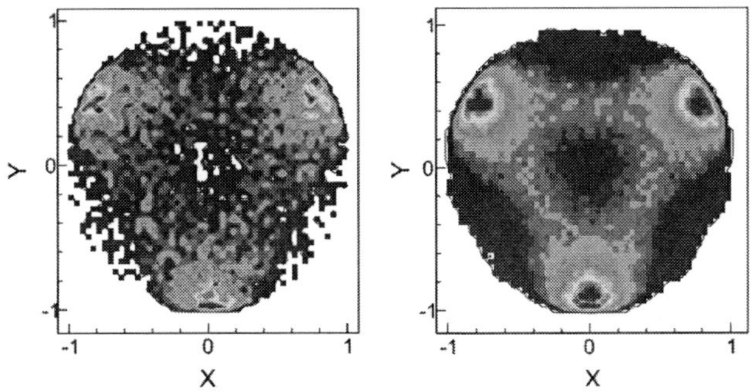

Figure 8. Peripheral collisions: Left panel: experimental energy Dalitz plot. Right panel: HFℓ Dalitz plot.

We would expect that the simultaneous decay should result in three equal kinetic energies, i.e. in an enhancement of the central part of the plot which is not the case of our data. On the contrary, it shows enhancements in regions where the energy of two α-particles are very close one to the other and the third α-energy is far from the two others. This can be considered as a signal of the predominance of the sequential decay, as confirmed by HFℓ predictions (see figure 8 right panel).
Other observables, proposed in [12] have been analyzed such as the mean energy $< E_\alpha >$, the max-

 DOI: 10.1051/epjconf/201612211002

imum energy normalized to the total energy ϵ_{max} and the deviation from the average center-of-mass energy, defined as $E_{rms} = \sqrt{< E_\alpha^2 > - < E_\alpha >^2}$. The results show a good agreement with HFℓ predictions, i.e. with a sequential decay [20].

3.4 The ^{16}O+^{12}C reaction

For the new performed experiment ^{16}O+^{12}C at three different beam energies, namely E_{beam} = 90, 110 and 130 MeV, the analysis has been focused up to now on the preliminary checks of collected data.

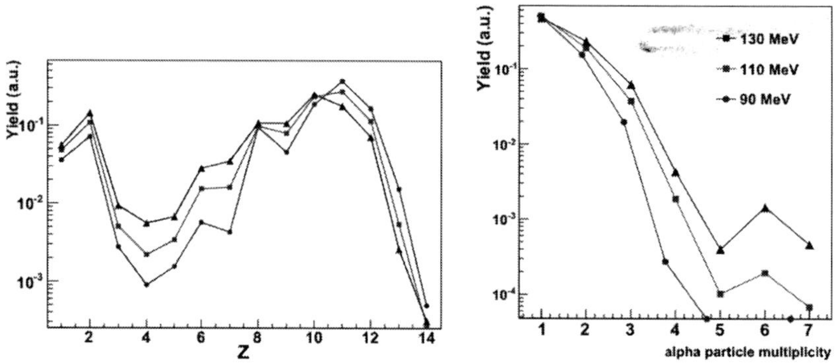

Figure 9. Preliminary results of inclusive charge distribution (left part) and alpha particle multiplicity (right part) of events selected for the reaction ^{16}O + ^{12}C at three different beam energies.

In the case of complete fusion, such reactions lead to a fused ^{28}Si* system respectively at 55, 62 and 74 MeV excitation energy. From our previous results on the decay of excited ^{24}Mg we expect cluster correlations to possibly persist in the ^{28}Si* system, even at such high excitation energies. Extrapolating the information from Ikeda diagrams, we also expect the cluster degree of freedom be differently expressed varying the E* of the system, thus leading to different decay patterns deviating from statistical expectations. For the three different beam energies, the charge distributions corresponding to the preliminary selection of the fusion-evaporation channel out of the entire dataset is shown in figure 9, together with the α multiplicity spectra detected at GARFIELD angles in coincidence with an evaporation residue in the RCo. As expected, we note a quick increase of the alpha multiplicity when the beam energy goes up.

4 Conclusions and perspectives

In this work we have discussed the clustering effects in nuclear reactions involving N=Z nuclei. In particular results from the measurements with a ^{12}C beam on a ^{12}C target at 95 MeV and the ^{14}N+^{10}B reaction at 80 MeV beam energy are presented.

We have compared experimental data for the two reactions among them and, moreover, each experiment with the results of HFℓ statistical calculations for the decay of the ^{24}Mg* source. The measured data are compatible with the expected behavior of a complete fusion-evaporation reaction, with some exception of specific channels corresponding to the emission α particles in coincidence with a Oxygen, Carbon or Neon residue. The experimental branching ratio excess for α particle emission has been quantified for both reactions, putting into evidence an effect due to the cluster nature of projectile and target in $^{12}C+^{12}C$ reaction but, at the same time, an indication of the persistence of cluster correlations also in the hot fused ^{24}Mg, as suggested by data from $^{14}N+^{10}B$ collisions.

To investigate the importance of discrete levels and the clustering effects in the CN decay, we have studied the decay of the well known Hoyle state in three α-particles. We have isolated this level in the 6-alpha decay channel in central collisions and, for the first time, we have compared its features with those of the same Hoyle level populated for the $^{12}C^*$ quasi-projectile. It has been debated if the decay mechanism is sequential, through a $^8Be_{g.s.}$ formation, or if a simultaneous decay in three α-particles could be contribute. Our results obtained comparing peripheral and central reactions with the formation of $^{12}C^*$ Hoyle state in the decay chain of $^{24}Mg^*$, give no indications of deviations from the sequential decay mechanisms. Further measurements are needed, increasing the energy and the number of nucleons involved, in order to investigate if *in medium* effects may distort the decay of the Hoyle state.

A natural extension of the same experimental scheme will be to compare the decay of a compound system formed in collisions between stable α-clustered reaction partners, or induced by an unstable light projectile on a chosen target in the future SPES facility at Legnaro National Laboratory. Together with the previously studied ^{24}Mg, the compound nucleus ^{28}Si can be formed with light radioactive beams provided by SPES.

References

[1] W. Von Oertzen 2010 Clusters in Nuclei, Lecture Notes in Physics 818 109-127.
[2] K. Ikeda, Prog. Theor. Phys. (Suppl.) extra number, 464 (1968).
[3] G. Baiocco et al., Phys. Rev. C 87,(2013) 054614.
[4] L. Morelli et al., J. Phys. G: Nucl. Part. Phys. 41 (2014) 075107.
[5] Y. Alhassid , G. F. Bertsch Phys. Rev. C 68 (2013) 044322;
[6] R. B. Wiringa, Phys. Rev. C 62 (2000) 014001.
[7] F. Hoyle The Astrophysical Journal, Supplement Series 1954 **1** 12.
[8] M. Freer et al. PR 107 (1994) R1751.
[9] J. Manfredi et al. PRC 85 (2012) 037603.
[10] O. S. Kirsebom et al. PRL 108 (2012) 202501.
[11] T. K. Rana et al. PRC 88 (2013) 021601.
[12] M. Itoh M et al. PRL 113 (2014) 102501.
[13] Ad. R. Raduta et al. PLB 705 (2011) 65.
[14] L. Quattrocchi et al. 2015 Contribution to Nucleus nucleus 2015 NN2015, Catania, Italy.
[15] M. Bruno et al., Eur. Phys. Journ. A 49 (2013) 128.
[16] G. Pasquali et al., Nucl. Instr. And Meth., A 570 (2007) 126.
[17] M. E. Ortiz et al Phys. Rev. C 25 (1982) 1436.
[18] L. Morelli et al., J. Phys. G: Nucl. Part. Phys. 41 (2014) 075108.
[19] L. Morelli et al., annual report 2014 LNL.
[20] L. Morelli et al 2014 J. Phys. G: Nucl. Part. Phys. to be published.

EPJ Web of Conferences **122**, 11003 (2016) DOI: 10.1051/epjconf/201612211003
*CNR*15*

α cluster structures in unbound states in ^{19}Ne

Reiji Otani[1], Masataka Iwasaki[1], and Makoto Ito[1,a]

[1] *Department of Pure and Applied Physics, Kansai University, 3-3-35 Yamatecho, Suita, Osaka 564-8680, Japan.*

Abstract. Cluster structures in ^{19}Ne are studied by the microscopic and macroscopic cluster models. In the microscopic calculation, the coupled-channels problem of (^{3}He+^{16}O) + (α+^{15}O) is solved, and the adiabatic energy surfaces, which are the series of the energy eigenvalues as a function of the He–O distance, are investigated. In the adiabatic energy curves, the several local minima are generated in the spatial region of the small core distance, where the neutron hole inside of the He or O nucleus is strongly coupled to the residual nuclei. The energy spectra, which are constructed from the strong coupling states, nicely reproduce the the low-lying energy levels in the ^{19}Ne nucleus. In the macroscopic approach, the α + ^{15}O potential is evaluated from the elastic scattering of the α + ^{15}N system, and the resonant levels of the α + ^{15}O system are calculated under the absorbing boundary condition. The potential model predicts the existence of the resonances above the α threshold, which has a weak-coupling scheme of the α particle and one hole inside of the ^{16}O nucleus. The extended microscopic calculations of (^{3}He+^{16}O) + (α+^{15}O) + (^{5}He+^{14}O) are performed in order to see the coupling effect of the $5p$-$2h$ configuration, which corresponds to the shell model limit of the ^{5}He + ^{14}O cluster configuration. The extended calculation suggests that the ^{5}He + ^{14}O configuration plays an important role on the formation of the $3/2^{+}$ resonance at 0.5 MeV with respect to the α threshold.

1 Introduction

The α cluster structures have been extensively studied for the so-called $4N$ nuclei with $N = Z$, such as ^{8}Be = 2α, ^{12}C = 3α, ^{16}O = α + ^{12}C, and ^{20}Ne = α + ^{16}O [1–3]. In current studies, the importance of the cluster degrees of freedom has been extended to the neutron-rich ($N > Z$) systems, which are obtained by adding extra neutrons to the $4N$ cluster systems. In the neutron-rich systems, such as ^{12}Be = $2\alpha + 4N$ [4], ^{16}C = $3\alpha + 4N$ [5] and ^{22}Ne = α + ^{16}O + $2N$ [6], various chemical-bonding-like structures are generated by the coupling of the cluster relative motion and the single particle motion of extra neutrons. In particular, the drastic structure changes among the chemical bonding-like states are pointed out in the continuum energy region of ^{12}Be [4].

On the contrary, cluster structures in the $4N$ systems with a hole in a cluster core, are also interesting research subjects [7–11]. For example, there is a recent study of ^{11}B = $\alpha + \alpha + t$, corresponding to a proton hole system of ^{12}C = 3α [7]. Pioneering work of the hole system is the study of the ^{19}F nucleus [8–11]. The ^{19}F nucleus is the one proton deficient system of ^{20}Ne, and the rotational bands

[a] e-mail: itomk@kansai-u.ac.jp

of α + ^{15}N and t + ^{16}O, which are calculated from the folding type potential, were clearly assigned to the energy levels observed in the α and t transfer reactions [8]. The energy levels in ^{19}F have also been analyzed in the more sophisticated models, such as the full microscopic α + ^{15}N cluster model on the basis of the generator coordinate method (GCM) [9, 10] and the orthogonalized condition model (OCM) of the (t + ^{16}O) + (α + ^{15}N) coupled-channels [11]. In the coupled-channels analysis, t and α cluster bands are clearly identified although a considerable mixture of these two configurations occurs.

In the previous analyses on ^{19}F [9, 11], the so-called weak coupling states of the proton hole are confirmed above the α threshold energy. For example, in the α + ^{15}N system, a proton hole inside of the ^{16}O cores loosely couples to the relative motion of the binary cluster cores. In a naive consideration, the weak coupling state of the hole in ^{16}O and the α particle is also expected in the neutron-deficient system, α + ^{15}O in ^{19}Ne [10, 11]. In fact, a formation of α + ^{15}O in ^{19}Ne was discussed from the perspective of ^{19}F – ^{19}Ne Coulomb displacement energy [8, 11]. However, detailed analysis of the energy levels over a wide energy region, which covers the unbound continuum states, the analysis of the resonance width above the α threshold and the excitation function of the resonant scattering, have not been undertaken yet because available experimental data are still limited. Therefore, a study of the α + ^{15}O cluster structure in ^{19}Ne, corresponding to the $N < Z$ system, is very interesting from the viewpoint of the systematics of the clustering phenomena in the $N \neq Z$ systems.

Recently, we have applied the potential model to the α + ^{15}O system [12] and predicted the existence of the resonances, which have the weak coupling feature, above the α threshold. The potential model is useful to understand the resonant states, in which the α clustering is well developed. However, the microscopic cluster model, in which the anti-symmetrization among the nucleons are completely performed, is very important in analyzing the energy levels from the low-lying states to the highly-excited states in a unified manner. In particular, the anti-symmetrization effect is essential for the low-lying states, which is difficult to handle in the potential model. In the present report, we apply the microscopic cluster model of (^{3}He + ^{16}O) + (α + ^{15}O) and mainly analyze the low-lying states in ^{19}Ne. The microscopic model and the potential model are combined to analyze the levels from the low-lying region to the highly-excited region.

There is another reason why we focus on the α + ^{15}O structure in the ^{19}Ne nucleus. The ^{15}O(α,γ)^{19}Ne reaction is known to play a crucial role in the advanced stages of astrophysical hydrogen burning [13]. In the ^{15}O(α,γ)^{19}Ne reaction, the most crucial resonance is known to arise through the $J^\pi = 3/2^+$ resonant level at 504 keV with respect to the α + ^{15}O threshold (E_x = 4.03 MeV). Unfortunately, direct measurement of the resonance at 504 keV is still difficult because of the small strength of the resonance and its energy position extremely close to the α decay threshold. The analyses in Ref. [14] have pointed out that the intrinsic structure of the resonance at 504 keV in ^{19}Ne is not the α + ^{15}O cluster structure, but the five particle–two hole ($5p$–$2h$) shell-model configuration with the ^{14}O$_{g.s.}$ core. The $5p$–$2h$ configuration has a large overlap with the shell model limit of the ^{5}He + ^{14}O configuration. Therefore, the coupling of ^{5}He + ^{14}O on the $J^\pi = 3/2^+$ state is essential in the cluster model approach. In this report, the result of the extended microscopic calculation of (^{3}He+^{16}O) + (α+^{15}O) + (^{5}He+^{16}O) in $J^\pi = 3/2^+$ will also be reported.

2 Framework

2.1 microscopic cluster model

We apply the generalized two-center cluster model (GTCM) for the calculations of the low-lying levels [4]. The GTCM is the extended model of the microscopic cluster model on the basis of the generator coordinate method (GCM) [15]. The application of GTCM to Be isotopes ($\alpha + \alpha + N + N$

+...) has already been published in Ref. [4], and we briefly explain the formulation of GTCM in the ^{19}Ne nucleus. The basis function for ^{19}Ne is given by

$$\Phi^{J^\pi K}(S) = \hat{P}^{J^\pi}_K \mathcal{A}\left\{\psi(^3\text{He})\psi(^{14}\text{O})\prod_{j=1}^{2}\varphi_j\left(m_j\right)\right\}_S . \quad (1)$$

The basis function is constructed on the basis of ^{3}He + ^{14}O cores, and the two neutrons are treated as the active valence-nucleons. The $\psi(^3\text{He})$ and $\psi(^{14}\text{O})$ represent the core wave functions of the ^{3}He and ^{14}O nuclei, respectively. The former part is described by the $(0s)^3$ configuration in the harmonic oscillator (HO) potential, while the latter part is constructed by the $(0s)^4(0p)^{10}$ configuration. Two centers are placed with the relative distance parameter S, which is treated as the generator coordinate. The single-particle wave function for the j-th valence neutrons ($j = 1, 2$) localized around the ^{3}He or ^{14}O clusters is given by an atomic orbital (AO) $\varphi\left(m_j\right)$, which is classified by a set of the AO quanta $m_j = (k, C, \tau)$ with the labels of the HO orbitals k, the core C, and the neutron spin τ ($= \uparrow$ or $\downarrow$). The HO orbitals around the core C ($C = {}^3$He or ^{14}O) are $k = 0s$ ($C = {}^3$He) or $k = 0p$ ($C = {}^{14}$O). The basis functions with the full anti-symmetrization $\mathcal{A}$ are projected to the eigenstate of the total spin J, its intrinsic angular projection K, and the total parity π by the projection operator $\hat{P}^{J^\pi}_K$.

The total wave function is finally given by taking the superposition over S, $\mathbf{m}$ and K as

$$\hat{\Psi}^{J^\pi}_\nu = \int dS \sum_{\mathbf{m},K} C^{J^\pi\nu}_{\mathbf{m}K}(S)\, \Phi^{J^\pi K}_{\mathbf{m}}(S) . \quad (2)$$

The coefficients $C^{J^\pi\nu}_{\mathbf{m}}(S)$ in Eq. (2) for the ν-th eigenstate are determined by solving a coupled channel GCM (Generator Coordinate Method) equation [15]

$$\left\langle \Phi^{J^\pi}_{\mathbf{m}}(S) \mid \hat{H} - E^{J^\pi}_\nu \mid \Psi^{J^\pi}_\nu \right\rangle = 0 . \quad (3)$$

Here $\hat{H}$ and $E^{J^\pi}_\nu$ denote the total Hamiltonian and the energy eigenvalues of the ν-th eigenstate for the J^π state, respectively. Before solving Eq. (3), we can investigate the structure changes with respect to the variation of the ^{3}He–^{14}O distance parameter. Namely, we solve the eigenvalue problem at a fixed S, such as

$$(\hat{H} - E^{J^\pi}_\mu(S))\, \Phi^{J^\pi\mu}_{AS}(S) = 0 \quad (4)$$

with

$$\Phi^{J^\pi\mu}_{AS}(S) = \sum_{\mathbf{m}K} D^{J^\pi\mu}_{\mathbf{m}K}(S)\, \Phi^{J^\pi\mu}_{\mathbf{m}K}(S) . \quad (5)$$

The μ-th eigenvalue $E^{J^\pi}_\mu(S)$ is a function of the relative distance-parameter S, and a sequence of $E^{J^\pi}_\mu(S)$ forms the energy surfaces. The energies $E^{J^\pi}_\mu(S)$ and wave functions $\Phi^{J^\pi\mu}_{AS}(S)$ correspond to the so-called "adiabatic energy surfaces (AESs)" and "adiabatic states (ASs)", respectively, in atomic physics [4]. The detailed properties of AESs and ASs are reported in Ref. [4].

As for the nucleon-nucleon (NN) interaction, we use the Volkov No.2 [16] and the G3RS [17] for the central and spin-orbit parts, respectively. The parameters in the NN interactions and the size parameter of HO are tuned to reproduce the ^{3}He and ^{4}He decay threshold of ^{19}Ne as much as possible. In the present calculation, the HO size parameter is set to $b = 1.6$ fm, while the strength of the Volkov No. 2 are taken to be $\pm$ 60 MeV without Bartlett and Heisenberg terms ($B = H = 0$). This parameter setting is the same as the previous microscopic calculation in ^{19}F [10]. We have found that the strength of Majorana term (M) must be varied depending on the parity: $M = 0.638$ for the positive parity and $M = 0.62$ for the negative parity. The feature of the parity dependence is also similar to the previous study of ^{19}F in Ref. [10]. The strength of the G3RS is taken to be $\pm$1500 MeV, which is the same order of the magnitude as the analysis of Be isotopes [4].

2.2 Macroscopic cluster model

The macroscopic potential model is also applied to calculate the unbound resonant states [12]. In the unbound resonances, the mean distance of α–^{15}O is expected to be enhanced, and two nuclei are weakly coupled. In the treatment of the weak coupling scheme, the potential model is quite useful. The Woods Saxon (WS) potential is assumed for the nuclear potential in α + ^{15}O, and its parameter set is fixed so as to reproduce the angular distribution of the α + ^{15}N elastic scattering.

In the determination of the parameters, the scattering problem of α + ^{15}N is solved in the range of E_α = 23.7 ~ 55 MeV. In the scattering calculation, the complex WS potential is employed, and we consider only the central part of the complex potential. The WS parameters of the real part is fixed over a whole energy range of the scattering calculation, while the parameters in the imaginary part are set to be energy dependent. The resultant WS potential reproduces the feature of the folding type potential, which used in the analysis of ^{20}Ne = α + ^{16}O [18].

The resonant levels of α + ^{15}O are calculated from the α + ^{15}N potential by adding the Coulomb and spin-orbit potentials. The strength of the spin-orbit potential is chosen to reproduce the energies of the bound levels of the $5/2^-$ and $3/2^-$ states, which are considered to be the pair levels in the spin-orbit splitting. Above the α decay threshold, the absorbing boundary condition (ABC) [19] is applied to identify the resonant levels, having the resonance energy E_R and the decay width Γ_R. The detailed explanations of the potential model plus ABC is shown in Ref. [12].

3 Results

3.1 Adiabatic energy curves

First, we solve Eq. (5) for the coupled channels of (^{3}He+^{16}O) + (α+^{15}O) and calculate the adiabatic energy surfaces (AESs). In Figure 1, the AESs with $J^\pi = 3/2^+$ are shown. In the AESs, the structure change from the two-body weak-coupling states to the one-body strong-coupling states occurs as the distance parameter S gets smaller. The vertical dashed line at S = 6.5 fm represents the boundary of the weak-coupling states and the strong-coupling states. Here, we call this boundary distance S_C.

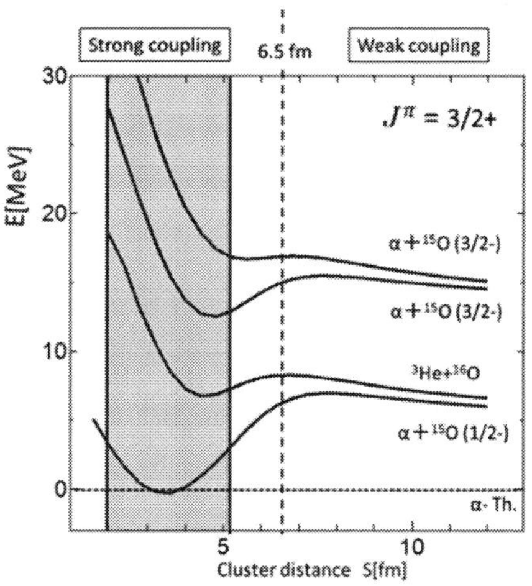

Figure 1. Adiabatic energy curves calculated from the coupled channels of (^{3}He+^{16}O) + (α+^{15}O). The abscissa and ordinate represent the He–O distance and the excitation energy measured from the α threshold energy (horizontal dotted line). See text for details.

At the outside region of S_C, the He and O nuclei slip to each other with the internal spins ($1/2^-$ or $3/2^-$ for ^{15}O and $1/2^+$ for ^{3}He) and the relative spin of L. In the particle-hole picture, the neutron hole in ^{16}O (α) weakly couples to the residual nucleus, α (^{16}O) . On the contrary, in the small distance of $S_C \leq 6.5$ fm, the effect of the nuclear interaction begins to be strong. The effect of the nuclear interaction leads to the strong mixture of the weak coupling states. The mixture of the weak coupling states corresponds the so-called strong coupling states, in which neutron holes are difficult to move inside of a nucleus freely, and the holes tend to be fixed to a certain spatial direction.

In the region of the strong coupling scheme, the local minima appear in the individual energy curves. This means that the energy levels respective to the energy minima are generated when we solve the coupled-channel GCM equation (3). In the present GCM calculation, the range of the distance parameter is set to $S = 2 \sim 5.2$ fm, which covers the local minima in the strong coupling region, as shown by the shaded area in Fig. 1. Therefore, the main components of the resultant energy levels are the strong coupling states, in which the neutron holes are tightly bound around the residual nuclei. Thus, the weak coupling states are excluded in the present computational space.

The weak coupling states are also possible to appear in the energy levels if the range of the distance parameter is extended in the GCM calculation. However, the weak coupling states are mainly appear as the resonant states in the continuum energy region and hence, the consideration of the scattering boundary condition is important in the careful analysis of the weak coupling states. The analysis of the the weak coupling states under the scattering boundary condition is a future subject [4].

3.2 Energy spectra

In Fig. 2, the energy spectra for the negative parties are plotted. The microscopic GCM spectra of the (^{3}He+^{16}O) + (α+^{15}O) at the middle position are compared with the experimental data plotted at the left side. The microscopic calculation reproduce the level ordering of the low-lying experimental spectra up to the α threshold (dashed line). The $7/2^-$ state becomes the bound state in the microscopic calculation, while the respective level in experiment, the spin of which is tentatively assigned, is resonant state just above the α threshold. This failure may be improved by tunning the parameters in the NN interaction and extending the range of the distance parameter S.

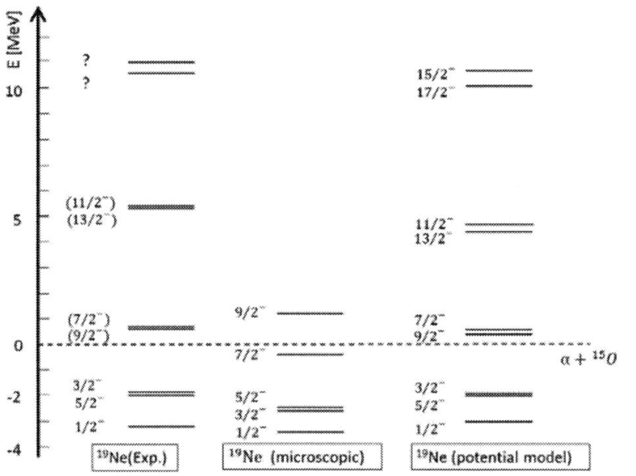

Figure 2. Energy levels of the negative parity states in ^{19}Ne. The left levels are the experimental data, while middle and right levels are the results of the microscopic and macroscopic models, respectively. The dotted line represents the α threshold energy, which is set to the zero energy.

The right side levels are the results of the potential model. In the potential model calculation, the low-lying bound levels are fairly reproduced. Furthermore, the potential model under the absorbing boundary condition (ABC) predicts the resonant levels above the α threshold. The resonant width is not shown for simplicity although the width are calculated in the ABC treatment. Unfortunately, the experimental information is still insufficient in the unbound region. Thus, the experimental investigation of the highly-excited states are strongly desired.

The result of the positive parity states are shown in Fig. 3. The microscopic GCM calculation with (^{3}He+^{16}O) + (α+^{15}O) is shown at the center position, while the experimental levels are plotted at the left position. The GCM calculation nicely reproduces the energy levels of the bound states below the α threshold although the energy position of the $7/2^{+}$ state is a little higher than the experimental observation. At the right side, the spectra obtained by the potential model are plotted in the unbound region above the α threshold, but the respective experiment is still insufficient. The resonance width and the excitation function of the resonant α scattering are predicted in Ref. [12]. Thus, the future experiments on the unbound resonances are very interesting.

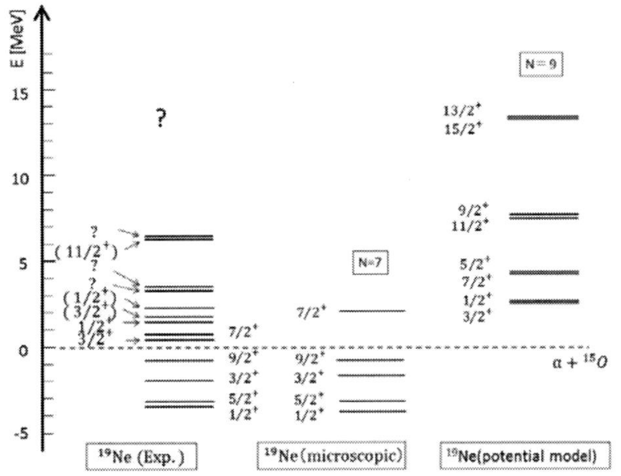

Figure 3. Same as Fig. 2 except for the positive parity states in ^{19}Ne.

The relative wave function of the α+^{15}O cluster configuration in the potential model (right levels) corresponds to the HO states with the total oscillator quantum number of $N = 9$, which is the same as the negative parity band in the ^{20}Ne = α + ^{16}O system [18]. On the contrary, the microscopic calculation is one lower nodal state with $N = 7$. In the potential model, the $3/2^{+}$ state becomes a ground state with $N = 7$, and its binding energy is about -11 MeV with respect to the α threshold. Therefore, the potential model does not reproduce the properties of the ground and low-lying levels below the α threshold, which are the member of the $N = 7$ state. This is because the He and O nuclei are strongly overlapped in the $N = 7$ state, where the anti-symmetrization effect among the nucleons becomes strong. In the potential model, the anti-symmetrization effect is not completely taken into account and hence, the binding energies of the $N = 7$ states are deeper than the energies of the experimental levels.

 DOI: 10.1051/epjconf/201612211003

The positive parity levels in Fig. 2 and the negative parity levels in Fig. 3 are belong to the $N = 8$ and $N = 9$ states, respectively, which correspond to the higher nodal state of $N = 7$. In these higher nodal states, the mean distance of the He and O clusters is much more enhanced than the distance in the $N = 7$ states, and the anti-symmetrization effect is not so effective. Therefore, the potential model nicely works in describing the experimental data corresponding to the higher nodal levels. The potential model is successful in handling the higher nodal states but the microscopic calculation is very important in the unified analysis of the energy spectra over a wide energy region.

3.3 Effect of the ^{5}He + ^{14}O configuration

In the result shown in Fig. 3, we can confirm that the $3/2^+$ state just above the α threshold is missing in both the microscopic and macroscopic calculations. This failure is the same as the results of the previous microscopic GCM calculation of ^{19}F = α + ^{15}N [10] and the coupled-channel OCM calculation in ^{19}F = (^{3}H+^{16}O) + (α+^{15}N) [11]. Therefore, the α and ^{3}He cluster configurations are considered to be minor component in the threshold $3/2^+$ state. This result is consistent to the previous experimental result, which points out the dominance of the $5p$–$2h$ configuration in the threshold $3/2^+$ state [14], because the α and ^{3}He cluster configurations have the large overlap with the $4p$–$1h$ and $3p$–$0h$ shell model configurations, respectively, in the $S = 0$ limit. Therefore, we should include other cluster configuration, which has a large overlap of the $5p$–$2h$ state.

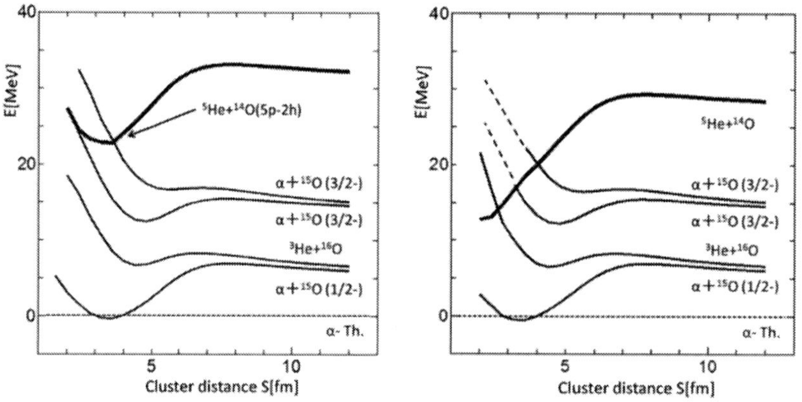

Figure 4. Adiabatic energy curves calculated by the coupled channels of (^{3}He+^{16}O) + (α+^{16}O) + (^{5}He+^{14}O). The left panel shows the truncated calculation, in which the $5p$–$2h$ configuration is mainly considered for ^{5}He + ^{14}O, while the full calculation is shown in the right panel. In both panels, the thick curves correspond to the solutions with the dominant ^{5}He+^{14}O channels. See text for details.

The ^{5}He + ^{14}O configurations are the possible configuration, which has a large overlap of the $5p$–$2h$ state. In order to improve the calculation for the $3/2^+$ state, we have also performed the extended microscopic calculation, which is the coupled-channels of (^{3}He+^{16}O) + (α+^{16}O) + (^{5}He+^{14}O), to include the $5p$–$2h$ configuration in the shell model limit ($S = 0$). The total number of the ^{5}He + ^{14}O configurations is 36. The computational process is divided into two stages. First, we truncate the model space of ^{5}He + ^{14}O to the specific 16 channels, which mainly change into the $5p$–$2h$ configurations in the $S = 0$ limit. The truncated $5p$–$2h$ channels are included in the extended calculation of (^{3}He+^{16}O) + (α+^{16}O) + (^{5}He+^{14}O). Secondly, the full model space is included in the calculation with the three partitions.

In Fig. 4, the adiabatic energies are shown for the truncated calculation (left panel) and the complete calculation (right panel). In the left panel, the thick curve shows the energy curve of the ^{5}He + ^{14}O configuration, while the thin curves show the adiabatic energies for the α and ^{3}He channels. In the thick curve, a local minimum appears around $S \sim 3$ fm, where the ^{5}He and ^{14}O nuclei strongly overlap to each other. This result means that the local minimum state corresponds to the $5p$–$2h$ states with the shell-model-like configuration.

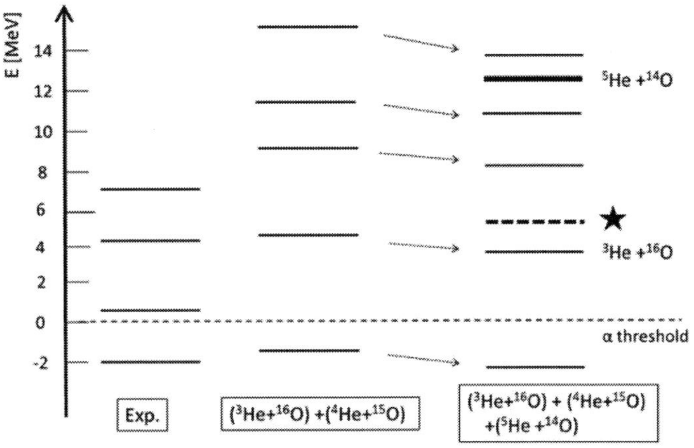

Figure 5. Energy spectra for the $3/2^+$ state. The left levels shows the experimental spectra of the $3/2^+$ state, while the middle and right spectra are the results of the microscopic calculation. The middle levels are the GCM solution of (^{3}He+^{16}O) + (α+^{16}O), while the right levels are the result of the (^{3}He+^{16}O) + (α+^{16}O) + (^{5}He+^{14}O) calculation. The arrows indicate the corresponding levels between the middle levels and the right levels. See text for details.

In the full calculation shown in the right panel, we find the further effect of other ^{5}He + ^{14}O channel. In the full calculation, the minimum point shifts to $S \leq 2$ fm, and the minimum energy of the full solution ($E_x \sim 13$ MeV with respect to the α threshold) is lower by about 10 MeV than the truncated calculation ($E_x \sim 23$ MeV). Therefore, the ^{5}He + ^{14}O correlation beyond the $5p$–$2h$ configuration is very important. However, the minimum energy in the full calculation is still higher by about 10 MeV than the threshold $3/2^+$ state ($E_x \sim 0.5$ MeV). On the contrary, the energy curves of α + ^{15}O and ^{3}He + ^{16}O are almost unchanged after the full configurations of ^{5}He + ^{14}O is included. This result means that the coupling effects of the ^{5}He + ^{14}O channels are weak for the α and ^{3}He channels. Therefore, we can consider that the ^{5}He + ^{14}O configurations are basically decoupled from the coupled-channels model space of the two partitions, such as (^{3}He+^{16}O) + (α+^{15}O).

In order to see the difference of the theoretical calculation and experimental $3/2^+$ levels more clearly, we have performed the GCM calculation for the (^{3}He+^{16}O) + (α+^{16}O) + (^{5}He+^{14}O) coupled-channels. The energy spectra are shown in Fig. 5. In the calculation of (^{3}He+^{16}O) + (α+^{16}O) (middle levels), there is only a bound $3/2^+$ state around the α threshold (dashed line), and no resonance appears around the threshold. On the contrary, a new levels with a dominant component of ^{5}He + ^{14}O appear around $E_x \sim 12$ MeV (thick level) in the extended calculation with the three partitions (right levels). However, the energy of the ^{5}He+^{14}O configuration is higher by about 10 MeV than the experimental $3/2^+$ state around the α threshold. The reason why the energy of the ^{5}He+^{14}O channel is too high is the incorrect reproduction of the threshold energy of the ^{5}He + ^{14}O channel in the theoretical calculation. The theoretical ^{5}He threshold is higher by about 7 MeV than the experimental threshold in the present setting of the nucleon-nucleon interaction. The ^{5}He nucleus is unbound with respect to the $\alpha + N$ threshold but this nucleus is handled by the simple HO wave function in the theoretical calculation. Therefore, the internal energy of the ^{5}He wave function is higher than the realistic ^{5}He nucleus.

Let us discuss the energy shift, which will arise from the correction of the threshold energy. If the ^{5}He threshold is correctly reproduced, the energy of ^{5}He+^{14}O will simply shift to the lower energy region by about 7 MeV, as shown by the dashed level with the star at the right level. Other levels will be unchanged even if the shift of the ^{5}He + ^{14}O level occurs. This is because of the de-coupling feature of ^{5}He + ^{14}O from the model space of the (^{3}He+^{16}O) + (α+^{15}O) coupled-channels. Therefore, the excitation energy of ^{5}He+^{14}O will be simply changed into $E_x \sim 5$ MeV with respect to the α threshold, (dashed level with the star). In this situation, the level repulsion is possible between the corrected ^{5}He+^{14}O state and the level at $E_x \sim 4$ MeV, which has a main component of ^{3}He + ^{16}O. Since the matrix element of the two neutron transfer is evaluated to be about $1 \sim 2$ MeV, the energy shift of the ^{3}He + ^{16}O state, which is generated from the perturbative coupling with ^{5}He + ^{14}O level, is estimated to be about -3 MeV. Therefore, there is a possibility that the $3/2^+$ state appear around the α threshold within the energy rang of about 1 MeV if the theoretical ^{5}He threshold is reproduced correctly. The reproduction of the ^{5}He threshold is important in the future calculation.

4 Summary and discussion

We have applied the microscopic and macroscopic cluster models to the ^{19}Ne nucleus and the energy levels are discussed. In the microscopic calculation, the coupled-channels calculation of (^{3}He+^{16}O) + (α+^{15}O) is performed on the basis of the generalized two-center cluster model (GTCM) [4]. In GTCM, the basis functions are constructed from the possible partitions of the valence neutrons around the ^{3}He and ^{14}O cores, and the anti-symmetrization among all the nucleons is completely taken into account. This model can produce the smooth connection of the weak coupling states and the strong coupling states in the adiabatic energy curves, which are the series of the energy eigenvalues as a function of the core distance.

In the small core distance, the strong coupling states are predominant, in which two nuclei are strongly overlapped with a specific geometrical configuration of the neutron hole. There are several local minima in the strong coupling region, and the strong coupling states are mainly included in the calculation of the generator coordinate method (GCM), in which the energy levels are calculated by superposing the distance parameter. The microscopic GCM calculation nicely reproduce the low-lying spectra below the α decay threshold. This result means the intrinsic character of the low-lying states are the strong coupling structure, in which the neutron hole is tightly bound inside of the α particle or the ^{16}O nucleus.

In the application of the macroscopic model, the α + ^{15}O potential is speculated from the elastic scattering of the mirror system, α + ^{15}N [12]. The resonant levels are calculated from the nuclear potential of α + ^{15}N by imposing the absorbing boundary condition [19]. The potential model predicts the existence of the several resonances in both of the positive and negative parity states. These resonances will be observed in the excitation function of the resonant α scattering [20]. Therefore, the experimental identification by the α resonant scattering is interesting in future studies.

The present calculations do not reproduce the $3/2^+$ state, which is considered to have the main component of the $5p$–$2h$ shell model configuration [14]. The $5p$–$2h$ configuration corresponds to the ^{5}He + ^{14}O configuration in the shell model limit and hence, we have performed the extended microscopic calculation with the ^{5}He + ^{14}O configuration. The extended calculation suggests the formation of a new $3/2^+$ state, which has the dominant component of the ^{5}He+^{14}O channel, but its excitation energy is much higher than the experimental data. Since this failure about the $3/2^+$ energy is due to the incorrect reproduction of the theoretical ^{5}He threshold, the parameters of the nucleon-nucleon interaction must be revised to reproduce the ^{5}He threshold.

The exact treatment of all the threshold energies will be quite difficult in the microscopic approach because the internal structure of ^{3,4}He and ^{5}He is different; specifically, the former nuclei are tightly bound, while the ^{5}He is unbound nucleus with the $\alpha + N$ structure. In order to overcome this threshold problem, the semi-microscopic cluster model, the orthogonalized condition model (OCM) [11], is considered to be better approach than the full microscopic approach. In the OCM calculation, the threshold energies of all the open channels are handled in a phenomenological manner, in which the theoretical threshold energies to be set to the experimental value. Thus, the application of the OCM calculation will be effective to solve the problem the $3/2^+$ state just above the α threshold.

In the present study, we have combined the microscopic and macroscopic cluster model for the analysis of the low-lying bound and the highly-excited resonant states, respectively. However, the unified calculation from the bound states to the continuum states is an ideal study of the nuclear structure. Since the macroscopic model is difficult to consider the anti-symmetrization effect correctly, which is prominent in the low-lying bound states, the microscopic calculation should be applied to such the unified calculation. The analysis by the microscopic calculations is now underway.

References

[1] K. Ikeda et al., Prog. Theor. Phys. Suppl. **68**, 1 (1986).
[2] H. Horiuchi et al., Suppl. Prog. Theor. Phys. **192**, 1 (2012), and references therein.
[3] M. Freer, Rep. Prog. Phys. **70**, 2149 (2007).
[4] M. Ito and K. Ikeda, Rep. Prog. Phys. **77**, 096301 (2014).
[5] N. Itagaki, T. Otsuka, K. Ikeda and S. Okabe, Phys. Rev. Lett. **92**, 142501 (2004).
[6] Masaaki Kimura, Phys. Rev. C**75**, 034312 (2007).
[7] T. Yamada and Y. Funaki, Phys. Rev. C**82**, 064315 (2010).
[8] B. Buck and A. A. Pilt, Nucl. Phys. A**280**, 133 (1977).
[9] F. Nemoto and H. Bando, Prog. Theor. Phys. **47**, 1210 (1972).
[10] P. Descouvemont and D. Baye, Nucl. Phys. A**463**, 629, (1986).
[11] T. Sakuda and F. Nemoto, Prog. Theor. Phys. **62**, 1274 (1979).
[12] R. Otani et al., Phys. Rev. C**90**, 034306 (2014).
[13] K. Langanke et al., Astrophys. Jour. **301**, 629 (1986).
[14] Z. Q. Mao et al., Phys. Rev. Lett. **74**, 3760 (1995).
[15] H. Horiuchi, Suppl. Prog. Theor. Phys. **62**, 90 (1977).
[16] A. B. Volkov, Nucl. Phys. **74**, 33 (1965).
[17] N. Yamaguchi, T. Kasahara, S. Nagata, and Y. Akaishi, Prog. Theor. Phys. **62**, 1018 (1979).
[18] Andrá T. Kruppa, and K. Kato, Prog. Theor. Phys. **84**, 1145 (1990).
[19] Y. Takenaka et al., Prog. Exp. Theor. **2014**, 113D04 (2014).
[20] H. Yamaguchi et al., Phys. Rev. C**87**, 034303 (2013); H. Yamaguchi et al., Phys. Rev. C**83**, 034306 (2011).

EPJ Web of Conferences **122**, 12001 (2016) DOI: 10.1051/epjconf/201612212001
*CNR*15*

Compound-nuclear reactions with unstable nuclei: Constraining theory through innovative experimental approaches*

J. E. Escher[1,a], A. P. Tonchev[1], J. T. Burke[1], P. Bedrossian[1], R. J. Casperson[1], N. Cooper[2], R. O. Hughes[1], P. Humby[2,3], R. S. Ilieva[3,4], S. Ota[1], N. Pietralla[5], N. D. Scielzo[1], and V. Werner[4,5]

[1] *Lawrence Livermore National Laboratory, Livermore, CA 94550, USA*
[2] *Department of Physics, University of Richmond, Richmond, Virginia 23173, USA*
[3] *Department of Physics, University of Surrey, Guildford, GU2 7XH, UK*
[4] *Wright Nuclear Structure Laboratory, Yale University, New Haven, CT 06520, USA*
[5] *Institut fuer Kernphysik, TU Darmstadt, Schlossgartenstrasse 9, 64289 Darmstadt, Germany*

Abstract. Cross sections for compound-nuclear reactions involving unstable targets are important for many applications, but can often not be measured directly. Several indirect methods have recently been proposed to determine neutron capture cross sections for unstable isotopes. We consider three approaches that aim at constraining statistical calculations of capture cross sections with data obtained from the decay of the compound nucleus relevant to the desired reaction. Each method produces this compound nucleus in a different manner (via a light-ion reaction, a photon-induced reaction, or β-decay) and requires additional ingredients to yield the sought-after cross section. We give a brief outline of the approaches and employ preliminary results from recent measurements to illustrate the methods. We discuss the main advantages and challenges of each approach.

1 Introduction

Compound-nuclear reactions play an important role in many applications. Their cross sections are required input for astrophysical models that describe stellar evolution and nucleosynthesis and for modeling processes that are relevant to generating energy. Neutron capture cross sections, for instance, are crucial for understanding the production of the elements from iron to uranium that we observe today. It is well known that nucleosynthesis of heavy elements beyond ^{56}Fe takes place primarily by neutron capture on lighter seed nuclei in the s (slow neutron capture) and r (rapid neutron capture) processes, with other processes contributing to the abundances of some specific isotopes. Near stability, neutron captures on s-process branch points, unstable nuclei with a lifetime long enough to allow the s process to proceed by either neutron capture or β decay, are of particular interest. The competition between capture and decay paths depends on environmental variables, such as neutron densities, temperatures, and pressure, as well as on nuclear properties. Measured s-process abundances, therefore, yield valuable insights into the detailed conditions of the astrophysical s process. Further away from stability, one finds isotopes for which the capture cross sections affect the isotopic abundance patterns predicted

[a] e-mail: escher1@llnl.gov

by r-process models. While the r process is known to take place in a high-temperature, high-flux environment, neither the exact path along the nuclear chart nor the astrophysical site where it takes place is known. Capture cross sections significantly affect the isotopic abundance patterns predicted by astrophysical models and need to be known to draw conclusions about stellar evolution and nucleosynthesis. Uncovering the origin of the heavy elements is one of the most important overarching questions in nuclear science [1].

However, due to the short half-live of many of the isotopes involved in the synthesis of the heavy elements, a large number of capture cross sections cannot be measured directly. Statistical reaction calculations typically use regional systematics to determine cross sections for isotopes at and near stability. In the absence of constraining auxiliary data, cross section uncertainties can be significant. Predictions of cross sections for nuclei away from stability, moreover, rely on extrapolations and become increasingly uncertain. An order of magnitude or two difference between different statistical calculations are not unusual in the latter case.

This contribution discusses indirect methods for constraining calculations of compound-nuclear reaction cross sections with particular focus on neutron-capture reactions. Specifically, we will focus on the production of the relevant compound nuclei (CN) via light-ion inelastic scattering and transfer reactions, photon-induced reactions, and nuclear β-decay. We will illustrate the methods with examples from recent experiments and discuss the challenges involved in extracting cross sections from the measurements.

2 The need for indirect measurements

2.1 Formalism for compound-nuclear capture reactions

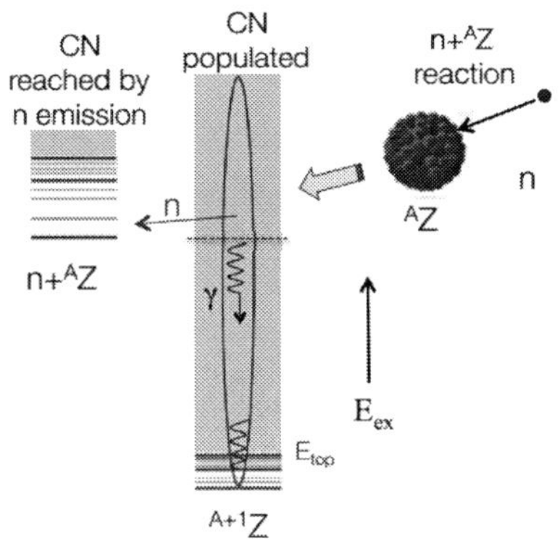

Figure 1. Schematic of a capture reaction. In a compound neutron capture process, the neutron and target nucleus ^{A}Z fuse to form the ^{A+1}Z compound nucleus, which subsequently decays via a γ-ray cascade. The capture cross section can be calculated using statistical Hauser-Feshbach theory, which requires nuclear level information, level densities, and γ-ray strength functions to properly describe the competition between the γ channel of interest and the competing particle emission channels (Here, we show only the competition between the γ and neutron channels.)

The formalism appropriate for describing compound-nucleus reactions is the statistical Hauser-Feshbach theory [2–4]. The average cross section per unit energy in the outgoing channel for reactions proceeding to an energy region in the final nucleus described by a level density is given by:

$$\frac{d\sigma_{\alpha\chi}^{HF}(E_a)}{dE_\chi} = \pi \lambda\!\!\!^{-}_{\alpha}^2 \sum_{J\pi} \omega_\alpha^J \sum_{\ell s \ell' s' I'} \frac{T^J_{\alpha\ell s} T^J_{\chi\ell' s'} \rho_{I'}(U') W_{\alpha\chi}(J)}{\sum'_{\chi''\ell''s''} T^J_{\chi''\ell''s''} + \sum_{\chi''\ell''s''I''} \int T^J_{\chi''\ell''s''}(E_{\chi''}) \rho_{I''}(U'') dE_{\chi''}}. \quad (1)$$

Here α denotes the entrance channel $a + A$ and χ represents the relevant exit channel $c + C$, E_a is the kinetic energy of the projectile, and $\lambda\!\!\!^{-}_\alpha$ is the reduced wavelength in the incident channel. The spin of the incident particle is i, the target spin is I, the channel spin is $\vec{s} = \vec{i} + \vec{I}$, and the compound-nucleus angular momentum and parity are $J\pi$. The statistical-weight factor ω_α^J is $(2J + 1)/[(2i + 1)(2I + 1)]$. Similarly, the spin of the outgoing particle is i', the spin of the residual nucleus is I', and the channel spin for χ is $\vec{s'} = \vec{i'} + \vec{I'}$. The quantities ℓ and ℓ' are the relative orbital angular momentum in the entrance and exit channels, respectively. The transmission coefficients are written as $T^J_{\alpha l s}$ and $\rho_{I'}(U')$ denotes the density of levels of spin I' at excitation energy U' in the residual nucleus. All energetically possible final channels χ'' have to be taken into account, thus the denominator includes contributions from decays to discrete levels in the residual nuclei (given by the first sum in the denominator, $\sum'$) as well as contributions from decays to regions described by a level density in the residual nuclei (given by the second sum in the denominator which involves an energy integral of transmission coefficients and level densities in the residual nuclei). Width fluctuation corrections $W_{\alpha\chi}$ are included in order to account for correlations between the incident and outgoing reaction channels [5, 6]. In writing Eq. 1, we have suppressed the parity quantum number except for that of the compound nucleus. In fact, the level density depends in principle on parity and all sums over quantum numbers must respect parity conservation.

For radiative neutron capture (α = n + ^{A}Z and $\chi = \gamma$ + $^{A+1}$Z), we usually need only the integral over the energy spectrum of primary γ rays emitted from the compound nucleus. (To determine cross sections for particular γ transitions or for the production of isomers, additional details of the γ-cascade need to be accounted for.) We integrate over all energies E_χ of the final-state channel and, in a first approximation, neglect the width fluctuation correlations. The primary effect of the correlations is an enhancement of the elastic scattering cross section. This allows us to write the cross section for the desired reaction as:

$$\sigma_{\alpha\chi}(E_n) \quad = \quad \sum_{J\pi} \sigma_\alpha^{CN}(E_{ex}, J, \pi) \; G_\chi^{CN}(E_{ex}, J, \pi)\,, \quad (2)$$

where $\sigma^{CN}(E_{ex}, J, \pi) = \sigma$ (n + ^{A}Z $\rightarrow$ $^{A+1}$Z*) denotes the cross section for forming the compound nucleus at excitation energy E_{ex} with angular-momentum and parity quantum numbers $J\pi$ and $G_\chi^{CN}(E_{ex}, J, \pi)$ is the branching ratio for the decay of this compound state into the desired exit channel χ. It contains transmission coefficients for the competing exit channels as well as the associated level densities and information on discrete levels. The kinetic energy E_n of the neutron is related to the excitation energy of the compound nucleus, E_{ex}, via $E_n = \frac{A+1}{A}(E_{ex} - S_n)$, where S_n is the energy required for separating a neutron from the compound nucleus $^{A+1}$Z. This factored form embodies the essential assumptions of the Hauser-Feshbach model, that formation and decay of the compound nucleus are independent processes, and that the total spin and parity of the compound system must be conserved.

2.2 Ingredients for calculating capture cross sections

Ingredients required to carry out Hauser-Feshbach cross-section calculations include nuclear binding energies, spins and parities of both ground and excited nuclear states, γ-branching ratios for these states, nuclear level densities and transmission coefficients for particles, photons, and fission. Much effort has been devoted to develop the requisite models and codes, and to formulate parameter recommendations [7]. Auxiliary experiments provide important constraints for these models and their parameters. For instance, neutron resonance measurements provide average level spacings and thus constraints for the level densities near the neutron separation energy. Average radiative widths provide information on the product of the level density and the γ-ray strength function, which is relevant to capture cross section calculations.

For neutron capture reactions involving nuclei near the valley of stability, the formation of the compound nucleus is reasonably well described, while a reliable description of the decay is more challenging, since the competition between all possible decay channels has to be properly accounted for in the calculated decay probabilities. Specifically, the γ-ray strength function for the CN ^{A+1}Z is needed and the level densities in both the target (^{A}Z) and compound nuclei are required, as well as information on the low-lying discrete states (energies, spins and parities) in the target nucleus.

For reactions involving unstable targets, the requisite neutron resonance spacings and average radiative capture widths are not available and the decay probabilities $G_{\chi}^{CN}(E_{ex}, J, \pi)$ become the primary source of uncertainty in calculations of capture cross sections. Evaluations typically rely on regional systematics in this case and a factor two (or more) difference between evaluated cross sections is not uncommon; for reactions involving nuclei far from stability the uncertainties increase to more than an order of magnitude. The indirect approaches discussed in this contribution aim at improving this situation by providing constraints for these $G_{\chi}^{CN}(E_{ex}, J, \pi)$.

3 CN production via inelastic scattering and transfer reactions

3.1 Method and illustrative example

In an approach that has become known as the 'surrogate method' [8], the compound nucleus $^{A+1}Z^*$ is produced by an inelastic scattering or transfer reaction $d + D \to b + {}^{A+1}Z^*$, and the desired decay channel $\chi({}^{A+1}Z^* \to c + C)$ is observed in coincidence with the outgoing particle b (see left panel of Fig. 2). The probability for forming $^{A+1}Z^*$ in the surrogate reaction (with specific values for E_{ex}, J, π) is $F_{\delta}^{CN}(E_{ex}, J, \pi, \theta_b)$, where δ refers to the entrance channel reaction $D(d, b)$ and θ_b is the angle of the outgoing direct-reaction particle b relative to the beam axis. The quantity

$$P_{\delta\chi}(E_{ex}, \theta_b) = \sum_{J,\pi} F_{\delta}^{CN}(E_{ex}, J, \pi, \theta_b)\ G_{\chi}^{CN}(E_{ex}, J, \pi)\ , \tag{3}$$

which gives the probability that the compound nucleus $^{A+1}Z^*$ was formed with energy E_{ex} and decayed into channel $\chi = \gamma + {}^{A+1}Z$, can be obtained experimentally, by measuring N_{δ}, the total number of surrogate events, and $N_{\delta\chi}$, the number of coincidences between the direct-reaction particle and the observable that identifies the relevant exit channel[1]: $P_{\delta\chi}^{exp}(E_{ex}, \theta_b) = N_{\delta\chi}(E_{ex}, \theta_b)/N_{\delta}(E_{ex}, \theta_b)\epsilon_{\delta}(E_{ex})$. Here, $\epsilon_{\delta}(E_{ex})$ denotes the efficiency for detecting the exit-channel χ (when in coincidence with the outgoing direct-reaction particle b). For example, a discrete γ-ray transition involving low-lying states in the decaying CN $^{A+1}Z^*$ can be used to identify the capture channel. In that case, $\epsilon_{\delta}(E_{ex})$ is the efficiency for detecting a γ-ray of that energy experimentally (preferably in a high-resolution Germanium detector).

To illustrate the idea, we discuss preliminary results from an experiment in the Zr region. The decay of compound nuclei in this region is known to be particularly sensitive to the spins and parities

[1] The angular dependence of $P_{\delta\chi}^{exp}(E_{ex}, \theta_b)$ arises from the fact that the population $F_{\delta}^{CN}(E_{ex}, J, \pi, \theta_b)$ of the compound nucleus depends on the angular-momentum transferred in the surrogate reaction, and hence on θ_b.

populated via a surrogate reaction [9–13]. Figure 2 gives a schematic representation of the surrogate ^{92}Zr(p,d) reaction, which was recently used to produce the CN ^{91}Zr at a range of excitation energies. This case is relevant to the ^{91}Zr(n,γ) reaction, for which directly-measured cross-section data exists. The experiment was carried out at the Texas A&M University Cyclotron Institute, where the K150 Cyclotron was employed to provide a 28.5-MeV proton beam. Particle-γ coincidence data was collected using the STARLiTeR array, a combination of a silicon telescope array and five HPGe clover detectors (see, e.g., Ref. [13]). The outgoing deuteron was detected at angles around 30 – 60 deg relative to the beam direction, and γ-rays associated with transitions between low-lying states (E_n < 2 MeV) were measured. The data in the right panel of Figure 2 gives the coincidence probability for the 1466 keV transition (from the $5/2^+$ state at 1.466 MeV to the $5/2^+$ ground state of ^{91}Zr) as a function of the excitation energy E_{ex} in the ^{91}Zr nucleus. E_{ex} was reconstructed from the measured energy and angle of the outgoing deuteron. When the CN ^{91}Zr is populated below the neutron threshold (S_n = 7.15 MeV), less than 10% of the decays proceed through this transition. The remaining γ-decay by-passes this transition, a fact that reflects the rich structure of this odd-A nucleus. As the excitation energy is increased beyond S_n, neutron emission begins to compete and the γ-probability drops quickly. The solid, dash-dotted, and dashed curves are the results of calculations that combine predictions for the weights $F_\delta^{CN}(E_{ex}, J, \pi, \theta_b)$ in Eq. 3 with a Hauser-Feshbach-type decay model for ^{91}Zr.

The distribution $F_\delta^{CN}(E_{ex}, J, \pi, \theta_b)$, which may be very different from the CN spin-parity populations following the absorption of a neutron in the desired reaction, has to be determined theoretically, so that the branching ratios $G_\chi^{CN}(E_{ex}, J, \pi)$ can be extracted from the measurements. In practice, the decay of the CN is modeled using a Hauser-Feshbach-type decay model and the $G_\chi^{CN}(E_{ex}, J, \pi)$ are obtained by adjusting parameters in the model to reproduce the measured probabilities $P_{\delta\chi}(E_{ex}, \theta_b)$. Subsequently, the sought-after cross section can be obtained by combining the calculated cross section $\sigma_\alpha^{CN}(E_{ex}, J, \pi)$ for the formation of $^{A+1}Z^*$ (from n+^AZ) with the extracted decay probabilities $G_\chi^{CN}(E_{ex}, J, \pi)$ for this state, see Eq. 2. At this point, the width fluctuation correlations can be included using standard correction factors [5, 6, 14].

The present case involves stable nuclei for which information on levels, level densities, and γ-ray strength functions is available, so the γ-cascade can be modeled and tests of the theory predicting the CN population $F_\delta^{CN}(E_{ex}, J, \pi, \theta_b)$ can be carried out. The ^{92}Zr(p,d) reaction involves the pickup of neutrons from inner shells of the ^{92}Zr nucleus. Neutron hole states $2p_{1/2}$, $2p_{3/2}$, $1f_{5/2}$, $1f_{7/2}$, etc., are involved in the production of ^{91}Zr near S_n. Their location and fragmentation as a function of ^{91}Zr excitation energy was obtained using the dispersive optical model approach of Mahaux and Sartor [15]. Cross sections for the one-neutron pickup process were calculated in the 1-step distorted-wave Born approximation and the relative weights of the final spins were determined as a function of $E_{ex}(^{91}\mathrm{Zr})$. The combination of these with the Hauser-Feshbach-type decay model results in predicted surrogate probabilities $P_{\delta\chi}^{exp}(E_{ex}, \theta_b)$ shown in the right panel of Fig. 2. We see that the probability calculated in the 1-step DWBA (solid red curve) does not agree well with the data. Contributions from two-step processes such as (p,p',d) and (p,d',d), in which the initial ^{92}Zr and the final ^{91}Zr are inelastically excited, are found to play an important role, as the dash-dotted and dashed curves show. Similar results are found for other γ-ray transitions in this nucleus (not shown). It becomes clear that in order to reproduce the surrogate data, structure information involving highly-excited regions of the ^{91}Zr nucleus is needed, as well as a detailed treatment of the surrogate reaction mechanism. The findings also underscore the value of benchmark experiments.

3.2 Advantages and challenges of CN production via light-ion reactions

An advantage of using inelastic scattering and transfer reactions to produce the CN of interest lies in the variety of projectile-target combinations that can be utilized. That makes it possible to reach a

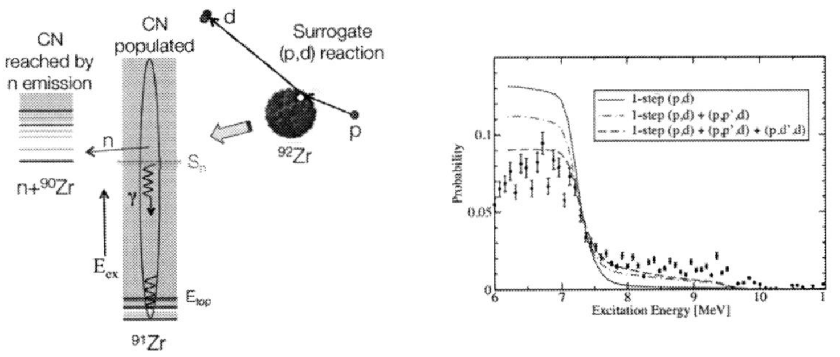

Figure 2. Strategy for obtaining cross section constraints from a (surrogate) light-ion inelastic scattering or transfer reaction. Left panel: Schematic representation of the surrogate reaction mechanism. The basic idea of the approach is to replace the first step of the desired reaction, $a + A$, by an alternative reaction, $d + D \rightarrow b + B^*$, that is experimentally easier to access yet populates the same compound nucleus. The subsequent decay of the compound nucleus into the relevant channel, $c + C$, can then be measured and used to extract the desired cross section. Here the desired reaction is ^{90}Zr(n,γ)^{91}Zr. The CN ^{91}Zr is produced via the ^{92}Zr(p,d)^{91}Zr one-nucleon pickup reaction. Neutron evaporation and γ emission compete in the decay of the CN. Characteristic γ-ray transitions in ^{91}Zr are measured in coincidence with the outgoing light ion. Right panel: Measured coincidence probability in the decay of ^{91}Zr. The probability of observing the 1466 keV transition in coincidence with the outgoing deuteron is given as function of E_{ex} (black data points with error bars). The curves show predicted coincidence probabilities based on one-step and two-step DWBA calculations. These results are preliminary.

large number of isotopes, in particular when the technique can be combined with radioactive beams in inverse-kinematics experiments [16]. In some cases the CN can be produced by multiple reactions which result in different spin-parity populations [12, 13], allowing for comparisons of cross sections obtained from different measurements and analyses. In addition, measurements of the outgoing light ion enable us to scan through a range of excitation energies in the CN and to obtain information both below and above the separation energy. This can be employed to probe different aspects of the CN decay.

To extract accurate neutron capture cross sections from surrogate data, we have to quantitatively account for the fact that the weights $F_{\delta}^{CN}(E_{\rm ex}, J, \pi, \theta_b)$, by which the decay probabilities $G_{\chi}^{CN}(E_{ex}, J, \pi)$ are multiplied in Eq. 3, are different from the relative formation cross sections $f_{\alpha}^{CN}(E_{ex}, J, \pi) \equiv \sigma_{\alpha}^{CN}(E_{ex}, J, \pi)/\sum_{J'\pi'} \sigma_{\alpha}^{CN}(E_{ex}, J', \pi')$ of the desired reaction, see Eq. 2. The decay probabilities $G_{\chi}^{CN}(E_{ex}, J, \pi)$ depend on the spins and parities of the decaying nucleus and that dependence is particularly strong in the Zr region, where the level density is low and neutron emission to discrete levels competes with γ emission [9, 10]. This is different from surrogate applications to (n,f) reactions. In the fission case the $G_{\chi=fission}^{CN}(E_{ex}, J, \pi)$ are much less sensitive to spin and parity and approximations can be employed which allow for analysis methods that do not require the knowledge of the CN spin-parity distributions [17–20].

Predicting these weights requires a framework for calculating cross sections of different reactions (stripping, pick-up, charge exchange, and inelastic scattering) to continuum states, for a variety of projectiles (p, d, t, α, etc.) and targets (spherical, deformed, and transitional). The ^{92}Zr example discussed showed that pickup reactions require information on the structure of the CN up to very high excitation energies (~10 MeV for capture). This can be a challenge, as both (quasi) single-particle and collective degrees of freedom can be relevant. Similar considerations apply to inelastic scattering, which involves (quasi) particle-hole excitations, and stripping reactions, which involve resonance states. The example also shows that details of the reaction mechanism need to be considered. Multi-step contributions can play a role and breakup (primarily for stripping reactions) needs to be accounted for.

4 CN production via photon-induced reactions

4.1 Method and illustrative example

Photon-induced reactions provide an additional mechanism for producing highly-excited nuclei, the decay of which is relevant to unknown capture cross sections. Inelastic scattering and photo-neutron reactions can be used to populate a nucleus of interest across a wide range of excitation energies, up to and beyond the neutron separation energy. Bremsstrahlung photons as well as quasi-monoenergetic γ-ray beams have been used for this purpose [21–29]. In a few experiments, neutrons emitted from the CN were detected and the (γ,n) cross sections were determined. Hauser-Feshbach calculations of these photo-neutron reactions were carried out and comparisons with the data were utilized to determine the most appropriate combination of level densities and γ-ray strength functions [28, 30]. These were subsequently employed to calculate the (n,γ) cross section of interest.

Alternatively, it is possible to measure the γ-ray strength function, an important ingredient for neutron-capture calculations, in photon-scattering experiments [22–24]. This approach makes it possible to determine the strength function below the particle emission threshold directly, without the need to model the reaction mechanism. As electric dipole transitions dominate, this approach provides access to the E1 strength function. Some setups also allow one to distinguish between different multipolarities [27, 29].

The idea of the approach is illustrated in Figure 3 for the γ + ^{76}Ge case. Nuclear resonance fluorescence (NRF) measurements of the nuclear dipole response of ^{76}Ge were recently carried out at the HIγS facility at TUNL at Duke University and at the DHIPS facility at TU Darmstadt, Germany: Werner and collaborators *et al.* [31, 32] combined results from a bremsstrahlung experiment with an experiment using quasi-monoenergetic, linearly polarized photons to measure spins and parities of 130 excited levels in ^{76}Ge and to determine the photoabsorption cross section for this nucleus in the energy range between $E_\gamma = 4$ MeV and $E_\gamma = 9$ MeV. The photoabsorption cross section is directly related to the γ-ray strength function $f(E_\gamma)$ of interest, $\sigma_{abs} = 3(\pi\hbar c)^2 E_\gamma f(E_\gamma)$. Here the assumption is made that the upward and downward strength functions are identical. We selected this example because it involves the same CN that has recently been produced via β-decay of ^{76}Ga and a comparison of the strength functions obtained from these very different experiments, using very different analysis techniques, can be carried out. The results will be discussed below.

4.2 Advantages and challenges of γ-induced reactions

The primary advantage of obtaining constraints for (n,γ) calculations from photon-induced reactions lies in the fact that this approach provides access to the γ-ray strength function in a model-independent manner (although it relies on a few assumptions, such as the validity of the Brink-Axel hypothesis). Well-established experimental techniques can be used to carry out the measurements. Recent improvements to the technique, such as the combination of an intense mono-energetic photon beam at the HIγS facility with γ-γ coincidence spectroscopy of the subsequent decays makes it possible to

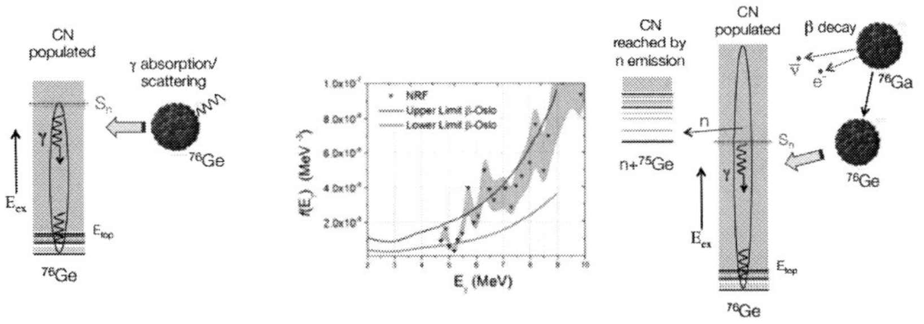

Figure 3. Strategies for obtaining cross section constraints from γ-induced reactions and nuclear β decay. Left panel: Schematic representation of the CN production via γ-induced reactions. Photons can be elastically or inelastically scattered or absorbed. Populating states below particle thresholds gives information on the γ-ray strength function, while producing the CN above S_n allows for a study of the competition between neutron and γ emission. Right panel: Schematic representation of the CN-production mechanism via β-decay. Center panel: Comparison of γ–ray strength functions extracted from from the photoabsorption cross section measured via nuclear resonance fluorescence experiments (γ absorption and emission) and a β–Oslo experiment, respectively. The former gives a more direct (model-independent) measurement of this quantity and thus provides a useful test of the latter method. The NRF results are preliminary.

study the detailed structure of the strength function in more detail than previously possible, in an energy regime that is important for neutron-capture calculations.

The fact that the spins populated in photon-induced reactions are determined by the ground state spin of the $^{A+1}$Z nucleus used in the experiment can be an advantage when photo-neutron data is used to constrain Hauser-Feshbach calculations. If both the target nucleus ^{A}Z and the compound nucleus $^{A+1}$Z of the desired ^{A}Z(n,γ) $^{A+1}$Z have similar spins, it can be expected that both reactions populate states of similar spins in the $^{A+1}$Z system, and thus both probe the competition of γ decay and neutron emission from states that are similar to each other. For cases with very different spins for the ^{A}Z and $^{A+1}$Z ground state, however, we can expect a spin-mismatch between the compound nuclei created in the different reactions. This is similar to the situation discussed above for light-ion reactions, although the effect of the electromagnetic interaction is well known and the spins and parities are accounted for in a straight-forward manner in the Hauser-Feshbach calculations.

The main drawback of using photon-induced reactions is the fact that it is limited to cases where the CN of interest has a stable ground state. It is therefore not possible to apply this strategy to nuclei that are far from stability. A limitation of solely considering the energy regime below the neutron threshold is that only the γ-ray strength function is obtained, while additional information is required to constrain the level densities of the nuclei involved.

5 CN production via β-decay

5.1 Method and illustrative example

For reactions involving compound nuclei on the neutron-rich side of the valley of stability, a third indirect approach can be considered: Nuclear β-decay from an unstable precursor nucleus $^{A+1}$(Z-1) can be used to produce the isotope of interest $^{A+1}$Z. This is schematically shown in the right panel

DOI: 10.1051/epjconf/201612212001

of Figure 3. The β-decay Q value determines the excitation energies that can be reached in the $^{A+1}$Z nucleus. Near stability, the available energy range may not reach up to the neutron separation energy. E.g., for the ^{76}Ga$\rightarrow^{76}$Ge example shown, states up to E_{ex} = 7.0 MeV can be populated, which is below the neutron threshold (S_n=9.4 MeV). With increasing neutron excess, the neutron binding energy decreases, the Q-value window increases, and consequently bound as well as neutron unbound states can be populated via β-decay.

Since allowed Gamov-Teller decay is the dominant mode, the spins populated in the $^{A+1}$Z nucleus lie within a narrow band (J^{CN} = J^{pre}-1, J^{pre}, J^{pre}+1, where J^{pre} is the ground state spin of the decaying precursor). When states above S_n are reached by β-decay, the subsequent competition between neutron and γ emission can be studied.

The ^{76}Ga$\rightarrow^{76}$Ge case was investigated recently by a group at the National Superconducting Cyclotron Laboratory at Michigan State University (NSCL/MSU) [33]. The experiment focused on measuring β-delayed γ emission with a total absorption γ spectrometer (TAGS). For γ emission in coincidence with β particles, the authors determined the matrix $P(E_\gamma, E_{ex})$, which gives the relative probabilities of primary transitions with energy E_γ in ^{76}Ge originating from energy bins at excitation energies E_{ex}. This matrix was obtained up to about 7 MeV and subsequently analyzed using the Oslo method [34–36].

The Oslo method was previously applied to γ-spectra measured in coincidence with charged-particle reactions. The analysis is aimed at extracting both level density and γ-ray strength function of a decaying nucleus from an experimental primary γ-ray matrix $P(E_\gamma, E_{ex})$. The analysis involves a series of steps in which the detector response function is accounted for, the primary γ-rays are isolated and the functional forms of the level density and γ-ray strength function are extracted [36]. Auxiliary data is required to determine the final physical solutions of these sought-after quantities. Typically, the level density is fixed by the requirement that it reproduces the cumulative number of levels at low energies and the neutron resonance spacing at the neutron separation energy, and the strength function is fixed by utilizing the average radiative width $< \Gamma_\gamma >$ at S_n. For the ^{76}Ge nucleus this information is not available and the authors used estimates based on microscopic calculations and systematics. The resulting level density and strength function were then used in a Hauser-Feshbach calculation of the ^{75}Ge(n,γ)^{76}Ge cross section. The cross section was found to be in agreement with standard Talys calculations. Unfortunately, no directly-measured data is available to benchmark the extracted capture cross section.

To better assess the promise of the β-Oslo approach, we thus compared the γ-ray strength function obtained in this approach with a more direct measurement of this quantity using the NRF approach discussed in the previous section. We find that, even though the NRF experiment yields a γ-ray strength function that has more structure and is slightly larger than the result from the β-Oslo method, both approaches give similar results (center panel of Figure 3). This additional, more direct, test of the β-Oslo method is encouraging. It would be useful to also carry out an independent test for the extracted level density.

5.2 Advantages and challenges of the β-decay method

A strong advantage of employing the β-decay approach is the fact that one can produce neutron-rich isotopes far from stability, with production rates that are potentially much higher than those obtained via light-ion inverse-kinematics reactions. The Q values for these nuclei make it possible to populate states both below and above the neutron separation energy. Since Gamov-Teller decay dominates, the spins populated in the intermediate nucleus lie in a narrow range and are determined by the ground state spin of the decaying precursor. Knowledge of the states populated is needed in order to infer information on the relative probabilities of the neutron and γ decay channels and to use that to constrain cross section calculations for capture reactions[2].

The Oslo analysis method seems to be less sensitive to the spin distribution of the γ-decaying nucleus than a γ-decay analysis that makes use of individual γ transitions. On the other hand, it has the drawback that additional, auxiliary information is required to normalize the level densities and γ-ray strength functions extracted. The data that is typically used to allow for this normalization, will most likely not be available for nuclei away from the valley of stability and alternative methods for obtaining the normalization factors need to be found.

6 Conclusions

Neutron-capture reactions on unstable isotopes are of interest to astrophysics and other applications. Indirect measurements play an important role in constraining calculations of the requisite cross sections. Several innovative methods for obtaining experimental constraints for the calculations are being explored. We have considered three approaches that produce the compound nuclei of interest via light-ion inelastic scattering and transfer reactions, photon-induced reactions, and nuclear β-decay, respectively. The constraints are obtained from measurements of the subsequent compound-nuclear decay. Each method has its unique strengths, as well as experimental and theoretical challenges that have to be overcome in order to provide reliable capture cross sections for nucleosynthesis modeling and other applications. Having several methods at one's disposal is important in order to validate the results obtained and to cover the reactions of interest.

Current investigations focus mostly on isotopes close to stability, where benchmarking is facilitated by the availability of auxiliary information and, in some cases, directly-measured cross sections. The longer-term goal is to develop methods that can be used for a wider range of reactions. Radioactive beams are now available at multiple facilities, such as Argonne National Laboratory (ANL) and the National Superconducting Cyclotron Laboratory at Michigan State University (NSCL/MSU). FRIB, which is under construction at MSU, will provide more intense beams of isotopes far from stability. The availability of these exotic beams make it possible to produce important isotopes using inelastic scattering and transfer reactions in inverse kinematics, as well as via β-decay, and to study of the subsequent competition between neutron and γ-ray emission.

Acknowledgments

This work was performed under the auspices of the U.S. Department of Energy by Lawrence Livermore National Laboratory under contract DE-AC52-07NA27344, with partial support through LLNL's LDRD (15-ERD-069) and ASC/PEM programs.

References

[1] Nuclear Science Advisory Committee, *Reaching for the Horizon: The 2015 Long Range Plan for Nuclear Science* (2015)
[2] W. Hauser, H. Feshbach, Phys. Rev. **87**, 366 (1952)
[3] E. Gadioli, P.E. Hodgson, *Pre-Equilibrium Nuclear Reactions* (Clarendon Press, Oxford, 1992)
[4] P. Fröbrich, R. Lipperheide, *Theory of Nuclear Reactions* (Clarendon Press, Oxford, 1996)

[2]An attempt to determine this competition by combining β-delayed γ data with β-delayed neutron data was recently published for the ^{87}Br→^{87}Kr, ^{88}Br→^{88}Kr, and ^{94}Rb→^{94}Zr cases [37]. The results of that experiment implied some unusual behavior of the strength function and need to be understood better.

[5] P.A. Moldauer, Phys. Rev. **123**, 968 (1961)
[6] S. Hilaire, C. Lagrange, A.J. Koning, Annals of Physics **306**, 209 (2003)
[7] R. Capote, M. Herman, P. Obložinský, P. Young, S. Goriely, T. Belgya, A. Ignatyuk, A. Koning, S. Hilaire, V. Plujko et al., Nuclear Data Sheets **110**, 3107 (2009)
[8] J.E. Escher, J.T. Burke, F.S. Dietrich, N.D. Scielzo, I.J. Thompson, W. Younes, Rev. Mod. Phys. **84**, 353 (2012)
[9] C. Forssén, F. Dietrich, J. Escher, R. Hoffman, K. Kelley, Phys. Rev. C **75**, 055807 (2007)
[10] J.E. Escher, F.S. Dietrich, Phys. Rev. C **81**, 024612 (2010)
[11] N.D. Scielzo, J.E. Escher, J.M. Allmond, M.S. Basunia, C.W. Beausang, L.A. Bernstein, D.L. Bleuel, J.T. Burke, R.M. Clark, F.S. Dietrich et al., Phys. Rev. C **81**, 034608 (2010)
[12] S. Ota, J. Burke, R. Casperson, J. Escher, R. Hughes, J. Ressler, N. Scielzo, I. Thompson, R. Austin, E. McCleskey et al., EPJ Web of Conferences **93**, 02001 (2015)
[13] S. Ota, J.T. Burke, R.J. Casperson, J.E. Escher, R.O. Hughes, J.J. Ressler, N.D. Scielzo, I.J. Thompson, R.A.E. Austin, B. Abromeit et al., Phys. Rev. C **92**, 054603 (2015)
[14] T. Kawano, P. Talou, H.A. Weidenmüller, Phys. Rev. C **92**, 044617 (2015)
[15] C. Mahaux, R. Sartor, Adv. Nucl. Phys. **20**, 1 (1991)
[16] J. Cizewski, R. Hatarik, K. Jones, S. Pain, J. Thomas, M. Johnson, D. Bardayan, J. Blackmon, M. Smith, R. Kozub, Nuclear Instruments and Methods in Physics Research Section B: Beam Interactions with Materials and Atoms **261**, 938 (2007)
[17] J.E. Escher, F.S. Dietrich, Phys. Rev. C **74**, 054601 (2006)
[18] J.T. Burke, L.A. Bernstein, J. Escher, L. Ahle, J.A. Church, F.S. Dietrich, K. Moody, E.B. Norman, L. Phair, P. Fallon et al., Phys. Rev. C **73**, 054604 (2006)
[19] G. Kessedjian, B. Jurado, M. Aiche, G. Barreau, A. Bidaud, S. Czajkowski, D. Dassié, B. Haas, L. Mathieu, L. Audouin et al., Physics Letters B **692**, 297 (2010)
[20] J.J. Ressler, J.T. Burke, J.E. Escher, C.T. Angell, M.S. Basunia, C.W. Beausang, L.A. Bernstein, D.L. Bleuel, R.J. Casperson, B.L. Goldblum et al., Phys. Rev. C **83**, 054610 (2011)
[21] R. Schwengner, R. Beyer, F. Dönau, E. Grosse, A. Hartmann, A. Junghans, S. Mallion, G. Rusev, K. Schilling, W. Schulze et al., Nuclear Instruments and Methods in Physics Research Section A: Accelerators, Spectrometers, Detectors and Associated Equipment **555**, 211 (2005)
[22] R. Schwengner, G. Rusev, N. Benouaret, R. Beyer, M. Erhard, E. Grosse, A.R. Junghans, J. Klug, K. Kosev, L. Kostov et al., Phys. Rev. C **76**, 034321 (2007)
[23] R. Schwengner, G. Rusev, N. Tsoneva, N. Benouaret, R. Beyer, M. Erhard, E. Grosse, A.R. Junghans, J. Klug, K. Kosev et al., Phys. Rev. C **78**, 064314 (2008)
[24] N. Benouaret, R. Schwengner, G. Rusev, F. Dönau, R. Beyer, M. Erhard, E. Grosse, A.R. Junghans, K. Kosev, C. Nair et al., Phys. Rev. C **79**, 014303 (2009)
[25] B.L. Berman, S.C. Fultz, Rev. Mod. Phys. **47**, 713 (1975)
[26] Kneissl, H. U Pitz, A. Zilges, Prog. Part. Nucl. Phys. **37**, 349 (1996)
[27] N. Pietralla, Z. Berant, V.N. Litvinenko, S. Hartman, F.F. Mikhailov, I.V. Pinayev, G. Swift, M.W. Ahmed, J.H. Kelley, S.O. Nelson et al., Phys. Rev. Lett. **88**, 012502 (2001)
[28] R. Raut, A.P. Tonchev, G. Rusev, W. Tornow, C. Iliadis, M. Lugaro, J. Buntain, S. Goriely, J.H. Kelley, R. Schwengner et al., Phys. Rev. Lett. **111**, 112501 (2013)
[29] B. Löher, V. Derya, T. Aumann, J. Beller, N. Cooper, M. Duchêne, J. Endres, E. Fiori, J. Isaak, J. Kelley et al., Nuclear Instruments and Methods in Physics Research Section A: Accelerators, Spectrometers, Detectors and Associated Equipment **723**, 136 (2013)

[30] A. Sauerwein, K. Sonnabend, M. Fritzsche, J. Glorius, E. Kwan, N. Pietralla, C. Romig, G. Rusev, D. Savran, L. Schnorrenberger et al., Phys. Rev. C **89**, 035803 (2014)

[31] R.S. Ilieva, P. Humby, N. Cooper et al. (2015), submitted

[32] V. Werner, *Dipole strength distributions from HIGS experiments*, in *Fifteenth International Symposium on Capture Gamma-Ray Spectroscopy and Related Topics (CGS15)* (2015)

[33] A. Spyrou, S.N. Liddick, A.C. Larsen, M. Guttormsen, K. Cooper, A.C. Dombos, D.J. Morrissey, F. Naqvi, G. Perdikakis, S.J. Quinn et al., Phys. Rev. Lett. **113**, 232502 (2014)

[34] M. Guttormsen, T. Ramsøy, J. Rekstad, Nuclear Instruments and Methods in Physics Research Section A: Accelerators, Spectrometers, Detectors and Associated Equipment **255**, 518 (1987)

[35] A. Schiller, L. Bergholt, M. Guttormsen, E. Melby, J. Rekstad, S. Siem, Nuclear Instruments and Methods in Physics Research Section A: Accelerators, Spectrometers, Detectors and Associated Equipment **447**, 498 (2000)

[36] A.C. Larsen, M. Guttormsen, M. Krtička, E. Běták, A. Bürger, A. Görgen, H.T. Nyhus, J. Rekstad, A. Schiller, S. Siem et al., Phys. Rev. C **83**, 034315 (2011)

[37] J.L. Tain, E. Valencia, A. Algora, J. Agramunt, B. Rubio, S. Rice, W. Gelletly, P. Regan, A.A. Zakari-Issoufou, M. Fallot et al., Phys. Rev. Lett. **115**, 062502 (2015)

EPJ Web of Conferences **122**, 12002 (2016) DOI: 10.1051/epjconf/201612212002
*CNR*15*

Calculations of Compound Nucleus Spin-Parity Distributions Populated via the (p,t) Reaction in Support of Surrogate Neutron Capture Measurements

J.Benstead[1,2,a], J.A.Tostevin[2], J.E.Escher[3], J.T.Burke[3], R.O.Hughes[3], S.Ota[3], R.J.Casperson[3], and I.J.Thompson[2,3]

[1] *AWE Aldermaston, Reading, Berkshire, RG7 4PR, United Kingdom*
[2] *University of Surrey, Guildford, Surrey, GU2 7XH, United Kingdom*
[3] *Lawrence Livermore National Laboratory, Livermore, California 94550, USA*

Abstract. The surrogate reaction method may be used to determine the cross section for neutron induced reactions not accessible through standard experimental techniques. This is achieved by creating the same compound nucleus as would be expected in the desired reaction, but through a different incident channel, generally a direct transfer reaction. So far, the surrogate technique has been applied with reasonable success to determine the fission cross section for a number of actinides, but has been less successful when applied to other reactions, e.g. neutron capture, due to a 'spin-parity mismatch'. This mismatch, between the spin and parity distributions of the excited levels of the compound nucleus populated in the desired and surrogate channels, leads to differing decay probabilities and hence reduces the validity of using the surrogate method to infer the cross section in the desired channel. A greater theoretical understanding of the expected distribution of levels excited in both the desired and surrogate channels is therefore required in order to attempt to address this mismatch and allow the method to be utilised with greater confidence. Two neutron transfer reactions, e.g. (p,t), which allow the technique to be utilised for isotopes further removed from the line of stability, are the subject of this study. Results are presented for the calculated distribution of compound nucleus states populated in ^{90}Zr, via the ^{92}Zr(p,t)^{90}Zr reaction, and are compared against measured data at an incident proton energy of 28.56 MeV.

1 Introduction

The majority of reactions of interest to the nuclear industry, and many relevant to astrophysics, involve the collision of an incident neutron with a target nucleus. Aside from elastic scattering, the reactions which may occur due to an incident neutron typically take place through an intermediate compound nucleus state, as shown in Figure 1.

Here the incident neutron is absorbed by, and shares its energy with, the target nucleus and forms an excited compound nucleus. This compound nucleus is unstable and will decay after some time to form the final products of the reaction. The basis of the surrogate reaction method is to exploit

[a] e-mail: james.benstead@awe.co.uk

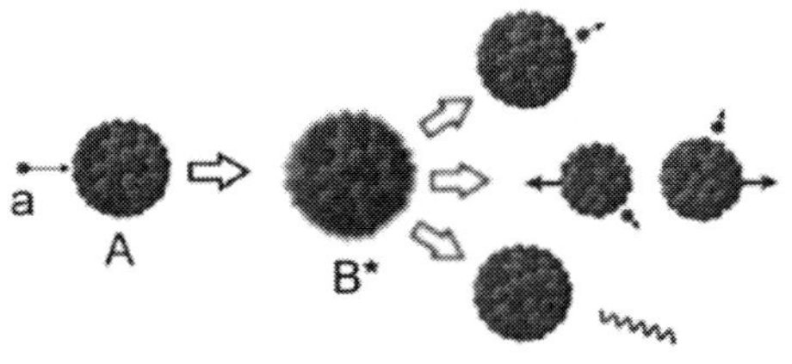

Figure 1. Illustration of a nuclear reaction proceeding through the compound nucleus stage. Here the projectile a collides with A to form B^*, which then decays to a number of possible products.

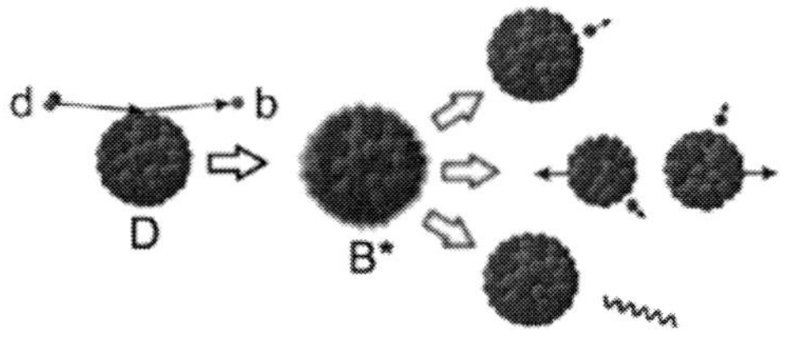

Figure 2. Illustration of the surrogate reaction method. Here, d interacts with D to form the compound nucleus B^*, which then decays in the same manner as would be expected through the desired reaction. The projectile d may simply be scattered inelastically, but most often loses or gains nucleons to form particle b.

the Bohr assumption, that the mode of decay of a compound nucleus is independent of the type of reaction from which it formed [1]. It is assumed that only the *spin distribution* of the states, in both energy and angular momentum, populated in the compound nucleus plays a role in determining the statistical likelihood of decays via each possible channel [2, 3].

In the surrogate reaction method, a suitable *surrogate* nucleus and reaction are sought, such that the same compound nucleus, and if possible its spin distribution, will be formed as is expected in the reaction of interest. The surrogate process is illustrated in Figure 2.

In the desired reaction case, one needs to determine the cross section $\sigma_{\alpha\chi}(E_a)$ for a reaction with incident channel α $(a + A)$ at an incident energy E_a and exit channel χ. The cross section $\sigma_{\alpha\chi}(E_a)$ is often split into two components,

$$\sigma_{\alpha\chi}(E_a) = \sum_{J,\pi} \sigma_{\alpha}^{B^*}(E_{\text{ex}}, J, \pi) G_{\chi}^{B^*}(E_{\text{ex}}, J, \pi) . \tag{1}$$

Here $\sigma_{\alpha}^{B^*}(E_{\text{ex}}, J, \pi)$ is the cross section for $(a + A)$ forming the compound nucleus B^* with an excitation energy E_{ex} in the state J^{π} and $G_{\chi}^{B^*}(E_{\text{ex}}, J, \pi)$ is the probability, or branching ratio, of B^* decaying to channel χ.

In a standard measurement $\sigma_{\alpha\chi}(E_a)$ is measured directly. In the surrogate approach however, $G_{\chi}^{B^*}(E_{ex}, J, \pi)$ is determined by indirect measurement and $\sigma_{\alpha}^{B^*}(E_{ex}, J, \pi)$ is calculated, generally via the optical model. The surrogate approach is therefore intrinsically a hybrid of theory and experiment.

Early surrogate experiments employed the Weisskopf-Ewing approximation [4], which assumes that the probability of decay to any channel χ is independent of the J^{π} components of the compound nucleus spin distribution, i.e. it is only dependent upon the excitation energy and $G_{\chi}^{B^*}(E_{ex}, J, \pi) \rightarrow G_{\chi}^{B^*}(E_{ex})$. The results of *(n,f)* surrogate measurements generally compare favourably with direct experimental data [2]. However, generally, for each isotope studied, this match breaks down at lower excitation energies and also, in some cases, above the threshold for second-chance fission [2].

The poorer performance of the surrogate method in certain energy regions is due to a so-called *spin mismatch* and the breakdown of the validity of the Weisskopf-Ewing approximation [5]. At low excitation energies the available levels which a nucleus may occupy are discretely spaced in energy, with each level possessing a unique J^{π} assignment. As the excitation energy increases, the density of levels increases until eventually a continuum of fragmented and overlapping states is reached. At lower excitation energies, a very particular spin distribution must be imparted to a target nucleus in order to populate the levels available with the same distribution strength as in the desired reaction channel. A surrogate reaction may quite easily populate a different spin distribution if the spin matching in the reaction used favours the transfer of very different values of J^{π}.

For fissile nuclei at higher excitation energies, the *(n,f)* channel is generally dominant and the distribution of fragments generated, which may be altered by the differing fission-decay channels taken, is not important. Rather, it is the integrated *(n,f)* cross section to all final products which is the value of concern. In other words, although the decay path taken from a higher excitation energy state may be different in the surrogate case compared to that of the desired reaction, it will still result in the production of fission fragments and hence the same *(n,f)* cross section.

The limitations of the surrogate method, due to spin mismatch, are greater for the case of the (n,γ) reaction. Surrogate (n,γ) studies have shown that, for the current surrogate approach employed, a more sophisticated application of theory is required to take into account the differences in spin distribution between the desired and surrogate cases [6].

Another assumption made during early applications of the surrogate method was that differences in both the type and relative strength of pre-equilibrium reactions could be ignored in both the desired and surrogate reaction channels. Pre-equilibrium reactions are sensitive to the nature of the reaction entrance channel, given that they may include single-particle (or hole) excitations which depend upon the configuration of the orbitals of the nucleus. The configuration of orbitals is dependent upon whether the incident particle has been captured, as in the case of a neutron-induced reaction, or has scattered and removed nucleons, which could be the case in a direct reaction. The effect of differing pre-equilibrium reactions on the validity of the surrogate approach depends upon the specific reactions, both desired and surrogate, being investigated.

In parallel to the efforts to develop additional theory to 'translate' from a measured surrogate spin distribution to the optimal/expected one, new types of surrogate reactions which may populate spin distributions closer to that expected in the desired reaction are being investigated. Although early studies employed the *(t,p)* reaction as a surrogate [7, 8], *(p,t)* has only been investigated for this purpose very recently [9, 10] and only in a preliminary manner.

In this study we have therefore developed a model for performing *(p,t)* calculations over a range of excitation energies in order to predict the observables of a typical surrogate measurement. The model has been designed to require only very limited generic prior information regarding the target nucleus to which it is applied, with the expectation that it can be applied relatively quickly to a range of isotopes in order to better inform the selection of candidate surrogate experiments, i.e. those in which

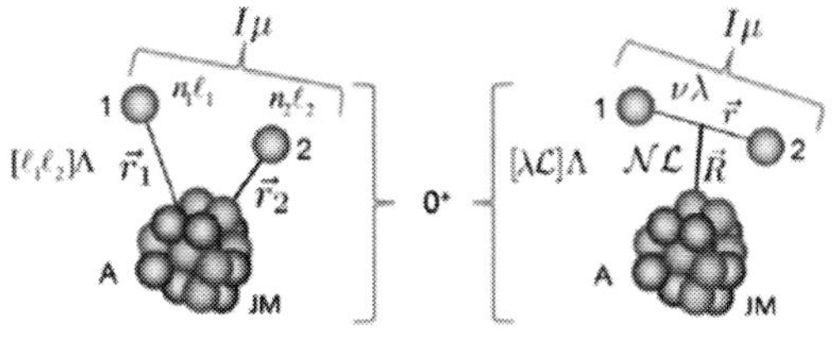

Figure 3. The various quantum numbers, angular momenta and their couplings present in the (A+2)-body target nucleus in the entrance channel in both the two neutron and di-neutron approaches.

the spin mismatch is minimised. This developed model has first been applied to recent measurements of the ^{92}Zr(p,t)^{90}Zr reaction.

2 Two-Neutron Transfer Model

In this section we briefly describe the (p,t) model developed for this study. In order to simplify the development of the model we begin by studying specific, important cases. We make the following assumptions:

- The target nucleus is even-even, i.e. it possesses even numbers of both protons and neutrons, is spherical, and in its ground state has a spin and parity assignment of $J^\pi = 0^+$.
- The two neutrons are transferred simultaneously, in one step, as a single spin-singlet 'di-neutron' object, with zero intrinsic angular momentum.

In order to perform calculations involving the transfer of a di-neutron, the quantum numbers and angular momenta of the two individual neutrons transferred must be translated into those of an equivalent single particle, as illustrated in Figure 3.

2.1 Formalism

Calculations of cross sections for the direct (p,t) transfer reaction on a mass $A + 2$ target nucleus, populating specific J$^\pi$, A-body final states, involve a number of components, see e.g. Glendenning [11]. The expression used for the cross section is

$$\frac{d\sigma}{d\Omega}(0^+ \to J^\pi) = |\sum_{N\Lambda S} G_{N\Lambda SJ} B_{N\Lambda SJ}(\vec{k}_p, \vec{k}_t)|^2 , \quad (2)$$

where $B_{N\Lambda SJ}$ is the (p,t) transition amplitude calculated via the Distorted Wave Born Approximation (DWBA) method and $G_{N\Lambda SJ}$, comprised of nuclear *structural* factors, is broadly analogous to the spectroscopic amplitude of single-nucleon transfer reactions. More specifically, G is given by the product of terms

$$G_{\gamma N\Lambda SJ} = \sum_{\gamma} g_\gamma \beta_{\gamma\Lambda SJ} \Omega_\nu \langle \nu\lambda, NL; \Lambda | n_1 l_1, n_2 l_2; \Lambda \rangle \ . \quad (3)$$

where $g_\gamma = \sqrt{2/(1+\delta_{ij})}$ is a symmetry factor dependent on the (like or unlike) pair of orbitals i, j occupied by the two transferred neutrons, $\langle \nu\lambda, NL; \Lambda | n_1 l_1, n_2 l_2; \Lambda \rangle$ is a *Moshinsky bracket* [12], $\beta_{\gamma\Lambda SJ}$

is the two-particle *parentage coefficient*, and Ω_n reflects the overlap of the two-nucleon relative motion wave functions between the initial and final states. Ω_n is assumed to be unity in the calculations presented here.

As we consider only even-even target nuclei and thus transfer from 0^+ ground states, the total and orbital angular momentum transferred are equal, $(\Lambda =)L = J$, and only *natural parity* final states with $\pi = (-1)^J$ are populated.

2.2 Structural factors

For the (p,t) pickup reaction the parentage coefficient, denoted by $\beta_{\gamma\Lambda SJ}$, measures the component in the (0^+) mass $A + 2$ target nucleus ground state of a specific J^π A-body residual nucleus final state plus two neutrons with quantum numbers $\gamma \equiv ([n_1 l_1 j_1], [n_2 l_2 j_2])$ coupled to $[\Lambda, S]J$. Explicit forms relevant to our cases are detailed by e.g. Glendenning [11].

The value of the Moshinsky bracket gives the amplitude (within the harmonic oscillator approximation) for the overlap between the wave functions of the two neutron single-particle orbitals, i.e. γ, in the target nucleus and the 0^+ di-neutron configuration, $[L, S = 0]J$. Here the $S = 0$ and relative s-wave restrictions are dictated, in the one-step approximation, by the $\langle \mathrm{p}|\mathrm{t}\rangle$ structure vertex with the outgoing triton.

2.3 Energy levels

The energies of the populated final states and the Q values for the individual transitions required for the DWBA transfer calculations are a necessary input. For these we used the experimental two-neutron separation energy S_{2n} of the target nucleus combined with Hartree-Fock calculations of the energies of the bound neutron single-particle states. The different degenerate J^π final states arising from the removal of neutrons from each pair of occupied neutron orbitals were split based on the phenomenological expectations, see e.g. Casten [13], from an attractive two-nucleon residual interaction.

The spherical Hartree-Fock [14] was used for these calculations. A variety of different Skyrme mean-field interactions were tested, but an inspection of previously identified levels, as shown in Figure 4, highlights a large number of levels with unassigned values of J^π, meaning determining a preferred potential is difficult. The SkX Skyrme model [15] was used for the current zirconium isotopes study.

2.4 DWBA calculations

The Surrey-version of the DWBA transfer code TWOFNR [16] was used for the (p,t) reaction cross section calculations. These cross sections showed little dependence on the choice of the proton optical model potential (OMP) selected. However, there was more significant sensitivity to the triton OMP used. Thus, calculations were performed using two available global triton OMPs [17, 18] to provide a first assessment of the uncertainty due to this physical input. In the results presented below, the proton OMP of Bechetti and Greenlees [19] was used with the triton OMP of Li *et al.* [17] unless otherwise stated.

2.5 Final-state energy spreading

There will be physical spreading, or fragmentation, of the strengths of the final states about their estimated energies (from the Hartree-Fock plus two-neutron residual interaction) [21]. We take this

 DOI: 10.1051/epjconf/201612212002

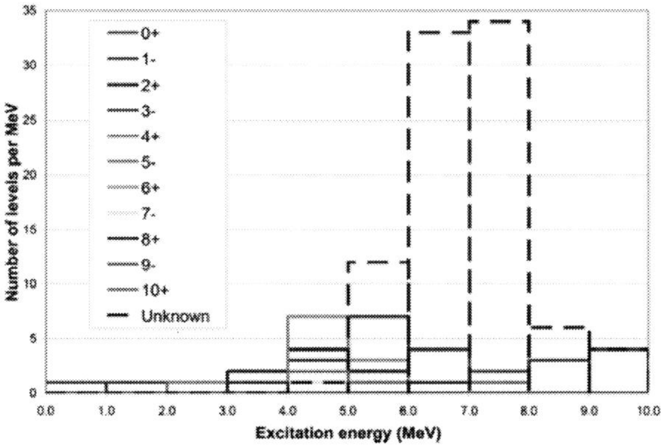

Figure 4. The density of levels of ^{90}Zr reported in the literature [20].

into account in the present calculations by the introduction of a parameterised spreading width, $\Gamma(E)$, for each state, dependent on its excitation energy above the Fermi energy of the residual nucleus E_F. We assume

$$\Gamma(E) = \frac{1}{2}\left[\frac{\epsilon_0(E-E_F)^2}{(E-E_F)^2+E_0^2} + \frac{\epsilon_1(E-E_F)^2}{(E-E_F)^2+E_1^2}\right], \tag{4}$$

where $\epsilon_0, \epsilon_1, E_0, E_1$ are chosen constants; taken here from Ref. [23] where they were used for ^{60}Ni .

This approach, apart from an additional factor of $\frac{1}{2}$, follows that of Brown and Rho [22]. The spreading of the strength of the transfer yield with excitation energy is distributed with a Breit-Wigner shaped form factor with a FWHM of $\Gamma(E)$. This method has been employed for related analyses of yield distributions following single-nucleon transfer reactions. The width calculated from Equation 4 is approximately linear with excitation energy for excited states below 25 MeV, and has a width $\Gamma(E) \approx 7$ MeV at $(E - E_F) = 25$ MeV.

3 Experimental Data

The experimental data compared to in this study were recorded in mid-2013. A 28.56 MeV proton beam was bombarded onto a thin $\approx$ 1mg/cm^2 isotopically purified ^{92}Zr foil. A large number of separate irradiations were conducted throughout the experiment with a total beam on target time of over 300,000 seconds (approximately 3.5 days). The average beam current was 1.56 nA.

3.1 STARS

The *S*ilicon *T*elescope *A*rray for *R*eaction *S*tudies (STARS) is an annular array of (*Micron Semiconductor* S2) silicon detectors used for the detection and identification of charged particles emitted from nuclear reactions [24].

In a typical setup, as illustrated in Figure 5, the STARS detector is placed within a scattering chamber behind a foil target. This foil is then bombared by incident charged particles and STARS located in the downstream position will detect the charged particles emitted, generally via direct reactions, into a range of scattering angles. The typical angular coverage of STARS is not typical and generally $\theta = 35°$ to $55°$. The STARS array is pixelated and can therefore record the angular distributions of emitted particles as well as their energies.

Measurements of the angular distribution of a state at a specific energy may be used to determine the J^{π} of that state. Unfortunately, in the measurements, the point of beam/target interaction was offset from the central value, leading to coarser resolution angular distributions to be observed for individual states.

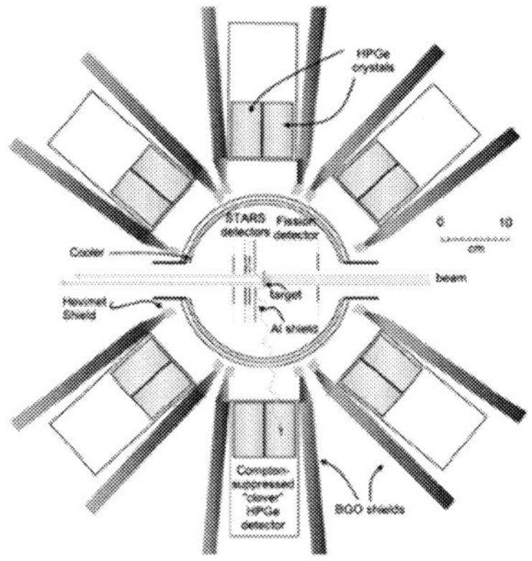

Figure 5. An overview of the STARLiTeR detector.

STARS is located at the K150 Superconducting Cyclotron Facility at Texas A&M University. STARS is fielded with an array of HPGe detectors referred to as the *Li*vermore *Te*xas *R*ichmond (LiTeR) array to form the STARLiTeR detector. The ability to detect γ-rays in coincidence with outgoing particle angular distributions allows for the identification of specific states populated in the excited residual nucleus and their decay pathways.

3.2 (*p,t*) measurement

The primary purpose of this ^{92}Zr foil irradiation was unfortunately not for studies of the ^{92}Zr(p,t)^{90}Zr reaction. Rather the experiment was tailored to study the ^{92}Zr(p,d)^{91}Zr reaction as a potential surrogate for the ^{90}Zr(n,γ)^{91}Zr reaction. This meant that the incident proton energy was optimised for measurements of the J^{π} distribution of ^{91}Zr at energies around its S_n value. Fortunately, given the ability of STARLiTeR to differentiate between outgoing deuterons and tritons, data for the ^{92}Zr(p,t)^{90}Zr reaction were also collected.

However, the energies of the states populated in ^{90}Zr are at energies below its S_n value, meaning that these data are unsuitable for direct use in a ^{89}Zr(n,γ)^{90}Zr surrogate study.

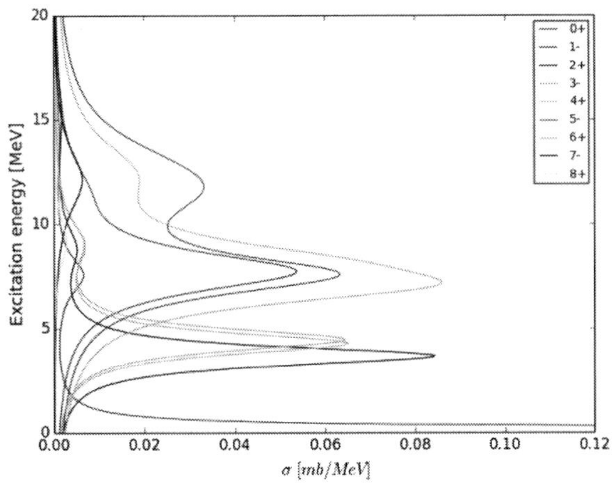

Figure 6. Calculated spin distribution for ^{90}Zr as a function of excitation energy following the ^{92}Zr(*p*,*t*) reaction.

4 Results and Analysis

4.1 Spin distribution

The model described above was used to calculate the J^π distribution for the excited states in ^{90}Zr expected to be populated following the ^{92}Zr(*p*,*t*)^{90}Zr reaction. This distribution is shown in Figure 6, with the individual levels summed over each value of J^π. The states have been spread in energy according to the method of Brown and Rho outlined in Section 2.5.

The excitation energies of these states may be translated into the energies of the tritons which would be emitted following their excitation (and are actually observed in the experiment). Figure 7 shows the normalised and summed J^π distribution as a function of outgoing triton, or rather product, kinetic energy, compared against the normalised measured data. The cross section is now only that for tritons incident on the inner radial ring of the STARS detector.

4.2 Identification of states

Figure 8 compares the calculated and measured data for the case of unspread levels (apart from spreading due to the experimental resolution). In this case, the comparison is for the triton distribution over the entire STARS detector. This is to avoid possible differences in the calculated positions of maxima and minima in angular distributions from masking the presence of a state. As may be seen in Figure 8, there is a mismatch in the positions of the calculated levels with those measured. However, it was found that this match does not improve through the use of either a different Skyrme potential in the Hartree-Fock level energy calculations or a more sophisticated shell model calculation. Rather the differences in the number of levels present and/or their energies may be due to fragmentation [21] and hence accounted for in a general sense via our energy spreading prescription described previously.

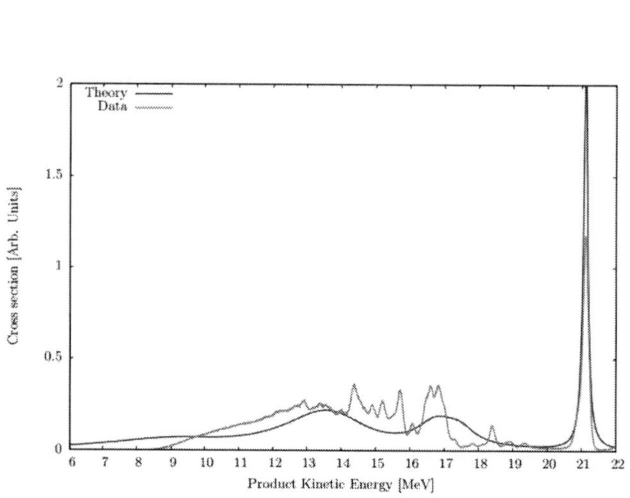

Figure 7. Comparison of theoretical results, using the half Brown and Rho spreading width, against experimental data for all rings of the detector.

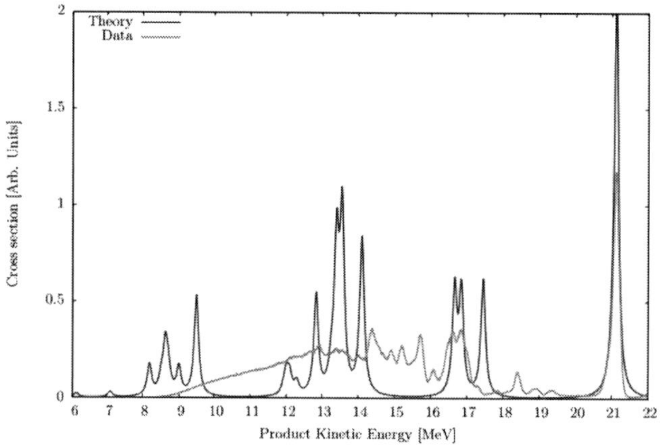

Figure 8. Theory vs data comparison for the triton spectrum integrated over all rings of the detector. In this case the theoretical states contributing to the total distribution are only spread by a width equal to that of the experimental resolution.

The angular distribution of tritons emitted due to the population of a specfic state will be related to its J^{π} value. By comparing the calculated angular distribution of a state with that of a measured state, at a coincident energy, we are able to better determine if we are calculating actual physical states.

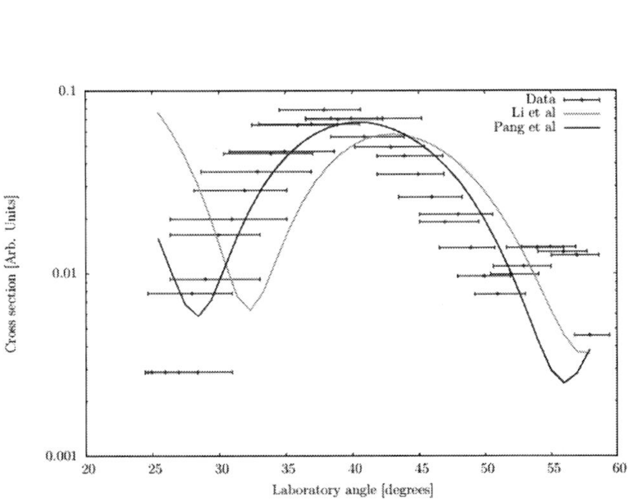

Figure 9. Comparison of the experimental angular distribution for the 0^+ ground state against the model predictions made using both the Li *et al.* and Pang *et al.* OMPs.

Figure 9 shows a comparison for the 0^+ ^{90}Zr ground state. It was found that the Li *et. al.* global triton OMP did not produce a good match against the experimental data. However, a calculation using the alternative Pang *et. al.* [18] OMP considered for this study produces a better match and perhaps indicates that further work to identify a preferential triton, and possibly proton, OMP should be performed.

4.3 Contaminants

It is possible that tritons may be observed which originate from (p,t) reactions on the other Zr isotopes present as contaminants in the ^{92}Zr target. Tritons emitted from the direct population of the ^{89}Zr and ^{88}Zr ground states will fall into our energy range of interest. The peak corresponding to the ^{89}Zr ground state will be at $E_t' = 17.85$ MeV and that corresponding to the ^{88}Zr ground state at $E_t' = 15.72$ MeV. An inspection of the data, combined with a knowledge of the estimated size of each peak relative to the ^{90}Zr ground state peak, suggests that the ^{88}Zr peak will be unobservable due to the background of tritons from the ^{92}Zr$(p,t)^{90}$Zr reaction in that region. However, the peak due to ^{89}Zr is possibly visible in the data above background at $E_t' = 17.85$ MeV.

4.4 Unnatural parity states

The model developed implicitly includes the assumption that unnatural-parity states are not populated. This is due to the two neutrons being transferred in a single step as a di-neutron, possessing an intrinsic spin of $S = 0$. Despite this assumption, the model still calculates the energies and J^{π} values of potential unnatural-parity states which would be populated via the transfer of various neutron pairs.

Table 1 lists the unnatural parity levels previously observed in ^{90}Zr [25], along with the two unassigned levels previously seen via the ^{92}Zr$(p,t)^{90}$Zr reaction, and compares these against the energies

Table 1. The excitation energies of unnatural-parity levels previously observed in ^{90}Zr [25]. Also shown are the energies of predicted potential unnatural-parity states along with the quantum numbers of the relevant transferred neutron pair.

Theory					Measured	
E^* [MeV]	J^π	N	$n_1l_1j_1$	$n_2l_2j_2$	E^* [MeV]	J^π
					4.238	?
					4.540	6^-
					4.580	1^+
4.613	3^+	2	$0g_{9/2}$	$1d_{5/2}$		
4.613	5^+	1	$0g_{9/2}$	$1d_{5/2}$		
					5.060	7^+
					5.068	?
					5.248	9^+
7.010	2^-		$1p_{1/2}$	$1d_{5/2}$		
					7.774	2^-
7.992	2^-		$0f_{5/2}$	$1d_{5/2}$		
7.992	4^-		$0f_{5/2}$	$1d_{5/2}$		
					8.032	2^-
8.510	2^-		$1p_{3/2}$	$1d_{5/2}$		

of the calculated unnatural parity states. It should be noted that transitions to unnatural states are not split by a residual interaction.

As can be seen in Table 1, the two predicted unnatural parity states at 4.613 MeV could be potential candidates for the unassigned levels at 4.238 MeV and 5.068 MeV. The two observed 2^- levels have predicted counterparts at energies which, although not the same, are close enough to provide a reasonable match given the limitations of our Hartree-Fock approach.

5 Summary and Outlook

A theoretical model has been developed to predict the excitation energy, spin and parity distributions of the residual nuclei following a (p,t) two-neutron transfer reaction. This model has been designed to be purely predictive and to require little or no detailed prior information on the target nucleus in question. The model developed has been applied to the case of 28.56 MeV protons incident on an isotopically enriched ^{92}Zr target, a case for which experimental data have recently been taken by a group from the Lawrence Livermore National Laboratory using the STARLiTeR detector at Texas A&M University.

5.1 Efficacy of the developed (*p,t*) model

The model developed makes a number of assumptions: the target nucleus is even-even, spherical; it is in its ground state with $J^\pi = 0^+$; and that the two neutrons are transferred simultaneously, in one-step, as a single spin-singlet di-neutron object, with zero intrinsic angular momentum. A detailed comparison between the model and the data for ^{92}Zr$(p,t)^{90}$Zr shows a reasonable match to the average trends of the observed spectra, but a breakdown when individual discrete states are scutinised in detail. In particular, the model fails to predict the population of a number of physical states observed in ^{90}Zr, suggesting a more sophisticated approach to the structural and/or dynamical components of the model is required.

5.2 Suitability of ^{92}Zr(*p,t*) as a surrogate for ^{89}Zr(*n*,γ)

Unfortunately, as the ^{92}Zr(*p,t*)^{90}Zr data used in this study were taken as part of an experiment tailored for (*p,d*), the energy of the proton beam was too low for the triton spectrum, at energies corresponding to excitation energies in the residual nucleus A at $\approx S_n^A$, to be observed. Therefore, the spin distribution calculated to be populated in this region cannot be compared to experimental data. Applying the model to a different Zr isotope for which experimental (p,t) data exist in the correct energy regime to act as an (n,γ) surrogate will be explored in a future study.

5.3 Outlook

The developed (*p,t*) model is sufficiently mature that it could be applied, with only minimal effort to any even-even, spherical, 0^+ target nucleus. It is hoped that this tool may prove of some use to the surrogate reaction community in helping to predict spin distributions populated via (*p,t*). However, it is acknowledged that an experiment resulting in a higher excitation energy of the residual compound nucleus is required in order to assess the model against data in the correct energy regime, i.e. at excitation energies around S_n.

Acknowledgments

This work was supported by the UK Ministry of Defence and was performed under the auspices of the US Department of Energy by Lawrence Livermore National Laboratory under contract DE-AC52-07NA27344. JAT acknowledges support of the Science and Technology Facilities Council (UK) grants ST/J000051 and ST/L005743. S.O. is supported by a JSPS Postdoctoral Fellowship for Research Abroad. British Crown Owned Copyright (2015)/AWE.

References

[1] J.M. Blatt and V.F. Weisskopf, *Theoretical nuclear physics* (John Wiley & Sons, 1952)
[2] J.E. Escher *et. al.*, Reviews of Modern Physics **84**, 353 (2012)
[3] G. Kessedjian *et. al.*, Phys. Lett. B **692**(5), 297 (2010)
[4] V.F. Weisskopf and D.H. Ewin, Phys. Rev **57**(6), 472 (1940)
[5] J.E. Escher and F.S. Dietrich, Phys. Rev. C **81**, 024612 (2010)
[6] N.D. Scielzo *et. al.*, Phys. Rev. C **81**, 034608 (2010)
[7] J.D. Cramer and H.C. Britt, Nucl. Sci. Eng. **41**, 177 (1970)
[8] H.C. Britt and J.B. Wilhelmy, Nucl. Sci. Eng. **72**, 2 (1979)
[9] R.O. Hughes *et. al.*, Phys. Rev. C **85**, 024613 (2012)
[10] R.O. Hughes *et. al.*, Phys. Rev. C **90**, 014304 (2014)
[11] N.K. Glendenning, Phys. Rev. B **137**, 102 (1965)
[12] M. Moshinsky, Nucl. Phys. **13**, 104 (1959)
[13] R.F. Casten, *Nuclear structure from a simple perspective* (Oxford University Press, second edition, 2000)
[14] B.A. Brown *et. al.*, Code DENS, provided as part of OXBASH, *Oxbash for Windows* (MSU-NSCL report number 1289, 2004)
[15] B.A. Brown, Phys. Rev. C **58**, 220 (1998)
[16] J.A. Tostevin, http://www.nucleartheory.net/NPG/code.htm (last accessed 6 October 2015)

[17] X. Li *et. al.*, Nucl. Phys. A **789**, 103 (2007)
[18] D.Y. Pang *et. al.*, Phys. Rev. C **79**, 024615 (2009)
[19] F.D. Becchetti and G.W. Greenlees, Phys. Rev. **182**(4), 1190 (1969)
[20] C.M. Baglin, Nuclear Data Sheets **113**(10), 2187 (2012)
[21] M.G.E. Brand *et. al.*, Nucl. Phys. A **531**(2), 253 (1991)
[22] G.E Brown and M. Rho, Nucl. Phys. A **372**(3), 397 (1981)
[23] M. Matoba *et. al.*, Nucl. Phys. A **581**, 21 (1995)
[24] S.R. Lesher *et. al.*, Nucl. Instr. and Meth. in Phys. Res. Section A: Accelerators, Spectrometers, Detectors and Associated Equipment **621**, 286 (2010)
[25] E. Browne, Nuclear Data Sheets **82**(3), 379 (1997)
[26] A.J. Koning *et. al.*, Proceedings of the International Conference on Nuclear Data for Science and Technology, April 22-27, 2007, Nice, France, pp211-214

EPJ Web of Conferences **122**, 12003 (2016) DOI: 10.1051/epjconf/201612212003
*CNR*15*

Exploring the spin states of ^{90}Zr populated by (*p,p'*), (*p,d*), and (*p,t*) reactions

S. Ota[1a], J.T. Burke[1], R.J. Casperson[1], J.E. Escher[1], R.O. Hughes[1], J.J. Ressler[1], N.D. Scielzo[1], I.J. Thompson[1], R.A.E. Austin[2], B. Abromeit[3], N.J. Foley[3], E. McCleskey[3], M. McCleskey[3], H. I. Park[3], A. Saastamoinen[3], and T. J. Ross[4]

[1]*Lawrence Livermore National Laboratory, Physics and Life Science Directorate, Livermore, CA 94551, USA*
[2]*St. Mary's College, Halifax, Nova Scotia, Canada*
[3]*Texas A&M University, Cyclotron Institute, College Station, TX 77840, USA*
[4]*University of Kentucky, Department of Chemistry, Lexington, KY 40506, USA*

Abstract. The ^{90}Zr nucleus was produced by three different reactions: ^{90}Zr(*p,p'*), ^{91}Zr(*p,d*), and ^{92}Zr(*p,t*), and the spin-parity (J^{π}) population of the ^{90}Zr states produced by these reactions was studied to investigate the surrogate reaction approach, which aims at indirectly determining cross sections for compound-nuclear reactions involving unstable targets such as ^{89}Zr(n,γ). Discrete γ-rays, associated with the de-excitation of ^{90}Zr and ^{89}Zr, were measured in coincidence with light ions at ^{90}Zr excitation energies extending above the neutron separation energy. Low-lying states populated by (*p,d*) and (*p,t*) reactions agreed well with the previous measurements. The measured γ transition systematics were used to gain insights into the J^{π} distribution of ^{90}Zr around the neutron separation energy and it was found that the (*p,p'*) reaction preferentially produces lower J states than (*p,d*) and (*p,t*) reactions in the studied energy region.

1 Introduction

Radiative neutron-capture (n,γ) cross sections at neutron energies of a few tens of keV to a few MeV are crucial inputs for nuclear-energy applications, astrophysical studies, and radiochemical applications. However, the cross sections for most short-lived isotopes remain poorly known due to their inaccessibility as target materials. To address this problem, various types of indirect methods to infer the (n,γ) cross sections have been investigated, such as the surrogate reaction method [1-4]. Also, approaches to obtain the γ-ray strength function to constrain the (n,γ) cross sections are being developed by taking advantage of mono-energetic γ-rays obtained from laser Compton back-scattering method [5], the Oslo method [6], and other techniques [7, 8].

The surrogate reaction method aims at determining neutron-induced reaction cross sections by accessing the compound nuclei of interest via an alternative reaction such as a transfer reaction or inelastic scattering involving stable beams and stable (or long-lived) targets. In a surrogate experiment, a nucleus produced via a direct reaction is assumed to subsequently equilibrate to form

[a] Corresponding author: ota2@llnl.gov

the same compound nucleus that is formed in the desired neutron-induced reaction. The outgoing particle from the surrogate reaction is detected in coincidence with an observable (e.g., a specific γ-ray transition) that is characteristic of the decay channel of interest and the measured coincidence probability is used to constrain the reaction cross section.

A primary difficulty in determining (n,γ) cross sections is the difference in the spin-parity (J^{π}) distributions of the compound nucleus created by the (n,γ) and the surrogate reactions. The γ-ray emission of the compound nucleus can be quite sensitive to the initial J^{π} distribution [9]. Therefore, in order to extract (n,γ) cross sections from surrogate data, it becomes necessary to take into account the J^{π} distribution of the decaying compound nucleus. When this distribution is known, the decay of the nucleus can be modelled and constraints on the desired (n,γ) cross sections are obtained by fitting the decay model to observables from the surrogate experiment. Details about the modelling based on the γ-ray decay probabilities can be found in [10].

With this background, we studied spin states in the nucleus ^{90}Zr populated by ^{90}Zr$(p,p')^{90}$Zr, ^{91}Zr$(p,d)^{90}$Zr, ^{92}Zr$(p,t)^{90}$Zr reactions. The ^{90}Zr nucleus is well suited for benchmarking the surrogate approach. The low level densities in the Zr mass region due to the presence of closed proton and neutron shells, leads to a competition between γ decay and neutron emission that is sensitive to the J^{π} distributions of the decaying compound nuclei and can be expected to be visible in the measured discrete γ-ray emission probabilities.

2 Experiment

The experiments were performed at the K150 Cyclotron facility at Texas A&M University. Enriched ^{90,91,92}Zr targets with thicknesses of 1.02, 1.01, and 0.960 mg/cm^2, respectively, were bombarded with a 28.56-MeV proton beam with an intensity of about 1.5 nA. Outgoing protons (for (p,p') reactions), deuterons (for (p,d)), and tritons (for (p,t)) were detected at forward angles. Measurements were run for 12, 36, and 84 hours for the experiments using the ^{90,91,92}Zr targets, respectively.

A schematic view of the experimental set up is shown in Fig. 1. The energy spectra and angular distribution of the light ions and prompt γ-rays were measured with the STARLiTeR detector system [11]. The STARLiTeR consists of three double-sided Micron S2 silicon detectors (referred to as ΔE, $E1$, and $E2$ hereafter) which are each segmented into 24 rings (one side) and 8 wedges (the other side), allowing the measurement of charged-particle scattering angles. The thickness of the ΔE, $E1$, and $E2$ detectors are 143 μm, 1000 μm, and 994 μm, respectively. The ΔE, $E1$, and $E2$ detectors were located at 19.2 mm, 23.4 mm, and 33.8 mm from the target and were used to identify charged-particles from (p,p'), (p,d), and (p,t) reactions covering angles between 25-58°.

For γ-ray detection, five Bismuth Germanate Oxide (BGO) Compton-suppressed High Purity Germanium (HPGe) clover detectors surrounded the silicon chamber. The energy resolution and the total absolute photopeak efficiency were measured using calibrated γ-ray sources placed at the target position before and after the experiments. The typical energy resolution was 2 keV for energies below 500 keV increasing to 5 keV at 3 MeV. The efficiency of the array was 3.8% at 150 keV, 1.5% at 500 keV, and 0.5% at 2 MeV after addback was applied. The γ-ray energies were measured up to about 5 MeV.

Coincident particle-γ events were identified based on the time difference between particle and prompt γ-ray signals. Further details on the detector arrays, data-taking system, and data analysis can be found in [4] and references therein.

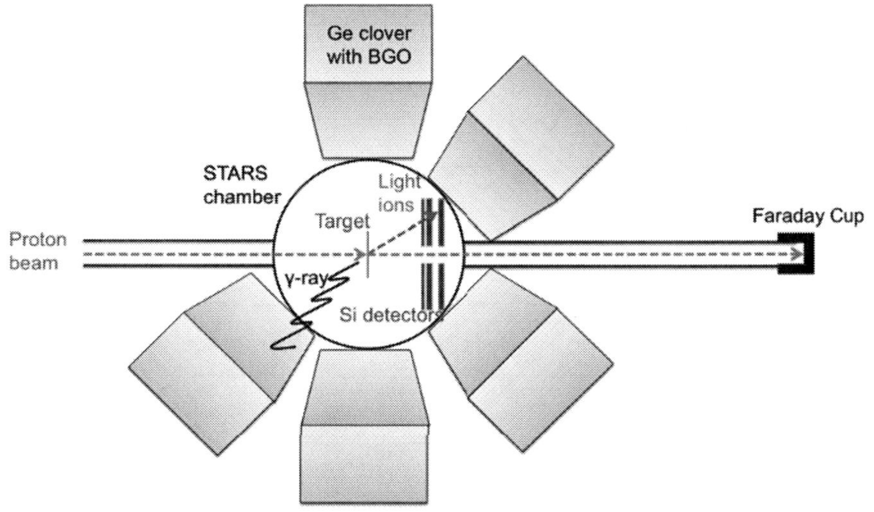

Figure 1. Schematic view of the experimental set up.

3 Spin population in low-lying states

3.1 Particle singles spectra

After particle identification using a conventional ΔE-range plot (see [3]), the detected particle energy was corrected for the recoil energy of the target nuclei and energy losses in the targets and dead layers of the Si detectors to obtain the total kinematic energy (E). The excitation energies (E_x) in ^{90}Zr can be determined from

$$E_x = E_b - E + Q, \tag{1}$$

where E_b, the beam energy, is 28.56 MeV, and Q-values for (p,p'), (p,d), and (p,t) reactions are 0, -4.969, and -7.346 MeV, respectively.

Fig. 2 shows the observed E distribution for ^{90}Zr(p,p'), ^{91}Zr(p,d), ^{92}Zr(p,t) reactions after correcting for backgrounds from Zr, C, and O contaminants. The full-width at half maximum (FWHM) for the ^{91}Zr(p,d) ground-state peak measured by the $\Delta E + E1 + E2$ detectors was about 300 keV. Similarly, the FWHM for the ^{92}Zr(p,t) ground-state peak measured by the $\Delta E + E1$ detectors was about 200 keV. This peak was observed at E = 23.6 MeV and 21.2 MeV for the ^{91}Zr(p,d) and ^{92}Zr(p,t) reactions, respectively, which agree well with the calculated values from Eq. (1).

In the ^{90}Zr(p,p') reaction, events with E > 21.5 MeV (E_x < 7 MeV) were mostly missed because these particles punch through the $E2$ detector. On the low-energy side, the events with E < 7.5 MeV (E_x > 21.0 MeV) were mostly cut off because these are stopped in the ΔE detector. Therefore, the energy range utilized in the ^{90}Zr(p,p') reactions is E_x = 7.0-21.0 MeV, which spans a 14-MeV energy region around the neutron separation energy of ^{90}Zr (S_n=11.97 MeV). Similarly, the energy ranges utilized in ^{91}Zr(p,d) and ^{92}Zr(p,t) reactions are E_x = 0-15 MeV and E_x = 0-11 MeV, respectively. These are wide enough to study excitations around S_n although the energy region above S_n is missed in the ^{92}Zr(p,t) reactions.

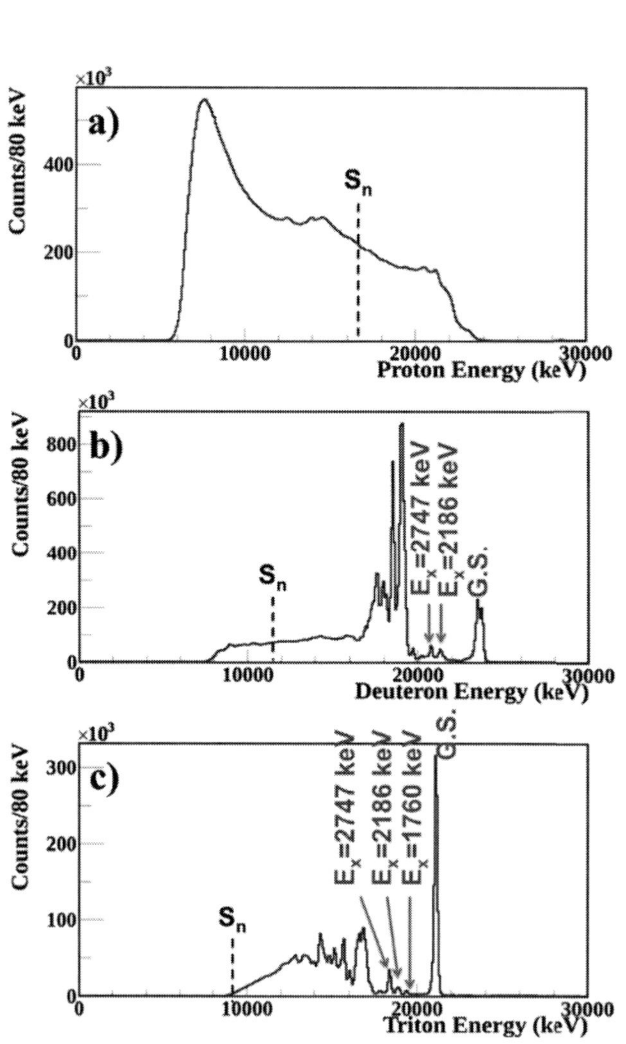

Figure 2. Particle singles spectra from a) ^{90}Zr(p,p')^{90}Zr, b) ^{91}Zr(p,d)^{90}Zr, c) ^{92}Zr(p,t)^{90}Zr reactions.

From Fig. 2 b) and c), we can see different ways of populating low-lying levels at $E_x < 3$ MeV in ^{91}Zr(p,d), and ^{92}Zr(p,t) reactions. The first excited state (E_x = 1760 keV (0^+)) is not observed in ^{91}Zr(p, d), but is observed in ^{92}Zr(p,t). This supports the result by [9] in which the spectroscopic factor of this state was estimated to be < 0.001 and therefore the cross section for the state should be extremely small in the ^{91}Zr(p,d) reaction. The second excited state (E_x = 2.186 MeV (2^+)) is clearly seen in both reactions. The third excited state (E_x = 2.319 MeV (5^-)) is not seen in the (p,d) data, which agrees with the measured spectrum in [12]. However, as discussed later, it was found that this state was just not well separated from the E_x = 2.186 MeV state in the (p,t) data. The mixed peaks of the fourth and the fifth excited states (E_x = 2.739 (4^-) and 2.747 MeV (3^-), respectively) are also observed in both reactions.

Several large peaks are observed around E_x = 4.0-6.5 MeV in Fig. 1 b) and at E_x = 4-8 MeV in Fig. 1 c). Although the limited energy resolution of the Si detectors do not allow for unambiguous structure information on these peaks, the particle-γ coincidence technique reveals the individual levels contributing to these peaks as described in following sections.

3.2 Particle-γ coincidence for (*p,d*) reaction

The particle-γ coincidence technique is of great help to identify the levels from which γ-rays are emitted with higher precision than is possible with the charged-particle measurements alone. Furthermore, since the γ-ray energies can be determined to ~ 1 keV, it is possible to assign these γ-rays to specific known levels at medium level density regions. A sophisticated ^{90}Zr(*n,n'*) experiment performed in the 2000's [13] reported more than 50 new levels and 170 γ-ray transitions below E_x = 5.8 MeV and greatly improved the information available on the excited states of ^{90}Zr. Here, the particle-γ coincidence technique was applied to study the peaks in Fig. 2 using the latest nuclear level information.

Fig. 3 a) shows the energy spectrum for deuterons in coincidence with γ-ray transitions from levels at E_x < 5.8 MeV in ^{90}Zr. As energies of the levels (E_i) which emit these γ transitions (i) are known from the literatures, the deuteron spectrum was created from these coincidence deuterons around the levels (typically collected from E_i ± 240 keV) taking into account the γ-ray branching ratios and detection efficiency. For instance, the γ-rays with the energies of 1656 and 3842 keV are emitted from the level of E_x =3842 keV (2^+) with the branching ratio of 0.145 and 0.855, respectively [13]. Then the deuterons in coincidence with the 1656 keV γ-ray are collected around E_x =3842 keV (E = 19.75 ± 0.24 MeV) correcting for the branching ratio (0.145) and the γ-ray detection efficiency. Similarly the deuterons in coincidence with 3842 keV γ-ray are collected at the same energy region and corrected for the branching ratio (0.855) and the efficiency. The sum of both events were considered to be the total deuteron singles count from E_x =3842 keV state (E = 19.75 MeV). The same procedure was performed for all the γ-ray transitions reported in [13].

To compare with the actual deuteron singles spectrum, the reconstructed deuteron spectrum was assumed to be Gaussian shape with the energy resolution (FWMH) of 200 keV as shown in Fig. 3 b). The low-lying excited state peaks such as E_x = 2186, 2747, 3308, and 3842 keV are well reproduced in both energy and intensity. The large peak at E_x = 4.5 MeV is well reproduced as well. It is remarkable agreement considering no angular correlation is taken into account in the reconstructed spectrum. The two large peaks at E_x = 5.5 MeV and E_x = 5.0 MeV are also well reproduced in energy, but intensities are both much reduced. These discrepancies in intensity come most likely from lack of data about discrete levels for excitation energies above 5 MeV. In fact, the reported number of levels above 5 MeV is only 32, which is clearly fewer than theoretically expected number of levels at this energy region.

Nevertheless, contributions to the two peaks at E_x = 5.5 MeV and 5.0 MeV are still explainable to some extent from the present data. Thus we can discuss some strongly populated levels for E_x = 4.0 - 5.8 MeV as follows. It turns out that the peak at E =19.0 MeV is comprised of E_x =4.541 (6^+, 26%), 4.454 (5^+ ,24%), 4.331 (3^+, 17%), 4.814 and 4.818 (3, and 3,4^+,11% for two), and 4.640 MeV (7,8, 10%) levels (percentage denotes contributions to the peak). Likewise, it was found that the peak at E = 18.5 MeV consists of E_x =5.060 (7^+, 56%), 4.992 (2 , 22%), 5.107 MeV (3,4^+, 15%) levels. It turns out that the peak at E =18.0 MeV dominantly consists of only 5.426 MeV (3) level although there remain uncertainties in the rest of the contributions as discussed above.

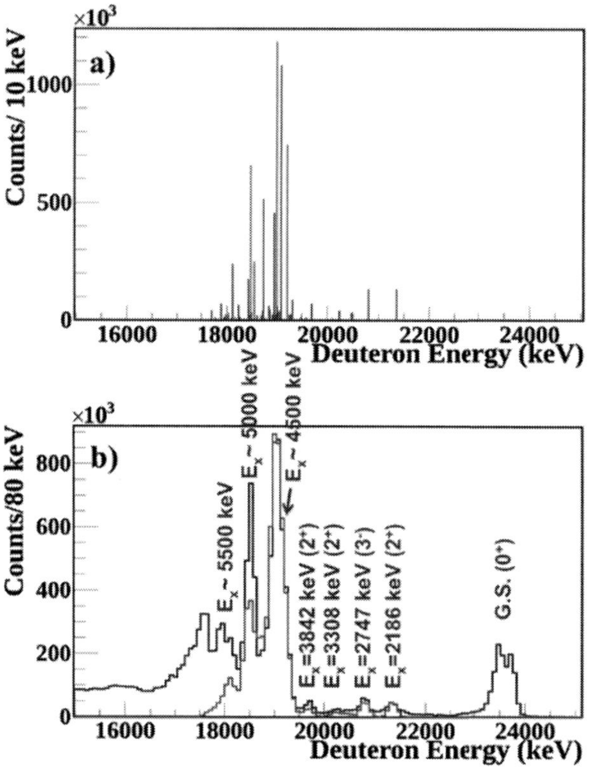

Figure 3. a) Estimated contributions from individual states to the ^{91}Zr(*p*,*d*) spectrum (see text for details). b) Simulated deuteron singles spectrum from the estimation (red) and actual deuteron singles spectrum (blue).

Ball and Fulmer [12] measured a deuteron spectrum from ^{91}Zr(*p*,*d*) reactions between the ground state and E_x = 6.5 MeV in ^{90}Zr at 30° with a 31 MeV proton beam. They used a magnetic spectrograph with a typical energy resolution of 18 keV. However, the possible ambiguities in energy calibration (on order of 10 keV) was pointed out in [13] and this made it difficult for them to compare the levels found through (*n*,*n'*) reactions with the levels reported by [12]. In the present work, the level energies were identified with 3-5 times better energy resolution than in the previous ^{91}Zr(*p*,*d*) measurement [12], it helps identify the states observed by [13]. The intense peak reported by [12] are E_x = 4.320, 4.443, 4.528, 5.050 MeV. These possibly correspond to E_x = 4.331, 4.454, 4.541, 5.060 MeV found in the present spectrum. Therefore, the present experiment agrees quite well with the past ^{91}Zr(*p*,*d*) experiment except for the ~10 keV calibration offset existing in their experiment.

3.3 Particle-γ coincidence for (p,t) spectrum

The same approach was taken for the ^{92}Zr(*p*,*t*) reaction. Fig. 4 a) shows a triton spectrum from the ^{92}Zr(*p*,*t*) reactions which is reconstructed by the same procedure as in the Fig. 3 a). Fig. 4 b) shows the simulated spectrum assuming the same energy resolution. As with the ^{91}Zr(*p*, *d*) case, the triton singles spectrum was reproduced well in energy although there exist some discrepancies in intensity.

Especially, the reduced intensity in the E_x = 2186 keV state is noticeable. This is possibly because the isomeric state of E_x = 2319 keV is not reconstructed in the present approach. The γ decay lifetime (τ = 809.2 ms) of the state is far longer than the particle-γ timing gate. In other words, the peak observed as the E_x = 2186 keV state is actually a doublet consisting of E_x = 2186 keV and 2319 keV states.

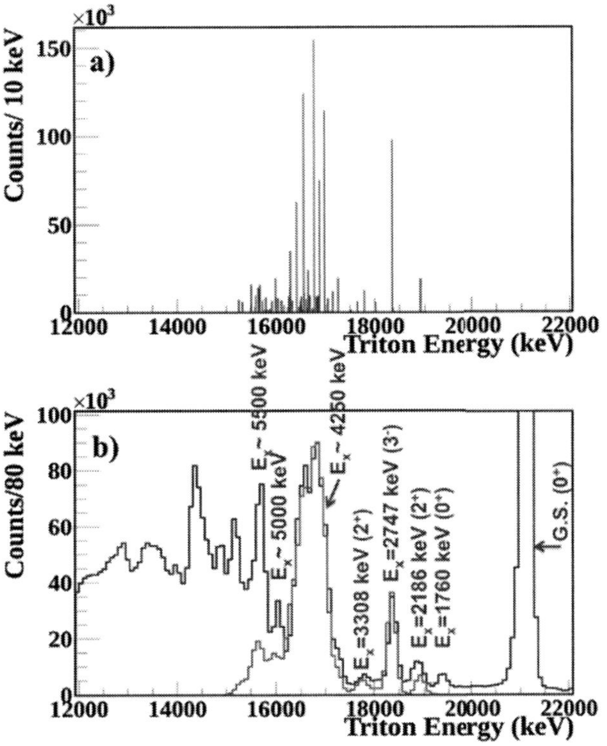

Figure 4. a) Estimated contributions from individual states to the ^{92}Zr(p,t) spectrum (see text for details). b) Simulated triton singles spectrum from the estimation (red) and actual triton singles spectrum (blue).

Figure 4 revealed the giant peak at E_x = 4.25 MeV consists of E_x = 4.331 (4^+, 18%), 4.541 (6^+, 18%), 4.124 (0^+, 16%), 4.229 (2^+, 10%), 4.681(2^+, 10%), 4.814 and 4.818 (3 and 3,4^+, 5% for two) MeV levels (percentages denote contributions to the peak). The peak at E_x = 5.0 MeV consists of E_x = 5.060 MeV (7^+) and the peak at 5.5 MeV consists of E_x =5.457 (4^+, 17%) and 5.513 MeV ((3,4), 15%). These peak structures can be compared with the ^{92}Zr(p,t) spectrum from [14]. Their spectrum was measured at 20° with 25 keV resolution using a 38 MeV proton beam. The intense peaks reported from the experiment are 4.125, 4.232, 4.335, 4.427, 4.543, 4.683, and 5.441 MeV. These definitely correspond to 4.124, 4.229, 4.331, 4.541, 4.681, and 5.457 MeV in the present spectrum. Despite the differences in measured angles and beam energy, the present spectrum agrees well with the past ^{92}Zr(p,t) experiment, and no significant mis-scaling in energy calibration is found from there.

3.4 Comparison of levels populated by ^{91}Zr(*p*,*d*) and ^{92}Zr(*p*,*t*) reactions

The states observed in either ^{91}Zr(*p*,*d*) or ^{92}Zr(*p*,*t*) reactions are summarized in Table 1 for comparison. Reaction dependence of populating these states is clearly observed and it can be seen that some states are populated by one reaction but not by the other. Although it is difficult to obtain a general trend about the reaction dependence from the present data, it seems that a high spin state such as $J = (7,8)\ \hbar$ is only populated by ^{91}Zr(*p*,*d*) and 0^+ states are only populated by ^{92}Zr(*p*,*t*) reaction. This may make sense considering the spin state of ^{91}Zr and ^{92}Zr ground states are $5/2^+$ (angular momentum: $L = 2\hbar$) and 0^+ ($L = 0\hbar$), respectively.

Table 1. The states of ^{90}Zr observed in either ^{91}Zr(*p*,*d*) or ^{92}Zr(*p*,*t*) reactions.
○ and × denote observed and unobserved (or possibly weakly populated), respectively.

E_x (keV)	J^π	^{91}Zr(*p*,*d*)	^{92}Zr(*p*,*t*)
1760	0^+	×	○
2319	5^-	×	○
4124	0^+	×	○
4229	2^+	×	○
4454	(5^+)	○	×
4640	7,8	○	×
4681	2^+	×	○
4992	2^-	○	×
5107	$3,4^+$	○	×
5426	3^-	○	×
5457	4^+	×	○
5513	$(3,4)^-$	×	○

4 Spin population in continuum region

4.1 γ-decay probability

In the highly-excited energy region (continuum region), we gain insights into the spins populated by the surrogate reactions by observing γ-decay probabilities, $P_i(E_x)$. $P_i(E_x)$ is a probability that the compound nucleus of interest (here ^{90}Zr), produced at a particular excitation energy E_x, decays via a specific γ-ray transition (i) that can be experimentally observed. This γ decay probability is given by

$$P_i\,(E_x) = (1 + a_{IC})\, N_{P\text{-}\gamma}(E_x, i)\,/\,(\,\varepsilon(E_\gamma)\, N_{singles}(E_x)), \tag{2}$$

where $N_{singles}$ gives the number of singles events (outgoing direct-reaction particle observed in detectors), $N_{P\text{-}\gamma}$ is the number of particle-γ coincidences observed for the transition of interest, ε denotes the γ-ray detection efficiency at the γ-ray energy E_γ, and a_{IC} gives the relevant internal conversion coefficient. $N_{singles}$ (E_x) is obtained from Fig. 1, while $N_{P\text{-}\gamma}$ (E_x) is the values obtained from the γ-ray spectrum gated on E_x by fitting a specific peak assuming a Gaussian shape. a_{IC} for the individual γ-rays were calculated using the software BRICC V. 2.0B [15].

4.2 Below the neutron separation energy

Fig. 5 shows $P_i(E_x)$ for the four discrete γ-ray transitions from ^{90}Zr low-lying states around E_x = 2-4 MeV as a function of excitation energies up to E_x = 20 MeV. The P_i (E_x) typically show some peaks at low excitation energies which correspond to direct population of levels which decay via these γ-ray transitions. In the region of 7 MeV $< E_x < S_n$, the level density is so high that the highly-excited nuclear system that is produced in the surrogate (transfer or inelastic scattering) reaction mixes with the surrounding states and equilibrates, i.e. the intermediate nuclear system becomes a compound nucleus. We observe that the P_i (E_x) are nearly constant in this high energy region. At excitation energies above S_n, the P_i (E_x) drop to nearly zero because of the competition from neutron emission.

While the γ-ray transitions from low J states such as J = 2 $\hbar$ and 4 $\hbar$ show a similar trend indicating that the P_i (E_x) are nearly independent of the reaction, the transitions from the higher J states such as J = 6 $\hbar$ and 8 $\hbar$ show significant reaction dependence and ^{90}Zr(p,p') shows notably smaller P_i (E_x) than the other two reactions. These suggest that ^{90}Zr(p,p') does not populate as many high J ($\geqq$ 6 $\hbar$) states as ^{91}Zr(p,d) and ^{92}Zr(p,t) reactions. More complete results and discussion about the γ-decay probabilities can be found in [4].

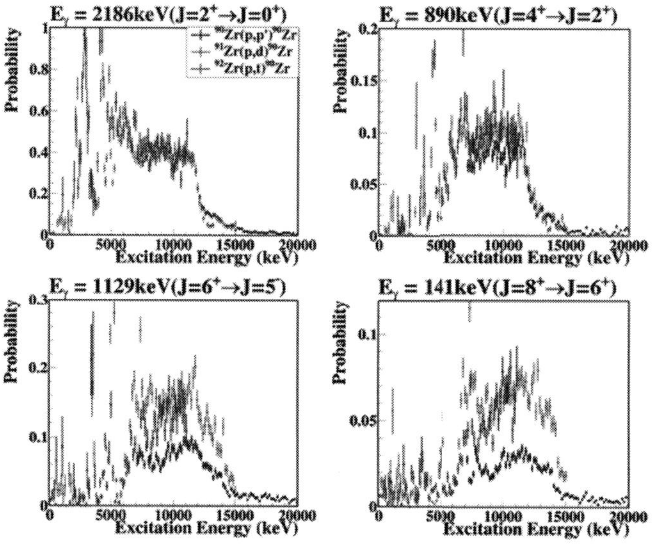

Figure 5. γ-decay probabilities for γ-ray transitions from different J states in ^{90}Zr (see [4] for more details).

4.3 Above the neutron separation energy

Figure 6 shows P_i (E_x) for the γ-ray transitions from ^{89}Zr produced by neutron evaporation from ^{90}Zr compound nucleus produced by ^{90}Zr(p,p'), ^{91}Zr(p,d), and ^{92}Zr(p,t) (i.e., by ^{90}Zr(p,p'n), ^{91}Zr(p,dn), and ^{92}Zr(p,tn)) as a function of E_x in ^{90}Zr (the ^{92}Zr(p,t) data did not extend above S_n). P_i (E_x) for γ-ray transitions in ^{89}Zr start to rise as the states become energetically accessible. The P_i (E_x) above S_n show J dependence as was observed below S_n in section 4.2. While the ^{90}Zr(p,p'n) plot shows lower P_i (E_x) than ^{91}Zr(p,dn) for high J states (J^π = 9/2$^+$ $\hbar$, 13/2$^-$ $\hbar$), the ^{90}Zr(p,p'n) plot shows higher P_i (E_x) than ^{91}Zr(p,dn) in γ-ray transtions from low J states such as 5/2$^-$ $\hbar$ and 3/2$^-$ $\hbar$. More complete results and a discussion of the γ-decay probabilities can be found in [4].

DOI: 10.1051/epjconf/201612212003

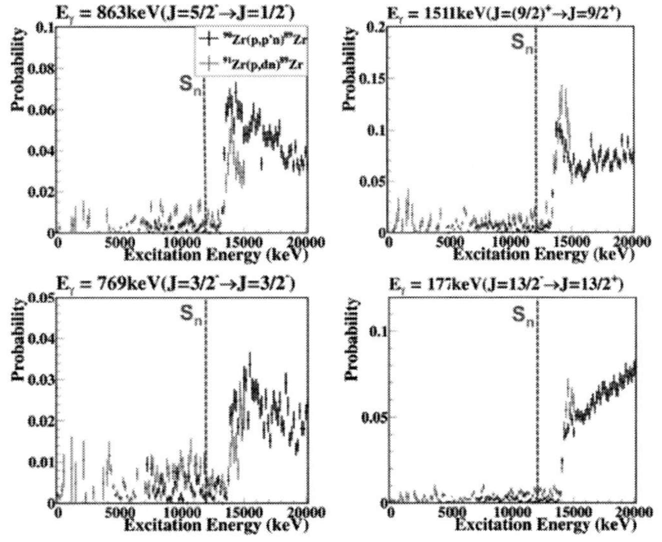

Figure 6. γ-decay probabilities for γ-ray transitions from different J states in ^{89}Zr (see [4] for more details).

5 Summary and Outlook

The ^{90}Zr nucleus spin distribution produced by three different reactions, namely *(p,p')*, *(p,d)*, and *(p,t)* reactions were studied. The outgoing particles were detected in coincidence with γ-rays emitted by the ^{89,90}Zr nuclei. These data, especially the spin populations around the neutron separation energy (S_n), can be used to understand the decay of the compound nucleus formed in ^{89}Zr(n,γ). The γ decay probabilities, P_i (E_x), are crucial to develop theoretical models for determining (n,γ) cross sections using the present surrogate reaction approach.

The low-lying states of ^{90}Zr at $E_x < 6$ MeV observed in the present *(p,d)*, and *(p,t)* experiments agreed well with the past measurements. Although it is difficult to obtain a general trend for the reaction dependence from the present data in the energy region, it seems s that a high spin state such as $J = (7{,}8)\ \hbar$ is only populated by ^{91}Zr*(p,d)* and 0^+ states are only populated by ^{92}Zr*(p,t)* reaction. This may make sense considering the spins of the ^{91}Zr and ^{92}Zr ground states are $5/2^+$ (angular momentum: $L = 2\hbar$) and 0^+ ($L = 0\hbar$), respectively.

We carried out measurements of P_i (E_x) for γ-rays emitted in the decay of ^{90}Zr nucleus produced near S_n. We observed that the ^{90}Zr*(p,p')* reaction produces notably fewer γ-rays from high J states than the ^{91}Zr*(p,d)* and ^{92}Zr*(p,t)* reactions. This suggests the inelastic scattering preferably populates lower J states in ^{90}Zr than the transfer reactions. This reaction dependence holds over several MeV above S_n.

Quantitative study regarding the J^π population of the ^{90}Zr compound nucleus and ^{89}Zr(n,γ) cross sections from the surrogate reaction approach will be forthcoming.

Acknowledgement

We express our thanks to the cyclotron staff at Texas A&M University. This work was performed under the auspices of the US Department of Energy by Lawrence Livermore National Laboratory under contract No. DE-AC52-07NA27344. One of the authors, S. O. is supported by a JSPS Postdoctoral Fellowship for Research Abroad. Partial support through the US Department of Energy's Topical Collaboration TORUS is acknowledged.

References

1. J. E. Escher, J. T. Burke, F. S. Dietrich, N. D. Scielzo, I. J. Thompson, and W. Younes, Rev. Mod. Phys. 84, 353 (2012)
2. N. D. Scielzo, J. E. Escher, J. M. Allmond, M. S. Basunia, C. W. Beausang, L. A. Bernstein, D. L. Bluel, J. T. Burke, R. M. Clark, F. S. Dietrich, P. Fallon, J. Gibelin, B. L. Goldblum, S. R. Lesher, M. A. McMahan, E. B. Norman, L. Phair, E. Rodriquez-Vieitez, S. A. Sheets, I. J. Thompson, and M. Wiedeking, Phys. Rev. C 81, 034608 (2010).
3. S. Ota, J. T. Burke, R. J. Casperson, J. E. Escher, R. O. Hughes, J. J. Ressler, N. D. Scielzo, I. J. Thompson, R. A. E. Austin, E. McCleskey, M. McCleskey, A. Saastamoinen, and T. Ross, EPJ Web Conf. 93, 02001 (2015).
4. S. Ota, J. T. Burke, R. J. Casperson, J. E. Escher, R. O. Hughes, J. J. Ressler, N. D. Scielzo, I. J. Thompson, R. A. E. Austin, B. Abromeit, N. J. Foley, E. McCleskey, M. McCleskey, H. I. Park, A. Saastamoinen, and T. J. Ross, Phys. Rev. C 92, 054603 (2015).
5. R. Schwengner, G. Rusev, N. Tsoneva, N. Benouaret, R. Beyer, M. Erhard, E. Grosse, A. R. Junghans, J. Klug, K. Kosev, H. Lenske, C. Nair, K. D. Schilling, and A. Wagner, Phys. Rev. C 78, 064314 (2008).
6. A. C. Larsen, M. Guttormsen, M. Krticka, E. Betak, A. Burger, A. Gorgen, H. T. Nyhus, J. Rekstad, A. Schiller, S. Siem, H. K. Toft, G. M. Tveten, A. V. Voinov and K. Wikan, Phys. Rev. C 83, 034315 (2011).
7. D. Savran, T. Aumann, A. Zilges, Prog. Part. Nucl. Phys. 70, 210-245 (2013).
8. L. Netterdon, A. Endres, S. Goriely, J. Mayer, P. Scholz, M. Spieker, A. Zilges, Phys. Lett. B 744, 358-362 (2015).
9. J. E. Escher and F. S. Dietrich, Phys. Rev. C 81, 024612 (2010).
10. J. E. Escher et al., EPJ Web Conf. (in this issue).
11. S. R. Lesher, L. Phair, L. A. Bernstein, D. L. Bleuel, J. T. Burke, J. A. Church, P. Fallon, J. Gibelin, N. D. Scielzo, and M. Wiedeking, Nucl. Instrum. Methods Phys Res., Sect. A 621, 286 (2010).
12. J. B. Ball and C. B. Fulmer, Phys. Rev. 172, 1199 (1968).
13. P. E. Garrett, W. Younes, J. A. Becker, L. A. Bernstein, E. M. Baum, D. P. DiPrete, R. A. Gatenby, E. L. Johnson, C. A. McGrath, S. W. Yates, M. Devlin, N. Fotiades, R. O. Nelson, and B. A. Brown, Phys. Rev. C 68, 024312 (2003).
14. J. B. Ball, R. L. Auble, and P. G. Roos, Phys. Rev. C 4, 196 (1971).
15. T. Kibedi, T. W. Burrows, M. B. Trzhaskovskaya, P. M. Davidson, and C. W. Nestor, Nucl. Instrum. Methods Phys. Res., Sect. A 589, 202 (2008).

First simultaneous measurement of fission and gamma probabilities of ^{237}U and ^{239}Np via surrogate reactions

P. Marini[1,a], Q. Ducasse[1,2], B. Jurado[1], M. Aiche[1], L. Mathieu[1], G. Barreau[1], S. Czajkowski,[1], I. Tsekhanovich[1], A. Moro[3], J. Lei[3], F. Giacoppo[4], A. Gorgen[4], Tornyi[4], L. Audouin[5], L. Tassan-Got[5], J. N. Wilson[5], F. Gunsing[6], M. Guttormsen[4], A. C. Larsen[4], M. Lebois[5], T. Renstrom[4], S. Rose[4], S. Siem[4], G. M. Tveten[4], M. Wiedeking[7], O. Serot[8], G. Boutoux[9], V. Méot[9], B. Morillon[9], D. Denis-Petit[9], O. Roig[9], S. Oberstedt[10], and A. Oberstedt[10]

[1]*CENBG, Chemin du Solarium B.P. 120, 33175 Gradignan, France*
[2]*CEA-Cadarache, DEN/DER/SPRC/LEPh, 13108 Saint Paul lez Durance, France*
[3]*University of Seville, Seville, Spain*
[4]*University of Oslo, Department of Physics, P.O. Box 1048, Blindern 0316 Oslo, Norway*
[5]*IPN Orsay, 15 rue G. Clémenceau, 91406 Orsay cedex, France*
[6]*CEA Saclay, DSM/Irfu, 91191 Gif-sur-Yvette cedex, France*
[7]*iThemba LABS, P.O. Box 722, 7129 Somerset West, South Africa*
[8]*CEA-Cadarache, DEN/DER/SPRC/LEPh, 13108 Saint Paul lez Durance, France*
[9]*CEA DAM DIF, 91297 Arpajon, France*
[10]*IRMM, Geel, Belgium*

Abstract. Fission and gamma decay probabilities of ^{237}U and ^{239}Np have been measured, for the first time simultaneously in dedicated experiments, via the surrogate reactions ^{238}U(^{3}He,^{4}He) and ^{238}U(^{3}He,d), respectively. While a good agreement between our data and neutron-induced data is found for fission probabilities, gamma decay probabilities are several times higher than the corresponding neutron-induced data for each studied nucleus. We study the role of the different spin distributions populated in the surrogate and neutron-induced reactions. The compound nucleus spin distribution populated in the surrogate reaction is extracted from the measured gamma-decay probabilities, and used as input parameter in the statistical model to predict fission probabilities to be compared to our data. A strong disagreement between our data and the prediction is obtained. Preliminary results from an additional dedicated experiment confirm the observed discrepancies, indicating the need of a better understanding of the formation and decay processes of the compound nucleus.

1 Introduction

Neutron-induced reaction cross sections of short-lived nuclei are relevant for fundamental nuclear physics, as well as for astrophysics and nuclear energy applications. Indeed these data play an important role in the understanding of the nucleosynthesis r- and s- processes, and for nuclear waste transmutation via fast neutrons. However the lifetimes of many relevant isotopes are often too short

[a]e-mail: marini@cenbg.in2p3.fr

for producing and/or handling a target, making the direct measurement of these cross sections very difficult or impossible.

The surrogate reaction method was proposed for the first time in the 70's by Cramer and Britt [1]. It is an indirect method which aims at determining compound nucleus reaction cross sections involving short lived and/or difficult-to-produce targets. The method is based on the Bohr hypothesis: the compound nucleus decay probability in a given channel is independent of the formation channel. Therefore the same compound nucleus A* formed in a neutron-induced reaction (n+(A-1)→A*) can be formed in a transfer reaction on a slightly different (but more accessible) target nucleus (b+Y→A*+c). In this case, by identifying the ejectile c one can determine the charge and mass (Z,A) of the decaying nucleus A, and the ejectile kinetic energy and emission angle provide its excitation energy E^*. The nucleus A can decay through different exit channels: fission, gamma emission, neutron emission, etc. Therefore the measurement of the number of coincidences between the ejectile and the decay products of interest, normalized to the total number of detected ejectiles (i.e. to the total number of nuclei A produced) allows one to extract the decay probability P^A_{decay} for the corresponding decay channel. The neutron-induced cross section for the nucleus A-1 can then be obtained as

$$\sigma^{A-1}_{decay} \simeq \sigma^A_{CN}(E_n)P^A_{decay}(E^*) \tag{1}$$

where $\sigma^A_{CN}(E_n)$ is the compound nucleus formation cross section via the (A-1)+n reaction and it is typically obtained by optical model calculations.

The method has the main advantages of allowing one to access short-lived nuclei, not otherwise accessible via direct measurements, and to simultaneously investigate several transfer channels on a broad excitation energy range. In addition this kind of measurements is performed with charged particles, whose beam intensity can be few orders of magnitude higher than the nowadays available neutron beams; and in a neutron-free environment, eliminating the issues related to neutron scattering typically associated to direct measurements. However, the equivalence of neutron-induced and surrogate reaction measurements relies on two hypotheses, which need to be tested. First, the formation of a compound nucleus must take place both in the neutron-induced and in the transfer reactions. This means that the formed nucleus loses memory of the entrance channel (except for the conserved quantities, i.e. energy and J^π) and its decay is independent of its formation. This assumption is justified by the high nuclear level density in the excitation energy region close to and above the neutron separation energy, which is the region of interest. The second one is that the decay probabilities of the compound nucleus are independent of its angular momentum and parity distributions -J^π - (the so-called Weisskopf-Ewing limit, see Ref. [2]), or that the J^π distributions populated in neutron-induced and transfer reactions are the same. Further details on the method and on the underlying assumptions can be found in Ref.[3].

Several measurements (e.g. Ref.[4]) showed a very good agreement of the fission cross sections obtained with the surrogate and direct methods for actinides. However, in recent experiments [5, 6] radiative capture cross sections on rare earths obtained in surrogate reactions were found to be higher up to a factor of 10 than the corresponding neutron-induced reaction data. These important discrepancies were attributed to the large differences in the angular momentum between the mother and the daughter nuclei around the neutron separation energy, which results in the hindering of the neutron emission channel and therefore in the increase of the gamma emission probability [6]. This effect is expected to be reduced when studying actinides, whose level densities are much higher than the rare earth one even at low excitation energies. However, a simultaneous measurement of fission and gamma emission probabilities of actinides was not performed up to now. In this work we report the results of two experiments where this simultaneous measurement was first performed with the aim of further investigating the validity of the assumption of the surrogate reaction method and therefore to pin down to which extent it can be applied to infer neutron-induced cross sections.

2 Experiments

The first measurement was performed in 2012 at the Oslo cyclotron. A deuteron and a ^{3}He beams at 15 and 24 MeV energies, respectively, were impinged on a 99.5% isotopically pure ^{238}U target of 260 μg/cm^2 thickness, deposited on a 40 μg/cm^2 C layer. The experimental setup coupled the CACTUS [7] NaI(Tl) array for gamma detection, the NIFF PPAC [8] for fission fragment detection and the SiRi silicon telescope array [9] for the ejectile detection and identification. A detailed description of the setup and of the data analysis can be found in [10–12].

A new dedicated experiment was performed in April 2015 at the Orsay TANDEM accelerator, with the aim of obtaining increased statistics, an increased precision in the excitation energy and in the fission and gamma emission probabilities. A 24 MeV ^{3}He beam was impinged on a 99.5% isotopically pure ^{238}U target of 260μg/cm^2 thickness, deposited on a 39μg/cm^2 C baking. The targets were produced at GSI and extreme attention was payed to reduce their oxidation.

The detection setup coupled dedicated detectors for the detection of gamma rays, fission fragments and light ejectiles. Gamma rays were detected by 4 C_6D_6 scintillators, placed at 90° polar angle around the target, assuring the detection over a broad energy range (up to 10 MeV). Their efficiency was determined with the EXtrapolated Efficiency Method [13]. For more details see Ref. [10, 14]. Six Ge detectors were also placed around the target to perform precise intensity measurements of selected gamma transitions. The fission fragments detector was constituted by 8 segmented photovoltaic cells. Five cells were arranged in a pentagone and covered the polar angles from 19° to 75° and the whole azimutal angle. The other three cells were placed upstream of the target and cover from 105° to 150° polar angle. The segmentation of the photovoltaic cells allows one to experimentally measure the fission fragments anisotropy, which affects the estimation of the fission detection efficiency. This will allow us to reduce the uncertainty on both the gamma decay and fission probabilities with respect to the previous experiment. Finally the ejectiles were detected by two double-side stripped ΔE(300μm)-E(2mm) silicon detectors, covering the polar angles from 116° to 159°. The silicon telescopes allow one to unambiguously identify the ejectile and measure its angle and kinetic energy. This information, combined with the reaction Q-value of the studied reaction, allows one to determine the excitation energy of the formed compound nucleus with a resolution of 50 – 80 keV. The experimental decay probability of the nucleus A* in the channel j (fission or gamma emission) can be obtained as:

$$P_j(E^*) = \frac{N^j_{coinc}(E^*)}{N_{singles}(E^*)\varepsilon^j(E^*)} \qquad (2)$$

where $N^j_{coinc}(E^*)$ is the number of ejectiles detected in coincidence with each decay channel product and $N_{singles}(E^*)$ is the total number of detected ejectiles, $\varepsilon^j(E^*)$ is the detection efficiency of the decay product.

3 Results

As mentionned, several nuclei can be accessed simultaneously during surrogate reaction measurements. In particular, in this experiment we measured the gamma decay and fission probabilities of ^{239}Np, ^{237}U and ^{238}Np via the ^{238}U(^{3}He,d) , ^{238}U(^{3}He,^{4}He) and ^{238}U(^{3}He,t) reactions, respectively. In a first moment we will focus on the (^{3}He,^{4}He) reaction channel (i.e. the decay of the excited ^{237}U), for which good quality neutron-induced cross section values are available.

DOI: 10.1051/epjconf/201612212004

In Fig.1a we present the preliminary fission probability obtained for this nucleus. The experimental data (crosses) are compared to results obtained in the Oslo experiment (squares) and to the evaluated neutron-induced data (full line) given by JENDL 4.0. The shown error bars account for the moment only for statistical uncertainties. The fission threshold is located around 6.1 MeV ^{237}U excitation energy. A very good agreement is found between the data obtained in the two experiments and the much higher statistics and excitation energy resolution of the Orsay experiment can be observed. An agreement between our experimental data and the neutron-induced data is observed for the fission threshold and the cross section values above the threshold. Similar agreements were found when analysing the other transfer reactions, although with slighty less statistics. In Fig. 1b the experimental gamma emission probability P_γ (full circles) of ^{237}U is shown. As expected the P_γ is equal to 1 below the neutron emission threshold of 5.1 MeV (we remind that the gamma emission is the only open channel below S_n since the nucleus is not fissile, and the proton separation energy is bigger than the neutron separation energy) and it significantly drops above this energy due to the competition with the neutron emission. Once more a good agreement with the data of the Oslo experiment is found. Our data are then compared to neutron-induced data (JENDL 4.0 - full line) and discrepancies a factor 3 at 6.4MeV are observed. In Fig. 1c we plot both the fission and gamma emission probabilities shown in Figs. 1a and 1b in the region where both decay channels are open simultaneously, and we compare the experimental data to the evaluated neutron induced data. Also in this excitation energy region we observe a good agreement with the neutron-induced data for the fission probability and a discrepancy of up to a factor 10 for the gamma emission probability. This seems to indicate that, while the fission probability is independent of the neutron emission hindering, and therefore independent of the compound nucleus populated J^π distribution, it is not the case for the gamma probability, which is strongly enhanced by the neutron emission hindering. However, calculations based on the statistical model with standard ingredients show a strong dependence of the fission probability on the spin.

To further investigate it, we compare the measured fission probability to the one calculated by the statistical model. Following the procedure described in [6] we extracted direct information on the populated J^π distribution from the experimental gamma decay probabilities, using the TALYS code [15]. Assuming a Gaussian angular momentum distribution, with no dependence on the excitation energy, the experimental gamma emission probability can be written as:

$$P_\gamma(E^*) = \sum_{J^\pi}[\frac{1}{2\sigma\sqrt{2\pi}}e^{-\frac{(J-\bar{J})^2}{2\sigma^2}}]G_\gamma(E^*, J^\pi) \quad (3)$$

where $G_\gamma(E^*,J^\pi)$ are the TALYS gamma decay probability. The unkonwn $\bar{J}$ and σ parameters, which correspond to the average and width of the spin distribution, are obtained by fitting the experimental data with Eq.3 in the compound-nucleus excitation energy region around 6MeV. The mean value of the surrogate spin distribution is around $6\hbar$ and the width is $2\hbar$. These values are higher than those obtained for the neutron-induced spin distribution, which is centered around $1\hbar$ with a width of about $0.5\hbar$. The surrogate spin distribution is now used as input parameter to the statistical model TALYS to determine the fission probability. The so-calculated fission probability is plotted in Fig.1a and compared to the experimental data. The calculated fission probability does reproduce neither the values nor the fission threshold obtained experimentally. In particular, we observe that the statistical model predicts a dependence of the fission threshold on the mean angular momentum of the compound nucleus, which increases as we increase the input mean spin of the compound nucleus. On the contrary, the agreement between the fission thresholds measured in surrogate and neutron-induced reactions observed experimentally indicates an independence of the fission probability to the compound nucleus angular momentum. Therefore, our experimental observations are not currently explained within a statistical model with standard ingredients.

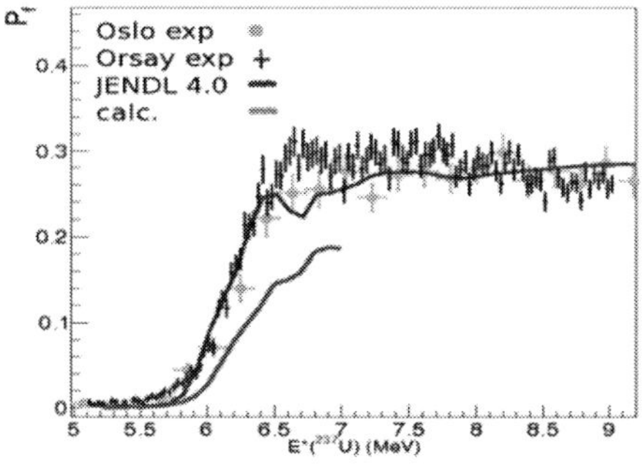

(a)

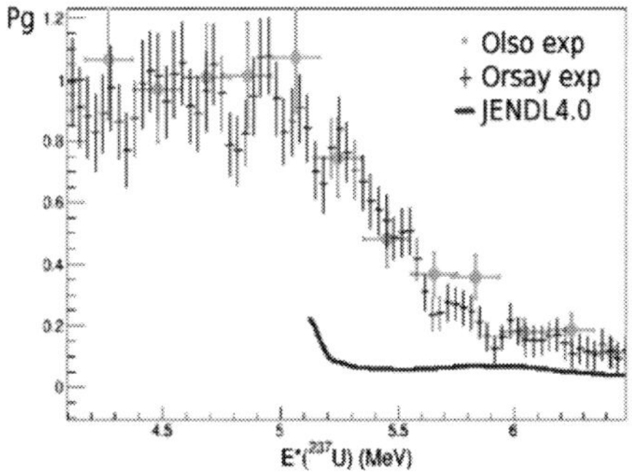

(b)

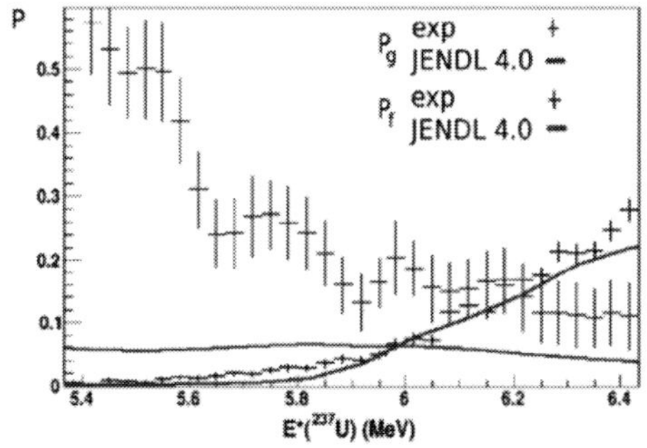

(c)

Figure 1: ^{237}U fission P_f (*a*) and gamma emission P_γ (*b*) probabilities as a function of the compound nucleus excitation energy (E*) obtained in the ^{238}U(^{3}He,^{4}He) reaction. Fig.*c* is a zoom of Figs. *a* and *b*, in the E* region where the fission and gamma emission channels are both open.

Similar results are obtained when studying the other transfer channels. In fig.2 we present the fission and gamma probabilities of ^{239}Np compound nucleus, obtained in the ^{238}U(^{3}He,d) transfer reaction. This is an interesting nucleus since the fission barrier is lower than the neutron separation energy, S_n. Therefore both fission and gamma decay channels are open below S_n and the sum of the corresponding decay probabilities should be 1. Below the fission barrier $P_\gamma = 1$ and it decreases around 5.5 MeV where the fission sets in. The fission probability increases up to an excitation energy of about 6.2 MeV, the S_n, and then it drops, due to the competition with the neutron emission. In fig 2 we also plot $P_f + P_\gamma$, whose good agreement with 1 below S_n gives us confidence in the procedure used in the analysis.

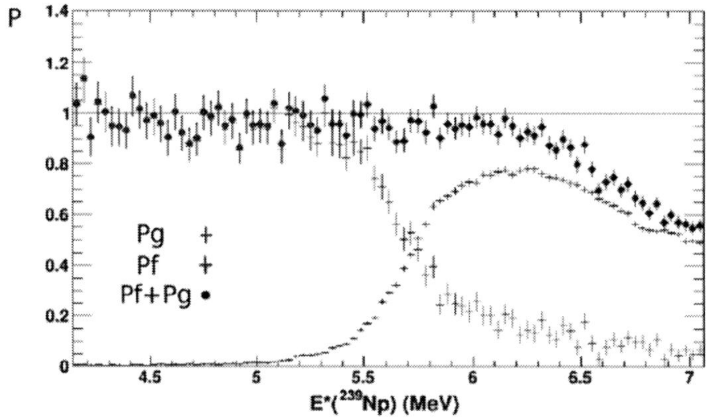

Figure 2: ^{239}Np P_f and P_γ probabilities as a function of the compound nucleus excitation energy (E^*) obtained in the ^{238}U(^{3}He,d) reaction and sum of P_f and P_γ.

4 Conclusions

In summary, we have performed two dedicated experiments to study the validity of the surrogate reaction method to extract neutron-induced reaction cross sections. It is the first time that transfer-induced gamma emission and fission probabilities of actinides are simultaneously measured. The comparison of our experimental data to those obtained in neutron-induced reactions shows a good agreement for the fission probability and a strong disagreement for the gamma emission probability for the same compound nucleus and excitation energy. This indicates a strong sensitivity of the gamma probability to the compound nucleus populated spin distribution at excitation energies slightly above the neutron separation energy. Indeed it was previously shown that the spin distribution populated in surrogate reaction is centered at higher values and it is broader than the one populated in neutron-induced reactions [6]. On the contrary we do not observe a dependence of the fission probability on the populated angular momentum distribution of the compound nucleus. We have compared these observations to the statistical model predictions. We have determined the spin distribution from a fit to the measured gamma emission probabilities via a statistical model calculation performed with the TALYS code. The so-obtained spin distribution is used as input parameter to deduce the fission probability. Statistical model calculations predict an influence of the angular momentum on the fission threshold. Such a dependence is not observed in the experimental data. Therefore our observations are nowadays not explained within a statistical model picture with standard ingredients. It is then crucial to better understand the formation and decay mechanisms of the compound nucleus in transfer reactions. Indeed, the surrogate reaction method allows one to access cross sections of short-lived nuclei, that cannot be directly measured.

Acknowledgement

We would like to thank the staff of the Oslo Cyclotron and the Orsay tandem for the support during the experiments and the GSI Target Laboratory for the production of high-quality ^{238}U targets.

We would also like to express our gratitude to A. Moro and J. Lei, and B. Morillon for performing deuteron break-up, and optical model and TALYS calculations, respectively, and for fruitful discussions. This work was supported by the University of Bordeaux, the European Commission within the 7th Framework Programme through Fission-2010-ERINDA (Project No 269499) and CHANDA (Project No 605203), by the European Atomic Energy Community's 7th Framework Programme under grant agreement number FP7- 249671 (ANDES) and by the French national research programs GEDEPEON and NEEDS.

References

[1] J. Cramer, H.C. Britt, Nucl. Sci. Eng. **41**, 177 (1970)
[2] V.F. Weisskopf, D.H. Ewing, Phys. Rev. **57**, 472 (1940)
[3] J.E. Escher, J.T. Burke, F.S. Dietrich, N.D. Scielzo, I.J. Thompson, W. Younes, Rev. Mod. Phys. **84**, 353 (2012), and refs. therein
[4] G. Kessedjian, J. B., M. Aiche, G. Barreau, A. Bidaud, S. Czajkowski, D. Dassié, B. Haas, L. Mathieu, L. Audouin et al., Phys. Lett. B **692**, 297 (2010)
[5] N.D. Scielzo, J.E. Escher, J.M. Allmond, M.S. Basunia, C.W. Beausang, L.A. Bernstein, D.L. Bleuel, J.T. Burke, R.M. Clark, F.S. Dietrich et al., Phys. Rev. C **81**, 034608 (2010)
[6] G. Boutoux, B. Jurado, V. Méot, O. Roig, L. Mathieu, M. Aïche, G. Barreau, N. Capellan, I. Companis, S. Czajkowski et al., Phys. Lett. B **712**, 319 (2012)
[7] M. Guttormsen, A. Atac, G. Løvhøiden, S. Messelt, T. Ramsøy, J. Rekstad, T. Thorsteinsen, T. Tveter, Z. Zelazny, Physica Scripta **T32**, 54 (1990)
[8] T. Tornyi, A. Görgen, M. Guttormsen, A. Larsen, S. Siem, A. Krasznahorkay, L. Csige, Nucl. Instr. and Meth. in Phys. Res. Sec. A **738**, 62 (2014)
[9] M. Guttormsen, A. Bürger, T. Hansen, N. Lietaer, Nucl. Instr. and Meth. in Phys. Res. Sec. A **648**, 168 (2011)
[10] Q. Ducasse, Ph.D. thesis, University of Bordeaux (2015)
[11] P. Marini, *Strength and limitations of the surrogate reaction method to acess neutron-induced cross sections of actinides*, in *NRM2015 Conference Proceedings* (2015)
[12] Q. Ducasse (2015), submitted to PRC
[13] G. Boutoux, B. Jurado, L. Mathieu, V. Méot, M. Aïche, G. Barreau, A. Blanc, I. Companis, O. Roig, C. Théroine, Nucl. Instr. and Meth. in Phys. Res. Sec. A **700**, 59 (2013)
[14] Q. Ducasse (2015), submitted to NIM A
[15] A.J. Koning, S. Hilaire, M. Duijvestijn, User Manual TALYS (2015)

Isospin influence on the decay modes of compound systems produced in the ^{78,86}Kr + ^{40,48}Ca at 10 AMeV

S. Pirrone[1], G. Politi[1,2], J.P. Wieleczko[5], E. De Filippo[1], B. Gnoffo[1], P. Russotto[1,2], M. Trimarchi[7,8], M. La Commara[3,4], M. Vigilante[3,4], G. Ademard[11], F. Amorini[6], L. Auditore[7,8], C. Beck[9], I. Berceanu[10], E. Bonnet[5], B. Borderie[11], G. Cardella[1], A. Chibihi[5], M. Colonna[6], A. D'Onofrio[4,12], J. D. Frankland[5], E. Geraci[1,2], E. Henry[13], E. La Guidara[1,14], G. Lanzalone[6,15], P. Lautesse[16], D. Lebhertz[5], N. Le Neindre[17], I. Lombardo[3,4], K. Mazurek[5], S. Norella[7,8], A. Pagano[1], E.V. Pagano[2,6], M. Papa[1], E. Piasecki[18], F. Porto[2,6], L. Quattrocchi[7,8], M. Quinlann[13], F. Rizzo[1,6], W. U. Schroeder[13], G. Spadaccini[3,4], A. Trifirò[7,8], J. Toke[13] and G. Verde[1,11]

[1]*INFN Sezione di Catania, Italy*
[2]*Dipartimento di Fisica e Astronomia, Università di Catania, Italy*
[3]*Dipartimento di Fisica, Università Federico II Napoli, Italy*
[4]*INFN Sezione di Napoli, Italy*
[5]*GANIL Caen,France*
[6]*INFN Laboratori Nazionali del Sud, Italy*
[7]*Dipartimento di Fisica, Università di Messina, Italy*
[8]*INFN Gruppo collegato di Messina, Italy*
[9]*IN2P3 - IPHC Strasbourg, France*
[10]*IPNE, Bucharest, Romania*
[11]*IN2P3 - IPN Orsay, France*
[12]*Dipartimento di Matematica e Fisica - Seconda Università di Napoli,Caserta, Italy*
[13]*University of Rochester, USA*
[14]*Centro Siciliano Fisica Nucleare e Struttura della Materia, Catania, Italy*
[15]*Università Kore, Enna, Italy*
[16]*IN2P3 - IPN Lyon, France*
[17]*IN2P3 - LPC Caen, France*
[18]*University of Warsaw, Poland*

Abstract. The study of the decay modes competition of the compound systems produced in the collisions ^{78}Kr+^{40}Ca and ^{86}Kr+^{48}Ca at 10AMeV is presented. In particular, the N/Z entrance channel influence on the decay paths of the compound systems, directly connected to the isospin influence, is investigated. The experiment was performed at the INFN Laboratori Nazionali del Sud (LNS) in Catania by using the 4π multi-detector CHIMERA. Charge, mass, angular distributions and kinematical features of the reaction products were studied. The analysis shows some differences in the contribution arising from the various reaction mechanisms for the neutron poor and neutron rich systems. Comparison with theoretical statistical and dynamical models are presented for the two systems. Besides a study of the influence of the energy on the entrance channel is performed for the ^{78}Kr+^{40}Ca reaction, by comparing the results of this experiment to those obtained for the same system at 5.5 AMeV with the INDRA device at GANIL.

[1] Corresponding author: sara.pirrone@ct.infn.it

1 Introduction

One of the principal aim in the field of the nuclear reactions is the study of the influence of the entrance channel N/Z ratio, directly connected to the isospin degree of freedom, on the reaction mechanism and on the formation of the fragments in the exit channel.

In this context, particular interest is focused on the heavy-ion reactions at low energy (E/A< 10-15 AMeV), where for medium-mass systems the fusion processes are predominant. Indeed, one expects that the N/Z ratio of the compound nucleus, plays an important role on its decay evaporation-process, providing crucial information on fundamental nuclear parameters as the level density, the fission barrier and the nuclear viscosity. Thus a program of detailed measurements of reaction cross-section for a large isotopic chain of compound nuclei, from neutron rich to neutron-poor, can provides careful information.

In this work, we present the up to date results of the ISODEC experiment, realized to study the competition among the various disintegration modes of ^{118,134}Ba compound nuclei produced in the reactions ^{78}Kr+^{40}Ca (neutron poor system) and ^{86}Kr+^{48}Ca (neutron rich system) at 10 AMeV. The experiment was performed at the INFN Laboratori Nazionali del Sud (LNS) in Catania by using the CHIMERA detector. It complements and improves the data obtained at 5.5 MeV/A for ^{78,82}Kr+^{40}Ca reactions [1], previously realized with beams delivered by GANIL facility and by using the INDRA detector.

The two investigated systems are different for 16 neutrons, the maximum difference that we can get by using stable nuclei; this allows to produce compound nuclei with similar both spin distribution and excitation energy in a large domain of N/Z (from 1.11 to 1.39). Such a set of data also will provide new constraint on sophisticated models attempting to describe statistical and/or dynamical properties [2] of excited nuclei.

2 Experimental set-up

In order to realize the presented scientific program, we need the measurements of several key observables such as the cross sections, multiplicities, angular and kinetic energy distributions of the various reaction products: Intermediate Mass Fragments - IMFs, Light Charge Particles - LCPs, Evaporation Residues - ER, and Fission Fragments - FF. The measurement of these observables requires a good isotopic resolution, a low energy thresholds for LCPs and IMFs, an high angular resolution and a broad angular acceptance.

The presence at LNS of CHIMERA [3], a second generation 4π multi detector, and the possibility to get beams with both good intensity and timing characteristics, strongly supported the necessity to perform the experiment in Catania.

Figure 1. CHIMERA in the scattering chamber at INFN-LNS

CHIMERA consists of 1192 detector telescopes, arranged on 9 rings in the forward part, that cover a polar angle from 1° to 30°, and 17 rings in spherical configuration, that cover from 30° to 176°. The geometrical efficiency is the 94% of the total solid angle. In figure 1, a picture of the detector in its own experimental scattering chamber at LNS is presented.

The single detection telescope consists of a silicon detector (Si, thickness about 300 μm) followed by a Caesium Iodine Thallium doped crystal, (CsI(Tl)), thickness from 3 cm to 12 cm), coupled to a photodiode.

The identification methods employed are:

ΔE-E for charge identification of particles punching through the Si detector and stopped in the CsI(Tl) with also mass identification for particles with Z<10;

E-TOF (Time of Flight) for mass identification, velocity and energy measurement of the particles stopped in the Si detector;

PSD (Pulse Shape Discrimination) in CsI(Tl), for isotopic identification of more energetic light charge particles;

PSD (Pulse Shape Discrimination) in Si detector, for charge identification of the particles stopped in the Silicon detector.

Besides, the CHIMERA multi detector is characterized by a low energetic detection threshold, that is less of 0,5 MeV/A for heavy ions and about 1 MeV/A for light particles.

Concerning the PSD in silicon detector [4, 5, 6], this was the last implementation made on the CHIMERA array and allows the use of this device to study reaction mechanism also in the low energy domain, extending the investigation dynamical range of the detector from the fusion to the multi fragmentation reactions.

In the experiment, self-supporting 1 mg/cm^2 thick ^{40}Ca and ^{48}Ca targets, prepared in collaboration between INFN-LNL and INFN-LNS Target Laboratories, were bombarded by ^{78,86}Kr beams at 10 AMeV, delivered by the LNS Superconductive Cyclotron, with typical beam current intensity of 800-1000 pA. The pulsed beam was delivered with a timing resolution in the range 800 ps - 1 ns. Inclusive and coincidence measurements were realized.

In figure 2 is reported as an example the Energy (E_{Si}) vs Rise Time (RT_{Si}) plot obtained by the PSD methods in silicon detector, for the n-poor system ^{78}Kr+^{40}Ca at 10 AMeV at θ=34°. Identification energy threshold of about 4.5 MeV/nucleon was obtained. Notice that in CHIMERA PSD identification of fragment are obtained mounting the silicon detectors in direct mode in order to allow good timing performances needed for the TOF identification.

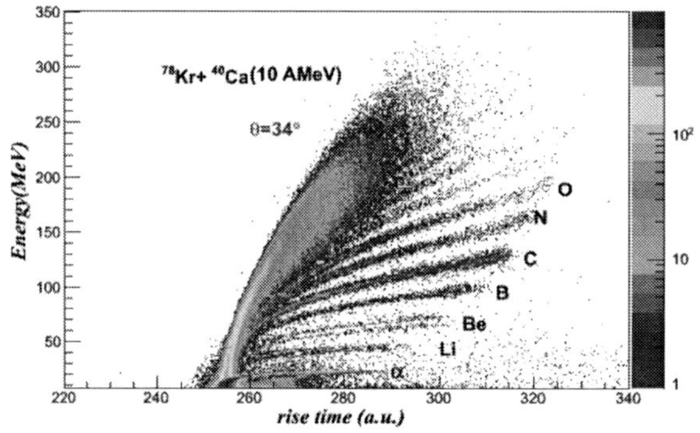

Figure 2. Energy-RiseTime plot for ^{78}Kr+^{40}Ca

These characteristics allowed the complete identification of LCP in a wide energy range, the complete identification in charge and mass of the IMF (3<Z<8) products, the charge identification up to Z =14-17 for fragments stopped in the Silicon, and up to about Z = 30 for the most energetic particles stopped in CsI.

In figure 3 is reported an example of typical matrix ΔE-E that shows the good identification capability of the array. We see a very good discrimination in charge and also in mass for the lighter nuclei, up to Carbon.

DOI: 10.1051/epjconf/201612213001

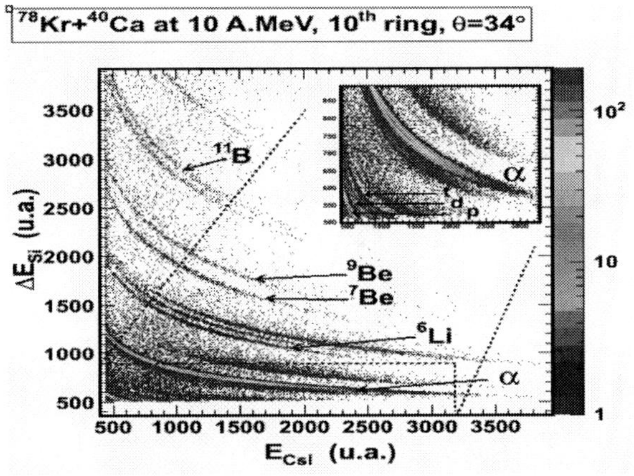

Figure 3. ΔE_{Si}-$E_{CsI(Tl)}$ plot for ^{78}Kr+^{40}Ca

In particular, in figure 4 is reported an example of isotopic identification for Carbon isotopes obtained at polar angle of 15° for the two systems [6].

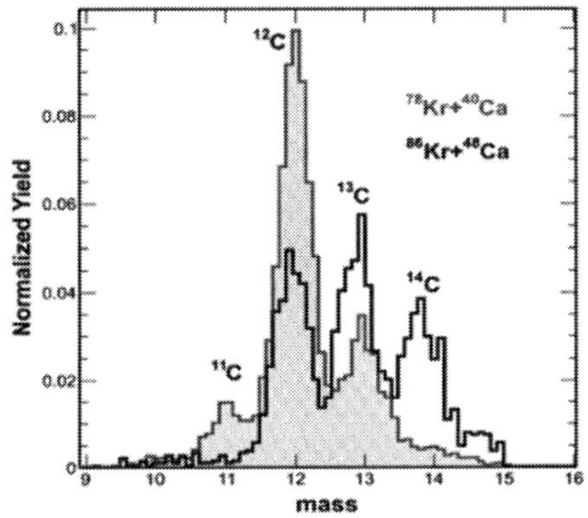

Figure 4. Carbon isotopes distributions at 15° for the two systems

3 Experimental Results

3.1 Mass distribution

In order to study the influence of the isospin degree of freedom it is essential to get the isotopic identification at least of part of the detected fragments. The capability of the CHIMERA detector allows us to get the mass distributions for the IMF $2 \leq Z \leq 8$.

In figure 5 the mass distribution for Z = 3, 4, 5 and for the two studied systems are reported.

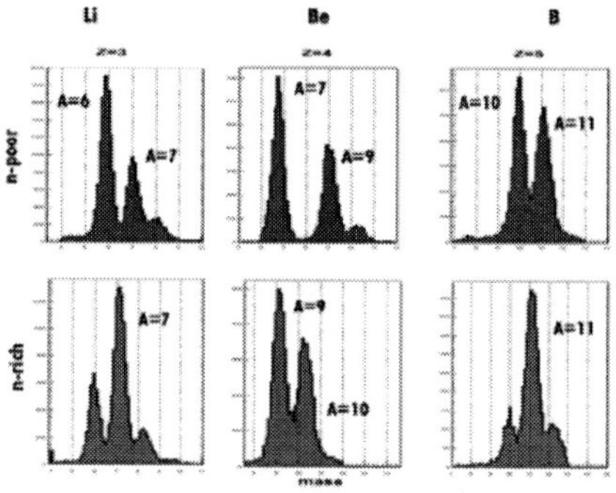

Figure 5. Mass distributions for Z=3,4,5, for the n-poor system and n rich system

One can see the different isotopic composition and relative enrichment in correspondence of the same Z, for the compared systems. In particular this effect is evident for the Be element, in which the isotopic composition goes from A=7, 9, in the n-poor system, to A= 9,10, in the n-rich one. These differences in the two systems show that a memory of the entrance channel is still present in this class of reaction products, and it could be due to an effect of the isospin influence on the reaction mechanism or, alternatively, on the structure of the emitted fragments.

3.2 Charge distribution

The influence of the isospin on the reaction mechanism and on the fragments production should appear also in the charge distribution.

A comparison between the fragments yields as a function of their charge for the studied systems is presented in Figure 6. The IMF cross section behaviour exhibits for both systems a strong even-odd effect, the staggering, due to a preferential production of fragments with even value of the atomic number. In agreement with other examples in literature [1, 8, 9] the staggering is more pronounced for the n-poor system respect to the n-rich one, in particular for Z<10 reaction products as it is shown in figure 6. This effect persists for higher Z, but with a smaller amplitude.

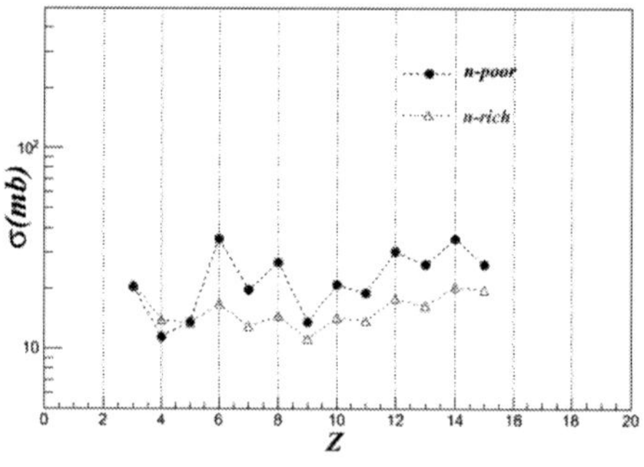

Figure 6: IMF Charge distributions, in red triangle for the $^{86}Kr+^{48}Ca$ reaction, in black circle for the $^{78}Kr+^{40}Ca$ reaction (color on line)

It seems that the neutron excess in the entrance channel affects the yields of the light fragments. An Influence can be originated also by structure effects [10] linked to the pairing forces, and it could be connected with the Symmetry Energy.

The charge distributions of the two systems were compared with the preliminary results of model calculations. In order to understand the origin of the different behaviour, we performed calculations with the dynamical DNS (Di Nuclear System) model for Z<10, and with the GEMINI ++ code, a Montecarlo Code basing on recent statistical model, for Z up to 16 .

Following the dynamical DNS (Di Nuclear System) model, a dinuclear system is formed in the beginning of the reaction, that can go towards a not fully equilibrated fast-fission or towards an equilibrated fissioning compound nucleus. A detailed description of the model can be found in [11,12].

The simulation of GEMINI++ code reproduces the decay of the compound nucleus following the formalism of the Hauser-Feshbach [13], taking in account the probability of asymmetric fission obtained by L. Moretto in an extension of the transitional state model [14,15].

The results of these comparisons are reported in figure 7.

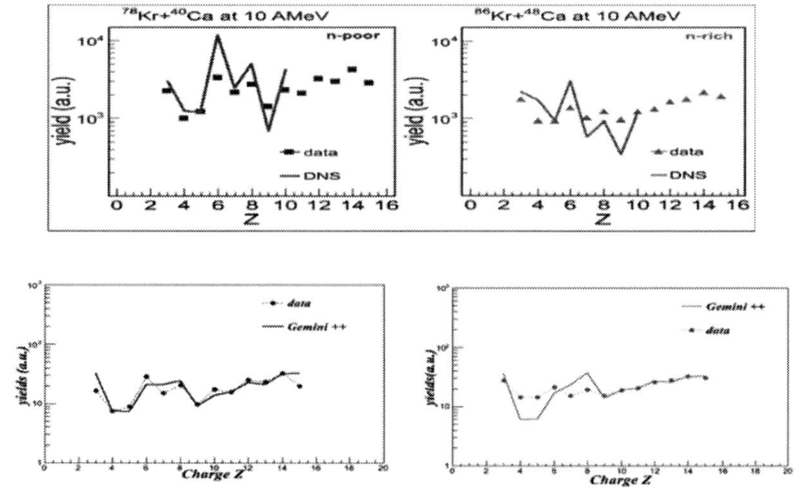

Figure7: top, comparison with DSN model; bottom, comparison with GEMINI++ model

The calculations seem to be in a slightly better agreement in the case of the neutron poor system in both cases. This behaviour could be the indication that the n-enrichment could favour new mechanism in the production of intermediate mass fragments, not considered in the used models, neither dynamics or statistical.

In both systems, for higher Z, the oscillation of the staggering of the yields is decreasing as the atomic number increases in agreement with the experimental findings.

3.3 Kinematical features

The good measurement of time of flight of the reaction products, allows to us to calculate the average velocity in the centre of mass $<v_{cm}>$ for Z from 3 to 17. The results are reported in Figure 8 at various laboratory angles for the neutron poor system. We note that for a given Z the average velocity in the centre of mass is constant within the experimental accuracy with respect to the emission angle.

This general behaviour observed for the emitted fragments, give us an indication of the high degree of relaxation in the production mechanism.

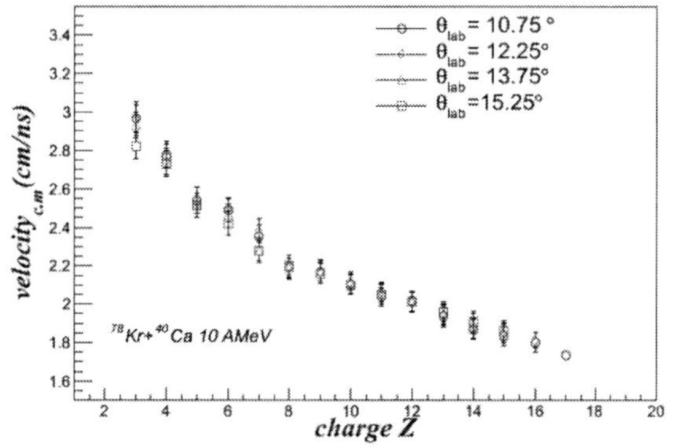

Figure 8. Average velocity distribution

As shown in figure 9, we also note that the mean values of the velocity of the IMF in the centre of mass frame show a quasi-linear decrease with the increase of the atomic number and they are well reproduced by the prediction of the fission systematic of Viola, with the Hinde's correction for the asymmetric process [16].

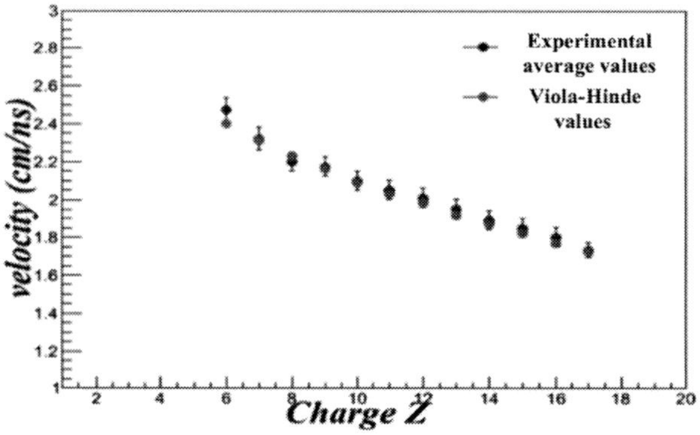

Figure 9 :Comparison of the mean values of the velocity, averaged over all angles, with the values of Viola, for the reaction $^{78}Kr+^{40}Ca$

3.4 Angular distribution

The study of the angular distribution of the emitted fragments allows us to get fundamental information to distinguish totally equilibrated processes, as the compound nucleus formation followed by evaporation or fission, from partially equilibrated process as the fast-fission or the DIC processes.

Some examples are given in the following. In figure 10, the angular distributions of fragments with Z from 12 to 32, for the neutron poor system, are compared to the *1/senθ* behaviour (dashed line), that is the behaviour of the isotropic emission typical of the fission fragments (FF) coming a fissioning compound nucleus.

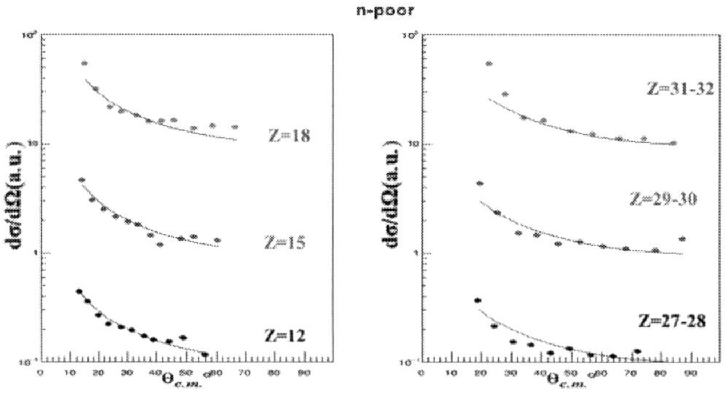

Figure 10 For the $^{78}Kr+^{40}Ca$ reaction. FFs angular distributions, in the center of mass frame, the solid line is $1/sin\theta_{cm}$

Besides, if we look at the angular distributions of the heavier reaction products in the laboratory frame; shown in figure 11, we see that they are very strongly forward peaked, as it is expected in the case of evaporation residue, (ER)

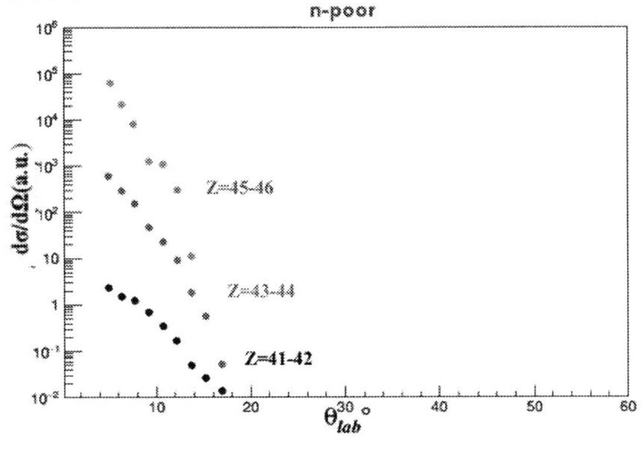

Figure 11 For the ^{78}Kr+^{40}Ca reaction. ER angular distributions, in the laboratory frame

Finally, starting from angular distributions for the different selected reaction products, with a suitable normalization, we obtain the cross sections for the different reaction mechanisms.

3.5 Global features

In order to investigate about the global features of the reactions complete events were selected by imposing the conditions of total detected mass M_{tot}>90 and total momentum P_{tot} >0.7P_{beam} [17,18]. By looking at the plot of the mass versus the parallel velocity for each reaction product, reported in figure 12, we can extract qualitative information about the competition among the different decay channels. In fact in this plot the different reaction products, coming by the Fusion-Fission and Quasi-Fission decay process or by Evaporation Residues one, can be easily localized. The results show that the *ER/FF* ratio is higher for the n-poor system compared to n-rich one.

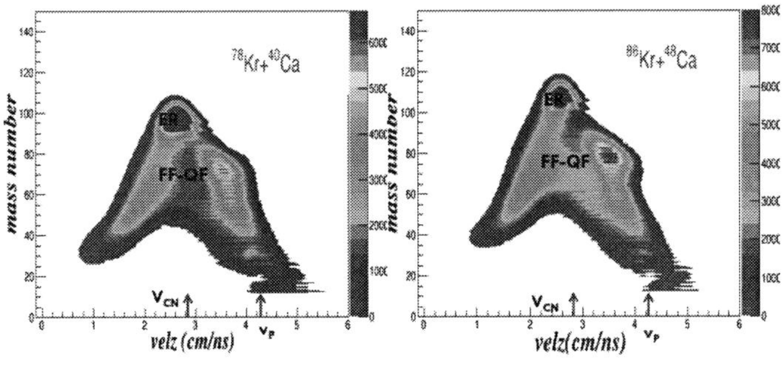

Figure12 Mass vs parallel velocity for the two systems.

4 Conclusions

Results of the study of the ^{78,86}Kr+^{40,48}Ca reactions at 10 AMeV incident energy are presented. Kinematical characteristics of the IMF seem to indicate an high degree of relaxation of the system formed in the collision before the breakup, as in typical fusion-fission like reaction mechanisms. Staggering effects are evident in the IMF charge distributions, as well as different isotopic composition and neutron enrichment for the reaction products in the two systems. Global feature analyses show differences between the two systems in the contribution arising from the various reaction mechanisms like fusion-fission, quasi-fission and fusion- evaporation.

The observed effects could be due to the role of the N/Z degree of freedom on the decay channels. Data analysis and more refined comparisons with theoretical models are in progress. A Letter of Intent has been presented for an experiment at radioactive ion beam facility SPES at INFN-LNL, with $^{88-94}$Kr + $^{40-48}$Ca at

10MeV/A, in order to study these effects in higher N/Z systems and further improve the knowledge in this field.

References

1. G. Ademard et al. Phys. Rev. C 83, 054619 (2011)
2. P.N. Nadtochy et al., Phys. Rev. C 75, 064614 (2007)
3. A. Pagano et al., Nucl. Phys. A734, 504 (2004)
4. G. Politi et al., IEEE Nuclear Science Symposium Conference Record N28, 1140 (2006);
5. M. Alderighi et al., IEEE Trans. Nucl. Sci. 52, 1624 (2005)
6. R. Bassini et al., IEEE Nuclear Science Symposium Conference Record N14, 173 (2006)
7. E. De Filippo, A. Pagano, EPJA681 32 (2014)
8. I. Lombardo et al., Phys. Rev. C 84, 024613 (2011)
9. G. Casini Phys.Rev. C 86, 011602(R) (2012)
10. M. D'Agostino et al., Nucl. Phys. A 861, 47 (2011)
11. Sh.A. Kalandarov, G.G. Adamian, N.V. Antonenko, W. Scheid, Phys. Rev. C 82, 044603 (2010)
12. A.S. Zubov, G.G. Adamian, N.V. Antonenko, S.P. Ivanova,W. Scheid, Phys. Rev. C 68,014616 (2003)
13. H.Hauser and H.Feshbach, Phys. Rev. 87 366 (1952)
14. L.G.Moretto Phys. Rev. A247, 221 (1975)
15. N.Bohr and J.A.Wheeler, Phys. Rev. 56, 426 (1939)
16. D.J.Hinde et al., Nucl.Phys. A472, 318 (1987)
17. G. Politi et al., EPJ Web of Conferences 21, 02003 (2012)
18. S. Pirrone et al., AIP Conf. Proc. 1524, 7 (2013)

EPJ Web of Conferences **122**, 13002 (2016) DOI: 10.1051/epjconf/201612213002
*CNR*15*

Diffracted Fringes of Compound Nucleus Levels

Kazumi Ideno
Yabe 636-1-234, Totsuka, Yokohama, Kanagawa 244-0004, Japan

Abstract. We investigate a relation between the energies of nuclear excited levels and its periodic phases for nuclei with A = 14 - 244 in the energy region up to several tens MeV. These levels include neutron and proton resonances, excited levels below neutron and proton separation energies and also vibrational and rotational bands in unstable nuclei. Here we use level periods less than 50 keV. We found that series of parabolic fringes appear in the plots of level energies vs. its periodic phases in various excitation modes. Distinguished fringes can be observed for neutron and proton resonances in nuclei with neutron or proton magic numbers: ^{37}Cl, ^{59}Ni, ^{61}Ni and ^{62}Ni. For neutron and proton resonances in a wide mass range of nuclei, parabolic fringes with the same periods and scales can be observed at the same incident energies. Each fringe is separated by a phase difference of $1/n$, where n is an integer. We interpret the parabolic fringes as a result of interference effects based on time; a quantized phase difference of $1/n$ can be connected to a discrete time delay of wave pulsations. These fringe spectra were compared between different nuclear excitation modes.

1 Introduction

A characteristic feature of compound nucleus is its complexity. For massive data of these nuclei, we tried to find out simple structure, e.g. periodic level structure for neutron resonances using a correlation function A_n [1-3]. Many resonance levels were found to be located with periods of ε = 3.06 eV in ^{180}Hf (^{179}Hf + n), ε = 4.37 eV in ^{178}Hf, ε = 5.44 eV in ^{124}Sb and ε = 17.6 eV in ^{169}Er. Later, Coceva et al. showed that the 4.37-eV periodicity in ^{178}Hf is contributed mainly by s-wave resonances with spin J = 3 [4]. This finding indicates that the level periodicity is relevant to symmetries such as spin. It was shown that these periodicities could not be expected from random-matrix statistical theory [5].

Detailed inspection of the 5.44-eV periodicity in ^{124}Sb revealed that most of the resonance energies below 1200 eV can be well approximated with integral multiples of 5.44 eV [6]. Observation of such level sequence indicates that periodic phases of level energies play an important role in the level occurrences. This finding led to more detailed examination of a relation between level energies and its periodic phases for neutron resonances. In the analysis, we used the above periods and also ones equal to its fractions. In the plots of level energies vs. its periodic phases, there appear fringe structures, which can be approximated with parabolic curves. We can observe parabolic fringes with the same scales in many different nuclei. These fringe structures in the eV-energy regions for medium and heavy nuclei are rather complicated. However, we found an illustrative example in a model spectrum of ^{236}U (^{235}U + n) [7, 8]. When we take a period of ε = 1.09 eV, all the observed levels between 1950 and 2250 eV are positioned on the same parabolic fringes with the same fringe centre.

Figure 1 shows the plots of level energies vs. periodic phases for *s*-wave levels with spin $J = 3$ and 4. All the more than five hundreds levels over this energy region are positioned on the parabolic fringe

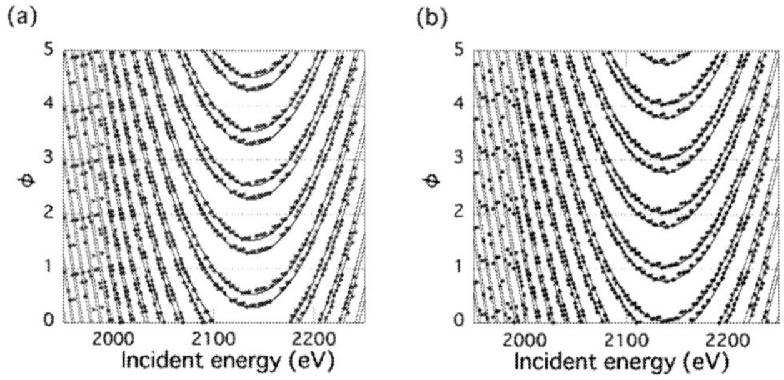

Figure 1. Plots of level energies vs. periodic phases with 1.09-eV period: spin $J = 3$ (a) and $J = 4$ (b).

curves, which are separated from each other by a phase difference of 1/4. There observed four multiple fringes between a phase difference of $\phi = 1$. First and second fringe curves select $J = 3$ levels, while third and fourth ones select $J = 4$ levels. If we use another period of 0.437 eV, we can observe different parabolic fringes; the levels with mixed spins of $J = 3$ and 4 are positioned on the same fringes where each fringe has the same fringe centre and is separated by a phase difference of 1/6. Since these level energies were obtained with a limited boundary condition [7], we call these resonance levels as a model spectrum. Later, the resonance data were revised by adopting a larger boundary condition and also by adjusting resonance parameters to fit fission cross-section measurements [9]. For the revised data [10], the fringe structure in Fig. 1 turned out to be obscured. However, the same fringes can be still traced [11].

Using nuclear data compilation [12], we performed the fringe analysis up to the energy regions of several tens MeV. We dealt with neutron and proton resonances and nuclear excited levels below neutron and separation energies. Considering a high precision of level energies and a large number of levels, neutron resonance data play a main role in the present analysis. In favourable cases, a number of observed levels in one nucleus amounts to several thousands.

2 Fringe analysis

Method of fringe analysis is very simple. Let divide *i*-th level energy E_i by a period ε, then we obtain

$$E_i = n_i\varepsilon + \eta_i, \tag{1}$$

where n_i is an integer and η_i a residual of E_i. We divide both sides of Eq. 1 by ε. Then we obtain

$$E_i/\varepsilon = n_i + \phi_i. \tag{2}$$

Here ϕ_i is in the range: $0 \leq \phi_i < 1$. We define ϕ_i as a periodic phase of level energy E_i in a periodic sequence starting from a point E_0. We usually take the starting point at zero excitation energy or zero incident energy. We treat equally such phases that differ from each other by integers: -1, 0, 1, 2... So that each level energy can have multiple numbers of phase: ... $\phi_i - 1$, ϕ_i, $\phi_i + 1$, $\phi_i + 2$...

We obtained diagrams of level energies vs. periodic phases using periods of ε. We extended the analysed energy region up to several tens MeV. We processed nuclei with $14 \leq A \leq 244$. In choosing level periods, we took notice that low-lying excited levels in odd As isotopes [13] and ^{71}Cu [14] can be well approximated by integral multiples of 16.5 keV. In the following analysis, we used the period of 16.5 keV and also its fractions such as 0.44, 0.55 and 1.65 keV.

3 Results and discussions

3.1 Neutron and proton resonances

Neutron resonance data are a promising target to perform the fringe analysis, since there observed a large number of levels with high resolution. For the present purpose, we mainly used the compiled data [12]; neutron and proton resonance levels are listed in units of excitation energy so that we can easily compare the results obtained from different excitation modes. Table 1 lists compound

Table 1. Data of typical compound nuclei, where series of parabolic fringes can be distinctively observed in the diagrams of level energies and its periodic phases. Data are from ref. [12].

Compound nucleus	Incident particle	Number of observed levels	Max. incident energy (*MeV*)	Max. excitation energy (*MeV*)
^{37}Cl	*p*	419	3.0	11.4
^{57}Fe	*n*	1242	9.3	16.9
^{60}Co	*n*	774	2.6	10.0
^{59}Ni	*n*	1989	9.8	18.8
^{61}Ni	*n*	2437	9.8	17.6
^{62}Ni	*n*	1712	9.8	20.4

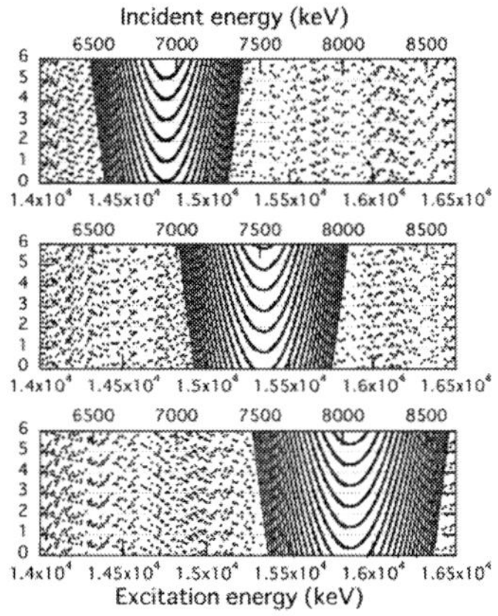

Figure 2. Plots of level energies vs. periodic phases with periods of ε = 1.32 keV (upper), 1.49 keV (centre) and 1.65 keV (lower).

nuclei, which can provide us with a starting point to understand systematically the fringe structures. Figure 2 shows the E-ϕ diagrams with periods of ε = 1.32, 1.49 and 1.65 keV for neutron resonance levels in a compound nucleus ^{61}Ni (^{60}Ni + n). It is seen that distinguished fringes appear at definite excitation energies and they are separated from each other by $\phi = 1$. As the level period decreases, the centre of the fringes shifts towards a lower-excitation energy and also the scale of the parabolic curves decreases. It is noted that all the observed ~220 levels between the 14.3- and 16.5-MeV excitation energies can be covered by the parabolic fringes with level periods changing continuously from 1.32 to 1.65 keV.

Fixing a level period, for example, at ε = 1.65 keV, we examined how the parabolic fringes behave when the excitation energies or incident neutron energies decrease. Figure 3 shows plots of

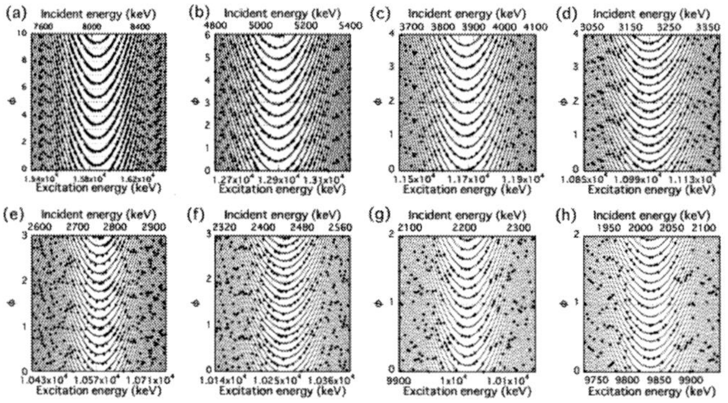

Figure 3. Plots of level energies vs. periodic phases with ε = 1.65 keV for ^{61}Ni. Multiplicity of the fringe increases continuously from $n = 1$ (a) to 8 (h) as the incident neutron energy decreases. Incident neutron energies at the fringe centres are E_f = 8.05 (a), 5.08 (b), 3.88 (c), 3.20 (d), 2.76 (e), 2.44 (f), 2.21 (g) and 2.02 MeV (h), where E_f denotes an incident neutron energy at a fringe centre.

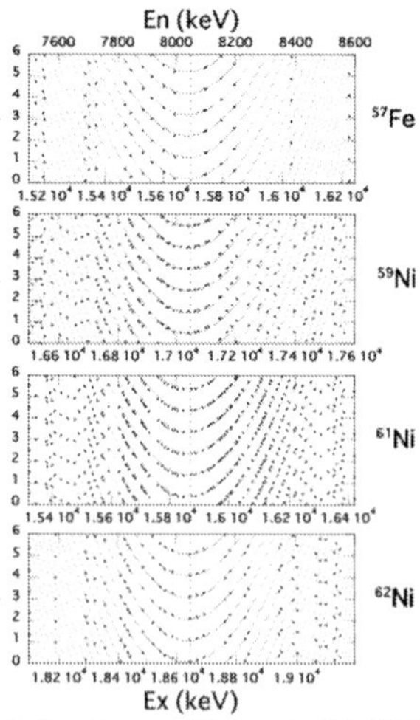

Figure 4. Plots of level energies vs. periodic phases with ε = 1.65 keV for ^{57}Fe, ^{59}Ni, ^{61}Ni and ^{62}Ni. Parabolic fringes with multiplicity of $n = 1$ in these nuclei have a fringe centre at the same incident energy of 8.05 MeV.

level energies vs. periodic phases for ^{61}Ni below an incident energy of 9.8 MeV. It is seen that at a definite excitation energy there appear parabolic fringes with a phase difference of $1/n$, n being an integer. Here, the phase difference of $1/n$ means that there observed n fringes between phases of ϕ and $\phi + 1$ at the same excitation energy. We call n as a multiplicity of the fringes. The multiplicity of the fringes increases from $n = 1$ to 8 as the excitation energy decreases. Fringes with multiplicity larger than $n = 8$ could not be resolved. Comparing observed parabolic fringes with each other among different nuclei, we found that fringe structures have a close similarity, especially among the nuclei listed in Table 1. In the nuclei ^{57}Fe, ^{59}Ni, ^{61}Ni and ^{62}Ni, there observed parabolic fringes, which have fringe centres at the same incident energies and have the same scales. Figure 4 shows these parabolic fringes with multiplicity of $n = 1$. Only difference among the parabolic fringes is about a bias of phase ϕ. Figure 5 shows plots of level energies vs. periodic phases with $\varepsilon = 0.55$ keV for ^{60}Co, where a sequence of fringes with multiplicities of $n = 1 \sim 3$ are observed below an incident energy of 1 MeV.

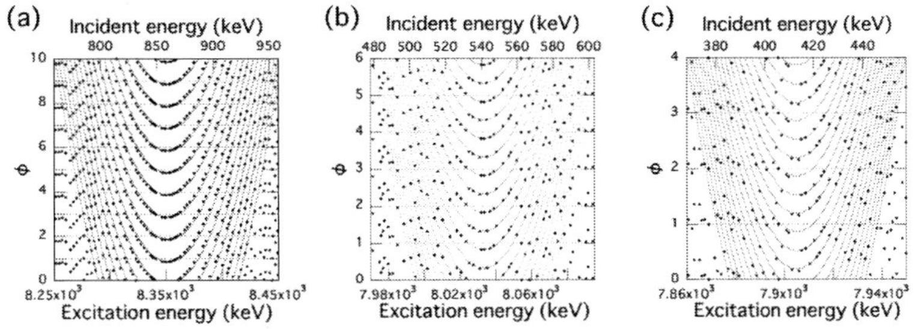

Figure 5. Plots of level energies vs. periodic phases with $\varepsilon = 0.55$ keV for ^{60}Co. Multiplicities of the parabolic fringes are $n = 1$ (a), 2 (b) and 3 (c), while the incident energies at the fringe centres are $E_f = 0.86$ (a), 0.54 (b) and 0.42 MeV (c), respectively.

Figure 6 shows plots of incident energies at fringe centres vs. level periods for ^{61}Ni and ^{60}Co.

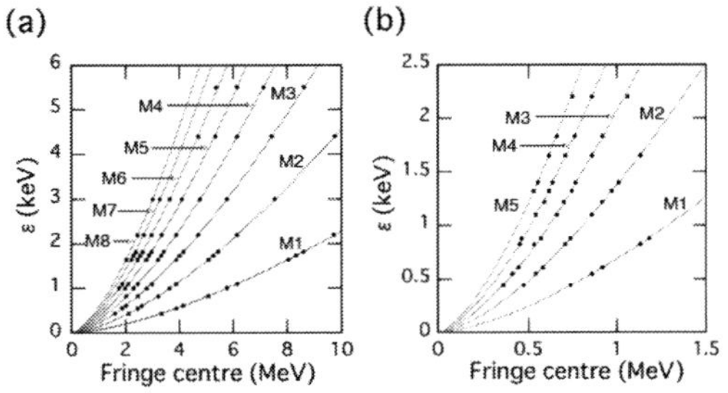

Figure 6. Plots of incident energies at fringe centres vs. level periods ε for ^{61}Ni (a) and ^{60}Co (b). Multiplicity of parabolic fringes is indicated by Mn.

As seen from the figure, there exists a simple relation between the energies of fringe centres and level periods. The two variables can be connected with following equation to a good accuracy:

$$\varepsilon = \alpha n E_n^{\frac{3}{2}}, \tag{3}$$

where α is a constant, n a multiplicity of the parabolic fringes and E_n an incident energy at the fringe centre. Equation 3 leads to a following relation:

$$E_n : E_1 = 1 : n^{\frac{2}{3}}. \quad (4)$$

Figure 7 shows plots of ratios E_n/E_1 for ^{61}Ni and ^{60}Co with level periods of $\varepsilon = 0.44 \sim 5.5$ keV. It is

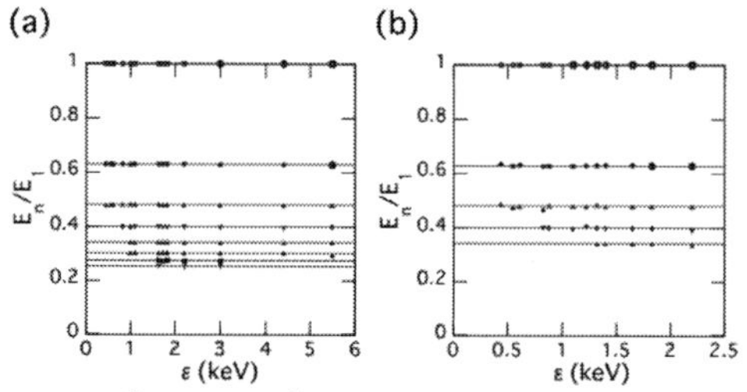

Figure 7. Plots of ratios of E_n to E_1 for ^{61}Ni (a) and ^{60}Co (b). Multiplicities of parabolic fringes range from $n = 1 \sim 8$ (a) and $1 \sim 5$ (b).

seen that the ratio of E_n to E_1 is same for different level periods and applicable both to ^{60}Co and ^{61}Ni. Table 2 lists the values of E_n/E_1 obtained from experiments and the ones predicted from formula 4.

Table 2. Ratios of E_n to E_1 for ^{60}Co and ^{61}Ni; n denotes multiplicity of the parabolic fringes.

	^{60}Co	^{61}Ni	
n	E_n/E_1 (Exp.)	E_n/E_1 (Exp.)	$1/n^{\frac{2}{3}}$ (Predicted)
1	1.000	1.000	1.000
2	0.630	0.631	0.630
3	0.480	0.482	0.481
4	0.397	0.398	0.397
5	0.343	0.343	0.342
6		0.303	0.303
7		0.274	0.273
8		0.250	0.250

The predicted values agree well with ones deduced from data. This fact indicates that we can use Eq. 4 to search for the locations of parabolic fringes with different multiplicities in the same sequence. Another important parameter characterizing the parabolic fringes is a scale factor of parabolic curves. This scale factor determines the energy range around the fringe centre where these fringes dominate. Periodic phases around the fringe centre can be expressed by parabolic curves as a function of level energy:

$$\phi_{ij} = k_n(E - E_n)^2 + \delta + i + \frac{j-1}{n};$$

$$0 \le \delta < 1;\ i = \ldots -1, 0, 1, 2, 3, \ldots;\ j = 1, 2, \ldots, n, \quad (5)$$

where n is a multiplicity of the fringes, ϕ_{ij} denotes periodic phase at a level energy of E, k_n a scale factor of the parabolas, E a level energy, E_n a level energy at the fringe centre and δ a constant. The constant δ corresponds to phase ϕ_{00} at $E = E_n$ with $i = 0$ and $j = 1$. The scale factor k_n can have positive or minus sign. Here we deal mainly with scale factors k_n with positive sign. All these fringes are parabolas of bottom-down type as shown in Figs.1 ~ 4. We found that a reciprocal of a scale factor k_n is proportional to an incident energy E_n at the fringe centre in the sequence of parabolic fringes:

$$1/k_n = \beta E_n, \tag{6}$$

where β is a constant. Figure 6 shows plots of $1/k_n$ vs. E_n for ^{61}Ni and ^{60}Co. It is seen that experimental

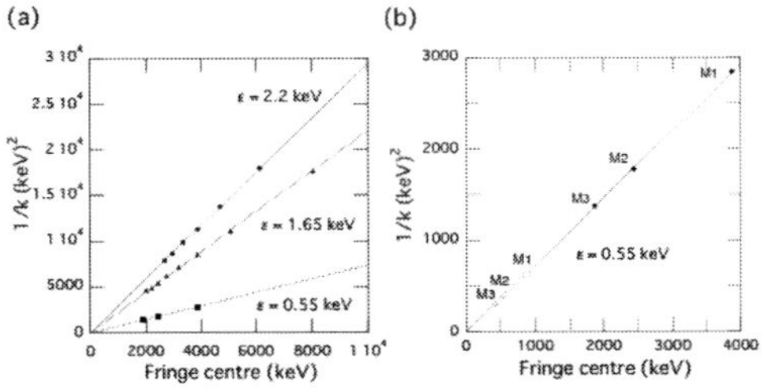

Figure 8. Plots of E_n vs. $1/k_n$ for ^{61}Ni (a) and ^{60}Co (b). (a): $n = 1 \sim 3$ ($\varepsilon = 0.55$ keV), $n = 1 \sim 8$ ($\varepsilon = 1.65$ keV) and $n = 2 \sim 7$ ($\varepsilon = 2.2$ keV); (b): upper three points (filled circles) with $n = 1 \sim 3$ (^{61}Ni) and lower three points (open circles) with $n = 1 \sim 3$ (^{60}Co).

data are on the linearly fitted lines with a good accuracy. A remarkable point in Fig. 8 (b) is that the same linear relation is commonly applicable to the sequences of parabolic fringes in ^{61}Ni and ^{60}Co.

We define a fringe radius r as a distance from a fringe centre at which periodic phase increases by one unit:

$$\phi_{OO}(r) = \phi_{OO}(0) + 1. \tag{7}$$

For the fringes with multiplicity of n, we denote the fringe radius by r_n. From Eq. 5, fringe radius r_n is connected with a scale factor k_n as follows:

$$r_n = 1/\,k_n^{\frac{1}{2}}. \tag{8}$$

Considering Eqs. 4, 6 and 8, we obtain following relation between fringe radii r_n and r_1 belonging to the same sequence:

$$r_n : r_1 = 1 : n^{\frac{2}{3}}. \tag{9}$$

For proton resonance levels in ^{37}Cl ($^{36}S + p$) (Table 1), we observed many parabolic fringes below an incident energy of 3.0 MeV. Plots of level energies vs. periodic phases for ^{37}Cl are shown in Fig. 9, where at an incident proton energy of 2.32 MeV, there appear distinguished parabolic fringes with multiplicities of n = 1, 2 and 3, each corresponding to level periods of ε = 0.55, 1.1 and 1.65 keV. These parabolic fringes shift towards lower energies as the level periods decrease. However, a dependence of ε on E_n is different from Eq. 3; $\varepsilon = fn(E_n + g)$, where f and g are positive constants. This indicates that another mechanism is operated to determine the positions of these parabolic fringes. We can observe similar parabolic fringes in ^{47}V at the same incident energy of E_1 = 2.32 MeV with ε = 0.55 keV.

We have described mainly parabolic fringes, which have sequence structure as shown in Figs. 3, 5. We can observe parabolic fringes over a wide mass range of nuclei. Many parabolic fringes have the same incident energies at the fringe centres and the same fringe radii. Here, an exception is a phase

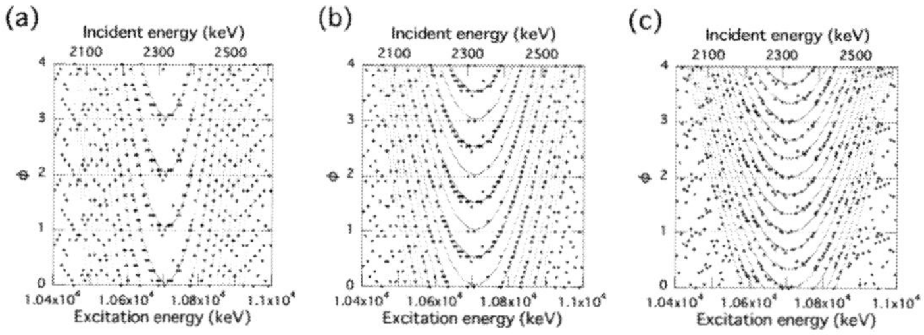

Figure 9. Plots of level energies vs. periodic phases with $\varepsilon = 0.55$ (a), 1.1 (b) and 1.65 keV (c) for ^{37}Cl. Observed parabolic fringes at the incident energy of 2.32 MeV have multiplicities of $n = 1$ (a), 2 (b) and 3 (c).

constant δ, which may differ from case to case. For neutron resonance levels in ^{52}V, we can observe a fringe sequence similar to the one in ^{61}Ni. Also, among nuclei such as ^{37}Cl, ^{41}Ca, ^{91,92}Zr and ^{208,209}Pb, we can identify fringe sequences similar to the one in ^{60}Co.

3.2 Nuclear excited levels below neutron and proton separation energies

We applied the fringe analysis to nuclear excited levels below neutron and proton separation energies. We searched for the same fringes as observed in neutron resonance levels. Here, we correspond nuclear excitation energies E_{exc} to incident neutron energies E_{inc}. Figure 10 shows plots of E_{exc} vs. ϕ with $\varepsilon = 1.65$ keV for ^{37}Cl, ^{61}Ni and ^{62}Ni. There appear fringes with multiplicity of $n = 1$, which are

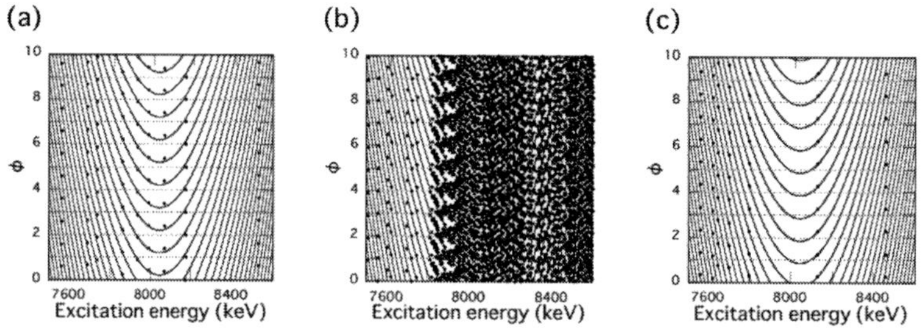

Figure 10. Plots of excitation energies vs. periodic phases for ^{37}Cl (a), ^{61}Ni (b) and ^{62}Ni (c). Neutron separation energy of ^{61}Ni is 7.82 MeV.

well reproduced by parabolic curves obtained from the 8.05-MeV fringes in ^{61}Ni. In Fig. 10 (b), dense fringes appear above $E_{exc} = 7.82$ MeV, which corresponds to a neutron separation energy of ^{61}Ni. Fringes at lower energies can be identified also by using the same parameters k_n and E_n determined from ^{61}Ni. The 8.05-MeV fringes can be observed in other nuclei such as ^{29}Si, ^{36}Cl, ^{41}K and ^{55}Fe.

We surveyed fringe spectra with the level period of 16.5 keV. Parabolic fringes can be resolved for nuclei with $N \geq 14$. These fringes have a fringe radius of $r_n = 0.5 \sim 1.5$ MeV and its fringe spectra are complex. However, there are cases where fringe spectra resemble each other in different nuclei, for example, in ^{46}V and ^{50}Mn.

3.3 Origin of parabolic fringes in the *E*- ϕ diagram

We interpret parabolic fringes in the E-ϕ diagram as a result of interference effects. We assume that periodic phase at E around a fringe centre E_n is a sum of periodic phases contributed from all points between E and E_n as a function of distance x from E_n as x/ε; finally, periodic phase at distance x from E_n is expressed by

$$\phi = eb\int_0^x \frac{x}{\varepsilon}\frac{dx}{\varepsilon} = eb\,\frac{x^2}{2\,\varepsilon^2}, \tag{10}$$

where e takes +1 in case of upward parabola and -1 in case of downward parabola; b is a constant determining a scale of the fringes. Phase at E is given by replacing x in Eq. 10 by $|E - E_n|$. Parabolic fringes with multiplicity of n are separated in phase by $\Delta\phi = 1/n$ from each other. If we connect a level period ε with a repetition cycle of $t = 2\pi\hbar/\varepsilon$, which is called Poincaré cycle [15], time delay of $2\pi(m - 1)\hbar/n\varepsilon$ can correspond to m-th fringe; here, $m = 1, 2,\ldots, n$. Time-based treatment of nuclear reaction is also made [16].

Downward parabolas can be actually observed in fringe spectra, but its role is not clear compared to upward parabolic fringes. Empirical rules on the correlation between scale factors and positions of parabolic fringes have been determined for typical examples, but we are far from understanding the whole picture of fringe structure.

4 Conclusion

Parabolic fringes with the same scale factors are observed at the same incident neutron energies below 10 MeV in nuclei such as ^{55}Fe, ^{57}Fe, ^{59}Ni, ^{61}Ni and ^{62}Ni. Spin assignments of these levels are limited to the energy region below 1 MeV. To study symmetry dependences of parabolic fringes, we need spin information on these levels at higher energies.

We interpret the parabolic fringes as a result of interference effects on periodic phases; a quantized phase difference of $1/n$ can be connected with a discrete time delay of wave pulsations caused by incident particles.

References

1. S. I. Sukhoruchkin, Sov. J. Nucl. Phys. **10**, 285 (1970)
2. K. Ideno and M. Ohkubo, J. Phys. Soc. Jpn. **30**, 620 (1971)
3. K. Ideno, J. Phys. Soc. Jpn., **37**, 581 (1974)
4. C. Coceva *et al.*, *Proc. Int. Conf. on Statistical Properties of Nuclei*, J. B. Garg, 447, Plenum, N.Y., 1972
5. K. Ideno, *Proc. Int. Conf. on Nuclear Data for Science and Technology*, 783, Mito, 1988
6. K. Ideno, *Proc. Int. Conf. on Neutrons in Research and Industry*, SPIE, Washington, vol. **2867**, 398, 1997
7. L. C. Leal *et al.*, Nucl. Sci. Eng. **109**, 1 (1991)
8. Landolt-Boernstein, New Series vols. **16/B**, Springer, Heidelberg, 1998
9. L. C. Leal *et al.*, Nucl. Sci. Eng. **131**, 220 (1999)
10. Landolt-Boernstein, New Series vols. **16/C**, Springer, Heidelberg, 2004
11. K. Ideno, **AIP915**, 819 (2008)
12. Landolt-Boernstein, New Series vols. **I/25, A, B**, Springer, 2012; **I/25C**, **D**, **E,** 2013
13. I. Hossain *et al.*, Phys. Rev. C **58**, 1318 (1998)
14. T. Ishii *et al.*, Phys. Rev. Lett., **84**, 39 (2000)
15. K. Izumo, Prog. Theor. Phys. **54**, 1378 (1975)
16. M. Ohkubo, Phys. Rev. C**53**, 1325 (1996); ibid., C**73**, 054609 (2006); C**87**, 014608 (2013)

Nuclear Poincaré cycle synchronizes with the incident de Broglie wave to predict regularity in neutron resonance energies

Makio Ohkubo[1,a]

[1] *N.Resonance Lab, 1663-39, Senba-cyo, Mito-shi, Ibaraki-ken 310-0851 Japan*

Abstract. In observed neutron resonances, long believed to be a form of quantum chaos, regular family structures are found in the s-wave resonances of many even-even nuclei in the tens keV to MeV region [M.Ohkubo, Phys. Rev. C **87**, 014608(2013)]. Resonance reactions take place when the incident de Broglie wave synchronizes with the Poincaré cycle of the compound nucleus, which is composed of several normal modes with periods that are time quantized by inverse Fermi energy. Based on the breathing model of the compound nucleus, neutron resonance energies in family structures are written by simple arithmetic expressions using S_n and small integers. Family structures in observed resonances of ^{40}Ca+n and ^{37}Cl+n are described as simple cases. A model for time quantization is discussed.

1 Introduction

In neutron-nucleus interactions, a compound nucleus (CN) is formed with excitation energy $E_x = S_n + E_n$, where S_n is neutron separation energy ~ 8 MeV, and E_n is the neutron kinetic energy in the center of mass system (CMS). In the compound nucleus, many degrees of freedom will be excited and mixed to form very complicated states or so-called chaos. Statistical distributions of many observed neutron resonance data are in good agreement with predictions of the Random Matrix Theory (RMT), therefore, the neutron resonance region is long believed to be a form of quantum chaos where no regularity in level spacings/energies is expected.

However, as is seen in every field of science, different methods of analysis extract different features of the system under scrutiny. Many facts contradicting RMT have been reported in observed resonance energies/spacings over the past five decades that relate the frequent appearance of special spacings by D_{ij} (spacings between two arbitrary levels) analysis or by Fourier-like analysis [1].

In order to understand the mechanism of neutron resonance reactions including special spacings/energies, we developed the "Breathing Model" of the compound nucleus, analogous to classical resonance phenomena where nuclear Poincaré cycles synchronize with the incident de Broglie wave at resonances. This model predicts resonance energies by simple arithmetic expressions with good accuracy. In section 2 of this article, the concept of the Breathing Model is described. In section 3, theoretical insight into time behaviors of normal modes uses time quantization by inverse Fermi energy. In section 4, simple cases of two normal modes found in observed resonance data are shown. In section 5, regularity in resonance energies are described. In section 6, a preliminary time quantization

[a] e-mail: makiohkubo@hi-ho.ne.jp

mechanism of a 'hot chain' is described. In section 7, transition from simple CN to complex CN are described. Discussions and conclusion are found in section 8.

2 Breathing model of the compound nucleus in resonance reactions

In the neutron-nucleus reaction, the wave packet of an incident neutron is divided into two components: (1) pass-by component without interaction; and (2) penetration component that penetrates the CN, excites many normal modes, and appears at the CN surface after every recurrence time τ_{rec} or the "Poincaré period" τ_{PP}. Interference takes place between the two components, and the observed cross-sections are the result of the interference between the two wave packet components. We define the initial "coalescent phase" as the wave crest of the pass-by component, the gather, and the flare up of penetration components on the CN surface (doorway) during a short time $\sim \tau_0$ (described below). This coalescent phase appears repeatedly with a time period τ_{rec} which depends on the ensemble of normal modes excited at resonance. This behaves as a scattering center for the pass-by component during the resonance lifetime $\sim 2\pi\hbar/\Gamma$, where Γ is the resonance width. At resonances, the incident de Broglie frequency synchronizes with the nuclear Poincaré cycle as illustrated in Fig. 1. If the mechanism above does not exist, an energy independent potential scattering cross-section ($\sim 10\ rmb = 10 \times 10^{-28}\mathrm{m}^2$) will be observed.

3 Theoretical insight

The penetration component excites the CN to energy $S_n(\sim 8$ MeV). As the lifetime of the CN formed by neutron resonance is relatively long, many excited degrees of freedom can be approximated by an oscillator ensemble rewritten as an ensemble of normal modes. Several combinations of normal modes are possible. The energy and time period of i-th mode are E_i and $\tau_i = 2\pi\hbar/E_i$, respectively. For M normal modes,the recurrence periods τ_{rec} are integer multiples of τ_i.

$$\tau_{rec} = n_i\tau_i, \qquad n_i\text{:integer}(i = 1, 2, ..., M), \tag{1}$$

where the influence of angular uncertainty $\sim$ 1 rad is negligibly small for the lifetime of the resonance. From Eq. (1), τ_i/τ_j $(i, j = 1, 2, ..., M)$ are integer ratios. Therefore, τ_i $(i = 1, 2, ..., M)$ are restricted to be integer multiples of unit time τ_0, or time quantization by τ_0. For unit time τ_0, it is plausible to use the inverse Fermi energy,

$$\tau_0 = 2\pi\hbar/G = 1.2 \times 10^{-22}\mathrm{s}, \tag{2}$$

where $G \sim 34.5$ MeV is almost the Fermi energy(the maximum energy of the nucleon in nuclear potential). Hence, the time period of the normal mode is $n_i\tau_0$, and its energy is G/n_i where n_i is an integer.

Although not all details are known, the Hamiltonian and wave functions of the CN described by M normal modes are formally written as

$$H = H_1 + ... + H_M, \tag{3}$$

$$\Psi(x, t) = \psi_1(x, t) \otimes ... \otimes \psi_M(x, t). \tag{4}$$

The time periodic recurrence of $\psi_i(x, t)$ and $\Psi(x, t)$ are

$$\psi_i(x_i, t) = \psi_i(x_i, t + n_i\tau_0), \tag{5}$$

$$\Psi(x, t) = \Psi(x, t + \tau_{rec}). \tag{6}$$

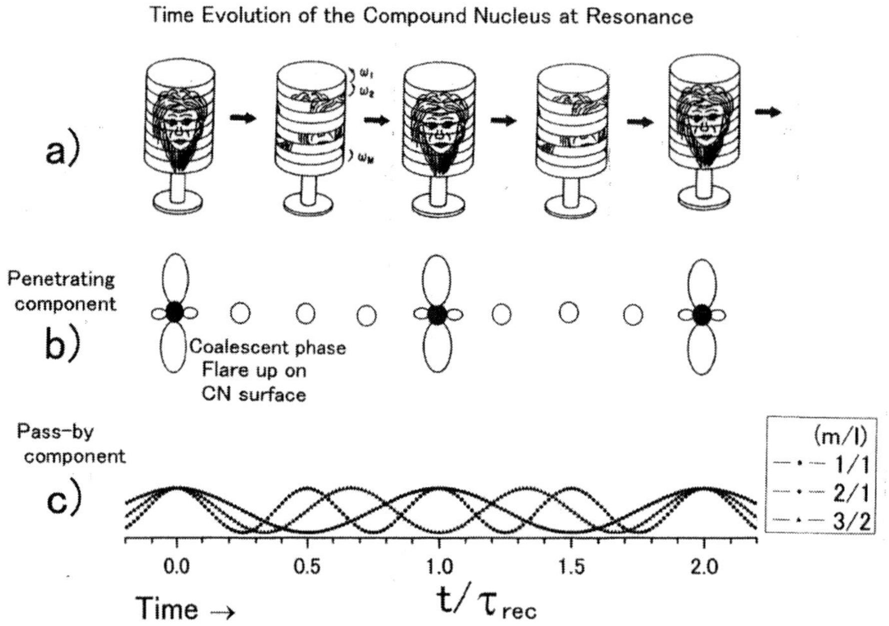

Figure 1. Time evolution of the compound nucleus at resonance. (a) A toy model is used to show recurrence of many oscillators rotating with arbitrary frequencies. (b) Penetration component, which excites CN and recurs (flare up on CN surface) at coalescent phases with the Poincaré periodicity, synchronizing with the pass-by component (c) at resonance.

The nuclear Poincaré period τ_{pp} or recurrence time τ_{rec} is defined as

$$\tau_{pp} = \tau_{rec} = LCM(n_1, ..., n_M)\tau_0, \quad (7)$$

where LCM is the least common multiple for a set of integers $(n_1, ..., n_M)$.

The recurrence energy E_{rec} is defined as

$$E_{rec} = \frac{G}{LCM(n_1, ..., n_M)}. \quad (8)$$

The excitation energy of the CN caused by the penetration component S_n is written as

$$S_n = E_1 + ... + E_M = G\left(\frac{1}{n_1} + ... + \frac{1}{n_M}\right) = G\sum_{i=1}^{M}\frac{1}{n_i}. \quad (9)$$

Neutron resonances take place when the period of the incident wave (pass-by component) $\tau_w(= 2\pi\hbar/E_n)$ are in simple integer ratios (k/m) of the nuclear Poincaré period τ_{rec}.

$$\tau_w = (k/m)LCM(n_1, ..., n_M)\tau_0. \quad (10)$$

That is, the possible resonance energies E_n in CMS are

$$E_n = (m/k)E_{rec}, \tag{11}$$

composing a family structure, where k, and m are small integers.

4 Simple cases

Despite expectations of complex features of CN, many simple cases are found in observed resonance data where only two normal modes are excited. The ratio S_n/E_n are in a simple form

$$\frac{S_n}{E_n} = (n_1 + n_2)\left(\frac{k}{m}\right). \tag{12}$$

and S_n is

$$S_n = \left(\frac{1}{n_1} + \frac{1}{n_2}\right)G. \tag{13}$$

In this case, many observed resonances dispose in vicinity of (IVO)

$$E_n = \left(\frac{m}{k}\right)E_{rec} = \left(\frac{m}{k}\right)\frac{G}{n_1 n_2} = \left(\frac{m}{k}\right)\frac{S_n}{n_1 + n_2}, \tag{14}$$

composing a family structure.

4.1 ^{40}Ca+n

In s-wave resonances in ^{40}Ca+n [2], we found that $S_n/E_n = 17(k/m)$ for many resonances (^{40}Ca (I)) which provided the clue to this problem [3]. Also, S_n is written as $S_n = 8362$ keV $= (17/70)G = (1/7 + 1/10)G$ where $G = 34434$ keV. Among 40 s-wave resonances observed in $0 \leqslant E_n^{obs} \leqslant 2500$ keV region, 15 resonances are IVO $E_n^C = (S_n/17)(m/k) = 491.9(m/k)$ keV with $(k, m \leqslant 10)$, where (m/k) are 1/3 , 3/7, 1/2, 7/10, 8/9, 1/1, 10/9, 7/6, 5/4, 4/3, 7/4, 2/1, 3/1, 4/1 and 9/2. For 10 resonances, the deviations $\delta = E_n^{obs} - E_n^c$ are within a region of 5 keV width.

The statistical probability of the appearance of the pseudofamily are calculated assuming a random distribution of $n_l = 40$ levels in the energy region $R = 2330$ keV [1]. Candidate cases (m/k) $(k, m \leqslant 10)$ are counted on a (k, m) plane to be $B = 53$. For a level placed at random in the region, the probability of being on a candidate region is $p = (\epsilon B)/R = 0.114$. For 40 levels placed at random, the expected number of integer ratios is $n_l p = 4.6$. The probability for 10 levels out of 40 being integer ratios is calculated by the binomial distribution: $P_r(10, 40, p) = {}_{40}C_{10}p^{10}(1-p)^{30} \approx 0.0083 = 0.83\%$, and the sum $P_s = \sum_{j=10}^{40} P_r(j, 40, p) \leqslant 1.2\%$.

In ^{40}Ca+n in the same energy region, another family coexists showing $S_n/E_n = 39(k/m)$. Many resonances dispose IVO $E_n^c = (S_n/39)(m/k) = 214(m/k)$ keV, called ^{40}Ca (II).

Similar regular family structures are found in s-wave resonances in target nuclei of ^{54}Cr,(I)(II), ^{64}Ni, ^{90}Zr, and ^{208}Pb.

For resonances including mixed J^π, similar family structures are found in target nuclei of ^{26}Mg, ^{37}Cl, ^{48}Ca, ^{50}Ti, and ^{58}Fe.

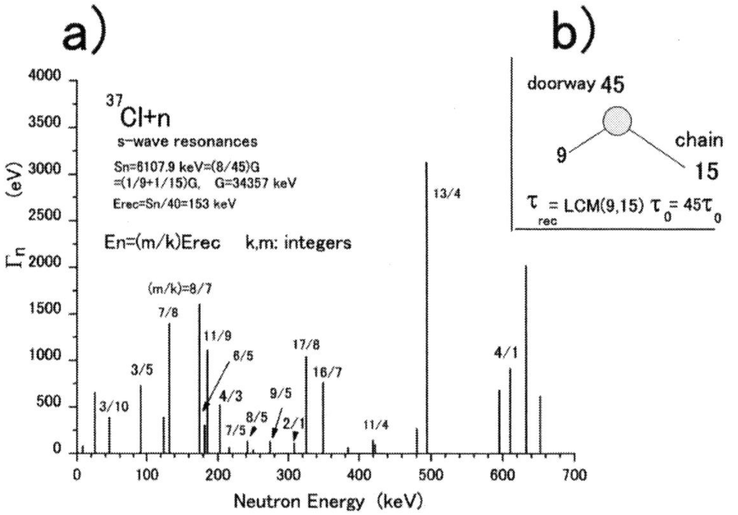

Figure 2. a) ^{37}Cl+n resonances. Neutron width Γ_n vs. incident neutron energy up to 700 keV [4]. S_n = 6107.9 keV can be written as $(8/45)G = (1/9 + 1/15)G$, where G = 34357 keV. To meet energy region requirements, a recurrence energy $E_{rec} = (1/225)G = S_n/40 = 152.7$ keV is used. b) Two normal modes (hot chains) of periods $9\tau_0$ and $15\tau_0$ couple to the doorway (circle) where coalescent phases appear with a time period of $45\tau_0$ repeatedly.

4.2 ^{37}Cl+n

As an example, the regular family structures of s-wave resonances in ^{37}Cl+n are described below where 1^+ and 2^+ resonances coexist. Original data from Sayer et al. [4] observed up to 1 MeV. The separation energy S_n is rewritten as S_n = 6107.9 keV = $(8/45)G = (1/9 + 1/15)G$ where G = 34357 keV. The recurrence energy is $E_{rec} = G/45 = S_n/8 = 763.5$ keV. In observed values S_n/E_n, a factor 40 appears for many resonances. Then $S_n = (40/225)G$ and one fifth harmonics of $E_{rec} = S_n/40 = 152.7$ keV are used. Many observed resonance energies E_n^{obs} dispose IVO E_n^c = $152.7(m/k)$ keV. Below $E_n \leqslant 500$ keV, 23 resonances are observed.

For 10 resonances, m/k are; 3/10, 3/5, 7/8, 8/7, 6/5, 4/3, 7/5, 8/5, 9/5, and 2/1 as illustrated in Fig.2a. Among these resonances, deviations $\delta = E_n^{obs} - E_n^c$ are within a width of 0.8 keV for 6 resonances. The probability of the appearance of these dispositions is calculated similar to the ^{40}Ca (I) described above. The parameters are: $n_l = 23$ in the region R = 500 keV, ϵ = 0.8 keV, B = 55, and $p = \epsilon B/R = 0.088$. $P_r = {}_{23}C_6 p^6(1-p)^{17} = 0.0097 = 0.97$ % and P_s are calculated to be 1.3 %.

5 Regularity in resonance energies

The probability of the appearance of the regular family structures P_s are quite small for several nuclei. In Table II of Ohkubo[1], P_s are listed for seven cases of four target nuclei, ^{40}Ca,(I)(II), ^{54}Cr,(I)(II), ^{64}Ni, and ^{90}Zr, where the average P_s is 1.3 %. In addition, P_s is 1.3 % for ^{37}Cl. We have investigated neutron resonances of about 40 light and magic nuclides. If we assume random dispositions of levels for all 40 nuclides, the expectation of it is less than one nuclide. However, it appears in 5 nuclides. The probability U of appearing in 5 nuclides among 40 nuclides with $P_s = 1.3\%$ is calculated by the binomial distribution as

$$U = {}_{40}C_5(0.013)^5(0.987)^{35} \approx 1.5 \times 10^{-4}. \tag{15}$$

The occurrence of a phenomenon with very small probability means a failure of the random hypothesis, and the existence of regular family structures are verified with the statistical significance level of $\sim 10^{-4}$. Though the random level ensemble substitutes for the RMT ensemble, this conclusion will not be changed for RMT ensemble.

6 Time quantization of normal modes

Time quantization of normal modes is inevitable for resonance reactions because of the finite time restriction of τ_{rec} as described in Eqs. (1)and (2). For the mechanism of time quantization of normal modes, we consider "hot chains" excited in CN. For example in ^{40}Ca(I), two normal modes of $7\tau_0$ and $10\tau_0$ are excited. These normal modes work as a time-delay-system for the penetration components. For the time-delay-system of $7\tau_0$, we consider a hot chain composed of seven elements. An element of the chain couples to the doorway, where the penetration component, called fire, enters and exits, and interference takes place with the pass-by component. The fire transfers to the neighbouring element in unit time τ_0, and the fire circulates in the chain in $7\tau_0$, repeatedly. Similar mechanisms work for the $10\tau_0$ chain. Fires in two chains gather and flare up at the doorway to form a scattering center for the pass-by component with a period $LCM(7, 10) = 70\tau_0$ repeatedly during the lifetime of the resonance.

7 From simple to complex

The simple cases above are for light or magic nuclei where neutron resonance level densities are not very high. Few normal modes couple to the doorway as illustrated in Fig. 2(b) and Fig. 3(a). For medium and heavy nuclei where level densities are high, many (M) normal modes couple to the doorway as illustrated in Fig. 3(b). M are estimated to be $M \leqslant 10$ for observed resonances [5]. Several different resonance families coexist in the same energy region and display complex features. For the resonances in the eV region, $LCM(n_1, ..., n_M)$ will be increased to $10^4 \sim 10^7$, and family structure analysis will be an interesting and difficult problem.

8 Discussion and Conclusion

The classical analogy of resonance phenomena is valid for neutron resonance reactions with some modification of physical concepts, i.e., the nuclear Poincaré cycle synchronizes with the incident de Broglie wave. Resonance families derived by arithmetic expressions in Eq. (11), agree well with the observed resonance energies. Time quantization of normal modes by τ_0 is essential to this work and simplifies the excitation modes in the energy region $\sim S_n$. Since τ_0 deviates about $\pm 1\%$ from family to family depending on the internuclear interactions, the value $G = 2\pi\hbar/\tau_0 \sim 34.5$ MeV deviates consequently.

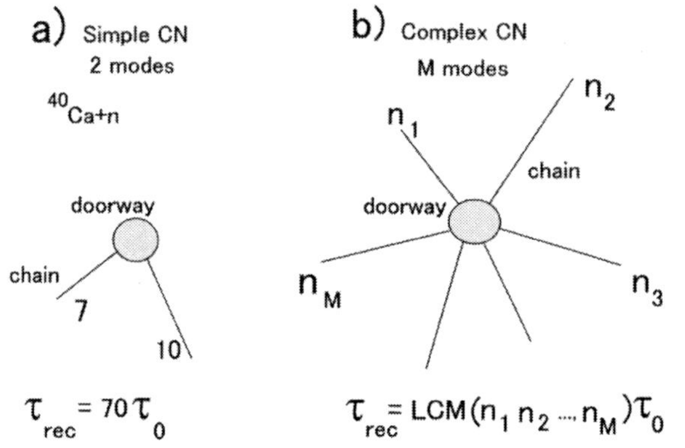

Figure 3. (a) Simple case of ^{40}Ca+n compound nucleus. Two normal modes (hot chains) of periods $7\tau_0$ and $10\tau_0$ couple to the doorway (center circle) where coalescent phases appear with time period $70\tau_0$ repeatedly. (b) Complex compound nucleus. Many ($M \leqslant 10$) normal modes couple to the doorway where coalescent phases appear with time period $\tau_{rec} = LCM(n_1, ..., n_M)\tau_0$, repeatedly.

Time behaviors of the compound nucleus are extracted by analysing energy structures of neutron resonance levels. This is mathematically analogous to determining crystal structures from neutron diffraction patterns using a neutron plane wave $\exp(i(kx - \omega t))$, where k is wave vector, x is a space coordinate, ω is frequency, and t is time. The crystal diffractions are in the (k, x) domain, scattering from atoms at space-periodic lattices, whereas the neutron resonances are in the (ω, t) domain, scattering from time-periodic coalescent phases on the CN surface.

The traditional idea of quantum chaos in the neutron resonance region is not correct. For a long time, it was surmised that regularities in resonance positions/spacings did not exist because of the success of the Random Matrix Theory, supported by many observed neutron resonance data analysed by traditional statistical methods. These methods, however, could not detect strong correlations of family structures as described in this article. To this point, it must be stressed that traditional statistical analyses have no ability to distinguish between chaos and regularity. Our results offer a new explanation for a regular system for this region.

On the other hand, rules on strengths and J^π of each resonance, which will depend on spatial components of wave functions, are not known at present. However, this approach seems to be promising to investigate nuclear dynamic structures in discrete energy regions.

References

[1] M.Ohkubo; *Phys.Rev.C* **87**, 014608(2013)

[2] R.Toepke; *KFK-2122*, Karlsruhe(1974)
[3] M.Ohkubo; *Phys.Rev.C* **80**, 024607(2009)
[4] R.O.Sayer,K.H.Guber, L.C.Leal and N.M.Larson; *Phys.Rev.C* **73**, 044603(2006)
[5] M.Ohkubo; *Phys.Rev.C* **53**,1325(1996)

AUTHOR INDEX

AUTHOR INDEX

AUTHOR INDEX

AUTHOR INDEX

AUTHOR INDEX